FOOD ANALYSIS:
Theory and Practice
Second Edition

Yeshajahu Pomeranz

Department of Food Science and Human Nutrition
Washington State University
Pullman, Washington

Clifton E. Meloan

Department of Chemistry
Kansas State University
Manhattan, Kansas

An **avi** Book
Published by Van Nostrand Reinhold
New York

An AVI Book
(AVI is an imprint of Van Nostrand Reinhold)
Copyright © 1987 by Van Nostrand Reinhold

Library of Congress Catalog Card Number 86-28824

ISBN 0-442-28316-4

Printed in the United States of America

Van Nostrand Reinhold
115 Fifth Avenue
New York, New York 10003

Van Nostrand Reinhold International Company Limited
11 New Fetter Lane
London EC4P 4EE, England

Van Nostrand Reinhold
480 La Trobe Street
Melbourne, Victoria 3000, Australia

Nelson Canada
1120 Birchmount Road
Scarborough, Ontario M1K 5G4, Canada

16   15   14   13   12   11   10   9   8   7   6   5   4   3

Library of Congress Cataloging-in-Publication Data

Pomeranz, Y. (Yeshajahu), 1922–
  Food analysis.

  Includes bibliographies and index.
    1. Food—Analysis.   I. Meloan, Clifton E.   II. Title.
TX541.P64   1987   664'.07   86-28824
ISBN 0-442-28316-4

Contents

Preface

This book has been designed for use as a text by undergraduate students majoring in food science and technology and as a survey of modern analytical methods and instruments for the worker and researcher in the field of food analysis.

In preparation of the manuscript, we followed the recommendations for subject coverage in a textbook of food analysis for students of food science and technology, as contained in the report of the Task Force on Food Analysis of the Institute of Food Technologists' Council Committee on Education.

Over 15 years have passed since the first edition of this book was published in 1971. Over those years it has become the most widely accepted text on food analysis. In the revised edition in 1978 we made a great number of small revisions and added new sections on high pressure liquid chromatography, affinity chromatography, immobilized enzymes, and infrared reflectance spectroscopy.

Tremendous advances have been made in analytical chemistry, in general, and food analysis, in particular. There have been fascinating developments in approaches, methodology, automation, and, foremost, in instrumentation. In preparing the third edition of the book we were faced with the challenge of adding new and important material and judiciously eliminating some of the old material if it was found obsolete or less important, doing all this without affecting the philosophy, structure, and totality of the book. This required a tremendous amount of work and some difficult decisions.

Practically every chapter from the old edition has been trimmed, revised, and updated—in most cases with a complete change in organization and contents. We believe the result is an entirely new book that retains the important points of the previous editions and updates them to make the book as relevant as possible. It does so in a single, compact volume that spans the broad spectrum from theory to practice in food analysis.

As in the previous editions, basic principles, rather than details of analytical methods, are stressed. Although the emphasis is on modern and sophisticated instruments and methods, we have also described the classical procedures that have been in use for many years. In chapters devoted to instrumentation, we have attempted to provide the background theory that is required for understanding the principles of instrumental assays. Included are diagrams and descriptions of typical instruments, information on their application and precision, and sample problems with detailed solutions. Whenever applicable, we have compared the instrumental and assay procedures to evaluate their usefulness and limitations. The chapters on instrumentation end with problem questions (and answers) to develop the subject matter covered in the text further.

In the sections devoted to applications, we describe the use of the basic instruments and discuss their potential and limitations in solving specific problems.

This book is not meant to replace standard methods of analysis; its main purpose is to explain the background and principles of those methods. By stressing fundamental principles, rather than detailed methodology, the book should provide a useful source both for the student and for the experienced researcher. To understand most of the material, the reader is expected to have studied general, organic, analytical, and food chemistry and have an appreciation of biochemistry. To allow for the student's limited training in physical chemistry, we have included the physicochemical principles, terminology, and detailed computations.

While we realize that some of the instrumental techniques that require expensive or more sophisticated biological methods may not be within the reach of many food chemists, we include them to stimulate the thinking of the searching student and the experienced researcher. We hope that these new procedures will contribute to an appreciation of the scope and potentialities of food analysis. Some of those techniques and procedures can yield excellent results with simplified adaptations—provided the principles are understood. We hope most readers will agree with our selection of methods, techniques, and approaches, and we would greatly appreciate comments from teachers, researchers, food analysts, and, most important, students in food science and technology.

We wish to thank the authors of articles, reviews, and books, as well as industrial companies and publishers for permission to reproduce material. Book reviewers in scientific and technical journals are thanked for their constructive criticism, as are colleagues and users of the previous two editions for the many useful suggestions for preparation of this second revised edition.

Y. Pomeranz
C. E. Meloan

1
GENERAL

1
Searching the Literature

Keeping informed of current developments is important to all professional workers, particularly those in a rapidly expanding area such as analytical chemistry. While no analyst can be familiar with all of the recent advances, he or she can learn how and where to look most effectively for the needed information. An attempt to survey the vast quantity of scientific literature may so frustrate inexperienced workers that they decide to ignore it altogether. Alternatively, they may sacrifice bench work to do practically endless reading and searching the growing literature.

In an effort to help analysts avoid either of these extremes, we outline the main sources of information and general approaches to a literature search in this chapter. By using an intelligent, systematic approach, analysts can be both "literature-conscious" and productive in the laboratory.

For over 300 years periodicals were the principal means for the exchange of scientific information. There are two basic types of scientific journals: (1) primary sources, which contain reports of original research and include detailed descriptions of the experimental procedures and the data derived from them and (2) secondary sources consisting of abstracts and reviews, which provide a condensed source of information.

Abstracts and indexes serve as keys to the contents of the periodical literature. Abstracts, which are short summaries of published articles, provide a quick way for readers to learn about current research and to decide which reports are most pertinent to their interests; indexes, which may list subjects, authors or formulas, assist in the search of published information. Most journals issue an index to each volume, generally annually, and some issue cumulative indexes covering several years. The most important compilations of abstracts and indexes cover a number of periodicals in a specific area, such as *Chemical Abstracts* in the area of chemistry. In recent years, considerable scientific and technical information has become available in audiovisual form. Listings of the audiovisual materials on foods and nutrition and of food and nutrition bibliographies available in the United States have been published by Oryx Press (Phoenix, AZ) for the National Agricultural Library, U.S. Department of Agriculture.

SEARCHING A NEW FIELD

General Background

If you wish to learn about a new field, the best source to consult at first is an advanced textbook. For general background information, several encyclopedias devoted to chemistry and technology are available. These include the technical and detailed Kirk-Othmer's *Encyclopedia of Chemical Technology* (Interscience Publishers) and Ullmann's *Enzyklopedie der Technischen Chemie* (Urban and Schwarzenberg Publishing) Ullmann's work is now available in English (Verlag Chemie); *The Encyclopedia of Science and Technology* (McGraw-Hill Book Co.) is less detailed but covers a wider field.

An excellent source of information is the single-volume *Merck Index of Chemicals and Drugs*. This compilation, revised every several years, contains formulas, preparations, and properties of over 10,000 chemicals.

In the identification and determination of food composition, the analyst often has to determine the physical properties of the substance. Relevant reference data can be obtained from various standard tables, handbooks, and some of the newer encyclopedias and dictionaries. Information on physical properties is often published by manufacturers of chemicals. In addition to industrial laboratories, the U.S. National Bureau of Standards publishes authoritative compilations of physicochemical data.

The *Handbuch der Organischen Chemie*, initiated by F. K. Beilstein, is the largest and most comprehensive source of information in organic chemistry. The 27 volumes of the main work published by the German Chemical Society included 200,000 entries covering the literature prior to 1910. Supplements are published periodically. The *Handbuch* deals with well-defined organic compounds and with natural materials of unknown structure. The extent of the coverage varies with the significance of the compound and the available information. Included are (whenever available) names, formulas, structures, history, occurrence, preparation, properties, technology, analysis, and reactions. In recent years, several new references on organic chemistry have been published, but the *Handbuch* continues to maintain its prominent position.

Reviews

Once you have acquired general background information in a new field, the next step is to consult a survey or a critical review of the current knowledge in the particular field. Monographs are basically comprehensive surveys of current knowledge on a specific subject. One of the best known is the *Advances in Chemistry Series* published by the American Chemical Society. Another monograph series is published in *Annals of the New York Academy of Science*. Both series are largely based on symposia organized by scientific societies. Other sources of review articles include annual review and special review periodicals, and special issues or selected parts of regular periodicals. Since 1959, *Chemical Abstracts* of the American Chemical Society has published annually a *Bibliography of Chemical Reviews*.

Since 1904, the Chemical Society (England) has published authoritative summaries of the previous years' important papers in *Annual Reports on the Progress of Chemistry*. A parallel series, *Reports on the Progress of Applied Chemistry*, is published by the Society of Chemical Industry. The reviews on foods are subdivided into the major areas and each section contains 100 to 200 references to original research papers, reviews, and proceedings of scientific conferences. Each year, the April issue of *Analytical Chemistry* is devoted to review papers; various aspects of food analysis are covered every second year. The reviews in *Analytical Chemistry* in areas such as biochemistry, clinical chemistry, water analysis, and various areas of inorganic, physical, and organic analyses also can provide useful background information for food analysts.

Among the most useful sources of current information, for the reader with sufficient background knowledge, are articles in annual publications such as *Advances in . . . , Annual Review of . . . , Progress in . . . ,* and *Methods in . . . .* Those pertinent to food analyses are listed at the end of this chapter.

Theses

Theses and dissertations contain comprehensive reviews, generally in a limited area. While most of the new information in theses is published in scientific journals, the literature reviews are shortened to reduce cost of publication. In many countries lists of higher education degree theses are published annually. Since 1938, theses from U.S. universities have been processed for microfilming by University Microfilms of Ann Arbor, Michigan. Abstracts of up to 600 words of such theses are published by *Dissertation Abstracts*. Theses of a chemical nature are listed in *Chemical Abstracts*, and entire theses can be purchased from University Microfilms. An annual list of U.S. Masters' Theses in the Pure and Applied Sciences has been published since 1955–1956 by the Thermophysical Properties' Research Center, Purdue University, Lafayette, Indiana.

Symposia

Programs of conferences and symposia are published in several journals. A comprehensive list is published in *Science*; more selective and limited lists are published in *Food Technology* and *Chemical & Engineering News*. In the food sciences, many such meetings are sponsored by government departments (e.g., various agencies of the U.S. Department of Agriculture). University presses, professional groups, commercial publishing houses, and private companies frequently reproduce the lectures and stimulating discussions of meetings in which prominent scientists present invited review papers in the area of their competence. Several renowned *Symposia on Foods* have been published as hardcover books by AVI Publishing Co.

Abstracts of papers presented at scientific meetings generally are published for members in the form of books by the sponsoring scientific society (e.g., American Chemical Society or Federation of Biological Sciences) or are included in periodicals (e.g., *Food Technology* for Institute of Food Technologists, *Cereal Foods World* for American Association of Cereal Chemists, and *Journal of the American Oil Chemists' Society* for AOCS).

Trade Publications

Although the purpose of most house organs and trade publications is to sell products, most such advertisements contain valuable information. The publications of many manufacturers are on a high scientific level. Some of these publications are the best sources on properties and applications of specialized equipment and chemicals. Several manufacturers periodically issue bibliographies and abstracts of technical–scientific articles in a specific area; some publish periodicals that reproduce pertinent articles from regular scientific journals; and some prepare detailed handbooks giving specifications, properties, and details of analytical procedures in selected areas (microbiology, enzymatic assay, electrophoresis, immunochemistry, automation in analytical chemistry). Industrial manuals are, of course, indispensable in installing, using, and servicing equipment. Several scientific journals periodically prepare lists of major commercial supply houses. A "Comprehensive Guide to Scientific Instruments" is published annually in *Science*.

Translations

Scientific journals continue to be the most important source of detailed information. Searching the literature becomes more difficult with the increased quantity of available literature and more publications in foreign languages. Many of the latter have a summary in English. In addition, active programs for translation from Slavic, Japanese, and Chinese literature are available.

Abstracts and Bibliographic Lists

Information on important publications is available in bibliographical lists or abstracting journals. *Chemical Abstracts* publishes over 200,000 abstracts a year, selected from about 10,000 journals in over 50 languages. The section on biochemistry covers various aspects of biology and chemistry of materials of plant, animal, and microbial origin. Included are informative abstracts of original scientific papers, patents, and some reviews; as well as lists of theses, monographs, books, reviews, and proceedings.

Several periodicals publish abstracts in a more restricted field. *Analytical Abstracts* and the abstract section of *Z. Analytische Chemie* are concerned primarily with analytical procedures. The abstract section in *Z. Lebensmitteluntersuchung Forschung* covers production, processing, storage, chemistry, and analyses of foods and agricultural products. The abstract sections in *J. American Oil Chemists' Society, J. Institute of Brewing*, and many others cover in a broad sense the respective areas. Food patents are abstracted in *Food Technology*. Some research organizations prepare excellent abstracts for member companies or for the general public. Information on cereals is abstracted in Germany by *Documenta Cerealia*.

Abstracts on current work (including theses) in East European countries may be found as an appendix to the journal *Die Nahrung*. Abstracts in the *Zentral Dokumentation Dienst Sozialistischer Länder-Nahrung und Ernahrung* are an excellent source of information on scientific work in foods and nutrition. Although some western periodicals are abstracted, the emphasis is on East European publications.

In addition to the comprehensive *Chemical Titles*, published by the

American Chemical Society, several publications provide more limited bibliographic lists. Selective lists are included in some periodicals such as *J. Chromatography* and *J. Lipid Research*. *Current Contents®* (Institute for Scientific Information, Philadelphia, PA) publishes seven weekly lists of publications in 6500 journals. The following four lists are of interest to food analysts:

Letter code	Coverage	Approximate number of journals covered
A	Agriculture, biology, environmental sciences	1000
E	Engineering, technology, applied sciences	770
L	Life sciences	1100
P	Physical, chemical, and life sciences	1300

Food analysts use primarily methods approved by various associations such as the Association of Official Analytical Chemists, American Association of Cereal Chemists, American Oil Chemists' Society, American Public Health Association, and American Society of Brewing Chemists (and their national or international counterparts). Most of the methods recommended by the United States and other organizations have been developed after years of collaborative testing and are considered reliable and official.

MODERN INFORMATION RETRIEVAL SYSTEMS

During the past few decades, computerized systems for storing and retrieving scientific information have been developed. An awareness of the systems available and how to use them properly can significantly reduce the time and effort involved in a "literature search" of a particular field.

Six papers presented at a symposium held during the thirty-fifth annual of the Institute of Food Technologists (IFT) in 1975 reviewed computer-based literature searches in industry, university and government (Burton 1976; Cuadra and Boyle 1976; Fisher 1976; Hopper 1976; Mann 1976; Stadelman 1976). The availability of databases on magnetic tape was described by Van Dyke and Ayer (1972) and Caponio and Moran (1975). Mermelstein (1977) described the available on-line retrieval systems as an alternative to manual searching, and Sze (1981) updated the area of information retrieval through computerized literature searching. Dialog Information Services, Inc. (Palo Alto, CA) has a broad information retrieval system that covers over 180 data bases. Several databases of particular interest to food scientists and technologists (available as of December 1984) are described in the remainder of this chapter.

1. *AGRICOLA*, 1970–present, 1,850,000 records (pieces of information), monthly updates (National Agricultural Library, Beltsville, MD). *AGRICOLA* (formerly *CAIN*) is the cataloging and indexing data base of the National Agricultural Library (NAL). This file provides coverage of journal and monographic literature on agriculture and related subjects. Since *AGRICOLA* represents the actual holdings of the National Agricultural Library, it provides substantial coverage of subject matter contained in a large library (see also Gilreath 1979; Turner 1983).

2. *BIOSIS PREVIEWS*, 1969–present, 4,100,000 records, biweekly updates (BioSciences Information Service, Philadelphia, PA). *BIOSIS PREVIEWS* contains citations from *Biological Abstracts* and *Biological Abstracts/RRM* (formerly *Bioresearch Index*), the major publications of the BioSciences Information Service. Together, these publications constitute the major English language service providing worldwide coverage of research in the life sciences. Over 9000 primary journals and monographs as well as symposia, reviews, preliminary reports, semi-popular journals, selected institutional and government reports, research communications, and other secondary sources provide citations on all aspects of the biosciences and medical research. Abstract searches are available for *Biological Abstracts* records from July, 1976, to the present.

3. *CA SEARCH*, 1967–present, 6,250,000 records, biweekly updates (Chemical Abstracts Service, Columbus, OH). The *CA SEARCH* database contains bibliographic data, keyword phrases, and index entries for documents covered by Chemical Abstracts Service. *CA SEARCH* is an expanded database that contains the basic bibliographic information in *Chemical Abstracts*. Index entries consist of CA General Subject Headings from a controlled vocabulary, the CAS Registry Numbers (a unique number assigned to each specific chemical compound), and modifying phrases. Additional keyword index phrases and cross-referenced CA General Subject Headings are also included.

4. *CAB ABSTRACTS*, 1972–present, 1,600,000 records, monthly updates (Commonwealth Agricultural Bureaux, Farnham Royal, Slough, England). *CAB ABSTRACTS* is a comprehensive file of agricultural and biological information containing records in the 26 main abstract journals published by Commonwealth Agricultural Bureaux. Over 8500 journals in 37 languages, as well as books, reports, and other publications, are scanned. In some instances less accessible literature is abstracted by sci-

entists working in other countries. About 130,000 items are selected for publication yearly; significant papers are abstracted, while less important works are reported with bibliographic details only.

5. *CRIS* (USDA), 40,000 records, monthly updates (U.S. Department of Agriculture, Washington, DC). *CRIS* (*Current Research Information System*) is a current-awareness database of research projects in agriculture and related sciences sponsored or conducted by USDA research agencies, state agricultural experiment stations, state forestry schools, and other cooperating state institutions. Currently active and recently completed projects within the last two years are included.

6. *DISSERTATION ABSTRACTS ONLINE*, 1861–present, 850,000 records, monthly updates (University Microfilms International, Ann Arbor, MI). *DISSERTATION ABSTRACTS ONLINE* is a subject, title, and author guide to dissertations accepted at accredited institutions in the United States. In addition, *DISSERTATION ABSTRACTS ONLINE* serves to disseminate citations for thousands of Canadian dissertations and an increasing number of papers accepted at institutions in other foreign countries.

7. *FOOD SCIENCE AND TECHNOLOGY ABSTRACTS*, 1969–present, 265,000 records, monthly updates (International Food Information Service, Reading, Berkshire, England). *FOOD SCIENCE AND TECHNOLOGY ABSTRACTS* (*FSTA*) provides access to research and new development literature in areas related to food science and technology. Allied disciplines such as agriculture, chemistry, biochemistry, and physics are also covered. Papers from related disciplines such as engineering and home economics are included when relevant to food science. *FSTA* provides indexing to over 1200 journals from over 50 countries, patents from 20 countries, and books in any language.

8. *FOODS ADLIBRA*, 1974–present, 75,000 records, monthly updates (Komp Information Services, Louisville, KY). *FOODS ADLIBRA* contains information on developments in food technology and packaging. New food products introduced since 1974 are covered, and nutritional and toxicology information is also included. Major significant research, such as technological advances in processing methods and packaging, are reported. Brief abstracts from over 250 trade periodicals constitute the bulk of the *FOODS ADLIBRA* database; over 500 highly technical research journals are scanned for additional information. Patents from the United States and some British patents are included. *FOODS ADLIBRA* is a useful source of information on government guidelines and regulations on the processing and packaging of foods. Marketing and management news and statistics as well as information on world food economics can also be found in this source.

9. *LIFE SCIENCES COLLECTION*, 1978–present, 490,000 records, monthly updates, (Cambridge Scientific Abstracts, Bethesda, MD). *LIFE*

SCIENCES COLLECTION contains abstracts of information in the fields of animal behavior, biochemistry, ecology, entomology, genetics, immunology, microbiology, toxicology, and virology. The worldwide coverage is of journal articles, books, conference proceedings, and report literature.

10. *NTIS*, 1964–present, 1,100,000 records, biweekly updates (National Technical Information Service, U.S. Department of Commerce, Springfield, VA). The *NTIS* database consists of government-sponsored research, development, and engineering plus analyses prepared by federal agencies, their contractors or grantees. This database is the means through which unclassified, publicly available, unlimited distribution reports are made available for sale. State and local government agencies are now beginning to contribute their reports to this file.

BIBLIOGRAPHY

Burton, H. D. (1976). Computer-based literature searching at the USDA/ARS. *Food Technol.* **30**(5), 70, 72.

Caponio, J. F., and Moran, L. (1975). CAIN: A computerized literature system for the agricultural sciences. *J. Chem. Inf. Comput. Sci.* **15**(3), 158–161.

Cuadra, C. A., and Boyle, H. (1976). On-line information services for food science and technology. *Food Technol.* **30**(5), 60, 62, 63.

Fisher, D. A. (1976). Keeping current through information services. *Food Technol.* **30**(5), 66, 68.

Gilreath, C. L. (1979). "Agricola Users' Guide. Agricultural Reviews and Manuals." ARM-H-7. Science Educ. Admin., U.S. Dept. Agric., Beltsville, MD.

Hopper, P. F. (1976). Information systems in industry. *Food Technol.* **30**(5), 74, 75, 76.

Mann, E. J. (1976). The international food information service. Past, present, and future. *Food Technol.* **30**(5), 54, 56, 58.

Mermelstein, N. H. (1977). Retrieving information from the food science literature. *Food Technol.* **31**(9), 46–48, 52–55.

Stadelman, W. J. (1976). Information systems in the university. *Food Technol.* **30**(5), 78, 84.

Sze, M. C. (1981). Meeting the information needs of food scientists through computerized literature searching. *Food Technol.* **35**(10), 92–97.

Turner, P. A. (1983). The use of mini and macrocomputers at the National Agricultural Library. *Agric. Libr. Inf. Notes* **9**(9), 1–5.

Van Dyke, V. J., and Ayer, N. L. (1972). Multipurpose cataloging and indexing system (CAIN) of the National Agricultural Library. *J. Libr. Autom.* **5**(1), 21–29.

Reviews

Advances in Agronomy
Advances in Analytical Chemistry and Instrumentation
Advances in Carbohydrate Chemistry
Advances in Cereal Science and Technology

Advances in Chemistry Series
Advances in Chromatography
Advances in Clinical Chemistry
Advances in Colloid Science
Advances in Comparative Biochemistry and Physiology
Advances in Enzymology and Related Subjects of Biochemistry
Advances in Food Research
Advances in Lipid Research
Advances in Protein Chemistry
Annual Reports on Progress of Chemistry
Annual Review of Biochemistry
Annual Review of Microbiology
Annual Review of Nutrition
Annual Review of Physiology
Bacteriological Reviews
Biological Reviews
Botanical Reviews
Chemical Reviews
Chromatographic Reviews
Critical Reviews in Foods and Nutrition
Methods of Biochemical Analysis
Methods in Enzymology
Nutrition Abstracts and Reviews
Physiological Reviews
Progress in the Chemistry of Fats and Other Lipids
Progress in Food and Nutrition Science
Progress in Lipid Research
Recent Advances in Food Science
Reports on the Progress of Applied Chemistry
Yearbook of Agriculture—U.S. Dept. Agriculture

Abstracts, Bibliographies, Indexes

Agricultural Index
Analytical Abstracts
Applied Science and Technology Index
Bibliographic Index
Bibliographic Current List of Papers, Reports, and Proceedings of Inter-
    national Meetings
Bibliography of Agriculture
Bibliography of Chemical Reviews
Biological Abstracts
Chemical Abstracts

Chemical Titles
Current Chemical Papers
Current Contents
World Bibliography of Bibliographies

Periodicals

Acta Chemica Scandinavia
Agricultural and Biological Chemistry (Tokyo)
Agronomy Journal
Alimenta
Analyst, The
Analytical Biochemistry
Analytical Chemistry
Analytica Chimica Acta
Angewandte Chemie
Annalen der Chemie
Annales des Falsifications et des Fraudes
Annals of the New York Academy of Sciences
Applied Microbiology
Applied Spectroscopy
Archives of Biochemistry and Biophysics
Australian Journal of Biological Sciences
Biochemical and Biophysical Research Communications
Biochemical Journal
Biochemistry
Biochimica et Biophysica Acta
Biotechnology and Bioengineering
British Journal of Nutrition
Bulletin de la Societe Chimique de France
Bulletin de la Societe de Chimie Biologique
Canadian Institute of Food Science and Technology Journal
Canadian Journal of Biochemistry and Physiology
Carbohydrate Research
Cereal Chemistry
Cereal Foods World
Chemische Berichte
Chemistry and Industry (London)
Chemistry and Physics of Lipids
Clinica Chimica Acta
Electroanalytical Chemistry
Endeavour
Enzymologia

Ernaehrungsforschung
European Journal of Biochemistry
Experientia
Federation Proceedings
Fette, Seifen, Anstrichmittel
Food Engineering
Food Manufacture
Food Microstructure
Food Science
Food Technology
Getreide, Mehl, und Brot
Helvetica Chimica Acta
Hoppe Seylers' Zeitschrift fur Physiologische Chemie
Industrial and Engineering Chemistry
Journal of Agricultural and Food Chemistry
Journal of Bacteriology
Journal of Biochemistry (Tokyo)
Journal of Biological Chemistry
Journal of Carbohydrate Chemistry
Journal of Cereal Science
Journal of Chromatography
Journal of Dairy Research
Journal of Dairy Science
Journal of Food Biochemistry
Journal of Food Science
Journal of Food Science and Technology
Journal of Food Technology
Journal of Gas Chromatography
Journal of General Microbiology
Journal of Histochemistry and Cytochemistry
Journal of Lipid Research
Journal of Milk and Food Technology
Journal of Molecular Biology
Journal of Nutrition
Journal of Texture Studies
Journal of the American Chemical Society
Journal of the American Oil Chemists' Society
Journal of the Association of Official Analytical Chemists
Journal of the Association of Public Analysts
Journal of the Chemical Society
Journal of the Institute of Brewing
Journal of the Science of Food and Agriculture
Laboratory Investigations

Laboratory Practice
Lebensmittelwissenschaft und Technologie
Lipids
Microchemical Journal
Microchimica Acta
Milling
Mitteilungen aus dem Gebiete der Lebensmitteluntersuchung und Hygiene
Muhle, Die
Nahrung, Die
Nature
Naturwissenschaften, Die
Nutrition Reports International
Phytochemistry
Plant Physiology
Poultry Science
Proceedings of the National Academy of Science, U.S.
Proceedings of the Nutritional Society of England and Scotland
Proceedings of the Society of Experimental Biology and Medicine
Qualitas Plantarum—Plant Foods for Human Nutrition
Science
Staerke, Die
Stain Technology
Transactions of the New York Academy of Science
Zeitschrift fur Analytische Chemie
Zeitschrift fur Ernaehrungswissenschaft
Zeitschrift fur Lebensmittel Untersuchung und Forschung
Zeitschrift fur Naturforschung

2
Sampling

The validity of the conclusions drawn from the analysis of a food depends, among other things, on the methods used in obtaining and preserving the sample. Sampling and any subsequent separations are the greatest sources of error in food analyses. An ideal sample should be identical in all of its intrinsic properties with the bulk of the material from which it is taken. In practice, a sample is satisfactory if the properties under investigation correspond to those of the bulk material within the limits set by the nature of the test.

According to Kratochvil and Taylor (1981), the major steps in sampling are (1) identification of the population from which the sample is to be obtained, (2) selection and obtaining of gross samples of the population, and (3) reduction of each gross sample to a laboratory-size sample suitable for analysis. It has been shown that if the analytical uncertainty is less than one-third of the sampling uncertainty, additional reduction of the analytical uncertainty is of little significance. For random errors, the overall standard deviation s_o is related to the standard deviation for the sampling operation, s_s, and to that of the remaining analytical operations, s_a, by the expression $s_o^2 = s_a^2 + s_s^2$.

If the sampling uncertainty is very large because of the heterogeneity of the population, examining a great number of samples by a rapid screening method of relatively low precision may be the best answer to reducing the total uncertainty. This principle was demonstrated for the determination of mycotoxins by Schuller *et al.* (1976). At very low levels of aflatoxin, the standard deviation of sampling is well over 80% of the total standard deviation, as one badly infected kernel (of peanuts, for example) can contaminate a relatively large lot of relatively sound kernels and raise the mycotoxin level above the acceptable level.

Kratochvil and Taylor (1981) distinguish several types of samples. *Random samples*, in which the bulk of the material is divided into a number of real or imaginary segments and sample increments (for a bulk sample), are selected according to a predetermined pattern. Sampling at evenly spaced intervals over the bulk is simple and widely used but subject to large bias. *Systematic samples* are collected to test changes in composition with time, temperature, spatial location, or treatment. The results are tested for statistically significant differences. *Representative samples* can be obtained only from truly homogeneous materials. Since these are seldom available, *composite samples* often are used instead. Obtained by special techniques designed to produce representative samples, they are most useful when the average composition of a lot is of main interest. A glossary of terms used in sampling is given in Table 2.1.

Benton-Jones and Steyn (1973) developed a comprehensive list of recommendations for sampling field and vegetable crops, fruits, and nuts. They indicate the stage of growth, plant part, and number of plants to sample, and describe in detail the preparative steps to be followed before the plant material is subjected to actual chemical analyses. In addition to information of the storage and transport of fresh plant material before cleaning and drying and on storage of tissue powder before analysis, they include procedures for (1) cleaning to remove tissue surface contamination, (2) drying to stop enzymatic reactions, (3) actual grinding to reduce particle size for analysis, and (4) drying to a constant weight. Soil and plant tissue preparation procedures were reviewed also by Lockman (1980), and automated sample preparation was described by Burns (1981).

SAMPLES

Samples for analysis should be large enough for all intended determinations. Homogeneous samples of 250 g (or ml) are generally sufficient. Samples of spices are often limited to 100 g, and those of fruits and vegetables increased to 1000 g. Samples should be packed and stored in such a way that no significant changes occur from the moment of sampling until the analysis is completed. The container should be identified clearly. Official and legal samples must be sealed in such a way that they cannot be opened without breaking the seal.

The quality control laboratory analyzes various types of samples (Pearson 1958). *Raw materials* are analyzed to determine whether the delivery approximates previous deliveries or if the material from a new supplier is up to the buying sample. *Process control samples* are generally analyzed

Table 2.1.   Glossary of Sampling Terms[a]

Term	Definition
Bulk sampling	Sampling of a material that does not consist of discrete, identifiable, constant units, but rather of arbitrary, irregular units
Composite	A sample composed of two or more increments
Gross sample	Also called bulk sample, lot sample; one or more increments of material taken from a larger quantity (lot) of material for assay or record purposes
Homogeneity	The degree to which a propety or substance is randomly distributed throughout a material; homogeneity depends on the size of the units under consideration; thus a mixture of two minerals may be inhomogeneous at the molecular or atomic level, but homogeneous at the particulate level
Increment	An individual portion of material collected by a single operation of a sampling device, from parts of a lot separated in time or space; increments may be either tested individually or combined (composited) and tested as a unit
Individuals	Conceivable constituent parts of the population
Laboratory sample	A sample, intended for testing or analysis, prepared from a gross sample or otherwise obtained; the laboratory sample must retain the composition of the gross sample; often reduction in particle size is necessary in the course of reducing the quantity
Lot	A quantity of bulk material of similar composition whose properties are under study
Population	A generic term denoting any finite or infinite collection of individual things, objects, or events in the broadest concept; an aggregate determined by some property that distinguishes things that do and do not belong
Reduction	The process of preparing one or more subsamples from a sample
Sample	A portion of a population or lot; may consist of an individual or groups of individuals
Segment	A specifically demarked portion of a lot, either actual or hypothetical
Strata	Segments of a lot that may vary with respect to the property under study
Subsample	A portion taken from a sample; a laboratory sample may be a subsample of a gross sample; similarly, a test portion may be a subsample of a laboratory sample
Test portion	Also called specimen, test specimen, test unit, aliquot; that quantity of a material of proper size for measurement of the property of interest; test portions may be taken from the gross sample directly, but often preliminary operations, such as mixing or further reduction in particle size, are necessary

[a] From Kratochvil and Taylor (1981).

by rapid in-plant tests (e.g., refractometer, hydrometer) as a guide to processing adjustments needed to produce an acceptable and uniform product. Periodic checks of *finished products* show whether the food meets legal requirements, is acceptable to the consumer, and has reasonable shelf life. *Buying samples* are submitted by suppliers of raw materials prior to delivery. Most *complaint samples* are submitted by customers. In a competitive market, information about products being sold by other manufacturers is of interest to management. The composition of *competitors' samples* is also valuable in developing new products.

According to Kramer and Twigg (1970), factors that determine selection of a sampling procedure include (1) *purpose of inspection*—acceptance or rejection, evaluation of average quality, and determination of uniformity; (2) *nature of lot*—size, division into sublots, and loading or stacking; (3) *nature of test material*—its homogeneity, unit size, previous history, and cost; and (4) *nature of test procedures*—significance, destructive or nondestructive assay procedures, and time and cost of analyses.

STATISTICAL CONCEPTS

Statistically sound sampling plans and procedures have been developed to provide the most efficient technique in terms of cost and information. Details of such procedures are readily available (Barlett and Wegener 1957; Bowman and Remmenga 1965; Bowker and Goode 1952; Bureau of Ordnance 1952; Cochran 1963; U.S. Dep. Defense 1963; Dodge and Romig 1944; Kramer and Twigg 1966; U.S. Dept. Agr. 1964). In this section, we summarize the basic statistical concepts involved in scientific sampling procedures.

In a series of n observations in which x is the value of a single observation and $\bar{x}$ the mean (average) of n observations, d is the deviation of a single observation from the mean $d = x - \bar{x}$. The standard deviation is $s_0 = (\sum d^2/n)^{1/2}$. If the total number of items is N, the true mean is $\mu = \sum x/N$, the true standard deviation is $\sigma = (\sum d^2/N)^{1/2}$, and $\bar{x} = \sum x/n$ is an estimate of the true mean. Generally, s_0 is smaller than σ, as it is derived from a smaller number of observations and is less likely to include all extreme values. To account for this, for n smaller than 10, s is calculated as $s = [\sum d^2/(n - 1)]^{1/2}$ and gives a better estimate of σ than s_0.

The dimensions of s are the same as of x. The deviation can also be expressed as $t = d/s$. This is useful in calculating the fraction of a sampled lot that varies from the mean by a certain value. It has been found that one standard deviation ($t = 1$) represents a deviation on both sides of the mean (total range of $2s$) within which 68.2% of all observations will occur for a population with a normal distribution provided a sufficiently large (>25) number of observations is made; 95.4% of all observations fall within a range of $t = 2$ (or $4s$); and 99.7% within a range of $t = 3$ (or $6s$). This means that the chance that a single observation will not exceed $3s$ is over 99%, and the chance that the error will not exceed s is 68%. The standard error $s_{\bar{x}}$ (standard deviation of the sample mean) decreases as the number of samples increases according to the relation $s_{\bar{x}} = s/n^{1/2} = [\sum d^2/n(n - 1)]^{1/2}$. Thus, if s is known, one can determine the number of samples (or sample increments) required for a desired standard error $s_{\bar{x}}$. As a result of the nonlinear relation between $s_{\bar{x}}$ and n, only seldom is it justified to increase n above 20. In practice, in 20 observations the range covered is about $3s$, i.e., three times the standard deviation.

From the equation $n = (ts/e)^2$ it is possible to calculate the number of observations n with the standard deviation s required to have a confidence that the error e of the observed mean will not exceed a certain magnitude. In that equation, t is the confidence or probability factor; when $t = 1$ the probability is 68%, when $t = 2$ it is 95%, etc. The above is true provided the analytical error is substantially smaller than the sampling error. Generally, a confidence of 95% ($t = 2$) is satisfactory; for a confidence of 99%, $t = 2.6$.

MANUAL SAMPLING

Samples are frequently taken manually. Apparently homogeneous materials such as single-phase liquids or well-mixed powders should be mixed thoroughly immediately before sampling. Small quantities of powders or solutions can be mixed by rotating and shaking in a closed container that has a volume at least twice that of the sample. Mixing may also be accomplished by pouring the material several times from one container to another. Laboratory samples of powders or ground materials may be obtained by quartering of thoroughly mixed samples, discarding two opposite quarters, remixing the remaining material, and repeating the process till the sample is reduced to a desired size. Sample dividers that mechanically mix and divide powdered or granular materials (Fig. 2.1) may be purchased from several apparatus supply houses.

Granular or powdered solids are generally sampled by probes and triers (Figs. 2.2 and 2.3). Liquids require thorough mixing before sampling. Partly or completely frozen, crystallized, or solidified fluids must be liquefied completely and mixed. If such mixing is practically unattainable, samples must be taken at various heights. Examples of liquid samplers are shown in Fig. 2.4. Milk must be thoroughly mixed because the fat rises to the top and the composition changes on standing. On the other

Fig. 2.1. Boerner sampler used to separate grain into laboratory-size samples. (Courtesy of Seedburo Equipment Co.)

Fig. 2.2. Spiral probe for sampling soybeans. (Courtesy of Seedburo Equipment Co.)

Fig. 2.3. Triers for sampling sacked rice (top) and cottonseed (bottom). (Courtesy of Seedburo Equipment Co.)

Fig. 2.4. Liquid sampler with graduated extension tube that telescopes inside the plunger and locks at any point to determine the sampling distance from the bottom. Valve lifts when the bottom of the tank is reached or the valve may be lifted by a cord attached to the valve plunger to collect samples from intermediate levels. (Courtesy of Seedburo Equipment Co.)

hand, excessive mixing of cream is inadvisable. Butter and hard cheese samples are generally taken with a stainless steel borer; soft cheeses are sampled by cutting out a representative segment. The greatest difficulty arises in sampling large fruits and vegetables. Often selection of a large number of individual units, to compensate for variation, is required.

CONTINUOUS SAMPLING

In continuous quality control operations, manual samplers have been replaced by mechanical samplers. There are three basic types of mechanical samplers (Johnson 1963). The *riffle cutter* is composed of equally spaced dividers designed to remove continuously a small fraction of the stream (Fig. 2.5). Usually, a riffle divides the stream equally and the sample is passed through the same riffle or successive riffles for further proportional reduction of the sample to a quantity convenient for analysis. This device is commonly used in laboratories in cutting and quartering a larger sample. The *circular* (or Vezin) sampler can be used for intermittent or continuous sampling and is suitable for both wet and dry materials. The cutter appears as a truncated wedge of a circle that passes through the falling stream once each revolution. Size of the segment or cutter opening determines the size of the sample. The sample is large (5–10%

Fig. 2.5. In the riffle-type sampler, material poured into the hopper is divided into two equal portions by two series of chutes, which discharge alternately in opposite directions into separate pans.

of the entire stream) and a secondary sampling is generally required. If the feed stream is relatively homogeneous, the sampler can be converted to an intermittent type engaged by a magnetic brake and timer. The most popular and least expensive sampling unit is the *straight-line* sampler, which can be operated either intermittently or continuously. In this type of sampler, the cutter moves in a straight line and at uniform speed across the entire stream.

The factors that affect the amount of sample taken by a sampler include the size of the cutter opening, the speed of the cutter as it travels through the stream, and the frequency of making a sample cut. Generally, to allow free entrance into the cutter, the minimum width of a cutter opening should be three times the size of the largest particles to be sampled. The cutter should move through the stream at a speed that gives the smallest possible sample without deflecting particles that should enter the cutter. The amount of sample taken per cut (Q) can be determined by the formula $Q = PW/L$ where P is the feed speed (lb/sec), W the cutter width (in.), and L the cutter speed (in./sec; generally about 30). Frequency of sampling depends on the uniformity and homogeneity of the sampled material. Mechanical sampling may be combined with high-speed analysis and automated processing adjustment. Sampling at high flow rates may require secondary and even tertiary subsampling to reduce the primary sample to a convenient quantity for analysis.

SAMPLING ERRORS

Sampling errors are caused by several factors. Lack of randomness in sample selection may result from both instrumental limitations or deficiencies and from human bias.

Manual methods of sampling powdered or granular materials are subject to numerous errors (Quackenbush and Rund 1967). Baker *et al.* (1967) studied the factors that bias sampling by triers. These factors include (1) *particle shape*—round particles flow into the sampler compartments more readily than angular particles of similar size; (2) *surface adhesiveness*—an uncoated hygroscopic material flows into the sampler compartment more readily than nonhygroscopic materials of similar shape and of either larger or smaller size; and (3) *differential downward movement* of particles (on the basis of size) when disturbed during sampling. Consequently, horizontal cores contain a higher proportion of smaller-sized particles than vertical cores. The latter is probably also the main bias factor affected by diameter and opening size of sampling tube.

Changes in composition may occur during or after sampling. Typical changes include gain or loss of water, loss of volatiles, physical inclusion of gases, reaction with container material or foreign matter in container, and damage to fruits or vegetables by mechanical injury leading to enhanced enzymatic or chemical changes.

The main problem arises, however, from the nonhomogeneity of many foods. Both macroheterogeneity (among various units of a lot) and microheterogeneity (within various parts of a unit) are common. The latter is especially important in determinations of vitamins and other minor components. For example, nicotinic acid and thiamine are concentrated in the aleurone and scutellum tissues of the wheat kernel; epidermis cells of grapes are rich in anthocyanin pigments; and essential oils in citrus fruits are mainly in cells of the flavedo layer. But major components also are distributed unevenly. Proteins, lipids, minerals, and crude fiber are higher in the outer layers than in the endosperm of cereal grains; variations in water, sugars, and organic acids are found in various tissues of fruits and vegetables; and the uneven distribution of fat in meat makes it imperative to express some analytical values on a fat-free basis.

Most of these difficulties are overcome by fine grinding and mixing of large samples. In some instances, however, attempts to homogenize a food sample are wrought with difficulties; in others, apparently homogeneous preparations have a tendency to segregate or stratify. Failure to recognize and appraise the variations in a sample may limit or even invalidate conclusions from analytical data.

To summarize, the aim of sampling is to secure a portion of the material that satisfactorily represents the whole. The more heterogeneous the material, the greater the difficulties and required efforts to obtain a truly representative sample.

BIBLIOGRAPHY

Baker, W. L., Gehrke, C. W., and Krause, G. F. (1967). Mechanisms of sampler bias. *J. Assoc. Off. Anal. Chem.* **50**, 407–413.

Barlett, R. P., and Wegener, J. B. (1957). Sampling plans developed by U.S. Dept. of Agr. for inspection of processed fruits and vegetables. *Food Technol.* **11**, 526–532.

Benton-Jones, J. Jr., and Steyn, W. J. A. (1973). Sampling, handling, and analyzing plant tissue samples. *In* "Soil Testing and Plant Analysis," rev. ed., pp. 249–270. Soil Science Society of America, Madison, WI.

Bowker, A. H., and Goode, H. P. (1952). "Sampling Inspection by Variables." McGraw-Hill, New York.

Bowman, F., and Remmenga, E. E. (1965). A sampling plan for determining quality characteristics of green vegetables. *Food Technol.* **19**, 617–619.

Burns, D. A. (1981). Automated sample preparation. *Anal. Chem.* **53**, 1402A, 1404A, 1406A, 1408A, 1410A, 1412A, 1414A, 1417A, 1418A.

Bureau of Ordnance. (1952). Sampling Procedures and Tables for Inspection Variables. NAVORD-OSTD-80. Bur. Ordnance, Dept. Navy. USGPO, Washington, DC.

Cochran, W. G. (1963). "Sampling Techniques," 2nd ed. Wiley & Sons, New York.

Dodge, H. F., and Romig, H. G. (1944). "Sampling Inspection Tables." Wiley & Sons, New York.

Johnson, N. L. (1963). Sampling devices. *Food Technol.* **17**, 1516–1520.

Kramer, A., and Twigg, B. A. (1970). "Fundamentals of Quality Control for the Food Industry," 3rd ed., Vol. 1. AVI Publishing Co., Westport, CT.

Kratochvil, B., and Taylor, J. K. (1981). Sampling for chemical analysis. *Anal. Chem.* **53**, 924A–926A, 928A, 930A, 934A, 936A, 938A.

Lockman, R. B. (1980). Review of soil and plant tissue preparation procedures. *J. Assoc. Off. Anal. Chem.* **63**, 766–769.

Pearson, D. (1958). Food analysis—techniques, interpretation and legal aspects. IV. The quality control chemist in the food industry. *Lab. Pract.* **7**, 92–94.

Quackenbush, F. W., and Rund, R. C. (1967). The continuing problem of sampling. *J. Assoc. Off. Anal. Chem.* **50**, 997–1006.

Schuller, P. L., Horwitz, W., and Stoloff, L. (1976). Review of aflatoxin methodology. A review of sampling plans and collaboratively studied methods of analysis for aflatoxins. *J. Assoc. Off. Anal. Chem.* **59**, 1315–1343.

U.S. Dep. Agric. (1964). U.S. standards for sampling plans for inspection by attributes—single and double sampling plans. *Fed. Regist. 29-FR-5870.*

U.S. Dep. Defense. (1963). "Sampling Procedures and Tables for Inspection of Attributes." *105D.* Mil. Std.

3
Preparation of Samples

The care, time, and effort devoted to the preparation of samples for analysis should be commensurate with the information required and the accuracy and precision of the analytical results desired. If the sample is not prepared properly for analysis, or if the components become altered during preparation, the results will be inaccurate regardless of the effort, the precision of the apparatus, and the techniques used in the analysis (Entenman 1961).

The purpose of sample preparation is to mix thoroughly a large sample in the laboratory. This apparently homogeneous sample must then be reduced in size and amount for subsequent analysis. Grier (1966) reported results of a survey based on questionnaires submitted to 200 plant scientists concerning the materials used in testing (from dry, stemmed hay to wet fruits); mills for size reduction of dry materials (Wiley mills, hammer mills, choppers); equipment for size reduction of wet material (food choppers, blenders, high-speed mixers); and ovens and lyophilizers for drying. The problems encountered by the analysts in the preparation of samples for analyses included difficulties in obtaining representative small samples from large samples; loss of plant material; removal of extraneous material from plants without removal of plant constituents; enzymatic changes before and during analyses; compositional changes during grinding; metal contamination during grinding; changes in unstable components (e.g., chlorophylls, unsaturated fatty acids); and special preparation problems in analyses of oil seed materials.

Both the nature of the food and the analyses to be performed must be considered in the selection of instruments for grinding. An analysis can sometimes be made directly on fresh or dried material. Total nitrogen by the Kjeldahl method and total ash determinations generally require little disintegration, other than that necessary to provide a homogeneous and representative sample (Pirie 1956). In practice, however, some disintegration is required, and for this a wide range of techniques and equipment is available.

For determination of moisture, total protein, and mineral contents, dry foods are generally ground to pass a 20-mesh (openings per linear inch) sieve. For assays involving extraction (lipids, carbohydrates, various forms of protein), samples are ground to pass a 40-mesh sieve. The advantages of using finer ground samples are outweighed by losses of total material and moisture, heating and undesirable chemical modifications, and even difficulties in assay (as in clogging of filter pores in extraction or filtration).

GRINDING DRY MATERIALS

Mechanical methods for grinding dry materials range from the simple pestle and mortar to elaborate and effective devices for grinding. For fine grinding of dry materials, power-driven hammer mills are widely used. To control the fineness of grinding, various screens through which the ground material must pass are inserted (Fig. 3.1). Hammer mills are rugged and efficient, and not easily damaged by stones and dirt, which may contaminate some samples. They are used to grind such materials as cereals, oil meals, and most foods that are reasonably dry and do not contain excessively high amounts of oil or fat.

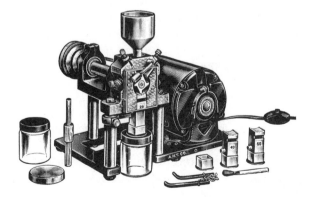

Fig. 3.1. Wiley Mill, intermediate, uses interchangeable sieve top-delivery tubes of 10–80 mesh per inch. (Courtesy A. H. Thomas Co.)

The fineness of grinding is largely affected by the composition of the food, especially the moisture content. Generally, foods are ground better after drying in a desiccator or vacuum oven. In some instances, however, pulverizing air-dried materials gives best results.

For small samples, especially if fine grinding is required, ball mills are used. The material is ground in a container half-filled with balls. The container rotates constantly and the balls exert an impact-grinding action. The main objection to such mills is that they require many hours (or even days) for satisfactory grinding.

Grinding of wet samples may result in significant losses of moisture and heat generation, accompanied by chemical changes. Grinding, especially of hard materials, may result in serious contamination from the abrasives of the mill. This may lead to serious errors, especially in determinations of mineral components. The contamination can be minimized, or at least recognized, if the working parts of the grinder are made of a resistant material (glass, ceramic, agate) or of a material that can be easily determined.

Chilled ball mills can be used to grind frozen materials without preliminary drying. Grinding of frozen foods reduces undesirable chemical changes (e.g., Maillard reaction involving interaction between sugars and amino acids) at room temperature. Clemments (1959) described a grinder for frozen, resilient plant tissues. Similar grinders can be used for grinding bread crumbs and samples of foods in which heat-labile components are to be studied.

GRINDING MOIST MATERIALS

For disintegration of moist materials, various fine-slicing devices are available. Some moist materials are disintegrated best by bowl cutters (for leafy vegetables and fleshy tubers and roots) or meat mincers (for fruits, roots, and meat products). Any of these devices can be either hand- or power-driven. In addition, power-driven pestles and mortars, especially if an abrasive (such as sand) is added, can give fine subdivision of moist materials.

The commercially available tissue grinders are used for small samples of soft material (Fig. 3.2). The sample is forced through the annulus between two concentric cylinders, one of which is driven mechanically. In addition to batch grinders, choppers and graters for continuous food disintegration are available.

For grinding dilute suspensions and most soft and pasty foods, various modifications of the Waring blendor can be used. In these instruments, knives rotating at up to about 25,000 rpm disintegrate a sample suspended

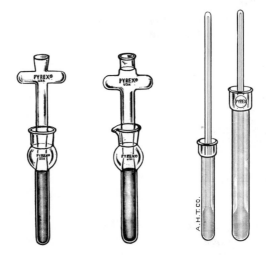

Fig. 3.2. Tissue grinders from left to right: Ten-Broeck plain; Ten-Broeck with pour-out spout; and homogenizers with piston-type, ground glass pestle. (Courtesy A. H. Thomas Co.)

in an extractant. Vessels to accommodate 10–2000 ml are available. Although blenders are ideal for routine disintegration of tissues for most analytical determinations, they cannot homogenize a thick slurry or highly texturized material because the rotating knives form a cavity in such materials. Blenders have been used successfully for disintegration (and sometimes simultaneous extraction) of such diverse materials as oil seeds, creamed cottage cheese (Perlmutter 1953), French dressing (Ratay 1953), and a great variety of animal and plant products (Jones and Ferguson 1951; Benton-Jones and Steyn 1973; Lockman 1980).

In several types of colloid mills, the dilute suspension is pumped through a controlled gap between smooth or slightly serrated surfaces until the particles have been sufficiently disintegrated by shear.

Sonic and supersonic vibrations have been adapted for the dispersion of foods. Attempts have been made to disintegrate a tissue by saturating it with gas under pressure and then allowing the gas to expand by suddenly releasing the pressure. Shaking suspensions with pure sand or small glass beads for wet grinding in a ball mill is often highly efficient. This principle is the basis of the Mickle disintegrator.

ENZYMIC AND CHEMICAL TREATMENT

In addition to mechanical methods, enzymic and chemical procedures can be used to disintegrate various types of materials. Pure cellulases are particularly useful in preparing materials of plant origin; proteases and carbohydrases can be used to solubilize high-molecular-weight components in many foods. Dimethylformamide, urea, pyridine, phenol, dimethyl sulfoxide, synthetic detergents, and reducing agents (for disulfide bond cleavage) are some examples of chemicals that are effective in dispersing or solubilizing foods or food components for analysis.

ENZYME INACTIVATION

One of the problems facing a food analyst is enzymatic modification following sampling or the preparation of samples for analyses. Generally, if total contents of a specified compound (e.g., minerals, carbohydrates, nitrogen) are determined, enzyme inactivation is not essential. If, however, various forms of a compound (sugars, free and bound forms of lipids, groups of proteins) are to be determined, the tissue must be treated in such a way that potentially troublesome enzymes are immediately and completely inactivated.

Whenever possible fresh material should be analyzed. To preserve the components in a living tissue in their original state, several methods of enzyme inactivation can be used. The treatment required to achieve enzyme inactivation varies widely with the size, consistency, and composition of the material; the enzymes present; and the intended analytical determinations. For example, fungal amylases are generally heat labile and can be inactivated at relatively low temperatures; some bacterial amylases are highly resistant and may survive bread baking temperatures. Extraction of chlorogenic acid from seeds or dry tissues requires heating to 90–100°C for 1 hr to inactivate polyphenolases. Juicy tissues should be heated for a time that will result in a temperature of 90–100°C for 5–10 min in the center of the sample.

Drying should be as rapid and at as low a temperature as possible. Such drying is greatly facilitated by spreading the sample over a wide area. Generally, drying at 60°C under vacuum is recommended. If the sample contains no heat-sensitive or volatile compounds, heating for several minutes at 70–80°C may be advisable. Such heating inactivates most enzymes and modifies cells to enhance the rate of their drying. During drying, enzymes and vitamins usually are destroyed, proteins and lipids are almost invariably modified, and some flavor components are volatilized. If drying is not done carefully, caramelization and sugar inversion in acid foods are likely to occur. Case-drying, as a result of formation of an impervious outside layer, is often quite troublesome.

The difficulties encountered in drying plant materials were illustrated by the investigations of Gausman et al. (1952). Unripe corn kernels were dried at three temperatures (44, 54, and 83°C). Drying at the higher temperatures, especially in samples containing high moisture levels (50% or above), lowered the apparent starch contents and increased the sugar contents. Pyridoxal and pantothenic acid levels were decreased, but apparent riboflavin and nicotinic acid levels were higher after drying at elevated than at room temperatures.

Some plant materials can be stored at -20 to $-30°C$ provided they can be cooled to the low temperature within 1 hr. However, plant acid phosphatases function at $-28°C$ in the frozen state and in 40% methanol solution. Tissues or extracts taken for studies of phosphate metabolism can be stored in the cold after drying (Bieleski 1964). Most foods are preserved best (even for long-term storage) by freeze drying. But freezing alone does not stop enzyme action. Actually, slow freezing disrupts tissues and often enhances enzymatic changes. Fresh foods, in which the enzymes have not been inactivated prior to freezing, are especially susceptible to enzymatic attack during and after thawing.

Some enzymes can be inactivated by inorganic compounds that cause irreversible enzyme poisoning; by a shift in pH, provided other compounds are not affected by high acidity or alkalinity; or by salting out, i.e., by high concentrations of ammonium sulfate. The most common methods of halting enzyme action in analyses of plant tissues include treatment with 80% methanol or ethanol, ice-cold 5 or 10% perchloric or trichloracetic acid, or a mixture of methanol–chloroform–2 M formic acid (12:5:3, by volume) (Bieleski 1964). Most commonly, the plant material is cut rapidly to enhance penetration and dropped into hot redistilled 95% ethanol containing enough precipitated calcium carbonate to neutralize the organic acids in the plant tissues. The amount of ethanol is such as to give a final concentration of 80%. This treatment generally preserves carbohydrates, but proteins may be modified during subsequent storage.

Processing (grinding, homogenization, mixing) of fresh tissues in air is likely to oxidize labile reduced compounds. In many instances, the oxidation is caused by enzymes that show limited or no net oxidative effects in intact tissues. Such changes are often eliminated by enzyme inactivation. Oxidation also may be reduced if tissues are ground in the presence of reducing agents such as bisulfite or dithiothreitol (Cleland 1964). In some instances, neither heating nor the addition of reducing agents eliminates undesirable oxidative changes. For example, up to 75% of the ascorbic acid in plant tissues may be oxidized during grinding unless proper techniques are used.

MINIMIZING LIPID CHANGES

Traditional methods of preparing samples may affect the composition of lipid extracts. Because the proportions of lipid classes in living tissue may change upon death, the samples must be chilled rapidly prior to extraction or frozen quickly for storage (Holman 1966). Incomplete extraction procedures leave more polar lipids than nonpolar components unextracted. In the determination of fatty acids, saponification of tissue minimizes losses of lipids that are extracted with difficulty but may decrease yields of polyunsaturated fatty acids that react with other tissue components in the presence of alkali. Prolonged saponification causes isomerization of polyunsaturated forms to conjugated forms.

Storage of lipids rich in polyunsaturated fatty acids presents several problems. Methylene-interrupted polyunsaturated acids, because of their activated methylene groups, are easily attacked by oxygen. The rate of oxidation depends on the degree of unsaturation. The maximum rate of oxidation for linoleate is approximately 20 times that of oleate. Each additional double bond in the molecule of polyunsaturated fatty acids increases the oxidation rate at least twofold. The oxidation rate of a sample is the weighted average of the concentration and rates of oxidation of the component fatty acids (Holman 1966; Frankel 1984).

Dry fat samples should be stored under nitrogen or dissolved in petroleum ether. Storage in ethyl ether is undesirable because it tends to form oxidative peroxides. Dilution with petroleum ether and flushing in a stream of pure nitrogen is a good practice for temporary protection. The rate of oxidation is temperature dependent. At $-20°C$, the rate of oxidation is about 1/16 that at room temperature. Addition of an antioxidant (e.g., propyl gallate or santoquin) is effective provided it does not interfere with the analytical determinations. The antioxidant 4-methyl-2,6-ditertiarybutylphenol is useful in chromatographic fractionations as it is easily separable from the common lipids (Wren and Szczepanowska 1964). Light (especially fluorescent lighting) activates oxidation. Polyunsaturated fatty acids undergo less deterioration when they are stored in frozen ($-20°C$) intact tissues than after they are extracted from tissues.

CONTROLLING OXIDATIVE AND MICROBIAL ATTACK

To minimize oxidative changes, preservation at low temperatures under nitrogen is recommended for most foods. Compositional changes (desiccation, moisture absorption, volatilization) in relatively dry foods are reduced by storage in hermetically closed containers at about 4°C. If such containers contain powdered materials and are opened frequently for removing samples, it is advisable to check whether moisture changes have taken plane. Frozen foods should be wrapped in plastic material or placed in airtight containers to reduce dehydration.

To reduce or eliminate microbial attack, several methods can be used. They include freezing, drying, and the use of preservatives or a combination of any of the three. The commonly used preservatives include sorbic acid or sorbate, sodium benzoate, sodium salicylate, tylosin, formaldehyde, mercuric chloride, toluene, and thymol. The most appropriate method or preservative depends on the nature of the food, expected contamination (natural or additives), storage period and conditions, and the analyses that are to be performed.

BIBLIOGRAPHY

Benton-Jones, J., Jr., and Steyn, W. J. A. (1973). Sampling, handling, and analyzing plant tissue samples. *In* "Soil Testing and Plant Analysis," rev. ed., pp. 249–270. Soil Science Society of America, Madison, WI.

Bieleski, R. L. (1964). The problem of halting enzyme action when extracting plant tissues. *Anal. Biochem.* **9**, 431–442.

Cleland, W. W. (1964). Dithiothreitol, a new protective reagent for SH groups. *Biochemistry* **3**, 480–482.

Clemments, R. L. (1959). A technique for the rapid grinding of frozen plant tissue. *J. Assoc. Off. Agric. Chem.* **42**, 216–217.

Entenman, C. (1961). The preparation of tissue lipid extracts. *J. Am. Oil Chem. Soc.* **38**, 534–538.

Frankel, E. N. (1984). Lipid oxidation: Mechanisms, products and biological significance. *Am. Oil Chem. Soc.* **61**, 1908–1917.

Gausman, H. W., *et al.* (1952). Some effects of artificial drying of corn grain. *Plant Physiol.* **27**, 794–802.

Grier, J. D. (1966). Preparation of plant material for analysis. *J. Assoc. Off. Anal. Chem.* **49**, 291–298.

Holman, R. T. (1966). Polyunsaturated acids; General introduction to polyunsaturated acids. *Prog. Chem. Fats Other Lipids* **9**, 3–12.

Jones, A. H., and Ferguson, W. E. (1951). Methods of preparing food and plant products with the Waring blendor and the limitations of that equipment. *Food Res.* **16**, 281–284.

Lockman, R. B. (1980). Review of soil and plant tissue preparation procedures. *J. Assoc. Off. Anal. Chem.* **63**, 766–769.

Perlmutter, S. H. (1953). Report on preparation of samples of creamed cottage cheese with the Waring blendor. *J. Assoc. Off. Agric. Chem.* **36**, 187–190.

Ratay, A. F. (1953). Report on preparation of sample and sampling of French dressing. *J. Assoc. Off. Agric. Chem.* **36**, 758–759.

Wren, J. J., and Szczepanowska, A. D. (1964). Chromatography of lipids in the presence of an antioxidant 4-methyl-2,6-di-tert-butylphenol. *J. Chromatogr.* **14**, 405–409.

4
Reporting Results and Reliability of Analyses

The basic purpose of an analytical assay is to determine the mass (weight) of a component in a sample. The numerical result of the assay is expressed as a weight percentage or in other units that are equivalent to the mass/mass ratio. The mass (weight) of a component in a food sample is calculated from a determination of a parameter whose magnitude is a function of the mass of the specific component in the sample.

Some properties are basically mass dependent. Absorption of light or other forms of radiant energy is a function of the number of molecules, atoms, or ions in the absorbing species. Although certain properties, such as specific gravity and refractive index, are not mass dependent, they can be used indirectly for mass determination. Thus, one can determine the concentration of ethanol in aqueous solutions by a density determination. Refractive index is used routinely to determine soluble solids (mainly sugars) in syrups and jams.

Some mass-dependent properties may be characteristic of several or even of a single component and may be used for selective and specific assays. Examples are light absorption, polarization, or radioactivity. Some properties have both a magnitude and specificity parameter (nuclear magnetic resonance and infrared spectroscopy). Such properties are of great analytical value because they provide selective determinations of a relatively large number of substances.

In this chapter, we describe conventional ways of expressing analytical results and discuss the significance of specificity, accuracy, precision, and sensitivity in assessing the reliability of analyses.

In recent years the metric SI system of units has gained worldwide

Table 4.1.   Base Units in the SI System

Quantity	Name	Symbol
Length	meter	m
Time	second	s
Mass	kilogram	kg
Electric current	ampere	A
Thermodynamic temperature	kelvin	K
Other temperature usages	degrees Celsius	°C
Luminous intensity	candela	cd
Amount of substance	mole	mol

Table 4.2.   Supplementary Units in the SI System

Quantity	Name	Symbol
Plane angle	radian	rad
Solid angle	steradian	sr

Table 4.3.   Derived Units with Special Names

Quantity	Name	Symbol	Equivalent to
Electric conductance	siemens	S	A/V
Electric capacitance	farad	F	C/V
Electric potential	volt	V	W/A
Electric resistance	ohm	Ω	V/A
Energy, work, quantity of heat	joule	J	N·m
Force	newton	N	$m \cdot kg/s^2$
Frequency	hertz	Hz	1/s
Illuminance	lux	lx	$cd \cdot sr/m^2$
Inductance	henry	H	Wb/A
Luminous flux	lumen	lm	cd·sr
Magnetic flux	weber	Wb	V·s
Magnetic flux density	tesla	T	Wb/m^2
Power, energy flux	watt	W	J/s
Pressure, stress	pascal	Pa	N/m^2
Quantity of electricity, electric charge	coulomb	C	A·s

Table 4.4. Multiple and Submultiple Units

Symbol	Prefix	Equivalent to
T	tera	10^{12}
G	giga	10^9
M	mega	10^6
k	kilo	10^3
h	hecto	10^2
da	deka	10^1
d	deci	10^{-1}
c	centi	10^{-2}
m	milli	10^{-3}
μ	micro	10^{-6}
n	nano	10^{-9}
p	pico	10^{-12}
f	femto	10^{-15}
a	atto	10^{-18}

acceptance. It has been recommended or required by the International Organization for Standardization (ISO), the International Union of Pure and Applied Chemistry (IUPAC), and the International Union of Pure and Applied Physics (IUPAP), as well as by an increasing number of scientific and professional organizations in the United States and by the industry and the trade. The SI system contains seven base units, two supplementary units, 15 derived units having special names, and 14 prefixes for multiple and submultiple units. All physical properties can be quantified by 38 names (Conant 1975). The units of the SI system are given in Tables 4.1–4.4.

REPORTING RESULTS

In reporting analytical results, both the reference basis and the units used to express the results must be considered. For example, analyses can be performed and the results reported on the edible portion only or on the whole food as purchased. Results can be reported on an as-is basis, on an air-dry basis, on a dry matter basis, or on an arbitrarily selected moisture basis (e.g., 14% in cereals).

To convert contents (%) of component Y from oven-dried (OD) to an as-received (AR) basis, or vice versa, the following formulas are used:

$$\%Y_{OD} = \frac{\%Y_{AR} \times 100}{(100 - \%loss_{OD}})$$

$$\%Y_{AR} = \frac{\%Y_{OD}(100 - \%loss_{OD})}{100}$$

To convert contents from an as-received basis to an arbitrary moisture basis, the following formula is used:

$$\%Y = \frac{\%Y_{AR}(100 - \text{arbitrary moisture }\%)}{100 - \% \text{ moisture}_{AR}}$$

To weigh out a sample on an arbitrary moisture (AM) basis, use the following:

$$\text{sample weight} = \frac{\% \text{ dry matter}_{AM}}{\% \text{ dry matter}_{AR}} \times \text{required sample weight}$$

To obtain % dry matter, subtract percentage of moisture from 100. If the moisture has been determined in two stages, air drying followed by oven drying, compute total moisture contents of sample as follows:

$$TM = A + \frac{(100 - A)B}{100}$$

where TM is the % total moisture, A the % moisture loss in air drying, and B the % moisture of air-dried sample as determined by oven drying.

Tables, nomograms, and calculators are available to simplify calculations in expressing results on a given basis, or for weighing samples on a fixed moisture basis (e.g., 20% in dried fruit). In view of the very wide range in moisture contents in various foods, analytical results are often meaningless unless the basis of expressing the results is known.

Expressing analytical results on an as-is basis is wrought with many difficulties. It is practically impossible to eliminate considerable desiccation of fresh plant material. In some instances, even if great pains are taken to reduce such losses, the results may still vary widely. For example, the moisture contents of leafy foods may vary by as much as 10% depending on the time of harvest (from early morning to late afternoon). Similarly, the moisture contents of bread crust and crumb change from the moment bread is removed from the oven as a result of moisture migration and evaporation. Absorption of water in baked or roasted low-moisture foods (crackers, coffee) is quite substantial. In most cases, storing air-dried foods in hermetically closed containers is least troublesome. Once the moisture contents of such foods are determined, samples can be used for analyses over a reasonable period.

The concentrations of major components are generally expressed on a percentage by weight or percentage by volume basis. For liquids and beverages, g per 100 ml is often reported. Minor components are calculated as mg (or mcg) per kg or liter; vitamins in mcg or international units per 100 g or 100 ml. Amounts of spray residues are often reported in ppm (parts per million).

In calculating the protein contents of a food, it is generally assumed the protein contains 16% nitrogen. To convert from organic nitrogen (generally determined by the Kjeldahl method, see Chapter 37) to protein, the factor of $6.25 = 100/16$ is used. In specific foods known to contain different concentrations of nitrogen in the protein, other conversion factors are used (5.7 in cereals, 6.38 in milk). Heidelbaugh *et al.* (1975) compared three methods for calculating the protein content of 68 foods: (1) multiplication of Kjeldahl nitrogen by 6.25; (2) multiplication of Kjeldahl nitrogen by factors ranging from 5.30 to 6.38 depending on the type of food; and (3) calculation on the basis of amino acid composition, determined by chemical analyses. Up to 40% differences in protein content were found depending on the calculation method. There were, however, only small differences in mixed diets representing typical menus.

If a food contains a mixture of carbohydrates, the sugars and starch are often expressed as dextrose. In lipid analyses (free fatty acids or total lipid contents) calculations are based on the assumption that oleic acid is the predominant component. Organic acids are calculated as citric, malic, lactic, or acetic acid depending on the main acid in the fruit or vegetable.

Mineral components can be expressed on an as-is basis or as % of total ash. In either case the results can be calculated as elements or as the highest valency oxide of the element.

Amino acid composition can be expressed in several ways: g amino acid per 100 g sample, or per 100 g protein, or per 100 g of amino acids. For the determination of molar distribution of amino acids in protein, g-moles of amino acid residue per 100 g-moles of amino acid are computed.

In trade and industry, empirical tests are often used. For example, fat acidity of cereal grains is often expressed as mg KOH required to neutralize the fatty acids in 100 g food. Acidity is often expressed for simplicity in milliliters of $N/10$ or N NaOH. The acidity of acid phosphates in baking powders is reported in industry as the number of parts of sodium bicarbonate that are required to neutralize 100 parts of the sample.

RELIABILITY OF ANALYSES

The reliability of an analytical method depends on its (1) specificity, (2) accuracy, (3) precision, and (4) sensitivity (Anastassiadis and Common 1968).

Specificity is affected primarily by the presence of interfering substances that yield a measurement of the same kind as the substance being determined. In many cases, the effects of the interfering substances can be accounted for. In calculating or measuring the contribution of several interfering substances, it is important to establish whether their effects are additive.

Accuracy of an analytical method is defined as the degree to which a mean estimate approaches a true estimate of an analyzed substance, after the effects of other substances have been allowed for by actual determination or calculation. In determining the accuracy of a method, we are basically interested in establishing the deviation of an analytical method from an ideal one. That deviation may be due to an inaccuracy inherent in the procedure; the effects of substances other than the analyzed one in the food sample; and alterations in the analyzed substance during the course of the analysis.

The accuracy of an analytical assay procedure can be determined in two ways. In the absolute method, a sample containing known amounts of the analyzed components is used. In the comparative method, results are compared with those obtained by other methods that have been established to give accurate and meaningful results.

The absolute method is often difficult or practically impossible to apply, especially for naturally occurring foods. In some cases, foods can be prepared by processing mixtures of pure compounds. If the mixtures are truly comparable in composition to natural foods, meaningful information is obtained.

Several indirect methods are available to determine the accuracy of analyses. While these methods are useful in revealing the presence of errors they cannot prove the absence of errors. When a complete analysis of a sample is made and each component is determined directly, a certain degree of accuracy is indicated if the sum of the components is close to 100. On the other hand, an apparently good summation can result from compensation of unrelated errors in the determination of individual components. A more serious error can result from compensation of errors that are related in such a way that a negative error in one component will cancel a positive error in another component. This may be particularly important in incomplete fractionations. For example, the sum of proteins separated according to differences in solubility may be close to 100%, yet the separation of individual components may be incomplete or of limited accuracy.

In the recovery method, known amounts of a pure substance are added to a series of samples of the material to be analyzed and the assay procedure is applied to those samples. The recoveries of the added amounts are then calculated. A satisfactory recovery is most useful in demonstrating absence of negative errors.

If the accuracy of an analytical method is affected by interference from substances that cannot be practically eliminated, a suitable correction can sometimes be applied. Such a correction is often quite complicated because the results may be affected by concentration of the interfering or assayed substance, or by their interaction in food processing or during the analysis.

Precision of a method is defined as the degree to which a determination of a substance yields an analytically true measurement of that substance. It is important to distinguish clearly between precision and accuracy. In industrial quality control, it often is unimportant whether analysis of numerous similar samples yields exactly accurate (i.e., true) information regarding the composition of the sample. The information may be useful provided the difference between the precise and accurate determination is consistent. The analysis that gives the actual composition (or in practice the most probable composition) is said to be the most accurate. For instance, direct and accurate determination of the bran content of wheat flour is both laborious and imprecise, but the bran content can be estimated directly from the amount of crude fiber in a flour. This estimation is based on the fairly constant ratio between crude fiber (determined by a precise, but not accurate, empirical method) and actual bran contents. Still simpler, is the estimation of bran content from total mineral content or reflectance color assay of a flour. Accuracy and precision are compared graphically in Fig. 4.1. The relative accuracy and precision of the crude fiber method for estimating bran content in wheat flour is represented by diagram (ii) in Fig. 4.1.

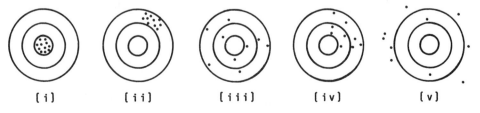

[i]   [ii]   [iii]   [iv]   [v]

Fig. 4.1. (i) Accuracy, good; precision, good. (ii) Accuracy, poor; precision, good. (iii) Accuracy, good; precision, poor. (iv) Accuracy, poor; precision, poor. (v) Random. (From Brown 1977 and Hislop 1980.)

To determine the precision of an analytical procedure and the confidence that can be placed on the results obtained by that procedure, statistical methods are used. The most basic concept in statistical evaluation is that any quantity calculated from a set of data is an *estimate* of an unknown *parameter* and that the estimate is sufficiently reliable. It is common to use English letters for estimates and Greek letters for true parameters (see Chapter 2 on sampling).

If n determinations $x_1, x_2, \ldots, x_n$ are made on a sample, the average $x = \sum x/n$ is an estimate of the unknown true value μ. The precision of the assay is given by the standard deviation σ:

$$\sigma = \pm [\sum (x - \mu)^2/n]^{1/2}$$

If the number of replicate determinations is small (<10), an estimate of the standard deviation (s) is given by

$$s = \pm [\sum (x - \bar{x})^2/(n - 1)]^{1/2}$$

The divisor $n - 1$ used to estimate s is termed the degrees of freedom and indicates that there are only $n - 1$ independent deviations from the mean. The standard deviation is the most useful parameter for measuring the variability of an analytical procedure.

If s is independent of x for a given concentration range, s can be computed from the results of replicate analyses on several samples of similar materials. In that case, the sums of the squares of the deviations of the replicates of each material are added, and the resultant total is divided by the number of degrees of freedom (the sum of the total number of determinations, n, minus the number of series of replicate determinations).

A complicating factor in determining the precision arises when the standard deviation varies with the concentration of the element present. Sometimes the range of concentration can be divided into intervals and the standard deviation given for each interval. If the standard deviation is approximately proportional to the amount present, precision can be expressed as a percentage by using the coefficient of variation (CV).

$$CV = \frac{\text{standard deviation}}{\text{amount present}} \times 100$$

If the data show a varying standard deviation, transformation of the data into other units in which the standard deviation is constant is often useful. Two widely used transformations are square roots or logarithms.

Chemical analyses are made for various purposes and the precision required may vary over a wide range. In the determination of atomic weights, an effort is made to keep the error below 1 part in 10^4 to 10^5. In most analytical work, the allowable error lies in the range 1 to 10 parts per 1000 for components comprising more than 1% of the sample. As a rule, analyses should not be made with a precision greater than required. Up to a point, precision is a function of time, labor, and overall cost (Youden 1959).

The precision of an analytical result depends on the least exact method used in obtaining the result. In expressing the result, the number of figures given should be such that the next to the last figure is certain and the last figure is highly probable yet not certain. Thus 10% and 10.00% denote widely varying precision (Paech 1956). The following is an example of how an average result computed from several determinations should be expressed. Assume the moisture content of sugar is determined in triplicate, and the following results are obtained: 1.032, 1.046, and 1.036%. The average is 1.038%. However, because the difference between 1.032 and 1.046 is larger than 0.010, the results should not be expressed with more than two figures after the decimal point. Thus, the average result should be reported as 1.04% (not 1.038%), indicating that the first figure after the decimal point is certain, and the second one is probable but uncertain.

The results of weighing, buret reading, and instrumental (including automatic) reading have limitations. Replication of analyses eliminates some of the errors resulting from sampling, from heterogeneity of sampled material, and from indeterminate—accidental or random—errors in the assay. Although repetition of an assay generally increases the precision of the analysis, it cannot improve its specificity and accuracy. If, however, reasonable specificity and accuracy have been established, the precision of the assay is an important criterion of its reliability.

Sensitivity of a method used in determining the amount of a given substance is defined as the ratio between the magnitude of instrumental response and the amount of that substance. Sensitivity is measured and expressed as the smallest measurable compositional difference between two samples. That difference becomes meaningful if it exceeds the variability of the method. In instrumental analysis, the signal to noise ratio should be at least $2:1$.

Sensitivity can be increased in two ways: (1) by increasing the response per unit of analyzed substance (e.g., in colorimetric assays by the use of color reagents that have a high specific absorbance; in gravimetric determinations by the use of organic reagents with a high molecular weight) and (2) by improving the discriminatory power of the instrument or operator (e.g., in gravimetry by using a microbalance; in spectrophotometry by using a photomultiplier with a high magnifying power) (Anastassiadis and Common 1968).

According to Horwitz (1982, 1983), the important components of reliability, which are listed in their order of importance for most purposes in food analyses, are as follows:

1. *Reproducibility*—total between-laboratory precision
2. *Repeatability*—within-laboratory precision
3. *Systematic error or bias*—deviation from the "true" value
4. *Specificity*—ability to measure what is intended to be measured
5. *Limit of reliable measurement*—the smallest increment that can be measured with a statistical degree of confidence

Typical analytical systematic errors (biases) are plotted in Fig. 4.2. Detection and determination of errors were described and discussed in detail by Cardone (1983A,B). Tolerances and errors are depicted in Fig. 4.3, in which the tolerance limits for the measured property are given by L_p and C_m indicates the uncertainty in the measurement. The values of L_p and C_m include estimates of the bounds for systematic errors or biases (B) and estimates of random errors (s, the estimate of standard deviation). C_m should be less than L_p. The confidence limits for $\bar{x}$, the mean of n replicate measurements, are

$$C_m = \pm[B + (t_s/\sqrt{n})]$$

where t is the so-called Student factor.

For regulatory purposes, reliability is paramount and reproducibility is the critical component (Horwitz 1982). The between-laboratory coefficient of variation CV is represented by

$$CV(\%) = 2^{(1 - 0.5 \log C)}$$

where C is the concentration expressed as powers of 10 (e.g., 1 ppm, or 10^{-6}, $C = -6$). The value of CV doubles for each decrease in concen-

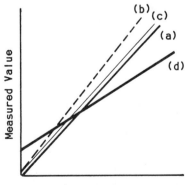

Fig. 4.2. Typical analytical systematic errors (bias). (a) Unbiased, (b) measurement-level related, (c) constant error, (d) combination of b and c. (From Taylor 1981.)

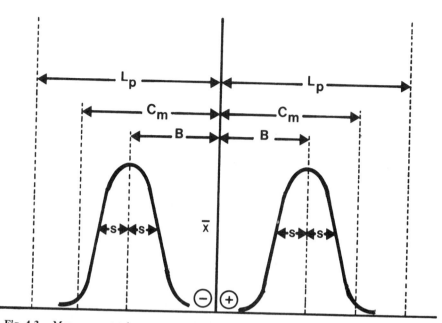

Fig. 4.3. Measurement tolerance and errors. See text for discussion. (From Taylor 1981.)

tration of two orders of magnitude. The between-laboratory coefficient of variation at 1 ppm is 16% (2^4). The within-laboratory CV should be one-half to two-thirds of the between-laboratory CV. The interlaboratory coefficient of variation as a function of concentration is illustrated in Table 4.5 and Fig. 4.4. The largest contributor to experimental errors in instrumental methods are systematic errors (Horwitz 1984), which are difficult to measure without interlaboratory comparisons. They can be reduced by incorporating reference physical constants and certified standards.

The precision characteristics of 18 analytical methods for trace elements subjected to interlaboratory collaborative studies over the last 10 years by the Association of Official Analytical Chemists were examined by Boyer *et al.* (1985). Removal of outliers and statistical calculations were standardized by the use of a computer program. Most of the studies, which represented a variety of analytes, matrices, and measurement techniques over a range of concentration of 100 g/kg to 10 µg/kg, were distributed about a curve defined by the equation

$$\% \, RSD_x = 2^{(1 \, - \, 0.5 \log C)}$$

where RSD_x is the among-laboratory standard deviation and C the concentration expressed as a decimal fraction (e.g., 1 ppm = 10^{-6}), irrespective of analyte, matrix, or measurement technique. The within-labo-

Table 4.5. Interlaboratory Coefficient of Variation (*CV*) as a Function of Concentration[a]

Analyte	Approx. concentration			Determinative methods	Approx. CV (%)
	Range	Unit	Mean (as fraction)		
Salt in foods	0.25–20	%	1×10^{-1}	potentiometric	$\sqrt{2} = 1.4$
Drug formulations	0.1–60	%	1×10^{-2}	chromatographic separations; spectrophotometric, automated, manual	$2 = 2$
Sulfonamides in feeds	0.01–0.05	%	2×10^{-4}	spectrophotometric	$2^2 = 4$
Pesticide residues	0.03–17	ppm	1×10^{-6}	gas chromatography	$2^4 = 16$
Trace elements			1×10^{-6}	atomic absorption	$2^4 = 16$
Aflatoxins B_1, B_2, G_1, G_2	2–200	ppb	1×10^{-8}	thin layer chromatography	$2^5 = 32$
Pesticide residues in total diet	1–100	ppb	1×10^{-8}	gas chromatography	$2^5 = 32$
Aflatoxin M_1	0.05–50	ppb	1×10^{-9}	thin layer chromatography	$2^{5.5} = 45$
Copper			0.5×10^{-6}	atomic absorption	22
Zinc			0.15×10^{-6}	atomic absorption	54
Lead			0.05×10^{-6}	voltametric	80
Cadmium			0.005×10^{-6}	voltametric	220

[a] From Horwitz *et al.* (1980).

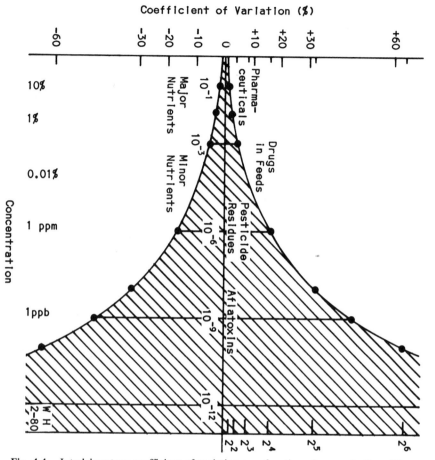

Fig. 4.4. Interlaboratory coefficient of variation as a function of concentration. (From Horwitz *et al.* 1980.)

ratory relative standard deviation RSD_0 was usually one-half to one-third of RSD_x. Positive deviations from this curve with decreasing concentration could be explained by heterogeneity of the material, free choice of analytical method, or concentrations below the limit of determination. The presence of more than 20% outlying laboratory results or RSD_x degenerating faster than the "normal" rate with decreasing concentration was taken by the authors to indicate that a particular method is inapplicable at or below the level generating the imprecise data.

Optimizing chemical laboratory performance was the subject of a symposium organized by the Association of Official Analytical Chemists (Garfield *et al.* 1980). The symposium covered a wide range of topics including design, criteria, and maintenance of quality assurance programs; reference standards; maintenance of records; and government regulations as they relate to good manufacturing practices and good laboratory practices.

BIBLIOGRAPHY

Anastassiadis, P. A., and Common, R. H. (1968). Some aspects of the reliability of chemical analyses. *Anal. Biochem.* **22**, 409–423.

Boyer, K. W., Horwitz, W., and Albert, R. (1985). Interlaboratory variability in trace element analysis. *Anal. Chem.* **57**, 454–459.

Brown, S. S., ed. (1977). "Clinical Chemistry and Chemical Toxicology of Metals." Elsevier/ North Holland, Amsterdam.

Cardone, M. J. (1983A). Detection and determination of error in analytical methodology. Part I. In the method verification program. *J. Assoc. Off. Anal. Chem.* **66**, 1257–1282.

Cardone, M. J. (1983B). Detection and determination of error in analytical methodology. Part II. Correction for corrigible systematic error in the course of real sample analysis. *J. Assoc. Off. Anal. Chem.* **66**, 1238–1294.

Conant, F. S. (1975). Using the SI system of units. *Rubber Chem. Technol.* **48**, 1–13.

Garfield, F. M., Palmer, N., and Schwartzman, G., eds. (1980). "Optimizing Chemical Laboratory Performance through the Application of Quality Assurance Principles." Association of Official Analytical Chemists, Washington, DC.

Heidelbaugh, N. D., Huber, C. S., Bednarczyk, J. F., Smith, M. C., Rambaut, P. C., and Wheeler, H. O. (1975). Comparison of three methods for calculating protein content of foods. *J. Agric. Food Chem.* **23**, 611–613.

Hislop, J. S. (1980). Choice of the analytical method. *In* "Trace Element Analytical Chemistry in Medicine and Biology" (P. Bratter and P. Schramel, eds.), pp. 747–767. de Gruyter, Berlin.

Horwitz, W. (1982). Evaluation of analytical methods for regulation of foods and drugs. *Anal. Chem.* **54**, 67A, 68A, 70A, 72A, 74A, 76A.

Horwitz, W. (1983). Today's chemical realities. *J. Assoc. Off. Anal. Chem.* **66**, 1295–1301.

Horwitz, W. (1984). Reduction in variability of instrumental methods. *J. Assoc. Off. Anal. Chem.* **67**, 1053–1057.

Horwitz, W., Kamps, L. R., and Boyer, K. W. (1980). Quality control. Quality assurance in the analysis of foods for trace constituents. *J. Assoc. Off. Anal. Chem.* **63**, 1344–1354.

Paech, K. (1956). General procedures and methods of preparing plant materials. *In* "Modern Methods of Plant Analysis," Vol. 1 (K. Paech and M. V. Tracey, eds.). Springer-Verlag, Berlin (German).

Taylor, J. K. (1981). Quality assurance of chemical measurements. *Anal. Chem.* **53**, 1588A, 1589A, 1591A.

Youden, W. J. (1959). Accuracy and precision: Evaluation and interpretation of analytical data. *In* "Treatise on Analytical Chemistry" (I. M. Kolthoff and P. J. Elwing, eds.). Interscience Encyclopedia, New York.

II
METHODS AND
INSTRUMENTATION

5
Theory of Spectroscopy

In this chapter, we discuss the theory of spectroscopy, i.e., the inter-action of radiation with matter, which is the basis for many common instrumental techniques of analysis. In succeeding chapters, the various types of spectroscopic analysis and the instruments used in each are considered in more detail. A knowledge of how these instruments work is a prerequisite for their maximum utilization. When an instrument manufacturer sells a piece of equipment, it is guaranteed to meet certain specification and performance levels. However, most instruments are capable of delivering far more than that, and if analysts understand how the equipment works and what it is supposed to do, they can usually extend its practical capabilities manyfold. It is also necessary to have an understanding of how an instrument operates in order to know whether the data obtained from it are correct.

Consider the light you are reading this book by. The radiation coming from that light consists of many wavelengths, ranging from very short to very long. Physicists have studied that radiation and have found that each ray behaves as if it has an electric component and a magnetic component acting at right angles to each other; they have called it *electromagnetic radiation*. (The term *light* is reserved only for the visible region. The terms *ultraviolet light* and *infrared light* are not correct and should be *ultraviolet* and *infrared radiation*.)

Several units and terms used to describe electromagnetic radiation are given in Table 5.1. To get some idea of the physical significance of those units consider the interatomic distances within a molecule. Those distances are measured in nanometers (e.g., the distance between the hy-

Table 5.1. Spectroscopy Units

Unit	Symbol	Relation to other units
Micrometer	μm	10^{-6} m
Nanometer (old millimi-cron, mμ)	nm	10^{-9} m; 1/1000 μm
Angstrom	Å	10^{-10} m; 1/10 nm
Wavenumber	cm^{-1}	v/c; $1/\lambda$ (see equation 5.1)
Electron volt	eV	23.06 kcal/mole or 8066 cm^{-1}
Erg	erg	6.24 × 10^{11} eV/mole

drogen and chlorine atoms in HCl is 0.127 nm). If the distance from New York to Los Angeles is 1 in., then on the same scale a nanometer would be about the width of your hand and a micrometer would be about the length of a football field.

There are two basic equations useful in spectroscopy. In the first,

$$\lambda v = c \qquad (5.1)$$

λ is the wavelength of the radiation in cm, v is the frequency of the radiation in cycles/sec, and c is a constant, the speed of light, equal to 3 × 10^{10} cm/sec. The second equation is

$$E = hv \qquad (5.2)$$

where E is the energy of the radiation in ergs, h is Planck's constant (6.62 × 10^{-27} erg-sec), and v is the frequency of the radiation, as in equation 5.1.

Suppose the rays of radiation from your desk lamp, plus those from any other source, are plotted in chart form as in Fig. 5.1, with the very short wavelengths at the top left and the very long wavelengths at the bottom right. The energy equivalents of various wavelengths are indicated along the top of the plot, and the effects of that radiation when it interacts with a molecule are indicated along the bottom.

The regions are established by the limitations of the instruments and the boundaries are continuously changing as the instruments improve. Notice that the term "instrument" was used, not machine. We drive to work in machines, but measurements are made with instruments.

ABSORPTION OF RADIATION

Consider a ray of radiation from your desk lamp shining on your shirt. The ray can either interact with the molecules in the shirt or pass through. What allows some rays to interact and some to pass through?

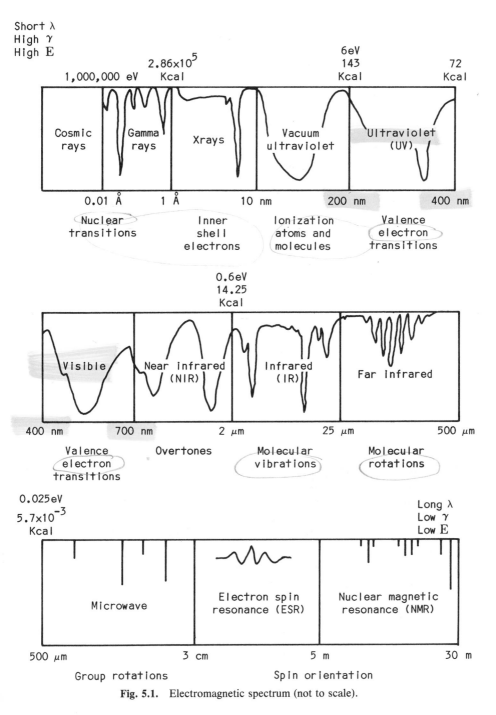

Fig. 5.1. Electromagnetic spectrum (not to scale).

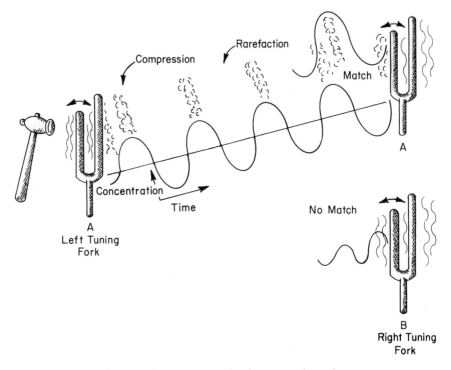

Fig. 5.2. Frequency matching for energy absorption.

Two main requirements must be met in order for a ray of radiation to be absorbed by a molecule: (1) the incident ray must have the same frequency as a rotational, vibrational, electronic, or nuclear frequency in the molecule, and (2) the molecule must have a permanent dipole or an induced dipole (in more common terms, there must be something for the absorbed energy to do—work must be done).

To visualize the first requirement, refer to Fig. 5.2 and recall your physics experiment with tuning forks. When the left tuning fork is struck with a hammer, the fork bends to the right compressing the air molecules against it. Those molecules, in turn, compress against other molecules further to the right (compression). When the left tuning fork returns to its original position, it leaves a reduced concentration of air molecules (rarefaction). If those concentrations are plotted vs. time, a wave pattern is produced.[1] When the compressed air reaches the tuning forks on the right side, both forks are bent to the right. Both forks will want to return

---

[1] Neither sound waves nor radiation waves *wiggle* through the air. Many students confuse the concentration vs. time plot with the actual physical process.

to their original positions. The top fork (A) will return quite easily because there are less molecules to oppose it since it is in phase with the rarefaction. The bottom (B) fork will return to its original position before the entire compression wave has passed it, and the fork will meet some resistance that will slow it down. The B fork again will move to the right because of its normal vibrational frequency. This time, however, it does so as a rarefaction passes, and the fork gets little push from the weak concentration of air molecules. When the fork moves again to the left, it collides directly with the second compression coming by and is stopped from vibrating. The net effect is that the right hand A fork has the same frequency as the left fork and vibrates, while the right B fork does not vibrate because it does not have a matching frequency.

A similar effect occurs when the electric and magnetic components of an electromagnetic ray of radiation interact with the electric and magnetic components of a molecule. If they match, the energy can be transferred (absorbed); if they do not match, the ray cannot interact and will pass on through the molecule.

The second requirement is somewhat harder to visualize. Suppose the frequencies match and the energy can be transferred. Where does it go and what does it do? It must be used up by doing work. Figure 5.3, which illustrates diagramatically the vibrations of a CS_2 molecule, may help to understand what happens.

Consider (1), a CS_2 molecule with the C and S atoms at normal distances. Since C is relatively more electropositive than S, the center of positive charge for the entire molecule acts as if it were at the center of the molecule. The two negative S atoms have the *center* of their negative charges also at the center of the molecule. Now let the molecule vibrate symmetrically; that is, the right-hand S atom moves away from the C atom the same distance that the left-hand S atom moves away (2). It takes work to move the left negative S away from the positive C, and it takes the same amount of work to move the right negative S away from the positive C. The net effect however is that no work has been done. Remember that *work* (w) equals *force* (f) $\times$ *distance* (d). In this case the distance to the right is exactly opposite the distance to the left, so the forces cancel each other. You can see this because the center of the negative charge did not move away from the center of the molecule. The same result occurs in (3), in which both S atoms move toward the C atom by the same amount. Since there is no *net* work to be done, the energy of the ray can not stay in the molecule, and the ray will pass through even though it had the same frequency as a vibration in the molecule.

In the vibration shown in (5), the work (force times distance) for the left S does not equal the force times distance for the right S and the effect

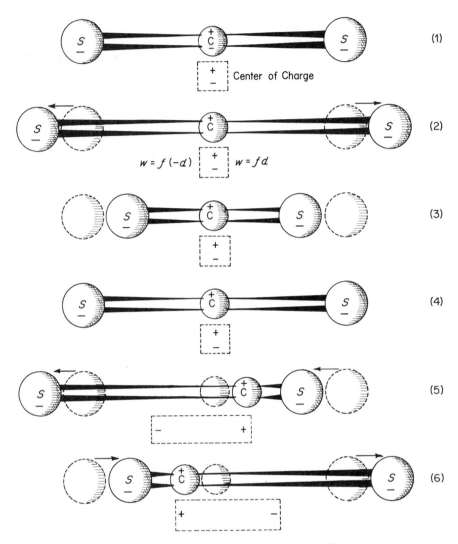

Fig. 5.3. Symmetric and asymmetric vibrations of a CS_2 molecule.

is that the center of negative charge moves to the left of the positive C. The reverse is true with (6). Since it takes work to separate the positive and negative center of charges, there is a place for the energy to go and the ray of radiation will be absorbed. In both (5) and (6), the center of charges consists of a positive and a negative end, a *dipole*. Some molecules have a dipole naturally; others can have an induced dipole when the electric and magnetic field of a ray of radiation approaches the molecule.

Now that we know what is required for the absorption of radiation, let us consider what happens when rays of different wavelengths and energies interact with matter. As indicated in Fig. 5.1, the effect of radiation on molecules depends on its wavelength. At long wavelengths (low energies), atom, group, and molecular rotations occur; at intermediate wavelengths and energies, molecular vibrations occur; and at longer wavelengths (higher energies) electronic transitions take place. Nuclear transitions are induced by very high energy radiation. In the following sections, the rotations, vibrations, and electronic transitions that are utilized in spectroscopic methods of analyses are explained.

ROTATIONS

When radiation having energies found in the nuclear magnetic resonance (NMR) region strikes a molecule, there is only enough energy to cause atom rotations. To illustrate what happens, refer to the diagram of acetic acid (CH_3COOH) in Fig. 5.4. As shown, the H atoms can rotate about the C—H and O—H bonds, and the O around the C=O bond, etc. Since there are three times as many C—H rotations possible as O—H, and since the H is bonded more weakly to the C than to the O, we would expect two wavelengths at which H absorption would occur, and all other things being equal one band would be three times stronger than the other.

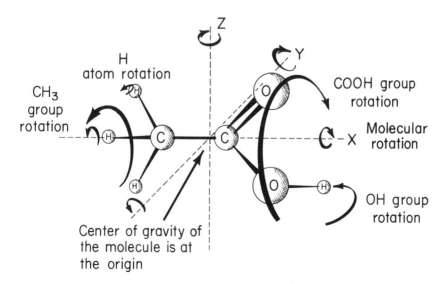

Fig. 5.4. Atom, group, and molecular rotations of acetic acid molecule.

Because the O rotation around the double bond is restrained, the C=O absorption occurs at a different wavelength and with reduced intensity than either the C—H or O—H absorption.

Since every compound has a different arrangement and number of atoms, the wavelengths of absorption will vary in intensity and distribution in the electromagnetic spectrum. This results in a *fingerprint* that is used to determine which atoms and groups are present, the shape of the molecule, and how much of each is present. The energies involved in the NMR region are of the order of 10^{-5} kcal/mole. Groups rotate with 0.01–0.1 kcal. If 0.1–2 kcal/mole of energy is added (far infrared region), there is enough energy to rotate the entire molecule.

There are three *fundamental rotations* for all nonlinear molecules. A molecule can rotate around the X, Y, or Z axis. Since most molecules are unsymmetrical, the rotations about each axis have a different frequency depending on the shape of the molecule. What happens if you strike a molecule at an angle and cause it to "tumble"? The "tumbling" motion can be resolved into three fundamental rotations. A "tumbling" molecule may be rotating 20% along the X axis, 70% along the Y axis, and 10% along the Z axis, the combined effect of which is "tumbling." The important idea is that regardless of the kind of "tumbling" a molecule does, it can always be resolved into a combination of the X, Y, and Z rotations. We cannot predict how each individual molecule will rotate, but we can experimentally show what the *average* of the billions of molecules in a sample will do. The spectra we see measure that average.

Only two axes of rotation are significant for linear molecules (Fig. 5.5). Hold a pencil horizontally in front of you and rotate it along its horizontal axis. The electric and magnetic fields produced by this motion are too weak to interact with a beam of radiation, so there is no absorption band at that frequency.

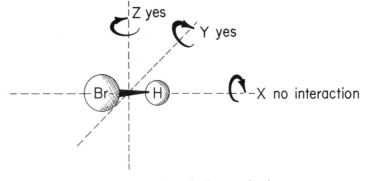

Fig. 5.5.   Rotation of a linear molecule.

VIBRATIONS

If 1–20 kcal of energy per mole is added to a molecule (infrared region), there is enough energy to cause the molecule to vibrate. There are many different ways a molecule can vibrate. As shown in Fig. 5.6, the atoms can compress or stretch along the axis of a bond. This is called a *stretching* vibration. The atoms can *bend symmetrically* or *asymmetrically*; if all the atoms in the molecule move, a *deformation* vibration occurs.

A *fundamental vibration* is a movement of atoms such that the center of gravity of the molecule does not change. This means that when any atom of a molecule moves (mass times distance) to the left, another atom or atoms must move to the right an equal mass times distance. Refer to Fig. 5.3, parts 5 and 6. Notice that when both S atoms move to the left, the C atom compensates by moving to the right.

Two rules give the theoretically possible number of fundamental vibrations:

$$\text{nonlinear molecules} \quad 3N - 6 \quad\quad (5.3)$$

$$\text{linear molecules} \quad 3N - 5 \quad\quad (5.4)$$

where N is the number of atoms in the molecule. Ethyl acetate has 14 atoms and is nonlinear; therefore, it has $(3 \times 14) - 6$ or 36 fundamental vibrations. Hydrobromic acid (Fig. 5.5) has 2 atoms and is linear; therefore, it has $(3 \times 2) - 5$ or 1 fundamental vibration.

The presence of a fundamental vibration does not mean that an absorption will occur. For the latter to take place, work must be done. Benzene, C_6H_6, has $(3 \times 12) - 6$ or 30 fundamental vibrations, but only four are active (i.e., absorb radiation) in the infrared. Since benzene has seven planes and a point of symmetry (Fig. 5.7), if something bends or stretches on one side of the molecule exactly like a bend or stretch on the opposite side, no net work is done and no absorption can take place.

Another factor that reduces the number of frequencies that actually

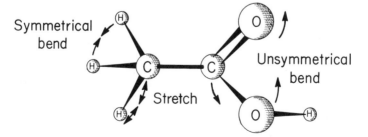

Fig. 5.6. Some molecular vibrations of acetic acid molecule.

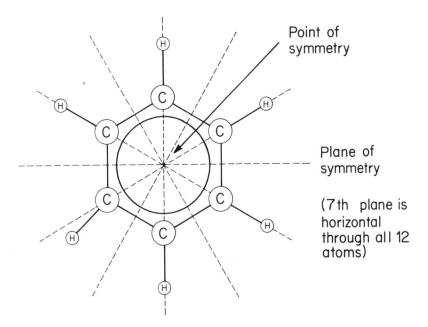

Fig. 5.7. Because of the symmetry of the benzene molecule, only four of its 30 fundamental vibrations absorb infrared radiation.

appear is *degeneracy*. For example, the CO_2 molecule has $(3 \times 3) - 5$ or 4 fundamental vibrations, but we see only two absorption bands in the infrared. Figure 5.8 shows the four fundamental vibrations. The two bending vibrations have the same frequency and will appear as only one band. Since the two vibrations have the same frequency, they are said to be twofold degenerate. If four vibrations had the same frequency, they would be fourfold degenerate. No absorption occurs at the symmetrical stretching frequency since no net work is done.

ELECTRONIC TRANSITIONS

If radiation of 10–100 kcal/mole is applied to a molecule (near infrared, visible, ultraviolet), electronic transitions may occur. That is, an electron in the molecule is raised to a higher energy level. If even more energy, generally greater than 100 kcal/mole, is added, there is enough energy not only to raise an electron to a higher energy level but to completely remove the electron from the molecule, producing an ion. This begins to occur in the ultraviolet region and continues into the X-ray, gamma ray, and cosmic ray regions. Because those ions react differently than the ions and

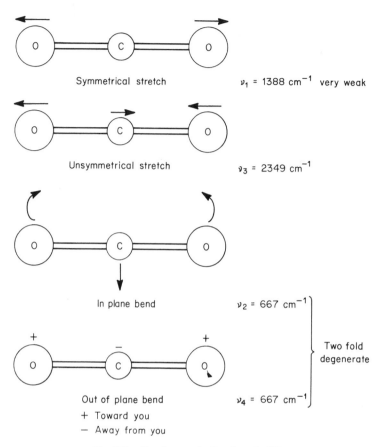

Symmetrical stretch $\nu_1 = 1388$ cm^{-1} very weak

Unsymmetrical stretch $\nu_3 = 2349$ cm^{-1}

In plane bend $\nu_2 = 667$ cm^{-1}

Out of plane bend $\nu_4 = 667$ cm^{-1}
+ Toward you
− Away from you

Two fold
degenerate

Fig. 5.8. Fundamental vibrations in CO_2.

molecules normally used in body processes, it is unsafe to stand in front of X-ray equipment very long; but it is perfectly safe to stand in front of a radio with its low energy waves. If several thousand kcal/mole is added to a molecule, then nuclear transitions occur, that is, the entire nucleus changes energy states.

MOLECULAR ENERGY STATES

All atoms and molecules contain energy. This energy takes many forms and varies greatly as to the amount. When an atom or a molecule has its lowest energy, it is said to be in the *ground state*. This is the state preferred by the molecule. If the molecule has more energy than the lowest energy, it is in an *excited state*. When a molecule absorbs energy in the form of radiation or by a collision and is raised from the ground state to an excited state, an *absorption spectrum* results. When the molecule gives up its excess energy and goes back to the ground state, an *emission spectrum* results.

There are many possible excited states, but there is not an infinite number. As illustrated in Fig. 5.9, a definite amount of energy (*quantum*) is required for each transition. According to the classical description involving a continuous number of energy levels, if a molecule absorbed energy, it would start to rotate at a higher frequency. If radiation of this new frequency was then absorbed, the molecular would rotate faster. This

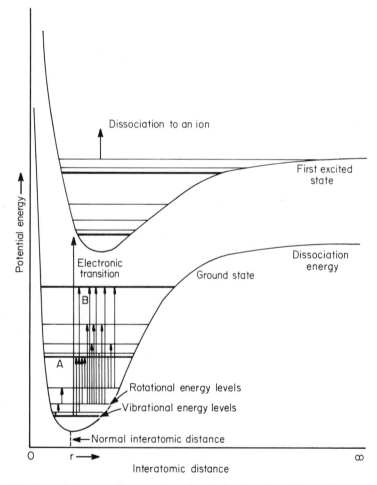

Fig. 5.9. Potential energy diagram of rotational, vibrational, and electronic transitions.

process could continue until the molecule was spinning so fast that it could tear itself apart. Experimentally we know that this does not happen.

According to our modern knowledge, a molecule can absorb only certain energies. If a molecule absorbs energy and rotates, it must get exactly the right energy to make the next jump before anything further will happen. Since the next jump requires twice the original energy, it is harder for the second step to take place. The third jump is even harder because this takes four times the original amount of energy. If the molecule gets an extra large amount of energy, rather than rotate at a destructive rate the molecule will start to vibrate and rotate slower (Fig. 5.9). This process can be repeated several times until the molecule vibrates with such an amplitude that it again risks destroying itself. Rather than destroy itself, an electron within the molecule moves to an excited state and uses up the excess energy.

It is important to note that a molecule does not vibrate faster with an absorption of energy but with a greater amplitude. Atoms within molecules vibrate between 10^{12} and 10^{15} times a second.

PROBLEMS

5.1. Ham radio operators who communicate worldwide operate in the 14.000–14.350 MHz (megahertz) frequency range. How long (in meters) in the wavelength associated with a frequency of 14.017 MHz? (Note: 1 MHz = 10^6 cycles/sec.) *Ans.* 21.4 m.

5.2. Channel 9 on CB radios is the distress frequency. It has a wavelength of 11.08 m. What is its frequency? *Ans.* 27.065 MHz.

5.3. When the Pioneer X space satellite left our solar system, it took 4.5 hr for its signal to return to earth. How far away (in meters and miles) was it at that time? (Note: 1 statute mile = 1609 m.) *Ans.* 4.86×10^{12}; 3.02×10^9 miles.

5.4. The flavor components of baked bread contain low-molecular-weight alcohols as part of their makeup. These compounds contain the —OH functional group, which absorbs radiation at 2.7 μm. What is the vibrational frequency of the —OH group? (Note: 1 μm = 1×10^{-4} cm.) *Ans.* 1.1×10^{14} sec^{-1}.

5.5. AM radio station KBUR in Burlington, Iowa, operates on a frequency of 1490 kHz. If you listened to music from a radio tuned to that frequency until you absorbed one mole of those photons, how many calories of energy would you receive? (Note: 1 cal = 4.18×10^7 ergs.) *Ans.* 1.42×10^{-4} cal.

5.6. Radiation from a mercury vapor lamp (254 nm) has been tested for controlling bacteria on meats. How many calories of energy would you receive if you absorbed one mole of photons from such a lamp? Notice the difference between standing in front of a radio and in front of a mercury vapor lamp. This is why safety signs should be posted when mercury lamps are on. *Ans.* 1.13×10^5 cal.

5.7.   The preservative sodium benzoate ($C_7H_5O_2Na$) is used in most soft drinks. How many fundamental vibrational frequencies are there for the benzoate ion? *Ans.* 36.

6
Visible and Ultraviolet Spectroscopy

The energies associated with the visible and ultraviolet regions of the electro-magnetic spectrum range from 40 to about 140 kcal/mole. Such energies are sufficient to cause electronic transitions within molecules and to ionize many substances. The electrons involved are the outer or bonding electrons, and the nomenclature used to describe the process is based on the bonds formed. When two or more atoms unite to form a molecule, the electrons involved can form sigma bonds (σ) on *pi* bonds (π); there also may be some unused or *nonbonding* (n) electrons.

Formaldehyde will be used to illustrate the various possibilities (Fig. 6.1). Because the various electrons are involved in bonding to different extents, their energies are different, as shown in Table 6.1. Figure 6.2 summarizes the molecular orbital picture for formaldehyde. Notice that the bonding orbitals have a large overlap with each other, whereas the antibonding orbitals have little overlap. In practice, interaction between sigma and pi electrons is small and the main transitions are $\sigma \rightarrow \sigma^*$ (vacuum UV), $\pi \rightarrow \pi^*$ (UV), $n \rightarrow \pi^*$ (near UV, visible), and $n \rightarrow \sigma^*$ (vacuum UV). It takes about 10^{-15} sec for one of those transitions to take place. The $n \rightarrow \pi^*$ transition is forbidden by selection rules, and therefore the intensity of these bands is weak, about 10^{-6} that of the $\pi \rightarrow \pi^*$. This means that most of the spectra observed in the UV and visible range is due to a π electron being raised to an excited state. The $n \rightarrow \pi^*$ transitions may be seen if nonpolar solvents are used. We can distinguish a $\pi \rightarrow \pi^*$ from a $n \rightarrow \pi^*$ band by obtaining spectra in both a nonpolar and a more polar solvent. If the band is due to a $\pi \rightarrow \pi^*$ transition, the band will shift to longer wavelengths (a *bathochromic* or red shift) as solvent po-

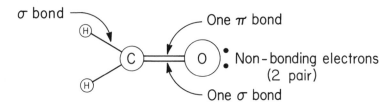

Fig. 6.1. *n*, σ, and π bonds in formaldehyde.

larity increases; if the band is due to a $n \rightarrow \pi^*$ transition, the band will shift to shorter wavelengths (a *hypsochromic* or blue shift). Figure 6.3 illustrates the terms used for spectral shifts.

The geometry of a molecule in an excited state may be different from its geometry in the ground state. Formaldehyde in the ground state is a flat molecule; in the excited state it is bent 20°and some of the bonds are longer (Fig. 6.4). This structural change can cause some rather drastic changes in chemical behavior. For example, phenol is 100,000 times more acid in the excited than in the ground state.

An extension of the electronic transition process is most dramatically illustrated in the visible region with molecular complexes. This is the process known as *charge transfer,* which Rao (1967) describes as follows:

The main spectral feature accompanying complex formation is the broad intense absorption band in the visible or ultraviolet region due to an electronic transition. Just as the excitation of an electron in an individual molecule by a quantum of radiation may be associated with intramolecular rearrangement of charge, similarly in the complex formed by the association of two molecular or ionic species, the excitation of an electron by a photon can involve a charge rearrangement in the complex. This rearrangement, according to the Mulliken theory, involves a transfer of an electron or part of it from one component of the complex to the other. The reaction may therefore be represented as follows for the alkali halide, MX.

$$M^+ + X^- + h\nu \rightarrow (M^+, X^-)^* \rightarrow M + X \tag{6.1}$$

Table 6.1. Electronic Energy Levels in a Molecule

Type		Symbol[a]	
Least stable	Antibonding	σ	σ*
Highest energy	Antibonding	π	π*
	Nonbonding	*n*	*n*
	Bonding	π	π
Most stable Lowest energy	Bonding	σ	σ
Energy			

[a] Asterisk is read as "star."

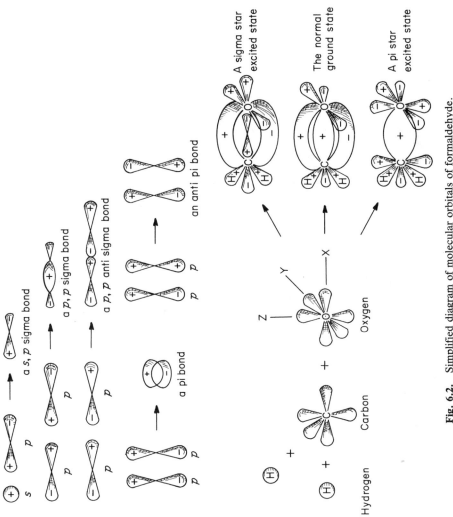

a s, p sigma bond

a p, p sigma bond

a p, p anti sigma bond

a pi bond

an anti pi bond

A sigma star
excited state

The normal
ground state

A pi star
excited state

Hydrogen + Carbon + Oxygen

Fig. 6.2. Simplified diagram of molecular orbitals of formaldehyde.

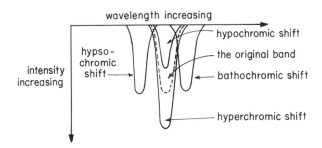

Fig. 6.3.   Spectral shift nomenclature.

where either or both the products M and X may be in the excited state. The primary excitation process is reversible and the excited complex $(M^+, X^-)^*$ will return to the ground level unless a secondary interaction can take place during its life span to give rise to the products M and X. It is the very short lifetime of the excited state (10^{-8} sec) which restricts the quantum efficiency of the reaction giving rise to color centers.

Figure 6.5 shows the position of the bands for the more common groups in the UV region. Because the bands are too broad (charge transfer) or too weak ($n \rightarrow \pi^*$), no similar chart in the visible region has been prepared.

Water is the most common solvent for the visible region. Methyl iso-butyl ketone and chloroform dissolve many organic compounds. In the ultraviolet region, cyclohexane is a good solvent for aromatic compounds; if a more polar solvent is needed, 95% ethanol is satisfactory. Other solvents that transmit down to 200 nm are acetonitrile, cyclopentane, hexane, heptane, methanol, 2,2,4-trimethyl pentane, and 2-methyl butane. Almost all of those compounds are saturated (sigma bonded) aliphatic organic materials. The transition for these solvents would be a $\sigma \rightarrow \sigma^*$, which requires more energy than is available in the UV region. Therefore, those materials are transparent in the UV region and do not interfere with the sample spectra.

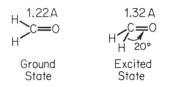

Fig. 6.4.   Geometry of formaldehyde in the ground and excited state.

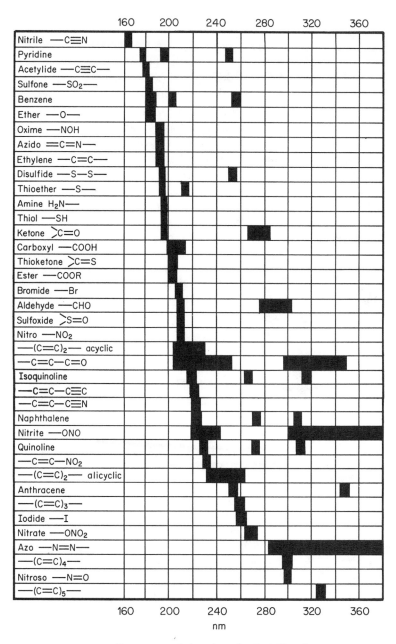

Fig. 6.5. Spectra correlation chart for the UV region.

INSTRUMENT COMPONENTS

All spectroscopic instruments contain the same basic components: a source, an attenuating device, a monochromator, a cell to hold the sample, a detector and amplifier, and a meter or recorder to observe the signal. The simplest and least expensive arrangement is a *single-beam* instrument (Fig. 6.6A). However, if the source voltage fluctuates during a scan or the solvent absorbs certain wavelengths, an error is introduced into the spectra. That error can be corrected in *double-beam* instruments by splitting the source radiation into two beams (Fig. 6.6B). Half of the time the beam passes through the reference cell containing everything but the sample, and the other half of the time the beam passes through the sample cell. Any source or solvent effects are registered by both the sample and the reference; therefore, if the detector measures the difference between the signals, the spectrum obtained is due entirely to the sample. The device used to split the source radiation into two beams is called a *chopper*, which is a rotating half disk with a mirror on one side. This disk turns at about 1000 rpm or faster. Single-beam instruments can be used for routine sample analysis, but double-beam instruments are required for good, fast, research work.

Sources

In the visible region, a tungsten lamp is a satisfactory radiation source. Only about 15% of the energy from a tungsten lamp is in the visible region, yet it can provide sufficient energy for most measurements in that region. Only a few percent of the radiation from a tungsten lamp is ultraviolet radiation. The most common sources for the UV region are the hydrogen discharge lamp, the mercury vapor lamp, and the deuterium (D_2) discharge lamp.

Attenuating Devices

An attenuator regulates the amount of radiation coming from the source. Since different wavelengths have different energies and since mirrors, lenses, cells, solvents, and monochromators all absorb a certain amount of radiation, some means must be provided to control the intensity of the radiation coming from the source. This can be done in several ways.

Some instruments have a sliding wedge that fits over the slit through which the source radiation passes. As the wedge is adjusted to cover the slit, a decreasing amount of radiation can get through.

The source also can be attenuated by varying the voltage to the lamp

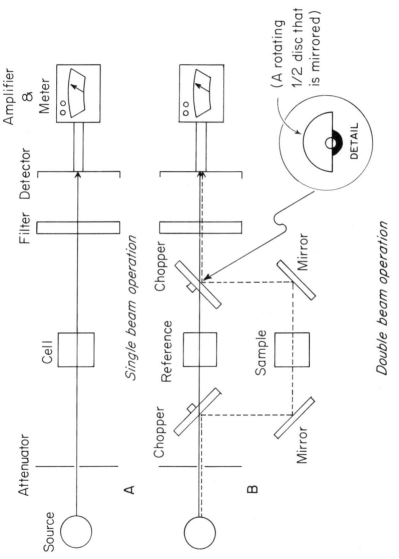

Fig. 6.6. Basic components of single- and double-beam instruments.

filament. Because the intensity of the radiation varies as

$$I = kV^n \qquad (6.2)$$

where I is the intensity, k the proportionality constant, V the voltage, and $3 \leq n \leq 4$, a small line voltage change will produce a large intensity change. That is why the meter needle moves continuously in many low-priced instruments. In more expensive instruments, voltage regulators are used to eliminate the fluctuation.

Cells

The cells may be either round or square. For measurements in the visible region, cells are usually made of glass. However, glass is not satisfactory in the UV region because it absorbs most of the radiation below 360 nm. Two good materials for the UV region are quartz and fused silica. Quartz has slightly better dispersing ability than fused silica, but fused silica transmits down to 185 nm, whereas quartz transmits only to 200 nm.

If Pyrex glass cells get mixed in with quartz cells, it is hard to tell them apart. They can be distinguished by placing the cells in a solution of trichloroethylene, which has a refractive index (1.475) equal to that of Pyrex (1.474). A Pyrex cell placed in this solution will disappear, but a quartz cell will remain visible.

Round cells should always be placed in the instrument in the same direction every time. This is necessary because round cells are seldom uniform in diameter and the glass thickness varies. Cells also should be filled nearly full. If they are filled only a little past the light path, reflection and stray-light errors may occur.

Monochromators

Before the various types of monochromators are discussed, we will explain their purpose. Suppose radiation of the wavelengths indicated in Fig. 6.7 (top) passes through a cell full of purple $KMnO_4$ solution. Figure 6.7 is a simplification (if not mechanically correct) of what happens. In order to obtain 100% transmittance (T), we put 20 units of each of the five wavelengths through the sample. However, the violet, blue, yellow, and red wavelengths are essentially unaffected by the permanganate solution and pass through; the green is absorbed, as indicated in the absorption spectrum shown in Fig. 6.8. Thus, in this case the meter needle reads 75%T, a 25%T change.

It is apparent that only the green radiation is of any analytical value. The other wavelengths simply "clutter up" the detector. The lower part

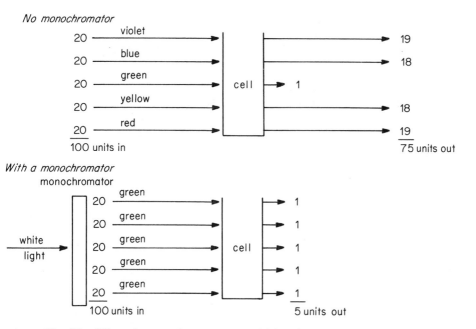

No monochromator

With a monochromator

Fig. 6.7. Effect of a monochromator on sensitivity of spectral measurements.

of Fig. 6.7 shows what happens when all but the green wavelengths are eliminated by use of a monochromator in the source beam. Again, 100 units of radiation enter the sample, but all the extraneous wavelengths are filtered out and only green gets through. The meter now reads 5%T. In this case, the change or deflection in *T* is 95% with a monochromator

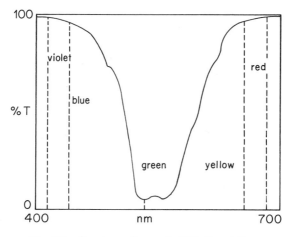

Fig. 6.8. Spectrum of a purple $KMnO_4$ solution.

but is only 25% without a monochromator. Thus we have almost a tenfold increase in sensitivity.

A *monochromator* is a device that will pass a limited number of wavelengths of radiation. The least expensive instruments use colored glass or solutions as monochromators. Actually their band pass widths may be as much as 150 nm, or half of the visible region. A much narrower band pass, 30–50 nm, can be obtained with an interference filter (Fig. 6.9). The filter consists of two pieces of glass, each mirrored on one side and separated by a spacer. Imagine it as a jelly sandwich: the bread is the glass, the butter the silver mirror, and the jelly the spacer. When radiation strikes one glass surface and passes through it, the radiation now travels to the other side where it strikes the mirrored surface at an angle. If it does not have enough energy to get through the mirrored surface, it is reflected back to the opposite mirror, a process that may be repeated several times. Along the way the wave may be reinforced by another wave of the same wavelength. If their crests coincide at the mirrored surface, their combined energies break through the mirrored surface. This is *constructive interference*. Waves of other wavelengths cannot get in phase at the mirrored surface because of the space limitations set up by the spacer and cancel each other out, a process known as *destructive interference*. The radiation emerging from the exit side of the filter will not only be λ but λ/2, λ/3, λ/4, etc., since those wavelengths can also be in phase in the space between the glass. These are shorter wavelengths and of higher energy than the desired wavelength, and can cause a great deal of difficulty if they reach the detector. Therefore a piece of glass or other absorber is used to remove them and act as a secondary filter.

A prism (Fig. 6.10) can produce even more nearly monochromatic radiation. A single prism is capable of providing band widths of 10–20 nm.

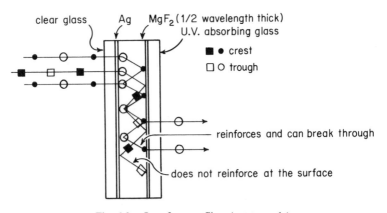

Fig. 6.9. Interference filter (not to scale).

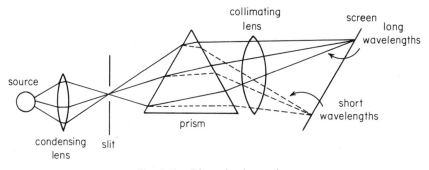

Fig. 6.10. Dispersion by a prism.

An advantage of a prism over an interference filter is that the prism provides a continuous spectrum over a much wider range. A disadvantage is that it has nonlinear dispersion. Because of this effect, two wavelengths separated by 5 nm at the short wavelength end of the spectrum are dispersed more than two wavelengths separated by 5 nm at the long wavelength end, as illustrated in Fig. 6.11. This makes it mechanically difficult to scan the wavelengths.

An entirely different approach for producing monochromatic radiation is to use a *diffraction grating*. A grating is made by evaporating a thin metallic film onto a glass surface and then ruling lines in it to make openings. A good grating has 15,000 parallel smooth lines to the inch. The main disadvantage of transmission-type grating is that the support absorbs much of the radiation. The *reflection grating* does not have this disadvantage. As shown in Fig. 6.12, λ_1 must travel further than λ_2 from the source to the viewer. If this distance difference is equal to one wavelength, constructive interference takes place. Equation 6.3 gives the relationship necessary:

$$n\lambda = d(\sin i + \sin \theta) \qquad (6.3)$$

where n is the order number. The other terms are defined in Fig. 6.12.

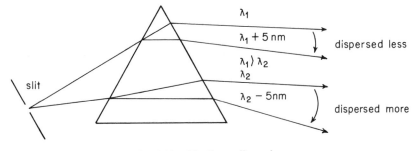

Fig. 6.11. Nonlinear dispersion.

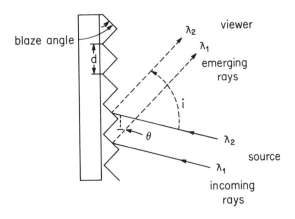

Fig. 6.12.   Reflection grating.

Detectors

The main type of detector for the UV and visible regions is the phototube. Figure 6.13 shows a single-stage phototube.

The incident radiation strikes the cathode, which is coated with a mixture of cesium oxide, silver oxide, and silver. The energy from one photon of radiation releases one to four electrons from the cathode surface. Those electrons are attracted to the anode by applying a positive potential of 50–100 V. This signal is amplified further until it is strong enough to move a recorder pen. The detection and much of the amplification can be com-

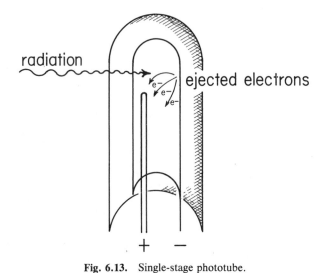

Fig. 6.13.   Single-stage phototube.

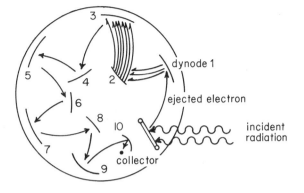

Fig. 6.14. Photomultipler tube.

bined into one operation by the 10-stage photomultiplier shown in Fig. 6.14. The initial process is the same as in a single-stage phototube, but the anode is another electron-emitting screen like the cathode of the single-stage phototube. Each emitted electron is accelerated by a 100-V potential applied to each *dynode*. The accelerated electron increases its en-

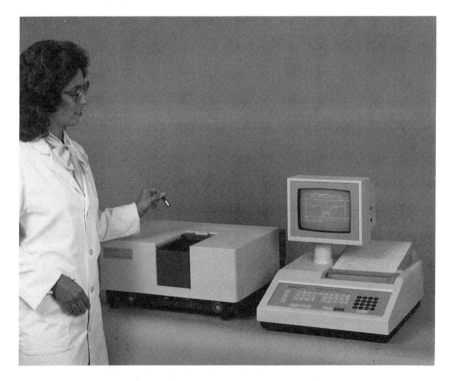

Fig. 6.15. An IBM ultraviolet–visible spectrophotometer. (Courtesy IBM, Danbury, CT.)

ergy to the point that when it strikes the first dynode it has enough energy
to release several more electrons. This process is repeated until dynode
10 is reached. By now, several million electrons have been released and
a corresponding amplification of the signal is obtained.

Spectrophotometers

Figure 6.15 is a photo of a modern ultraviolet–visible spectrophotometer. It is a double-beam instrument with a holographically ruled single-
grating monochromator. It has a spectral range of 195–900 nm with a 0.1-
nm resolution and a photometric range of −0.3 to 3.0 absorbance units.
A digital interface gives the microprocessor control over the optics, all
automated features, and displays. In normal operation, spectra can be
generated by depressing a single key. This instrument has automatic base-
line correction, spectral subtraction, axis expansion, smoothing, differentiation, and multiwavelength measurements. If something malfunc-
tions, the self-diagnostics built into the microprocessor determine where
the problem exists. In 1983 this instrument cost $13,000.

QUANTITATIVE ANALYSIS

Ultraviolet and visible spectral measurements can be used for quantitative analysis because the amount of radiation transmitted by a solution
depends on its concentration and pathlength. In this section we describe
these relationships and present the relevant equations. Selection of the
best wavelength for measurement and the range of intensity also are
discussed.

Consider the glass cell in Fig. 6.16A having a thickness b and filled
with a solution of concentration c. The radiation entering the cell has an
intensity I_0, and the emerging radiation has a value of I. The ratio of I/I_0
is called *transmittance* (T), and $I/I_0 \times 100$ = percentage transmittance
($\%T$). Since radiation I_0 has not entered the cell and none has been absorbed, it can be set arbitrarily at 100%. Assume that I is 50%. If the
thickness of the cell is doubled (Fig. 6.16B), I drops to 25% or one-half
of 50, just as 50 was one-half of 100. If the cell path is further increased,
the intensity keeps decreasing by the same factor.

The curved line in Fig. 6.17 shows the plot of I vs. cell thickness. The
differential equation describing this curve can be rearranged and integrated to give

$$I/I_0 = e^{-kb} \tag{6.4}$$

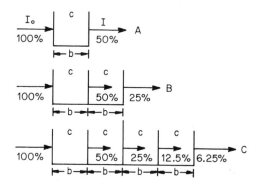

Fig. 6.16. Effect of path length (b) on transmittance.

By the same line of reasoning, a curve of $\%T$ vs. concentration c, similar to Fig. 6.17, can be plotted and the following equation derived:

$$I/I_0 = e^{-k_1 c} \qquad (6.5)$$

Equations 6.4 and 6.5 can be combined to give

$$I/I_0 = e^{k_{12} bc} \qquad (6.6)$$

To be useful as an analytical tool, spectral measurements should double in value if the concentration doubles. This is not the case with equation 6.6. A new term, *absorbance* (A), is defined as

$$A = \log 1/T = \log I_0/I \qquad (6.7)$$

Notice that this involves common logs (log), not natural logs (ln). After multiplying the log of both sides of equation 6.6 by 2.3 (to convert natural

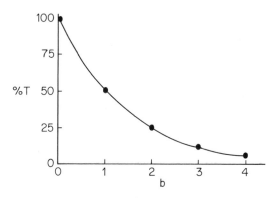

Fig. 6.17. Variation of $\%T$ with cell thickness (b).

log to common log), we have

$$\log I/I_0 = -2.3\, k_2 bc \qquad (6.8)$$

Multiplying by -1 to get rid of the minus sign and combining $2.3\, k_2$ into a new constant a yields

$$\log I_0/I = abc \qquad (6.9)$$

By substituting A into equation 6.9, we obtain

$$A = abc \qquad (6.10)$$

where a is the absorptivity, b the cell thickness in cm, and c the concentration in any convenient units (usually mg/ml or μg/ml). However, if the work is to be published, c must be in moles/liter and then a is changed to ϵ (molar absorption). Equation 6.10 is known as *Beer's law*.

To convert A to %T, equation 6.7 is rewritten as

$$\log I_0 - \log I = abc = A \qquad (6.11)$$

I_0 is always set at 100%, so $\log I_0$ is 2. Log I is the %T read on the instrument. Therefore,

$$2 - \log \%T = A \qquad (6.12)$$

Example: Caffeine in coffee and tea can be determined by its absorption at 276 nm. The caffeine is first extracted with dilute NH_4OH and then separated from other materials on a Celite column eluted with $CHCl_3$. If a standard sample containing 0.5 mg caffeine/50 ml of $CHCl_3$ had a reading of 80%T, calculate the concentration of caffeine in an unknown sample solution with a %T of 60. 10-mm cells were used:

For the standard, $A =$	For the unknown $A =$
$2 - \log \%T$	$2 - \log \%T$
$2 - \log 80$	$2 - \log 60$
$2 - 1.903$	$2 - 1.778$
0.097	0.222

To determine the absorptivity a, employ Beer's law (equation 6.10) using the experimental information obtained from the standard:

$$A = abc$$

$$0.097 = a \times 1\ \text{cm} \times 0.5\ \text{mg}/50\ \text{ml}$$

and

$$a = 0.0194$$

The a thus obtained can only be used when concentrations are expressed in mg/50 ml. Determine the caffeine concentration of the unknown sample as follows:

$$A = abc$$

$$0.222 = 0.0194 \times 1 \text{ cm} \times c$$

and

$$c = 1.14 \text{ mg caffeine/50 ml of CHCl}_3$$

Selecting the Wavelength and Transmittance for Measurement

In general, the best wavelength at which to make spectral measurements is the one that transmits the least radiation. The reason for this is illustrated by the sample spectra in Fig. 6.18. The top spectrum was obtained at a sample concentration of 1 mmole; the bottom one at a concentration of 20 mmole. If a measurement is made at about 415 nm, a 10%T change is obtained for a twentyfold concentration change, but if the measurement is made at about 515 nm, an 80%T change is obtained for the same twentyfold concentration change. Thus, the sensitivity of measurements is greatest at the wavelength that transmits the least radiation (i.e., absorbs the most).

The accuracy of spectral measurements is affected by %T, as shown

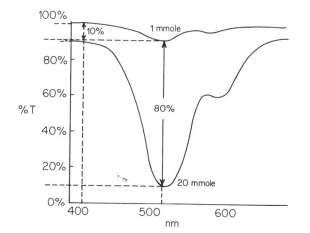

Fig. 6.18. Sample spectra illustrating effect of wavelength on sensitivity of measurements.

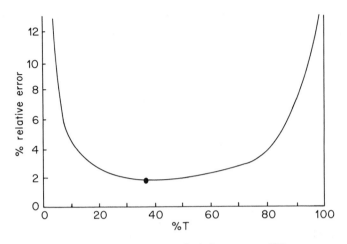

Fig. 6.19. Relationship of relative error to %*T*.

in the plot of % relative error vs. %*T* in Fig. 6.19. For a solution that obeys Beer's law, the least error occurs at 36.7%*T*. Thus, the most accurate readings are obtained when %*T* is near this value. However, readings between 20 and 80%*T* are also reasonably accurate. In contrast, very low and very high readings have a large error and should not be used for analytical determinations. The %*T* can be adjusted by changing either the concentration of the sample solution (by diluting or concentrating it) or the path length of the cell.

RECENT APPLICATIONS

Propoxur, a nonsystemic carbamate insecticide used on field crops, fruits, and vegetables, can be determined by a simple colorimetric method (Kalapanda *et al.* 1983). A diazonium salt is prepared with 4,4′-diaminodiphenyl sulfone and the resulting dye is measured at 500 nm. Carbaryl in grains can also be determined by a diazonium salt method (Appaiah *et al.* 1982) in which the pesticide is hydrolyzed with alcoholic KOH to 1-naphthol then coupled with 4-aminophenazone.

Ott and Gunther (1983) developed a rapid method for field screening citrus fruits for contamination by parathion. The method, which is sensitive to 5 ppm on a 1-g rind sample, can be used by a grower to determine when the pesticide level has decreased to a level safe enough to harvest the crop.

The usual method for determining vitamin E has been color formation with iron. In an improved method developed by Contreras-Guzman *et al.*

(1982), Cu^{2+} is reduced to Cu^+ by vitamin E and the Cu^+ then chelated with either cuproin or bathocuproin. The color is much more stable and just as sensitive as that formed in the iron method.

Caramel is often added to whiskey and its amount is expressed as a color intensity unit (CIU) that is $1000A$ at 525 nm. Strunk et $al.$ (1982) and 20 laboratories agreed on a uniform method for determining caramel in whiskey.

Galoux et $al.$ (1982) developed a rapid method for determining 3-amino-1,2,4-triazole in grain and meat. It involves a modified Storherr and Burke separation followed by color development with N-(1-naphthyl)ethylene diammonium dichloride and measurement at 455 nm. The method is sensitive to 0.05 ppm.

The reaction of benzophenone with the $P = S$ group of thiophosphates serves as the basis for the spectrophotometric determination of these compounds in technical materials and formulations. According to Sane and Kamat (1982), recoveries are 99–100% and the method covers the range 35–350 μg/ml at 599 nm.

Ultraviolet spectrophotometry was used by Byler et $al.$ (1983) to investigate the reaction between nitrite ion and cysteine to determine if nitroso compounds were formed. They found that S-nitrosothiols were formed at concentrations of nitrite and pH levels usually employed for food preservation. The authors speculated that this reaction might be an enzyme inhibitory reaction that kills bacteria on the foods.

PROBLEMS

6.1. What is the absorbance of a solution that has a $\%T$ of 58? $Ans.$ 0.236.

6.2. The color intensity of whiskey is expressed in CIUs, defined as $1000 \times A$ at 525 nm by the method of Strunk et $al.$ (1982). Whiskeys usually range from 29 to 374 CIU. A bourbon had an absorbance reading of 0.165 at 525 nm. What is the color intensity of this liquor? $Ans.$ 165.

6.3. Slits were first used in the light path in an absorption spectrophotometer by E. K. Vierordt in 1873. Why use a slit? Why not use a square or circle? Hint: Place two toothpicks on top of each other (lines of radiation) and then move the top one to the side until you can tell that there are two toothpicks. Repeat with two coins and compare the results.

6.4. An analyst, using the method of Galoux et $al.$ (1982), found that the minimum concentration of 3-amino triazole (MW = 84) that could be determined in meat products was 0.05 ppm. At this concentration, the absorbance of the colored reactin product was 0.02 with 1.00-cm cells. What is the molar absorptivity (ϵ) of this reaction? $Ans.$ 33,600.

6.5. The method of Contreras-Guzman and Strong (1982) for determining vitamin E (avg. MW = 416) uses Cu^{2+} and urea as a catalyst for color development. A vitamin E solution of 1.2×10^{-5} M had an absorbance of 0.445 in 10-mm cells at 478 nm. What a should be used if c is to be measured in μg/ml? *Ans.* 0.089.

6.6. Iron in a biscuit sample is to be determined by forming the tris(1,10-phenanthroline) iron (III) chelate. A 0.50-μg/ml standard produced an absorbance of 0.34 at 515 nm with 1.00-cm cells. Calculate the %T and the a for this system. *Ans.* %T = 45.7; a = 0.68 ml/μg cm^{-1}.

6.7. Butylated hydroxyanisole (BHA, FW = 180) is often added to cereals as a fat antioxidant. The current law permits no more than 200 ppm BHA to be present. Assume the molar absorptivity is 8000 (10-mm cells) and the density of fat is 0.90. What must be the weight of a cereal sample to obtain a %T reading of 37% if 200 ppm BHA is present? *Ans.* 48.4 g.

BIBLIOGRAPHY

Appaiah, K. M., Ramakrishna, R., Subbarad, K. R., and Kapur, O. (1983). Spectrophotometric determination of carbaryl in grains. *J. Assoc. Off. Anal Chem.* **65**, 32–34.

Byler, D. M., Gosser, D. K., and Susi, H. (1983). Spectrophotometric estimation of the extent of *S*-nitrosothiol formation by nitrite action on sulfhydryl groups. *J. Agric. Food Chem.* **31**, 523–527.

Contreras-Guzman, E. S., and Strong, F. C., III. (1982). Determination of tocopherols (vitamin E) by reduction of cupric ion. *J. Assoc. Off. Anal. Chem.* **65**, 1215–1221.

Galoux, M., VanDamme, J. C., and Bernes, A. (1982). Colorimetric determination of 3-amino-1,2,4-triazole in grains and meat. *J. Assoc. Off. Anal. Chem.* **65**, 24–27.

Kalapanda, M. A., Omprakash, K., and Krishnarajpet, V. N. (1983). Colorimetric determination of propoxur and its residues in vegetables. *J. Assoc. Off. Anal. Chem.* **66**, 105–107.

Ott, D. E., and Gunther, F. A. (1983). Colorimetric method for field screening above tolerance parathion residues on and in citrus fruit. *J. Assoc. Off. Anal. Chem.* **66**, 108–110.

Rao, C. N. R. (1967). "Ultraviolet and Visible Spectroscopy." Butterworths, London.

Sane, R. T., and Kamat, S. S. (1982). Simple colorimetric method for determination of organothiophosphate pesticides in technical materials and formulations. *J. Assoc. Off. Anal. Chem.* **65**, 40–42.

Strunk, D. H., Timmel, B. M., Hamman, J. W., and Andreasen, A. A. (1982). Spectrophotometric determination of color intensity of whiskey: Collaborative study. *J. Assoc. Off. Anal. Chem.* **65**, 224–226.

7
Measurement of Color

Color and discoloration of many foods are important quality attributes in marketing. Although they do not necessarily reflect nutritional, flavor, or functional values, they relate to consumer preferences based on the appearance of the product. Color characteristics of foods can result from both pigmented and originally nonpigmented compounds.

The broad area of color in foods may be divided into two general subtopics: (1) addition of approved synthetic colors to achieve a desired appearance and (2) determination of natural pigments (Francis 1963). Acceptability, uniformity, and reproducibility are the most important factors in the use of synthetic colors. Natural pigments in foods are determined as an index of economic value (i.e., in grading) or to control color in processing and storage.

Color is often used to determine the ripeness of fruit (green color in tomatoes or peaches; white color in lima beans). The color of potato chips is largely controlled by the reducing sugars content, by storage conditions, and by subsequent processing. The yellow color of egg yolk and of the skin of chicken is a direct function of the amount of pigment present. The color of poultry meat also can be influenced by the amount of finish put on the bird. Inclusion of xanthophyll pigments or antioxidants in poultry feed can influence skin and shank color. The brown discoloration in frozen turkeys is independent of the pigment quantity, but is related to freezing rate. Color is one of the more prominent visible characteristics of raw and cured meats.

Color measurement was reviewed by Francis and Clydesdale (1975), Hunter (1979), Francis (1983), and Clydesdale (1984).

OPTICAL ASPECTS

Color arises from the presence of light in greater intensities at some wavelengths than at others. In practice, it is limited to the band of the spectrum from 380 to 770 nm, the part of the electromagnetic spectrum that is visible to the human eye. Precise measurement of color in foods can be made, and the color expressed in terms of internationally accepted units, provided the measurement is not complicated by appearance factors such as gloss, mottling, and texture (Kramer and Twigg 1970). Color is an appearance property attributable to the spectral distribution of light; gloss, transparency, haziness, and turbidity are properties of materials attributable to the geometric manner in which light is reflected and transmitted.

From the point of view of optics, color is the stimulus that results from the detection of light after it has interacted with an object. Thus, three factors are involved: a light source, an object, and a receiver–detector. For standard comparisons, three standard illuminants (light sources) have been established by the International Commission on Illumination (called CIE from the initials of the French name, Commission Internationale de l'Eclairage): Illuminant A—incandescent lamp at 2848 K; Illuminant B— noon-sunlight lamp at 5000 K; and Illuminant C—average daylight at 6740 K.

The light may be reflected, transmitted, absorbed, or refracted by an illuminated object. If practically all radiant energy in the visible range is reflected from an opaque surface, the object appears white. If the light through the entire visible spectrum is absorbed in part, the object appears to be gray. If it is absorbed almost completely the object is black.

The term *lightness* (or *value*) refers to the relation between reflected and absorbed light, without regard to specific wavelength. *Hue* is the aspect of color that we describe by words as green, blue, yellow, or red. This perception of color results from differences in absorption of radiant energy at various wavelengths. Thus, if the shorter wavelengths of 400– 500 nm are reflected to a greater extent than other wavelengths, the color is described as blue. Maximum reflection in the medium wave length range results in a green or yellow color; and maximum reflection at longer wavelengths (600–700 nm) indicates red objects. The term *chroma* (also called *saturation* or *purity* refers to reflection at a given wavelength and indicates how much a color differs from gray. It describes, for example, how a red brick differs from a red tomato if the two are of the same lightness and hue. Visual perception of color can be described by the three variables hue, value, and chroma arranged in a cylindrical coordinate system, as shown in Fig. 7.1. If the light is reflected from a surface evenly at all angles, we have the impression of a product with a flat, dull, or diffuse

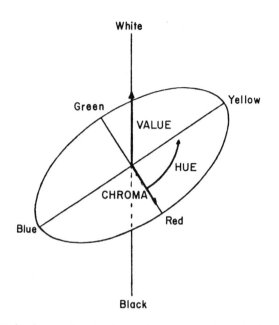

Fig. 7.1. The visual perception of color can be represented in a simple case by the three variables hue, value, and chroma arranged in a cylindrical coordinate system. (From Bernhart 1969.)

appearance. If the reflection is stronger at a specific angle or in a beam, we observe gloss or sheen as a result of *specular* or directional reflectance.

As mentioned previously, in practice it is also important to know the effect of physical characteristics of a surface on its resultant reflectance. Color, as complex as it is, is only one aspect of the appearance of an object. Gloss, texture, relative opacity, and surface uniformity are also important. Thus, colors of objects with different appearances can be matched only for a given set of viewing conditions. It is virtually impossible, for instance, to match the appearance of a flat surface to a textured one.

The preparation of a sample for color measurements is dictated by the specific requirements for which it is evaluated (Mackinney and Chichester 1954; Little and Mackinney 1969). This requirement ranges from the absolute restriction that the sample be intact and unaltered to a wide latitude in sample alteration.

CIE COLOR SYSTEM

The standard color measurement system against which other systems should be compared is that proposed by the CIE in 1931. The CIE system specifies a color by three quantities—X, Y, and Z—called *tristimulus values*. These values represent the amounts of three primary colors— red, green, and violet—that are required for a standard observer to get a match. If each of the tristimulus values is divided by the sum of the three, the resulting values x, y, and z, called *chromaticity coordinates,* give the proportion of the total stimulus attributable to each primary. Since the sum of the three is unity, the values x and y alone can be used to describe a color. A two-dimensional plot of x vs. y forms a chromaticity diagram. The third dimension, of lightness or darkness, is defined by the Z tristimulus value. The chromaticity diagram of the CIE system is shown in Fig. 7.2.

Several recording spectrophotometers draw a continuous curve of the light reflected by a sample for the visible spectrum range. The essential components of such a reflectance spectrophotometer are shown in Fig. 7.3. The CIE has recommended a standard method of reducing by calculation the spectrophotometric curve across the entire range of visible wavelengths to three numbers that adequately and accurately describe the color. For this purpose, CIE has established three curves, called color mixture curves, that represent the color vision of the standard observer under standard light conditions. For a color specification, the reflectance of a sample at wavelength intervals of 10 nm is multiplied by the respective

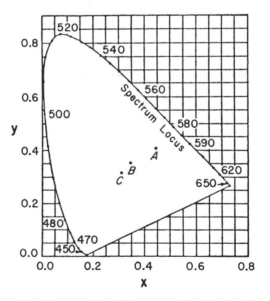

Fig. 7.2. Chromaticity diagram of the CIE system. All real colors lie within the horse-shoe-shaped locus marked with the wavelengths of the spectrum colors. (From Bernhardt 1969.)

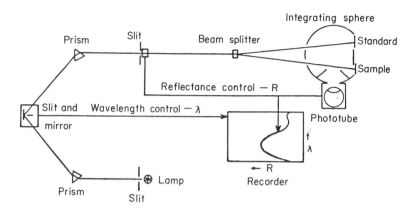

Fig. 7.3. The essential components of a reflectance spectrophotometer. A white-coated integrating sphere is used to collect the light reflected from the surface of the sample and direct it to the phototube. (From Bernhardt 1969.)

values of X, Y, and Z for a standard light source. This shows how much energy of each wave length of the illuminant source is reflected from the sample. Next, the products of X, Y, and Z are totaled for each wavelength. Then x, y, and z are determined from the folowing ratios:

$$x = \frac{X}{X + Y + Z}, \quad y = \frac{Y}{X + Y + Z}, \quad z = 1 - (x + y) \quad (7.1)$$

This calculation is quite laborious and has been replaced by tristimulus integrators attached to the spectrophotometer.

An ideal sample for measuring reflectance is flat, homogeneously pigmented, opaque, and light-diffusing. Practically all foods are irregular in shape, texture, particle size, and surface characteristics; are both light-transmitting and light-diffusing; and show inhomogeneous pigment distribution. Ideal samples for transmittance measurements are clear and moderately light-absorbing. In practice, these criteria are seldom met.

Compressing a dry powdered sample (e.g., flour) to a pellet gives an approximately ideal reflecting surface. Some products (e.g., coffee beans) can be ground and compressed, but the procedure is difficult to reproduce. Some dried foods (lyophilized egg whites or yolks, instant coffee, sweet potato flakes, gelatin desserts) can be pressed into wafers between thin disks of Teflon (Berardi *et al.* 1966). The surface reflectance characteristics of thin layers of high-moisture homogenized samples are affected by their translucency, depth of layer, and background. Resolution of spectral reflectance curves of homogenized tuna samples is greatly improved by treating the surface of the sample with sodium dithionate (Little and Mackinney 1969).

SIMPLIFIED TRISTIMULUS COLOR SYSTEMS

Color identification with a recording spectrophotometer is costly and time-consuming. More economical and practical, though less accurate, tristimulus systems are available for routine purposes.

In the *Munsell system,* color is determined by using scales of hue, value, and chroma. Hue is defined on the Munsell color tree in the circumferential direction by five main and five intermediate hues for decimal notation. Chroma (the degree of saturation) is defined by 10 or more steps in the radial direction, and value (the degree of lightness) by 10 steps in the vertical direction. In the Munsell system, three or four overlapping disks are used (Nickerson 1946)—one each for hue and chroma, and one or two to provide adjustment for value. The disks are adjusted until the color obtained by rapid spinning of the disks matches the color of the tested object. The results may be reported as such, as a ratio of the two chromatic disks, or may be transformed to the CIE system values by calculation (Essau 1958) or by the use of chromaticity diagrams on which Munsell notations are superimposed (Billmeyer 1951).

Nickerson (1946) described an instrument that used a Munsell disk colorimeter in which the angle, intensity, and type of light source are controlled. To reduce the time required for color matching, the lens is rotated instead of rotating the disks. A colorimeter with disks of specified Munsell notations is also available.

The tristimulus photoelectric colorimeter developed by Hunter (1952) is relatively inexpensive, rugged, and well-adapted to routine tests. The instrument consists of three separate circuits, filters, and photocells (Fig. 7.4). The Hunter R_d (diffuse reflectance) or L (lightness) values are directly comparable to Y in the CIE system or value in the Munsell system. The Hunter a value denotes redness or greenness; the Hunter b value measures yellowness or blueness (Fig. 7.5). The a values are functions of X and Y, the b values of Z and Y. The a and b values provide information equivalent to that of the hue and chroma dimensions of the Munsell system. Hunter and CIE values are interconvertible by calculation (Hunter 1958). Similarly, Hunter values can be converted to Munsell notations mathematically (Davis and Gould 1955) or by use of charts (Billmeyer 1951).

The *Lovibond Tintometer* (Fig. 7.6) is a visual colorimeter used widely in the oil industry. The instrument has a set of permanent glass color filters in the three primary colors—red, yellow, and blue. The colors are calibrated on a decimal scale in units of equal depth throughout each scale. The oil sample is placed in a glass cell and the filters are introduced into the optical system until a color match is obtained under specified con-

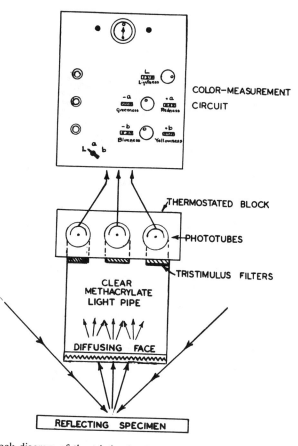

Fig. 7.4. Block diagram of the tristimulus Hunter colorimeter. (Courtesy Hunter Associates Laboratory, Inc.)

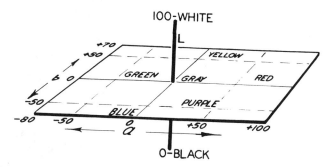

Fig. 7.5. The axes of L, a, b color values of the Hunter colorimeter. (Courtesy Hunter Associates Laboratory, Inc.)

Fig. 7.6. The Lovibond Tintometer is commonly used to measure color of oils. (Courtesy Lovibond of America, Inc.)

ditions of illumination and viewing. Color of oils is measured with transmitted light. Either transmitted light for molten waxes and fats or reflected light against a white background for solid samples is used. Attachments to the Lovibond Tintometer provide means of converting the readings from Lovibond units into trichromatic coordinate values. In addition, modifications for determining the color of liquid, solid, powder, or paste materials are available.

ABRIDGED METHODS

The most accurate measures of color generally require a function involving all three coordinates. In some cases, however, the work and computations may not be justified by the small increase in accuracy. In some instances, color can be determined satisfactorily by abridged methods instead of determining three dimensions of color.

In frozen lima beans, the Hunter L value (or Y of the CIE system) correlates well with panel scores. The effect of color on grade score of lima beans can be determined directly from a simple table that compares the Hunter L value with scores for three grades (Kramer and Hart 1954). In contrast, all three Hunter values (L, a, and b) must be determined with applesauce, since no single value or combination of two values correlates as well with consumer evaluations as multiple correlation including all three values. For determining U.S. grade for color of applesauce, nomographs are available.

The color of raw tomatoes can be measured with the Agtron E colorimeter from the reflectance ratio at 546 and 640 nm of cut surfaces of two halves of a tomato (Smith and Huggins 1952) or from determination of a single reflectance value on the Agtron F colorimeter (Smith 1953).

An approximation of the color of tomato juice also can be obtained from a reading of percent reflectance at one wavelength on the Agtron F (Gould 1954). The correlation with panel scores is improved if two color attributes (or their ratio) such as a (redness) and b (yellowness) are determined (Kramer 1954); these values can be determined with the Gardner color and color difference meter. A still more accurate figure representing the color can be obtained by including a function of all three color parameters. Such a score developed by Yeatman *et al.* (1960) for grading purposes is described by the function

$$\frac{aL}{\sqrt{a^2L + b^2L}} \frac{1}{L} \tag{7.2}$$

An instrument called the "tomato colorimeter" was developed by Hunter and Yeatman (1961) to measure the color of tomato juice in terms of this equation.

The brightness, or tristimulus y, is a useful index of the degree of roasting, development of flavor, and amount of extractable solids in coffee (Francis 1963). For fine control of color development, however, roasting to a given color is inadequate. The time–temperature relations in roasting must be controlled, and the relation between visual appearance and tristimulus reading must be measured. Whenever an abridged method is used, it is always advisable to select the appropriate parameter for the abridged procedure after tristimulus color data of the product have been correlated with panel scores.

Aqueous solutions of pure sucrose are clear and water-white. Such solutions may display a slight coloration and haze due to traces of impurities. Since the coloration is low, it is difficult to express in terms of tristimulus coordinates or related systems. The color is, therefore, commonly expressed in terms of an absorption index at a specified wavelength (Bernhardt 1969). When the solution has a haze, the transmitted light is attenuated through scattering and the absorption index cannot be determined in a single measurement. The scattering (turbidity) is the result of differences in the refractive indices of the solution and the trace impurities suspended in the solution. Total attenuation index of sugar solutions measure the contributions of absorption and turbidity. Total attenuation can be determined by measuring light transmitted by the colored and scattering solution. The turbidity can be determined independently of absorption by measuring scattered light. The absorption index (light attenuation due to color) can be calculated from the difference between the total attenuation and scattering. In determining the attenuation index through transmittance measurements, serious errors may occur if some of the scattered light is permitted to enter the photodetector. The scatter error is reduced in a special photometer designed by Bernhardt *et al.* (1962).

Simple color-grading equipment using predetermined fixed color ranges is available. For rapid grading of raw materials (e.g., lima beans, maple syrup, honey) standard colored plastic blocks, visually equivalent to the desired color, are used.

Little (1973) described the development of model systems that varied in a known and controllable fashion for color dimension; the determination of conditions for analyzing samples for physical and chemical properties that govern changes in perceived color; and the definition and construction of relevant scales. The results were then used to demonstrate the relationship between visual evaluations and physically determined quantities. The method was applied to five foods:

(1) canned tuna—opaque, irregularly pigmented with a labile pigment system and high specific absorption;

(2) applesauce—translucent, moderate light-scattering, and low specific absorption;

(3) milk—highly turbid, moderately light-transmitting, and low absorption;

(4) white wine—highly transparent with practically no turbidity and very low absorption, and

(5) red and rose wines—moderate to low transmittance with moderate to high absorption and very low turbidity.

The analytical methods included, as appropriate, reflectance measurements under oxidizing and reducing conditions, thin-layer transflectometry, and calculation of Kubelka–Munk optical constants. The same conceptual model was used for different products, but the procedural details differed depending on the constraints imposed by the physical and chemical characteristics of the foods.

METHODS BASED ON PIGMENT CONTENT

Methods for extracting pigments and measuring their concentration at specified wavelengths in a photoelectric colorimeter or spectrophotometer have been reviewed by Kramer (1965). Such methods are satisfactory provided the color impression of a food correlates well with the amount of the extracted pigment. In many foods, however, color as seen by the eye is not closely correlated with the concentration of extracted pigment. In addition, this approach places undue emphasis on total pigment content, whereas in practice, color of the outer surface is responsible for consumers' color impressions.

For example, the value of spectrophotometric determination of pigments in meat extracts is limited. The published methods measure the proportion of various forms of myoglobin in the total sample, whereas generally only the surface color is evaluated by the consumer. There are, however, many instances in which transmittance of light at a specified wavelength of a pigment extract is a satisfactory measure of color.

A translucent solution transmits its own color and absorbs the complimentary color. Red solutions absorb most in the green portion of the spectrum, yellow solutions in the blue-violet, and blue solutions in the orange-red. The major water-soluble red pigment of beets is betanin. A purple solution of beet juice transmits the purple color but absorbs the complimentary color. Consequently, at 525 nm (green) transmittance is lowest and absorption is highest. At 700 nm (red) practically all the light is transmitted. The greater the pigment concentration, the lower the percent transmittance of the complimentary color. In most analytical assays, the transmittance of the complimentary color is measured (Kramer and Smith 1946).

Beta-carotene is important in sweet potatoes as a nutritional component (precursor of vitamin A) and as a contributor to their pleasant orange color. Correlation between beta-carotene absorption and color is high, provided the effects of moisture content, pithy breakdown, and texture are accounted for. Generally beta-carotene comprises up to 90% of the total carotenoid pigments in sweet potatoes. If the pigment composition varies widely, both the type and quantity of each major pigment must be determined and weighted for their contribution to the color (Francis 1969).

BIBLIOGRAPHY

Anon. (1966). Color matching scores. *Chem. Eng. News* **50**, 51–52, 55–56.

Berardi, L. C., Martinez, W. H., Bondreaux, G. J., and Frampton, V. L. (1966). Rapid reproducible procedure for wafers of dried food. *Food Technol.* **20**, 124.

Bernhardt, W. O. (1969). Color and turbidity in solutions. *Food Technol.* **23**(1), 30–31.

Bernhardt, W. O., Eis, F. G., and McGinnis, R. A. (1962). The sphere photometer. *J. Am. Soc. Sugar Beet Technol.* **12**(2), 106.

Billmeyer, F. W. (1951). "Nomographs for Converting Hunter Color Values to C. I. E. Values." E. I. duPont de Nemours and Co., Wilmington, DE.

Billmeyer, F. W. (1967). Optical aspects of color. *Opt. Spectra* **1**(2), 59, 61–63.

Clydesdale, F. M. (1984). Color measurement. *In* "Food Analysis: Principles and Techniques," Vol. 1 (D. W. Gruenwedel and J. R. Whitaker, eds.), pp. 95–150. Marcel Dekker, New York and Basel.

Commission International de L'Eclairage. (1931). "Proceedings of Eighth Session." Cambridge, England. Sept. 19–29. Cambridge Univ. Press.

Davis, R. B., and Gould, W. A. (1955). A proposed method for converting Hunter color difference meter readings to Munsell hue, values, and chroma notations corrected for Munsell values. *Food Technol.* **9**, 536–540.

Essau, P. (1958). Procedures for conversion of color data for one system into another. *Food Technol.* **2**, 167–168.

Francis, F. J. (1963). Color control. *Food Technol.* **17**(5), 38–42, 44–45.

Francis, F. J. (1969). Pigment content and color in fruits and vegetables. *Food Technol.* **23**(1), 32–36.

Francis, F. J. (1983). Colorimetry of foods. *In* "Physical Properties of Foods" (M. Peleg and E. B. Bagley, eds.), pp. 105–123. AVI Publishing Co., Westport, CT.

Francis, F. J., and Clydesdale, F. M. (1975). "Food Colorimetry: Theory and Applications." AVI Publishing Co., Westport, CT.

Gould, W. A. (1954). Simplified color instrument now available to industry. *Food Packer* **35**(1), 33.

Hunter, R. S. (1952). "Photoelectric Tristimulus Colorimetry with Three Filters. Circ. *C 429.* U.S. Dept. Comm. Natl. Bur. Std. (U.S.).

Hunter, R. S. (1958). Photoelectric color difference meter. *J. Opt. Soc. Am.* **48**(12), 985–995.

Hunter, R. S. (1979). How does it look to you? *Ind. Res. Dev.* **8**, 68–71, 72.

Hunter, R. S., and Yeatman, J. N. (1961). Direct-reading tomato colorimeter. *J. Opt. Soc. Am.* **51**, 552.

Kramer, A. (1954). Color dimensions of interest to the consumer. Quartermaster Food and Container Institute, Symposium—Color in Foods. Chicago, IL.

Kramer, A. (1965). Evaluation of quality of fruits and vegetables. *In* "Food Quality: Effects of Production Practices and Processing" (G. W. Irving, Jr., and S. R. Hoover, eds.). Publ. 77. Am. Assoc. for the Advancement of Science, Washington, DC.

Kramer, A., and Hart, W. J. (1954). Recommendations on procedures for determining grades of raw, canned, and frozen lima beans. *Food Technol.* **8**, 55–62.

Kramer, A., and Smith, H. R. (1946). Preliminary investigation and measurement of color in canned foods. *Food Res.* **11**, 14–31.

Kramer, A., and Twigg, B. A. (1970). "Quality Control for the Food Industry," Vol. 1, 3rd ed. AVI Publishing Co., Westport, CT.

Little, A. C. (1973). Color evaluation of foods—Correlations of objective facts with subjective impressions. *In* "Sensory Evaluation of Appearance of Materials," pp. 109–127. ASTM STP *545.* Am. Soc. Testing Materials, Philadelphia, PA.

Little, A. C., and Mackinney, G. (1969). Colorimetry of foods—the sample as a problem. *Food Technol.* **23**(1), 25–28.

Mackinney, G., and Chichester, C. O. (1954). The color problem in foods. *Adv. Food Res.* **5**, 301–551.

Mackinney, G., and Little, A. (1962). "Color of Foods." AVI Publishing Co., Westport, CT.

Nickerson, D. (1946). Color measurement and its application to the grading of agricultural products. U.S. Dep. Agric. Misc. Publ. *580.*

Smith, T. J. (1953). Tomato grading by electronics. *Food Eng.* **25**(9), 53.

Smith, T. J., and Huggins, R. A. (1952). Tomato classification by spectrophotometry. *Electronics* **25**(1), 92.

Yeatman, J. N., Sidwell, A. P., and Norris, K. H. (1960). Derivation of a new formula for computing raw tomato juice color from objective color measurement. *Food Technol.* **14**, 16.

8
Fluorimetry

The phenomenon of *luminescence* was first studied by Stokes in about 1852. Stokes noticed that when fluorspar was placed in the sun it seemed to *glow*. The orange jackets of highway workers that appear very bright even on cloudy days, road signs that "glow in the night," and the afterglow when a television set is turned off are modern examples of this phenomenon. Luminescence is a general term to describe systems that can be made to glow. Such systems can be classified according to the glow-producing mechanism. The two major divisions of present-day analytical importance are *fluorescence* and *phosphorescence.*

In general, compounds that fluoresce or phosphoresce contain either an electron-donating group (amines, alcohols, and hetero atoms) or multiple conjugated double bonds (aromatic rings). Notice that these groups contain either nonbonding or pi electrons. The presence of groups that tend to withdraw electrons (e.g., carboxyl, azo, the halides, and nitro groups) usually destroy fluorescence. Figure 8.1 shows the energy levels involved in fluorescence and phosphorescence.

Consider a molecule having a pair of nonbonding electrons in which one electron is spinning opposed to the other. Such a state is called a *singlet state.* It generally requires radiation energy in the ultraviolet region to raise one of these electrons to an excited state (1 to 2 in Fig. 8.1). This process happens so fast (about 10^{-15} sec) that the molecule does not have time to change its basic geometry. The electron may immediately lose all of the energy it just acquired and return to its original energy level, a process known as *resonance fluorescence* (2 to 1 in Fig. 8.1).

Vibrational relaxation is the process by which a molecule in the excited

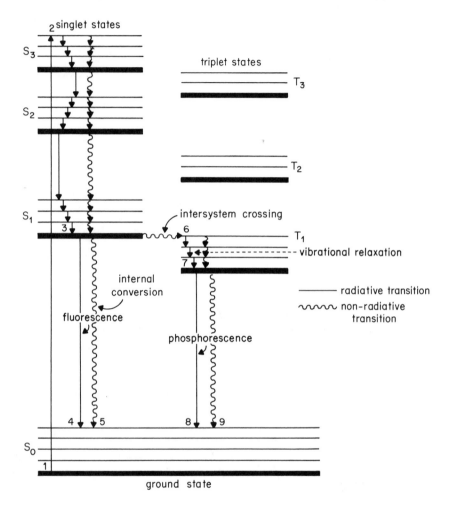

Fig. 8.1.   Molecular energy levels involved in fluorescence and phosphorescence.

state loses energy by dropping back to the lowest vibrational level in the excited state (2 to 3 in Fig. 8.1). In solution this excess energy can be taken up by the solvent molecules and no radiation is given off. The entire process takes from 10^{-11} to 10^{-13} sec. This process is so efficient that molecules in solution, generally only emit radiation from the lowest vibrational level of an excited state. The lifetime of molecules in this lowest level of the first excited state is from 10^{-7} to 10^{-9} sec.

If the molecule now returns to the ground state with the emission of radiation (3 to 4 in Fig. 8.1), the process is called fluorescence. Note that the energy emitted is less than the amount of energy originally absorbed.

This emitted energy is usually in the visible or near infrared region. Our eyes cannot detect the ultraviolet radiation that started the process, but they can detect the lower energy visible radiation that ends the process; this explains why certain substances tend to glow. Physically, fluorescence is regarded as an instantaneous process, that lasts about 10^{-8} sec.

Intersystem crossing is the process of changing from a singlet state (opposed electron spins) to a triplet state (matched electron spins) as shown in 3 to 6 in Fig. 8.1. This process can be very fast (about 10^{-8} sec) and is enhanced by the presence of heavy atoms within the molecule. Fast as it is, intersystem crossing is still much slower than vibrational relaxation and as a result, intersystem crossing usually occurs after the molecule has reached the lowest vibrational level in the singlet excited state. After intersystem crossing has taken place, the molecule immediately (10^{-13} sec) goes to the lowest vibrational level of the triplet state by vibrational relaxation. The lifetime of this lowest triplet state varies from 10^{-4} to 10 sec, which is very long compared with that of the lowest excited singlet state. If the molecule drops from the triplet state to the ground state with the emission of a quantum of radiation, the process is called phosphorescence (7 to 8 in Fig. 8.1). Physically this is a slow process, and phosphorescence is usually associated with compounds that tend to have an afterglow, that is they glow after the source of exciting radiation is shut off.

Because the triplet state lifetime is relatively long, there is a greater chance for the molecule to lose its energy through a radiationless transition either by vibrational coupling or by collision. The latter is so common that few systems will phosphoresce at room temperature and they must be cooled to liquid N_2 temperature (77°K) before the collisions are reduced to such a degree that phosphorescence can occur. Largely because of this, fluorescence is more easily measured than phosphorescence. The term *fluorimetry* refers to the measurement of fluorescent radiation; an instrument for such measurements is a *fluorimeter*.

INSTRUMENT COMPONENTS

The basic components of a fluorimeter are similar to those of an ultraviolet spectrophotometer. The radiation source must be in the ultraviolet range. Hydrogen, deuterium, and mercury lamps are commonly used; recently, xenon arcs have become more popular as fluorimeter sources. Quartz or fused silica optics and cells are required. Since the fluorescent radiation is in the visible region, a multiplying phototube can be used for the detector.

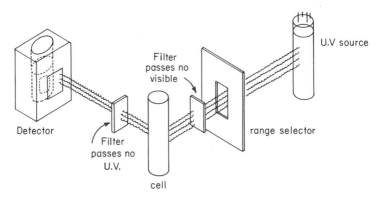

Fig. 8.2.   Diagram of a Turner fluorimeter. (Courtesy of Turner, Inc.)

The main difference between a fluorimeter and an ultraviolet spectro-photometer is that the detector is at right angles to the incident radiation in a fluorimeter (Fig. 8.2) and in line with the incident radiation in an ultraviolet instrument (refer to Fig. 6.6). Fluorescent radiation is emitted in all directions, whereas transmitted radiation is emitted only in the same direction as the incident radiation. This difference is utilized in the design of a fluorimeter, which measures the fluorescent radiation that *comes from the side of the sample*. It is possible to measure the amount of fluorescence that *comes out of the sample in the same direction* as the incident radiation. This requires a very good monochromator to separate the excitation radiation from the fluorescent radiation, and can be done accurately only with great difficulty. It is comparable to weighing a ship and then the ship and the captain to determine the weight of the captain. However, if the detector is placed at a right angle to the source radiation, only the fluorescent radiation is measured and much more accurate results can be obtained.

EXPERIMENTAL PRECAUTIONS

Some compounds are destroyed by the intense ultraviolet radiation that is used to excite the molecules. Consequently, a slow analyst may find it difficult to get reproducible fluorimetric measurements.

The adjustment of pH is very important in fluorimetry. If the non-bonding electrons are protonated, they are no longer nonbonding and the fluorescent spectra can be shifted considerably or even destroyed.

Temperature is another factor that must be controlled because most fluorescing molecules lose their energy by collision rather than by fluorescing. A good way to reduce molecular collision is to solidify the material by freezing it. If a system does not fluoresce or does so weakly, place the sample and the cell into a dry ice–acetone bath or, better yet, into liquid nitrogen, which freezes the sample in a few seconds and in many cases increases the fluorescence intensity 10 to 1000 times. The aluminum 8-hydroxyquinoline system can be changed quite noticeably just by cooling it with the refrigerated water from a drinking fountain. With liquid nitrogen, readings must be taken within a few seconds after the cell is removed from the freezing solution because the cold cell will freeze moisture from the air onto its surface and ruin the analysis. This can be reduced considerably by blowing dry air through the cell compartment.

At high sample concentrations, the emitted fluorescent radiation from one molecule has just the right energy to be absorbed by a second molecule. This radiation absorbed by the second molecule may then be lost by further collision or it may be reradiated. If it is reradiated, the original ray has gone through two molecules; yet, a detector would count it as having come from only one molecule and low results would be obtained. Such *self-absorption* is a very serious problem, and only solutions in which the value of *abc* in equation 8.1 is below 0.01 will follow Beer's law.

Quenching is a process in which foreign substances lower or eliminate the fluorescence. Usually only a trace is necessary to ruin a determination (Hercules 1966).

QUANTITATIVE ANALYSIS

The fluorescence intensity is proportional to the initial radiation times *abc*. Equations similar to those in Chapter 6 can be derived to give

$$\log\left(\frac{F_0}{F_0 - F}\right) = abc \qquad (8.1)$$

where F_0 is 100 (usually set by a quinine standard), F the fluorescence measured, b the cell length, and c the concentration, and a the proportionality constant. Note that it is $\log[F_0/(F_0 - F)]$ that is equal to *abc*. Equation 8.1 can be written as

$$\log F_0 - \log(F_0 - F) = abc \qquad (8.2)$$

where F_0 is usually set at 100 using a 1-ppm quinine standard and log of 100 is 2. We then get

$$2 - \log(100 - F) = abc \qquad (8.3)$$

For very dilute solutions the determination of an unknown is basically done as in visible or ultraviolet spectroscopy. The method of *standard addition,* however, is much faster and is often preferred. The standard addition method involves measuring the fluorescence of the unknown, adding a small known amount of the material being determined to the original sample, and measuring the fluorescence again. From the increase in fluorescence due to the added standard, the original concentration can be computed. If the calibration plot is linear, using the method of standard addition is simple. If, however, the calibration plot yields a curve, some calculation is necessary. The following example will show how this calculation is done.

Example: Cholesterol is determined in egg noodles by treating the product with a base, extracting it with ether, irradiating it at 546 nm, and measuring the fluorescence at 577 nm. Assume that 2 g of such a product (10% moisture) was treated as described, extracting both the cholesterol and small amounts of other steroids. The extract was treated to oxidize cholesterol to a nonfluorescing compound. The solution was diluted to 100 ml; 25 ml was placed in a fluorimeter cell (previously calibrated with quinine); and a reading of 7% was recorded. The oxidized cholesterol in the fluorimeter cell was reduced back to a fluorescing species, producing a reading of 60%. When a 24-ml aliquot of the original sample solution was added to 1 ml of a standard solution of cholesterol containing 0.1 mg/ ml, an instrument reading of 90% was obtained. What is the % cholesterol in egg noodles?

Substituting the reading for the standard plus the unknown into equation 8.3, we get (uncorrected for background)

$$abc = 2 - \log(100-90) = 2 - 1.0 = 1.0$$

In a case like this, when it is assumed that abc is greater than 0.01, the background reading cannot just simply be subtracted from the sample reading because it is the $\log F_0/(F_0 - F)$ that is equal to abc, and to subtract one fluorescence reading from another would be similar to subtracting %T in colorimetry, which is not permissible.

Twenty-five milliliters of the sample solution produced a background fluorescence of 7%. By using 24 ml in the standard and 25 ml for the background reading, the effect of sample size is very small.

For the background,

$$abc = 2 - \log(100 - 7) = 2 - 1.97 = 0.03$$

Therefore, $1.0 - 0.03 = 0.97 = \log[F_0/(F_0 - F)]$ corrected for background.

The value of 0.97 is due to the standard plus the unknown and so the unknown must now be subtracted. The problem now becomes: What fluorescence reading would be produced by 24 ml of cholesterol solution if 25 ml had a reading of 60%? Since a and b of abc remain constant, $\log[F_0/(F_0 - F)]$ should be in the ratio 24/25:

$$\frac{25}{\log[100/(100 - 60)]} = \frac{24}{\log[100/(100 - F)]}$$

Let $\log[100/(100 - F)] = X$. Then

$$\frac{25}{0.398} = \frac{24}{X}$$

and solving for X,

$$X = 0.382 \quad \text{(uncorrected for background)}$$

$$= 0.387 - 0.03 = 0.35 \quad \text{(corrected for background)}$$

When the unknown 0.35 is subtracted from the unknown plus the standard, 0.97, then the difference 0.62 must be due to the 0.1 mg cholesterol added, $0.62/0.1 = 0.35/R$, and solving for R,

$$R = 0.057 \text{ mg cholesterol/25 ml sample solution}$$

The cholesterol percentage on a dry-matter basis is then

$$\frac{0.000057 \times 4 \times 100}{1.8} = 0.013\%$$

RECENT APPLICATIONS

Fluorescence equations were developed by Kollig (1983) to determine chlorophyll and pheophytin in plant samples under various environmental conditions.

By far the most important recent application of fluorescence in the food industry is associated with the fluorescence detector used in high-performance liquid chromatography. The details of this detector will be described in Chapter 21. This method has been used to determine polynuclear aromatic compounds in fish and crustaceans that may have been contaminated with crude oil from spills. Several of these materials are known carcinogens in animals. Detection down to 0.1–0.2 ppb for benz(a)pyrene is now possible with a fluorescent detector (Joe et al. 1982).

Moye et al. (1983) used a post-column fluorogenic labeling procedure to determine glycophosphate herbicides in fruits and vegetables. Labeling the herbicide with o-phthaldehyde mercaptoethanol, they could detect 5 ppb of aminomethyl phosphoric acid.

Kaykaty and Weiss (1983) detected residues of the coccidiostat Lasalocid in poultry from 5 ppb to 5 ppm with an 84% recovery. They excited samples at 310 nm and measured fluorescence at 430 nm. Lasalocid is also used as a component of feedlot feeds for cattle, as it increases feeding efficiency.

PROBLEMS

8.1. Eosin is a dye used to stain muscle tissue. Solutions of eosin were compared in a fluorimeter and the following values were obtained:

Eosin (μg)	Reading	Eosin (μg)	Reading
1	10	6	53
2	19	7	60
3	27	8	66
4	37	9	71
5	45		

Plot eosin (μg) vs. fluorimeter reading. Over what range is the relationship $F = KI_0 abc$ valid?   *Ans.* Up to about 4.5 μg.

8.2.   Anthracene is used as a starting material to make several food-coloring dyes. Pure anthracene has a violet fluorescence, but tetracene impurities impart a greenish-yellow fluorescence. If 1 mg of tetracene dissolved in 1 liter of ethanol had a F of 80 at 540 nm and a 95-mg of impure anthracene dissolved in 1 liter of ethanol had a F of 5.5 at 540 nm, what is the percentage of tetracene in the anthracene? (Assume that the value of abc is less than 0.01, so the simple fluorescence equation can be used.)   *Ans.* 0.037%.

8.3.   A sample of bologna is to be checked for its vitamin B_1 content. A 7-g sample is extracted with HCl, treated with the enzyme phosphatase, and diluted to 50 ml. A 10-ml aliquot is treated with ferricyanide and then diluted to 25 ml. A second 10-ml aliquot is diluted as a blank. A standard solution containing 0.1 μg/ml is treated in the same manner and a blank prepared. The following readings were obtained:

Fluorescence reading	Solution
4	Sample blank
5	Standard blank
48	Sample, oxidized
82	Standard, oxidized

Calculate the amount of B_1 in an average slice (20 g) of this bologna. *Ans.* 13.2 μg.

BIBLIOGRAPHY

Hercules, D. (1966). "Fluorescence and Phosphorescence Analysis." Interscience Publishers, New York.

Joe, F. L., Jr., Salemma, J., and Fazio, T. (1982). High performance liquid chromatography with fluorescence and ultraviolet detection of polynuclear aromatic compounds. *J. Assoc. Off. Anal. Chem.* **65**(6), 1395–1402.

Kaykaty, M., and Weiss, G. (1983). Lasalocid determination in animal blood by HPLC fluorescence detectors. *J. Agric. Food Chem.* **31**, 81–84.

Kollig, H. (1983). Derivation of fluorometric chlorophyll and pheophytin equations. *J. Assoc. Off. Anal. Chem.* **66**(3), 592–593.

Moye, H. A., Miles, C. J., and Scherer, S. J. (1983). A simplified HPLC residue procedure for the determination of glycophosphate herbicides in fruits and vegetables employing post column fluorogenic labeling. *J. Agric. Food Chem.* **31**, 69–72.

9
Infrared Spectroscopy

Infrared spectroscopy is believed to have started with an experiment reported by Herschel in 1800. Herschel placed a thermometer at successive points in a daylight spectrum and noticed an unusually large heating effect in the region immediately beyond the red region of the visible spectrum, hence the term *infrared*. The next century was spent in developing detectors, in attempting to understand infrared sources; and in making empirical observations on structure–spectra relationships. Physicists made the early significant contributions, particularly the molecular vibration–rotation theories developed before World War I. When it was shown that infrared could be used for functional group analysis just prior to World War II, chemists became interested in this type of spectroscopy. Today, Fourier transform infrared instruments offer the highest resolution and speed. These will be discussed later.

In infrared work two units of wavelength are commonly used, the micrometer (old micron) and the wavenumber. Rotational frequencies are of the order of 10^{11}–10^{13} hertz (Hz, 1 Hz = 1 cycle/sec); vibrational frequencies are 10^{13}–10^{15} Hz. In order to have more manageable numbers and still maintain a proportional relationship between energy and frequency, the wavenumber concept was developed.

Recall equation 5.1, $\lambda\nu = c$, where λ is the wavelength in cm, ν the frequency in cycles/sec, and c the speed of light (3×10^{10} cm/sec). This can be rearranged to give

$$\frac{\nu}{c} = \frac{1}{\lambda} \tag{9.1}$$

the wavenumber. The symbol for wavenumber is $\bar{v}$, with units cm^{-1}. Since $\bar{v}$ is simply v divided by the speed of light, its numerical value is much smaller (20–5000) and more convenient to work with than the values of v.

By multiplying the numerator and denominator of equation 5.2, $E = hv$, where E is the energy and h is Plank's constant, to give

$$E = hv \frac{c}{c} = h\bar{v}c \qquad (9.1a)$$

we see that the wavenumber is proportional to energy. Since λ is often given in μm, not cm, equation 9.1b is used to convert wavelength to wavenumber:

$$\bar{v} \, (cm^{-1}) = 10,000/\lambda \, (\mu m) \qquad (9.1b)$$

As an example of this conversion, the frequency in wavenumbers corresponding to a wavelength of 4.62 μm is calculated as follows:

$$\bar{v} = \frac{10,000}{4.62} = 2165 \, cm^{-1}$$

INSTRUMENTATION

Sources

In infrared spectroscopy the source must provide a continuous spectrum of radiation at a high intensity. As explained earlier, a molecule absorbs radiation if the radiation matches a frequency of some transition within the molecule. If a molecule can absorb radiation, it can also emit it and thereby become a source for that wavelength. Since a continuous spectrum contains all wavelengths, the source compound must be able to absorb all wavelengths in order to emit all wavelengths. If a compound absorbs all of the wavelengths striking its surface, it will be black and is known as the *ideal black body radiator*.

Figure 9.1 is a plot of intensity of emitted radiation at various temperatures vs. wavelength for an ideal black body radiator: Notice that as the temperature of the source increases, the intensity of the emitted radiation increases. Equation 9.2 shows the relationship

$$I_{\lambda max} = 1.3T^5 \times 10^{-15} \qquad (9.2)$$

where I is the intensity in watts/cm^2 and T the temperature (degrees Kelvin) of the source. This equation indicates that if the absolute temperature of the source is doubled, its emission intensity increases 2^5 or 32 times.

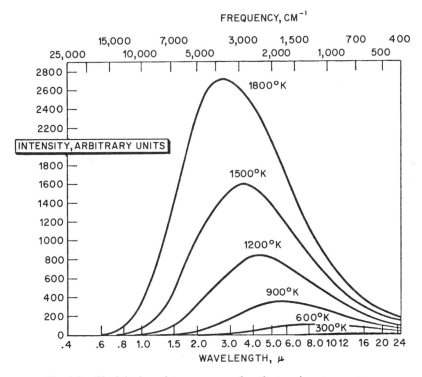

Fig. 9.1. Black body emittance vs. wavelength at various temperatures.

Therefore one way to increase sensitivity would be to increase the source temperature. However, this cannot be done at random because, as can also be observed from Fig. 9.1, as the temperature of the source increases, the wavelength of maximum intensity shifts to a shorter wavelength.

Rayleigh has shown that short wavelength radiation can be scattered easily, according to the relationship

$$T = k(cld^3)/(d^4 + \alpha\lambda^4) \tag{9.3}$$

where T is the amount of scattering. Notice that with λ^4 in the denominator, a very small shift in radiation to a shorter wavelength will cause a very large increase in the amount of radiation that can be scattered. This short-wavelength radiation has a high energy since $\Delta E = h\nu$. Even if a small amount of this high-energy radiation reaches the detector, it will overwhelm the low-energy signal that should be detected. Great care must be taken with infrared instruments to ensure that this stray radiation does not reach the detector. This is done by controlling the source temperature and/or by using filters and baffles.

A source temperature of about 1500 K is generally used in commercial infrared instruments although experimental research instruments may use 6000 K carbon arcs. Three common types of infrared sources are the Nernst glower, a globar, and nichrome wire.

Nernst Glower. A hollow rod is made of a fused mixture of the oxides of zirconium (90 parts), yttrium (7 parts), and either erbium or thorium (3 parts). The rod is usually about 3 mm in diameter and about 5–6 cm long. Sometimes a molybdenum strip is attached to each end of the electrode as an expansion spring to prevent the joint from cracking when the instrument is repeatedly turned on and off. The glower has a negative coefficient of resistance and must be preheated before it will conduct. If the glower is water-jacketed, 1–2% stability can be maintained; if in addition it is also enclosed, 0.1% stability can be achieved.

The glower has an intense radiation up to about 10 μm and its normal lifetime is up to several hundred hours. This source produces about 200 times more radiation at 1–2 μm than it does between 14 and 15 μm.

Globar. A globar is made from carborundum (SiC) and is about 6 mm in diameter and 5 cm long. The usual operating temperature is about 1400°C with 2000°C, the point where SiC vaporizes, being the upper limit. This source conducts at room temperature and requires no preheating. A globar has many advantages: it is sturdy, it is more stable than a Nernst glower, it is better than a Nernst glower beyond 10 μm, and its radiation matches that of a blackbody.

Disadvantages of a globar are that it has a relatively short life; overheats at temperatures in excess of 1200°C and must be cooled; requires a lot of power; and contains a binding material that evaporates gradually and absorbs some of the radiation. In addition, the operating voltage gradually changes because the binding material is evaporating and this requires a more complex electronic control of the source.

Nichrome Wire. A closely wound spiral of nichrome wire, heated to 1500 K by the passage of an electric current, is another type of infrared source. Its main advantage is a positive temperature coefficient of resistance. Both the glower and the globar have negative coefficients; that is, their temperature is lowest and the resistance highest at the ends of the source where the electrical connections are made. This results in a high dissipation of energy at the connections and causes their eventual arcing and subsequent burnout. The image of the wire spiral is not uniform and this is eliminated by enclosing it in a ceramic cylinder.

Cell and Prism Materials

If a material is to be transparent in the infrared region and thus serve as cell and prism material, its molecules must have frequencies of motion different from that of the incident radiation and the material to be examined. Most materials examined by infrared are organic compounds, which are held together mainly by covalent bonds. Cell materials should thus be made from something different than covalently bonded compounds. Ironically bonded systems, especially the alkali or alkaline earth halides, are the most common cell materials. Several of these materials are shown in Table 9.1.

As indicated in Fig. 5.1, the entire infrared region is considered to extend from 700 nm to about 500 μm. Why then is there interest in materials that only transmit to 5–7 μm? The reason is that much of the information that can be obtained in the far infrared and microwave regions can be obtained between 2 and 7 μm if the monochromator has sufficient dispersion to resolve the vibrational–rotational interactions in this region. If this can be done, the very expensive and difficult to operate far infrared and microwave instruments are not necessary.

In general, the alkali halides have several disadvantages as cell and prism materials. They are water soluble, cleave and scratch easily, and

Table 9.1. Infrared Transmitting Materials Commonly Used

Material	Usual cutoff wavelength (μm)	Refractive index
Glass	2.5	
SiO_2 (quartz)	3.5	1.45
TiO_2 (rutile)	5.0	2.6
LiF	5.5	1.38
Al_2O_3 (ruby)	5.5	1.77
MgO (periclase)	7.0	1.74
Irtran 1	7.5	1.36
CaF_2	8.5	1.43
Irtran 3	10	1.39
BaF_2	11	1.6
Si	11	3.4
Irtran 2 (ZnS)	14	2.28
NaCl (halite)	15	1.54
Irtran 4	20	2.42
KCl (sylvite)	20	1.49
AgCl	22	2.07
KBr	25	1.53
KRS-6 (TlCl–TlBr 40%)	30	2.18
CsBr	35	1.69
KRS-5 (TlI–TlBr 42%)	39	2.63
KI	40	1.67
CsI	50	
Diamond	250	2.41
Polyethylene	>250	

have a high temperature coefficient, so close temperature control is necessary to secure mechanical stability and constant dispersion. Silver chloride is not water soluble but is quite sensitive to visible radiation. Probably the best salt material at the present time is the eutectic mixture of thalium bromide (42%) and thalium iodide (58%), known as KRS-5. It is easily recognized because of its orange color. It is toxic, hard to polish, and stains readily; but its solubility in water is low and its high refractive index (2.63) make it a favorite for multiple internal reflection attachments, which will be discussed in a later section.

Although diffraction gratings are replacing prisms almost completely in infrared spectrophotometers, the need for a good cell material still remains. Today, the *Irtran* series of materials provides the best combination of low water solubility, high refractive index, and the ability to withstand large temperature changes. For example, Irtran-2 shows no change when exposed to the following: water at 23°C for 336 hr; 6% NaCl solution for 168 hr; 0.5 N HNO$_3$ for 24 hr; 0.5 N NH$_4$OH for 24 hr; and ethyl ether, acetone, chloroform, or benzene for 24 hr. It can withstand temperatures of more than 800°C and can be polished to good optical tolerances. Samples of 0.08-in. material have survived water quench tests from 165°C to ambient temperature.

Types of Cells

The purpose of the cell is to hold the compounds while they are being examined with the instrument. The cell windows are made from the materials discussed in the previous section, and extreme care must be used to keep the cells in good condition. Cells are usually stored in desiccators to reduce moisture problems, and they should not be exposed to large differences in temperature lest they crack. Cells are generally carried downward, that is, with the hand on top, so body heat does not warm the cell thus changing its optical properties. The high solubility of most cell materials makes it mandatory that the solvents and compounds placed in the cells are as dry as possible. Carbon tetrachloride and carbon disulfide are favorite solvents (Note: CS$_2$ reacts with primary amines). Cell thickness varies from a few hundredths of a millimeter for liquid samples to several meters for those used with gases. As a result, a variety of cells are required for the normal range of problems encountered.

Sandwich Cell. The sandwich cell is probably the most common type. A thickness of 0.01–1.0 mm is normal.

Wedge Cell. In spectral measurements of solutions in double-beam instruments, the spectrum of the solvent interferes unless it is blanked out. This is difficult to do in the infrared region because the cell thickness is so small that it is hard to find two perfectly matched cells, and even then the amount of solvent in the reference would be more than in the sample cell. What is needed is a continuously variable cell thickness for the reference side. The wedge cell, which contains a wedge-shaped spacer, is used for this purpose. For example, in a sample cell of 0.025-mm thickness, a wedge varying from 0.01 to 0.08 mm is used. This cell is moved horizontally until the major solvent band is just cancelled out. It can be done quite quickly.

Gas Cell. For fairly pure compounds, a 2- to 10-cm gas cell is used. For trace analysis, it is customary to use a multiple-pass cell, since cells larger than about 10 cm do not fit into most instruments.

Solid Samples—KBr Pellet Methods

The infrared spectra of many solids can be obtained from a solution of the compound, but solvent bands may interfere and many compounds are not soluble in the solvents commonly used. To overcome these difficulties, solid samples are used. However, to simply place small crystals between the cell windows does not work satisfactorily because too much radiation is scattered. In order to reduce scattering, the sample particles must be about 2 μm in diameter; if the particles are larger, they must be surrounded by a medium of similar refractive index.

A technique previously used to prepare solid samples was the mull method. The procedure involves placing 2–5 mg of the sample into a small mortar, adding 4 or 5 drops of Nujol (mineral oil), and grinding the mixture into a paste. This paste is spread between the cell windows. The mineral oil reduces the scattering considerably. The main disadvantage is that the —CH_2— and —CH_3 bands of the mineral oil interfere with the sample, and that solvents, including mineral oil, often interact with the sample to distort the spectra.

A technique that solves most of these problems is the KBr pellet method, which was developed by Stinson and O'Donnell (1952). In this procedure, the compound is mixed with optically pure KBr and pressed into a disk-shaped pellet, which is placed in the sample beam of the instrument. Normal pellets are about the size of a dime. If only qualitative results are desired, or if an internal standard is used, then the rapid and inexpensive procedure illustrated in Fig. 9.2 can be used.

To prepare KBr pellets by the rapid method, proceed as follows. Grind 0.1 g of sample in a small agate mortar until it is very fine. Add 2 g of dry KBr and grind the two together until they appear to be completely mixed. (A "wiggle bug" is convenient for this.) With a cork borer, cut a hole about 1 cm in diameter in a piece of filter paper or stiff cardboard.

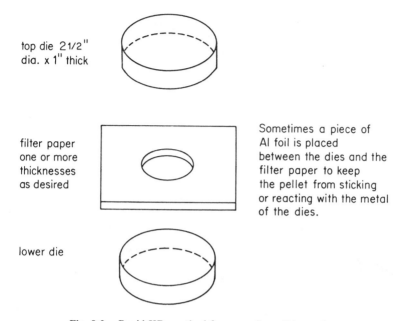

top die 2 1/2" dia. x 1" thick

filter paper one or more thicknesses as desired

Sometimes a piece of Al foil is placed between the dies and the filter paper to keep the pellet from sticking or reacting with the metal of the dies.

lower die

Fig. 9.2. Rapid KBr method for preparing solid samples.

The pellet thickness can be varied by the number of pieces of cardboard used. Place a 5-cm square of aluminum foil on the lower die, place the cardboard on top, add 0.2–0.4 g of the pellet mixture, place another piece of aluminum foil on top, and finally add the top die. Place this in a press (any machinist can make an A frame about 18 in. high and a heavy-duty car or light truck hydraulic jack can be used), and apply about 80,000 psig for about 1 min and then release the pressure. The pellet will usually crack in several places at this time because trapped air is being released. Now reapply the pressure and remelt the KBr. The pellet formed will be imbedded in the paper and can easily be handled; in fact, it can be mounted directly in the sample holder if the original piece of cardboard was cut correctly. The pellet should be transparent. A milky pellet indicates that not enough pressure was applied, while a brownish or wet pellet indicates that the KBr is old and needs replacing. A wet pellet is indicated by 30–40% transmittance at the 2 μm part of the spectra.

An even simpler and less expensive method for preparing pellets is to use the minipress shown in Fig. 9.3. It consists of a big nut with a bolt in each end. The ends of the bolts are polished and hardened. Pressure is applied by two wrenches and the sample is left in the center of the nut when the two bolts are removed. The nut is then placed in the beam of the instrument. The only disadvantage is that the sample must be de-

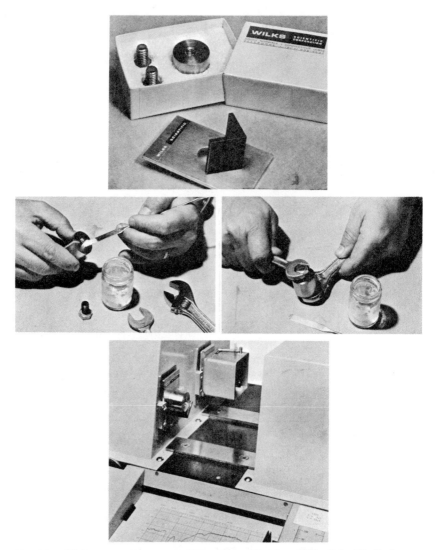

Fig. 9.3.   Minipress used to prepare KBr pellets. (Courtesy Wilks Scientific Co.)

stroyed before a new sample can be made thus preventing accumulation of a "file" of samples for a particular project.

When KBr pellets are used, it is necessary to reduce the reference beam intensity to compensate for the KBr. Rather than make another pellet, which is hard to duplicate, various attenuation devices have been made. The two shown in Fig. 9.4 are the most common commercial attenuators; the top one can be made quite easily by an individual with a small piece of metal and a scrap of venetian blind type window screen.

Fig. 9.4. External beam attenuators. (Courtesy Wilks Scientific Co.)

Detectors

The detectors used for the visible and ultraviolet regions require too much energy to respond to be useful in the infrared. To understand this, suppose someone lit a match on a dark night several blocks away from you. Your eyes would immediately detect the visible radiation, but your body would not feel the heat given off. An infrared detector must detect this heat. If 0.5% analytical accuracy is to be achieved, then temperature differences of 5×10^{-5}°C must be measured. Infrared detectors are therefore constructed differently than those for the visible and ultraviolet regions. The three most common detectors found in commercial instruments are thermocouples, bolometers, and pneumatic devices.

Thermocouples. Figure 9.5 shows the basic idea of how a thermocouple works. A thermocouple is formed when two different metals are placed together. Two such junctions are necessary, one to serve as the detector and one to serve as a reference. Radiant energy is focused on the hot junction; because of the different characteristics of the two metals, one side of the junction warms up more than the other side. This temperature (energy) difference causes electrons to leave one metal faster than the other metal, and the net result is that at the junction a small voltage is produced proportional to the amount of heat striking the junction. The cold junction serves as a reference.

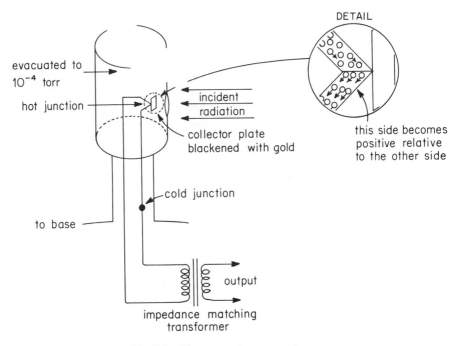

Fig. 9.5. Thermocouple construction.

Blackening the absorbing surface of the hot junction improves the efficiency of the absorption of the radiation. If gas molecules from the surrounding air strike the junction, they can cool it, thereby decreasing its sensitivity. Therefore, the junction is placed in a small cavity and evacuated. The combination of surface blackening and evacuation can improve the sensitivity 10 to 20 times.

Bolometers. While a thermocouple works fairly well as an infrared detector, more sensitive devices often are needed. A bolometer, which measures a difference in current, is such a device that is based on the familiar Wheatstone bridge circuit shown in Fig. 9.6(A).

A commercial bolometer (Fig. 9.6B) contains a ribbon of electrical conducting material (thermocouple, thermistor, wire filament) through which a steady current is passed. When radiation strikes the sensing element, the electrical resistance changes and the corresponding current change is measured. Because of the Wheatstone bridge circuit in the system, any infrared radiation changes due to variations in external conditions that change the temperature are canceled by the reference side of the detector; thus, only the radiation from the sample is measured.

The sensitive element in a bolometer may be a thermocouple, a piece

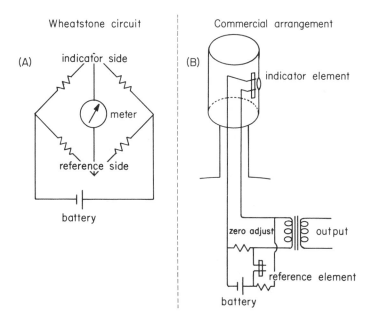

Fig. 9.6. Bolometer construction.

of metal (nickel is popular), a thermistor (a thin piece of Mn, Ni, or Co oxides), or a superconductor. A superconductor is a substance that loses its normal resistance when cooled to very low temperatures (Fig. 9.7). Notice that the resistance change per unit of temperature change is much greater in the superconducting transition region than in the normal region. Superconducting bolometers are 20–30 times more sensitive than normal bolometers. Tantalum nitride and niobium nitride are two compounds that exhibit this effect. However, liquid helium and/or liquid hydrogen must be used to attain sufficiently low temperatures.

Pneumatic Detectors. Developed by Golay (1947), the pneumatic detector uses an entirely different approach than other detectors. It was mentioned earlier that when striking a match at night visibility detection is more sensitive than infrared detection. What Golay did in effect was to make a visible detector that was controlled by infrared radiation. Figure 9.8 schematically shows this detector.

The detector consists of a pneumatic chamber with a capillary tube covered with a flexible cover that has a mirrored surface on the outside. This chamber is filled with xenon. Inside the chamber is a metallized membrane of very low thermal capacity (Al on collodion) to absorb the incident radiation. The infrared radiation is absorbed by the inner membrane, which in turn heats the xenon. As the gas in the chamber expands

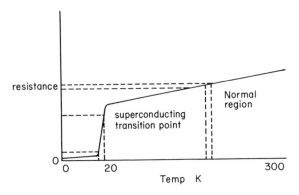

Fig. 9.7. Plot of resistance vs. temperature for a superconducting material.

and contracts it moves the membrane at the end of the capillary. A faint beam of light is directed at the membrane and the amount the light moves is detected with a photocell.

A Golay detector is very sensitive, usually being surpassed only by superconducting bolometers. It is capable of covering the infrared region out to 700 μm. Its short wavelength sensitivity falls when the wavelength is comparable to the film thickness, and its long wavelength sensitivity falls when the wavelength is comparable to the diameter of the receiving element.

Multiple Internal Reflection

Multiple internal reflection (MIR) or attenuated total reflection (ATR), developed by Fahrenfort (1961), is used particularly in the infrared spectral region for the analysis of a variety of materials. Not since the introduction of the KBr technique has there been a development of equal importance in the field of infrared spectroscopy. The technique is partic-

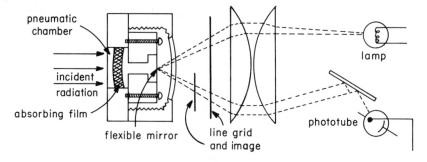

Fig. 9.8. Golay pneumatic detector.

incident radiation   reflected radiation

Fig. 9.9.   Reflected radiation.

ularly valuable if it is undesirable or inconvenient to impose on the specimen the preparative procedures required for normal transmission work, or where the region of interest is a thin layer adjacent to a surface.

Figure 9.9A shows how the reflection of radiation is commonly depicted. Under certain conditions this has been found to be incorrect; what really happens is that the incident beam penetrates a few molecular layers before it is reflected (Fig. 9.9B). The amount of reflection that occurs is roughly inversely proportional to the ratio of the index of refraction of the plate and that of the medium in contact with it. Since air has a low refractive index (1.0), a radiation beam being totally reflected from the face of the prism in contact with air suffers practically no change. If a material with an index equal to or greater than that of the plate is placed in contact with a plate, all of the energy will pass into the second medium.

Further, the index of refraction of a material undergoes a rapid change at wavelengths where the material absorbs. Hence, when such a material is brought in contact with a prism (see Fig. 9.10), the internally reflected beam will lose energy at these absorbed wavelengths (where the index is high), so that a plot of the reflected energy will produce a curve that is very similar to, but not identical to, a conventional transmission spectrum. The amount of apparent penetration of energy into the sample depends on the prism's refractive index (high index—low penetration) and the angle of incidence of the beam of radiation (high angle—high penetration).

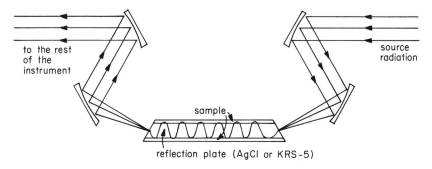

to the rest of the instrument

source radiation

sample

reflection plate (AgCl or KRS-5)

Fig. 9.10.   Diagram of light path for multiple internal reflection measurements. (Courtesy Wilks Scientific Co.)

The actual thickness of the sample has no effect on the spectrum—and herein lies the real importance of the MIR technique, as the spectroscopist no longer must prepare a thin film for the infrared energy to pass through. An equivalent spectrum can be obtained by merely placing the sample in direct contact with the reflecting face of the prism. Figure 9.10 shows how a multiple internal reflection (MIR) spectrum is obtained.

KRS-5 is usually preferred as the prism material because it has a high refractive index (2.63) which provides a better refractive index ratio.

QUALITATIVE ANALYSIS

Absorption bands in the shorter wavelength region (2–10 μm) are thought to result mainly from the stretching and bending vibrations of individual bonds, and are therefore considered characteristic of the diatomic structural units of functional groups. Bands in the longer wave-

Table 9.2. Preliminary Analysis of an IR Spectrum

(1) Absorption at 2.5 to 3.2[a]	OH, NH compounds
plus	
a. 5.7 to 6.1	acids
b. 5.9 to 6.7	amides (usually two bands)
c. 7.5 to 10.0	—O— compounds
d. about 15.0	primary amines (broad)
(2) Sharp absorption at 3.2 to 3.33	Olefins, aromatics
plus	
a. 5.0 to 6.0	benzenoid patterns (weak)
b. 5.9 to 6.1	olefins
c. 6.1 to 6.9	aromatics (two bands)
d. 11.0 to 15.0	aromatics (several strong bands)
(3) Sharp absorption band at 3.33 to 3.55	Aliphatics
plus	
a. 6.7 to 7.0	—CH$_2$—, —CH$_3$
b. 7.1 to 7.4	—CH$_3$
c. 13.3 to 13.9	—(CH$_2$)$_4$—
(4) Two weak bands at 3.4 to 3.7	Aldehydes
plus	
a. 5.7 to 6.1	aldehydes and ketones
(5) Absorption at 4.0 to 5.0	Acetylenes, nitriles
(6) Strong sharp bands at 5.4 to 5.8	Esters, acyl halides (1 peak), anhydrides (2 peaks)
plus	
a. 7.5 to 10.0	—O— compounds
(7) Strong sharp bands at 5.7 to 6.1	Aldehydes, ketones, acids
(8) Strong bands at 7.5 to 10.0	—O— compounds
(9) Strong bands at 11.0 to 15.0	Aromatics, chlorides

[a] All band wavelengths in μm.

length region (7–25 μm) appear to be caused by more complex vibrations of polyatomic units and of the molecule as a whole.

It should be noted that infrared spectra determined in different solvents are not always identical. If a compound is pure, it will have a moderate number of bands and they will be sharp. Generally, impurities must be present at concentrations of 1% or higher before they can be detected successfully.

Because many functional groups have characteristic patterns of infrared absorption bands, infrared spectra are useful for qualitative analysis, i.e., the identification of compounds. Even if an exact identification cannot be made based on the infrared spectrum, it can provide clues to a compound's identity. Table 9.2 and Fig. 9.11 contain information concerning infrared absorption by various groups and classes of compounds.

QUANTITATIVE ANALYSIS

Quantitative analysis using infrared radiation is extremely difficult on an absolute basis because it is hard to determine cell thickness accurately and because wavelength settings are hard to reproduce. However, some good empirical methods have been developed that make quantitative analysis quite practical.

Since most infrared work is done in cells 0.01–0.25 mm thick, this thickness must be known to 0.0001–0.0025 mm in order to get even 1% accuracy. There are two methods for determining cell thickness (or spacing) to this accuracy. One method is based on an interference pattern similar to that of the interference filter, and the other on use of a standard absorber. In the first method, the spectrum of a freshly polished, empty cell is determined. The spectrum in Fig. 9.12 is typical for an instrument linear in micrometers. The cell thickness can be determined from such a spectrum by the equation

$$b = \frac{n(\lambda_1 \times \lambda_2)}{2(\lambda_1 - \lambda_2)} \tag{9.4}$$

where b is the spacing in micrometers, n the number of bands between λ_1 and λ_2, and λ_1 and λ_2 the wavelengths in μm between which the bands are counted. Note the manner in which n is counted.

For example, the spacing of a cell in millimeters is calculated as follows if the empty cell produces a spectrum between 3 and 5 μm with peaks at 3.05, 3.44, 3.90, 4.31, and 4.72 μm:

$$b = \frac{4(4.72 \times 3.05)}{2(4.72 - 3.05)} = 17.24 \quad \mu m = 0.0172 \quad mm$$

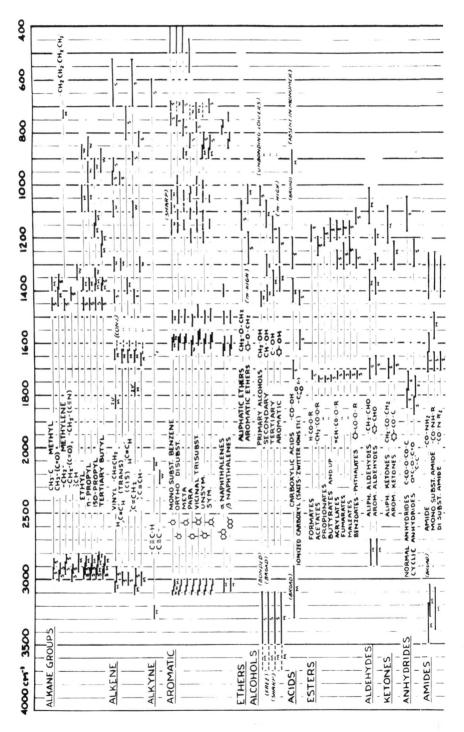

122

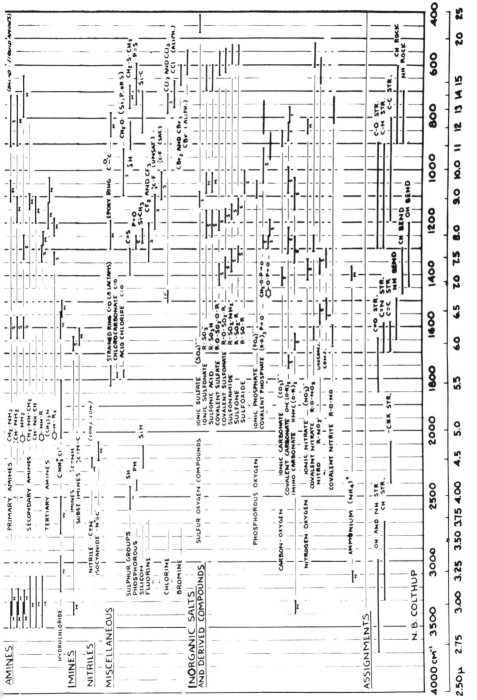

Fig. 9.11. Spectra–structure correlations. (Courtesy Stamford Research Labs., American Cyanamid Co.)

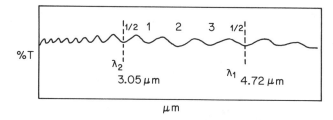

Fig. 9.12. Interference pattern of an empty cell.

The second method of determining cell thickness—the standard absorber method—is used for calibrating cells that are not smooth enough to give interference fringes. Benzene is an excellent standard absorber for such purposes; its 1960 cm^{-1} (5.05 μm) band may be used for calibrating cells that are less than 0.1 mm in pathlength and the 845 cm^{-1} (11.8 μm) band may be used for calibrating cells 0.1 mm or greater in pathlength. Cell thickness is calculated from the absorbance of standard benzene as follows:

(1) For the 1960 cm^{-1} band,

$$\text{cell thickness (mm)} = 0.1 \times \text{absorbance} \qquad (9.5)$$

(2) For the 845 cm^{-1} band,

$$\text{cell thickness (mm)} = \text{absorbance}/0.24 \qquad (9.6)$$

The second problem in using infrared spectroscopy for quantitative analysis involves determining the absorptivity (a). This is difficult because it is almost impossible to reproduce slit widths and wavelength settings with sufficient accuracy to ensure that an a determined by one operator will be the same as an a determined by another, especially if different

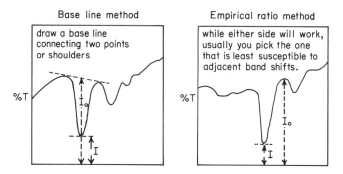

Fig. 9.13. Two empirical methods for determining absorptivity in the infrared region.

instruments are used. Because of these difficulties, several empirical methods have been developed. The two useful ones are the *base line method* and the *empirical ratio method*. These methods are illustrated in Fig. 9.13. Plot the log of I_0/I vs. concentration to establish a calibration curve.

INFRARED AS A GAS CHROMATOGRAPHIC DETECTOR

An infrared spectrophotometer is an excellent detector in combination with a gas chromatograph to identify the resolved components. Because materials can emerge from a gas chromatograph very rapidly, it is necessary to have either a few-second scan rate or a means of bypassing the unwanted peaks. Also, the sample must be kept hot and detection limits sufficient to see the small amount of sample, which is even further diluted by the carrier gas, must be provided. Infrared instruments with a few-second scan rate are available but are quite expensive. Figure 9.14 shows a method to do this, as developed by the Wilks Co. The idea is to provide a long path in the infrared beam for the sample to pass through. The tube is gold-plated to provide low absorption losses, yet be chemically inert.

FOURIER TRANSFORM INFRARED (FT-IR) SPECTROSCOPY

The instruments described so far have involved what are known as *dispersive* techniques; that is, the various wavelengths of radiation from the source are dispersed by a monochromator and then passed through the sample and detected one at a time. Not only is the information from those wavelengths not being measured at the time lost, but the process is slow. An alternative approach is to have all of the wavelengths arrive at the detector at the same time and by a mathematical treatment of the data, called *Fourier transform,* sort out the results from the individual signals. Modern computers permit the difficult mathematics to be done in a few seconds with the result that spectra can be obtained much faster, with smaller samples, and at longer wavelengths where much less energy is available.

This type of instrument requires an *interferometer.* The first interferometer was built by A. A. Michelson over 100 years ago. This apparatus splits the radiation from the source into two beams. The radiation from each beam is reflected by mirrors and returned along the same paths to

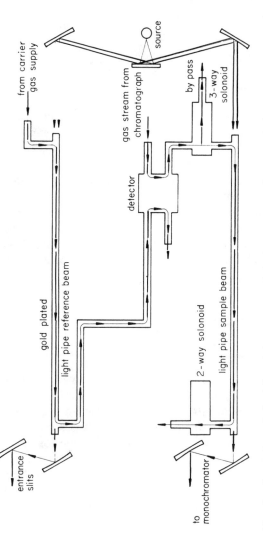

Fig. 9.14. Gas flow schematic for infrared detector coupled with a gas chromatograph. (Courtesy Wilks Scientific Co.)

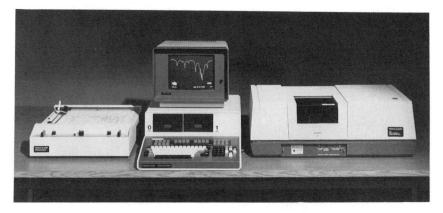

Fig. 9.15. Perkin-Elmer Model 1800 FT-IR. (Courtesy Perkin-Elmer Co., Norwalk, CT.)

recombine. If a sample is placed in the path of one beam then the radiation from the two beams can either interfere constructively or destructively depending on the phase difference of the two beams; the resulting pattern is an *interferogram,* which is a measurement of the energies as a function of optical path difference. An additional feature is to make one mirror of the interferometer movable and measure the amount of this movement with a laser. By doing this it is possible to obtain wavelength measurements to a few hundredths of a wavenumber.

Interferograms do not look at all like normal infrared spectra and must be treated by Fourier transform mathematics to produce an "infrared spectrum." Figure 9.15 is a photo of a modern FT-IR instrument, a Perkin-Elmer Model 1800, and Fig. 9.16 is a diagram of the optical path. This instrument can cover the range from 4500 to 450 cm^{-1} in 25 sec and requires only a 1.5-μm sample.

Three main advantages of FT-IR are conventionally cited, although several other aspects of interferometry often lead to useful improvements in the quality of spectra. The most important, although often grossly overestimated, advantage to FT-IR is *multiplexing,* known as Fellgett's advantage. As all frequencies are measured simultaneously in interferometry, Fellgett's advantage can be used either to acquire spectra much faster or to improve the signal-to-noise ratio for a given resolution. For instance, between 4000 and 400 cm^{-1} there are 3600 spectral elements of 1-cm^{-1} resolution. Assuming that all other factors are equal, an interferometer can acquire the spectrum in 1/3600 the time required of a dispersive spectrometer. Alternatively, because the noise in IR detectors, which is essentially random, increases as the square root of the number of scans N and the signal increases proportionally to the number of scans, addition of the 3600 scans by the interferometer will improve the signal-

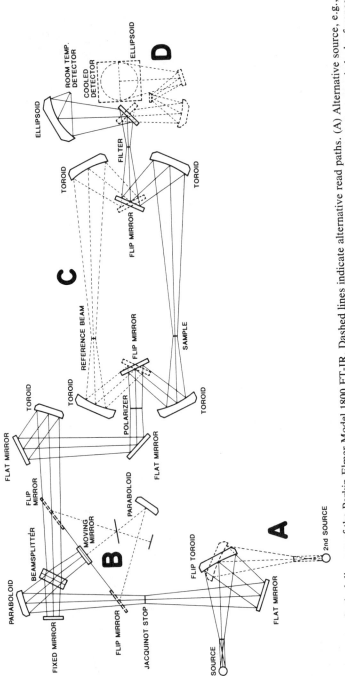

Fig. 9.16. Optical diagram of the Perkin-Elmer Model 1800 FT-IR. Dashed lines indicate alternative read paths. (A) Alternative source, e.g., mercury arc for far-IR or heated sample for emission studies. (B) Second interferometer, e.g., Mylar for far-IR. (C) Software switched reference beam. (D) Alternative detector, e.g., liquid N2 cooled MCT. (Courtesy Perkin-Elmer Company, Norwalk, CT.)

to-noise ratio by $N/N^{1/2} = N^{1/2}$, or by a factor of 60. It can be seen that as the number of spectral resolution elements is decreased, either by sacrificing resolution or by investigating a smaller region of the spectrum, Fellgett's advantage is correspondingly reduced.

The *throughput,* étendue, or Jacquinot's advantage is a measure of the relative amounts of energy allowed to reach the detector. It is a mistake to compare slit areas with apertures; the actual amount of this advantage is more closely approximated by comparing the area of the collecting mirrors in an interferometer with the area of the grating in a dispersive instrument, or comparing the f-numbers of the cones of radiation. This advantage for an interferometer theoretically can be from 5 to 30 times although in practice it is somewhat less.

The Connes advantage makes use of the high precision of lasers to very precisely measure frequencies. There are no drawbacks to laser precision except for alignment problems, and most Fourier transform software also includes calibration routines.

INFRARED REFLECTANCE SPECTROSCOPY

Extensive investigations have led to the development of instruments that utilize infrared reflectance to measure moisture, protein, oil, fiber, and other major components of foods and agricultural products. Typical spectra for protein, oil, starch, and water are given in Fig. 9.17.

The reflectance spectrum of a sample of ground beef (Fig. 9.18) is the summation of the spectra of its components water, carbohydrates, proteins, and oil. Several methods of treating the data can be used to predict the composition from reflectance spectra:

1. Log $(1/R)$ is determined at up to eight selected wavelengths for each component.

2. The difference in log $(1/R)$ at up to four selected pairs of wavelength points is measured. The wavelengths are selected at a maximum absorption wavelength for the measured component and at a nearby wavelength at which the absorption is a minimum.

3. The function $(R_{\lambda_1} - R_{\lambda_2})/(R_{\lambda_1} + R_{\lambda_2})$, i.e., the slope of the reflectance curve in the region of an absorption band, varies linearly with concentration of a component; λ_1 and λ_2 are close wavelengths, near to, but not at a peak absorption point for that component.

4. The second derivative of the log $(1/R)$ at a peak absorption point.

All four data treatments involve developing multiple correlations for calibration and use of those calibrations to determine protein, oil, or water content.

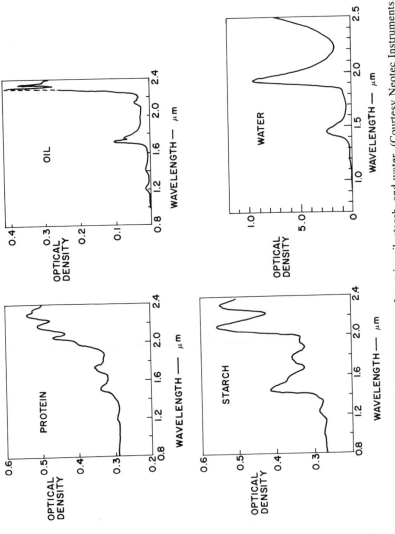

Fig. 9.17. Near-infrared reflectance spectra of protein, oil, starch, and water. (Courtesy Neotec Instruments.)

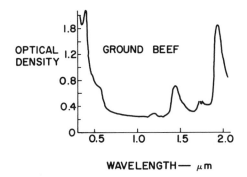

Fig. 9.18. Near-infrared reflectance spectrum of ground beef. (Courtesy Neotec Instruments.)

There is a good linear relationship between chemical composition and reflectance in the near infrared (NIR). However, reflectance spectra are sensitive to extremes in particle size (reflectance increases as particle size decreases). A reflectance change of up to 50% can occur as particle size is reduced; the second derivative technique has the least sensitivity to particle size effects.

As it is undesirable to extrapolate beyond the compositional limits of samples used for calibration and standardization, it is necessary to obtain samples that cover the entire expected range of composition. Finally, temperature changes of the instrument can cause both wavelength and reflectance sensitivity changes.

Instruments

Several companies now market near infrared reflectance instruments to measure the composition of foods and agricultural products. These companies include Technicon Instruments Corp. (Tarrytown, NY), Gardner Neotec Instrument Division, Pacific Scientific Co. (Silver Spring, MD), Dickey-John Corp. (Auburn, IL), and Percon Corp. (Stockholm, Sweden). Two common NIR reflectance instruments are the Neotec grain quality analyzer and the Dickey-John infra-analyzer. Reflectance instruments for scanning the whole NIR region and for analyses of small samples (e.g., single kernels) also are available. For additional information on NIR reflectance instruments and the applications of this technique, see Norris (1974), Trevis (1974), Williams (1975), Pomeranz and Moore (1975), Wetzel (1983), and Norris and Williams (1984).

In the Neotec grain quality analyzer, the light beam passes through a *tilting* filter system that changes angle in relation to a tungsten light source (Fig. 9.19). In this manner, more than 300 measurement points are scanned

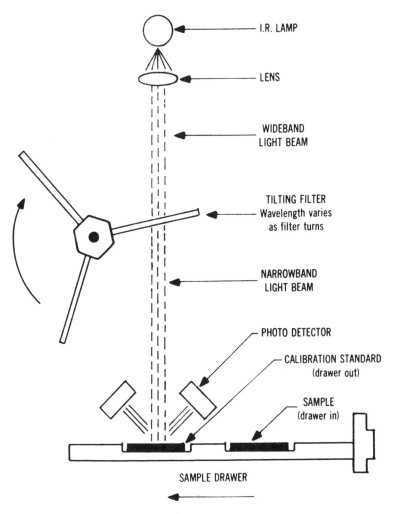

Fig. 9.19.   Diagram of the grain quality analyzer. (Courtesy Neotec Instruments.)

10 times every sec using only three narrow-bandpass infrared filters. At all 300 points, the amount of IR light reflected by the grain sample is measured by a photodetector. Data is automatically fed into a built-in computer, which solves third order equations and displays protein, oil, and water content on the digital readout.

The Dickey-John infra-analyzer utilizes the difference in reflectance of six narrow wave bands of energy in the near infrared spectrum from a ground sample (Fig. 9.20). The reflected energy is detected by a sensitive photocell, sampling the reflected energy level from each of the six dif-

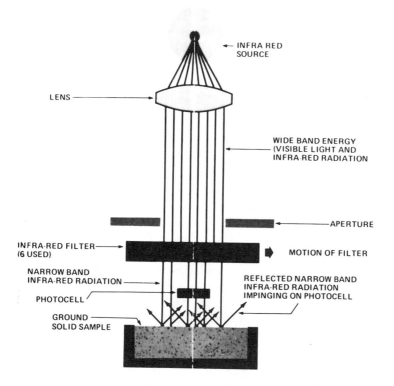

Fig. 9.20. Diagram of the Infra-Analyzer. (Courtesy Technicon Instruments.)

ferent wave bands many times per second. This signal output is amplified and channeled thru a synchronizing device and then is processed thru the computing electronic circuitry.

Other Applications and Limitations

Bengtsson and Larsson (1984) discussed the limited value of NIR reflectance methods to predict the nutritive value of forages. The application of NIR reflectance spectrophotometry to the nondestructive analysis of foods, feeds, and various agricultural products was reviewed by Polesello and Giangiacomo (1983). Robertson and Barton (1984) found NIR reflectance more precise for determination of oil than of water in sunflower seeds. Several authors have reported on the use of NIR reflectance to determine protein in wheat (Osborne 1983), for protein testing in pulse breeding programs (Williams *et al.* 1978), to determine protein in pea flour (Davies and Wright 1984), and to analyze dried milk and cheese powders (Ereifej and Markakis 1983). Weaver (1984) reported NIR reflectance analytical calibrations for the determination of moisture and fat in milk and milk powder, for moisture in soft cheese and butter, and for fat in cream.

Osborne *et al.* (1984A) used NIR reflectance analysis to measure protein, fat, and moisture content of sliced white bread baked by a pilot-scale version of a typical U.K. commercial process. One advantage of their procedure was that bread could be analyzed without the need to dry and powder the samples. Standard deviations between NIR reflectance and standard procedures were 0.20% for protein, 0.18% for fat, and 0.51% for moisture. Although NIR reflectance offers a quick and simple method of quality control (fat and sucrose content) in dry cake mixes, it is less accurate than other methods (Osborne *et al.* 1983). Similarly, NIR reflectance was of limited value in the compositional analysis of biscuits and biscuit doughs; only the calibration for fat was sufficiently accurate to detect metering errors of ±5% relative to the total amount of fat in the recipe (Osborne *et al.* 1984B). NIR reflectance also was of limited value in quality control of malting barley (Kato and Sanada 1983) and provided no useful information on breadmaking quality of wheat, beyond that which may be predicted from protein content (Osborne 1984). The use of NIR reflectance to determine the nutritionally limiting amino acid lysine in cereal grains (Rubenthaler and Bruinsma 1978; Gill *et al.* 1979) was extended by Williams *et al.* (1984) to other amino acids in wheat and barley.

RECENT APPLICATIONS

Mills *et al.* (1982) used an interference filter to isolate the region from 3.4 to 3.5 μm (the CH stretch), replacing the carbonyl band at 5.9 μm of the ester linkage in fats formerly used, and were able to estimate the fat content in aqueous fat emulsions. Dioxathin is an acaricide used in a cattle dip and residues of this material can sometimes be found in meat from these animals. Luguru and Saidi (1982) were able to use a combination of the cis isomer band at 1099 cm^{-1} and the trans isomer band at 1200 cm^{-1} to determine this compound to the 500-ppm level.

Bjarna (1982) developed a multicomponent (protein, fat, water) analysis of meat products using a combination of infrared frequencies. Mills *et al.* (1983) reported that the infrared method for determining fat in meat products after treatment by the Mojonnier method was sufficiently accurate to act as a standard method. On-line detection of the cis and trans isomers of permethrin that had been separated by high-pressure liquid chromatography (HPLC) was done by Papadoulou-Mourkidou *et al.* (1983).

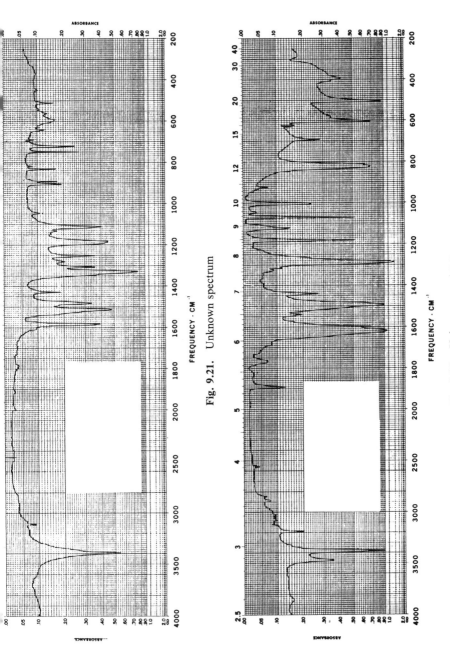

Fig. 9.21. Unknown spectrum

Fig. 9.22. Unknown spectrum

A near infrared diffuse reflectance procedure for the compositional analysis of nonfat dry milk was developed by Bair et al. (1984). They used a combination of bands to determine casein (2190 nm), water (1940 nm), fat and casein (2310 nm), and casein plus lactose (2100 nm).

PROBLEMS

9.1. Convert the following units as indicated:
5.0 μm to angstroms   Ans. 20,000.
5.9 μm to wavenumbers   Ans. 1695.
800 cm^{-1} to μm   Ans. 12.5.
15.3 μm to cm^{-1} Ans. 654.

9.2. An empty absorption cell was placed in an infrared spectrophotometer and a spectrum was recorded over the range 8–10.5 μm. The spectrum showed maxima at 8.10, 8.51, 9.00, 9.52, and 10.10 μm. Calculate the spacing between the plates in this cell in mm. Ans. 0.082 mm.

9.3. If an infrared cell is reported to be 0.020 mm thick, how many fringes (bands) would you expect to find in its spectrum between 4 and 10 μm? Ans. 6.

9.4. Refering to Fig. 9.13 and remembering that absorbance $A = \log I_0/I$, calculate the A for the 2.95 μm band in Fig. 9.21. Ans. 0.77.

9.5. Using the data in Fig. 9.21 and an organic qualitative analysis book, e.g., Shriner and Fuson, identify the following compound. It produces a brown color with ceric nitrate and forms a brown precipitate with ferric hydroxide. Its empirical formula is $C_6H_4ClNO_3$. Ans. 2-chloro-4-nitrophenol (Sadtler Spectra No. 8471).

9.6. A compound gives a positive Hinsberg test and the resulting compound is soluble in alkali. It is known to contain Br but gives a negative NaI test. Its density is 1.8 g/ml. Use Fig. 9.22 and an organic qualitative analysis book (such as Shriner and Fuson) to identify this material. Ans. p-bromoaniline (Sadtler Spectra No. 8481).

BIBLIOGRAPHY

Bair, R. J., Frank, J. F., and Lowenstein, M. (1983). Compositional analysis of nonfat dry milk by using near infrared diffuse reflectance spectroscopy. *J. Assoc. Off. Anal. Chem.* **66**, 858–864.

Bengtsson, S., and Karlsson, K. (1984). Prediction of the nutritive value of forages by near infrared reflectance photometry. *J. Sci. Food Agric.* **35**, 951–958.

Bjarna, E. (1982). Multicomponent analysis of meat products by infrared spectrophotometry: Collaborative study. *J. Assoc. Off. Anal. Chem.* **65**, 696–700.

Colthup, N. B. (1950). Spectra–structure correlations in the infrared region. *J. Opt. Soc. Am.* **40**, 397–400.

Davies, A. M. C., and Wright, D. J. (1984). Determination of protein in pea flour by near infrared analysis. *J. Sci. Food Agric.* **35**, 1034–1039.

Ereifej, K. I., and Markakis, P. (1983). Analysis of dried milk and cheese powders by near infrared reflectance spectroscopy. *In* "Instrumental Analysis of Foods: Recent Report. Proc. Symp. Int. Flavor Conf.," pp. 237–242. Academic Press, Orlando FL.

Fahrenfort, J. (1961). Attenuated total reflectance. *Spectrochim. Acta.* **17**, 698–709.

Gill, A. A., Starr, C., and Smith, D. B. (1979). Lysine and nitrogen measurements by infrared reflectance analysis as an aid to barley breeding. *J. Agric. Sci.* **93**, 727–729.

Golay, M. (1947). A pneumatic infrared detector. *Rev. Sci. Instrum.* **18**, 357–362.

Kato, T., and Sanada, M. (1983). Use of near IR spectroscopy in the quality control and breeding of barley and hops (in Japanese). *Shokuhin Kogyo* **26**, 52–55. [*Chem. Abstr.* 1984, **101**, 53182u.]

Luguru, S. M., and Saidi, A. B. (1982). Infrared spectrophotometric determination of dioxathin in cattle dip water. *J. Assoc. Off. Anal. Chem.* **65**, 930–932.

Mills, B. L., and van de Voort, F. R. (1982). Evaluation of the CH stretch measurement for the estimation of fat in aqueous fat emulsions using infrared spectroscopy. *J. Assoc. Off. Anal. Chem.* **65**, 1357–1361.

Mills, B. L., van de Voort, F. R., and Usborne, W. R. (1983). Mojonnier method as reference for infrared determination of fat in meat products. *J. Assoc. Off. Anal. Chem.* **66**, 1048–1050.

Norris, K. H. (1964). Reports on design and development of a new moisture meter. *Agric. Eng.* **45**, 370–372.

Norris, K. H., and Williams, P. C. (1984). Optimization of mathematical treatments of raw near-infrared signal in the measurement of protein in hard red spring wheat. I. Influence of particle size. *Cereal Chem.* **61**, 158–165.

Osborne, B. G. (1983). Investigation of the performance of an improved calibration for the determination of protein in UK home-grown wheat by near infrared reflectance analysis. *J. Sci. Food Agric.* **34**, 1441–1443.

Osborne, B. G. (1984). Investigations into the use of near infrared reflectance spectroscopy for the quality assessment of wheat with respect to its potential for bread baking. *J. Sci. Food Agric.* **35**, 106–110.

Osborne, B. G., Fearn, T., and Randall, P. G. (1983). Measurement of fat and sucrose in dry cake mixes by near infrared reflectance spectroscopy. *J. Food Technol.* **18**, 651–656.

Osborne, B. G., Barrett, G. M., Canvain, S. P., and Fearn, T. (1984A). The determination of protein, fat and moisture in bread by near infrared reflectance spectroscopy. *J. Sci. Food Agric.* **35**, 940–945.

Osborne, B. G., Fearn, T., Miller, A. R., and Douglas, S. (1984B). Application of near infrared reflectance spectroscopy to the compositional analysis of biscuits and biscuit doughs. *J. Sci. Food Agric.* **35**, 99–105.

Papadoulou-Mourkidou, E., Iwata, Y., and Gunther, F. A. (1983). Application of HPLC system with an on-line infrared detector to the residue analysis of permethrin. *J. Agr. Food Chem.* **31**, 629–633.

Polesello, A., and Giangiacomo, R. (1983). Application of near infrared spectrophotometry to the nondestructive analysis of foods. A review of experimental results. *Crit. Rev. Food Sci. Nutr.* **18**, 203–230.

Pomeranz, Y., and Moore, R. B. (1975). Reliability of several methods for protein determination in wheat. *Baker's Dig.* **49**, 44–48, 58.

Robertson, J. A., and Barton, F. E. (1984). Oil and water analysis of sunflower seed by near-infrared reflectance spectroscopy. *J. Am. Oil Chem. Soc.* **61**, 543–547.

Rubenthaler, G. L., and Bruinsma, B. L. (1979). Lysine estimation in cereals by near infrared reflectance. *Crop Sci.* **18**, 1039–1040.

Stinson, M. M., and O'Donnel, M. J. (1952). The infrared and ultraviolet absorption spectra of cytosine and isocytosine in the solid state. *J. Am. Chem. Soc.* **74,** 1805–1808.

Trevis, J. E. (1974). Seven automated instruments. *Cereal Sci. Today* **19,** 182–198.

Weaver, R. M. V. (1984). Near infrared reflectance analysis applied to dairy products. *Spec. Publ., R. Soc. Chem.* **49,** 91–102.

Wetzel, D. L. (1983). Near-infrared reflectance analysis; sleeper among spectroscopic techniques. *Anal. Chem.* **55,** 1165A, 1166A, 1170A, 1172A, 1174A, 1176A.

Williams, P. C. (1975). Application of near infrared reflectance spectroscopy to analysis of cereal grains and oilseeds. *Cereal Chem.* **52,** 561–576.

Williams, P. C., Stevenson, S. G., Starkey, P. M., and Hawtin, G. C. (1978). The application of near infrared reflectance spectroscopy to protein testing in pulse breeding programs. *J. Sci. Food Agric.* **29,** 285–292.

Williams, P. C., Preston, K. R., Norris, K. H., and Starkey, P. M. (1984). Determination of amino acids in wheat and barley by near infrared reflectance spectroscopy. *J. Food Sci.* **49,** 17–20.

10
Flame Photometry, Atomic Absorption, and Inductively Coupled Plasmas

The next several chapters will be devoted to methods for determining the various elements in foods. These include elements that are nutrients as well as those that are toxic, those that are present naturally, and those that become contaminants during processing or storage. Because researchers are finding that the relative ratios of combinations of elements are important, it is no longer adequate to determine the concentration of one element. For example, copper poisoning can be cured by adding 1/6 the amount of molybdenum, and Zn:Cu ratios must be kept below 14:1 to reduce the risk of heart attack. Figure 10.1 (Thompson 1970) shows several interrelationships between elements that have been demonstrated in test animals.

Some of the more common elements that have been or are currently being used in packaging materials or allied products in concentrations greater than 100 ppm (0.010%) are Sb, As, Ba, Cd, Cr, Co, Cu, Fe, Pb, Mn, Hg, Mo, Ni, Sr, Sn, Ti, Zn, Zr, Ce, La, Al, and V. In addition, if anything goes wrong with the product in a container, the coating is one of the first items to be suspected. This means that any element present that affects the quality of the product must be monitored at some time or another. In beer, for example, aluminum, sodium, calcium, magnesium, potassium, copper, zinc, and iron must be monitored at the ppm level (0.0001%). Thus, sodium, calcium, magnesium, and potassium must be added to the original list and the required levels of detection reduced 100-fold.

The current climate of fear and suspicion of chemicals, as well as the sincere interest of basic research scientists concerning the pathways of

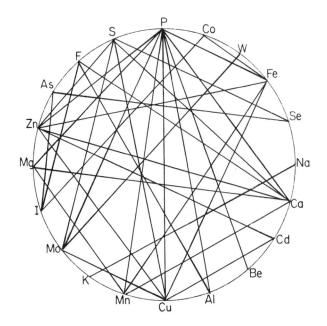

Fig. 10.1.   Mineral interrelationships in animals. (From Thompson 1970.)

trace elements, requires that we be prepared to examine almost the entire periodic table at the ppb level 0.0000001%. The techniques described in this chapter (flame photometry, atomic absorption, and inductively coupled plasmas), x-ray dispersion, and nuclear methods are those most commonly used to detect low levels of trace elements and for multiple-element determinations.

FLAME PHOTOMETRY

Almost everyone who has taken a course in chemistry has placed various salts on a spatula and held them in the flame of a Bunsen burner and watched the colors form. If you do this systematically, you find that only the alkalies, alkaline earths, and a few other elements give off radiation that appears colored.

Figure 10.2 is a simplified energy diagram that may help explain why some elements give off radiation that appears colored and others do not. The heat energy from the flame can raise an electron in some atoms (elements X and Y) from the ground state to an excited state. If the excited electron drops back to the ground state from its original level in one jump, the radiation given off is called a *resonance line*. If the electron loses its energy in steps, some of the energy may be given off in the visible and

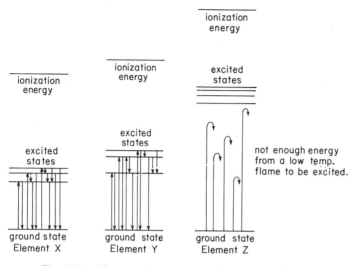

Fig. 10.2. Diagram of energy states of atoms in a flame.

infrared regions or it may lose its energy entirely by radiationless colli-sions. With the alkali and alkaline earth metals the energy required to excite the atoms is relatively low and there is enough energy in a gas–air flame (900–1200°C) to do this. However, with other elements the en-ergy required to get to the excited state is greater than a Bunsen burner can provide (element Z).

Each metal has a different set of energy levels because each has a different nuclear charge and a different number of electrons. Therefore the wavelengths of emitted radiation will be different for each element, and a measure of the position of these wavelengths when they are dis-persed can be used to identify which atoms are present. If the intensity of this radiation is measured, then quantitative analysis is also possible.

A flame photometer is designed to provide a flame whose temperature is hot enough to excite as many elements as possible, to determine which wavelengths are given off, and to measure their intensities. It is possible to determine between 50 and 60 elements using present-day instruments. Figure 10.3 is a block diagram of a flame photometer. A flame photometer is, in essence, a UV–visible spectrophotometer in which the cell and source have been combined.

Inferences

Flame photometry interferences can generally be considered to be due to ionization, background emission, self-absorption, chemical interac-tions, and spectral overlap.

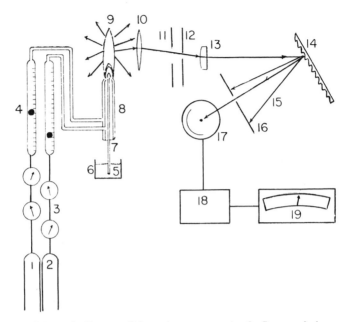

Fig. 10.3.   Schematic diagram of the major components of a flame emission spectrophotometer. (1) H_2, C_2H_2, (2) O_2, air, or N_2O, (3) pressure gauges, (4) flow meters, (5) sample
solution, (6) sample holder, (7) capillary, (8) nebulizer, (9) flame, (10) condensing lens, (11)
mask, (12) entrance slits, (13) collimating lens, (14) diffraction grating, (15) diffracted beam,
(16) exit slits, (17) phototube detector, (18) amplifier, (19) readout meter or recorder.

Ionization causes a decrease in intensity and occurs generally at low
concentrations. At low concentrations there are few atoms present, they
absorb all the energy, and they can be ionized. As the concentration
increases, the energy is spread over more atoms, and less ions are produced. If ionization is a problem, addition of a small amount of an easily
ionizable metal, such as one of the alkalies, is recommended. The added
metal takes up excess energy and the material sought will be ionized less.

Background emission comes from the flame and from other elements
in the sample. This can be handled as follows: Measure the emission due
to the flame without any sample present and set the instrument at zero.
If other elements are present, prepare a synthetic blank and determine
the radiation at the peak intensity. Then move off to one side of the peak
emission line and take another (the lowest) reading. If it is not zero, then
background emission is present.

Self-absorption occurs when radiation emitted from one atom is just
the right energy to be absorbed by another atom. The net effect is that
several atoms can be involved in absorbing and emitting, yet the detector

sees only the final process and indicates a much lower concentration than is actually present. This is a very serious problem, and the process is so efficient that advantage is taken of it in atomic absorption spectroscopy.

Spectral interferences are caused by two or more elements having emitted radiation whose wavelengths overlap. A good monochromator will generally reduce this considerably. Most of the remaining spectral interference is due to the emission of compounds. However, compound emission is generally in the green-red region, whereas atom spectra is in the violet-ultraviolet region. Therefore, if spectral interference is a problem, switch to another wavelength, preferably toward the UV region.

Chemical interference is caused by anions being present that complex the cations, thus, effectively lowering the metal concentration. Phosphates and aluminates are particularly troublesome. Chelating agents are generally used to reduce this type of interference. The chelate, although it ties up the metal also, is an organic material and burns away in the flame, releasing the metal.

Processes within the Flame

The process involved in flame photometry and atomic absorption is actually a very inefficient one, and those attractive colors you see are due to a relatively few atoms. Figure 10.4 shows a representation of some of the steps involved with sodium chloride as the test material.

Step 1. The sodium chloride solution is aspirated into the flame by flame gases such as oxygen–hydrogen, oxygen–acetylene, air–acetylene, nitrous oxide–acetylene, and nitric oxide–acetylene. The solution is broken into drops, about 95% of which are too big to be useful because the solvent cannot be evaporated fast enough. This portion of the sample passes through the flame without further action.

Step 2. The solvent is removed and the solid salt is dispersed in the flame gases. The heat energy from the combustion and the collisions between molecules produces atoms, ions, and free radicals.

Step 3. Neutral atoms of the metal element are produced. Since neutral atoms require the smallest amount of energy to raise them to an excited state, any process that interferes with this process decreases the overall sensitivity of the measurement. Compounds may form with the combustion gases and their products, as well as with other elements in the sample. Although the number of neutral atoms produced varies considerably, efficiencies of 10–15% are more likely than 80–90%.

Step 4. Absorbed radiation raises a neutral atom to an excited state. This process may not take very much energy, but it is a very inefficient.

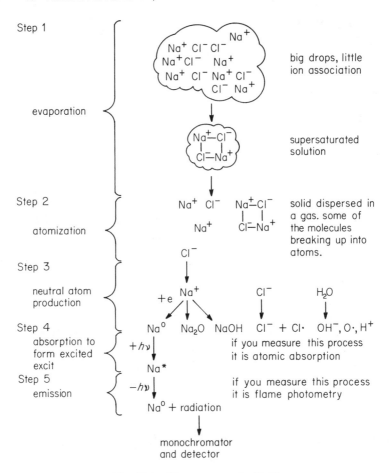

Fig. 10.4.   Flame process for NaCl.

Equation 10.1 can be solved to find out how many of the neutral atoms are eventually excited.

$$N_{ex} = N_g \frac{g_{ex}}{g_g} e^{-\Delta E/kT} \tag{10.1}$$

where N_{ex} is the number of neutral atoms excited, N_g the number of neutral atoms in the ground state, g_{ex}/g_g the statistical weight factor for the particular energy levels (for sodium it is 2 and for zinc it is 3), ΔE the energy to excite the atom, k the Boltzmann constant, and T the flame temperature in K. Table 10.1 shows the results of this calculation for sodium and zinc. Notice that at 3000 K only 1 out of 58,800 neutral sodium atoms and 1 out of 558 trillion zinc atoms is excited. Zinc cannot be

Table 10.1. Values of N_{ex}/N_g for Sodium and Zinc

Element	2000 K	3000 K	4000 K
Na	9.86×10^{-6}	5.88×10^{-4}	4.44×10^{-3}
Zn	7.29×10^{-15}	5.58×10^{-10}	1.48×10^{-7}

determined with a flame photometer, but can be determined easily by atomic absorption. Sodium can be determined in the 0.01-ppm region by flame photometry.

Step 5. The excited atom can return to the ground state either by losing its energy by molecular collision or by giving off a ray of radiation. It is unfortunate that we cannot control the direction in which the ray of radiation will come off. Since it can be radiated in all directions, most of it is lost and only that amount that can be collected by mirrors and lenses can be used.

In addition to the above losses, there is a high dilution loss. In order to vaporize 2 ml of a solution, about 10 liters of gas is required. If only 0.1 ml (95% is lost because of large drops) of the sample is used, then this represents a dilution of 10,000. The original 10 liters of gas is further expanded to about 80 liters when heated to the flame temperature. The net result is that there are many places where an improvement in sensitivity can be made.

Equation 10.1 indicates that sensitivity can be improved by lowering the ΔE or raising the temperature. Neutral atoms have the lowest ΔE, so it would be logical to make more neutral atoms. It has been found that if alkali metals are added in small amounts to the sample, more neutral atoms of the element desired are formed. An electrical discharge across the flame has a similar effect.

As shown in Table 10.1, if the temperature is 4000 K, then 1 atom of sodium in 4000 is excited, a substantial improvement over the 1 in 58,000 at 3000 K. The problem is how to produce a hotter flame. Theory predicts that the flame temperature for an oxygen–hydrogen flame should be much hotter than the normally measured 2800 K. However, the combustion product, water, dissociates appreciably at 2800 K and dissipates much of the energy, thus lowering the flame temperature. A combustible mixture is needed that will produce products that will not dissociate. An oxygen–cyanogen $(CN)_2$ flame produces CO_2 and N_2, both of which dissociate little. As a result, a flame temperature of 5000 K is possible. However, cyanogen is quite expensive and very toxic, so its use has been limited.

We have seen that there are many interferences in flame photometry. Moreover, the emission of radiation (Step 5), which is measured in flame photometry, is very inefficient. Those limitations led to a new approach pioneered by Walsh (1955).

ATOMIC ABSORPTION SPECTROSCOPY

Recall from the previous discussion on NaCl that at 3000 K only 1 out of 58,800 neutral atoms are excited. Recall also that a major interference in fluorescence and flame photometry is self-absorption, that is, wavelengths emitted from one atom are easily absorbed by another atom of the same kind. This phenomenon is utilized in atomic absorption (AA) spectroscopy to increase the efficiency at which neutral atoms are excited.

An atomic absorption instrument (Fig. 10.5) contains a source of radiation whose wavelengths are exactly the same as those required to excite the atoms in the flame, thereby providing a much more efficient way

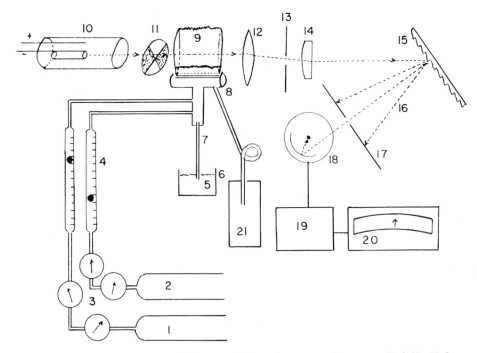

Fig. 10.5. Basic components of an atomic absorption spectrophotometer. (1) C_2H_2, N_2O, or H_2, (2) air, (3) pressure gauges, (4) flow meters, (5) sample solution, (6) sample holder, (7) capillary, (8) nebulizer, (9) slot flame, (10) hollow cathode, (11) chopper, (12) condensing lens, (13) entrance slits, (14) collimating lens, (15) diffraction grating, (16) diffracted beam, (17) exit slits, (18) phototube, (19) amplifier, (20) readout meter or recorder, (21) gas trap and waste.

for the neutral atoms to be excited. Notice in Fig. 10.5 that a *chopper* and a *hollow cathode* have been added to a regular flame photometer. The hollow cathode is used to produce a high intensity of radiation similar to that absorbed by an element to be determined in the flame. The chopper is used to produce a pulsating signal for easier amplification and to provide a means to distinguish hollow cathode radiation from flame radiation.

In flame photometry, the heat energy from the flame and the energy from molecular collisions are absorbed. This is not very efficient because many "energies" are available and only a few will excite the neutral atoms. This is where the hollow cathode lamp is effective. Its purpose is to provide the "right kind" of radiation. If magnesium is to be determined, then a magnesium hollow cathode is used; if calcium, then a calcium hollow cathode is adequate.

The main difference between flame photometry and atomic absorption is that in flame photometry the radiation emitted from the flame is measured, and in atomic absorption the decrease in the intensity of the radiation from the hollow cathode due to the absorption by the atoms in the flame is measured.

At the present time the following elements *cannot* be determined directly by AA spectroscopy: Fr, Ra, Ac, Tc, Os, C, N, P, O, S, Po, all halogens, all inert gases, and Pm in the rare earths. However, it is believed that in the near future it should be possible to determine Ra, Os, C, N, P, S, and the halogens except At.

Hollow Cathode

A hollow cathode, diagrammed in Fig. 10.6, operates as follows. Electrons given off by the cathode accelerate toward the anode. Along the way they ionize argon atoms, which are then attracted to the cathode.

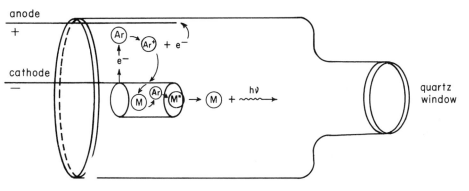

Fig. 10.6. Operation of a normal hollow cathode.

When the relatively large, highly accelerated argon ions smash into the cathode, their energy is sufficient to knock several metal atoms off of the cathode surface. These atoms may then be excited by collision with the filler gas ions. When the excited atoms give off their energy, the radiation emitted is characteristic of the metal the cathode was made from. As an example, a copper cathode will give off radiation that copper atoms in the flame can readily absorb and that no other element will absorb. Therefore, atomic absorption spectroscopy is almost completely free of spectral interference.

As a source of radiation for AA, a hollow cathode has two advantages: it provides high-intensity radiation and decreases the *Doppler effect*. If an atom is moving in the same direction as the ray of radiation it emits, then the atom's velocity is added to the velocity of the emitted ray changing its frequency. The reverse is true when the atom moves away from the emitted ray. This is the Doppler effect and it is undesirable because the atoms now give off a band of energies rather than the single energy desired. This decreases the efficiency of absorption and also causes greater spectral interference. The shape of the hollow cathode provides an electric and magnetic field that reduces the Doppler effect.

Chopper

The atoms in the flames of an AA instrument still emit radiation, and since this includes the same wavelength as that being absorbed, an error is introduced. This is eliminated by the use of a chopper. The chopper cuts the beam from the hollow cathode, producing an alternating pulse of energy, while the energy emitted by the atoms in the flame is continuous. It is possible electronically to distinguish these different signals by the detector system.

Burners

The first AA instruments used the same type of burners as flame photometers. These are seldom used now because, being the turbulent-flow type, they give high background noise and the number of atoms in the light path is small. In order to increase the sensitivity of the system, mirrors can successfully be used with some elements. Changing the flame profile is sometimes useful even if a mirror system is not used. This may be done by moving the burner a few millimeters and may improve the sensitivity severalfold (Fig. 10.7).

To increase sensitivity without mirrors, several burners have been placed in a row, thereby providing a longer sample path. This increases

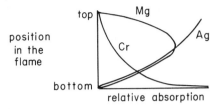

Fig. 10.7. Effect of position in the flame on absorption for magnesium, silver, and chromium.

sensitivity, but it is difficult to align all the burners and the resultant noise is disturbing. A uniform flame that burns quietly and that has a long pathlength is attained in the laminar flow, elongated flame, or slot burner (Fig. 10.8).

The long flame in a slot burner provides a longer path for absorbing radiation. It is a quiet flame because the big solvent droplets that produce the turbulence are eliminated by draining. A preheater is sometimes used to evaporate large amounts of solvent. Care must be taken with this type of burner because the fuel is premixed before it gets to the flame; when sample aspiration is stopped, a slight negative pressure develops, drawing the flame inside the chamber and producing an explosion. Air is used rather than oxygen for combustions because the flame speed is slower and the explosion hazard is reduced. With an air–acetylene mixture it is

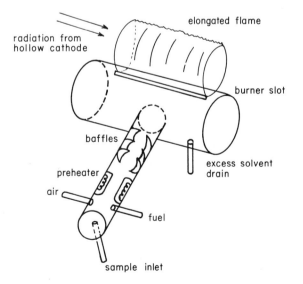

Fig. 10.8. Components of a slot burner.

Table 10.2. Comparison of Flame Mixtures, Combustion Temperatures, and Flame Speed in Atomic Absorption

Mixture	Combustion temperature (°C)	Flame speed (cm/sec)
Air–propane	1925	82
Air–acetylene	2300	160
50% O_2, 50% N_2	2815	640
O_2–acetylene	3060	1130
N_2O–acetylene	2955	180
NO–acetylene	3080	90

possible to aspirate 70% perchloric acid without an explosion in a slot burner. The burner head should be coated with Desicote to keep it from dissolving.

Nitrous oxide can be used instead of air. Nitrous oxide does not form oxides with the aspirated metals nearly as readily as air or oxygen, so more atoms are reduced. In addition, a nitrous oxide flame is several hundred degrees hotter than an air–acetylene flame; because of this, flame compounds are destroyed to a greater extent in a nitrous oxide flame even though flame speed is slow. Table 10.2 compares flame mixtures, temperatures, and speeds. Since nitrous oxide is an anesthetic, a good exhaust system must be provided.

ELECTROTHERMAL PROCESS (CARBON ROD FURNACE)

A typical flame or atomic absorption instrument requires a sample flow rate of 2 ml/min and 45–60 sec to obtain a stable signal. If the sample could be evaporated faster, then more atoms would be in the beam and less sample would be required. In a carbon rod furnace, now called the electrothermal process because materials other than carbon can be used, a 1–10-μl sample in a hollow carbon or ceramic rod is heated to 3000°C in 3 sec. This results in lowering the detection limits for most elements by 1000 to 1 million times.

Figure 10.9 is a diagram of one type of carbon rod furnace. The ends of the rod are connected to electrical terminals and 2–12 V and 10–500 A are applied. The furnace can be brought to white heat within 2 sec. Argon is used to sweep out oxygen, which would cause the carbon to burn, and cooling water is necessary to keep the rest of the instrument from getting too hot. The sample is injected with a microliter syringe into the top hole. A four-step sequence then follows: (1) drying, (2) ashing, (3) atomization,

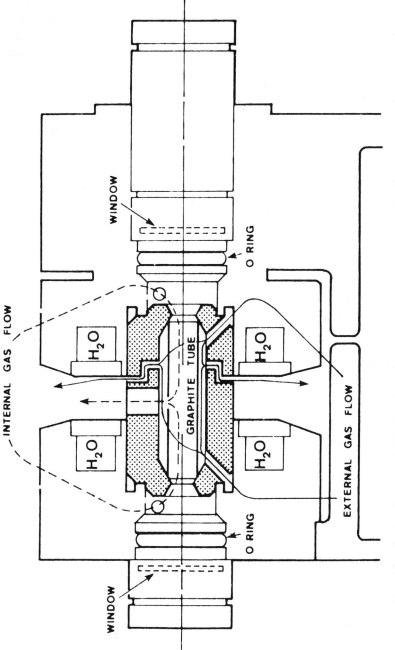

Fig. 10.9. Schematic diagram of Perkin-Elmer carbon rod furnace. (Courtesy Perkin-Elmer Corp., Norwalk, CT.)

and (4) cooling. The times and temperatures of each step can be computer controlled. The first step typically lasts 20 sec at 20 A (200°C). Ashing is then done to remove organic matter that forms a smoke, blocking the light path and causing very high results. This step requires an intermediate temperature and time, typically 30 sec at 40 A (400°C). The sample is then atomized, the atoms ascending into the beam where their absorbance is measured. A 3-sec burn at a temperature of 1950°C requires 125 A. The current is shut off and a 30-sec cooling period is allowed.

Manual injection of the sample usually does not provide precise measurements. However, with an automatic sample injector the precision obtained can be within 1%.

INDUCTIVELY COUPLED PLASMA (ICP) EMISSION SPECTROSCOPY

A disadvantage with atomic absorption spectroscopy is that only one element can be determined at a time. With flame photometry or emission spectroscopy several elements can be determined simultaneously, but the detection limits are quite high for many elements. An examination of equation 10.1 indicates that an increase in flame temperature will provide more excited atoms, thus increasing the sensitivity of the process. One of the first methods to increase source temperature was to replace the flame with an arc formed between two carbon rods. Temperatures of 5000 K could easily be obtained. A better way to obtain even higher temperatures is to use a plasma torch. The carbon electrodes of a conventional emission spectrograph are replaced by the plasma torch.

A *plasma* is a flame or electrical discharge that has an unusually high concentration (>1%) of positive and negative ions present. These flames can be made to have very high temperatures, from a few thousand degrees to 50,000 K; temperatures of 9000–10,000 K are typical in current analytical instruments. Figure 10.10 illustrates the basic design of the plasma burner; Figs. 10.11–10.13 will be used to explain the plasma-forming process.

Plasma Formation

According to Fassel and Kniseley (1974), a plasma can be formed by placing a quartz tube about 2.5 cm in diameter inside a coil connected to a high-frequency generator operating typically at 4–50 MHz with power levels of 2–5 kW. Argon is passed through the tube, and the plasma is started by using a Tesla coil to generate the first free electrons.

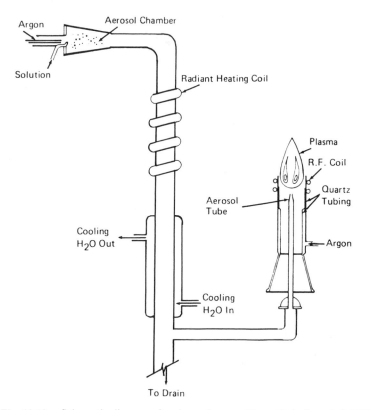

Fig. 10.10.   Schematic diagram of a plasma burner. (From Kniseley *et al.* 1973.)

Let us now examine the events leading to the formation of the plasma. The high-frequency currents flowing in the induction coil generate oscillating magnetic fields whose lines of force are axially oriented inside of the quartz tube and follow elliptical closed paths outside of the coil, as shown in Figure 10.11. The induced axial magnetic fields in turn induce the seed electrons and ions to flow in closed annular paths inside the quartz tube space. This flow of electrons—the eddy current—is analogous in behavior to the current flow in a short-circuited secondary of a transformer. Because the radio frequency field is an alternating field, the induced magnetic fields vary in both direction and strength. This causes the electrons to be accelerated each half cycle. The accelerated electrons and ions meet resistance to their flow, Joule or ohmic heating is a natural consequence, and additional ionization occurs. The steps just discussed lead to the almost instantaneous formation of a plasma. The overall appearance is a three-zone flame: the inner core, which is very intense, brilliant white, and nontransparent; a second cone about 1–3 cm above

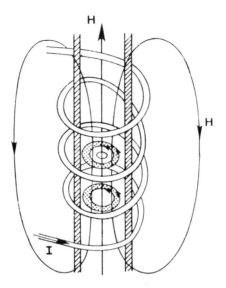

Fig. 10.11. Magnetic fields and eddy currents generated by an induction coil. (From Fassel and Kniseley 1974.)

the induction coil, which is bright but somewhat transparent; and the tail flame, which is barely visible until sample is added to the plasma. The second region is used for analytical work.

The plasma formed in this way attains a gas temperature of 9000–10000 K. At these temperatures the plasma must be thermally isolated, otherwise the quartz tube would melt under sustained operation. This isolation is achieved by Reed's vortex stabilization technique (Fig. 10.12), which utilizes a flow of argon introduced tangentially. Figure 10.12 also shows the complete assembly of concentric tubes and argon flow patterns for sustaining the plasma after it has formed. The tangential flow of argon, which is typically about 10 liter/min, streams upward tangentially, cooling the inside walls of the outermost quartz tube and keeping the plasma away from the wall. The plasma itself is anchored near the exit end of the concentric tube arrangement.

In addition to the vortex stabilization flow of argon, another lower velocity flow of about 1 liter/min transports the sample as an aerosol, a powder, or a thermally generated vapor to the plasma.

Because of the high temperatures of the plasma and the resulting sharp temperature gradient, it was initially difficult to get the sample actually into the flame. For several years this was a major drawback to the ICP technique, and the observed detection limits were never as low as believed possible. This is illustrated in Fig. 10.13 (left); that is, the sample simply goes around the outside of the plasma and little reacts. Fassel and Kni-

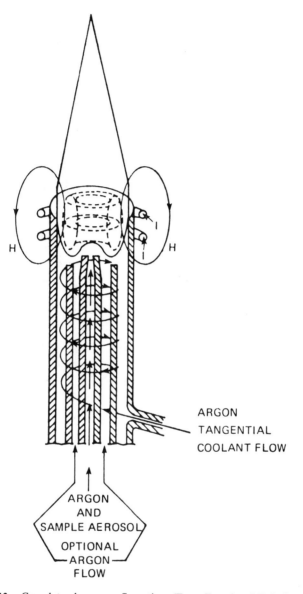

Fig. 10.12.   Complete plasma configuration. (From Fassel and Kniseley 1974.)

seley (1974) found that if the frequency of the induction coil is increased to about 30 MHz, the plasma becomes doughnut shaped (Fig. 10.13 right) and the sample can be injected into the center of the doughnut. Although the temperature of this region is not the maximum, it is about 7000 K, which is more than twice that of a normal hot flame. The net result was

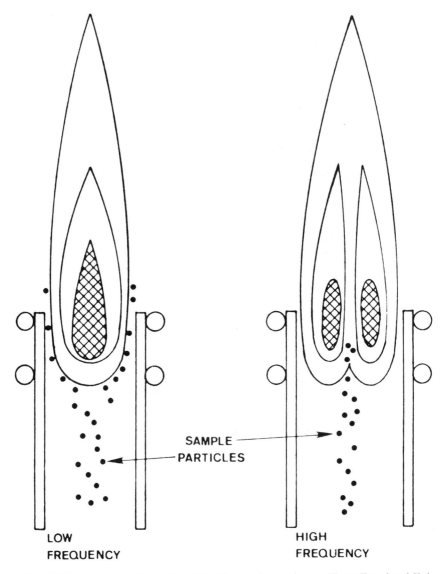

Fig. 10.13.   Sample particle paths with different plasma shapes. (From Fassel and Kniseley 1974.)

a general lowering of the detection limits by two or three orders of magnitude, and ICP emission spectroscopy was taken from the ppm range into the ppb range.

Figure 10.14 is a diagram of the optical path of a current ICP apparatus. Such an instrument can determine 50 elements in a liquid sample within 2 min, most at the ppb level.

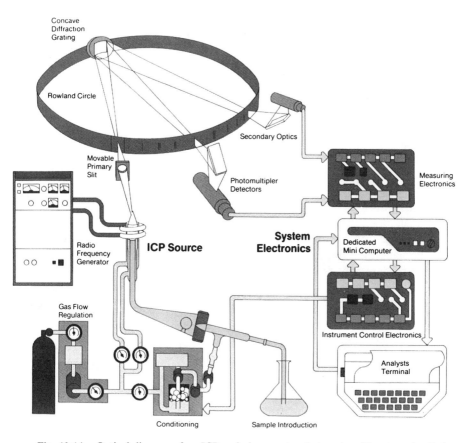

Fig. 10.14. Optical diagram of an ICP emission spectrophotometer. (Courtesy Applied Research Laboratories, Inc.)

Advantages

The advantages of ICP emission spectroscopy have been described by Fassel and Kniseley (1974) as follows:

Typical residence times of the sample in the plasma before reaching the observation height are in the 2.5-msec range. The combination of high temperatures and relatively long residence times should lead to complete solute vaporization and a high, if not total, degree of atomization of the analyte species in the core of the plasma. Once the free atoms are formed they occur in a chemically inert environment, as opposed to the violently reactive surroundings in combustion flames. Thus their lifetime, on the average, should be longer than in flames.

The plasma system described above possesses other unique advantages. First, after the free atoms are formed, they flow upstream in a narrow cylindrical radiating channel so it is easier to use all of the available radiation. Second, at the normal height of observation, the central axial channel containing the relatively high number density of analyte free atoms has a rather uniform temperature profile with very few low temperature atoms surrounding it. Thus self absorption is held to a minimum. This means that with good detectors, calibrating curves covering at least five orders of magnitude are possible thus eliminating the usual multiple dilutions required previously.

The other advantages of these plasmas worthy of note are (a) no electrodes are used, hence contamination from the electrodes normally used in other plasmas or arcs is eliminated; and (b) the plasma operates on nonexplosive noble gases, hence the systems can be used in locations where combustibles are not allowed.

DETECTION LIMITS

A distinction is made here between detection limits and sensitivity. Sensitivity is usually defined as the amount of an element that will produce a 1% change in the absorption reading, whereas detection limits are defined as the amount of material that will produce a signal twice as large as the background noise. Atomic absorption is capable of detecting 0.001 μg/ml for some systems, with 1 μg/ml being most common. Electrothermal methods vary from 0.01 μg/ml for zinc to 40 μg/ml for palladium and platinum. ICP varies from 0.02 μg/ml for strontium to 300 μg/ml for tin.

RECENT APPLICATIONS

To determine the amount of lead and cadmium that leaches from foodware, a 4% solution of acetic acid is allowed to stand in contact with the tableware for 24 hr. The extracted metals are then determined by AA or ICP. Gould et al. (1983) found that heating the solution to 120°C for 2 hr increased the amount of material leached from the tableware by a factor of 10 and permitted determination by conventional AA.

Kumpulainen et al. (1983) used electrothermal methods to determine selenium in foods by using 0.5% nickel as a matrix modifier to prevent selenium losses during the ashing step. Levels to 1 ppm can be determined with 98% recovery.

Cary and Rutzke (1983) developed an electrothermal method to determine chromium in plant tissues. The chromium is concentrated by coprecipitation with ferric hydroxide, and the redissolved iron is removed by extraction with MIBK before the determination.

BIBLIOGRAPHY

Cary, E. E., and Rutzke, M. (1983). Electrothermal atomic absorption determination of Cr in plant tissues. J. Assoc. Off. Anal. Chem. 66, 850–852.

Fassel, V. A., and Kniseley, R. N. (1974). "Inductively coupled plasma-optical emission spectroscopy." Iowa State University, Ames, IA.

Gould, J. H., Butler, S. W., and Steele, E. A. (1983). Release of Pb and Cd: Comparison of two hot leach methods with a room temperature method, using specially glazed ceramic. *J. Assoc. Off. Anal. Chem.* **66**, 1112–1116.

Knieseley, R. N., Fassel, V. A., and Butler, C. C. (1973). Application of inductively coupled plasma excitation sources to the determination of trace metals in microliter volumes of biological fluids. *Clin. Chem.* **19**, 807–812.

Kumpulainen, J., Raittila, A., Lehto, J., and Koivistainen, P. (1983). Electrochemical atomic absorption spectrometric determination of selenium in foods and diets. *J. Assoc. Off. Anal. Chem.* **66**, 1129–1135.

Thompson, D. J. (1970). "Trace Elements in Animal Nutrition," 3rd ed. International Minerals and Chemical Corp., Skokie, IL.

Walsh, A. (1955). The application of atomic absorption spectra to chemical analysis. *Spectrochim. Acta* **7**, 108–112.

11
X-Ray Methods

Roentgen discovered X rays in 1895 and within a few years reported several of their basic properties. Among these properties are the following:

(1) X rays travel in straight lines and with the speed of light;

(2) X rays are not charged particles;

(3) the generation of X rays may be accomplished by impinging a beam of high-energy electrons onto a target, the higher atomic weight targets being the most efficient sources for X rays;

(4) X rays effect a change in photographic emulsions;

(5) electrical charges are dissipated when exposed to X rays; and

(6) X rays induce fluorescence in many materials such as calcium tungstate and zinc sulfide.

In 1912, when it was found that a copper sulfate and zinc blend would diffract X rays, the analytical significance of X rays was first realized.

Today, many types of X-ray analysis have been developed including X-ray absorption, diffraction, fluorescence, emission, absorption edge, low-angle scattering, k-capture, soft X rays, and, most recently, the electron probe and nondispersive techniques. In direct quantitative analysis, X-ray fluorescence and nondispersion X-ray methods are now the most important techniques.

X-ray absorption and fluorescence are atomic properties that are only slightly affected by the way an atom is bonded to another atom. The relationship between the mass absorption coefficient and wavelength is simple. Elaborate separation schemes are not necessary. These factors combine to make X-ray techniques fast, simple, accurate, and in many cases, quite sensitive analytical tools.

PRODUCTION AND DETECTION OF X RAYS

Figure 11.1 will be used to help explain how X rays are produced. Suppose an electron traveling at a very high velocity strikes an atom. Since it has a high energy, the electron can overcome the negative charge due to the electrons around the nucleus and penetrate into the inner levels of the atom. During this process, the electron will lose some of its energy because it requires work to push through the negatively charged field. However, if the electron has sufficient energy so that it is not repelled, it may transfer a part of its remaining energy to an electron in the K shell around the nucleus, causing it to be ejected. This creates a vacancy in a very low energy level that electrons from other orbitals will try to fill. Since L-shell electrons are the most readily available, they usually succeed in filling this vacancy. When the L electron drops from its higher energy (less stable position) to the lower energy (more stable position), it loses a large amount of energy. When calculated in terms of wavelengths by equations $E = h\nu$ and $\lambda\nu = c$, the wavelengths are only a few angstroms in length.

The vacancy in the K shell could have been filled by an electron from the M shell, a less likely process but a more energetic one if it happens. Since in both cases a K-shell vacancy is filled, we call the X rays produced K X rays; the one filled from the closest shell is a K_α and the other a K_β X ray.

If the original entering electron penetrates only as far as the L shell, then the X ray would be called an L X ray and there would be L_α and L_β X rays.

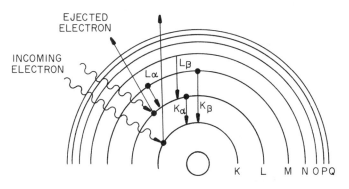

Fig. 11.1. Electronic transitions involved in production of X rays.

Suppose a K electron is ejected and the vacancy is filled by an L electron. The L vacancy is in turn filled by an M electron, etc. When does the process stop? Theoretically it can go on until the outermost orbital is reached. But in practice it stops rather quickly. The efficiency of X-ray production is quite low. Only about 0.01–0.1% of an electron beam current produces X rays. This means that there is a very large number of electrons coming into the orbitals from the outside to fill the vacancies and stop the process.

Because each element has a different arrangement of electrons about its nucleus and thus a different set of energies, X rays of characteristic wavelengths and energies are emitted by each element. Figure 11.2 is a plot of X-ray intensity vs. wavelength. Notice that the intensity of K_β is about 10–15% that of K_α, but its energy is greater. The large amount of low-intensity continuous radiation in Fig. 11.2 is caused by the other 99.9% of the electrons that do not produce X rays, but still enter the electrical field of the atom and are slowed down (braking or Bremsstrahlung) by radiative collision with the orbital electrons.

X-Ray Tubes

The purpose of an X-ray tube is to provide a beam of electrons with sufficient energy to produce X rays. Figure 11.3 is a diagram of an X-ray tube. A high voltage is placed across the filament and an electron cloud is formed around it. When 5000–100,000 V is applied to the target (anode), the electrons leave the cathode and hit the target, producing X rays characteristic of the target material. These X rays are emitted out of the tube

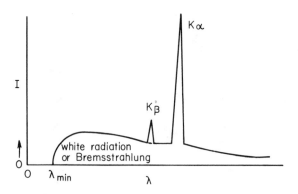

Fig. 11.2.   Typical X-ray emission spectrum.

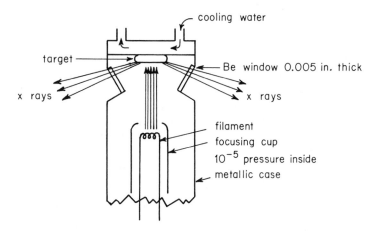

Fig. 11.3. Schematic diagram of an X-ray tube.

through a thin Be window. Much of the energy is expended as heat and the target must be cooled, usually by water.

Detectors. Figure 11.4 illustrates the various types of detectors usually employed in X-ray instruments. The detailed operation of these will be explained in Chapter 19 on radioactivity.

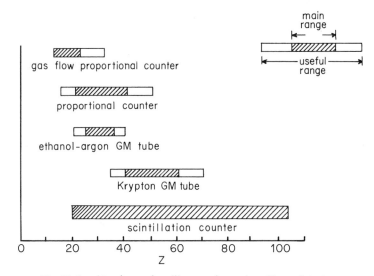

Fig. 11.4. Atomic number (2) range for various X-ray detectors.

X-RAY DIFFRACTION

X-ray diffraction is not a routine analytical technique, but a knowledge of the principle involved is necessary for an understanding of X-ray fluorescence. As noted already, X rays travel in straight lines and are very penetrating. In addition, as electromagnetic radiation, X rays may be diffracted. In the case of X rays, the right order of spacings for diffraction exists in crystals, in which the atomic or ionic distances are of the order of a few angstroms.

When an X ray strikes a layer of atoms in a crystal it can be diffracted as shown in Fig. 11.5. If another X ray (λ_2) strikes the layer below the first layer, and the total distance it travels is an even number of wavelengths behind the first wavelength (λ_1), then λ_2 can add to λ_1, since they are both in phase. If this is repeated for several layers, the emerging X-ray beam is strong enough to be detected by a photographic film or other X-ray detector.

X-ray diffraction is now used primarily to examine the crystalline structure of materials and to determine bond lengths and angles, expansion coefficients, and atomic and molecular weights. In food analysis it has found only limited application.

X-RAY FLUORESCENCE

Figure 11.6 shows one arrangement for an X-ray fluorescence instrument. The X rays from an X-ray tube are used as the source of high energy to produce secondary X rays in the sample. The sample is placed in an

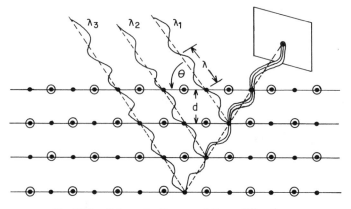

Fig. 11.5.   Schematic diagram of X-ray diffraction.

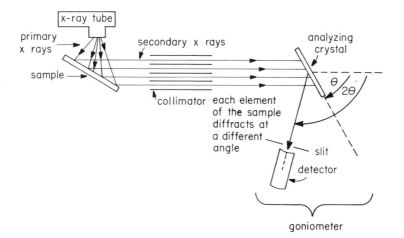

Fig. 11.6. Main components of an X-ray fluorescence instrument that utilizes flat-crystal reflection method.

aluminum or plastic holder. The secondary X rays from the sample are characteristic of the elements in the sample. These X rays are collimated with a set of capillary tubes, usually made from nickel, and directed upon a crystal of known spacing d such as LiF. Recall the Bragg equation, $n\lambda = 2d \sin \theta$. We know d, and if θ is measured, then λ, the characteristic X rays, and therefore the elements present in the sample can be determined. Since the angle θ depends on the wavelength, the analyzer crystal must be rotated to determine the various elements. As the crystal rotates through an angle θ, the detector must rotate through an angle 2θ. The crystal, the detector, and the mechanical parts necessary to provide the rotation is known as a *goniometer*.

The collimator greatly reduces the intensity of the final signal at the detector. One arrangement to eliminate the collimator is to use a curved crystal. The crystal used is a thin piece of mica. By bending the crystal, there is a focusing of the wavelength to a point in front of the detector.

X-Ray Sources

The choice of source in X-ray fluorescence is determined by the elements being examined. This is somewhat of a peculiarity limited to X rays and may be explained by referring to Fig. 11.7.

As the wavelength decreases, the mass absorption coefficient (μ/ρ) decreases. When the wavelength slightly exceeds the wavelength of a characteristic emission line for the sample, there is a rapid increase in the absorption of X rays. "As a general rule, the closer the characteristic

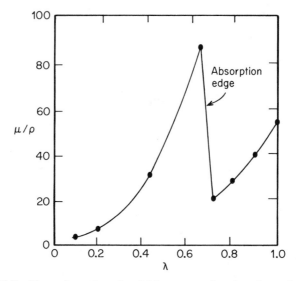

Fig. 11.7.   X-ray absorption of molybdenum as a function of wavelength.

tube radiation approaches the absorption edge value on the short wave-length side, the more efficient is the X-ray excitation process'' (Birks 1960). Each element has 1 K, 3 L, and 5 M absorption edges.

   For biochemical assays, chromium, molybdenum, and wolfram are the tubes of choice. Most are operated at 50 kV and 30–45 mA. At 60 kV there is barely enough energy to excite the barium K_α ($Z = 56$) line, and usually less intense L lines are used for the heavier elements.

Collimators

   Collimators may be either capillary tubes or parallel plates, with the latter the most common at the present. Generally they are made out of nickel. They range in size from 2 to 10 cm in length with a separation of 0.12 to 0.5 mm. The closer the spacing and the longer the collimator, the higher the resolution but the poorer the detection limits.

Analyzing Crystals

   The Bragg equation places some severe restrictions on d. As a result, a variety of different spacings must be available if all elements are to be examined with maximum sensitivity. Table 11.1 lists several crystal materials and the lowest atomic number element that can be determined with each.

Table 11.1. Selected Analyzing Crystals for X-Ray Fluorescence

Material	2d Spacing (Å)	K_α	L_α
LiF	4.028	K(19)	In(49)
NaCl	5.039	S(16)	Mo(42)
Quartz	8.50	Si(14)	Rb(37)
Ethylene diamine tartrate (EDDT)	8.808	Al(13)	Br(35)
Calcium sulfate	15.12	Na(11)	Ni(28)
Mica	19.8	F(9)	Mn(25)
Potassium acid phthalate (KAP)	26.0	O(8)	V(23)
Half K salt of cyclohexane, 1,2-di acid	31.2		
Tetradecano amide	54		
Dioctadecylterephthalate	84		
Dioctadecyladipate	90		

The reflectivity of the last four materials in Table 11.1 is extremely high, which means that the detection limits for the lighter elements can be lowered. Those materials are still being evaluated.

The use of oriented soap films that can be made easily into curved surfaces is a recent innovation that permits d values from 25 to 100 Å.

Detectors

The most common detectors in X-ray fluorescence instruments are gas-flow proportional counters, thallium-activated NaI scintillation counters, or the xenon proportional counter. They are discussed in detail in Chapter 19.

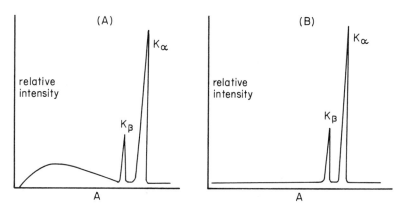

Fig. 11.8. Comparison of X-ray emission and fluorescence spectra. (A) Chromium target at 50 kV; (B) chromium excited by X rays from a wolfram tube (fluorescence) operated at 50 kV.

It should be noted that whereas the characteristic spectrum of an X-ray tube has superimposed on it the white radiation, the fluorescent spectrum does not have this disadvantage. This is shown in Fig. 11.8.

NONDISPERSIVE X-RAY SPECTROSCOPY

In the section on Fourier transform principles in Chapter 9, the advantages of nondispersion of the radiation were explained. The same principles apply to X rays. Since each element emits X rays of definite energies, it should be possible to have the X rays emitted from several elements arrive at a detector at the same time and then be separated based on their respective energy levels.

Recent advances in electronics have permitted the construction of *pulse height analayzers* with sufficient accuracy to make it possible to collect the emitted X rays from an irradiated sample and separate them without using a dispersing crystal. This technique, known as *nondispersive X-ray spectroscopy,* can be used to detect several elements at one time. Nondispersive instruments can be made small enough to be portable, and have sufficient sensitivity for most applications.

Pulse Height Analyzers

A differential pulse height analyzer is a device to determine the number of pulses of any given energy of either gamma rays or X rays. As diagrammed in Fig. 11.9, any X ray with sufficient energy to produce an

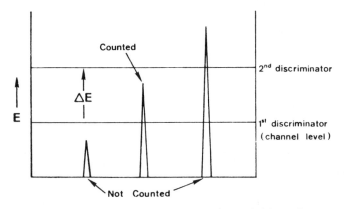

Fig. 11.9.   Operation of a grid potential in pulse height analyzer.

electrical pulse that is larger than the potential set on the first discriminator and less than the potential on the second discriminator will be counted. The energy range between the discriminators is known as a *window*. If the pulse is not large enough or if it is too large, it will not be counted. For example, if iron, chromium, and tin are present in a mixture and chromium has an X ray with energy to fit into the window, then chromium X rays will be counted and those from iron and tin not. For multiple-element analysis a series of windows are required, as shown in Fig. 11.10.

In practice it is fairly common to tie the two discriminators together with the window ΔE set at some given value. Then the channel level is varied through the energy region selected, as shown in Fig. 11.10. Note that ΔE is constant, but the channel is varied. In a multichannel analyzer, several hundred channels operate at once rather than moving one over a range.

Figure 11.11 shows a diagram of a nondispersive X-ray instrument. The Applied Research Laboratories Model N940 nondispersive analyzer (Fig. 11.12) is an example of this type of instrument. It has a radioactive source, weighs about 70 kg, and is in effect "portable."

A spectrum is obtained by exciting the sample with an isotope X-ray source (Table 11.2); the spectrum is then analyzed by an energy-responsive proportional counter, usually referred to as a detector. The use of a nondispersive system allows a close-coupled source/detector geometry, which eliminates the need for vacuum or helium purged path.

Detector pulses, after amplification, are fed to a pulse height analyzer (PHA) for effective element separation, then to a counter, and finally to a four-digit light-emitting diode (LED) display unit. When the atomic numbers of the wanted elements are very close together, one or more easily changed filters can be inserted to give further physical discrimination.

Analysis times with this type of instrument are relatively short, often only a few seconds; since the method is nondestructive, it is possible to obtain many repeat measurements from the same sample. An electronic timer for determination of the analysis period is provided.

Table 11.2. Isotope Sources for
Nondispersive X-Ray Instruments

Isotope	Typical activity (mCi)	Half-life (yr)
^{55}Fe	20	2.7
^{241}Am	3	470
^{238}Pu	30	86

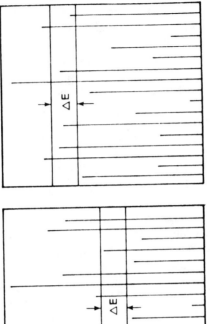

Fig. 11.10. Use of discriminators in obtaining differential spectra.

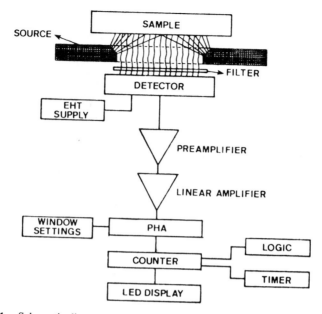

Fig. 11.11. Schematic diagram of a nondispersive X-ray instrument. (Courtesy Applied Research Laboratories, Sunland, CA.)

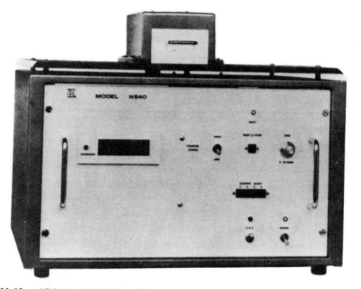

Fig. 11.12. ARL Model N940 nondispersive X-ray analyzer. (Courtesy Applied Research Laboratories, Sunland, CA.)

Table 11.3. Element Coverage of Source/Detector Combinations in Nondispersive X-Ray Instruments

	Si	P	S	Cl	K	Ca	Sc	Ti	V	Cr	Mn	Fe	Co	Ni	Cu	Zn	As	Sr	Mo	Ag	Sn	Ba
K spectra	Si	P	S	Cl	K	Ca	Sc	Ti	V	Cr	Mn	Fe	Co	Ni	Cu	Zn	As	Sr	Mo	Ag	Sn	Ba
L spectra	Sr	Mo		Sn				Ba							W		Pb	U				

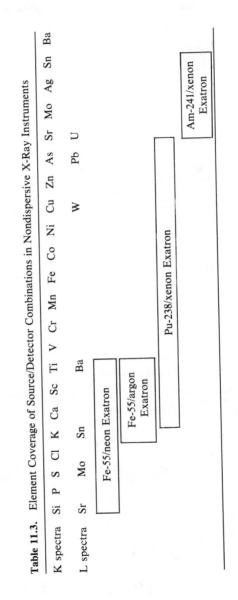

Fe-55/neon Exatron

Fe-55/argon Exatron

Pu-238/xenon Exatron

Am-241/xenon Exatron

Sources

The sources are annular in shape so that the excited radiation from the sample passes through the center of the source to the detector. This arrangement permits extremely close coupling, with a pathlength of less than 10 mm, and gives adequate sensitivity for light elements. Solid, liquid, and powder samples are equally acceptable; the maximum sample size is approximately 50 mm in diameter × 45 mm high. Polyethylene sample holders, specially designed for liquid and powder samples, are available. As there is no vacuum differential, extremely thin windows can be used; a 6 μm thick Mylar film is typical.

Detectors

Very high detection efficiency are provided by gas-filled Exatron proportional detectors. These detectors produce output pulses proportional in amplitude to the energy of the X ray that produces the pulse. The detector is placed to receive as much of the fluorescent radiation as possible, and its output thus consists of a train of pulses whose heights are related to the energies of the incoming X rays. Detectors are available containing neon, argon, or xenon gas. Table 11.3 shows the various detector/source combinations that can be used in the model N940 X-ray analyzer and the element range covered. The details of how these detectors work are discussed in Chapter 19.

QUANTITATIVE ANALYSIS

X-ray spectroscopy can be used for quantitative analysis because the intensity of the radiation from a given element is nearly proportional to the amount of the element present in a sample. Since there are small deviations from linearity, it is the usual practice to prepare calibration curves.

In addition, one must correct the intensities, or intensity ratios, for absorption of the particular radiation by the sample. A method developed by Cuttitta and Rose (1968) is quite effective in reducing matrix problems.

RECENT APPLICATIONS

Raptis et al. (1980) used X-ray fluorescence to determine selenium in rice, wheat, and fish. The method, with absolute detection limits of 70 ng, is not as sensitive as the normal AA hydride method but is more accurate and precise. The anticaking agent sodium aluminosilicate was detected in coffee creamers by Ogawa *et al.* (1980) using X-ray fluorescence ratios of Na, Al, and Si. Pickford (1981) has reviewed several applications of X-ray fluorescence.

BIBLIOGRAPHY

Birks, L. S. (1960). The electron probe: An added dimension in chemical analysis. *Anal. Chem.* **32**(9), 19A–28A.

Cuttitta, F., and Rose, H. J. (1968). Slope-ratio technique for the determination of trace elements by X-ray spectroscopy. A new approach to matrix problems. *Appl. Spectroc.* **22**, 321–324.

Ogawa, S., Tonogai, Y., Ito, Y., and Iwaida, M. (1980). Detection and determination of sodium aluminosilicate in commercial coffee-cream powder (in Japanese). *Shokuhin Eiseigaku Zasshi* **21**(2), 148–149. [*Anal. Abstr.* **41**, 4F49.]

Pickford, C. J. (1981). Chemical aspects of trace constituents of diet. II. Sources of, and analytical advances in, trace inorganic constituents in food. *Chem. Soc. Rev.* **10**, 245–254. [*Anal. Abstr.* 1982, **42**, 3F2.]

Raptis, S. E., Wegschneider, W., Knapp, G., and Toelg, G. (1980). X-ray fluorescence determination of trace selenium in organic and biological matrices. *Anal. Chem.* **52**, 1292–1296.

12
Potentiometry

Because most food analysts are relatively unfamiliar with electroanalytical techniques and instruments, potentiometric methods are not commonly used in food analysis. The purpose of this chapter is to explain the major principles of potentiometry and to give some examples of potential and useful applications.

The heart of any electroanalytical technique is the electrode. When an electrode is placed in a solution, it tends to send its ions into the solution (electrolytic solution pressure), and the ions in solution tend to react with the electrode (activity). These two factors, the electrode pressure and the activity of the solution ions, combine to produce what is called an *electrode potential*. At a given external pressure and temperature, the potential is a constant that is characteristic of the metal.

We distinguish two classes of electrodes, *reference electrodes* and *indicator electrodes*. The potential of a reference electrode does not change significantly with changes in the surrounding solution, whereas the potential of an indicator electrode changes as the concentration of one of the ions in the solution changes.

No one has succeeded in devising a method for measuring an absolute single electrode potential, and a second electrode is always necessary to complete the circuit. However, relative single-electrode potentials can be obtained quite easily. By international agreement the potential of the reaction

$$2H^+ + 2e^- \leftrightharpoons H_2 \qquad (12.1)$$

is given a value of 0.00 V and all other electrodes are compared to it.

Table 12.1. Reduction Potentials

Oxidized form	Reduced form	E^0 (V)
$F_2 + 2e^-$	$= 2F^-$	$+2.85$
$Pb^{4+} + 2e^-$	$= Pb^{2+}$	$+1.69$
$MnO_4^- + 8H^+ + 5e^-$	$= Mn^{2+} + 4H_2O$	$+1.52$
$ClO_3^- + 6H^+ + 6e^-$	$= Cl^- + 3H_2O$	$+1.45$
$Cr_2O_7^{2-} + 14H^+ + 6e^-$	$= 2Cr^{3+} + 7H_2O$	$+1.36$
$ClO_4^- + 8H^+ + 8e^-$	$= Cl^- + 4H_2O$	$+1.34$
$O_2 + 4H^+ + 4e^-$	$= 2H_2O$	$+1.229$
$NO_3^- + H_2O + 2e^-$	$= NO_2^- + 2OH^-$	$+1.01$
$Ag^+ + e^-$	$= Ag$	$+0.7995$
$Fe^{3+} + e^-$	$= Fe^{2+}$	$+0.771$
$O_2 + 2H^+ + 2e^-$	$= H_2O_2$	$+0.682$
$Hg_2SO_4 + 2e^-$	$= 2Hg + SO_4^{2-}$	$+0.615$
$Fe(CN)_6^{3-} + e^-$	$= Fe(CN)_6^{4-}$	$+0.36$
$Cu^{2+} + 2e^-$	$= Cu$	$+0.345$
$Hg_2Cl_2 + 2e^-$	$= 2Hg + 2Cl^-$	$+0.268$
$AgCl + e^-$	$= Ag + Cl^-$	$+0.2222$
$Sb_2O_3 + 6H^+ + 6e^-$	$= 2Sb + 3H_2O$	$+0.152$
$Sn^{4+} + 2e^-$	$= Sn^{2+}$	$+0.15$
$2H^+ + 2e^-$	$= H_2$	0.000
$Fe^{3+} + 3e^-$	$= Fe$	-0.036
$Pb^{2+} + 2e^-$	$= Pb$	-0.126
$Sn^{2+} + 2e^-$	$= Sn$	-0.140
$Ni^{2+} + 2e^-$	$= Ni$	-0.25
$Tl^+ + e^-$	$= Tl$	-0.336
$Cd^{2+} + 2e^-$	$= Cd$	-0.4020
$U^{4+} + e^-$	$= U^{3+}$	-0.50
$Cr^{3+} + 3e^-$	$= Cr$	-0.71
$Zn^{2+} + 2e^-$	$= Zn$	-0.762
$Al^{3+} + 3e^-$	$= Al$	-1.67
$Th^{4+} + 4e^-$	$= Th$	-2.06
$Na^+ + e^-$	$= Na$	-2.71
$K^+ + e^-$	$= K$	-2.92
$Li^+ + e^-$	$= Li$	-3.06

Table 12.1 shows several of common electrode systems and their potentials. The term E^0 refers to the potential of the sytem when all the species present are at unit activity.

The terms *activity, activity coefficient,* and *concentration* are quite confusing to many students. An example to illustrate the difference may help. Suppose there are 20 students in class (the concentration C) and that 10 of them did today's assignment (the activity a). The fraction (activity coefficient, γ) that are working is 0.5. The general equation is

$$a = \gamma C \qquad (12.2)$$

E^0 therefore is an ideal situation that is not likely to occur often in practice. The E^0 must be corrected for the actual system; this can be done by applying the Nernst equation.

$$E_{actual} = E^0 + \frac{2.3RT}{nF} \log \frac{a_{ox}}{a_{red}} \tag{12.3}$$

where R is 8.316 joules/mole-degree, T the temperature in K, n the number of electrons involved, F the Faraday (96,500 coulombs), a_{ox} the actual activity of the oxidized species, and a_{red} the actual activity of the reduced species. (At 25°C, $2.3RT/F = 0.0591$.)

Concentrations are much easier to work with than activities, and fortunately at low concentrations the activity so closely approaches the concentration that actual concentrations can be used in the Nernst equation. Using the permanaganate system in Table 12.1 as an example, we have

$$E_{actual} = 1.52 + \frac{0.0591}{5} \log \frac{[MnO_4^-][H^+]^8}{[Mn^{2+}][H_2O]^4}$$

If an electrode reaction is written as a reduction like the reactions in Table 12.1, a minus $(-)$ sign indicates the metal is a better reducing agent than H_2, and a plus $(+)$ sign indicates that the metal is a better oxidizing agent than H_2. The signs also indicate the polarity of the electrode if it is used in a battery.

The difference between the equilibrium electrode potential and the actual electrode potential is called *overvoltage*. This occurs because some electrodes, because of temperature, current density, electrode crystal structure, or surface area, do not arrive at an equilibrium value rapidly. Excess energy required to force the equilibrium is provided by an increase in the applied voltage; it is more negative for the cathode and more positive for the anode. A few examples are shown in Table 12.2.

Table 12.2. Hydrogen Overvoltage on Several Metals[a] (1.0 M H_2SO_4)

Metal	Current density (amp/cm^2)	
	0.01	0.10
Platinized platinum	0.035	0.055
Smoth platinum		0.39
Gold	0.4	1.0
Iron	0.56	0.82
Silver	0.3	0.90
Mercury	1.2	1.3

[a] From Lingang (1958).

Overvoltage is very important in analytical procedures. Water will form hydrogen at 1.23 V. This means that under ideal conditions, a metal that requires 1.3 V to be reduced could not be determined, because the water in the system would decompose first. However, as a result of electrode overvoltage, it may be possible to plate out the metal, providing its overvoltage is significantly different from that of hydrogen.

REFERENCE ELECTRODES

The major reference electrodes are the standard hydrogen electrode (SHE), the saturated calomel (SCE), and the silver–silver chloride electrode. The standard hydrogen electrode is the international standard, but is seldom used for routine work because more convenient electrodes and reliable calibration buffers are available. The SHE will not be discussed here.

Calomel Electrode

The calomel electrode is the most widely used reference electrode. Its operation is based upon the reversible reaction

$$Hg_2Cl_{2(s)} + 2e^- = 2Hg + 2Cl^- \tag{12.4}$$

The values of E_{25}^0 with different salt bridges are 0.2444, saturated KCl (common U. S. electrode); 0.2501, 3.5 N KCl (common European electrode); and 0.3356, 0.1 N KCl. Figure 12.1 shows one arrangement of this electrode.

The calomel electrode is unstable above 80°C and should be replaced by a Ag–AgCl electrode at high temperature.

In order to make an electrical connection between the electrode and the sample solution, some means must be provided that will produce a negligible potential and be reproducible. This can be done by a *salt bridge*. The saturated KCl salt bridge is the most common type. To understand why, refer to Fig. 12.2, which is a schematic diagram of an enlarged section of a liquid–liquid junction.

Consider the A^+ and B^- ions. Initially there are none of these in the sample solution and a large concentration in the salt bridge. These ions, following the laws of thermodynamics, will diffuse from the region of greater concentration (more order, lower entropy) to that of lower concentration (less order, more entropy). However, the smaller A^+ ions can diffuse faster than the larger B^- ions with the result that a small charge difference or potential is established. The same thing can happen with

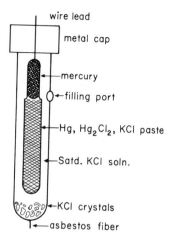

Fig. 12.1. Saturated calomel electrode is the most common reference electrode.

the C^+ and D^{2-} ions from the sample diffusing into the salt bridge. The net result is called the *liquid–liquid junction potential* and can amount to 20–30 mV. This is a large value compared with the few tenths of a millivolt variation that is necessary for good electrode potential measurements.

However, if the hydrated $+$ ion and the $-$ ion are the same size, they will diffuse at the same rate and no junction potential will be established. Since K^+ and Cl^- are two ions with approximately equal mobilities, KCl is the usual salt bridge material. This choice considerably reduces half of the junction potential. We are not able to choose our sample solutions so that the ions diffuse at the same rate. However, diffusion by the sample

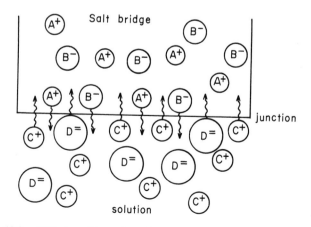

Fig. 12.2. Diffusion of ions gives rise to liquid–liquid junction potentials.

ions is greatly reduced if the salt solution in the bridge is saturated with
KCl. In this case, because there are millions of ions coming out of the
bridge compared with a few sample ions, the sample ions find it hard to
diffuse against this stream of ions. Although the solution diffusion is not
eliminated, it is reduced to the point where the junction potential is less
than 1 mV.

A solution 3 N in KCl or 1N in KNO_3 has been found to be the best
salt bridge concentration. For ions that have Cl^- interference (Pb, Hg,
Ag), a $NaNO_3$ salt bridge is preferred. The necessity of having a stream
of ions coming from the salt bridge makes the depth of immersion of the
electrode into the sample solution quite important. A pressure head of
about 2 cm is required to ensure that the KCl salt solution will stream
out of the fiber or porous disk at the rate of 0.1 ml/day (Fig. 12.3).

Silver–Silver Chloride Electrode

The Ag–AgCl electrode, which is very reproducible, is based on the
reversible reaction

$$AgCl_{(s)} + e^- = Ag_{(s)} + Cl^- \tag{12.5}$$

For this electrode, $E^0 + E_j = 0.1992$ (0.1 N HCl + saturated KCl) or
0.1981 (in buffer solutions).

Figure 12.4 shows a Ag–AgCl elecrode. The electrode is prepared by
electroplating a layer of silver on a platinum wire, then converting the
surface silver to AgCl by electrolysis in HCl. An alternative method uses
thermal decomposition of Ag_2O and $AgClO_4$.

The solution surrounding the electrode should be saturated with KCl

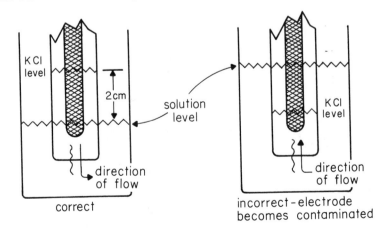

Fig. 12.3. Depth of immersion for calomel electrodes.

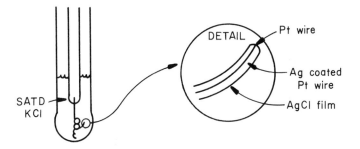

Fig. 12.4. Silver–silver chloride electrode is used as internal reference electrode in glass electrodes.

and AgCl. The electrode is very sensitive to Br^- and O_2 interference, and requires *about 30 hr to reach equilibrium after it is initially prepared.* For most determinations this electrode is relatively free of interfering side reactions and is the internal reference electrode in glass electrodes.

INDICATOR ELECTRODES

Glass Electrode

The most common electrode in use today for the measurement of hydrogen ion concentration is the glass electrode. The relationship between pH and potential is derived as follows. The half cell reaction $H_2 = 2H^+ + 2e^-$ can be substituted into equation 12.3 to give

$$E = E^0 + \frac{0.0591}{2} \log \frac{(a_{H^+})^2}{(a_{H_2})} \tag{12.6}$$

Since $\log(a_{H^+})^2 = 2 \log(a_{H^+})$ and $(a_{H_2}) = 1$ at atmospheric pressure,

$$E = E^0 + \frac{0.0591}{2} \times 2 \times \log(a_{H^+}) \tag{12.7}$$

Recalling that

$$pH = -\log(a_{H^+}) \tag{12.8}$$

and substituting this into equation 12.7, we get

$$E = E^0 - 0.0591 \; pH \tag{12.9}$$

Thus, pH is linearly related to the electrode potential. This is why most pH meters have both a millivolt and a pH scale. A schematic diagram of a glass electrode is shown in Fig. 12.5.

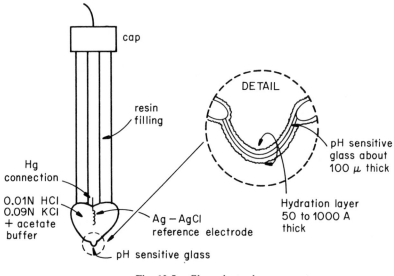

Fig. 12.5.   Glass electrode.

The following discussion on the operation of the glass electrode is from a presentation by Rechnitz (1967):

If a cross section is taken through a membrane of a functioning glass electrode, one finds that the membranes structure is distinctive and consists of several discrete regions and interfaces. This structure can be represented by the simplified schematic:

internal solution	hydrated gel layer	dry glass layer	hydrated gel layer	external solution

The operation of glass electrodes must be examined in terms of these various layers and of the processes that take place at the interfaces. Additional "intermediate" layers may also exist on either side of the dry glass layer.

In practice, the dry glass layer constitutes the bulk of the membrane thickness, and the hydrated layers vary in depth from 50 to 1000 Å for the most useful electrodes, depending on the hygroscopicity of the glass. When the dry electrode is first immersed in an aqueous medium, the formation of the hydrated (external) layer causes some swelling of the membrane. Thereafter, a constant dissolution of the hydrated layer takes place with the accompanying further hydration of dry glass so as to maintain the thickness of the hydrated layer at some roughly constant, steady state value. The rate of dissolution of the hydrated layer depends on the composition of the glass and also on the nature of the sample solution. The rate of dissolution largely determines the practical lifetime of the electrode; lifetimes vary from a few weeks to several years. (Storing the electrode in aqueous solution when not in use increases its life about 30%. Repeated leaching and drying hasten the destruction of the silicon–oxygen network by the extracted lye.)

The hydrated layer must be present on the external electrode surface for glass electrodes to function properly in solutions containing water, probably because the movement of ions in glass is aided by the hydration of the glass. For example, the diffusion coefficients of univalent cations in hydrated glass, about 5×10^{-11} cm^2/sec, are about 1000 times greater than in dry glass.

Schwabe and Dahms using radioactive tracers to study the hydrated layer found that:

"(1) The hydrated glass surface undergoes cation exchange according to $Na_{soln}^+ + H_{glass}^+ = Na_{glass}^+ + H_{soln}^+$ when the membranes are in contact with a solution containing Na^+. Anions are not exchanged. The pH dependence and mass balance of the cation exchange process rules out the self exchange of Na^+ alone.

"(2) Greater quantities of metal ion are taken up than are required for simple monolayer coverage of the glass, indicating that metal ions diffuse into the hydrated layer.

"(3) The takeup of metal ions by the glass as a function of pH can be correlated with the electrode properties of the glass. Furthermore, the selectivity order of ion exchange for a series of alkali metal ions is the same as the potentiometric selectivity order of the glass electrode."

The overall potential at the glass electrode consists of the algebraic sum of a phase boundary (ion exchange) contribution and a diffusion potential contribution.

An understanding of the true mode of the functioning of glass electrodes was delayed for many years by the uncritical acceptance by chemists of the attractive, but erroneous, hypothesis that hydrogen ions selectively penetrate the glass membrane to yield the electrode potential. Although it is true that hydrogen ions (not hydronium ions) undergo exchange across the solution-hydrated layer interface, these ions do not penetrate the glass membrane under normal circumstances. Schwabe and Dahms elegantly demonstrated this with their coulometric experiment involving prolonged electrolyses on glass electrode bulbs filled with tritium labeled sample solutions. If hydrogen ions penetrate the membrane, some appreciable fraction of the total quantity of electricity passed during the electrolysis should be accounted for by the transport of T^+, that is, tritium should be found on the nonlabeled side of the membrane. In fact, the quantity of tritium found on the nonlabeled side of the membrane never exceeded the natural tritium content of the outside solution even after 20 hr of intensive electrolysis at elevated temperatures.

Of course, some small but finite current must flow during the potentiometric measurement and, thus, there must be transport across the glass membrane system. Charge can be transferred across the solution–hydrated layer interface by ion exchange and, within the hydrated layer, by diffusion. But, how then is charge carried through the dry glass portion of the membrane to complete the circuit? The absence of a Hall effect in electrode glasses indicates that electronic conduction can be ruled out (even in glass electrodes with semiconductor additives), and all available experimental evidence points to the fact that the current is carried by anionic mechanism. When an $Na_2O–Al_2O_3–SiO_2$ glass is electrolyzed under completely anhydrous conditions, sufficient amounts of metallic sodium are always formed at the cathode to account for all of the current passed in the electrolysis. The current is, in fact, carried entirely by the cationic species of lowest charge available in any given glass. No single sodium (for sodium silicate glass) moves through the entire thickness of the dry glass membrane but, rather, the charge is transported by an interstitial mechanism where each charge carrier needs to move only a few atomic diameters before passing on its energy to another carrier.

Ion-Selective Electrodes

The glass electrode is generally thought of as being only of value in measuring pH, but if the composition of the glass membrane is changed, this type of electrode can be quite sensitive for other cations. Electrodes for Na, K, NH_4, Ag, Li, Cs, and Rb (all univalent ions) are commercially available.

For example, a glass consisting of 72.2% SiO_2, 6.4% CaO, and 21.4% Na_2O (mole %) is the usual glass for H^+ determinations, but if the glass composition is changed to 71% SiO_2, 11% Na_2O, and 18% Al_2O_3, the electrode is 2800 times more sensitive to Na than K. If the electrode composition is 68% SiO_2, 27% Na_2O, and 5% Al_2O_3, the electrode is K sensitive ($K_{K^+/Na^+} = 20$). An electrode made from a glass containing 52.1% SiO_2, 19.1% Al_2O_3, and 28.8% Na_2O is 100,000 times more sensitive to Ag^+ than H^+.

Precipitate-Impregnated Membrane Electrodes. According to Rechnitz (1967), "the success of glass membranes as cation-selective electrodes rests largely on the fact that the hydrated glass lattice contains anionic 'sites' that are attractive to cations of appropriate charge to size ratio." If a similar exchange process could be set up at a membrane material having cationic sites, it would in principle be possible to prepare anion-selective electrodes. Actually, it is quite easy to construct ion-exchange membranes that permit exchange of either cations or anions; the difficulty is that those membranes show insufficient selectivity among anions and cations of a given charge to be satisfactory as practical electrodes. The problem then is one of finding an anion-exchange material that will display appreciable selectivity among anions of the same charge, and also have suitable properties to permit its processing into a membrane electrode. Thus far, silicone rubber has been found to be the most effective.

An inert, semiflexible matrix (silicone rubber) is used to hold an active precipitate phase (AgI for an I^- electrode) in place. Such membranes are called heterogeneous or precipitate-impregnated membranes. Fisher and Babcock (1958) used radioactive tracer materials to show that the electrode potential is determined by the electrical charge on the surface of the inorganic precipitate particles, and that the current is carried by the transport of the counter ion through the membrane.

The main advantage of membrane-type electrodes over the older metal–metal halide electrodes is their insensitivity to redox interferences and surface poisoning. Figure 12.6 is a schematic diagram of a precipitate-impregnated electrode.

At the present time electrodes sensitive to chloride, bromide, iodide, sulfide, sulfate, phosphate, and hydroxide ion have been developed.

Solid-State Electrodes. The active membrane portion of a solid-state electrode consists of a single inorganic crystal doped with a rare earth (Fig. 12.7). For example, the Orion fluoride electrode is crystalline lanthanum fluoride that has been doped with europium(II) to lower its electrical resistance and facilitate ionic charge transport. This electrode has

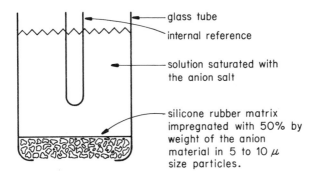

glass tube

internal reference

solution saturated with the anion salt

silicone rubber matrix impregnated with 50% by weight of the anion material in 5 to 10 μ size particles.

Fig. 12.6. Precipitate-impregnated electrodes sensitive to a number of anions are available.

approximately a tenfold selectivity for fluoride over hydroxide, and at least a 1000-fold selectivity for fluoride over chloride, bromide, iodide, hydrogen carbonate, nitrate, sulfate, and monohydrogen phosphate. The fluoride electrode does not require preconditioning or soaking prior to use.

Solid-state electrodes are available to detect the anions Cl^-, Br^-, I^-, and S^{2-} and the cations Cd^{2+}, Co^{2+}, Pb^{2+}, and As^{3+}.

Liquid–Liquid Membrane Electrodes. The range of selective ion-exchange materials could be greatly extended if such materials could be used in electrodes in their liquid state. Liquid ion-exchange materials that possess high selectivity for specific ions may be tailored by appropriate chemical adjustment of the exchanger on the molecular level.

The main problems hindering the development of successful liquid–liquid electrodes are mechanical. It is necessary, for example, that the liquid ion exchanger be in electrolytic contact with the sample solution, yet actual mixing of the liquid phases must be minimal.

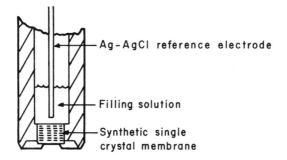

Ag–AgCl reference electrode

Filling solution

Synthetic single crystal membrane

Fig. 12.7. Solid-state electrode.

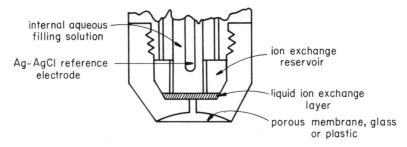

Fig. 12.8.   Liquid–liquid membrane electrode.

At the present time electrodes of this type are available for Ca^{2+}, Mg^{2+}, Cu^{2+}, Cl^-, ClO_3^-, and NO_3^-. Figure 12.8 shows a liquid–liquid membrane electrode.

Enzyme Electrodes

Enzyme electrodes are used to determine certain uncharged molecules. This type of electrode contains an enzyme that can convert the molecule to be measured into an ion that can be detected with a conventional electrode. For example, urea can be determined by coating the end of an ammonium ion electrode with uricase imbedded in a gel. As urea in a sample penetrates the gel, it reacts with the enzyme uricase, which con-

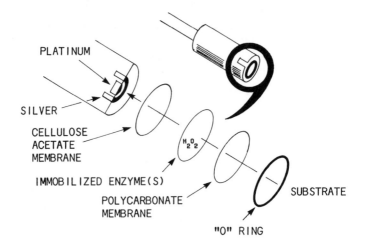

Fig. 12.9.   Components of an enzyme electrode that detects glucose, sucrose, lactose, and ethanol. (From Mason 1983.)

verts part of the urea to ammonia. The ammonia in the presence of water becomes ammonium ion, which is then detected by an ammonium ion electrode.

An enzyme electrode can be prepared by anyone although the commercial models are more rugged. To prepare an electrode, a piece of nylon stocking is placed over the end of the electrode and a thin film of polyacrylamide gel containing the specific enzyme is polymerized onto this network. This gel is then covered with a film of clear plastic such as Saran wrap. Both the stocking and the plastic membrane can be held in place with a rubber band placed around the neck of the electrode. In this manner any electrode can be prepared, limited only by the enzyme and detector electrode. Commercial electrodes use immobilized enzymes, which is clearly a preferred method.

A commercial enzyme electrode to determine glucose, sucrose, lactose, and ethanol is diagrammed in Fig. 12.9. The hydrogen peroxide (H_2O_2) liberated by the enzymatic reactions is detected with a platinum anode. This electrode utilizes the following reactions for the different substrates:

Glucose:

$$\beta\text{-D-glucose} + O_2 \xrightarrow[\text{oxidase}]{\text{glucose}} H_2O_2 + \text{gluconic acid}$$

Sucrose:

$$\text{Sucrose} + H_2O \xrightarrow{\text{invertase}} \text{fructose} + \alpha\text{-D-glucose}$$

$$\alpha\text{-D-glucose} \longrightarrow \beta\text{-D-glucose}$$

$$\beta\text{-D-glucose} + O_2 \xrightarrow[\text{oxidase}]{\text{glucose}} H_2O_2 + \text{gluconic acid}$$

Lactose:

$$\text{Lactose} + O_2 \xrightarrow[\text{oxidase}]{\text{galactose}} H_2O_2 + \text{oxidation product}$$

Ethanol:

$$\text{Ethanol} + O_2 \xrightarrow[\text{oxidase}]{\text{alcohol}} H_2O_2 + \text{acetaldehyde}$$

$H_2O_2:$

$$H_2O_2 \xrightarrow[\text{anode}]{\text{Pt}} 2H^+ + O_2 + 2e^-$$

Bimetallic Electrodes

A very sensitive, yet inexpensive and simple electrode pair consists of two wires, usually of different composition. Their performance is based on the fact that one of the electrodes reaches equilibrium with the solution before the other one does. The main advantages of electrodes of this type are that they can be made very tiny and they can be bent into any shape

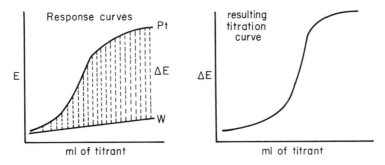

Fig. 12.10. Response and titration curves for a Pt–W bimetallic electrode.

desired. Nickel and platinum are preferred for organic oxidation–reduction titrations. A platinum–platinum rhodium pair is satisfactory for acid–base titrations. Two cases and the resulting titration curves are shown in Fig. 12.10 and 12.11.

POTENTIOMETRY

The finest electrodes in the world are of no value unless the potentials they generate are measured and recorded accurately. The methods of potentiometry consist of measuring the potential difference between an indicator electrode and a reference electrode at various intervals during the progress of a titration. An electrode potential is an equilibrium quantity, and during its measurement the equilibrium must not be disturbed or the potential obtained will be incorrect.

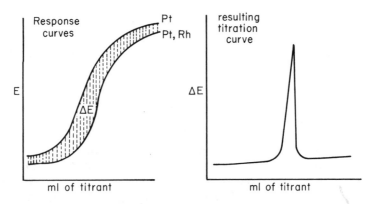

Fig. 12.11. Response and titration curves for a Pt/Pt–Rh bimetallic electrode.

Most potential-measuring devices, such as voltmeters, require considerable current for operation. When this current is passed through the electrode, either metal is plated out from the solution or gas is evolved. In either case, the equilibrium at the electrode changes and the measured potential is not that of the solution but that next to the electrode. This may be quite different from the solution potential and usually is continuously changing. The net result is that the measurement is erratic and a more refined method of measuring potential is necessary.

In the null method for measuring potential, an equal but opposite potential is applied to the electrolytic cell. This does not alter the electrode process, but it does reduce the current flow to zero, thus maintaining the equilibrium. When the solution potential balances the external potential, a galvanometer placed in the circuit will show no deflection. The cell potential can then be determined by measuring the external potential. The apparatus for doing this is called a *potentiometer*. A diagram of a student potentiometer is shown in Fig. 12.12.

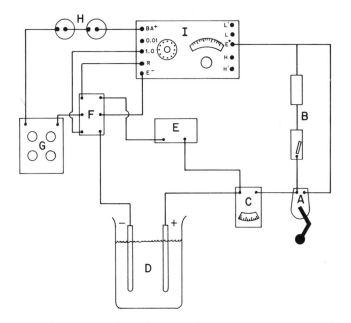

Fig. 12.12. Components of a student potentiometer. (A) Tap key, which is depressed only for an instant as a long depression will change the cell equilibrium, (B) resistances for coarse and fine adjustment, (C) galvanometer, which can be used with the metal electrodes but not with glass electrodes, (D) titration cell, (E) standard cell to calibrate the instrument, (F) double-pole double-throw switch, (G) decade resistance box for a coarse adjustment to balance the potential of the bucking batteries (H), (H) 1.5-V batteries to provide the potential in oposition to the cell voltage, (I) potentiometer, which contains a precision slide wire resistor and a series of 0.1 V steps to cover the range 0–1.5 V.

Direction of Redox Reaction

A question often encountered is in what direction does an oxidation–reduction reaction proceed? Suppose you had a solution of U^{3+} and wanted to oxidize it to U^{4+} with Sn^{4+}. Is the reaction feasible? To answer this question, proceed as follows: (1) write the half cell *reduction* reactions (Table 12.1 or equivalent) with the most positive reaction on top, and (2) proceed in a clockwise direction *around* the reactions to get the reaction direction. This is shown below for the Sn^{4+}/U^{3+} system.

$$Sn^{4+} + 2\,e^- = Sn^{2+} + 0.04\text{ V}\ \Big)\ \text{direction of}$$
$$U^{4+} + 1\,e^- = U^{3+} - 0.05\text{ V}\ \Big/\ \text{the reaction}$$

Notice that the system will go as desired. A solution of Sn^{2+} could not be oxidized to Sn^{4+} with a solution of U^{4+}—the reverse reaction—unless some additional external potential was applied.

Complex Ions

Measurements of metal ions is complicated by the presence of complexing agents. The net effect of a complexing agent is to lower the concentration of the metal ion and completely alter the measured potential. This is of real importance in food analysis when trace metals are to be determined and complexing agents are present.

Example: A food sample was suspected of containing cyanide. The sample was extracted with water, made 0.01 M in $ZnCl_2$, and diluted to 1 liter. The potential of a metallic zinc electrode and a 1 M calomel electrode pair dipping into the solution was -1.342 V. Knowing that the dissociation constant of $Zn(CN)_4^{2-}$ is 8.9×10^{-17}, calculate the concentration of CN^- in the solution.

$$Zn^{2+} + 2e^- = Zn, \qquad E_1^0 = -0.763\text{ V}$$

$$Hg_2Cl_2 + 2e^- = 2Hg + 2Cl \qquad E_2^0 = +0.282\text{ V}$$

The reaction of the electrode pair is

$$Zn + Hg_2Cl_2 = Zn^{2+} + 2Hg + 2Cl^-$$

$$E_{cell} = E_2^0 - E_1^0 + \frac{0.0591}{2}\log\frac{[Zn^{2+}][Hg]^2[Cl^-]^2}{[Zn][Hg_2Cl_2]}$$

Since all species except Zn^{2+} are in their standard states, their activities are equal to unity, which gives the following:

$$E_{cell} = E_2^0 - E_1^0 + \frac{0.0591}{2} \log[Zn^{2+}]$$

$$-1.342 = 0.282 - [-0.763] + \frac{0.0591}{2} \log[Zn^{2+}]$$

and

$$\log[Zn^{2+}] = -(0.594)/(0.0591) = -10.05$$

$$[Zn^{2+}] = 8.9 \times 10^{-11} \text{ moles/liter}$$

Since we started with 0.01 M Zn^{2+}, this means that the Zn^{2+} concentration has been greatly reduced by the presence of a complexing agent, which we are assuming is CN^-.

$$Zn(CN)_4^{2-} = Zn^{2+} + 4CN^-$$

$$K_d = \frac{[Zn^{2+}][CN^-]^4}{[Zn(CN)_4^{2-}]} = 8.9 \times 10^{-17}$$

$$8.9 \times 10^{-17} = \frac{[8.9 \times 10^{-11}][CN^-]^4}{[0.01 - 8.9 \times 100^{-11}]}$$

(ignore 8.9×10^{-11} because it is small compared to 0.01). Therefore, $[CN^-] = 0.01$ M. However, you must remember that the 0.01 M $Zn(CN)_4^{2-}$ would require 0.04 M CN^- to form the complex. The $[CN^-]$ measured is actually that which is in excess of the amount necessary to make the complex. The total CN^- present is therefore 0.04 + 0.01 or 0.05 M.

pH Meters

The electrical resistance of the glass electrode is very high, being of the order of 5×10^5 ohms in very thin-walled and fragile membranes, and as high as 2×10^8 ohms in rugged commercial products. The measurement of the potential of the glass cannot be made with an ordinary potentiometer because the current that passes through the glass membrane under the very low potential of the cell is insufficient to move the galvanometer needle. Vacuum tube amplifying devices capable of registering minute amounts of current were first substituted for the galvanometer, and these permitted the measurement of the potential across the glass membrane. Tubes have now been replaced by transistors. Essentially, the commercial pH meter is an ordinary potentiometer in which the galvanometer has been replaced by a transistor device for indicating the potential difference. Transistors require only minute current for their operations and they may be left connected to the cell.

Nonaqueous Titrations

Water has some major limitations as a titrating medium for organic compounds. Not only are relatively few organic compounds soluble in water, but most of those that are soluble are too weakly basic or acidic to be titrated accurately. It has long been known that the strength of basic compounds, such as amines, is much greater in glacial acetic acid than in water and that acidic compounds such as sulfonamides and phenols are more acidic in a basic solvent such as dimethylformamide.

To understand this phenomenon, refer to Fig. 12.13A, a diagram of base B in water. The H^+ from the titrant can react with the base to form BH^+ or with the water to form H_3O^+. Because the electrons in weak bases, such as amines, are not as readily available as the electrons in water ($K_a = 10^{-7}$), the proton will react with the water, not completely but sufficiently to ruin the stoichiometry of the base titration. However, if the base is placed in acetic acid ($K_a = 1.8 \times 10^{-5}$), the H^+ from the titrant will prefer to react with the more readily available electrons of B: and the titration will go to completion.

A convenient electrode pair for nonaqueous titrations is silver–silver chloride and glass, the latter being the indicator electrode as it is when pH determinations are made in aqueous solutions. The silver–silver chloride electrode is preferred to a saturated calomel electrode or a mercury-mercurous sulfate electrode, because it does not require an aqueous salt bridge. The glass electrode must be soaked in water every hour or so when used in nonaqueous solvents in order to provide the hydrated layer necessary for the electrode to work.

Voltage measurements in nonaqueous systems should be done with a pH meter rather than the student-type potentiometer if a glass electrode

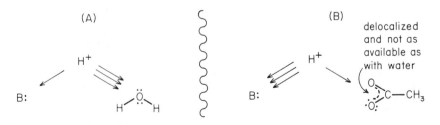

Fig. 12.13.   Titration of base B: in water (A) and acetic acid (B).

is used. However, since pH meters are calibrated for the glass and calomel electrodes in aqueous solution, the emf produced by the glass and silver–silver chloride electrodes in either acidic or basic media differs markedly from the true value. If the measured value is not corrected, it should be reported as *apparent pH*.

RECENT APPLICATIONS

Mason (1983) determined glucose, sucrose, lactose, and ethanol in foods and beverages using an immobilized enzyme electrode. Contamination of spices with uric acid was correlated to insect infestation by Brown *et al.* (1982) who monitored oxygen released from a uricase reaction. A glucose-sensing electrode using glucose oxidase, in conjunction with a glassy carbon electrode, has been described by Matsumoto *et al.* (1980) who used it to measure glucose in orange juice.

The ammonia content of milk was determined by Soederhjelm and Lindquist (1980) using an ammonia electrode; they reported that the ammonia content may be used as an indicator of biological aging. Fluoride in plants was determined by Eyde (1982) on alkaline-fused ash solutions with a fluoride electrode, and fluoride in salt samples was determined by Kokubu *et al.* (1980) after concentrating the samples on a Zr (IV) loaded cation-exchange column. Sodium hypobromite has been used in a potentiometric titration for nitrogen in meat products by Schillak (1980).

PROBLEMS

12.1. What is the pH of a solution at 25°C if a potential of 0.642 V is observed between a hydrogen electrode and a SCE? *Ans.* 6.72.

12.2. What potential would be shown by an antimony electrode (Table 12.1) relative to a SCE in a solution at pH 6.3? *Ans.* 0.768 V.

12.3. Calculate the potential in the following cell. A Pt wire dips into a solution that is 0.05 M in Sn^{4+} and 0.025 M in Sn^{2+}. This solution is connected by a salt bridge to another solution that is 0.086 M in Hg^{2+} and 0.107 M in Hg^+. A Pt electrode dips into this latter soluiion. *Ans.* 0.124 V.

12.4. A 0.2 M solution of HF is placed in a polyethylene beaker. To this is added an equal volume of 0.03 M Al^{3+}. A calomel electrode and an Al wire, connected to a potentiometer, are placed in the solution. What is the potential? (Note: $AlF_6^{3-} = Al^{3+} + 6\ F^-$ with $K_d = 1.4 \times 10^{-20}$.) *Ans.* 1.519 V.

BIBLIOGRAPHY

Brown, S. M., Abbott, S., and Guarino, P. A. (1982). Screening procedure for uric acid as indicator of infestation in spices. *J. Assoc. Off. Anal. Chem.* **65**, 270–272.

Edye, B. (1982). Determination of fluoride in plant material with an ion selective electrode. *Fresenius' Z. Anal. Chem.* **311**, 19–22. [*Anal. Abstr.* **43**, 4G5.]

Fisher, R. B., and Babcock, R. F. (1958). Effects on aging of reagent soutions on the particle size of precipitates—electrodes consisting of membranes of precipitates. *Diss. Abstr.* **19**, 428.

Kokubu, N., Hayasida, Y., Kobayashi, T., and Yamasaki, A. (1980). Determination of fluoride content in various sodium chloride samples by zirconium(IV) loaded cation exchange resin column inorganic affinity chromatography. *Denki Tsushin Daigaku Gakuho* **31**, 113–116. [*Anal. Abstr.* 1982, **42**, 3B208.]

Lingane, J. J. (1958). "Electroanalytical Chemistry," 2nd ed. Wiley (Interscience), New York.

Mason M. (1983). Determination of glucose, sucrose, lactose, and ethanol in foods and beverages, using immobilized enzyme electrodes. *J. Assoc. Off. Anal. Chem.* **66**, 981–984.

Matsumoto, K., Yoshioka, S., Nomura, T., and Osajima, Y. (1980). Glucose electrode using glassy carbon as base electrode. *Kyushu Daigaku Nogakubu Gakugei Zasshi* **34**, 115–121. [*Chem. Abstr.* 1981, **94**, 2968m.]

Rechintz, G. A. 1967. Ion selective electrodes. *Chem. Eng. News,* June 12, 146–158.

Schillak, R. (1980). Determination of total nitrogen in fresh animal tissue by digestion with hydrogen peroxide and potentiometric titration of ammonium nitrogen with sodium hypobromite solution. *Chem. Anal. (Warsaw)* **25**, 181–190. [*Anal Abstr.* 1981, **40**, 2F19.]

Soederhjelm, P., and Lindquist, B. (1980). The ammonia content of milk as an indicator of its biological deterioration or aging. *Milchwissenschaft* **35**, 541–543. [*Chem. Abstr.* **93**, 202766k.]

13
Coulometry

Coulometric titrations are determinations in which the titrant or reagent is electrically generated at the surface of an electrode (working electrode) immersed in a solution with a second electrode (counter electrode). Because the titrant is used almost at the instant it is generated at the electrode, coulometric methods can be designed to employ titrants that cannot be easily used in burets, such as chlorine, bromine, iodine, titanium (II), copper (I), and chromium (II). Submicrogram quantities of materials can be determined. The instrumentation is generally quite simple and of low to moderate cost. The analyses are rapid and there are no dilution corrections. The electrodes can be placed in remote places and the system is easy to automate. These advantages combine to make coulometry a potentially powerful analytical technique.

Coulometric titrations are based on Faraday's law, which simply states that the passage of 1 farad of electricity (96,494 coulombs) will oxidize or reduce 1 g-equivalent of the substance under consideration. Mathematically this can be stated as

$$g = itM/nF \tag{13.1}$$

where g is the weight of the substance in g, i the current in amps, t the time in sec, M the formula weight of the substance, n the number of electrons transferred per atom, and F is 96,494 (Faraday's constant).

It is necessary that only one reaction take place at the electrodes and that it proceed with 100% current efficiency; that is, every electron reacts only with the system being investigated. In order to maintain 100% current efficiency, low currents are generally used, and the electrolysis products formed at the counter electrode must not be allowed to diffuse to the working electrode. A porous barrier is generally used to obstruct diffusion, yet allow the passage of current by the supporting electrolyte.

An alternate method (Lingane 1958) is to add a material to the system that *can be electrolyzed* at the counter electrode, and whose reaction products are inert to the electrolysis at the working electrode should they migrate there. Hydroxylamine and hydrazine are two such compounds.

$$2NH_2OH = N_2 + 2H^+ + 2H_2O + 2e^- \tag{13.2}$$

$$NH_2NH_2 = N_2 + 4H^+ + 4e^- \tag{13.3}$$

The H^+ formed will not be reduced if a mercury electrode is used because of its overvoltage and of course N_2 is inert.

Example: If a current of 65.80 mA flows for 290.8 sec through a solution of cadmium nitrate, how many milligrams of cadmium will be deposited at the cathode?

$$\frac{65.80 \times 10^{-3} \, (\text{amp}) \times 290.8 \, (\text{sec}) \times 112.41 \, (\text{g})}{2 \, (\text{electrons}) \times 96,494} \times \frac{1000 \, \text{mg}}{\text{g}} = 11.15 \, \text{mg}$$

From the knowledge of the time of the current flow and the magnitude of the current, one may determine the number of gram-equivalents of titrant generated and therefore the number of gram-equivalents of the substance being titrated.

Two processes are normally employed: constant current or controlled potential. The constant current process will be discussed first.

CONSTANT-CURRENT COULOMETRY

In this constant-current coulometry, the current i is maintained constant and the time t required for the material in the sample to completely react is determined. Since $it = q$ (coulombs) and 96,494 coulombs = 1 equivalent, t, the percent composition of the sample can be quickly determined. For simple systems this is by far the easiest method. However, if more than one species is present, extreme care must be taken because of *concentration overvoltage* and *activation overvoltage*, both of which reduce the current efficiency to less than 100%.

Concentration overvoltage arises in the following manner. Recall the Nernst equation,

$$E_{\text{actual}} = E^\circ + \frac{0.0591}{n} \log \frac{[\text{ox}]}{[\text{red}]}$$

The E_{actual} depends on the concentration of the ionic species *at the surface* of the electrode. This may be considerably different than the bulk concentration, as shown in Fig. 13.1. Suppose 1000 electrons/sec reach the electrode surface from a constant-current supply and 2000 sample ions arrive each second at the electrode surface to react. Since there are more sample ions than electrons, all of the electrons are used up and there is 100% current efficiency. Now suppose that after awhile the concentration of sample ions drops to 500 ions/sec. The constant-current supply, however, will continue putting 1000 electrons/sec on the electrode surface; these electrons must be used up. Since there are not enough sample ions for the electrons provided, the electrons must be used somewhere else. Other ions present in the sample will react, and we can tell from the Nernst equation what happens. From the Nernst equation it can be seen that the potential will change as the concentration changes. Different elements react at different potentials and other elements will react as well as the sample ions. This means that 100% current efficiency is no longer obtained and the analysis is in error. What is needed to prevent this is vigorous stirring and low currents.

Activation overvoltage is a deviation of the electrode potential from its Nernst equation value due to the passage of *electrolytic current* (not the electrode exchange current) in the solution itself. This shifts the cathode to a more negative value and the anode to a more positive value. For small currents the effect is linear with an increase in current; for large currents, the shift is linear with the logarithm of the current. Again low currents are required.

Example: Suppose that 5.00 ml of platinum(IV), added to an electrolyte of sodium bromide containing stannic chloride, was titrated with electrogenerated stannous ion, using platinum and gold electrodes. The endpoint was detected potentiometrically after 297.3 sec at a constant current. The *iR* drop across a standard 10.00-ohm resistor was 0.0965 V. What is

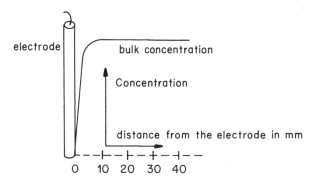

Fig. 13.1. Variation in concentration of ions with distance from electrode.

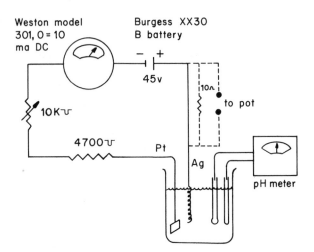

Fig. 13.2. Schematic diagram of a simple coulometric titration system.

the normality of the platinum(IV) solution? $E = iR$ or $i = E/R = 0.0965/$
$10.00 = 0.00965$ amp:

$$\frac{0.00965 \times 297.3}{96,494} = 2.973 \times 10^{-5} \text{ g-equiv}$$

$$\frac{2.973 \times 10^{-5} \text{ g-equiv}}{5.00 \text{ ml}} \times \frac{1000 \text{ ml}}{1.0 \text{ liter}} = \frac{0.00595 \text{ g-equiv}}{\text{liter}} = 0.00595 \text{ } N$$

A simple inexpensive circuit, which is suitable for many titrations, is
shown in Fig. 13.2. The electrodes shown are used conveniently in the
coulometric titration of a number of acids; however, other electrodes
could be used as well. If the current measured does not have to be known
too accurately, a 0 to 10-mA ammeter is suitable. Alternatively, if more
accurate current values are needed, a 10.00-ohm standard resistor may
be inserted, in series, into the circuit and the potential drop across this
resistance determined by employing a potentiometer and using Ohm's
law, $E = iR$.

CONTROLLED-POTENTIAL COULOMETRY

The main problem with constant-current coulometry in complex systems is that the currents required to maintain 100% current efficiency are
so small that the analysis can take several hours.

An improvement in this respect is accomplished by *controlled-potential*

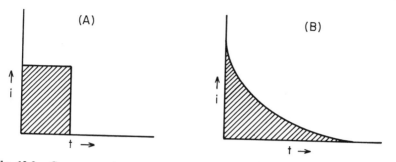

Fig. 13.3. Current vs. time plots for constant-current (A) and constant-potential (B) coulometry.

coulometry. The system requires three electrodes; two that are the same as in constant-current coulometry and a third electrode to monitor the cathode potential. The third electrode is placed near the cathode and draws very little current, so it does not disturb the system appreciably. If the cathode potential varies from a pre-determined value, then the system is automatically corrected back to the proper value. The device to maintain this constant potential is known as a *potentiostat.* Since the potential is controlled, only one element will be determined at a time. However, the current will decrease as the concentration of the element decreases. Figure 13.3 compares the constant-current and constant-potential processes.

When the current has decreased to 0.1% of the initial current, the deposition may be regarded as completed. In some cases the current decreases to a level above 0.1% of the original value, and remains unchanged. This residual current is due to oxygen or to slow evolution of hydrogen. A usual method for determining the time for electrolysis is the *10 half-life* concept. The half time, or half life, is determined by noting the current at the start, and then measuring the amount of time it takes to reduce this value by one half. Table 13.1 shows the extent of deposition associated with various half times.

Table 13.1. Fraction Deposited per Number of Half Times

Fraction deposited	Number of half times
0.50	1
0.90	3.33
0.99	6.67
0.999	10.0
0.9999	13.33

USE OF COULOMETRY WITH GAS CHROMATOGRAPH

One of the most common applications of coulometry in food analysis is as a detector for a gas chromatograph such as that manufactured by the Dohrman Co. Basically the idea is as follows. The volatile food components, or extracts, are separated by the chromatograph and passed through a quartz combustion tube heated to 800°C. The material is decomposed to simple compounds such as CO_2, H_2O, NO_2, SO_2, Cl_2, and Br_2. Those materials are then passed into a coulometric titration cell and titrated by a constant-current process, the selectivity depending on the type of electrodes and the supporting solution. Figure 13.4 is a diagram and a photograph of the Dohrman sulfur cell.

The following formulas taken from the Dohrman technical bulletin show how the calculations are made.

$$\text{microequivalents} = \frac{q \times 10^6}{96,494 \text{ coulombs/equiv}} \qquad (13.4)$$

$$q = it = \left(\frac{\text{mV/in.}}{R(\Omega)}\right) 10^{-3} \times \frac{60 \text{ sec}}{\text{min}} \times \frac{\text{min}}{\text{in.}} \times A \qquad (13.5)$$

$$= V/R \times 6.0 \times 10^{-2} \times S \times A \qquad (13.6)$$

where V is the mV/in. recorder sensitivity, R the ohms resistance of the range ohms switch, S the chart speed of the recorder in min/in., and A the area under the peak in square inches. Substituting equation 13.6 into equation 13.4 gives

$$\text{microequivalents} = \frac{V \times S \times 6.0 \times 10^{-2} \times 10^6 \times A}{R \times 96,494} \qquad (13.7)$$

The quantity X (in μg) of a chlorinated compound can be determined by applying the following formula:

$$X = A \frac{\text{min}}{\text{in.}} \frac{\text{mV}}{\text{in.}} \frac{35.45 \text{ g}}{\text{equiv}} \frac{60 \text{ sec}}{\text{min}} \frac{10^6 \text{ }\mu g}{g} \frac{10^{-3} \text{ V}}{\text{mV}} \times 10^2$$

$$\times \left[\text{recorder input resistance} \right. \qquad (13.8)$$

$$\times \left. \left(\text{range} \times \%Cl \times 96,494 \frac{\text{coulombs}}{\text{equiv}} \right) (\text{ohms}) \right]$$

For a sensitivity of 0.1 mV/in. and a chart speed of 0.5 in./min, equation

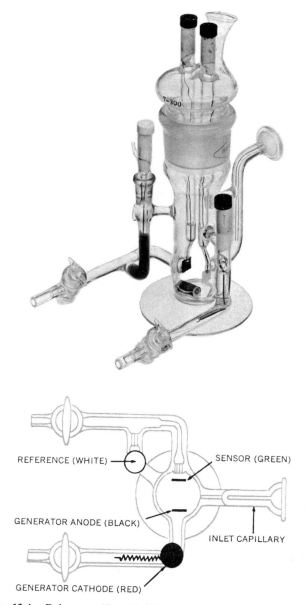

REFERENCE (WHITE)

SENSOR (GREEN)

GENERATOR ANODE (BLACK)

INLET CAPILLARY

GENERATOR CATHODE (RED)

Fig. 13.4. Dohrman sulfur cell. (Courtesy Dohrman Instrument Co.)

13.8 becomes

$$Cl\ (\mu g) = \frac{A \times 442}{range \times \%Cl} \qquad (13.9)$$

$$S\ (\mu g) = \frac{A \times 199.2}{range \times \%S} \qquad (13.10)$$

$$N\ (\mu g) = \frac{A \times 174.2}{range \times \%N} \qquad (13.11)$$

Determination of Halogen

The halogenated compound is converted to Cl_2, Br_2, or I_2 during the combustion process. The electrolytic cell contains 70% acetic acid in water. The reference electrode is Ag–AgOAC (satd).

A typical cell reaction, using chloride as an example, is $Ag^+ + Cl^-$ = AgCl. Depletion of silver in the electrolyte is detected by the reference/sensor pair and is replaced electrically at the generator electrodes by a current from the microcoulometer. The reaction at the generator electrode is $Ag = Ag^+ + e^-$.

Determination of Sulfur

The sulfur compound is converted to SO_2 at temperatures above 800°C. The electrolyte is 0.04% acetic acid, 0.05% KI. The generator and sensor electrodes are Pt and the reference electrode is $Pt:I_3^-$.

A typical cell reaction, using SO_2 as an example, is $I_3^- + SO_2 + H_2O$ = $SO_3 + 3I^- + 2H^+$. The triiodide depletion in the electrolyte is detected by the reference/sensor electrodes and is replaced at the generator electrodes. The reaction at the generator anode is $3I^- = I_3^- + 2e^-$. Any system that will react with iodide (I_3^-) can be determined.

Determination of Nitrogen

The nitrogen is converted to ammonia by combustion with H_2 over a nickel catalyst. The electrolyte in the cell is 0.04% Na_2SO_4, the generator and sensor electrodes are platinum, and the reference electrode is lead in saturated $PbSO_4$. Any acid or base can be titrated with this system. BaO removes acid gases, so bases can be determined.

A typical cell reaction, using ammonia as an example, is as follows:

$$NH_3 + H_2O = NH_4OH$$

$$NH_4OH + H^+ = NH_4^+ + H_2O$$

Depletion of the hydrogen ion in the electrolyte is detected by the reference/sensor electrodes and is replaced as follows:

$$\tfrac{1}{2}H_2 = H^+ + e^-$$
$$H_2O + e^- = OH^- + \tfrac{1}{2}H_2$$

As little as 500 pg of N_2 can be determined.

RECENT APPLICATIONS

The Dohrman detector for gas chromatography continues to be the dominant application of the coulometric principle in the analysis of food residues. The detector is particularly useful for the determination of pesticide residues containing sulfur (D. Mansky, personal communication, 1985).

The coulometric titration of proteins in milk products has been described by Brauner et al. (1981) using electrogenerated bromate. Coulometric titration of H_2SO_4, resulting from combustion of samples in oxygen and treatment with H_2O_2, is the basis of a method developed by Xiang et al. (1982) for the microdetermination of sulfur in organic compounds. Glucose has been determined by generation of excess iodine, addition of excess sodium thiosulfate, and coulometric bipotentiometric back titration with iodine (Markova et al. 1982).

Table 13.2. Systems for Electrogenerating Various Titrants

Substance determined	Electrogenerated titrant
Phenols	Bromine
Phosphites	Bromine
Olefins	Bromine
Cyclic β diketones	Bromine
Sulfur compounds	Iodine
Proteins	Hypobromite
Organic chlorides and bromides	Silver(I)
Alkaloids and organic bases	Silver(I)
Chromate, dichromate, oxalate, molybdate, tungstate, ferrocyanide	Lead(II)
Phosphate	Bismuth(III)
Organic acids in sugars	OH^-
Organic nitrogen compounds	H^+
Aliphatic amides	Oxidation at platinum electrode
Catechol	Oxidation at platinum electrode
Nitroso compounds	Reduction to amines
Nitro compounds	Reduction in dimethyl sulfoxide

New equipment has been developed by Damokos (1982) and by Last (1982). The former developed a universal apparatus for coulometric and other titrations and the latter designed a digital coulostatic coulometer. The possibilities for the future are almost unlimited. Table 13.2 shows some of the recent systems for electrogenerating various titrants. Almost all of these can be applied to food analyses either directly or indirectly.

PROBLEMS

13.1. The following measurements were made in a coulometric titration of arsenic(III) ions extracted from 50 lb of Jonathan apples. The titrant was electrogenerated bromine.

iR drop across the resistance: 0.6748 V

calibrated resistance: 89.52 ohms

generation time: 6.36 min

Calculate the amount of arsenic (ppm) present on the apple skins. *Ans.* 0.16 ppm.

13.2. In an electrolytic determination of bromide ion in 50.00 ml of a solution, the quantity of electricity, as determined from a mechanical current–time integrator, was 131.5 coulombs. Calculate the weight of bromide ions in the original solution. Calculate the potential of the silver electrode that should be employed throughout the electrolysis.

$$K_{sp}AgBr = 5.0 \times 10^{-13}, \quad AgBr + e^- = Ag + Br^-$$

$$E° = +0.073 \text{ V}$$

Ans. (a) 5.45×10^{-2} gm; (b) -0.653 V.

13.3. A particular timer employed in constant-current coulometric titrations is marked in units of 0.01 min. If it is desired to have each unit (0.01 min) correspond conveniently to 1 microequivalent, what value of the current in milliamperes would be required from the power supply? *Ans.* 160 mA.

13.4. In the determination of copper in a brass cooking pan, the copper was coulometrically determined by using thioglycollic acid. The endpoint was detected amperometrically using two mercury electrodes. A 2.705-mg aliquot of the brass was titrated at a constant current of 10.66 mA. The time of the analysis was 247.3 sec. Determine the copper content, in weight percent. *Ans.* 31.4%.

13.5. Atrazine, $C_2H_{14}CIN_5$ (16.41% Cl), is extracted from 2 lb of cranberries with 100 ml of hexane. A 10-μl sample is injected into a gas chromatograph and the Cl determined coulometrically. The chart speed is 0.5 in./min, the range resistance is 256 ohms, the peak area is 1.8 in.^{2}, and the recorder sensitivity is 0.1 mV/in. Calculate the concentration of atrazine in the cranberries in ppm. *Ans*. 2.08 ppm.

13.6. Simetryne, $C_8H_{15}N_5S$ (15.03% S), is extracted from 100 lb of potatoes with 2 liters of hexane which is then concentrated to 100 ml. A 10-μl sample is injected into a gas chromatograph and the sulfur determined coulometrically. The chart speed is 1 in./min., the recorder sensitivity is 0.05 mV/in., and the range resistance is 1280 ohms. The peak area is 0.15 in.^{2}. Calculate the simetryne (ppb) in the potatoes. *Ans*. 0.34 ppb.

BIBLIOGRAPHY

Brauner, J., Ficnar, J., and Kubin, J. (1981). Coulometric titration of protein in milk. *Prum. Potravin* **32**(8), 455–457. [*Anal. Abstr.* 1982, **42**, 2F29.]

Damokos, T., and Toth, E. (1983). Cell for coulometric titrations with GDTA, *Magy. Kem. Foly.* (*Hungary*) **89**, 565–6.

Last, T. A. (1982). Coulostatic coulometer with digital counter. *Anal. Chem.* **54**, 2327–2332.

Lingane, J. J. (1958). "Electroanalytical Chemistry," 2nd ed. Interscience Publishers, New York.

Markova, T. R., Turyan, Y. I., Strizhov, N. K., and Korobko, L. V. (1982). Coulometric determination of glucose. *Zv. Vyssh. Uchebn. Zaved., Pishch Teknol.* **5**, 143–145. [*Chem. Abstr.* **98**, 154727c.]

Xiang, L., Chen, M., and Gao, L. (1982). Coulometric microdetermination of sulfur in organic compounds. *Youji Huaxue* **3**, 193–195. [*Chem. Abstr.* **97**, 119838x.]

14
Conductivity

The most common uses of the principles of conductivity in the analysis of foods are in detectors for ion chromatography, high-performance liquid chromatography, and gas chromatography. Minor uses involve the detection of adulteration of liquids and occasionally the titration of weak acids and bases.

In conductivity measurements, use is made of two parallel platinum foil (1-cm^2) electrodes, about 1 cm apart (Fig. 14.1). When a voltage is applied to two electrodes in a solution containing ions, the anions move toward the anode and the cations toward the cathode. The amount of current that will pass between these two electrodes is a function of the concentration, temperature, and kind of ions. The polarity of the electrodes is usually changed about 1000 times a second in order to prevent electroplating and electrode polarization. The advantages of conductivity-type detectors are that they measure cell resistance without inducing electrolysis and they respond to all types of ions.

The conductivity K of a material is usually thought of as the conductance of a 1-cm cube of the solution. A more useful value, *equivalent conductance*, is the conductance per unit concentration of charge-carrying constituent. The equivalent conductance is defined as

$$\Lambda = 1000K/C \qquad (14.1)$$

where C is the concentration expressed as normality and K the *specific conductance* in mhos (the reciprocal of the resistance). The equivalent conductance Λ varies with concentration, and the value of Λ at infinite dilution is Λ_0. Equivalent conductivities are more commonly broken into

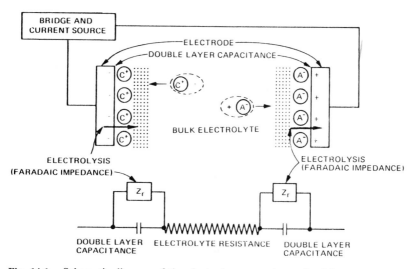

Fig. 14.1. Schematic diagram of the electrode process in conductivity measurements. (From Braunstein and Robbins 1971.)

ionic conductivities $(\lambda +, \lambda -)$, which can be added together to form *molar conductivities*:

$$\Lambda_0 = \sum(\lambda +) + \sum(\lambda -) \tag{14.2}$$

Values of ionic conductivities at infinite dilution at 25°C for a number of ions are given in Table 14.1.

Table 14.1. Ionic Conductances at 25°C in Water

Ion	λ_+	Ion	λ_-
H_3O^+	350	OH^-	197.5
Li^+	39	NH_4^+	80
Na^+	50	Cl^-	76
K^+	74	Br^-	78
NH_4^+	74	I^-	76.5
Ag^+	62.5	NO_3^-	71
$1/2\ Mg^{2+}$	55	HCO_3^-	44
$1/2\ Ca^{2+}$	60	CH_3COO^-	41
$1/2\ Sr^{2+}$	60	ClO_4^-	67
$1/2\ Ba^{2+}$	64	$1/2\ CO_3^{2-}$	70
$1/2\ Cu^{2+}$	56	$1/2\ C_2O_4^{2-}$	24
$1/2\ Zn^{2+}$	55	$1/2\ SO_4^{2-}$	80
$1/2\ Pb^{2+}$	73	$1/2\ CrO_4^{2-}$	82
$1/2\ Fe^{2+}$	54	$1/3\ PO_4^{3-}$	80
$1/3\ Fe^{3+}$	68	$1/3\ Fe(CN)_6^{3-}$	101
$1/3\ La^{3+}$	70	$1/2\ Fe(CN)_6^{4-}$	111

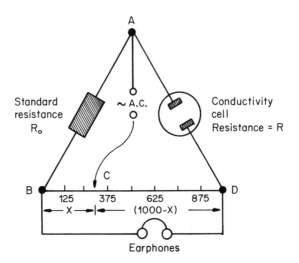

Fig. 14.2. Schematic circuit diagram for the measurement of conductance.

The instrument generally used to measure conductivity is a Wheatstone bridge circuit arranged as shown in Fig. 14.2. A standard resistance R_0 is set to approximately match the resistance of the conductivity cell, R. By movement of the slide wire contact C, a point will be found where the noise heard in the earphones is minimized. Then,

$$\frac{\text{cell resistance}}{\text{std resistance}} = \frac{R}{R_0} = \frac{\overline{CD}}{\overline{BC}} = \frac{(1000 - X)}{X} \tag{14.3}$$

where $\overline{BD} = 1000$ divisions and X is the number of divisions. Thus, the measurement of X as a function of the volume of titrant added permits us to prepare conductometric titration plots.

ACID–BASE TITRATIONS

Consider the case of the titration of 100 ml of 0.10 N NaOH with 1.0 N HCl. From Table 14.1 we see that the OH$^-$ has a large conductance, the Na$^+$ a fairly small conductance. As the HCl titrant is added, the reaction

$$OH^- + H^+ = H_2O$$

proceeds, so that the OH$^-$ concentration decreases, and hence the conductance of the solution decreases (Fig. 14.3).

After the equivalence point is reached, the added H$^+$ is in excess, and

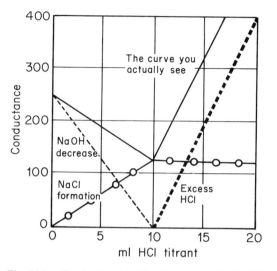

Fig. 14.3. Conductometric titration of NaOH with HCl.

as this H^+ has a very large conductance, the conductivity of the solution again rises sharply.

Note that the NaOH contribution to the conductance decreases extremely rapidly while the NaCl contribution increases slowly, thereby decreasing the slope of total conductance. Following the equivalence point, the NaCl contribution to the conductance remains constant while the HCl contribution rises sharply. Thus the total conductance of the solution rises sharply following the endpoint.

With weak acids, weak bases, or both, the shapes of the titration curves may differ from that in Fig. 14.3; however, the principle in constructing these curves is the same as with strong acids and bases. Several titration curves involving weak acids and bases are shown in Fig. 14.4. Because weak acids and bases are only slightly ionized, they do not themselves

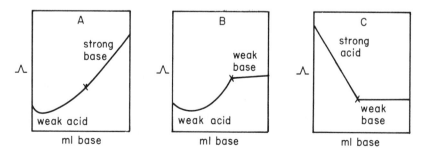

Fig. 14.4. Conductometric titration curves involving weak acids and bases.

contribute much to the total conductance of the solution, although the salt formed still exerts an influence. Note in Fig. 14.4 that a weak base is a better titrating agent than a strong base for a weak acid.

The conductance (G) of a solution depends on the concentration of the ions present. As a titration proceeds and the solution is diluted, the conductance decreases due to this dilution effect and the endpoint is less distinct than it could be. To remedy this, two approaches are generally used. The first is to use a titrant ten times stronger than the titrated material so the dilution is held to a minimum. If this fails, then a correction for the dilution is made using the equation

$$G_{actual} = \left(\frac{V + v}{V}\right) G_{obs} \qquad (14.4)$$

where V is the initial volume and v the volume of titrant added.

Example: Assume that 20 ml of NaOH was titrated to 50% of the equivalence point by adding 5.85 ml of 1.00 N HCl. The observed conductance was 29.0 ohm^{-1} cm^2. What is the actual conductance?

$$G_{actual} = \left(\frac{20 + 5.85}{20}\right) \times 29.0 = 37.4 \text{ ohms}^{-1} \text{ cm}^2$$

OXIDATION–REDUCTION SYSTEMS

Conductometric titrations involving redox systems are very difficult because the equivalent conductances of the ions involved are quite similar (e.g., $Fe^{2+} = 54$, $Fe^{3+} = 68$).

CONDUCTIVITY DETECTORS FOR CHROMATOGRAPHY

High-performance Liquid Chromatography and Ion Chromatography

The same kind of conductivity detector can also be used for high-performance liquid chromatography (HPLC). The same style of detector can be used for ion chromatography. The principles of both HPLC and ion chromatograph, and the use of conductivity detectors, are discussed in Chapter 21.

Gas Chromatography

The first reasonably successful conductivity detector for gas chromatography was the one developed by Coulson (1965). The sample is separated by a gas chromatograph and combusted in a short, high-temperature furnace; the combustion gases then are mixed with a stream of deionized water, and the conductivity of this solution measured. The sensitivity depends upon producing either H_3O^+ or OH^- ions, which have the highest conductivities. This detector works well with halogens, sulfur, and nitrogen (convert to NH_3 with H_2), and is as sensitive as thermal conductivity cells for carbon.

According to Coulson, electrolytic conductivity gas chromatography requires less cleanup of food extracts, soil extracts, and water extracts than is required for microcoulometric gas chromatography, because smaller aliquots of samples may be analyzed. Smaller samples injected into the chromatograph result in longer column life and sharper peaks due to more rapid volatilization of the pesticide residues from other plant extractives. Sharper peaks result in better resolution and more positive identification of unknown peaks. The greater sensitivity of electrolytic conductivity detection makes possible routine analysis at the low ppb level.

The Coulson detector had the disadvantage that electroplating often occurred and space charges built up, both causing changes in sensitivities and unstable base lines. The problem was solved to a large extent by Hall (1974) who applied alternating current to the electrodes rather than the direct current used by Coulson. A diagram of a Hall detector used with a Tracor instrument is shown in Fig. 14.5.

Operation of a Hall detector, which is a differential conductivity cell, is described as follows in the Tracor instrument manual:

> The gaseous reaction products formed in the reactor are directed to the cell [Fig. 14.6] and enter through the gas inlet. The conductivity cell is constructed of stainless steel and inert plastic. The differential conductivity cell contains the reference conductivity cell, the gas–liquid contactor, the gas–liquid separator and the analytical conductivity cell.
>
> The conductivity solvent enters through the solvent inlet and flows through the reference conductivity cell that is formed by the top and outer electrode assemblies. The solvent then flows into the gas–liquid contactor where it is mixed with the gaseous reaction products entering through the gas inlet. The heterogeneous gas–liquid mixture formed in the gas–liquid contactor is separated into gas and liquid phases in the gas–liquid separator. The gas phase exits through the hollow bottom electrode. The liquid phase flows between the outer wall of the bottom electrode and the inner wall of the gas–liquid separator. The cavity formed by these surfaces constitutes the analytical conductivity cell. The liquid phase exits into the hollow bottom electrode through a small hole in the wall of this electrode. At this point the gas and liquid phases are recombined and returned to the solvent reservoir.

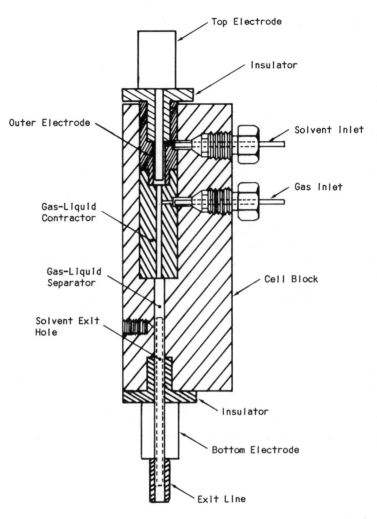

Fig. 14.5. Diagram of a Hall detector. (Courtesy Tracor Instruments Inc.)

RECENT APPLICATIONS

Masoud and Cha (1982) used a conductivity detector on an HPLC to detect the antioxidants BHA and BHT. Detection of estrogenic hormone traces in meat with an electrochemical detector scanning HPLC effluent was demonstrated by Frischkorn *et al.* (1980), who proportioned lithium electrolyte into the mobile phase. An HPLC separation with electrochemical detection gave Alawi and Ruessel (1981) detection limits of 10 ppb in milk for sulfonamides.

Sugars in sugar refinery products have been determined by HPLC using refractometric and conductometric determination (Charles 1981). Sinalbin and sinigrin, components of mustard oil and horseradish, were separated by isotachophoresis and then detected conductometrically by Klein (1982). Frank *et al.* (1982) developed an automatic method for determining the oxidation stability of oil and fatty products; in their procedure the volatile reaction products are captured in a water trap and detected conductometrically.

Okayama *et al.* (1981) determined overall organic acids by conductivity measurements after adding excess imidazole or 2-aminopyridine. The Hall detector is used continuously at the FDA Regional Laboratory in Kansas City to monitor nitrogen-containing pesticide and herbicide residues in foods (K. Griffith, personal communication).

BIBLIOGRAPHY

Alawi, M. A., and Ruessel, H. A. (1981). Determination of sulfonamides in milk by HPLC with electrochemical detection. *Fresenius Z. Anal. Chem.* **307**(5), 382–384. [*Chem. Abstr.* **95**, 148771.]

Braunstein, J., and Robbins, G. D. (1971). Electrolytic conductance measurements and capacitive balance. *J. Chem. Educ.* **48**, 52–57.

Charles, D. F. (1981). Analysis of sugars and organic acids. Cane sugar refinery experience with liquid chromatography on a sulfonic acid cation exchange resin. *Int. Sugar J.* **83**, 169–172, 195–199. [*Anal. Abstr.* 1982, **42**, 1F22.]

Coulson, D. M. (1965). Electrolytic conductivity detector for gas chromatography. *J. Gas Chromatogr.* **3**, 134.

Frank, J., Geil, J. V., and Freaso, R. (1982). Automatic determination of oxidation stability of oil and fatty products. *Food Technol.* **36**, 71–76.

Frischkorn, C. G. B., Smyth, M. R., Frischkorn, H. E., and Golimowski, J. (1980). Determination of traces of oestrogenic growth promoting hormones in meat be high performance liquid chromatography with voltammetric detection. *Fresenius Z. Anal. Chem.* **300**(5) 407–412. [*Anal. Abstr.* **39**, 5F40.]

Hall, R. C. (1974). A highly sensitive and selective microelectrolytic conductivity detector for gas chromatography. *J. Chromatogr. Sci.* **12**, 152–160.

Klein, H. Z. (1982). The quantitative determination of *p*-hydroxybenzyl glucosinolate (sinalbin) and allyl glucosinolate (sinigrin) using isotachophoresis in seeds of *Sinapis alba*, *Brassica juncea and Brassica aigra*. *Z. Acker-Pflanzenbau* **150**(5), 349–355. [*Chem. Abstr.* **96**, 67331j.]

Masoud, A. N., and Cha, Y. N. (1982). Simultaneous use of fluorescence, ultraviolet, and electrochemical detection in high performance liquid chromatography—Separation and identification of phenolic anti-oxidants and related compounds. *J. High Resolut. Chromatogr. Commun.* **5**, 299–305. [*Chem. Abstr.* **97**, 108604s.]

Okayama, K., Matsomoto, K., Yamamoto, M., and Osajima, H. (1981). Sourness of food and its electrochemical measurement. VI. Application of an organic weak base (imidazole or 2-aminopyridine) in aqueous solution for electrochemical measurement of the organic acid content. *Anal. Abstr.* **41**, 179–182.

Small, H., Stevens, T. S., and Bauman, W. C. (1975). Novel ion exchange chromatographic method using conductimetric detection. *Anal. Chem.* **47**, 1801–1809.

15
Electrophoresis

In a strict sense, electrophoresis refers to the movement of charged colloidal particles and macromolecular ions under the influence of an electric field. Electrophoresis in food analysis is, generally, restricted to proteins. The charge on the protein depends on the pH of the solution. Every protein has an isoelectric point (pI) at which the net charge is zero and at which it has a zero mobility. At pH values below pI, the protein migrates as a cation, the mobility increasing with decreasing pH. At pH values above the pI, the protein migrates as an anion, the mobility increasing with increasing pH.

Differences in migration velocities provide a sensitive means of separating components of a mixture that is otherwise difficult to fractionate. Electrophoresis is often the only method available for the quantitative analysis and fractionation of biological fluids, and for the characterization of their purified components. In addition it is a powerful tool for the detection and characterization of macromolecular interactions.

Electrophoretic methods can be divided into moving boundary electrophoresis and zone electrophoresis. Zone electrophoresis has by far the widest range of applications in food analysis and its variations will be discussed.

ZONE ELECTROPHORESIS

Zone electrophoresis on solid supports and in gels (Fig. 15.1) is the method of choice in situations where separations, rather than precise

214

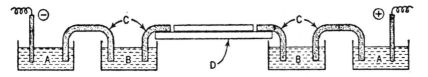

Fig. 15.1. Diagram of apparature for a zone electrophoresis. A—salt solution; B—buffer; C—filter paper soaked in buffer; and D—supported gel or paper.

physicochemical measurements, are desired. In addition, zonal methods are suitable for the analysis of small quantities of material by fairly simple procedures.

In conventional zone electrophoresis (Fig. 15.2), the sample components are separated based on their differences in net charge, size and shape. The separation takes place at a constant pH and ionic strength. Polyacrylamide gel, cellulose, or granulated gels are used as stabilizing media. The sample is applied as a narrow zone on top of the stabilizing gel and when the electric field is applied, the sample components will migrate into the gel. Separation in the stabilizing media takes place because of different mobilities of the sample components. The separated zones migrate one after the other out of the gel into the funnel-shaped elution chamber where they are then flushed out by a continuous stream of elution buffer.

The important advantages of zone electrophoresis are the following: (1) simple and inexpensive apparatus that permits simultaneous analysis of several samples in a relatively routine procedure, (2) simple procedures for visualization of zones and for isolation of fractions, (3) improved resolution by combining electromigration with molecular sieving (e.g., in starch and polyacrylamide gels), and (4) adaptability to either large-scale preparative or microanalytical separations. In addition, zone electrophoresis can be used for the investigation of low-molecular-weight substances that are difficult to analyze by the moving boundary methods. These advantages are gained, however, at the sacrifice of accuracy and precision, particularly with respect to the determination of mobilities from migration rates in solid media.

Among the many supporting media that have been used for zone electrophoresis, the most important ones are paper, cellulose acetate, and agar, starch, and polyacrylamide gels. In starch gel and polyacrylamide gel electrophoresis, molecular sieving is utilized to great advantage. Because size, shape, and electrophoretic mobility determine relative mobility in the gels, the degree of resolution is much improved. In addition, strongly dissociating agents (such as concentrated urea solutions) can be incorporated into the starch gel to permit separation of protein subunits, which tend to aggregate in aqueous solutions.

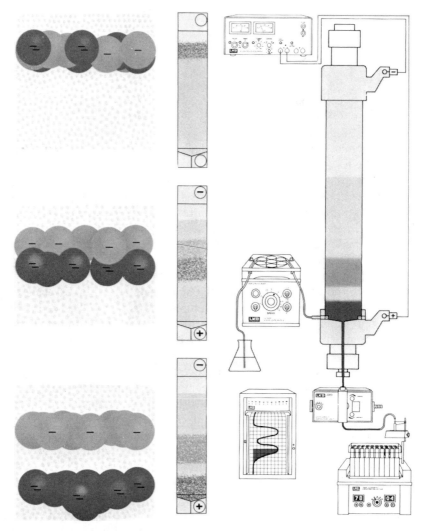

Fig. 15.2. Conventional zone electrophoresis apparatus and diagram of separation process. (Courtesy of LKB Instruments.)

Paper electrophoresis was first used for protein separations by von Klobusitzki and Konig (1939). In paper electrophoresis, a narrow band of protein solution is deposited on a support strip that is soaked in buffer. A constant voltage gradient, applied through the two ends of the paper, separates the proteins into bands according to their electrophoretic mobilities. To overcome the partial sorption of the proteins on the paper, cellulose acetate strips were developed. The latter have the additional advantage that they can be made transparent by treatment with paraffin oil or 5–15% acetic acid in ethanol at the end of electrophoresis, and then stained to reveal the proteins. The dark-stained protein bands stand out better against the transparent background.

Instead of paper, some investigators prefer open-pored gel-like agar or agarose at a concentration of about 1% (Wieme 1963). With the gel methods, the sample is deposited by pipetting a measured amount of protein in a slot in the gel. Agar gels are negatively charged and show a strong electroosmotic transport toward the cathode; consequently, the sample should not be deposited too far from the middle of the gel plate. Because agarose gels are neutral, this problem does not arise with them. In addition, the migration of many proteins, and particularly lipoproteins, is limited by their interaction with agar but not with agarose. The use of starch as a support was introduced by Smithies (1955) and of polyacrylamide by Barka (1961) and by Ornstein and Davis (1962).

After the electrophoretic run, the supporting medium (paper, cellulose acetate, or gel) is "fixed" in alcohol or acid, or heated to insolubilize the proteins. The supporting medium is then stained with a protein stain. Cellulose acetate strips are generally stained with a Ponceau-S solution in 5% trichloracetic acid; thus, the fixing and staining operations are combined into one step. The strips are washed with 4% acetic acid containing tap water and can be stored for reference, photographed, or scanned with a densitometer. Strips stained with Amido Black are washed for several hours in 10% acetic acid or methanol. For lipoprotein staining, Sudan Black is used most. The strips are generally washed with 50% ethanol–water mixtures. Glycoproteins can be stained with PAS (periodic acid–Schiff reagent containing fuchsin). Replacing fuchsin by pararosaniline considerably improves the stability of the reagent. After separation and staining, separated bands can be eluted and the intensity of the stain measured spectrophotometrically.

For scanning separated bands, several instruments are available. The absorbance of unstained bands at 205 nm provides a measure of the peptide bond, but measurements at this low wavelength require rather expensive equipment. Ultraviolet scanning at 280 nm has the drawback that the absorbance of various proteins at that wavelength depends on their tryptophan and tyrosine contents, which varies widely among proteins.

Scanning of separated and stained bands, followed by integration, provides information on the relative distribution of the proteins. The interpretation of the quantitative results must be made with caution. Generally, there is some overlapping of proteins and the extent of staining of individual protein bands (on an equiprotein basis) varies widely.

PAPER ELECTROPHORESIS

The basic process of paper electrophoresis is as follows: The material to be separated into its components is either spotted or streaked onto the center of a strip of paper that is saturated with a buffer solution. Each end of the paper dips into a container holding several hundred milliliters of additional buffer solution. An electrode is placed in each container, one being connected to the positive terminal of a DC source and the other to the negative terminal. A potential of about 5–10 V/cm is applied, and if any charged particles are present in the material, they will move toward the terminal of opposite charge.

There are four main techniques involving the use of paper as a stabilizer: the sandwich, ridgepole, solvent immersion, and horizontal strip techniques. Schematic diagrams of the last two of these are shown in Fig. 15.3 and 15.4.

In the sandwich technique, a paper strip saturated with buffer and containing a spot of the specimen to be analyzed is placed between glass plates and connected to electrode vessels. The edges of the glass plates are usually closed with silicone grease to retard evaporation. If clamps are used to press the plates together, great care must be taken to ensure that the pressure is uniform, otherwise erratic migration will occur due to potential changes in the buffer system.

The solvent immersion technique (Fig. 15.3) consists of immersing a paper strip soaked in buffer solution into a nonconducting, immiscible solvent such as carbon tetrachloride or chlorobenzene. Chlorobenzene is preferred because its density is nearly the same as the buffer-saturated paper, and this keeps the paper from floating or sinking. Thus, more reproducible results can be obtained. Two major advantages of this technique are that buffer evaporation is held to a minimum and the immersion solvent acts as a coolant to reduce thermal migration within the paper, which can be a serious problem if high currents are used. The electrode vessels are made quite large so that they will hold a large amount of buffer.

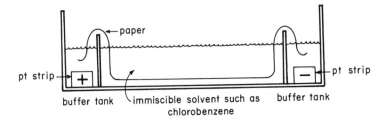

Fig. 15.3. Solvent immersion paper electrophoresis.

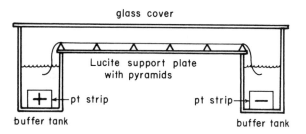

Fig. 15.4. Horizontal strip paper electrophoresis.

This is necessary to reduce pH changes caused by the electrode reaction at the ends of the filter paper.

In the horizontal suspension technique (Fig. 15.4), the paper strip is held in a taut horizontal position. The ends of the paper dip into electrode vessels. The entire system is placed in a box or chamber to maintain constant temperature. In some cases helium is placed in the chamber to act as a heat sink. This is one of the best techniques for mobility studies because the paper is horizontal and the effects of movement of the supporting solutions are minimized.

CALCULATION OF MOBILITY

In order for a particle to migrate in an electric field, it must possess a net electrostatic charge. This charge is an integral multiple of 4.8×10^{-10} esu. Many compounds that do not normally have a charge can be separated electrophoretically provided some charging process is used. Charging processes that have been found to be successful are reactions with acids and bases, dissociation into ions by polar solvents, hydrogen bonding, chemical reactions, polarization, and ion pair formation.

The widest application of electrophoresis is in separations for the isolation and purification of small amounts of compounds. The best method of referring to such separations and of comparing one system to another has been the *mobility* of the compounds involved. Correct mobilities are therefore important and anyone using electrophoresis should be familiar with the various factors influencing the mobility and know what corrections should be made.

Consider a particle placed on a strip of paper that is saturated with a buffer and has a potential applied to it. The force F exerted on the particle is equal to the charge of the particle Q times the field strength E.

$$F = QE \qquad (15.1)$$

At first glance it would appear that the particle would move toward one end of the paper with an increasing velocity. However, it turns out that the velocity is constant. The reason for this is that as the particle moves through the buffer it meets a retarding force caused by the viscosity of the solvent; this retarding force increases linearly with the particle acceleration, thus maintaining the velocity of the particle essentially constant. Stokes has shown that this opposing force for a sphere moving in a viscous medium can be expressed

$$F_s = 6\pi r \eta v \tag{15.2}$$

where F_s is the viscous retarding force, r the radius of the particle in cm, η the viscosity of the medium in poises, and v the electrophoretic velocity in cm/sec. When these forces are equal

$$F_s = EQ = 6\pi r \eta v \tag{15.3}$$

The mobility of a particle, U, is defined as

$$U = v/E \tag{15.4}$$

where U is in cm^2 volt^{-1} sec^{-1} and E is in volts/cm. Substituting equation 15.3 into equation 15.4, we get

$$U = Q/6\pi \eta r \tag{15.5}$$

Example: Calculate the mobility of the Ba^{2+} ion in a 0.0625 M solution of $BaCl_2$. The radius of Ba^{2+} in solution is 2.78 Å and the viscosity of this solution is 10.310 millipoises.

Referring to equation 15.5:

$$Q = 2 \times 4.8 \times 10^{-10} \text{ esu}$$

$$r = 2.78 \times 10^{-8} \text{ cm}$$

$$\eta = 1.03 \times 10^{-2} \text{ poises}$$

Since mobility has the units cm^2 volt^{-1} sec^{-1}, the factor 300 is necessary to convert from practical volts to esu's.

$$U = \frac{2 \times 4.8 \times 10^{-10}}{6 \times 3.14 \times 2.78 \times 10^{-8} \times 1.03 \times 10^{-2} \times 3 \times 10^2}$$

$$= 5.9 \times 10^{-4} \text{ cm}^2 \text{ volt}^{-1} \text{ sec}^{-1}$$

The mobility calculated in this way with equation 15.5 cannot be verified experimentally because there is an additional retarding effect on the mobility due to the negative chloride ions moving in the opposite direction from the Ba^{2+} ions. This has a tendency to decrease the mobility of the Ba^{2+}, and is related directly to the ionic strength of the solution. By incorporating the Debye–Huckel equation, the following correction term is obtained:

$$\frac{1}{1 + rA\sqrt{\mu}}$$ (15.6)

where μ is the ionic strength, A a constant (0.233×10^8 for water at 25°C), and r the radius of the particle in cm. The ionic strength $\mu = \frac{1}{2}\sum cZ^2$, where c is the concentration in moles/liter and Z the valence of the ion.

Example: Referring to the previous example, what is the mobility of the Ba^{2+} ion when the other ions are considered?

$$U = \frac{Q}{6\pi r\eta} \times \frac{1}{1 + rA\sqrt{\mu}}$$

$$\sqrt{\mu} = [\tfrac{1}{2}(0.0625 \times 2^2) + (2 \times 0.0625 \times 1^2)]^{1/2}$$

$$\mu = 0.433$$

The correction term then is

$$\frac{1}{1 + 2.78 \times 10^{-8} \times 0.233 \times 10^8 \times 0.433} = 0.781$$

Multiplying the mobility obtained in the previous example by 0.781 gives

$$U = 5.9 \times 10^{-4} \times 0.781 = 4.62 \times 10^{-4} \quad cm^2\, volt^{-1}\, sec^{-1}$$

The preceding equations were developed from theory; however, in practice a relationship for calculating mobilities that involves more readily obtainable measurements is desirable. Recalling that the units involved in equation 15.4 were volts/cm and cm/sec, the equation can be rewritten as equation 15.7, in which needed parameters are easily obtained.

$$U = p_m l_m / Vt$$ (15.7)

where U is the mobility in $cm^2\, volt^{-1}\, sec^{-1}$, p_m the distance the ion moves in cm, l_m the distance between the electrodes in cm, V the voltage in volts (not to be confused with E, which is volts/cm), and t the time in sec. For some electrophoretic cells, the distance l_m is hard to obtain accurately, so volts/cm is determined from

$$E = i/qk$$ (15.8)

where E is the field strength in volts/cm, i the current in amperes, q the cross-sectional area of the cell in cm^2, and k the specific conductivity of the solution in $ohm^{-1}cm^{-1}$. The value of k is determined from

$$k = C/R$$ (15.9)

where R is the resistance of the solution in the cell in ohms and C the cell constant in cm^{-1}. The cell constant is determined by the classical method of Jones and Bradshaw (1933) in which they found that if 7.457 g of dry KCl were added to 1 liter of water at 20°C the resulting conductivity was 0.007138 $ohm^{-1}cm^{-1}$ at 0°C.

Example: A 3% solution of egg albumin buffered at a pH of 7.2 was placed in an electrophoresis cell having a diameter of 1.2 cm. A current of 20 mA was used and it took 208 min for the egg albumin to move 5.87 cm toward the positive electrode. If the cell constant was 18.3 cm^{-1} and the resistance of the system was 7690 ohms at 0°C, calculate the mobility of the material.

Step (1). Calculate the specific conductivity using equation 15.9.

$$k = \frac{18.3 \text{ cm}^{-1}}{7690 \text{ ohms}} = 0.00238 \text{ ohm}^{-1}\text{cm}^{-1}$$

Step (2). Calculate E using equation 15.8.

$$E = \frac{0.020}{(3.14 \times 0.6^2)(0.00238)} = 7.43 \text{ volts/cm}$$

Step (3). Calculate U using equation 15.7. Remember that the l_m has been incorporated into the E value since $E = V/l_m$.

$$U = \frac{-5.87}{(208 \times 60)(7.43)} = -6.3 \times 10^{-5} \text{ cm}^2 \text{ volt}^{-1} \text{ sec}^{-1}$$

The sign is negative because the ion was negative as indicated by motion toward the positive electrode.

Heat Production

A problem that often arises in determining mobilities is heat production. The current passing through a cell develops heat and this not only causes convection currents, which change the mobilities, but increases the rate of evaporation of the buffer solution when paper electrophoresis is employed. This loss of buffer causes a change in mobility due to changes in the potential and current; it also causes a capillary action in the paper since the drier paper will take up solution from the buffer tank. The usual maximum amount of heat that can be tolerated is 0.15 W/cm^3. The heat produced can be calculated from the equation.

$$H = i^2/q^2k \tag{15.10}$$

where H is the heat developed in W/cm^3 and i, q, and k are as previously defined.

Example: Determine the heat developed in the solution described in the previous example.

$$H = \frac{(0.020)^2}{(1.13)^2(0.00238)} = 0.132 \text{ W/cm}^3$$

Correction for Electro-Osmosis

It has been found that if water is placed in a capillary tube and positive and negative electrodes are placed in contact with the water, the water will migrate toward the negative electrode indicating that the water is positively charged. This migration can be altered by placing other ions in the system, and in fact can be reversed. This movement of the solution in an electric field is called *electro-osmosis*. Consequently, the mobilities must be further corrected because the compound migrating in the electrophoretic system can be held back or speeded up by the movement of the buffer solution. This is similar to the action of a tail wind or head wind on an airplane.

To determine the electro-osmosis correction, a material with no inherent mobility is selected. Therefore, if it moves, it does so because of the solution movement carrying it along. This movement can either be added or subtracted from the mobility of the compound under investigation to correct for the electro-osmosis. Dextran, which has a mobility of only -0.16×10^{-5}, is commonly used to determine the correction factor for proteins. The corrected mobility is given by:

$$U_{cpd} = \frac{P_{cpd} \pm P_{dex}}{Et} \tag{15.11}$$

where P_{cpd} is the distance the compound moved and P_{dex} the distance the dextran moved. If the dextran moves opposite to the compound, then P_{dex} is added ($+$) to P_{cpd}; if it moves in the same direction, then P_{dex} is subtracted ($-$).

Migration Pathlength Correction

It has been found that the mobility of a compound determined in free solution is different from that determined using paper as a support. This is believed to be due to the fact that the particle in the paper must wind in and out among the fibers of the paper, thus traveling a longer path than that which is actually measured. Figure 15.5 illustrates the problem involved.

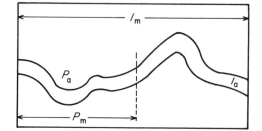

Fig. 15.5. Particle migration in paper.

Referring to Fig. 15.5, we can write the following:

$$l_a = l_m(l_a/l_m) \tag{15.12}$$

$$p_a = p_m(l_a/l_m) \tag{15.13}$$

where l_a and p_a refer to a paper system and l_m and p_m to a solution system. Rearranging equation 15.7 gives

$$p_m = UVt/l_m \tag{15.14}$$

However, we know this is not correct and should be expressed in terms of p_a and l_a as

$$p_a = UVt/l_a \tag{15.15}$$

In order to get this into a workable equation, the relationships shown in equations 15.12 and 15.13 are substituted for p_a and l_a.

$$p_m = \frac{UVt}{l_m} \left(\frac{l_m}{l_a}\right)^2 \tag{15.16}$$

The value of $(l_m/l_a)^2$ can be obtained by conductivity measurements as follows:

$$(l_m/l_a)^2 = R_o v_t/R v_b \tag{15.17}$$

where R_o is the resistance, in ohms, of the electrolyte occupying a volume equal to the buffer in the paper, R the resistance, in ohms, of the paper saturated with buffer, v_t the total volume of the paper and the buffer in it, and v_b the volume of the buffer in the paper.

Another way of determining the ratio is to obtain the mobility in the paper and also in free solution. Any difference will be due to a different pathlength. Typical values of (l_m/l_a) are 0.5–0.7 for most papers.

GEL ELECTROPHORESIS

This is a relatively new development of a general technique called gel electrophoresis. In moving boundary electrophoresis it is difficult to remove a particular fraction without stirring and remixing the solution. To overcome this problem many different types of supporting materials (e.g., paper) were tried. If starch gel is the support material, the separated components can be removed easily by slicing the gel.

Gel electrophoresis, however, often takes a long time, in part because the sample must be added in a broad band. The technique of *disk electrophoresis* is a way of concentrating the sample in a very narrow "disk" before the actual separation occurs. (The separated components are also disk-shaped.) This means that the separation can be done much faster and better. For example, where a filter paper technique yields seven protein fractions of a serum, a gel gives 30 components. Although it is true that a narrow initial sample zone is one reason for good separations, another property of a gel is that it acts like a sieve. The theory of disk electrophoresis technique was developed primarily by Ornstein (1964), and the actual analytical technique was developed by Davis (1964).

In cases where the ionic mobilities are quite similar, paper electrophoresis may not resolve the components. However, a gel is a latticed network whose pores can be made of molecular size and act as a sieve to further separate compounds based on differences in size and shape.

The gel usually used in disk electrophoresis is polyacrylamide, made by polymerizing acrylamide with N,N-methylene bis-acrylamide (Bis). The polymerization is catalyzed by riboflavin and/or N,N,N,N-tetra-methylenediamine (Temed) and by light; a lamp is used to hasten the process once the solutions are placed in the cell. The pore size can be varied by the ratios of the chemicals used. In a recent technique, a continuous gradient of pores down the column was used. These polyacrylamide gels are relatively inert and have almost no ionic side groups.

A disk cell contains three sections: a large-pore sample-concentration gel, a large-pore spacer gel to remove the materials used to concentrate the sample, and a small-pore gel used to separate the components (Fig. 15.6). The operation of a disk electrophoresis apparatus has been described by Davis (1964):

> The concentration step is achieved by introducing a mixture of sample ions into an electrophoretic column near the boundary of two ions whose sign is like that of the sample ions at a given pH. One ion is faster (chloride), the other, an ion of a weak acid or base (glycine), slower than all of the sample ions at this pH. The counter ion (Tris, hydroxymethyl aminomethane), the ion of sign opposite that of the slow, fast, and sample ions, serves as a buffer to set initially and then to regulate the various pHs behind the moving

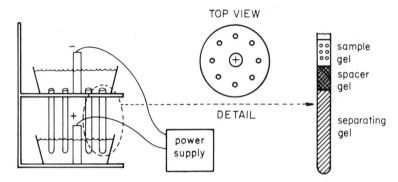

Fig. 15.6.   Disk electrophoresis apparatus.

anion boundary. Tris is used with proteins. The electrical polarity is set so that the fast ion is situated ahead (in the direction of migration) of the sample and slow ions. Application of a voltage results in the segregation from one another and stacking of the constituent ions of the sample into contiguous zones in order of their relative mobilities, the entire sample being sandwiched between the fast and slow ions.

The sample thickness can be very thin. Bromphenol blue is added as a marker dye to show when the migration down the entire cell is complete.

The next step is to remove the slow ion from the sample. The spacer segment is used for this by employing a different pore size and a different pH, so that the ion of the weak acid or base now continuously overtakes and passes through the sample species, establishing a comparatively uniform voltage gradient in which electrophoretic separation of the sample occurs.

The third step is to actually electrophoretically separate the sample components. A small-pore gel is used for this. After the separation, the gel is removed from the cell and transferred to a larger-diameter cell, and dye is added to react with the separated compounds. The unbound dye is removed from the gel by electrophoresis.

Sodium Dodecyl Sulfonate Disk Electrophoresis

A technique used primarily to determine the molecular weight of proteins involves adding a surface active agent such as sodium dodecyl sulfonate (SDS) to the sample. Being both a potent protein denaturant and solubilizing agent, SDS cancels differences in intrinsic charge of the proteins and usually converts them to a rodlike shape whose lengths vary with the molecular weight. A series of standards are separated at the same time and the proteins visualized with Coomassie Brilliant Blue R-250. A fairly linear relationship is obtained for molecular weights of 15,000–70,000. Proteins of higher molecular weights do not readily pass through the gel, and a nonlinearity occurs with molecular weights of 15,000 down to 6000. Below 6000, all of the proteins form essentially the same size particles, and if separation is desired, urea must be added to disperse the particles.

Isotachophoresis

Isotachophoresis is an electrophoretic technique by which sample components are separated based on differences in their net mobility. The

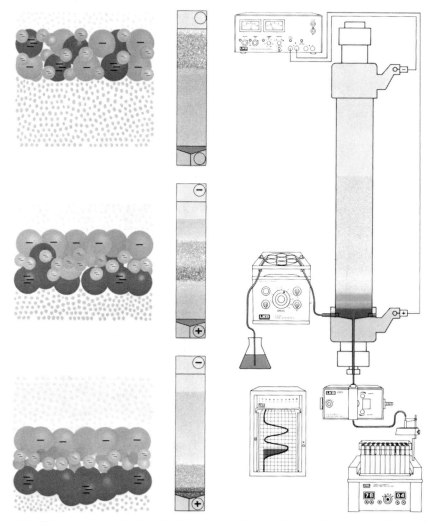

Fig. 15.7.   Apparatus for isotachophoresis and diagram of separation process. (Courtesy of LKB Instruments.)

separation is performed in a polyacrylamide gel utilizing a discontinuous buffer system (Fig. 15.7). The sample components and the "spacer ions" are dissolved in the terminating electrolyte, and the leading electrolyte is initially applied throughout the polyacrylamide gel. When the electric field is applied, separation of the samples takes place between the leading and terminating electrolytes during migration. When segration is completed, the separated zones and the spacer ions migrate with the same velocity in immediate contact with each other into the funnel-shaped elution chamber where they are then flushed out by a continuous stream of elution buffer.

Isotachophoresis has a high resolving power because of its concentrating effect. The zone boundaries are actively sharpened during the experiment, which prevents diffusional broadening.

ISOELECTRIC FOCUSING

Consider the amino acid NH_2—R—COOH, which in neutral solution would exist as NH_3^+—R—COO$^-$. It may react with an acid or base as follows:

$$NH_3^+—R—COO^- + H^+ = NH_3^+—R—COOH \quad \text{(a positive ion)}$$

$$NH_3^+—R—COO^- + OH^- = NH_2—R—COO^- + H_2O \quad \text{(a negative ion)}$$

If placed in an electrophoresis cell, the amino acid would migrate toward the anode in basic systems and toward the cathode in acid systems. As mentioned before, there should be a pH at which the net charge is zero, and the amino acid would show no movement. This point where no movement occurs is called the *isoelectric point*; it is important as it affects separations. Consider the situation where the pH is such that part of the amino acid ions are positive and part are negative. Here we should get a very diffuse band and probably no separation. What is necessary, therefore, to get sharp resolution is to have the compound be separated as far away from the isoelectric point as possible, so that all of the ions are of the same charge.

Electrofocusing or *isoelectric focusing* is the name given to the phenomenon occurring to ampholytes in a pH gradient influenced by an electric field (Haglund 1967). The gradient is obtained by imposing a dc potential on an electrolyte system in which the pH steadily increases from the anode to the cathode. Provided the pH gradient is stable, ampholytes (such as proteins or peptides) present in the electrolyte system are repelled by both electrodes, and each ampholyte species collects at a place in the gradient where the pH of the gradient is equal to the isoelectric point of that species. Though the theory of electrofocusing has been known for a long time, it has only recently become a practical method of protein fractionation. This has been mainly due to commercial availability of carrier ampholytes for use in electrofocusing, and of improved instruments for separation.

The carrier ampholytes are mixtures of aliphatic polyamino–polycarboxylic acids with molecular weights of 300–600. The pK values of the amino acid carboxylic groups determine the pI of each ampholyte. When preparing a column, the carrier ampholytes are dissolved in a density gradient of a nonionic compound (e.g., sucrose) to obtain a convection-free medium. The sample can be layered at some chosen level in the column or evenly distributed throughout the whole volume. The pH gradient is formed by the carrier ampholytes when a voltage (200–1200 V, dc, yielding a current up to 10 mA) is applied across the mixture. The proteins in the sample migrate to the point where they are electrically neutral (i.e., pH = pI) (see Fig. 15.8). To reduce convection, the column is thermostated during operation. The time required for electrofocusing normally ranges from 24 to 72 hr. At the end of the run, the contents of the column are drained (without remixing the gradient-stabilized and separated substances) through a capillary to a fraction collector. The effluent is detected by a flow analyzer to identify the separated sharp zones.

The two main applications of isoelectric focusing in biochemistry are (1) the analytical separation of high-molecular-weight ampholytes, especially proteins, according to their isoelectric points (a separation of two species requires a difference in isoelectric point of only 0.02 pH units) and (2) characterization of proteins by their isoelectric points in a single, simple experiment, compared to 16 runs required in free electrophoresis. In the pH range 5 to 8, the precision and reproducibility of isoelectric point determination in electrofocusing are about 0.01 pH unit.

As any other method, electrofocusing has several limitations. Thus, lipoproteins are denatured at the isoelectric point, and some proteins are only slightly soluble at their isoelectric point, especially in a salt-free system. Addition of urea increases protein solubility and may be useful in some cases. Increasing the ampholyte concentration, and thus the ionic strength, can also increase the solubility of some proteins.

Gel electrofocusing is a microanalytical modification of isoelectric focusing in a sucrose density gradient (Wrigley 1968). Carrier ampholytes and the protein sample are set in a polyacrylamide gel using conventional disk electrophoresis apparatus. Fractionation takes 1–3 hr. Protein zones are detected by precipitation in the gel with 5% trichloracetic acid or with protein dyes after removal of carrier ampholytes.

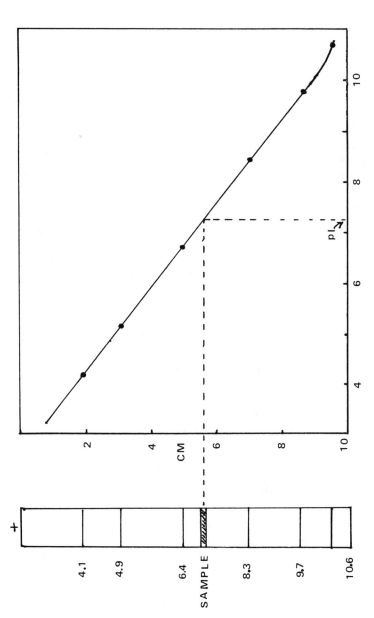

Fig. 15.8. Example of the use of pI marker proteins. Marker chemical: Acetylated cytochrome C at pI 4.1, 4.9, 6.4, 8.3, 9.7; cytochrome C at pI 10.6. (Courtesy U.S. Biochemical Corp. Cleveland, OH.)

PREPARATIVE ELECTROPHORESIS

Electrophoresis in horizontal thick slabs of buffer soaked into an appropriate medium (mainly potato starch) has been for many years the simplest method of preparative electrophoresis. By this method, batches of 5–50 ml liquid (or more in cooled columns) can be separated in 10–20 hr. A commercial, continuous fractionation and elution electrophoresis apparatus (Fig. 15.9), adapted from the design of Jovin *et al.* (1964), is available. Samples containing up to 200 mg proteins in 160 ml can be separated within 10 hr. The instrument is capable of utilizing several supporting media, though it has been used primarily for polyacrylamide gel electrophoresis.

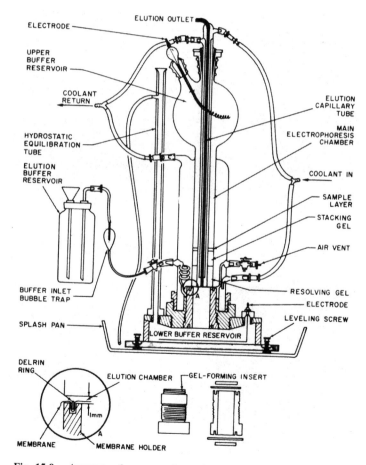

Fig. 15.9. Apparatus for preparative polyacrylamide gel electrophoresis.

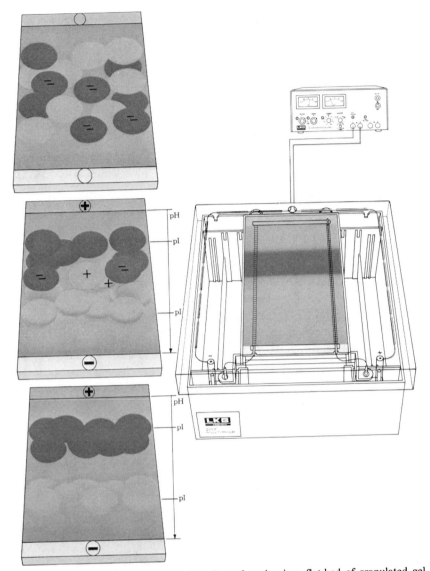

Fig. 15.10. Apparatus for preparative electrofocusing in a flat-bed of granulated gel. (Courtesy of LKB Instruments.)

Electrofocusing in a flatbed of granulated gel also can be for preparative purposes (Fig. 15.10). In this case the sample can be applied as a zone or throughout the whole gel bed. After the separation is completed, the gel bed is divided into sections by a fractionating grid and the sample components are eluted from the granulated gel.

RECENT APPLICATIONS

General methods for determining polysaccharide thickening agents in foods has been described by Pechanek *et al.* (1982). Isoelectric focusing has been used by Macgregor and Ballance (1980) to separate and then determine, by using the amylopectin B-limit dextrin plate technique, alpha-amylase enzymes in germinated barley. Ascorbate oxidase in wheat flour has been isolated and characterized by Pfeilsticker and Roeung (1980) using thin layer isoelectric focusing. Rennet enzymes have been fractionated and identified by Reimerdes and Geuer (1980) by electrophoresis techniques. Cliffe and Law (1979), using a modified starch gel electrophoretic technique, detected peptidase activity in starter cultures.

A rapid detection procedure for cow milk in goat milk has been described by Mitchell and Middleton (1980) using electrophoresis on cellulose acetate. Cow milk in sheep and goat milk has been detected by Kruse *et al.* (1982) by the use of isoelectric focusing on polyacrylamide gels containing urea. Protein patterns obtained from cooked and frozen crab meat have been shown by Krzynowek and Wiggin (1981) to differ among genera by the use of polyacrylamide gel isoelectric focusing.

Isotachophoresis has been used to determine L-ascorbic acid and dehydroascorbic acid (Rubach and Breyer 1980) and to determine histamine in fish by a capillary modification (Rubach *et al.* 1981).

Proteins of vegetable and animal origin have been differentiated by Ring *et al.* (1982) using disk polyacrylamide gel electrophoresis. The sodium dodecyl sulfate gel electrophoresis technique was used by Rizvi *et al.* (1980) to separate soya, egg albumin, and wheat gluten. Electrophoresis has been used by Armstrong *et al.* (1982) to determine isolated soybean protein in raw and pasteurized meat products and by Valas-Gellei (1981) to determine milk and soy proteins in meat products. Khatamian and Meloan (1983) used disk electrophoresis to detect adulteration of beef by other meats such as horse and deer.

PROBLEMS

15.1. A 0.2 M solution of $MgCl_2$ is prepared. What is the mobility of the chloride ion if its solution radius is 1.93 Å and the viscosity of the solution is 8.54 millipoises? What is the mobility if ionic strength is considered? *Ans.* -1.24×10^{-5} cm^2 volt^{-1} sec^{-1}; -1.07×10^{-5} cm^2 volt^{-1} sec^{-1}.

15.2. What is the solution radius of the acetate ion if a 0.125 M solution of sodium acetate showed a mobility of -2.49×10^{-4} cm^2 volt^{-1} sec^{-1}.

The viscosity is 9.633 millipoises. *Ans.* 1.69 Å.

15.3 The electrophoretic fractionation of a pig embryo plasma produced a spot 58 mm away from the origin 13 hr and 36 min after a current of 9 mA had been applied. The cell constant is 15.2 cm^{-1} for the 8-mm cell, and a resistance of 2550 ohms was obtained when a 0.02 M sodium phosphate 0.15 M NaCl buffer was used. Calculate the mobility and the heat evolved. *Ans.* 2.0×10^{-5} cm^2 $volt^{-1}$ sec^{-1}; 5.3×10^{-2} watts/cm^3.

15.4. The red blood cells of a guinea pig were found to migrate 38 mm in 86.5 min when placed in 0.15 M phosphate buffer. A current of 12 mA was used. The cell was 1.0 cm in diameter and had a constant of 18.9 cm^{-1}; the resistance was 8140 ohms when the buffer was in the cell. Calculate the mobility and the heat evolved. *Ans.* 1.12×10^{-4} cm^2 $volt^{-1}$ sec^{-1}; 1.0×10^{-1} watts/cm^3.

BIBLIOGRAPHY

Armstrong, D. J., Richert, S. H., and Reimann, S. M. (1982). The determination of isolated soybean protein in raw and pasteurized meat products. *J. Food Technol.* **17**, 327–337.

Barka, T. (1961). Studies on acid phosphates. II. Chromatographic separation of acid phosphates of rat liver. *J. Histochem. Cytochem.* **9**, 564–571.

Cliffe, A. J., and Law, B. A. (1979). An electrophoretic study of peptidases in starter streptococci and in cheddar cheese. *J. Appl. Bacteriol.* **47**, 65–73.

Davis, B. (1964). Disc electrophoresis. II. Method and application to human serum proteins. *Ann. N.Y. Acad. Sci.* **121**, 404–427.

Haglund, H. (1967). Isoelectric focusing in natural pH gradients—A technique of growing importance for fractionation and characterization of proteins. *Sci. Tools* **14**(2), 17–24.

Jones, G., and Bradshaw, B. C. (1933). The measurement of the conductance of electrolytes. V. A redetermination of the conductance of standard KCl solutions in absolute units. *J. Chem. Soc.* **55**, 1780–1800.

Jovin, J., Chirambach, A., and Naughton, M. A. (1964). An apparatus for preparative temperature regulated polyacrylamide gel electrophoresis. *Anal. Biochem.* **9**, 351–369.

Khatamian, N., and Meloan, C. E. (1983). The use of disc electrophoresis to detect adulteration of beef by other animal meats. *Anal. Lett.,* **16**(A15), 1127–1132.

Krause, I., Belitz, H. D., and Kaiser, K. P. (1982). Detection of cow milk in sheep and goat milk or cheese by isoelectric focusing on thin layers of polyacrylamide gels containing urea. *Z. Lebensm. Unters. Forsch.* **174**, 195–199.

Krzynowek, J., and Wiggin, K. (1981). Fish and other marine products: Generic identification of cooked and frozen crabmeat by thin layer polyacrylamide gel isoelectric focusing: Collaborative study. *J. Assoc. Off. Anal. Chem.* **64**, 670–673.

Macgregor, A. W., and Ballance, D. J. (1980). Quantitative determination of alpha-amylase enzymes in germinated barley after separation by isoelectric focusing. *J. Inst. Brew., London* **86**, 131–133. [*Anal. Abstr.* 1981, **40**, 4F45.]

Mitchell, G. E., and Middleton, G. (1980). Rapid detection of cows milk in goats milk. *Aust. J. Dairy Technol.* **35**, 15–16. [*Anal. Abstr.* 1981, **41**, 4F46.]

Ornstein, L. (1964). Disc electrophoresis. I. Background and theory. *Ann. N.Y. Acad. Sci.* **121**, 321–349.

Ornstein, L., and Davis, B. J. (1962). "Disc Electrophoresis." Eastman Kodak Co., Rochester, NY.

Pechanek, U., Blaicher, G., Pfannhauser, W., and Woidich, H. (1982). Thickening agents: Electrophoretic method for qualitative and quantitative analysis of gelling and thickening agents. *J. Assoc. Off. Anal. Chem.* **65**, 745–752.

Pfeilsticker, K., and Roeung, S. Z. (1980). Characterization of L-ascorbic acid oxidase (EC 1.10.3.3) from wheat flour. *Z. Lebensm. Unters. Forsch.* **171**, 425.

Reimerdes, E. H., and Geuer, M. M. (1980). The application of conventional techniques for the identification of milk coagulating enzymes in rennet and rennet substitutes. *Kiel. Milchwirtsch. Forschungsber.* **32**, 15–36. [*Anal. Abstr.* 1981, **41**, 2F47.]

Ring, C., Weigert, P., and Hellmannsberger, L. (1982). Differention of protein of vegetables and animal origin by disc polyacrylamide gel electrophoresis. *Fleischwirtschaft* **62**(5), 648–650. [*Chem. Abstr.* **97**, 22219h.]

Rizvi, S. S. H., Josephson, R. V., Blaisdel, W. J., and Harper, W. J. (1980). Separation of soy-spun fiber, egg albumin, and wheat gluten blend by sodium dodecyl sulfate gel electrophoresis. *J. Food. Sci.* **45**, 958–961.

Rubach, K., and Breyer, C. (1980). Isotachophoretic determination of L(+)-ascrobic acid and dehydroascorbic acid in diluted citrus juice concentrates. *Dtsch. Lebensm-Rundsch.* **76**, 228–231.

Rubach, K., Offizorz, P., and Breyer, C. (1981). Determination of histamine in fish and canned fish by capillary isotachophoresis. *Z. Lebensm. Unters. Forsch.* **172**, 351–354.

Smithies, O. (1955). Zone electrophoresis in starch gels; group variations in the serum of normal human adults. *Biochem J.* **61**, 629–638.

Valas-Gellei, A. (1981). Quantitative determination of milk and soy proteins in meat products. *Acta Aliment.* **10**(3), 187–199. [*Chem. Abstr.* **96**, 5019e.]

von Klobusitzki, D., and Konig, P. (1939). Biochemische studien uber die gifte der schlangengattung. *Arch. Exp. Path. Pharmakol.* **192**, 271–278.

Wieme, R. J. (1963). "Agar Gel Electrophoresis." Elsevier Publishing Co., New York.

Wrigley, C. (1968). Gel electrofocusing—A technique for analyzing multiple protein samples by electrofocusing. *Sci. Tools* **15**(2), 17–23.

16
Voltammetry
(Polarography)

The dropping mercury electrode opened the door to modern electroanalytical chemistry. Today, there are many variations of the original polarographic technique, such as ac, pulse, cyclic, and derivative polarography, chronopotentiometry, chronocoulometry, and both anodic and cathodic stripping techniques. For many years polarography in all of its forms was not well received by practicing food chemists because its detection limits were only about 10^{-5} moles/liter and complex mixtures could not be handled. Polarography could not compete with atomic absorption for metal analyses or with chromatography for organic compound analyses. The development of stripping techniques lowered the detection limits for most metals 1000-fold, and the use of pulsing and differential pulsing techniques lowered the detection limits for organic compounds.

It is not necessary to learn the details of all the voltammetry techniques. If you learn the basic principles of dc voltammetry, then the other techniques are readily understandable. Therefore, the discussion in this chapter is devoted primarily to dc voltammetry. The two variations of this basic technique most commonly used by food analysts, anodic stripping and pulsing, are described as well.

If the world of electroanalytical chemistry were limited to potentiometry, most of the measurements and information we are now able to obtain would be impossible. In a potentiometric system, the system controls us because we merely measure what is going on in the solution naturally. However, if we force the issue—that is, make something happen that would not ordinarily happen—we gain control of the situation and can obtain a wealth of new information. Remember—*voltage* controls what reacts and *current* controls how much reacts.

At the turn of this century, although it was well known that voltammetry had great possibilities, it was also found that reliable measurements could seldom be made. The problem was that the electrodes were not reproducible and the results depended to a large extent on the past history of the electrode.

This problem was solved in 1922 by Heyrovsky who used a dropping mercury electrode (Fig. 16.1). This consists of a very fine glass capillary filled with extremely pure mercury. As the mercury comes from the capillary, it forms a small drop which is continuously expanding, exposing a new and reproducible surface to the solution it is immersed in.

The advantages of the dropping mercury electrode are (1) the surface area is reproducible, (2) the new surface eliminates poisoning effects, (3) the high overpotential of hydrogen on mercury renders possible the deposition of substances difficult to reduce, (4) mercury forms amalgams with many elements and this reduces their potential, (5) diffusion currents assume a steady current immediately and are reproducible, (6) the range is from $+0.4$ to -1.8 V referred to a normal H_2 electrode (above $+0.4$ V, Hg dissolves and at -1.8 V, H_2 is evolved), and (7) the lower limit of detection with conventional polarography is 10^{-5} moles/liter.

Heyrovsky coined the term *polarography* because the method is based on the polarization of the cathode. We now prefer to call the process *voltammetry*, since the resulting curves are actually a plot of current vs. voltage.

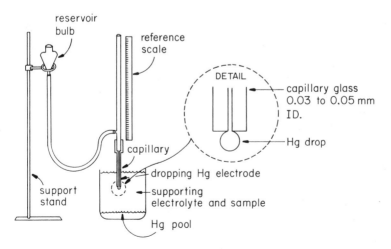

Fig. 16.1. Dropping mercury electrode.

A polarograph works as follows. The dropping mercury electrode (DME) is placed in the sample solution and adjusted so that about 20 drops a minute are obtained. This electrode is usually made the cathode (−). The anode (+) may be the mercury pool that forms at the bottom of the cell (not very reliable), a saturated calomel electrode (quite reliable), or any other type of reference electrode. The cathode is made the controlling electrode by making its area much smaller than that of the anode. A small electrode is more easily polarized and is the controlling electrode. Since the anode has a larger area and the current is small (microvolts at most), the concentration polarization at this electrode is usually negligible and its potential is regarded as constant. *The solution is not stirred.* The reason for this will be discussed later. A voltage, usually starting at zero and increasing negatively, is applied to the drop. When the drop voltage reaches the value necessary for a species in the sample to be reduced, a current will begin to flow due to the transfer of electrons from the mercury drop to the metal ions (the polarized electrode is depolarized). The current that results is then plotted vs. the voltage applied. Figure 16.2 indicates the basic electrical diagram.

Figure 16.3 shows a hypothetical voltamogram that would be expected in this system. When the voltage (E) is zero, the electrode has no attraction for any of the ions present, and no current (i) should flow. As we increase the voltage, we reach a point where the most easily reduced ion will react, and a current begins to flow. Now the current will reach a limit depending on how many ions are in the solution that can diffuse to the electrode. The greater the concentration, the greater the current; this relationship is the basis of quantitative voltammetry. If the voltage is again increased, we reach the point where the next ion can be reduced. However, the first ion is still being reduced so the current produced is the sum of both ions. The height of this wave will be proportional to the number of ions of the second type in the solution. If the voltage is further increased, a point will be reached at which the water solvent starts to decompose and to liberate hydrogen. Since the water concentration is large, it will produce a wave that will go off scale and completely obscure any metal ions that might have been reduced at a more negative voltage.

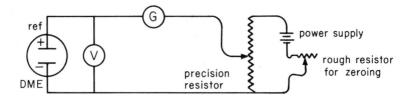

Fig. 16.2.   Electrical diagram for a simple dc polarograph.

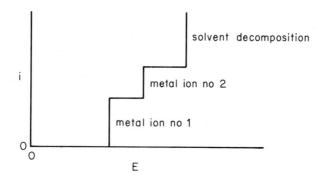

Fig. 16.3. Hypothetical voltamogram of current (i) vs. voltage (E).

Figure 16.3 shows a hypothetical result. What we actually get is shown in Fig. 16.4 for the reduction of lead using a very poor technique. In region 1 we had expected no current, yet there is a slight current that increases slightly with voltage. This is called the *residual current*; it is the current necessary to charge the drop, forming what is known as the *Helmholtz double layer*, more electrons being necessary as the voltage increases. In region 2 we might expect a straight line, but we get a saw-toothed signal. This is because the mercury drop continuously increases in size as it leaves the capillary. At the start the drop is small and few ions can diffuse into it, so the current is small. As the drop increases in size, more ions can diffuse into it and the current increases. When the drop falls off, the current will not go back to zero because there is still a small mercury

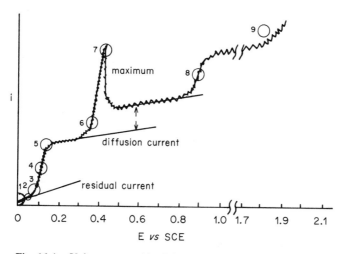

Fig. 16.4. Voltamogram of lead determined by a poor technique.

surface exposed, and also the speed of the pen is slow enough to dampen out part of the oscillation. The net effect is a saw-toothed pattern.

In the regions 3 and 5 we get a rounding of the curve rather than a sharp break. The reason for this is that the energies of ions are not all the same but follow a Gaussian distribution; thus, some are of low energy and some are of high energy. The high-energy ions will react with the electrode a bit sooner than the main bulk of the ions so an *early* current is obtained. The wave increases normally (region 4) until we approach the upper limit where the less energetic ions are not reduced as easily as the *normal* ions, and more voltage must be applied.

What is the source of the wave in region 4? Lead is supposed to be reduced at about -0.4 V versus a SCE, but the observed value is only -0.1 V. It turns out that oxygen in the solution is very easily reduced and the reduction occurs in two steps:

$$O_2 + 2H^+ + 2e^- = H_2O_2, \qquad E^\circ = -0.05 \text{ V} \qquad (16.1)$$

$$H_2O_2 + 2H^+ + 2e^- = 2H_2O, \qquad E^\circ = -0.9 \text{ V} \qquad (16.2)$$

Reduction of oxygen is a serious problem; it produces most of the current and would take up most of the chart paper, thereby reducing the utility of the method, if not controlled. Thus, oxygen must be removed. This is done by bubbling nitrogen through the sample solution for about 5 min before an analysis is made.

Region 6 is the normal lead wave. The peak (region 7) is called a *maximum*. Its exact cause is not completely known, but it can be eliminated

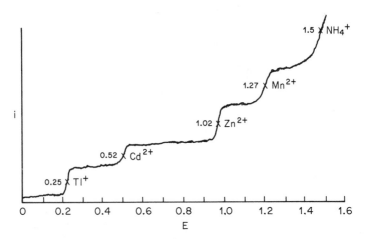

Fig. 16.5.   Voltamogram of a mixture of four metals with $(NH_4)_2SO_4$ as the supporting electrolyte.

by adding a *suppressor* to the sample. Suppressors are surface active agents such as gelatin or Triton X-100 (a synthetic detergent), or such compounds as bromcresol green. Only a few drops of dilute solution (0.002–0.01%) of the suppressor should be added or the entire wave can be eliminated. Region 8 is the second oxygen wave, and region 9 is the water solvent reduction.

Figure 16.5 shows a voltamogram of a mixture of four metals in an aqueous solution using $(NH_4)_2SO_4$ as the supporting electrolyte.

MIGRATION CURRENT

It was noted earlier that the sample solution should not be stirred. This is because the wave height is proportional to the sample concentration only if ions arrive at the cathode by one process, in this case by diffusion; stirring would upset the diffusion process. However, there are many many ions in solution and the cations, seeing the negative potential of the cathode, will move toward the cathode and the anions will move toward the anode. This current is not due to diffusion and is called the *migration current*. Can you see that a species such as $Cd(CN)_4^{2-}$ would actually cause a decrease in the Cd^{2+} current under these conditions?

To reduce that effect, a *supporting electrolyte* is added. This is a solution about 1000 times more concentrated than the solution to be measured, and it is made out of a salt that is very difficult to reduce (Na, K, Li). The effect of a supporting electrolyte is illustrated in Fig. 16.6. Since there are so many more K^+ ions than C^+ ions, when a potential is applied to the DME, the K^+ ions migrate to the electrode and surround it, but

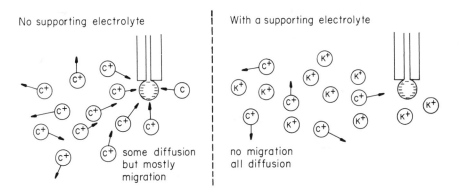

Fig. 16.6. A supporting electrolyte (K^+) reduces the migration of C^+ ions so that most of the observed current is due to diffusion of C^+ ions.

they cannot be reduced because the voltage (-0.60) is not high enough to reduce the K^+ ions. The C^+ ions other than those in proximity of the electrode are prevented from migration and reach the electrode because of normal diffusion.

DIFFUSION CURRENT

The experimental results of polarography were obtained long before a mathematical interpretation was forthcoming. The first successful equation that accounted for the reduction of the ions, the diffusion toward a spherical drop whose area was continuously changing and whose surface is moving toward the bulk solution, was developed by P. Ilkovic in 1934.

$$i_d = 607nD^{1/2}Cm^{2/3}t^{1/6} \tag{16.3}$$

where i_d is the average current in μA (708 is used when i_d is the maximum current), t the drop time in sec, m the flow rate of mercury in mg/sec, D the diffusion coefficient of the ion under study in cm^2/sec, C the concentration in millimoles/liter, and n the number of electrons taking place in the reaction at the dropping mercury electrode.

Temperature does not enter into this equation directly, but the diffusion coefficient changes 1–2%/°C at room temperature. Therefore for accurate work, temperature control to a few tenths of a degree is necessary.

Example: A 5×10^{-4} M solution of $BaCl_2$ in 0.1 M $(CH_3)_4NCl$ has a $E_{1/2}$ of -1.94 V vs SCE and shows an average diffusion current of 4.0 μA. The dropping rate was found to be 24 drops per min, and when 20 drops were collected, they weighed 0.0750 g. Calculate the diffusion coefficient.

$n = 2$ (since $Ba^{2+} + 2e^- = Ba$)

$C = 0.5$ millimoles/liter

$t = 60$ sec/24 drops $= 2.5$ sec/drop

$m = 0.0750$ g/20 drops $= 3.75$ mg/drop; $\times$ 24 drops/60 sec $= 1.5$ mg/sec

Substitute these values into equation 16.3:

$$4.0 = 607 \times 2 \times D^{1/2} \times 0.5 \times (1.5)^{2/3} \times (2.5)^{1/6}$$

Use logarithms to evaluate the fractional powers:

$\log(1.5)^{2/3} = 0.1761 \times \frac{2}{3} = 0.1174;$ antilog is 1.31

$\log(2.5)^{1/6} = 0.3979 \times \frac{1}{6} = 0.0663;$ antilog is 1.165

Thus

$$D^{1/2} = \frac{4.0}{607 \times 2 \times 0.5 \times 1.31 \times 1.165} = 0.0043$$

$$D = 1.9 \times 10^{-5} \text{ cm}^2/\text{sec}$$

QUANTITATIVE ANALYSIS

Polarography can be used for quantitative analysis since the diffusion current is proportional to the concentration of ions in a sample solution, as indicated in equation 16.3. Two approaches—the calibration curve and the standard addition method—are common in polarographic analyses.

Calibration Curve Method

The most convenient analytical method when a large number of similar samples are to be analyzed is the calibration curve method. Voltamograms are obtained for three to five solutions of varying concentrations and the diffusion current, corrected for residual currrent, is plotted vs. concentration. Figures 16.7 and 16.8 indicate how data are obtained and the corresponding calibration curve prepared; E_{de} is the potential of the dropping electrode.

Once such a calibration curve is prepared, the various unknowns can readily be determined. Care must be taken to ensure that the temperature and drop time are the same in all cases.

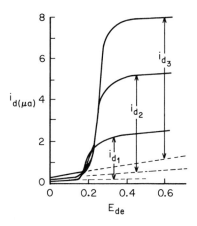

Fig. 16.7. Voltamograms of standard solutions at three concentrations.

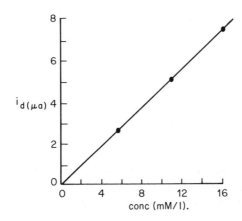

Fig. 16.8. Calibration curve prepared from data in Fig. 16.7.

Standard Addition Method

When only a few samples are to be analyzed, the method of standard addition appears to be the most convenient. Two voltamograms are required: one for the unknown, and one for the unknown plus an added known amount of the same material. Figure 16.9 illustrates how experimental data are obtained. The volume of the solution changes upon the addition of the standard, and this produces a corresponding change in the

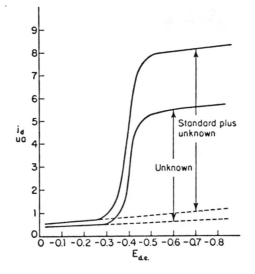

Fig. 16.9. The standard addition method.

concentration. The result is that a correction for this dilution must be made. Equation 16.4 shows how this is done.

$$C_{unk} = \frac{i_d v C_{std}}{\Delta i (V + v) + i_d v} \tag{16.4}$$

where i_d is the diffusion current of the unknown, Δi the increase in the diffusion current due to the added standard, V the original volume, $v =$ the added volume; C_{unk} the concentration of the original solution, and C_{std} the concentration of the added standard.

Example: A 20.0-g sample of grapefruit sections was stored in a new experimental tin can. This sample was homogenized, diluted to 250 ml, and centrifuged. A 25-ml aliquot was taken for polarographic analysis and a current of 24.9 μA was observed. When 5.0 ml of a standard solution containing 6.0×10^{-4} M Sn was added, the diffusion current rose to 28.3 μA. Calculate the % tin in the grapefruit (the atomic weight of tin is 118.70):

$$C_{unk} = \frac{24.9 \times 5 \times 6.0 \times 10^{-4}}{(28.3 - 24.9)(25 + 5) + 24.9 \times 5}$$

$$= 3.3 \times 10^{-4} \, M$$

$$118.70 \times 3.3 \times 10^{-4} \times \frac{250 \text{ ml}}{1000 \text{ ml}} = 0.0098 \text{ g of Sn}$$

$$\frac{0.0098 \text{ g} \times 100}{30.0 \text{ g}} = 0.032\%$$

ANODIC AND CATHODIC STRIPPING

In ordinary voltammetry with a mercury drop electrode, the diffusion current becomes too small to measure when the concentration of the metal ions reach about 10^{-5} M. In anodic stripping a single mercury drop is allowed to hang at the end of the capillary and the voltage applied is set so the metal desired will be reduced on the drop's surface and collect there. The solution is stirred so that more metal ion can get to the drop and be reduced. This is the concentration step and may proceed for several minutes, usually 2 to 5.

The potential on the drop is now reversed and the drop made the anode. Oxidation takes place in the drop with a corresponding movement of electrons and a measureable current. The reduced metal in the drop is now rapidly oxidized and returns (stripped from the drop) to the solution as ions. However, the local concentration of the ions at the surface of the drop is quite high and a current proportional to the concentration is produced. This is usually 1000 to 10,000 times greater than the forward process, and the detection limits are lowered accordingly. Precisions of $\pm 15\%$ have been reported for samples as dilute as 10^{-10} M and from ± 2 to $\pm 5\%$ for samples of 10^{-9} M or greater. Analyses with several metals simultaneously present are possible by controlling the stripping potential.

Cathodic stripping is the reverse of anodic stripping; that is, anions are first oxidized at the drop, then stripped by reducing them back to ions.

PULSE AND DIFFERENTIAL PULSE TECHNIQUES

The following description is abstracted from a preprint by M. D. Hawley. The growth of the mercury drop, as well as its periodic removal, constantly cause the electrode area to change. When either the potential of an electrode or its area is altered, current (non-Faradaic, since it does not arise from the electron-transfer reaction) must flow to charge the electrode surface. Although this non-Faradaic current is relatively small, when the scan rate is slow and the concentration of the electroactive species is large, it becomes an appreciable fraction of the total current if the concentration of the electroactive species is less than 10^{-4} M. In order to extend the detection limit below 10^{-5} M, phase sensitive methods are usually required to separate the Faradaic current from the non-Faradaic current.

In pulse polarography a square wave, typically 10–100 mV in magnitude, is superimposed upon a voltage that varies linearly. The non-Faradaic current arising from each step in potential decreases exponentially, while the decrease in the Faradaic current for a diffusion-controlled, reversible process is proportional to $t^{1/2}$. Thus, if the cell time constant is less than 1 msec and the current is sampled a few milliseconds after the potential step, the non-Faradaic current will have decayed nearly to zero, while the Faradaic current will remain relatively large. The next sampling of the current is made exactly one-half cycle later. As long as the electrode reaction is reversible chemically, a differential current will be observed in the vicinity of $E_{1/2}$ that will be proportional to the concentration of the electroactive species.

It should be noted that the plot of the differential current vs. the applied potential is very nearly symmetrical about $E_{1/2}$. Maximum sensitivity is attained for those redox systems in which electron transfer is rapid and the product of the electrode reaction is stable. Detection limits for $Cd(2+)$ and $Pb(2+)$ are reported to be less than 5×10^{-8} M.

Figure 16.10 illustrates normal pulse polarography. In the top of the figure, the sweep rate will be a function of both the increase in step size,

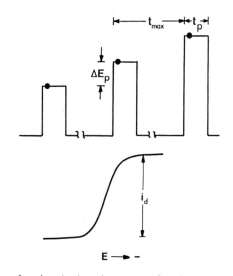

Fig. 16.10. Forcing function (top) and response function (bottom) in normal pulse polarography.

E_p, and the drop life of the mercury drop, t_{max}. Typically, 0.5 mV $< E_p$ < 4 mV and 1 msec $< t_{max} < 4$ msec, while the pulse length t is in the range 10 msec $< t_p < 100$ msec. The current is sampled as soon as the charging current due to the potential pulse has decayed to an insignificantly small fraction of the total current. The points at which the current is sampled are indicated by the black dots on the input waveform. The mercury drop is dislodged mechanically by means of a solenoid-operated hammer when the potential pulse is terminated. The curve at the bottom of Fig. 16.10 shows the response function. $E_{1/2}$ is the potential at which $i = i_d/2$. The recorded current i_d at any point is the difference in current at potential E_p and $E_{initial}$.

Figure 16.11 (top) shows the forcing function of differential pulse polarography. The sweep rate is a function of both step size E_s and the drop life of the mercury drop, t_{max}. Typically, 0.5 mV $< E_s < 4$ mV; 1 sec $< t_{max} < 4$ sec; and 5 mV $< E_p < 100$ mV. The current is sampled shortly before and after the application of the potential pulse; the points at which the current is sampled are indicated by the black dots. The duration of the pulse, t_p, normally is in the range 10 msec $< t_p < 100$ msec. In the response function (Fig. 16.11, bottom), the maximum differential current i_{dif} occurs when $E \simeq E_{1/2}$. Since the ratio of the concentrations of the oxidized and reduced forms of the reversible redox couple is then determined by the Nernst equation, substantial differential currents are observed only in the vicinity of $E_{1/2}$.

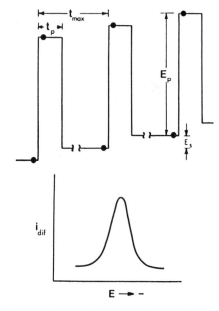

Fig. 16.11.   Forcing function (top) and response function (bottom) in differential pulse polarography.

RECENT APPLICATIONS

The electrochemical behavior of trichothecene toxins was related to the unsaturated alpha, beta keto group by Palmisano *et al.* (1981) who measured deoxynivalenol (Vomitoxin) by differential pulse polarography (DPP). Wood and Downing (1980) described a modified pulse polarographic method that could detect nicarbazin in chicken at the 0.1-ppm regulatory level.

Food colors have been determined by Fogg and Bhanot (1980) by dc voltammetry, and the behavior of some azo dyes, including the optimum conditions for DPP, was reported by Hart and Smyth (1980). Tartrazine (Yellow No. 5) can be determined in foods by a polarographic method developed by Kobori and Kawakami (1982). Fogg and Whetstone (1982) used DPP to study the degradation of Red 10B in foods.

A polarographic method has been used by Takahashi *et al.* (1982) to determine ascorbic acid oxidase inhibitor in fruits and vegetables. The determination of diacetyl and acetoin by DPP after separation by steam distillation was reported by Lechner *et al.* (1982).

A polarographic method for mercapto groups has been used by Lechner and Klostermeyer (1981) to detect whey powder in skimmed milk power. Polarographic suppression of the oxygen peaks during the determination of fructose has been found by Marek *et al.* (1980) to be a means of detecting the addition of invert sugar to honey.

A study of the anodic stripping method for determining lead and cadmium after dry ashing was reported by Capar *et al.* (1982). Mrowetz (1981) determined lead, cadmium, and copper by anodic stripping after dissolution of an oxide ash in sulfuric acid and then added HCl to determine tin. Selenocystine, cystine, and cysteine have been determined by Grier and Andrews (1981) by cathode stripping voltammetry.

PROBLEMS

16.1. A method is to be developed for determining traces of a thallium chloride rat poison on grapes. A 5×10^{-3} M solution had a diffusion current of 2.28 μA. If 25 sec were required for 8 drops of Hg to fall and the total weight of the Hg is 0.2453 g, calculate the diffusion constant for this system. *Ans.* 1.85×10^{-4} cm^2/sec.

16.2. Pyrazine (2e$^-$ change) in coffee aroma is to be determined. If the diffusion coefficient is 0.98×10^{-5} cm^2/sec and the same capillary was used as in problem 16.1, calculate the concentration of pyrazine using the Ilkovic equation. *Ans.* 0.356 millimoles/liter.

16.3. Arsenic(III) produces reduction waves when 1 N H$_2$SO$_4$ is used as the supporting electrolyte. A 20-g sample of a banana peel was examined for arsenic and produced a wave of 41.7 μA. Arsenic in concentrations of 150, 250, 350, and 500 μg produced wave heights of 19.3, 32.1, 45.0, and 64.3 μA, respectively, when treated in a manner similar to the sample. Calculate the % As in the sample. *Ans.* 0.0016%.

16.4. Potassium antimonyl tartrate in 0.4 N sodium tartrate has a half wave potential of -0.8 V vs. SCE. Several solutions of this compound were prepared and the following results obtained:

Conc. (mM)	μA
1.0	0.94
3.2	3.10
5.6	5.40
7.4	7.12

A 0.0407-g sample dissolved in 25ml of solution containing Sb had an i_d of 5.72 μA. Calculate the % Sb in the sample. *Ans.* 43.7%.

16.5. Potassium can be determined polarographically by forming potassium tetraphenyl borate(III), which is dissolved in N,N-dimethylformamide. Tetrabutylammonium iodide is used as the supporting electrolyte. The potassium impurities of the sodium salt were investigated by dissolving 0.3 g of the salt in 20 ml of the solvent, adding tetraphenyl borate and the supporting electrolyte, and diluting to 50 ml. A diffusion current of 5.32 μA was obtained. Ten milliliters of a standard solution containing 6.0 × 10^{-4} M K was added and a diffusion current of 6.4 μA was produced. Calculate the % K in the sample. *Ans.* 0.0017%.

BIBLIOGRAPHY

Caper, S. G., Gajan, R. J., Madzsar, E., Albert, R. H., Sanders, M., and Zyren, J. (1982). Determination of lead and cadmium in foods by anodic stripping voltammetry. II. Collaborative study. *J. Assoc. Off. Anal. Chem.* **65**, 978–991.

Fogg, A. G., and Bhanot, D. (1980). Effects of tetraphenylphosphonium chloride on d.c. and differential-pulse polarograms of synthetic food coloring matters. *Analyst* (*London*) **105**, 234–240.

Fogg, A. G., and Whetstone, M. R. (1982). Differential-pulse polarographic determination of Red 10B formed from the permitted food color Red 2B. *Analyst* (*London*) **107**, 455–459.

Grier, R. A., and Andrews, R. W. (1981). Cathodic stripping voltammetry of selenocystine, cystine, and cysteine in dilute aqueous acid. *Anal. Chem. Acta* **124**, 333–339.

Hart, J. P., and Smyth, W. F. (1980). A spectral study of some azo dyes of biological importance. *Spectrochim. Acta* **36A**(3), 279–284. [*Chem. Abstr.* **93**, 9546k.]

Heyrovsky, J. (1923). Electrolysis with a dropping Hg electrode. I. *Phil. Mag.* **45**, 305–315.

Kobori, S., and Kawakami, S. (1982). Polarographic studies on additives in foods. VI. Polarographic reduction wave of tartrazine (in Japanese). *Ulsunomiya Daigaku Kyoibugakubw Kiyo, Dia-Z-bu* **31**, 117–123. [*Chem. Abstr.* **97**, 125771M.]

Lechner, E., and Klostermeyer, H. (1981). Detection of the adulteration of skim-milk powder by whey powder (polarographic methods) (in German). *Mitchwissenschaft* **36**(5), 267–270. [*Chem. Abstr.* **95**, 23032n.]

Lechner, E., Kunder, I., and Klostermeyer, H. (1982). Polarographic determination of diacetyl and acetoin in butter (in German). *Z. Lebensm. Unters. Forsch* **173**(5), 372–375. [*Chem. Abstr.* **96**, 18758.]

Marek, M., Bacilek, J., and Jary, J. (1980). Polarographic determination of the adulteration of honey with inverted sugar. *J. Apic. Res.* **19**(4), 255–260. [*Chem. Abstr.* **94**, 172969e.]

Mrowetz, G. (1981). Inverse-polarographic determination of trace elements (cadium, lead, copper, tin) in evaporated milk (in German). *Milchwissenschaft* **36**(8), 479–481. [*Chem. Abstr.* **95**, 113539.]

Palimsano, F., Visconti, A., Bottalico, A., Lerario, P., and Zambonin, P. G. (1981). Differential-pulse polarography of trichothecene toxins: Detection of deoxynivalenol in corn. *Analyst* (*London*) **106**, 992–998.

Takahashi, K., Ohnishi, K., and Asao, T. (1982). Studies on ascorbic acid oxidase inhibitor. I. On the polarographical analysis of enzymic activity (in Japanese). *Mukogawa Joshi Diagaku Kiyo, Shokumotsu-hen* **27**, 65–70. [*Chem. Abstr.* **96**, 120971.]

Wood, J. S., and Downing, G. V. (1980). Modified pulse polarographic determination of nicarbazin in chicken tissue at the 0.1-ppm level. *J. Agri. Food Chem.* **28**(2), 452–454.

17
Mass Spectroscopy

Mass spectrometry has long been useful in the identification and determination of the composition of mixtures, particularly those organic in nature. Techniques are now available to use a mass spectrometer directly as a detector for the components separated by a gas chromatograph. Thus, an understanding of the basic principles of mass spectroscopy is essential for all analysts.

The basic idea of mass spectroscopy is to produce ions (only positive ions will be discussed here) by bombarding organic molecules with high-energy electrons, then accelerating these ions in a definite direction so that they can be separated according to their mass or velocity. The separated ions are then detected and their intensity measured.

COMPONENTS OF A MASS SPECTROMETER

There are several types of mass spectrometers available commercially. The three most common types are the electromagnetic (Fig. 17.1), the time of flight (Fig. 17.2), and the quadrupole (Fig. 17.3). Each spectrometer has three major components: an ion source, analyzer, and detector.

Ion Source

Consider the molecule $CH_3CH_2CH_2SH$ (*n*-propyl mercaptan). If a high-energy electron strikes this molecule and transfers its energy to it, the

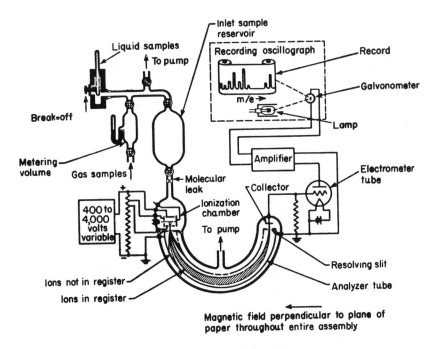

Fig. 17.1.   Electromagnetic mass spectrometer. (Courtesy Consolidated Electrodynamics Co.)

following ions may be produced (m/e = mass/charge):

$C_3H_8S + e^-$

$C_3H_5^+$ (m/e = 41) 4.7%

CHS^+ (m/e = 45) 11.2%

CH_2S^+ (m/e = 46) 42.8%

CH_3S^+ (m/e = 47) 3.4%

$C_3H_5S^+$ (m/e = 73) 0.9%

$C_3H_6S^+$ (m/e = 74) 21.2% (parent ion molecule)

negative neutral and all other + ions 15.9%

The percent composition obtained depends upon the energy of the impinging electrons. If a very low energy electron strikes a molecule, it may break only one bond. If the energy of the impinging electron is increased, different bonds can be broken. A plot of the fraction of the ions produced vs. the electron energy is called a *clastogram*.

A clastogram of *n*-propyl mercaptan is shown in Fig. 17.4. Notice that the ratio of the ions reaches a fairly constant value on the right of the

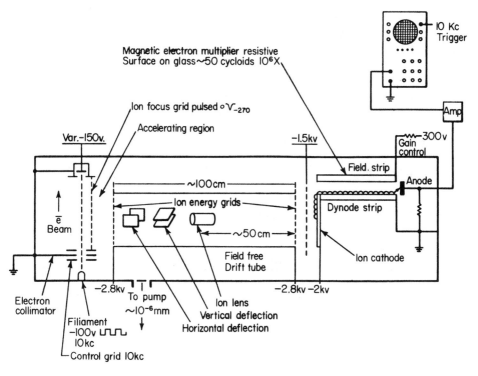

Fig. 17.2. Time of flight mass spectrometer.

diagram. This corresponds to 50–70 V applied to the ion source, and is the useful range for analysis.

Figure 17.5 shows an ion source in more detail. Electrons emitted from the filament are accelerated between E_1 and E_2 by a potential of about 70 V. When an electron strikes a molecule (M), positive ions (+), negative ions (−), electrons (e⁻), and neutral molecules may be formed. The pos-

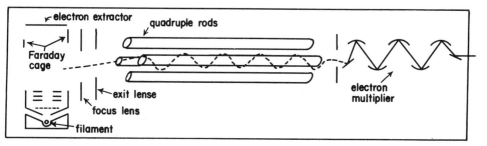

Fig. 17.3. Quadrupole mass spectrometer. (Courtesy Electronics Associates, Inc.)

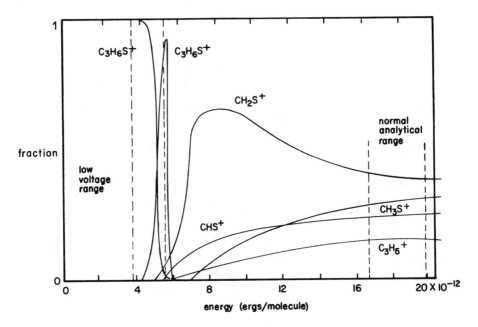

Fig. 17.4. Clastogram of *n*-propyl mercaptan, $CH_3CH_2CH_2SH$.

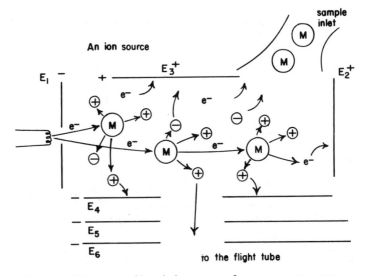

Fig. 17.5. Movement of ions in ion source of a mass spectrometer.

itive species are separated from the negative species by E_3 and E_4, which have a small potential of a few volts across them. The positive ions are then accelerated down the tube E_5, E_6, etc., each having several hundred volts applied to them. A total accelerating voltage of 2000 V is normal.

Analyzers

The ions accelerated in the ion source are then separated. In an electromagnetic spectrometer, the separation of ions of different mass to charge (m/e) ratios is accomplished by a magnetic field at right angles to the flight path of the ions. The lighter ions are deviated more than the heavier ions, so separation occurs. (Note: a vacuum of about 10^{-6} torr is always applied. The reason for this is that once the positive ions are produced, you do not want to lose them prematurely by collision with the other molecules and becoming neutralized.)

After formation of ions in an electron beam, the ions are accelerated through a potential drop V. In so doing, the ions receive kinetic energy T equal to

$$T = mv^2/2 = eV \qquad (17.1)$$

where m is the mass of the ion in grams, v the velocity of the ion in cm/sec, e the charge on the ion (4.8×10^{-10} esu), and V the voltage used to accelerate the ion in erg/esu.

Upon entering a magnetic field H, the ions will take paths that are arcs of a circle with a radius R equal to

$$R = mv/eH \qquad (17.2)$$

where H is in gauss (G) and R in cm. Combining equations 17.1 and 17.2 and rearranging gives

$$m/e = H^2 R^2/2V \qquad (17.3)$$

The angle of deflection (radius) is usually fixed for a given analyzer tube, so that to focus ions of given (m/e) values on the detector system, either H or V must be varied.

Example: Consider an electromagnet-type mass spectrometer having $V = 2000$ volts and $R = 7.00$ in. What must be the magnetic field to focus the CO_2^+ ion ($m/e = 44$) on the detector? Use equation 17.3. Here, R is in cm and V is in erg/esu. Since 300 practical volts = 1 erg/esu and 1 in. = 2.54 cm, we get

$$V = 2000/300 = 6.667 \text{ erg/esu}, \qquad R = 7.00 \times 2.54 = 15.78 \text{ cm}$$

Rearranging equation 17.3 and substituting the known values gives the following:

$$H^2 = 2V(m/e)/R^2$$

$$= \frac{2 \times 6.667 \times (44/6.02 \times 10^{23})}{(15.78)^2 \times 4.80 \times 10^{-10}} \times 9 \times 10^{20}$$

$$= 7.33 \times 10^6$$

$$H = 2708 \text{ gauss}$$

(9×10^{20} is the factor to convert esu^2 to emu^2).

A time of flight instrument does not have magnets, and the flight path of accelerated ions is straight instead of curved. Even without a magnetic analyzer, it is still possible to analyze the various ions on the basis of equation 17.1. Since all ions fall through the same potential drop V, all ions have the same kinetic energy T but different velocities. From equation 17.1 we see that velocity is inversely proportional to the square root of the mass, $v = (2T/m)^{-1/2}$. Since velocity v equals distance d divided by time t, we can derive

$$t = d(m/e2V)^{1/2} \qquad (17.4)$$

where t is the flight time in sec and d the flight length in cm.

Example: A time of flight mass spectrometer has a flight length of 93.0 cm and an accelerating potential of 2530 V. How long would it take for a CH_3OH^+ ion to traverse the spectrometer and how far behind a CH_2OH^+ ion would it be?

CH_2OH^+ (at. wt = 31)

$$t = 93.0 \times (31 \times 300 \times 10^{-16}/2 \times 2.530 \times 4.80 \times 6.02)^{1/2}$$

$$= 7.415 \ \mu\text{sec}$$

CH_3OH^+ (at. wt = 32)

$$t = 93.0 \times (32 \times 300 \times 10^{-16}/2 \times 2.530 \times 4.80 \times 6.02)^{1/2}$$

$$= 7.534 \ \mu\text{sec}$$

$$\Delta t = 0.12 \ \mu\text{sec}$$

The electric quadrupole mass spectrometer is basically different from the previous instruments. A schematic diagram of the positioning of the four long, parallel electrodes is shown in Fig. 17.6. Although the electrodes are shown to have a uniform hyperbolic cross section, cylindrical rods can be used if they are carefully spaced. The opposite pairs of electrodes are connected electrically.

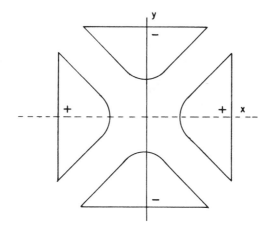

Fig. 17.6. Cross-sectional view of electrodes in a quadrupole mass spectrometer.

Both a dc voltage U and an rf voltage, $V_0 \cos \omega t$, are applied to the quadrupole array. Ions are injected from a conventional ion source into the rf field in the z direction. The ions travel with a constant velocity. Figure 17.7 shows what happens to ions that are too heavy or too light for the particular field strength.

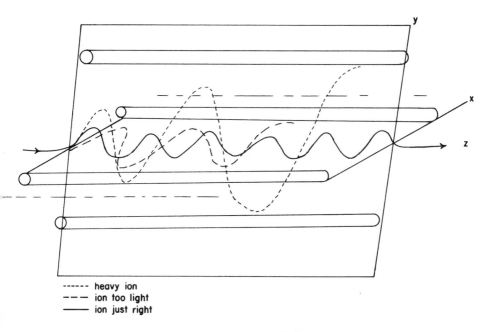

Fig. 17.7. Quadrupole oscillations.

Detectors

The detector in a mass spectrometer is an electrode onto which the positive ions fall. Connected in series between this electrode and ground is a resistor. Electrons from ground rush to neutralize the positive charge on the collector. Current across the resistor causes a potential drop that is proportional to the current flow (Ohm's law), so that by measuring the voltage the number of ions of each type can be determined.

As early as 1943, A. A. Cohen used an electron multiplier in a mass spectrometer (Fig. 17.8). The principle is the same as was described for photo-multiplier tubes in Chapter 6. The detector has a high sensitivity and a rapid response. Multiplication factors of 10^5 to 10^6 are usual.

The Wiley magnetic electron multiplier employs crossed magnetic and electric fields to control the electron trajectories (Wiley 1956). Figure 17.9 is a diagram of this detector. As the individual groups of ions arrive at the end of the field free flight tube (time of flight mass spectrometer), they collide with the plane ion cathode of the magnetic electron multiplier. The plane ion cathode is used because it eliminates ion transit time variations encountered with a curved ion cathode. Each ionization produces a group of electrons, and because of the crossed magnetic and electric fields present, the electrons follow a cycloidal path down the dynode strips of the multiplier. In this manner a current gain of the order of 10^6 is obtained.

LOW-VOLTAGE MASS SPECTROMETRY

At present, the primary use for a mass spectrometer by the food analyst is as a detector for a gas chromatograph. The cracking pattern tells what is present, and the height of the peaks can be used to determine how much is present. In order to simplify the analysis, a low ionization voltage is sometimes used. As the voltage of the electron beam is decreased, the

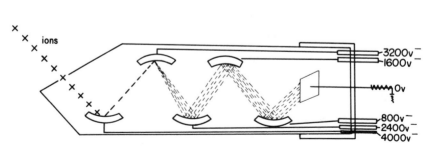

Fig. 17.8. Cohen-type electron multiplier.

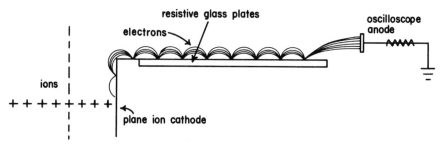

Fig. 17.9. Wiley electron multiplier.

amount of fragmentation is also decreased, particularly below about 25 V. However, there is also an accompanying reduction in sensitivity of ion detection. In practice, an intermediate value of the ionizing voltage is usually selected such that the fragmentation will be held to a minimum commensurate with the best sensitivity possible.

Figure 17.10 gives some indication of how much simpler the spectrum of propylene becomes at a lower voltage.

QUANTITATIVE ANALYSIS

There are times when the single peak coming from the gas chromatograph is really a mixture of two or more components. The height of the peaks in a mass spectrum is directly proportional to the pressure of the sample in the ion source. If the sample is a mixture, then the heights will be proportional to the partial pressures.

There are two methods by which the concentrations of components in a mixture can be calculated: (1) the linear simultaneous equations method

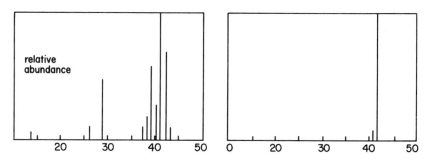

Fig. 17.10. Mass spectra of propylene with a source voltage of 70 V (left) and 12 V (right).

and (2) the technique of subtractions. The subtraction technique, which is the simpler of the two, is discussed in this section.

In mass spectroscopy, the partial pressures of the components of a mixture are additive. The basic idea of the subtraction technique is to take the unknown spectrum and then determine the spectrum for each compound known or suspected to be in the mixture. Since all of these spectra may have been determined at different pressures, it is necessary to convert them all to one common pressure, which is the pressure of the unknown. Each known spectrum is then subtracted from the unknown spectrum, in turn, until the unknown spectrum is reduced to zero.

Example: The data in Table 17.1 will be used to explain the subtraction technique. Notice that the peak with the greatest intensity is arbitrarily set at 100. All other peaks are proportional to the reference peak and the numbers in the tables show these ratios.

Let 1 be ethyl alcohol; 2, methyl alcohol; 3, methyl aldehyde; and 4, ethyl aldehyde. From Dalton's law, $P_1 + P_2 + P_3 + P_4 = P_t$, and we know that the total pressure P_t is related to the number of moles of gas. We may write this in terms of mole fraction of the components: $N_1 + N_2 + N_3 + N_4 = N_t = 1$. Let a be a factor that correlates P_t and N_t, i.e., $P_t a = N_t = 1$. Thus $a = 1/P_t$ and $P_1 a = N_1$; $P_2 a = N_2$; etc.

Let us first select the largest peak in the mass spectrum of the unknown that can arise from only *one* component of the mixture. In our example mixture, $m/e = 46$ can arise only from ethyl alcohol; $m/e = 47$ is too small to be reliable.

We see from Table 17.2 that pure ethyl alcohol has an intensity of 16.12 at $m/e = 46$, whereas the unknown has a value of 3.96 at the same m/e. (Note: it is not necessary for you to determine the spectra of each compound you suspect in the sample. Tables of spectra are commercially available and they can be used in exactly the same way as shown in this

Table 17.1. Mass Spectrum of Mixture of Unknown Composition

m/e	Ion abundance	m/e	Ion abundance
24	0.71	33	0.14
25	2.39	40	0.43
26	5.61	41	1.81
27	7.51	42	4.36
28	17.7	43	12.5
29	100.0	44	18.3
30	43.4	45	9.13
31	37.9	46	3.96
32	8.51	47	0.11

Table 17.2. Relative Ion Abundance in Mass Spectra of Known Compounds

m/e	Ethyl alcohol	Methyl aldehyde	Methyl alcohol	Ethyl aldehyde
24	0.41			1.55
25	2.05			4.77
26	8.31			9.08
27	23.2		0.15	4.50
28	5.61	30.9	6.35	2.67
29	23.9	100.0	64.7	100.0
30	6.05	88.5	0.80	1.14
31	100.00	1.91	100.0	
32	1.27		66.7	
33	0.07		1.02	
40	0.23			0.95
41	1.14			3.88
42	3.03			9.16
43	7.98			26.7
44	1.76			45.6
45	5.1			1.24
46	16.12			
47	0.47			

example.) This means that the two are at different pressures and we must express them on a common basis. This can be done by multiplying 16.12 by a factor to convert it to 3.96. Thus, $16.12 \, P_1 = 3.96$ and $P_1 = 0.2455$ in this case. Now, multiply all the peaks in the ethyl alcohol spectrum by 0.2455 to reduce it to the same pressure as it would have in the unknown. The results of this calculation are shown in Table 17.3.

This new reduced spectrum is now subtracted from the original unknown spectrum to give a new unknown spectrum without the ethyl alcohol component (Table 17.4). Notice that the $m/e = 46$ and 47 peaks have been reduced to zero.

Table 17.3. Ethyl Alcohol Spectrum × 0.2455

m/e	Ion abundance	m/e	Ion abundance
24	0.10	33	0.02
25	0.50	40	0.055
26	2.04	41	0.28
27	5.70	42	0.75
28	1.38	43	1.96
29	5.87	44	0.43
30	1.48	45	8.62
31	24.6	46	3.96
32	0.31	47	0.11

Table 17.4. Spectrum of Unknown Mixture Minus
the Ethyl Alcohol Spectrum

m/e	Ion abundance	m/e	Ion abundance
24	0.61	33	0.12
25	1.89	40	0.37
26	3.57	41	1.53
27	1.81	42	3.62
28	16.3	43	10.5
29	94.1	44	17.9
30	41.9	45	0.51
31	13.3	46	0.00 out
32	8.20	47	0.00 out

Now let us repeat this process, this time eliminating ethyl aldehyde, since it is the only remaining component that has any peaks above m/e = 40. The peak at m/e = 44 is the most reliable because it is the largest and no other compounds have a similar m/e. The factor to correct for differences in pressure is 45.6 P_4 = 17.9 (use the value in the spectrum in Table 17.4), and P_4 = 0.393. The new ethyl aldehyde spectrum is shown in Table 17.5. This spectrum is subtracted from that in Table 17.4 to give a new unknown spectrum, minus both the ethyl alcohol and the ethyl aldehyde, as shown in Table 17.6.

There are now only six peaks left, and they must be due to methyl alcohol and methyl aldehyde. Since methyl aldehyde does not have an m/e = 32 or 33 peak, these peaks are eliminated using the m/e = 32 peak for methyl alcohol. The factor is 66.7 P_2 = 8.2, and P_2 = 0.1230. The new methyl alcohol spectrum is Table 17.7. The new unknown spectrum minus ethyl alcohol, methyl alcohol, and ethyl aldehyde is Table 17.8.

Finally, 100 P_3 = 46.8, so P_3 = 0.468. The new methyl aldehyde spectrum is Table 17.9.

The final unknown spectrum is Table 17.10. Since all of the m/e values

Table 17.5. Ethyl Aldehyde Spectrum × 0.393

m/e	Ion abundance	m/e	Ion abundance
24	0.61	40	0.37
25	1.88	41	1.53
26	3.57	42	3.60
27	1.77	43	10.5
28	1.05	44	17.9
29	39.3	45	0.49
30	0.45		

Table 17.6.   Original Spectrum Minus Ethyl
Alcohol and Ethyl Aldehyde Spectra

m/e	Ion abundance	m/e	Ion abundance
24	0.00 out	33	0.12
25	0.02 out	40	0.00 out
26	0.00 out	41	0.01 out
27	0.04 out	42	0.02 out
28	15.3	43	0.00 out
29	54.8	44	0.00 out
30	41.5	45	0.02 out
31	13.3		
32	8.20		

Table 17.7.   Methyl Alcohol Spectrum × 0.123

m/e	Ion abundance	m/e	Ion abundance
27	0.02	31	12.3
28	0.78	32	8.20
29	7.96	33	0.12
30	0.10		

Table 17.8.   Original Spectrum Minus Spectra of
Three Components

m/e	Ion abundance	m/e	Ion abundance
28	14.5	31	1.0
29	46.8	32	0.0 out
30	41.4	33	0.0 out

Table 17.9.   Methyl Aldehyde Spectrum × 0.468

m/e	Ion abundance	m/e	Ion abundance
28	14.5	30	41.4
29	46.8	31	0.9

Table 17.10.   Final Spectrum of Unknown Mixture

m/e	Ion abundance	m/e	Ion abundance
28	0.0 out	30	0.0 out
29	0.0 out	31	0.1 out

have been reduced to zero or <0.1, this means that all of the sample has been accounted for and there is no other component left. Now $P_1 + P_2 + P_3 + P_4 = P_t$:

$$P_1 = 0.2455, \quad P_2 = 0.393, \quad P_3 = 0.123, \quad P_4 = 0.468$$

$$P_t = 1.2295$$

Therefore, $a = 1/P_t = 1/1.230 = 0.813$:

$N_1 = 0.2455 \times 0.813 = 0.200 = 20.0$ mole % ethyl alcohol

$N_2 = 0.123 \times 0.813 = 0.100 = 10.0$ mole % methyl alcohol

$N_3 = 0.468 \times 0.813 = 0.381 = 38.1$ mole % methyl aldehyde

$N_4 = 0.393 \times 0.813 = 0.319 = 31.9$ mole % ethyl aldehyde

COMBINING MASS SPECTROMETRY WITH GAS CHROMATOGRAPHY

A gas chromatograph can separate complex mixtures quantitatively into individual unidentified components, and the mass spectrometer is excellent for the identification of those compounds. However, combining these two instruments poses some problems. The components from a chromatograph are at atmospheric pressure and diluted with a large amount of carrier gas flowing at 50–100 ml/min. The mass spectrometer operates at 10^{-5} to 10^{-6} torr, and requires a relatively small sample.

Early attempts consisted of placing a cold trap at the chromatograph outlet and transferring the trapped components to the mass spectrometer. This procedure is quite slow, and the entire process must be repeated for each component.

What is needed is a means of removing the carrier gas, thereby concentrating the sample, reducing the pressure 100,000,000-fold in less than a minute, and scanning each component as it emerges from the chromatograph. Although rapid-scan spectrometers are available, interfacing them with a gas chromatograph is a problem. Several ways have been proposed, but the two most commonly used are the fritted glass tube (Watson and Bieman 1964, 1965), and the jet separator.

Fritted Glass

One type of interface consists of an ultrafine fritted glass tube surrounded by a metal chamber vacuum jacket (Fig. 17.11). A vacuum pump

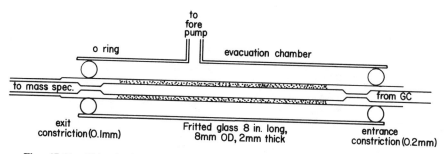

Fig. 17.11. Fritted glass interface for connecting a mass spectrometer to a gas chromatograph.

is attached to the metal chamber and the entire system is heated to 275–300°C. When the sample and the carrier gas emerge from the chromatograph, the low-molecular-weight helium atoms effuse through the fritted glass much faster than the higher-molecular-weight sample. The separated helium is pumped out of the system, and the sample plus some residual helium go to the mass spectrometer. The entrance constriction is required to maintain effusion. The mean free path of the gas must be 10 times the diameter of the pores through which it passes. Since the pores in the fritted tube are approximately 1 μm wide, the pressure inside the tube should be a few torr. The entrance constriction maintains this pressure drop. The exit constriction controls the sensitivity of the apparatus.

Jet Separator

Figure 17.12 is a diagram to illustrate the principle of the jet separator, and Figure 17.13 is a cross-sectional diagram of a commercial jet separator

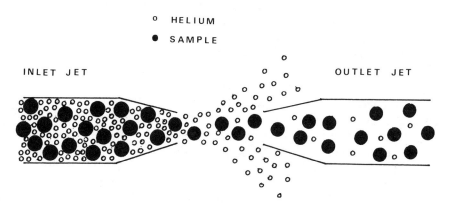

Fig. 17.12. Diagram of how a jet separator operates.

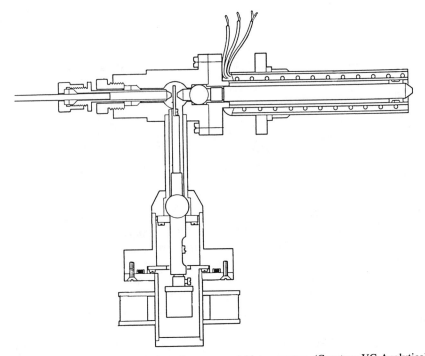

Fig. 17.13. Cross-sectional view of a commercial jet separator. (Courtesy VG Analytical Co.)

made in England, the VG 7070E high-resolution mass spectrometer. A jet separator operates as follows: The effluent from the gas chromatograph column passes into the inlet jet, which has a small diameter (0.1-mm) hole in the nozzel. The low-molecular-weight carrier gas (helium) tends to be ejected from the nozzle in the form of a cone with a large included angle, whereas the heavier sample vapor emerges in almost a straight line. Most of the sample, therefore, passes into the somewhat larger nozzle (0.25-mm diameter) in the outlet jet, which is positioned about 0.3 mm from the inlet jet. The lighter carrier gas is pumped away by the inlet system rotary pump connected to the jet body.

To reduce the possibility of sample decomposition in the separator, both the inlet and outlet jets are made of glass, although the body of the separator is made of stainless steel. Consequently, when the separator is connected to the gas chromatograph column with glass-lined stainless steel tubing, the sample is in contact only with glass, except possibly for a short length of metal tube fitted to some types of column.

The separator is fitted with a cump valve, the blade of which passes between the two jets, sealing the oriface in the outlet jet. The inlet jet is not sealed, so that flow from the column can continue unimpeded when this valve is closed, although no sample will enter the mass spectrometer. This valve can be used to prevent large solvent peaks entering the source or when conditioning new columns. The valve can also be used to stop the flow into the source to reduce the source pressure and the intensity of background bleed peaks when other inlet systems are in use.

RECENT APPLICATIONS

Brumley *et al.* (1981 A,B) confirmed aflatoxins B_1 and M_1 by negative ion chemical ionization mass spectrometry. (CI/MS), and monitored 12 ions and calculated ratios in a low-resolution GC/MS method for 2,3,7,8-tetrachlorodibenzodioxin in fish. Beck and Holmstedt (1981) derivatized 1,2,3,4-tetrahydro-1-methyl-β-carboline with pentafluoropropionic anhydride after extraction and measured it by monitoring single ions using a capillary CI/GC/MS. The mass spectrometric determination of triglycerides of selected fats by the direct chemical ionization method is described by Sebedio and Ackman (1981).

Chemical ionization MS has been used to identify positions of double bonds in polyunsaturated fatty acids by Susuki *et al.* (1981) and for the detection of hydroxy fatty acids in biological samples using both positive and negative CI/MS in combination with GC by Stan and Scheutwinkel-Reich (1980).

PROBLEMS

17.1.   Assume that in a time of flight mass spectrometer, with a flight path of 0.850 m, the accelerating voltage is 2800 V. What is the flight time in microseconds for the m/e = 84 and 86 of chloroform?   *Ans.* 10.60 and 10.33 μsec.

17.2.   In an observation of the *n*-butane spectrum with a conventional electromagnetic spectrometer, it was observed that the m/e = 29 peak was focused on the collector when an accelerating voltage of 2850 V and a magnetic field of 2430 gauss were employed. (a) If the magnetic field is held at 2430 gauss, what must be the accelerating voltage to focus the m/e = 27 ion on the collector? (b) If the accelerating voltage is held at 2950 V, what must be the magnetic field to focus m/e = 25 on the collector?   *Ans.* 3065 V; 2095 gauss.

17.3.   You are given the following mass spectrometric information for 1-butanol (A), 2-butanone (B), cyclobutante (C), 3-methyl-1-butyne, and an unknown mixture of the four compounds:

m/e	A	B	C	D	Unknown
26	7.9	5.3	22.6	5.8	18.0
27	55.7	16.2	42.0	36.8	90.0
28	17.7	3.0	100.0	2.5	35.5
31	100.0	0.6	—	1.2	48.0
39	18.9	2.4	19.8	29.3	42.0
41	62.8	1.7	89.3	17.1	52.5
43	59.6	100.0	0.1	0.8	97.0
53	1.2	0.6	4.7	100.0	100.0
56	86.2	0.2	62.2	...	54.5
57	6.6	6.1	2.7	...	7.9
67	...	...	...	61.6	57.4
72	0.16	16.5	...	...	12.2

Determine the mole fraction composition of the unknown using the subtraction technique. *Ans.* D = 40, B = 30, A = 20, and C = 10.

17.4. You are given the following mass spectrometric data for 1-propanol (A), 2-propanol (B), propanal (C), ethane thiol (D), and an unknown mixture of the four compounds:

m/e	A	B	C	D	Unknown
28	6.1	1.8	68.9	41.5	36.1
29	15.9	11.3	100.0	93.5	75.1
31	100.0	6.4	3.2	0.1	100.0
34	...	...	...	24.1	9.3
45	4.2	100.0	0.1	22.4	51.4
47	...	0.2	...	78.6	30.5
58	0.5	0.2	38.9	10.8	12.2
59	9.2	3.5	1.4	7.7	13.5
60	6.4	0.4	...	1.5	7.0
62	...	...	...	100.0	39.7

Determine the mole fraction composition of the unknown using the subtraction method. *Ans.* A = 50, B = 20, C = 10, and D = 20.

BIBLIOGRAPHY

Beck, O., and Holmstedt, B. (1981). Analysis of 1,2,3,4-tetrahydro-1-methyl-β-carboline in alcoholic beverages. *Food Cosmet. Toxicol.* **19**(2), 173–177. [*Anal. Abstr.* 1982, **43**, 4F53.]

Brumley, W. C., Nesheim, S., Trucksess, M. W., Trucksess, E. W., Dreifuss, P. A., Roach, J. A. G., Andrzejewski, D., Eppley, R. M., Pohland, A. E., Thorpe, C. W., and Sphon, J. A. (1981 A). Negative ion chemical ionization mass spectrometry of aflatoxins and related mycotoxins. *Anal. Chem.* **53**, 2003–2006.

Brumley, W. C., Roach, J. A. G., Sphon, J. A., Dreifuss, P. A., Andrejewski, D., Neiman,

R. A., and Firestone, D. (1981 B). Low resolution multiple ion detection gas chromato-graphic-mass spectrometric comparison of six extraction–cleanup methods for determining 2,3,7,8-tetrachlorodibenzo-*p*-dioxin in fish. *J. Agri. Food. Chem.* **29**, 1040–1046.

Cohen, A. A. (1943). The isotopes of cerium and rhodium. *Phys. Rev.* **63**, 219–223.

Schulte, E., Hoehn, M., and Rapp, U. (1981). Mass spectrometric determination of triglyceride patterns of fats by the direct chemical ionization technique (DCI). *Fresenius' Z. Anal. Chem.* **307**, 115–119. [*Anal. Abstr.* 1981, **41**, 6F75.]

Sebedio, J. L., and Ackman, R. G. (1981). Chromarods-S modified with silver nitrate for the quantitation of isomeric unsaturated fatty acids. *J. Chromatogr. Sci.* **19**, 552–557.

Stan, H. J., and Scheutwinkel-Reich, M. (1980). Detection of hydroxy fatty acids in biological samples using capillary gas chromatography in combination with positive and negative chemical ionization mass spectrometry. *Lipids* **15**(12), 1044–1050.

Suzuki, M., Ariga, T., and Sekine, M. (1981). Identification of double bond position in polyunsaturated fatty acids by chemical ionization mass spectrometry. *Anal. Chem.* **53**, 985–988.

Watson, J. T., and Bieman, K. (1964). High resolution mass spectra of compounds emerging from a gas chromatograph. *Anal. Chem.* **36**, 1135–1137.

Watson, J. T., and Bieman, K. (1965). Direct reading of high resolution mass spectra of gas chromatographic effluents. *Anal. Chem.* **37**, 844–851.

Wiley, W. C. (1956). Bendix time-of-flight mass spectrometer. *Science* **124**, 817–820.

18
Nuclear Magnetic Resonance

Nuclear magnetic resonance (NMR) was first observed in 1945, although the concepts of nuclear spin and magnetic moments upon which the technique is based go back to about 1925.

The NMR spectrometer is composed of the following basic units: magnet, radio-frequency (rf) oscillator, and rf detector. A simple schematic diagram of the equipment is shown in Fig. 18.1.

Three fields, all at right angles to each other, are required for NMR. The first, a stationary magnetic field (X axis), is required to produce a strong external field that will cause the axis of the spinning nuclei to tilt as they spin. This can be visualized by considering a spinning nucleus as a toy gyroscope. A gyroscope will spin about its axis of rotation, but when placed in an external field such as the earth's magnetic field, it will tilt and start a second rotation. This second rotation is called *precession*. In NMR, the magnet employed depends upon the type of analysis desired, but is usually of the order of 10,000 gauss, and must provide a very stable, homogeneous field.

The second field (Z axis), is an alternating radio-frequency field (60 MHz for Hydrogen). If the second field oscillates at the same frequency as the precessing nuclei, then energy can be transferred from the oscillator to the nuclei, and it does work by changing the angle of precession. If the nuclei get out of phase with the oscillator, they can lose their energy and return to a lower angle of precession (see Fig. 18.2). This causes the nuclei to "wobble." It must be understood that any one nucleus does not wobble continuously, but that some nuclei are being raised to high preces-

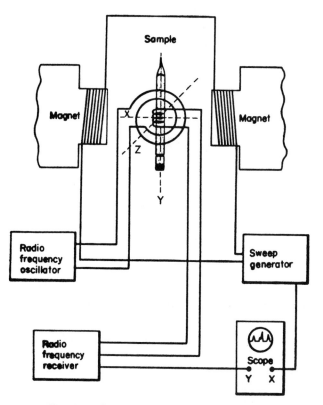

Fig. 18.1. Schematic of an NMR spectrometer.

sion levels while other nuclei are dropping back to lower precession levels, the net effect being that about 7 out of 1,000,000 nuclei "wobble."

The third field (Y axis) is used to detect this "wobble." It consists of a small coil of wire that has a voltage induced in it when the "wobbling" nuclei cut its lines of force. The small voltage generated is amplified, and the signal is displayed by a recorder.

The rf oscillator frequency can be varied to match the existing frequency of the precessing nuclei or the rf frequency can be fixed and the magnetic field varied thereby changing the precessional frequency to match the rf oscillator. In practice it is easier to change the magnetic field strength. The field may be swept at different rates, but usually the rate is about 5–10 milligauss/min. The spectrum obtained is a plot of signal strength on the Y axis against the magnetic field on the X axis.

The sample volume used may be varied but is commonly 0.01–0.5 ml. This small sample volume is a great advantage, as is the fact that NMR

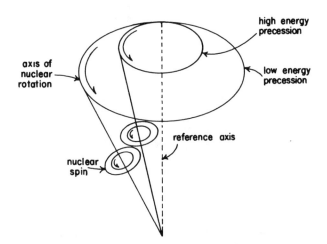

Fig. 18.2. Precession–energy relationship.

is totally nondestructive. For solids, about 50 mg is usually dissolved in about 0.3 ml of a solvent. Carbon tetrachloride or carbon disulfide is tried first because they have no interfering hydrogens. Deuterated solvents are then used, with water the last choice. The samples are placed in small-diameter tubes, generally about 5 mm od.

It is also usual to add directly to the sample, or place in a concentric sample tube, a known material to serve as the reference for the analysis. Tetramethyl silane is commonly used.

An additional technique, which is helpful in obtaining well-resolved spectra, is to spin the sample while it is in the magnetic field. This averages out the field inhomogeneities perpendicular to the direction of spinning.

INTERPRETATION OF NMR SPECTRA

Since the nuclei of atoms in different compounds are usually in different magnetic environments, they require different applied magnetic fields to reach resonance with an oscillator at a fixed frequency. Thus NMR spectra are like fingerprints of molecules.

To illustrate some of the information available from a proton spectrum, let us examine the ethyl acetate spectrum shown in Fig. 18.3. The spectrum has three groups of peaks with the following areas in arbitrary units: 93:140:145 for the quadruplet:singlet:triplet. Since the structural formula

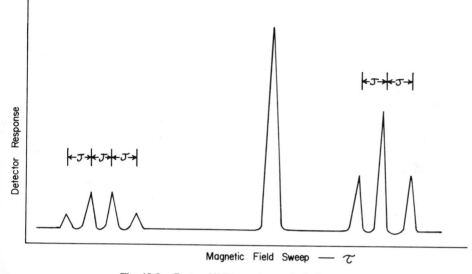

Fig. 18.3. Proton NMR spectrum of ethyl acetate.

of ethyl acetate is

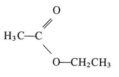

we would expect three groups of peaks, due to the three nonequivalent groups of protons: the CH_2 in the ethyl group, the CH_3 in the ethyl group, and the CH_3 in the acetate. The areas, proportional to the number of protons, would be in the ratio of 2:3:3. If we let the average of 140 and 145, 142.5, be equivalent to 3, we find our experimentally measured areas to be in the ratio of 1.96:2.95:3.05, which is very nearly 2:3:3, as anticipated. To identify the given lines within the groups, we see the singlet and the triplet must be the CH_3 groups, and that the quadruplet must be the CH_2 group using the above ratios.

Looking now at the spin–spin interactions of the protons in the CH_3 of the ethyl group of the ester with the neighboring CH_2 group we see that

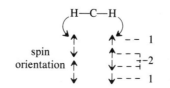

(where ↑ and ↓ are +1/2 and −1/2 spins, respectively) and thus we could expect a triplet, that is, three lines due to ↑↑ ↓↓ and ↓↑ coupling. The last of these occurs twice and therefore is expected to be twice as intense (i.e., an intensity ratio of 1:2:1). Thus we readily identify the triplet as the CH_3 from the ethyl group.

The nearby singlet is now expected also to be a CH_3 group. But there is only one line. This is because there are no nearby neighboring protons in the molecule to allow spin interactions with the protons of this CH_3 group; therefore, both from the intensity and spin considerations, we conclude that the singlet is the CH_3 group from the acetate portion of the molecule.

We are left with the fact that the quadruplet must be due to the CH_2 group. Again looking at the spin-spin interactions with the protons of the neighboring CH_3 group, we see that

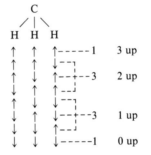

and therefore a quadruplet is confirmed, even to the prediction that the intensity ratios in the quadruplet be 1:3:3:1.

The following general rules about spin-spin interactions are helpful in calculating the number of lines expected and their approximate intensities: (1) Nuclei of the same group do not interact to cause observable splitting. (2) The multiplicity of the band from a group of equivalent nuclei is determined by the neighboring groups of equivalent nuclei. The neighboring group will cause a multiplicity of $s = (2nI + 1)$ in that group, where n is the number of equivalent nuclei of spin I. (3) The intensities are symmetric about the midpoint of the group and for the case of nuclei of $I = \frac{1}{2}$, their intensities are given by the coefficients in the expansion of $(r + 1)^{s-1}$.

Coupling Constants and Chemical Shifts

We have discussed that the NMR spectrum of ethyl acetate can be considered as three groups of peaks, one a singlet, another a triplet, and the third a quadruplet. It is of interest now to observe that the spin-spin splitting of a peak such as the triplet or quadruplet is a fixed amount. In Fig. 18.3, we see that these are in fact equal in both the CH_2 and CH_3 groups of the ethyl portion of ethyl acetate. This splitting is designated by J, the *coupling constant*, and is independent of the magnetic field. Some representative values of J are given in Table 18.1. The coupling constant in ethyl acetate is about 7 cps, and falls within the range of values given for CH_3CH_2X in Table 18.1. Thus, one quickly sees how values of coupling constants may aid in stucture determinations.

Another significant aid in structure determination is the *chemical shift*, δ, which is a function of electron density about the nucleus. The chemical shift is related to the distance from the center of a group of lines to a reference line. Some common reference substances, which can be used also as solvents and which produce only a single reference line, are water, benzene, and cyclohexane. The separation of the sample and reference lines is ΔH, and thus the separation of the lines is field-dependent.

Three ways are commonly employed to determine values of δ. (1) Use simply a mixture of the sample and the reference. Usually this is not very

Table 18.1. Coupling Constants

Type of compound	J (in cps)
CH_3X	12
CH_3CH_2X	5–8
H⧵C=C⧸H	17.5–18.5
⧸C=C⧵ with H and H	8.5–10.5
C=C with H and H	1.5–2.0
$HC\equiv CH$	9
(benzene ring) H, H	5–8

satisfactory, since the value may be concentration dependent. (2) Determine δ as a function of the concentration in the reference, and extrapolate to infinite dilution. (3) Place the sample into one tube and then place the reference into another tube which is then inserted concentrically into the sample tube.

Approximate values of δ are given in Table 18.2 for a number of different types of compounds.

For NMR proton work, δ has both $+$ and $-$ values. The use of δ values is somewhat inconvenient. The use of τ values,

$$\tau = 10.00 = (H_{SiMe_4} - H_S) \times 10^6/H_{SiMe_4} \qquad (18.1)$$

where $(H_{SiMe_4} - H_S)$ is the shift using tetramethylsilane as the reference and is expressed in ppm, allows us to express chemical shifts as positive values for all but the most acidic of protons. Increasing values of τ signify greater shielding of the proton. δ values relative to water, such as those

Table 18.2. Chemical Shift Values

Group	δ	Group	δ
—SO₃H	-6.5	Ar 　＼ 　　N—H 　／ Ar	$+2.0$
—COOH	-6.5	Ar—SH	$+2.0$
$\begin{array}{c}\quad O\\ \quad \parallel\\ —C—C\\ \quad\ \ \diagdown\\ \qquad\ \ H\end{array}$	-4.7	$\equiv$C—H	$+2.3$
—CHO	-3.3	CH₃—N	$+2.5$
$\begin{array}{c}\ \ O\\ \ \ \parallel\\ —C\\ \ \diagdown\\ \quad NH_2\end{array}$	-2.9	$\begin{array}{c}\quad\mid\\ =C—CH_3\end{array}$	$+3.3$
Ar—OH	-2.4	Ar—CH₃	$+3.4$
Ar—H	-2.0	$\equiv$C—CH₃	$+3.4$
$=$CH₂	-0.5	—CH₂—(cyclic)	$+3.5$
C$=$CH—C	-0.5	R—SH	$+3.5$
R—OH	-0.1	R—NH₂	$+3.7$
HOH	0.0	$\begin{array}{c}—C—CH_3\\ R\end{array}$	$+4.0$
AR—NH₂	$+1.5$	R 　＼ 　　N—H 　／ R	$+4.4$
—O—CH₃	$+1.5$		
C—CH₃—X	$+2.0$		

given in Table 18.2, can be readily converted to τ values by adding a constant of about 5.2 ppm.

WIDE LINE NMR

The spectra considered in the previous section were obtained from samples in solution. In such spectra, line widths are a few milligauss in width because the molecules are free to tumble in all directions and the fields produced by hydrogen nuclei, for example, tend to average out very rapidly. As a result, the line width is determined by the inhomogeneity of the applied magnetic field.

However, in a solid material, the nuclei are more or less fixed in position with respect to their neighbors, and because of this fixed spatial orientation, any given nucleus may be in a completely different magnetic field than that of its nearest neighbor. This means that different resonance frequencies are needed for each nucleus and the absorption band, that is the line width, will be quite large, in some cases several hundred gauss.

Figure 18.4 shows hydrogen in a solid (starch) and in the water surrounding it. This difference in band widths can be used to determine moisture in the presence of solids containing hydrogen. This analysis can be done in about 30 sec with an absolute error of 0.1%.

SUPERCONDUCTING MAGNETS

It was discovered many years ago that certain metals and alloys would lose practically all of their electrical resistance if they were cooled to a

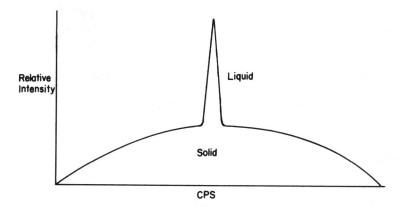

Fig. 18.4.   Solid and liquid proton NMR bands obtained from water in starch.

sufficiently low temperature; they became superconductors. Usually this requires liquid helium or hydrogen as the coolant. A few years ago alloys were found that could produce very high magnetic fields at liquid helium temperatures. The NMR instruments that have resulted from this combine the advantages of both the temporary and permanent magnet instruments previously used.

For example, electromagnet NMR spectrometers could be made to have higher magnetic fields than those with permanent magnets, so larger shifts could be obtained. This permitted several elements to be studied (H, F, P, N) and increased sample sizes, thus increasing the sensitivity. The disadvantages were that the magnets required a large amount of electrical power and as a result became hot. Cooling water had to be forced around the magnets in large volumes (15 liter/min), which made it difficult to stabilize the signal to much better than 10 Hz.

Permanent magnet instruments do not require large amounts of power (only that needed to measure the signal), do not get hot, and do not require cooling. Their stability is typically 1 Hz. However, the total magnetic field strength is not as high and therefore the splittings are not as large as with an electromagnet, usually limiting research to hydrogen. Such instruments weigh only about 0.5 ton compared with 4–5 tons for electromagnet instruments. Although permanent magnet NMR spectrometers are not as sensitive as the electromagnet type, they are sufficiently stable that students can use them and their price is reasonable. Such instruments were commonly purchased by most universities for routine hydrogen work.

The superconducting magnets allow one to produce a very high magnetic field with comparatively low power requirements, and the heating is more readily controllable. The result is a stability of about 1 Hz, similar to that of permanent magnets. The high fields obtainable in a superconducting NMR instrument allow splittings even better than those with a room-temperature electromagnet instrument. A modern superconducting NMR spectrometer can be used to determine almost any element in the periodic table.

Liquid nitrogen ($0.20/liter) is used to keep the liquid helium ($3.50/liter) cool. A double Dewar flask is used. The newer instruments have boil-off rates of helium of only about 0.04 liter/hr.

RECENT APPLICATIONS

A pulsed NMR method for determining lipids in food products in which up to 4% of water does not interfere is described by Lambelet (1980).

Robertson and Windham (1981) proposed a NMR method for determining the oil content of sunflower seed. Tulloch (1982) describes a method using ^{13}C NMR to nondestructively investigate the conjugated acids in seed oil triglycerides.

Low-temperature NMR has been used to determine unfreezable water in protein suspensions (Hays and Fennema 1982), and NMR spectroscopy has also been applied to studying the internal movement and binding of water in raw and roasted coffee and the grounds (Satanilla and Fritsch, 1981). O'Neil *et al.* (1980) measured phytate in cereals by Fourier transform NMR spectroscopy of phosphorus-31.

Proteins in fish or meat meals have been determined by Wright *et al.* (1980) by low-resolution pulsed NMR and by Rutar *et al.* (1980) and O'-Donnell *et al.* (1981) by magic angle sample spinning ^{13}C NMR.

BIBLIOGRAPHY

Hays, D. L., and Fennema, O. (1982). Methodology for determining unfreezable water in protein suspensions by low-temperature NMR. *Arch. Biochem. Biophys. 213*, 1–6. [*Chem. Abstr.* **96**, 33450g.]

Lambelet, P. 1980. Determination of lipids in foods by pulsed nuclear magnetic resonance (NMR). *Mitt. Geb. Lebenmittelunters. Hyg.* **71**, 119–123.

O'Donnell, D. J., Ackerman, J. J., and Maciel, G. E. (1981). Comparative study of whole seed protein and starch content via cross polarization. Magic angle spinning carbon-13 nuclear magnetic resonance spectroscopy. *J. Agric. Food Chem.* **29**, 514–518.

O'Neil, I. K., Sargent, M., and Trimble, M. L. (1980). Determination of phytate in foods by phosphorus-31 Fourier transform nuclear magnetic resonance spectrometry. *Anal. Chem.* **52**, 1288–1291.

Robertson, J. A., and Windham, W. R. (1981). Comparative study of methods of determining oil content of sunflower seeds. *J. Am. Oil Chem. Soc.* **58**, 993–996. [*Anal. Abstr. 1982* **43**, 2f70.]

Rutar, V., Blinc, R., and Ehrenberg, L. (1980). Protein content determination in solid organic materials by proton enhanced magic angle sample spinning C-13 NMR. *J. Magn. Reson.* **40**, 225–227. *Anal. Abstr. 1981* **41**, 4F29.]

Santanilla, J. D., Fritsch, G., and Mueller-Warmuth, W. Z. (1981). NMR studies on the internal movement and binding of water in raw and roasted coffee. *Z. Lebens. Unters. Forsch.* 172–173.

Saraf, D. N., and Fatt, I. (1967). Effect of electrolytes on moisture determination by nuclear magnetic resonance. *Nature* **214**, 1219–1220.

Tiers, G. (1958). Proton nuclear resonance spectroscopy. I. Reliable shielding values by "internal referencing" with tetramethylsilane. *J. Phys. Chem.* **62**, 1151–1152.

Tulloch, A. P. (1982). ^{13}C Nuclear magnetic resonance spectroscopic analysis of seed oils containing conjugated unsaturated acids. *Lipids* **17**, 544–550.

Wright, R. G., Milward, R. C., and Coles, B. A. (1980). Rapid protein analysis by low-resolution pulsed NMR. *Food Technol.* (*Chicago*) **34**, 47–52. [*CA* **94**, 638766.]

19
Radioactivity, Counting Techniques, and Radioimmunoassay

Radioactivity is a general term applied to the emission of high-energy particles and electromagnetic radiation emanating from the unstable and excited nuclei of atoms. Table 19.1 summarizes the characteristics of some of the different radiations.

Alpha particles are helium nuclei moving at high speeds (on the average, 1×10^9 cm/sec) emitted from unstable nuclei having large atomic numbers. All alpha particles from a given isotope have the same energy, nearly identical penetration ranges, which are very short, and produce about 25,000 ion pairs/cm while they last.

Beta particles are distinguishable from simple electrons only by the fact that they originate in the nucleus and are usually moving at high speed. A beta spectrum is continuous, having energies varying from a few thousand electron volts to several million electron volts. An average beta particle produces about 60 ion pairs/cm.

Gamma rays originate in a nucleus that has been left in an excited state because of some previous distintegration or interaction. The fact that gamma rays have discrete energies is evidence that the nucleus exists in various energy levels. Gamma rays have high penetrating power but 1/10 to 1/100 of the ionizing power of beta rays.

Table 19.2 summarizes the common units of radioactivity measurement.

RADIOACTIVE DECAY

Many equations have been developed to describe the various parts of the radioactive process, but the one equation that most people come in

Table 19.1. Radiation Characteristics

Radiation	Type	Charge	Typical energy range	Pathlength		Primary mechanism of energy loss
				Air	Solid	
Alpha	Particles	+2	5–9 Mev	3–5 cm	25–40 μm	Ionization excitation
Beta	Particles	–1	0–4 Mev	0–10 cm	0–1 mm	Ionization excitation
Neutron	Particles	0	0–10 Mev	0–100 m	cm	Elastic collision with nuclei
X ray	Electro-magnetic radiation	0	ev–100 Kev	μm–10 m	cm	Photoelectric effect
Gamma ray	Electro-magnetic radiation	0	10 Kev–30 Mev	cm–100 m	mm–10 cm	Photoelectric effect Compton effect Pair production

Table 19.2. Units of Measurement

Name	Symbol	Magnitude
Curie	Ci	3.7×10^{10} dis/sec
rad	rad	
Roentgen	R	1.6×10^{12} ion-pairs/cm of air
Specific activity		dis/g/sec

contact with is the half-life equation:

$$t_{1/2} = 0.693/\lambda \tag{19.1}$$

where λ is the radioactive decay constant ($\sec^{-1}$). If A is the count rate and A_0 is the number of counts at time zero, then equation 19.1 becomes

$$A = A_0 \exp(-0.693t/t_{1/2}) \tag{19.2}$$

Example: A sample of irradiated meat containing ^{35}S was known to contain 9.50 mCi originally. After 1 yr and 237 days how many disintegrations per minute (dpm, dis/min) occur in the sample?

$$t_{1/2} = 87.1 \text{ days for } ^{35}\text{S}$$

$$
\begin{aligned}
A &= 9.50 \exp(-0.693 \times (365 + 237)/87.1) \\
&= 9.50 \times 0.00835 \text{ mCi} \\
&= 0.0794 \text{ mCi} \\
&= 0.0794 \times 2.2 \times 10^6 \text{ dpm} = 176,000 \text{ dpm}
\end{aligned}
$$

COUNTING DEVICES

Although the radiations mentioned previously have high energies in bulk, individually these energies are not sufficient, by a factor of about one million, to permit direct observation and measurement. Detection and measurements are therefore done indirectly by utilizing the effects or interactions produced by these radiations as they traverse matter.

The four most common detectors in use today and their general characteristics are shown in Table 19.3. In recent years, semiconductor detectors also have been developed.

Geiger–Muller Counter

The Geiger-Muller (GM) tube is very sensitive to alpha and beta particles (98% efficient) compared to gamma rays (2% efficient). The GM

Table 19.3. Characteristics of Radioactivity Counters

Detector	Sensitive medium	Detector multiplication	Output signal (V)	Resolving time (sec)	Efficiency Electron	Efficiency X	Efficiency Gamma
Ionization chamber	Gas	1	10^{-6} to 10^{-3}	10^{-6} to 10^{-3}	Low	Low	Low
Proportional counter	Gas	10^2 to 10^4	10^{-4} to 10	10^{-6}	High	Medium	Low
Geiger counter	Gas	10^7	0.1 to 10	10^{-4} to 10^{-3}	High	Medium	Low
Scintillation counter	Solid liquid	10^6	10^{-2} to 10	10^{-2} to 10^{-6}	Medium	High	Very high

counter is relatively inexpensive and simple to operate, but it does not discriminate between types of radiation and it has a finite lifetime. It is steadily being replaced by proportional and scintillation counters.

Suppose a ray of radiation comes through the mica window of a GM tube and strikes an argon atom. The argon atom is ionized to produce a positive argon ion and an electron. The positive ion moves toward the cathode about 1000 times slower than the electron moves toward the anode. The electron, attached by the high potential of the anode, is rapidly accelerated. In fact, it has sufficient energy so that if it collides with an argon atom another ion and electron can be produced. Now there are two electrons accelerating toward the anode. These can produce 4, 8, 16, etc., electrons. The net result, called the *Townsend avalanche,* is that thousands of electrons reach the anode. When these electrons reach the anode, a small current is produced and the pulse signal is measured.

In addition, some of the electrons striking the anode may have a high enough energy to knock electrons from the anode. These electrons will immediately be reattracted to the anode and they in turn can knock other electrons loose. This is known as the *photon spread.*

The total time it takes for this signal to build up is known as the rise time t_r and is usually 2–5 μsec.

What happens to the positive ions during this time? They are slowly moving toward the cathode as a positive space charge. If they strike the cathode with their full energy, then more photoelectrons will be generated, more than the tube can handle. The net result is that the counter will burn out. What is needed is something to dissipate this energy and that from the photon spread. Molecules, with their many energy levels, are used for this purpose; ethanol and chlorine are favorites. The ionized argon atoms will transfer their energy to these molecules. The molecules may then form ions or free radicals or simply absorb the energy, but the net effect is that the energy is now so spread out that the cathode has little affinity for these particles, and no photoelectrons are produced. Since there is a limit to the amount of quenching gas that can be added to this type of counter, the counter will work only as long as quenching gas is present.

What happens if a second ray of radiation enters the counter before the first ray has completed its reaction? If the second ray ionizes an argon atom at a point between the cathode and the positive space charge, then the electron produced will not see the anode but will recombine with the argon ion instead. The net result is that a ray of radiation entered the counter but was not counted so the counter was dead.

Now consider a ray entering the counter between the positive space charge and the anode. The electron produced will see the anode and be attracted to it. However, it does not have as much room to operate in as the original pulse and may not be detected. Again the counter is dead. Dead times t_d vary but usually are between 80 and 100 μsec. This does not mean that after 100 μsec the counter is completely ready to go again. It means only that the next signal produced can be detected, but its amplitude will be very weak. The time it takes for the counter to completely recover is known as the recovery time t_{rec}, and this varies from 200 to 300 μsec. The counter dead time affects the counting rate and as such introduces an error into the measurement.

Proportional Counters

Geiger–Muller tubes are limited to about 15,000 counts per minute (cpm) because of their long dead time. If the voltage applied to the anode of a GM tube is reduced to a value at which the anode can collect electrons but not form the photon spread, then the output pulse is proportional to the energy of the initial ionization, since the number of secondary electrons now depends only on the number of primary ion pairs produced initially. A device operated in this manner is called a proportional counter.

The dead times of proportional counters are very short, of the order of 1 μsec, and therefore they can count up to 200,000 cpm. The pulse signal is much weaker than that with a GM tube so a much better amplification system is needed. You cannot make a proportional counter out of a GM tube by simply lowering the anode voltage.

Proportional counters can operate at atmospheric pressures, so the quenching gas can be added continuously. Since this can be endless, a proportional counter can count indefinitely. Proportional counters are very good for alpha and beta particles, and because of their ionization efficiency, these counters can easily distinguish alpha particles from beta particles.

Scintillation Counters

The basic principle behind the operation of the scintillation counter is that an energetic particle incident upon a luminescent material (a phosphor) excites the material; the photons created in the process of de-excitation are collected at the photocathode of the photomultiplier tube where the photons cause the ejection of electrons. The electrons ejected from the photocathode are then caused to impinge upon other electrodes, each approximately 100 V higher in potential. In the acceleration from dynode to dynode, more electrons are ejected and a large amplification is obtained.

It would be expected that the greater the energy of the incident particle, the greater the number of electrons that would be produced. This indicates that the scintillation counter could be used to obtain the energy of the particle. In a scintillation counter, advantage is not taken of the proportional properties; however, a scintillation spectrometer uses these proportional characteristics of the scintillation process to good advantage.

The *liquid scintillation counter* is particularly convenient for the counting of very low energy beta emitters, such as tritium, carbon-14, and sulfur-35, although it is not restricted to this use. As the name implies, the scintillator is not a crystal, such as NaI(Tl), nor a plastic, but a liquid solution. The solution is prepared by combining a good solvent, a scintillator, and a primary solute.

The pulses of light emitted by the scintillating solution (caused by the particles in the radioactive decay process) are observed with a photomultiplier tube and counted. It is common practice to cool the photomultipliers to lower their noise, which is caused in part by thermal emission from the photocathode.

Many materials have been tried as liquid scintillators. One of these contains 100 g of naphthalene, 7 g of 2,5-diphenyloxazole (PPO), and 50 mg of 1,4-bis(2,5-phenyloxazoyl)benzene (POPOP) per liter of pure 1,4-dioxane. The advantage of this liquid scintillator is its ability to dissolve many aqueous solutions. Here the PPO is the primary solute and POPOP is the secondary solute. A number of other good scintillator systems are known. In general, the alkylbenzenes are the best solvents.

Semiconductor Detectors

Germanium, a semiconducting material, can be made sensitive to gamma radiation by placing small amounts of an impurity, such as lithium, into the germanium crystal structure. When a gamma ray interacts in the crystal, an electrical charge is produced. The charge is collected and amplified to produce a voltage pulse whose height is proportional to the amount of energy deposited in the crystal by the gamma ray.

A single crystal of semiconductor material, such as silicon or germanium, will not make a suitable counter because of the dc current in the crystal. Random variations of this current may produce pulses similar to the radiation-induced pulses. To reduce this current, a p-n junction in reverse bias is used (Fig. 19.1). In thermal equilibrium, the conduction electrons contributed by the donor atoms (the impurity atoms in an n-type semiconductor) predominate in the n region. Here the conduction electrons neutralize the space charge of the donor atoms. Similarly, the holes contributed by the acceptor atoms (the impurity atoms in the p-type semiconductor) neutralize the space charge of the acceptor atoms in the

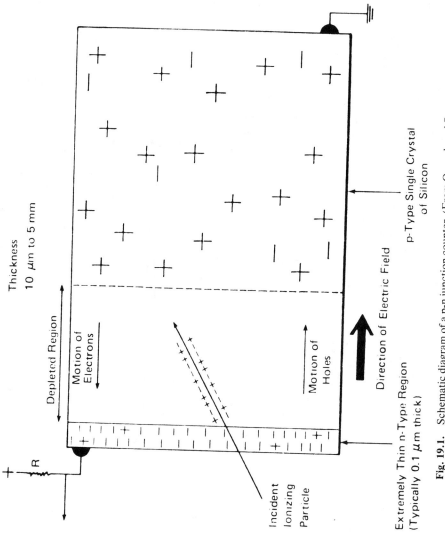

Thickness
10 μm to 5 mm

Depleted Region

Motion of Electrons

Motion of Holes

Direction of Electric Field

p-Type Single Crystal of Silicon

Extremely Thin n-Type Region
(Typically 0.1 μm thick)

Incident Ionizing Particle

R

Fig. 19.1. Schematic diagram of a p-n junction counter. (From Ouseph and Swartz 1974.)

p region. In addition, at the junction between the two regions, diffusion allows electrons to move into the p region and holes into the n region. In this process the electrons leave behind positively charged donor ions and the holes leave negatively charged acceptor ions. These ions are fixed and form a double layer. The electric field of the double layer essentially prevents further diffusion across the junction. In reverse bias, the holes accumulate closer to the negative electrode and the electrons to the positive electrode. The region in the middle is practically free of charge and is called the *depletion region*. Because of the concentration of the charges close to the electrodes, the potential drop is essentially confined to the depletion region. Consequently, if radiation enters the depletion space, electron–hole pairs, which are subjected to practically no potential difference, will rarely be collected at the electrodes. Therefore, the sensitive area of the counter is the depletion region.

Germanium (lithium) semiconducting detectors *must* be stored and op-

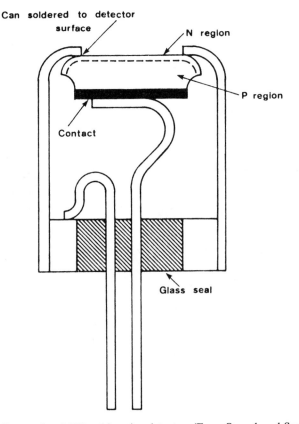

Fig. 19.2. Encapsulated diffused junction detector. (From Ouseph and Swartz 1974.)

erated at low temperatures, usually liquid nitrogen temperature, because of the high mobility of the lithium ions in germanium at room temperature. This cooling requirement decreases the portability of Ge(Li) detectors.

A variation of the detector just described is the *diffused junction detector*. This type of detector is often produced by diffusing a high concentration of donor impurities into a p-type material. Silicon is usually used as the base material. Single crystals of high resistance are sliced into 1-mm pieces, and phosphorous is diffused into one surface of these slices. A common method is to coat one side with phosphorous pentoxide dissolved in glycol and to heat the slice at 800°C in dry nitrogen for 0.5 hr. The phosphorous diffuses into the base material, while the glycol leaves the base material covered with a black residual deposit. After diffusion, proper electric connections are made to the p and n sides. A typical design of a diffused junction detector is shown in Fig. 19.2.

ISOTOPIC DILUTION

The technique of isotopic dilution is a simple and extremely useful method of quantitative analysis. The method involves three steps: (1) adding a small amount of a radiolabeled compound (of known specific activity) that is chemically identical to the unknown component of the unknown mixture being studied, (2) separating a small but pure fraction of the compound, and (3) determining its specific activity. It is then only a matter of calculation to find the amount of the component originally present in the unknown mixture.

Let us look at this technique in a more quantitative manner. Let A be the number of grams of unknown present. There is originally no activity in this material, so the specific activity S_A is zero. Now we shall add A^* grams of the radiolabeled component, chemically identical to that being analyzed. The total activity added is Q^* and the specific activity is $S_{A^*} = Q^*/A^*$. After the addition is complete, and thorough mixing has occurred, the total number of grams present is $(A + A^*)$, and the total activity is still Q^*.

Now let us separate A' grams of the pure component. Its specific activity would be $S_{A'} = Q'/A'$. The fraction separated comprises $A'/(A + A^*)$ of the total, and the activity is Q'/Q^* of the total.

$$A = A^* \left(\frac{S_{A^*}}{S_{A'}} - 1 \right) \tag{19.3}$$

Example: A sample of 15.0 g of liver contained an unknown amount of glycine. Ten milligrams of glycine with a specific activity of 300,000 dps/g was added to the sample and thoroughly mixed. Then 12.5 mg of pure glycine was separated and counted. The count rate was 250 dps. What was the percent by weight of glycine in the original sample?

$$S_{A'} = 250/0.0125 = 20,000 \text{ dps/g}$$

$$S_{A*} = 300,000 \text{ dps/g}$$

$$S_{A*}/S_{A'} = 300,000/20,000 = 15$$

$$A = A*(15 - 1) = 10(14) = 140 \text{ mg}$$

and therefore

$$0.140 \times 100/15.0 = 0.93\%$$

Example: To 200 ml of orange juice containing ethanol was added 5 μl of ethanol with a specific activity of 5×10^5 cpm/ml. The radiotracer was tritium. After thorough mixing, a 1.00-ml sample of ethanol was isolated from the mixture. Fifty microliters of the separated ethanol was found to have a counting rate of 200 cpm. What was the percent by volume of ethanol in the orange juice?

$$S_{A'} = 200/0.050 = 4000 \text{ cpm/ml}$$

$$S_{A*} = 500,000 \text{ cpm/ml}$$

$$S_{A*}/S_{A'} = R = 500,000/4,000 = 125$$

$125 \gg 1$, so that

$$A = RA* = 125 \times 0.005 \text{ ml} = 0.625 \text{ ml}$$

and therefore

$$0.625 \times 100/200 = 0.31\%$$

It is to be emphasized in this discussion that there are three "musts" to be observed when using isotopic dilution: (1) the tracer must be identical to the component of the mixture being investigated, (2) there must be no exchange of the tracer with any component of the mixture, and (3) there must be no side reactions with other components of the mixture.

NEUTRON ACTIVATION ANALYSIS

The previous section dealt with the use of naturally occurring radioactive materials, but what about those systems that do not have available natural radioactive isotopes or to which a natural radioactive isotope cannot be added. In such systems, nuclear transmutations must be produced by the interaction of the nuclei of matter with highly energetic charged particles (protons, deuterons, alpha particles), by neutrons, or by energetic photons. The production of radioactive isotopes generally is dependent upon such artificial nuclear transmutations.

A polonium–beryllium source is commonly used in the laboratory for low-level irradiations. The neutrons are produced by having an alpha particle that comes from polonium bombard a beryllium nucleus to produce a neutron and carbon-12. The reaction results in the production of neutrons of an average energy of about 4 MeV, which is far too great to produce the desired (n, γ) reaction with any reasonable efficiency. It is necessary therefore to slow down the neutrons to thermal energies (ca 0.025 meV) using water as a moderator. Collisions of the energetic neutrons with the hydrogen atoms in the water molecules result in a rapid decrease in the energy of the neutrons. One can consider then that the water moderator becomes filled with a neutron gas, the maximum density of which occurs at the source. The material to be irradiated is suspended in the water as close as is feasible to the neutron source.

Using cadmium in meat as an example, a neutron activation analysis simply involves irradiating the meat to produce radioactive cadmium, which is then counted. One problem is that while the sample is prepared and counted, the radioactive material may decay so rapidly that unless a correction is made a sizable error is present. In addition, one must know how long to irradiate a sample to make it radioactive enough to count? The rest of this section explains how these measurements and corrections are made.

Radioactive decay obeys a statistical law identical with that encountered in monomolecular and first-order chemical kinetics and in optical absorption phenomena. Such a law states that the rate of decay is proportional to the total number of radioactive atoms present.

The half-life of the radioisotope being produced in a nuclear transformation has a definite bearing on the total time over which one can efficiently produce transmutations. The rate of decay depends only upon the number of radioactive atoms present. As one produces transmutations, the rate of decay of the produced isotope will gradually increase. The number of radioactive atoms will stop increasing when their rate of production at any instant is equal to their rate of decay. If R is the rate of production at any instant by the source, the rate of growth of the number of radioactive atoms (N^*) is given by

$$dN^*/dt = R - \lambda N^* \tag{19.4}$$

Integration of equation 19.4 leads to

$$N^* = (Rt_{1/2}/0.693)(1 - 2^{-t/t_{1/2}}) \tag{19.5}$$

where t is the irradiation time and $t_{1/2}$, the half-life of the isotope being produced.

Two special cases need to be considered: production of isotopes with short and with long half-lives, relative to the time of bombardment. From equation 19.5 it is seen that for an irradiation time $t = t_{1/2}$, 50% saturation is reached; for $t = 2t_{1/2}$, 75% saturation; and for $t = 5t_{1/2}$, 97% saturation. Thus bombardment times greater than 4 or 5 half-lives of the radioisotope being produced are a waste of time.

A word should be said concerning the rate of production R of the radioisotope at any instant by the source. This production rate is dependent upon the neutron flux Φ in neutrons/cm^2/sec, upon the number of target nuclei N_t, and upon the nuclear cross section for the activation σ_{act}, in cm^2. Thus R is given by

$$R = \Phi\sigma_{act}N_t \tag{19.6}$$

Substituting equation 19.6 into 19.5 gives

$$N^* = \frac{\Phi\sigma_{act}N_t t_{1/2}}{0.693}(1 - 2^{-t/t_{1/2}}) \tag{19.7}$$

For the two cases discussed above, we obtain the following forms of equation 19.4:

for $t \gg t_{1/2}$,

$$N^* = \Phi\sigma_{act}N_t t_{1/2}/0.693 \tag{19.8}$$

for $t_{1/2} \gg t$,

$$N^* = \Phi\sigma_{act}N_t t/0.693 \tag{19.9}$$

Example: Let us assume that a sample containing 1.50 g of phosphorus is bombarded in a nuclear pile at a flux of 5×10^{11} neutrons/cm^2/sec for 1 week. The (n,γ) reaction produces phosphorus-32 with a half-life of 14.3 days. The cross section for the (n,γ) reaction is 0.19 barns (1 barn = 10^{-24} cm^2). Phosphorus-31 occurs to the extent of 100% in nature. What is the resultant activity?

Using equation 19.7, we have

$$N^* = \frac{\Phi\sigma N_t t_{1/2}}{0.693}[1 - \exp(-0.693t/t_{1/2})]$$

but

$$-dN^*/dt = \lambda N^*$$

so that

$$-dN^*/dt = \Phi\sigma N_t[1 - \exp(0.693/t_{1/2})]$$
$$= \frac{(5 \times 10^{11})(0.19 \times 10^{-24})(1.50)(6.02)}{31.0} \times 10^{23}(1 - 2^{-7.0/14.3})$$
$$= 2.76 \times 10^9 \times (1 - 0.71)$$
$$= 8.0 \times 10^8 \text{ dps}$$

$$\frac{8.0 \times 10^8 \text{ dps}}{3.7 \times 10^7 \text{ dps/mCi}} = 22 \text{ mCi}$$

(6.02×10^{23} is the number of atoms per mole.)

Example: If a sample of 10.0 g of ammonium nitrate is irradiated for 14.0 days in a flux of 1×10^{12} neutrons/cm²/sec, how many curies of carbon-14 will be produced? $\sigma_{act} = 1.75$ barns for the (n,q) reaction. $t_{1/2}$ of $^{14}C = 5770$ yr. ^{14}N is 99.635% abundant in nature. Since $t_{1/2} \gg t$, we may use equation 19.9:

$$\frac{-dN^*}{dt} = \lambda N^* = 0.693\Phi\sigma_{act}N_T(t/t_{1/2})$$
$$= \frac{(0.693)(1 \times 10^{12})(1.75 \times 10^{-24})(10.0)(2)(6.02 \times 10^{23})(14.0)(0.996)}{80 \times 365 \times 5770}$$
$$= 1.21 \times 10^6 \text{ dps}$$

$$\frac{1.21 \times 10^6 \text{ dps}}{3.7 \times 10^4 \text{ dps/}\mu\text{Ci}} = 33 \text{ }\mu\text{Ci}$$

Example: A sample of a metal (at. wt = 100) was found to contain a very small amount of copper impurity. A 1.00-g sample of the metal was subjected to neutron bombardment with a flux of 5×10^{11} neutrons/cm²/ sec for 3 hr. ^{63}Cu is 69.1%, and ^{65}Cu is 30.9% in nature. The cross section for the (n,γ) reaction on ^{63}Cu is 3.7 barns. ^{66}Cu is also produced. ^{64}Cu $t_{1/2} = 12.8$ hr, and ^{66}Cu $t_{1/2} = 5$ min. After the sample was removed from the pile it was allowed to stand for 24 hr while being worked up. A 0.10- g equivalent sample of the irradiated material was counted with a counter in which 25 cpm corresponded to 1000 dpm. The sample was found to count at the rate of 3750 cpm. What is the percent Cu impurity in the original metal? Assume no other radioisotopes are produced.

$$3760 \times \frac{1000}{25} = 150,000 \text{ dpm/0.10 g metal}$$

or 1,500,000 dpm/g. But 24 hr lapsed from the irradiation, and so the activity actually produced was (equation 19.2)

$$A = 1.50 \times 10^6 = A_0 e^{-0.693 \times 24.0/12.8} = A_0 e^{-1.300}$$

$$A_0 = 1.50 \times 10^6 e^{1.30} = 1.5 \times 10^6 \times 3.67 = 5.50 \times 10^6 \text{ dpm}$$

or

$$5.50 \times 10^6/60 = 9.18 \times 10^4 \text{ dps}$$

That is

$$-dN^*/dt = 9.18 \times 10^4 \text{ dps}$$

but

$$\frac{-dN^*}{dt} = \Phi\sigma_{act}N_t[1 - \exp(0.693t/t_{1/2})]$$

$$9.18 \times 10^4 = 5 \times 10^{11} \times 3.7 \times 10^{-24}N_t$$

$$\times [1 - \exp(-0.693 \times 3.00/12.8)]$$

$$= 1.85 \times 10^{-12} \times N_t(0.15)$$

$$N_t = 3.3 \times 10^{17}$$

That is, the number of ^{63}Cu target atoms was 3.3×10^{17}. The total number of copper atoms was

$$3.3 \times 10^{17}/0.691 = 4.8 \times 10^{17}$$

$$4.8 \times 10^{17} \times \frac{1}{6 \times 10^{23}} \times \frac{63.54}{1} = 5.05 \times 10^{-5} \text{ g Cu}$$

$$\frac{5.05 \times 10^{-5} \text{ g}}{1.00 \text{ g total}} \times 100 = 0.005\% \text{ Cu impurity}$$

RADIOIMMUNOASSAY

Although radioimmunoassay (RIA) is not a new technique, it is very difficult to do the first time it is used with a new compound, and it has only been recently that a variety of methods have been developed for food analyses. The following explanation provides a general description of the RIA technique.

Immunoassay is a method for detecting small amounts of biological components based on the formation of a precipitate when an antigen, in a narrow trough on the side of an electrophoresis strip, is allowed to diffuse to an antibody, which is the separated "spot." However, for many compounds (e.g., hormones and vitamins) that are present at ng or pg/ml levels, the amount of antibody present is too small to produce a precipitate. In such cases, a radioactive antigen (Ag*) can be prepared and reacted with the antibody (Ab) to produce a radioactive product (Ag*Ab), which is then separated and counted. This procedure is called radioimmunoassay. If the equilibrium constant for the reaction between Ag* and Ab is known quite accurately, then the amount of antibody present can be determined. Radioimmunoassay is a very sensitive method, capable of detecting at the ppb level and in some cases even less. This is not a simple technique for beginners.

The standard antigen must have the same antigen properties as the unknown. The object is to add a small radioactive fragment to the protein acting as the antigen. The two most common isotopes used are ^{125}I (0.8–2.0 mCi) and ^{131}I (2–5 mCi). Iodine-125 has a 2-month half-life, is 95% abundant, and has less overall activity than ^{131}I but it is easy to count; Iodine-131 has an 8-day half-life and is about 25–35% abundant, but it produces less gamma-counting efficiency than ^{125}I.

With hormones, the desirable reaction site to be labeled is the tyrosine portion. If the hormone does not contain a tyrosine, then labeled tyrosine is added to the molecule. This is done by employing chloramine-T to slowly release HOCl, which then reacts with NaI* to form HOI*. The HOI* is then reacted with tyrosine to place an OI* group in the 4 position. The sequence of reactions is shown below:

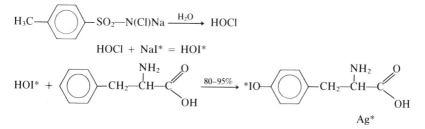

$$HOCl + NaI^* = HOI^*$$

RECENT APPLICATIONS

One effect of dietary fiber on human health was assessed by measuring iron binding in whole wheat flour and white flour at human gastric pH using ^{59}Fe (Dintzis and Watson 1984). The metabolic fate of stevioside, a new artificial sweetener that is supposed to have low intestinal absorption, was determined by ^{14}C measurements by Dunbar et al. (1984). The amount of absorption of ^{14}C labeled monensin, an antibiotic used to control coccidioses in chickens, was studied by Davison (1984). Zulalian *et al.* (1984) used ^{14}C-labeled sulfamethazine to determine its fate in swine and cattle. The use of liquid scintillation counting of ^{14}C has been suggested by Martin *et al.* (1981) for differentiating synthetic ethanol from ethanol produced by fermentation.

Radioimmunoassays for T-2 toxin in milk and in corn and wheat were reported by Lee and Chu (1981A,B). Veal was analyzed for stilbene derivatives by RIA by Brunn *et al.* (1982). A RIA procedure for 17-β-estradiol in milk and plasma was described by Glencross *et al.* (1981). A radioisotopic method using denatured ^{125}I-labeled human serum has been suggested by Kas and Rauch (1982) for the determination of low levels of proteolytic activity in foodstuffs. Mansell and Weiler (1980) reported a RIA method for the determination of limonin in citrus plants.

Patti *et al.* (1980) fused shellfish samples with Na_2O_2, precipitated the hydroxides and, after ion exchange, measured technicium-99 by beta counting. Allegrini *et al.* (1981) looked at total diet composites for iodine content by neutron activation analysis. McCurdy *et al.* (1980) coprecipitated iodine-131 with CuI in assaying milk for that element. A new method for determining the natural ^{14}C content of foods was published by Mueller and Fischer (1980) who absorbed CO_2 from combustion in NaOH, formed $BaCO_3$ liberated CO_2 into 1-aminopropane-2-ol, and then counted the resulting 1,3-bis-(2-hydroxypropyl) urea by liquid scintillation.

PROBLEMS

19.1. A breakfast cereal is known to contain thiamine. A 0.500-mg sample of ^{14}C-labeled thiamine, having a specific activity of 1.00 μCi/g, was added to 500 g of the cereal and mixed thoroughly by crushing and blending. A 0.40-mg sample of pure thiamine isolated from the mixture was found to have a disintegration rate of 5750 disintegrations per hour after all corrections for background, etc., were made. What is the milligram thiamine content per ounce of this cereal? *Ans.* 0.187 mg/oz.

19.2. It is desired to determine the aureomycin content of a fermentation broth. A 1.00-kg sample of the broth was subjected to isotope dilution analysis by adding 2.00 mg of ^{14}C-labeled aureomycin with a specific activity of 850 cpm/mg. After thorough mixing, a sample of aureomycin was separated and purified. A 0.65-mg sample of the purified aureomycin had a count rate of 118 cpm after correction for background. What is the percent by weight of aureomycin in the fermentation broth? *Ans.* 0.156%.

19.3. Following irradiation in a nuclear pile, 14.2 μcuries of carbon-14 activity had been induced in a 5.003-g sample of ammonium nitrate. Then 2.51 g of methylamine hydrochloride was added with thorough mixing to 0.1047 g of the irradiated NH_4NO_3 and dissolved in 50.0 ml of distilled water. Following a standard procedure, the phenyl isothiocyanate derivative of methylamine (ϕ—NH—CS—NHCH$_3$) was prepared from the mixture; 0.1528 g of this derivative was placed in a liquid scintillator and the carbon-14 was counted. A count rate of 122.3 cpm was observed; since the counting efficiency was 4.9%, the true count rate was 2508 dpm. What percentage of the produced carbon-14 activity is found in the form of methylamine? *Ans.* 1.4%.

BIBLIOGRAPHY

Allegrini, M., Boyer, K. W., and Tanner, J. T. (1981). Neutron activation analysis of total diet food composites for iodine. *J. Assoc. Off. Anal. Chem.* **64**, 1111–1115.

Brunn, H., Stojanowic, V., Flemming, R., Klein, H., Shirbini, A., and Becht, A. (1982). Determination of stilbene derivatives in veal products by radio-immunological and combined gas chromatographic and mass spectrometric studies. *Fleischwirtschaft* **62**, 1009–1010. [*Chem. Abstr.* **97**, 143226s.]

Davison, K. L. (1984). Monensin absorption and metabolism in calves and chickens. *J. Agric. Food Chem.* **32**, 1273–1277.

Dintzis, F. R., and Watson, P. R. (1984). Iron binding of wheat bran at human gastric pH. *J. Agric. Food Chem.* **32**(6), 1331–1360.

Dumbar, G. E., Bunes, L. A., Dietrich, P. S., and Stephenson, R. A. (1984). Diterpenoid sweeteners: Synthesis and sensory evaluation of biologically stable analogues of stevioside. *J. Agric. Food Chem.* **32**, 1321–1325.

Glencross, R. G., Abeywardene, S. A., Corney, S. J., and Morris, H. S. (1981). The use of 17β-estradiol antiserum covalently coupled to sepharose to extract from biological fluids. *J. Chromatogr.* **223**, 193–197.

Kas, J., and Rauch, P. (1982). Radioisotropic method for determination of low and very low proteolytic activities in foodstuffs. *Z Lebensm. Unters. Forsch.* **174**, 290–293.

Lee, S., and Chu, F. S. (1981A). Radioimmunoassay of T-2 toxin in corn and wheat. *J. Assoc. Off. Anal. Chem.* **64**, 156–161.

Lee, S., and Chu, F. S. (1981B). Radioimmunoassay of T-2 toxin in biological fluids. *J. Assoc. Off. Anal. Chem.* **64**, 684–688.

Mansell, R. L., and Weiler, E. W. (1980). Radioimmunoassay for the determination of limonin in citrus. *Phytochemistry* **19**, 1403–1407. [*Anal. Abstr.* 1981, **41**, 6F54.]

Martin, G. E., Noakes, J. E., Alfonso, F. C., and Figert, D. M. (1981). Liquid scintillation counting of C-14 for differentiation of synthetic ethanol from ethanol of fermentation. *J. Assoc. Off. Anal. Chem.* **64**, 1142–1144.

McCurdy, D. E., Mellor, R. A., Lambdin, R. W., and McClain, M. E. (1980). The use of cuprous iodide as a precipitation matrix in the radiochemical determination of iodine-131 in milk. *Health Phys.* **38**, 203–213. [*Anal. Abstr.* 1981, **40**, 6F22.]

Mueller, H., and Fischer, E. (1980). New method for determination of natural carbon-14 in biological material. *Z. Anal. Chem.* **302**, 199–202. [*Anal. Abstr.* 1981, **40**, 3F36.]

Ouseph, P. J. and Swartz, M. (1974). Topics in chemical instrumentation. *J. Chem. Educ.* **51,** A139–145.

Patti, F., Cappellini, L., and Jeanmaire, L. (1980). Determination of technetium-99 in marine biological samples: Choice of a method. *Anal. Abstr.* 1982, **42,** 1D47.

Zulalian, J., Stout, S. J., Bobeoch, C. N., Lucas, L. M., Miller, P., and Orloski, E. J. (1984). A study of the absorption, excretion, metabolism and residues in tissues in rats fed C-14 labeled sulfamethazine. *J. Agri. Food Chem.* **32,** 1434–1440.

20

Column Chromatography, Size Exclusion, Ion Exchange

No other separation process can match chromatography for simplicity, efficiency, and range of applications. Combined, these attributes make chromatography one of the great achievements of analytical chemistry.

Although the exact origin of the basic process may never be known, the work of Tswett who separated plant pigments by column techniques in 1906 is regarded as the first systematic study. Table 20.1 lists some of the types of chromatography currently used. The basic techniques, their major subdivisions, and the physical principles are discussed in this and the next several chapters.

In this chapter we discuss those column techniques that are normally done at or near atmospheric pressure. The techniques are classified based on the type of material used for the column packings. Those involving primarily inorganic material packings include displacement (or adsorption) and partition, chromatography. Those involving organic polymers are known as size-exclusion (gel-permeation and gel-filtration) and affinity (including immobilized-enzyme) chromatography. A third type is ion-exchange chromatography.

DISPLACEMENT AND PARTITION CHROMATOGRAPHY

Displacement Chromatography

The basic apparatus in displacement chromatography is a glass column a few centimeters in diameter and 10–20 cm in length, packed tightly with

Table 20.1.   Types of Chromatography

Stationary phase	Mobile phase	Type	Physical principle
Solid	Gas	Gas–solid	Adsorption
Solid	Liquid	Column, thin-layer, paper	Adsorption, partition, ion exchange, and gel permeation
Liquid	Liquid	Column, thin-layer, paper	Partition
Liquid	Gas	Gas–liquid	Partition

an inert filling. The solution, containing the compounds to be separated, is added to the top of the column in as thin a band as possible (Fig. 20.1A). Let us assume that both A and B are adsorbed strongly to the column packing, but B is adsorbed more strongly than A. A solvent, S, is added that is adsorbed more strongly to the column packing than either A or B.

When the solvent molecules enter the top of the column, they displace both A and B, leaving them momentarily free to migrate down the column. Since, in this case, B is more strongly adsorbed than A, the molecules of B will on the average be readsorbed more readily on the next available

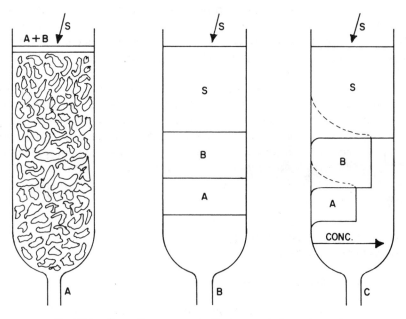

Fig. 20.1.   Separation process in displacement chromatography.

site, thus forcing the A molecules to go further down the column before being adsorbed. If more S is added to the top of the column, a continuing displacement takes place; that is, S displaces B, and B displaces A. The results are shown in Fig. 20.1B.

Figure 20.1C is a plot of the concentration of each component in 20.1B at its distance down the column. Notice that the bands actually overlap, and as a result this system is not used for quantitative analysis but is quite useful for preparative work.

Let us look at Fig. 20.1C a little closer. Notice that the concentrations per column length are different. The surface of the inert filling is really not inert but, because of crystal imperfections, impurities, and surface geometry, has places that are more reactive than the rest of the surface. These are called *reactive sites*. As the A molecules proceed down the column, they will be adsorbed on the reactive sites until all available sites are filled. The column is then *saturated,* so we reach a concentration plateau.

There is a small region ahead of the main concentration. This is due in part to the fact that all of the A molecules do not have the same energy for adsorption. There are some A molecules that have a lower energy than most of the rest; when they arrive at an active site, they do not have enough energy to react. Therefore they proceed down the column a bit further than the main body of A molecules. On the other hand there are some very active A molecules and can compete successfully with B molecules for the sites. These molecules provide the *tail.*

The B molecules are more energetic, as a group, then the A molecules, so reactive sites that would not interest an A molecule are quite appealing to a B molecule. Therefore the column saturation for B molecules is a bit higher than for the A molecules. It is even higher for the S molecules.

In displacement chromatography, also known as adsorption chromatography, the sample components are adsorbed to a granulated gel matrix by noncovalent bonds (such as hydrogen bonds), nonpolar interactions, and Van der Waal's forces. In the desorption step a gradient containing a nonpolar solvent exchanges with and displaces the sample components one at a time. Fig. 20.2 shows a typical apparatus for displacement (absorption) chromatography and a separation.

Partition Chromatography

The same equipment is used for partition chromatography as for displacement chromatography with one exception. The inert packing is coated with a film of a high-boiling liquid that is strongly adsorbed to the particle surface. This is called the *stationary phase*. A second liquid, the *mobile phase,* is then passed through the column. This solution is selected

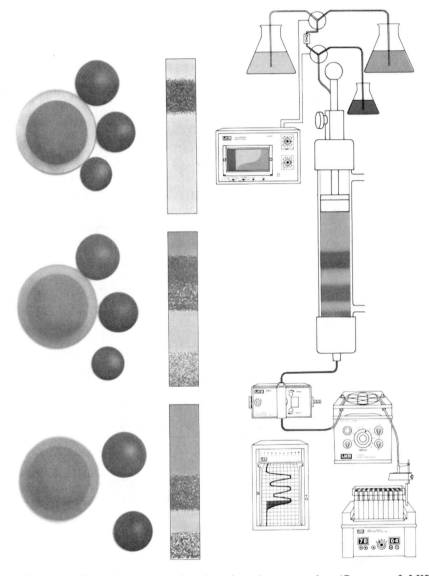

Fig. 20.2. Separation process in adsorption chromatography. (Courtesy of LKB Instruments.)

so that (1) it will not dissolve or mix appreciably with the stationary phase and (2) it will not adsorb to the inert phase as did the stationary phase. The partition process is illustrated in Fig. 20.3.

When a solution of A and B is placed at the top of the column (Fig. 20.3B) and the mobile phase S is added, the compounds proceed slowly

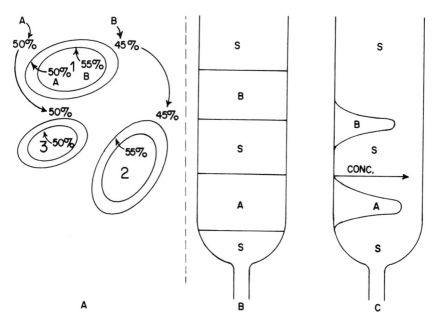

Fig. 20.3. Partition process.

down the column. As an example let us assume that we have 100,000 molecules each of A and B. These molecules are crowded together in the mobile phase and as they pass the stationary phase some of each dissolve in it. Because of a difference in the molecular structures of A and B, 55% of the B molecules will dissolve in the stationary phase, whereas only 50% of the A molecules will dissolve.

The molecules remaining in the mobile phase are carried down to the next particle where the process is repeated again, 55 and 50% dissolving, and 45 and 50% staying in the mobile phase. What about the particles still retained in particle 1? Fresh solvent is now surrounding this particle and the molecules, crowded in the stationary phase, dissolve into the mobile phase in the 50:50 and 55:45 ratios. These molecules can then proceed to particle 2 or 3 where the entire process is repeated.

Notice that at no time is there a complete and immediate separation, but that every time the partitioning takes place a few more of the A molecules get ahead of the B molecules. If we continue the process enough times, the two compounds will be completely separated from each other, although each component will be mixed with the eluting agent.

The concentration profile in Fig. 20.3C shows that the concentration of the band is not uniform as it was in displacement analysis, but that the complete separation of A and B is possible. Therefore partition analysis is preferred for quantitative analysis.

Tailing

There are times when the concentration profile looks like Fig. 20.4 rather than 20.3C. This phenomenon is called *tailing* and is very troublesome because it makes quantitative analysis difficult and sometimes impossible.

In order to explain the shape of the curve, we need to know the results of an entirely different experiment, the Langmuir adsorption isotherm. To illustrate this, one adds some dilute acetic acid to a few grams of charcoal in a flask. After thoroughly shaking the mixture, the charcoal is filtered off, and the acid is titrated to determine how much acid was adsorbed on the charcoal. When this is done for several concentrations of acid and the data are plotted, the result looks like Fig. 20.5.

The significance of the isotherm is that at low concentrations of acid, the acid is adsorbed very strongly to the charcoal (steep slope), but as the acid concentration increases, the acid does not adsorb as strongly (shallow slope).

Refer to Fig. 20.4. We see that there is a slight leading edge. Earlier we said that this was due to the less energetic particles getting ahead of the main body. This is true in part but notice that the molecules did not get very far ahead. When those leading molecules get ahead of the main body, their concentration is low and now they become tightly adsorbed. The main body coming along behind forms a normal curve.

What about those more energetic molecules? They are strongly adsorbed, but as long as they are with the main body of the band where the concentration is high they move slowly along. However, if the main body

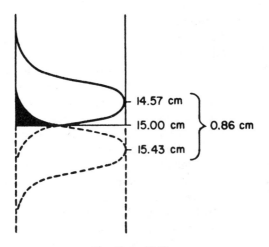

Fig. 20.4.   Tailing.

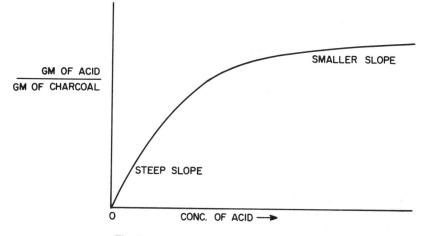

Fig. 20.5. Langmuir adsorption isotherm.

of the band passes over them, then they are in a low concentration region and again they are adsorbed strongly. Finally there are no more molecules to come along and displace them, so a tail is formed.

Apparently these tailing molecules are being adsorbed on the "inert" surface. What must be done is either to make the inert surface actually inert, or to use a more polar mobile phase to desorb the adsorbed molecules. The former can be done by acid washing or silinizing the column to cover the active sites. The latter can be done by a process called gradient elution whereby you start out with one solvent and then slowly mix into it another, more polar solvent. The net effect is to slowly change the polarity of the mobile phase.

R_f Value

In later discussions the term R_f *value* will be mentioned. This is a parameter that measures how far a material moves with respect to how far the solvent moves at the same time. This ratio is defined as

$$R_f = \frac{\text{distance the solute moves}}{\text{distance the solvent moves}} \tag{20.1}$$

A problem arises when one has to determine how far the solute has moved. Where do you measure the distance from? Since most bands have some tailing it is almost impossible to find the center of the band or the center of the concentration. What is easy to find is the front edge and this is the place to make the measurement. To remind us that it is the front edge, an f subscript is used.

Table 20.2.   Adsorbents in Decreasing Order of
Adsorptivity

Fullers earth (aluminum silicate)	Calcium phosphate
Charcoal	Potassium carbonate
Activated alumina	Sodium carbonate
Magnesium silicate (Florisil)	Talc
Silica gel	Inulin
Calcium oxide	Starch
Magnesium oxide	Powdered sugar
Calcium carbonate	

Solvents and Adsorbents

Unfortunately there are no formulas to calculate the proper solvent or adsorbent for the separation of a given compound. Tables 20.2 and 20.3 show some of the materials that have been found to work.

It should be emphasized that you are not limited to a single column packing or to the chemical characteristics of a given inert phase. Columns may be packed in segments, and they may be made acidic or basic, as desired. As an example of this, the separation of noscapine and pheniramine in cough syrup will be considered. Figure 20.6 shows how the columns are packed.

Column A is placed over Column B, and 100 ml of ether is passed through the columns. This removes the flavor components from the syrup, which would interfere later with the spectrophotometric determination. The middle tosic acid layer is a safety trap to ensure that the basic noscapine and pheniramine do not leave the top column during the ether wash. The $NaHCO_3$ is used to keep tosic acid from washing off the column, since tosic acid will also interfere with the spectrophotometric determination.

Then 150 ml of $CHCl_3$ is passed through both columns. Noscapine is

Table 20.3.   Solvents in Increasing Order of Polarity

Petroleum ether (30–50°C)	Toluene
Petroleum ether (50–70°C)	Esters of organic acids
Petroleum ether (50–100°C)	1,2-dichloroethane, dichloromethane, chloroform
Carbon tetrachloride	Alcohols
Cyclohexane	Water (varies with pH and salt concentration)
Carbon disulfide	Pyridine
Ether	Organic acids
Acetone	Mixtures of acids and bases with water, alcohol,
Benzene	or pyridine

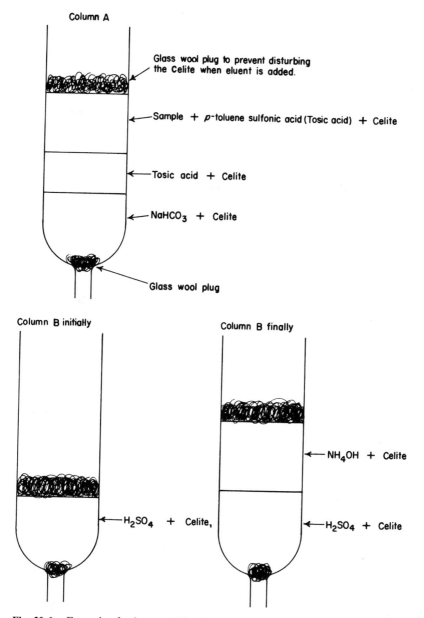

Column A

Glass wool plug to prevent disturbing the Celite when eluent is added.

Sample + p-toluene sulfonic acid (Tosic acid) + Celite

Tosic acid + Celite

NaHCO₃ + Celite

Glass wool plug

Column B initially

Column B finally

NH₄OH + Celite

H₂SO₄ + Celite,

H₂SO₄ + Celite

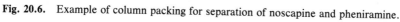

Fig. 20.6. Example of column packing for separation of noscapine and pheniramine.

completely removed from both columns and is in the eluate. Pheniramine is washed off of the weak tosic acid column but is retained on the stronger acid, H_2SO_4, column. Column A is discarded. In order to remove the pheniramine, column B must be made basic, so a NH_4OH + Celite plug is added. When $CHCl_3$ is added, the NH_4OH is washed over the H_2SO_4 making this part of the column basic, and the pheniramine is now washed off and can be determined.

Generally, the sample is added in as narrow a band as possible at the top of the column. In this case, this was not necesssary since the separation was so easy.

SIZE-EXCLUSION CHROMATOGRAPHY

In 1959, the Pharmacia Co. of Sweden found that dextran could be polymerized with epichlorohydrin to form a gellike polymer that would separate molecules on the basis of their size. The technique became known as *gel filtration*. The gel was quite fragile, would expand and contract considerably as the ionic strength of the eluting liquid changed, and was rather unstable to strong acid or base eluents. However, gel filtration with this polymer was an excellent method for separating compounds with molecular weights up to 40,000,000, for determining the mol wt distribution of polymers; for preparative fractionation of polymers, and the purification of biological samples; for determining complex equilibria constants; and for desalting. A few years later, John Moore of Dow Chemical Co. developed a styrene polymer that was less fragile, less expandable, and more resistant to pH changes than the dextran polymer. These materials also separated molecules based on size, and the technique became known as gel-permeation chromatography (GPC). Together, these techniques are now known as *size-exclusion* chromatography. Both systems are quite useful.

Probably the easiest way to understand size-exclusion chromatography is to first take a look at Fig. 20.7. Molecules of various size are drawn on the photo, and it can be seen that the large molecules cannot fit at all and would be washed out of the column immediately. Figure 20.8 is a pictorial representation of the separation process and a typical apparatus for gel filtration.

The molecular weight of the smallest compound that does not fit into the gel is called the *exclusion limit,* and the volume required to elute the large molecules is called the *void volume*, V_o. Smaller compounds can permeate the gel. In general, the smaller a compound, the further it per-

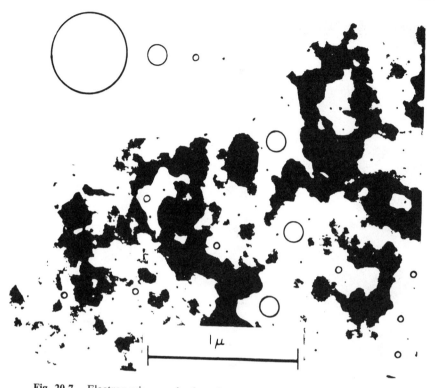

Fig. 20.7. Electron micrograph of a gel particle matrix. (From Cantow 1967.)

meates the gel and the more solvent is required to wash it out of the gel. If the compounds are extremely small, then their molecular size compared to the pores is so small that they behave the same and require essentially the same amount of solvent to elute them. This volume is called the *total permeation volume*, V_t. Figure 20.9 illustrates these terms. V_i is the *interstitial volume*.

A series of columns, each fitted with a different exclusion limit gel, can be used to separate a multiple-component mixture. A general rule is that compounds that differ by 10% in molecular weight can be separated in the same column.

As indicated previously, the main difference between gel filtration and gel permeation is the type of polymer gel used. Gel filtration utilizes dextran polymers and aqueous salt solutions for the mobile phase. Because these gels swell considerably and collapse rather easily, large-pore gels must be flushed from the bottom to top in order to prevent the gel from collapsing and plugging the column. The polymers used in gel permeation

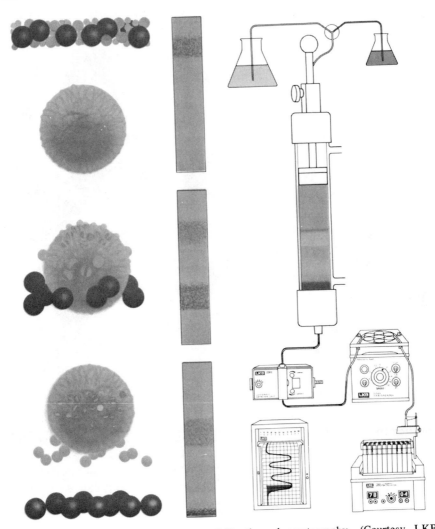

Fig. 20.8. Separation process in gel-filtration chromatography. (Courtesy LKB Instruments.)

are highly crosslinked styrene–divinyl benzene (Styragel, Poragel) or something similar. Biogel-P for example is a polyacrylamide polymer. These beads can be used with higher pressures, some like Hydrogel can withstand pressures up to 3000 psi. Organic solvents can be used; therefore, a wide variety of compounds can be separated. Some commonly used solvents are cyclohexane, chloroform, methylene chloride, dime-

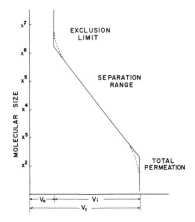

Fig. 20.9. Terms used in gel-permeation chromatography.

Table 20.4. Selected Characteristics of Biogel-P, Porasil, and Styragel[a]

Material	Range	Exclusion limit
Biogel		
P2	$10^2–1.8 \times 10^3$	4×10^3
P4	$8 \times 10^2–4 \times 10^3$	8×10^3
P6	$1 \times 10^3–6 \times 10^3$	
P30	$2.5 \times 10^3–4.0 \times 10^4$	
P100	$5 \times 10^3–1 \times 10^5$	
P200	$3 \times 10^4–2 \times 10^5$	
P300	$6 \times 10^4–4 \times 10^5$	
Porasil		
A		4×10^4
B		2×10^5
C		4×10^5
D		1×10^6
E		1.5×10^6
F		$>4 \times 10^6$
Styragel		
10^7A	$5 \times 10^5–5 \times 10^8$	5×10^8
10^6A	$10^5–5 \times 10^7$	5×10^7
10^5A	$5 \times 10^4–2 \times 10^6$	2×10^6
10^4A	$10^3–7 \times 10^5$	7×10^5
10^3A	$500–5 \times 10^4$	5×10^4
10^2A	$100–8 \times 10^3$	8×10^3

[a] Courtesy Waters Associates, Milford, MA.

thylformamide, and xylene. Table 20.4 shows a few characteristics of three gels.

Gel-permeation chromatography is one of the best methods to remove fats and waxes from food extracts when pesticide residues are to be determined. More than 98% of the fats can be removed in one pass through a 30-cm column. This is a slow process, in many cases requiring several hours, because the flow rates are so low. This system therefore lends itself to a method of automatically controlling the amount of eluent and the eluting time. Figure 20.10 is a photo of a commercial apparatus that can handle 12 samples at one time.

Fig. 20.10. Commercial instrument for automatic separation of fats from food extracts. (Courtesy Biochemistry Laboratories, Inc., Columbia, MO.)

AFFINITY CHROMATOGRAPHY (INCLUDING IMMOBILIZED ENZYMES)

The term *affinity chromatography* was first coined by Cuatrecasas *et al.* in 1968. (*See also* Cuatrecasas and Anfinsen 1971.) The potential applications of the technique are practically unlimited. They include purification, analysis, and characterization of biological substances; elucidation of mechanisms that govern biochemical systems; and immobilization of enzymes. Until the advent of affinity chromatography, practically all methods for purification of biological substances depended on gross chemical and/or physical differences between the substances that were to be separated. In gel filtration, separation is based on differences in molecular size and shape; in ion-exchange chromatography separations are based on differences in electrical properties. In affinity chromatography, separations are based on specific interactions between interacting pairs of substances, i.e., a macromolecule and its substrate, cofactor, allosteric effector, or inhibitor (Guilford 1973). In principle, a ligand is attached covalently to a water-insoluble matrix that has been tailor-cut to adsorb from a mixture only the component(s) with a specific affinity for the ligand. All other components pass through the adsorbent unrestrained. The adsorbed molecules can then be eluted after some change in conditions, e.g., a solution of free ligand or by changes in pH and/or ionic strength.

A simplified diagram of the process involved in affinity chromatography is given in Fig. 20.11. The mixture of proteins containing the desired enzyme (E) is applied to the column that supports the ligand (L). Proteins that have no affinity for the ligand are washed through with the buffer used for application. The desired enzyme is recovered by washing the column with a buffer of a different acidity. Another representation of the

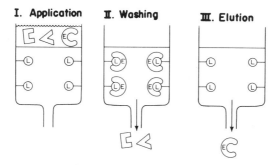

Fig. 20.11.   Diagram of separation process in affinity chromatography. (From Royer 1974.)

separation process is shown in Fig. 20.12. The term affinity chromatography is somewhat imprecise as it is more characteristic of an extraction (purification) using an adsorbent, rather than chromatography. The technique parallels, in principle, the use of insolubilized antigens as immunosorbents in the purification of antibodies.

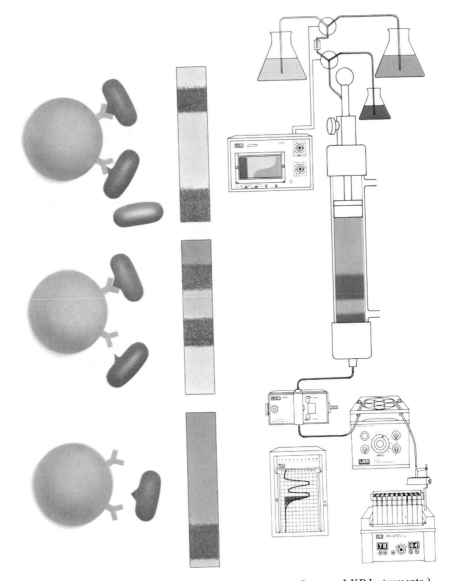

Fig. 20.12. Separation process in affinity chromatography. (Courtesy LKB Instruments.)

Affinity chromatography has several advantages over other methods for separation: (a) specificity; (b) as a small proportion of the total protein is adsorbed from a crude mixture, only a small amount of adsorbent is required; (c) the adsorbed material is rapidly separated from proteolytic enzymes and can be stabilized by ligand binding at the "active" site; and (d) the adsorbent can be regenerated many times.

The Matrix

A suitable system for affinity chromatography must meet several requirements. The matrix to which the ligands are attached must be capable of mild chemical modification without undergoing major physical changes, be practically free of residues capable of nonspecific interactions, have a loose lattice structure of sufficient porosity to allow unimpeded access of macromolecules to bound ligands, and be hydrophilic enough to permit interaction between two phases.

According to Royer (1974), the characteristics of the ideal matrix material are adequate chemical functionality for reaction with various groups on proteins and ligands, low cost, resistance to microbial attack, dimensional stability, durability, hydrophilicity (some proteins denature at hydrocarbon–water interfaces), capacity to regenerate, high capacity for enzyme or ligand, and accessibility to solvent.

Many materials have been used as matrices. The organic materials include carbohydrates (cellulose, agarose, starch), vinyl polymers (polyacrylamide and others), polymers of amino acids and their derivatives, amine-containing resins, Nylon, Dacron, etc. (Royer 1974). The inorganic materials include glass (porous and solid beads), metals (nickel screens), colloidal silica, and aluminas. Conjugates of silicas or porous glass with organic polymers are an example of an inorganic-organic combination.

Beaded agarose, polyacrylamide, and glass are used most widely because they have most of the required characteristics listed previously, can be used with certain nonaqueous systems, and are available commercially. The disadvantages of agarose are its high cost and susceptibility to microbial attack. The fibrous, heterogeneous cellulose has a low porosity, poor flow rate, and has a significant proportion of carboxy residues. Polystyrene and related polymers are unsuitable because they are highly lipophilic and because they strongly and nonspecifically adsorb many proteins (Guilford 1973). A comparison of the properties of several affinity matrices is given in Table 20.5.

The *ligand* must interact specifically and reversibly with the molecule to be purified. The ligand must be suitable for coupling to a matrix with minimal modification of the ligand structure which is essential for binding. Interactions that involve dissociation constants above 10^{-3} mole/liter are

Table 20.5.   Affinity Chromatography Matrices[a]

Matrix	Nonspecific adsorption	Binding capacity	Chemical stability	Mechanical and hydrodynamic properties
Sephadex, PAA	Excellent	Low	Excellent	Excellent
Cellulose	Excellent	Low	Excellent	Fair
Silica	Poor	Moderate	Excellent	Excellent
Sepharose	Excellent	High	Excellent	Excellent

[a] From Gelb (1973).

likely to be too weak. Almost all affinity chromatography systems involve a covalent bond between ligand and matrix. Before use, the ligand–matrix complex must be washed with several cycles of buffers of as wide a range of pH and ionic strength as the stability of the covalent bonds will allow. Detergents are sometimes used. Subsequent adequate washing of gels is essential to remove noncovalently bound ligand. (For details, see Anon. 1974.)

The amount of ligand on the matrix represents the maximum theoretical binding capacity. Since the matrix does not have an ideal porosity, only a fraction of the coupled molecules is accessible for binding. In addition, an adsorbed macromolecule may mask adjacent ligands and make them inaccessible. Three variations for fixing ligands to matrices are available: (a) direct linkage of ligand to carrier, (b) extension of ligand by an "arm" which is subsequently attached to the matrix, and (c) derivatization of the carrier with a spacer, the free end of which is functionally suitable for attaching the ligand (Guilford 1973).

Methods of covalent linking involving activation of polysaccharide matrices are shown in Figs. 20.13 and 20.14. Activation of polysaccharides with cyanogen bromide at pH 11 is one of the most widely used methods. The active form, presumably an imino-carbonate, bound to Sepharose is available commercially.

Although the activation of Sepharose with CNBr is easy to perform, CNBr is at best an unpleasant material with which to work and, if handled improperly, may even be dangerous. In many instances, it may be preferable to use CNBr-activated Sepharose 4B, which has been activated, stabilized with lactose and dextran, and lyophilized. The following general method (from Gelb 1973) may be used as a guide for the coupling of ligands containing primary amino groups to CNBr-activated Sepharose 4B:

> One gram of gel containing lactose and dextran is swollen for 10 min in 100 ml of 1 mM HCl, placed on a glass filter and aspirated almost to dryness. An additional 100 ml of 1 mM HCl is used to wash the gel and it is again aspirated to a semidry cake. Approximately 10 mg of the substance to be coupled is dissolved in 5 ml of 0.1 M NaHCO$_3$ containing

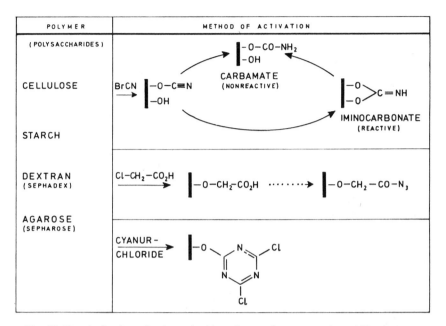

Fig. 20.13. Activation of polysaccharide polymers for enzyme immobilization. (From Nelboeck and Jaworek 1975.)

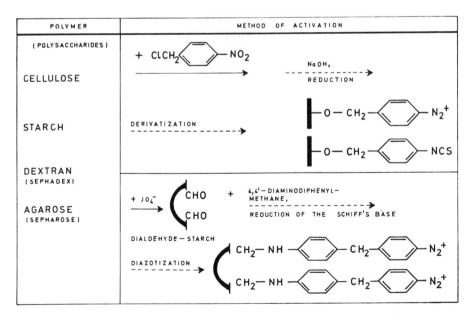

Fig. 20.14. Activation of polysaccharide polymers for enzyme immobilization. (From Nelboeck and Jaworek 1975.)

0.5 *M* NaCl. The coupling buffer containing the substance to be immobilized is added to the previously prepared gel and allowed to react for 2 hr at room temperature or overnight at 4°C. During this incubation period, the gel slurry should be agitated periodically. The use of magnetic stirring devices should be avoided, as this may cause fragmentation of the gel beads. Unbound material is removed by washing with 0.1 *M* NaHCO₃ containing 0.5 *M* NaCl and any remaining active groups are reacted with 1 *M* ethanolamine at pH 8 for 1–2 hr. Alternate washing with 0.1 *M* acetate buffer, 1 *M* NaCl, pH 4 and 0.1 *M* borate buffer, 1 *M* NaCl, pH 8 serves to remove any noncovalently adsorbed protein. Usually, three cycles with these washing buffers are sufficient.

Sepharose reacts primarily with primary amino groups of proteins. Another common matrix activation method involves azide-activated carboxymethyl cellulose. This results in binding of NH_2—, —SH, —OH, and histidyl residues. Activation of polysaccharides via triazine derivatives yields alkylating links. Replacing halogen atoms by other functional links significantly enhances variability of the method (Fig. 20.13).

Aldehyde–starch is a commercially available substrate for derivatization. Transformation with methylenedianiline into a Schiff's base and its reduction yields a product that can undergo diazotization. The diazonium groups are highly reactive to amino, tyrosine, and histidine (and less reactive to —SH and tryptophan) residues in proteins (Fig. 20.14).

Polyacrylamide matrices are based on a cross-linked copolymer of acrylamide *N,N'*-methylenebisacrylamide, which comprises a hydrocarbon framework carrying carboxamide side chains resistant to hydrolysis at pH 1–10 (Guilford 1973). This material is available commercially in spherical beads of various sizes and porosities. Several methods are available for modifying the structure with groups suitable for covalent bonding and preparation of biospecific adsorbents. Methods involving azide formation result in gel shrinkage and reduced porosity. This limits usefulness of polyacrylamide for affinity chromatography of large proteins. Polyacrylamide can also be made functional by treatment with glutaraldehyde to which ligands are attached as Schiff's bases.

Porous glass is an attractive support material for several reasons. Glass is resistant to changes in pH and solvent, mechanical damage, and microbial attack; it is rigid, regenerable, and provides a hydrophilic environment. On the other hand, porous glass is expensive, has less binding capacity than agarose, and erodes at basic pH. Coating with zirconium oxides improves stability but is costly. Activation of the porous glass surface starts with silanization with γ-amino propyltriethoxysilane (Fig. 20.15). The amino groups of the resulting alkylamine glass may be reacted directly with the carboxyl groups of proteins (by using a carbodiimide) or ligands (by using dicyclohexyl carbodiimide). An arylamine derivative may be prepared from the amino alkyl derivative by *p*-nitrobenzoylation followed by reduction. The derivative can be diazotized and reacted with proteins or ligands through phenolic, imidazole, or amino groups.

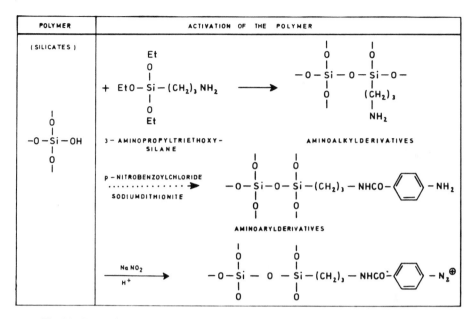

Fig. 20.15. Activation of porous glass (silicate) matrix for enzyme immobilization. (From Nelboeck and Jaworek 1975.)

Use of nylon matrices involves controlled acid hydrolysis of polyca-prolactam chains to yield an "etched" product (Fig. 20.16). Both the free carboxyl and amino groups can be derivatized. Adsorption involves formation of a Schiff base between the amino groups of the matrix and primary amino groups of the protein.

Another group of matrices involves copolymerization of an inactive monomer with a functionalized monomer capable of copolymerization. This approach is the basis of numerous matrices, including some that involve covalent binding of enzymes to copolymers with anhydride groups. The functional monomers can be maleic acid; the inactive co-monomer, ethylene. Functionalized maleic acid can be in the form of the anhydride, azide, acid chloride, hydrazide, etc. Ethylene can be replaced by acrylamide or vinyl ether. The principle of covalent immobilization of enzymes through functionalization of monomers is illustrated in Fig. 20.17.

Matrix Modification and Spacers

Because of steric considerations, it may be necessary to separate a small ligand from the backbone of the matrix by a "spacer" to enhance binding capacity. The effect of spacer groups on the accessibility of bind-

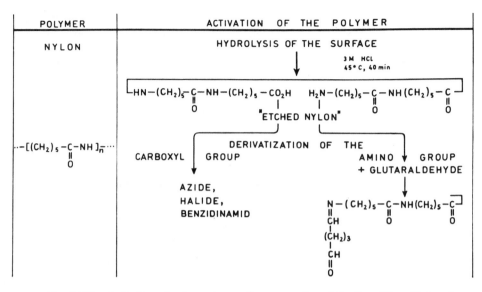

Fig. 20.16. Activation of nylon polymers for enzyme immobilization. (From Nelboeck and Jaworek 1975.)

COMONOMER (INACTIVE)	COMONOMER (FUNCTIONALIZED)	CROSSLINK	ACTIVATED POLYMER

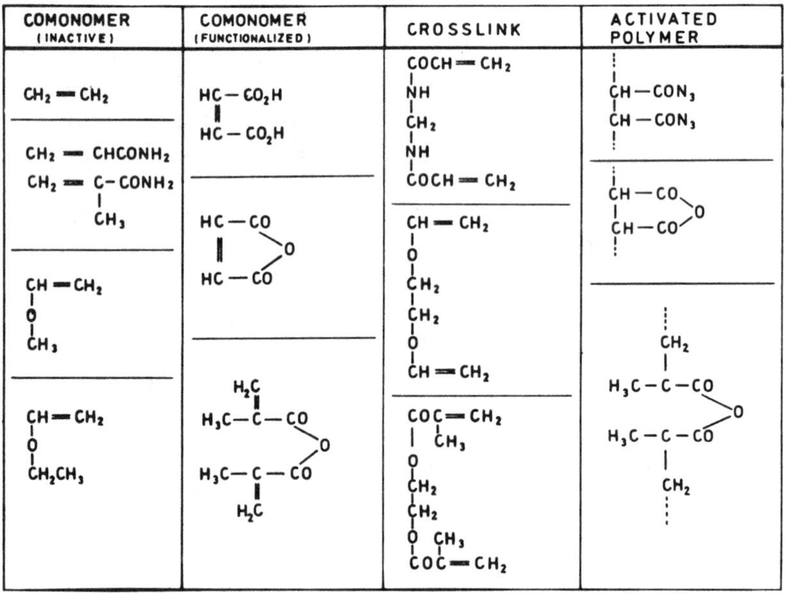

Fig. 20.17. Covalent immobilization of enzymes by functionalization of monomers. (From Nelboeck and Jaworek 1975.)

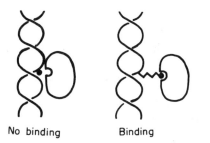

No binding   Binding

Fig. 20.18. Effect of spacer groups on the accessibility of binding sites. *Left*—Matrix with no spacer, i.e., ligand directly coupled to CNBr-activated Sepharose 4B. *Right*—Matrix with spacer, i.e., ligand coupled to either AH- or CH-Sepharose 4B. (From Gelb 1973.)

ing sites is illustrated in Fig. 20.18. In addition, some ligands have no NH_2 group for direct coupling. Activated agarose and polyacrylamide can be modified with bifunctional reagents of the general structure NH_2—R—X, where X is a functional group and R, chemically inert, determines rigidity, hydrophilicity, and maximum length. The disadvantage of some of the derivatives is lack of selectivity of the end product (Guilford 1973).

Immobilized Enzymes

The development of techniques for preparing immobilized enzymes that retain catalytic activity along with greatly improved stability has resulted in extensive use of bound enzymes (in place of free enzymes) in assay methods (Anon. 1974).

According to recommendations of the Meeting on Enzyme Engineering held in Henniker, NH, August 1971 (Nelboeck and Jaworek 1975), immobilized enzymes can be classified as follows:

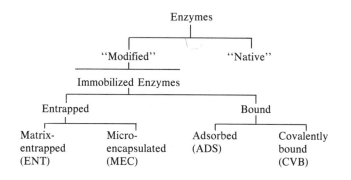

Whereas entrapped enzymes are isolated from large molecules that cannot penetrate into the matrix, bound enzymes may be exposed to molecules of all sizes. Immobilization of enzymes by affinity chromatography involves covalent binding.

The advantages of immobilized enzymes have been described as follows (Anon. 1974):

> Enzymes which are immobilized by covalent linkage to an insoluble carrier matrix have a number of unique advantages over the free enzymes. These are due to increased stability of the enzyme and to the ease with which it can be removed from the reaction mixture. Stability is increased both to thermal denaturation and, in the case of proteases, to autodigestion; the bound enzyme can thus be stored for long periods without loss of activity. As the enzyme can be removed from the reaction mixture quickly by filtration or centrifugation, the reaction can be controlled without the addition of inhibitors. The reaction product is obtained without contamination, and the recovered enzyme can be reused many times. Since the immobilized enzyme can be packed in a column and the reaction can be controlled without the loss of enzyme's activity, continuous flow reactions, either degrading or synthetic, can be carried out. Enzyme reactions of this type are not possible with conventional soluble enzymes.

When an enzyme is immobilized, several changes in the enzyme's apparent behavior may occur. The optimal pH may shift, depending on the carrier. There is, generally, an increase in K_m; the increase is related to the change in the substrate and/or carrier, diffusion effects, and (possibly) tertiary changes in enzyme configuration. In many cases, the rate of thermal inactivation and denaturation of an immobilized enzyme is less than that of the free enzyme. This does not necessarily imply superior operational stability, as such stability depends also on carrier durability, organic inhibitors, metal inhibitors, etc. (Weetal 1974).

Immobilized enzymes are generally less active than the corresponding native enzymes. Enzymes that act on low-molecular-weight substrates have after binding a residual activity of 40–80%; bound hydrolases retain only 5–40% of their activity for high-molecular-weight substrates. This partial loss of activity is offset by repeated use. Some of the reasons for modified activity include steric hindrance (resulting from attachment of proteases, amylases, and nucleases to several points on the support), electrostatic interactions between the support and similarly or oppositely charged enzyme substrates, and diffusion effects.

ION-EXCHANGE CHROMATOGRAPHY

The equipment used in ion-exchange separations is similar to that used in column chromatography and for most separations can be identical. Figure 20.19 is a diagram of one of the more elaborate columns. The

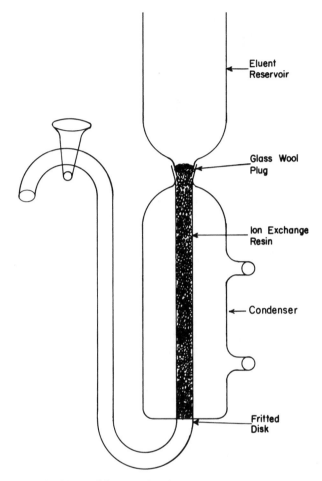

Fig. 20.19. Diagram of typical ion-exchange column.

apparatus contains a water jacket and a means to prevent the column from going dry.

It has been said that ion exchange goes back to the biblical time when Moses was leading the Israelites safely through the wilderness and made bitter water potable by stirring it with a tree. The oxidized cellulose may have exchanged with the bitter electrolytes of the water.

Be that as it may, H. S. Thompson and T. Way in 1850 are credited with the first systematic study. Thompson wanted to know why $(NH_4)_2SO_4$ and KCl did not wash out of soils. When $(NH_4)_2SO_4$ was added at the top of the column, $CaSO_4$ was eluted. When KCl was added, $CaCl_2$ was eluted. In every case the cation was exchanged.

Way was Thompson's student and continued the work. He arrived at the following conclusions, which remain essentially unchanged to this day:

(1) The exchange of Ca^{2+} and NH_4^+ in soils noted by Thompson was verified.

(2) Exchange of ions in soils involved the exchange of equivalent quantities.

(3) Certain ions were more readily exchanged than others.

(4) The extent of exchange increased with concentration, reaching a leveling-off value.

(5) The temperature coefficient for the rate of exchange was lower than that of a true chemical reaction.

(6) The aluminum silicates present in soils were responsible for the exchange.

(7) Heat treatment destroyed the exchange capacity.

(8) Exchange materials could be synthesized from soluble silicates and alums.

(9) Exchange of ions differed from true physical adsorption.

In modern ion-exchange chromatography, components with different net charges are separated when a gradient of increasing ionic strength is used as the eluant. The gel matrix can carry either positive or negative groups. In the case of a positively charged gel matrix, sample components with negative net charges will be adsorbed. During desorption the negatively charged sample components are exchanged by the negative ions from the salt gradient. Each sample component is then desorbed at a specific ionic strength and continuously eluted from the column. An ion-exchange separation is represented in Fig. 20.20.

Ion-Exchanger Materials

$Na_2Al_2Si_3O_{10}$ is a natural zeolite in which 2 out of 10 atoms carry a negative charge. These crystals resemble open sponges with charges sticking out all over. The following equation involving another zeolite shows which atoms exchange:

$$Na_2O \cdot Al_2O_3 \cdot 4SiO_2 \cdot 2H_2O + 2KCl$$
$$= K_2O \cdot Al_2O_3 \cdot 4SiO_2 \cdot 2H_2O + 2NaCl$$

(20.2)

Inorganic exchangers are limited in their capacity and they are destroyed by strong acids and bases.

Adams and Holmes (1935) prepared the first organic resins. These are more stable, have a larger capacity, and can be tailor-made to increase

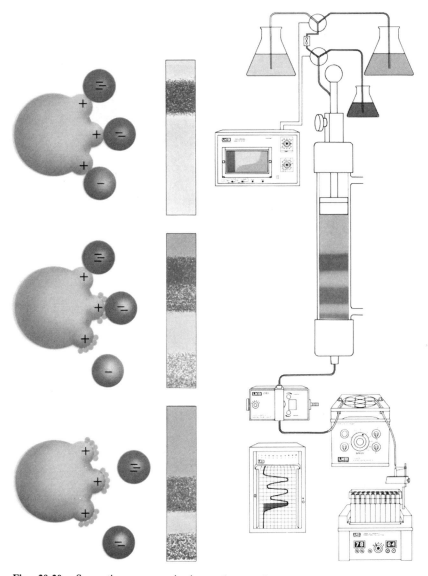

Fig. 20.20. Separation process in ion-exchange chromatography. (Courtesy LKB Instruments.)

selectivity. The structural equation shows how the resins are made:

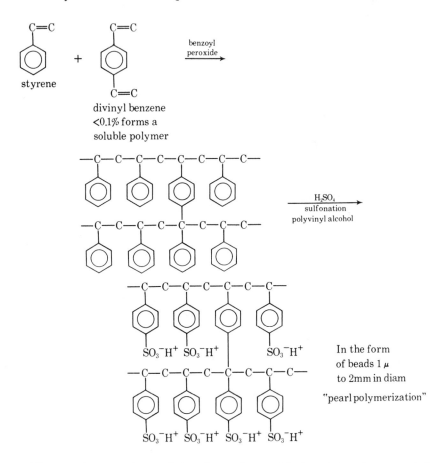

The reactive group containing the exchangeable ion determines a resin's characteristics. The amount of divinyl benzene determines how porous the bead will be.

Some of the common exchange groups used in ion-exchange resins are listed in Table 20.6.

Ion-Exchange Affinity

1. At low concentrations and ordinary temperatures, the extent of exchange increases with increasing valency of the exchanging ion

$$Na^+ < Ca^{2+} < Al^{3+} < Th^{4+}$$

Table 20.6. Typical Exchange Groups

Cation exchangers		Anion exchangers
—SO$_2$H		—N(CH$_3$)$_3^+$ Cl$^-$
—COOH	Strong	(CH$_3$Cl)$_2^-$
		$-\overset{+}{N}\underset{\diagdown}{\diagup}$
		CH$_2$OH
—CH$_2$SO$_3$H		
—OH		—NR$_2$
—SH	Weak	—NHR
—PO$_2$H$_2$		—NH$_2$

2. At low concentrations and ordinary temperatures and constant valence, the extent of exchange increases with increasing atomic number of the exchanging ion.

$$Li^+ < Na^+ < K^+$$

$$= NH_4^+ < Rb^+ < Cs^+ < Ag^+ < Be^{2+} < Mn^{2+} < Mg^{2+}$$

$$= Zn^{2+} < Cu^{2+} = Ni^{2+} < Co^{2+} < Ca^{2+} < Sr^{2+} < Ba^{2+}$$

3. At high concentrations the differences in the exchange "potential" of ions of different valences (Na$^+$, Ca^{2+}) diminishes, and in some cases reverses.

4. At high temperatures, in nonaqueous media, or at high concentrations, the exchange potentials of ions of similar charge become quite similar and even reverse.

5. The relative exchange potentials of various ions may be approximated from their activity coefficients. The higher the activity coefficient, the greater the exchange.

6. The exchange potential of the H$^+$ and OH$^-$ ions varies considerably with the nature of the functional group, and depends upon the strength of the acid or base formed between the functional group and either the hydrogen or hydroxyl ion. The stronger the acid or base, the lower the exchange potential.

7. For a weak base exchanger, the exchange follows the order

$$OH^- > SO_4^{2-} > CrO_4^{2-} > \text{citrate} > \text{tartrate} > NO_3^- > \text{arsenate}$$

$$PO_4^{3-} > MoO_4^- \quad \text{acetate} = I^- = Br^- > Cl^- > F^-$$

Complex and Chelate Resins

In an effort to improve the selectivity of ion-exchange resins, various complexing and chelating agents have been built into the resin structure. Below are examples of some of these:

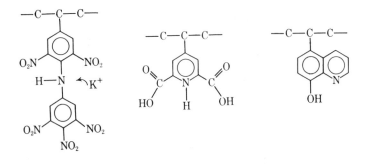

Ion-Retardation Resins

Bio-Rad AG 11A8 is an example of an ion-retardation resin (Fig. 20.21). This resin is used primarily for the desalting of biological fluids. It is made by polymerizing acrylic acid inside Dowex 1. The result is a spherical resin bead containing paired anion and cation exchange sites.

Retardation is due to the resin's paired anion- and cation-exchange sites, which attract mobile ions and associate weakly with them. Anions and cations are absorbed in equivalent amounts, but can be eluted with water. Organic molecules, even ionic species such as acidic and basic amino acids, are usually not absorbed by the resin.

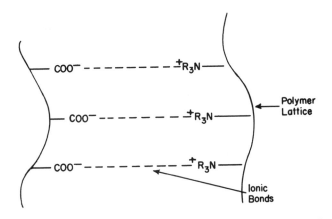

Fig. 20.21.   Structure of Bio-Rad AG 11A8, an ion-retardation resin.

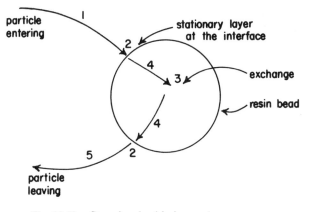

Fig. 20.22. Steps involved in ion-exchange process.

The resin removes both acids and salts from solutions of amino acids, polypeptides, proteins, enzymes, nucleic acids, and sugars.

Ion-Exchange Kinetics

It is believed that ion exchange occurs *within* the resin bead because of its small outer surface and rather high capacity (2–10 meq/g). The ion-exchange process has been broken into five possible steps (Fig. 20.22) in an attempt to determine which step controls the rate of the exchange.

Steps 1 and 5 are not rate determining if the solution is agitated, such as happens when the solvent moves through a column. Step 3, the actual exchange, is believed to be instantaneous and therefore is not a limiting factor. Step 4, the migration through the bead to the exchange site, depends on the degree of cross-linking and the concentration of the solution. If the solution concentration is greater than 0.1 M, then this step controls the rate of exchange. Step 2, the passage from the outer solution into the bead matrix, is believed to control the rate if the solution concentration is less than 0.001 M.

For concentration ranges between 0.001 and 0.1 M, a combination of steps 2 and 4 controls the rate.

SELECTING A CHROMATOGRAPHIC TECHNIQUE

Fractionation According to Classes of Compounds

Adsorption chromatography primarily separates nonpolar aliphatic and aromatic compounds according to the type and number of functional groups. The functional groups can be arranged in order of increasing adsorption affinity (and decreasing elution ease) as follows:

1. Chlorides: —Cl
2. Hydrocarbons: —H
3. Ethers: —OCH₃
4. Nitro compounds: —NO₂
5. Tertiary amines:

6. Esters:

7. O-acetyl compounds:

8. Primary amines: —NH₂
9. N-acetyl compounds:

10. Alcohols: —OH
11. Amides:

12. Acids:

If two or more constituents are present in a molecule, their influence on the rate of migration is roughly additive, but steric hindrances may be of greater importance.

Ion-exchange chromatography is particularly suitable for fractionation of compounds having ionized groups, e.g., sugar phosphates from sugars. Great differences in the net ionic charges of various classes lead, in turn, to class separations.

Partition chromatography is useful in fractionating hydrophilic components of widely different polarities into classes.

Reversed-phase partition chromatography is one of the liquid–liquid systems in which the stationary liquid phase is nonpolar and the mobile phase is polar. If the substances being separated are only sparingly soluble in water, they merely move with the solvent system and no separation results. In such a case it may be advantageous to impregnate paper with a nonaqueous medium to act as the stationary phase. In normal chromatography the stationary phase, being aqueous, is more polar than the mobile phase. The technique where the mobile phase is more polar is called *reversed-phase chromatography*. The mobile phase is not necessarily water, though mixtures used normally contain some water. A number of substances have been employed as supports for the stationary phase. The greatest problem is to get evenly impregnated papers. Silicone-impregnated papers are available commercially.

Reversed-phase partition chromatography is useful in separating lipophilic compounds into classes. Polar substances migrate ahead of less polar constituents of a mixture; that is, the separation pattern is the reverse of that observed in adsorption or partition chromatography.

Fractionation within Homologous Series

Adsorption chromatography is best for separating the first five to seven members of a homologous series, or their simple derivatives, according to differences in molecular weight. Mixtures of homologues with more than eight carbon atoms migrate as classes, ahead of the shorter-chain compounds.

Ion exchange is suitable for fractionating mixtures of the lowest members of homologous series of ionic compounds.

Partition chromatography is used in separating hydrophilic homologues. As in adsorption chromatography, the higher members migrate ahead of the homologues of shorter chain lengths.

Reverse partition is best for efficient separation of homologous compounds with more than ten carbon atoms.

Fractionation According to Degree of Unsaturation

Adsorption chromatographic separations are based primarily on the capacity to form complexes with $AgNO_3$.

Ion exchange does not resolve compounds differing only in the number of double bonds.

Partition and reverse-phase partition chromatography can be used for separations of compounds varying in degree of unsaturation, but the separations are not very sharp. To overcome that limitation, derivatives of the compounds that are to be separated can be prepared and fractionated. Thus, treatment with mercuric-acetate yields acetoxymercurimethoxy derivatives that can be more easily separated. After chromatography, the derivatives are treated with hydrochloric acid to obtain groups of original compounds.

RECENT APPLICATIONS

Sakuri *et al.* (1979) determined BHA and BHT antioxidants in fish by a pentane–CH_3CN partition column, with a Florisil cleanup and GC separation. Minicolumn techniques for screening for mycotoxins were described by Holaday (1981). McKinney (1981) employed a silica SepPAK as a first-stage cleanup of a toluene–CH_3CN solution of cottonseed extract and then went to a minicolumn. Havery *et al.* (1982) reported a fast method for determining volatile amines in milk consisting of CH_2Cl_2 elution from a Celite column. A minicolumn of aluminum oxide was used by Gorin and Heidema (1980) to clean up onion extracts before HPLC separation. Waniska and Kinesella (1980) found that adsorption chromatography was the best method to separate oligosaccharides. The flavanols L-epicatechin and D-catechin have been isolated from grapes by Baranowski and Nagel (1981) using column chromatography. Kundu and Deb (1981) developed a column chromatographic method for determining unsaponifiable matter in fats and oils. Wiggins *et al.* (1982) determined vitamins A and D in foods by HPLC after a preliminary separation on a dry alumina column.

A cleanup for bis(2-ethylhexyl) phthalate traces in fish oil that used

Biobead and alumina columns was described by Burns *et al.* (1981). An automatic method for reducing carbohydrates using a reaction with blue tetrazolium after gel chromatography was described by Robin and Tollier (1981). Barth (1980) developed a method for determining the molecular weight distribution of pectins using gel chromatography. Caramel-type food colors have been separated into their color components by gel-permeation chromatography by Hellwig *et al.* (1981). Schulte (1982) presents a simple apparatus for the rapid analysis of polymerized triglycerides by gel chromatography with Biobeads S-X and methylene chloride as the eluent. Nagayama *et al.* (1982) extracted bromate from bread, adsorbed it on DEAE-Sephadex A25 and determined it by thin-layer chromatography after elution. High-resolution gel chromatography has been shown by Rutschmann *et al.* (1982) to separate protein fractions on a preparative scale. Barth (1982) has separated hydrolyzed plant proteins with a high-performance size-exclusion technique. Davison (1984) isolated the metabolites of monesin on a DEAE Sephadex column eluted first with water then with 1 *M* KBr.

Nose *et al.* (1982) reported a method for the simultaneous detection of synthetic antimicrobials in meat products based on selective ion exchange. Three cation-exchange columns connected in series were used by Schmidt *et al.* (1981) for the separation of malto-, xylo-, and cello-oligosaccharides. Byrne *et al.* (1981) reported a method for the determination of β-sitosterol and β-sitosterol-D-glucoside in whiskey by HPLC after collecting them on a short Amberlite XAD-2 column. Sen-Jones *et al.* (1982) evaluated a procedure for separating elements of importance in foods by ion exchange before direct or hydride generation ICP determination. Maher (1981) used ion exchange to separate inorganic from methylated arsenic species in clams. Inui *et al.* (1982) cleaned up acid digests of tea and liver by ion exchange before determining selenium in a graphite tube furnace. Droz *et al.* (1982) used cation exchange for sample cleanup before separation of juice and wine acids by reversed-phase HPLC. The use of ion-exchange resins has been suggested by Satyanarayana *et al.* (1981) as an efficient means of extracting proteins from oilseed flours.

BIBLIOGRAPHY

Adams, B. A., and Holmes, E. L. (1935). Adsorptive properties of synthetic resins. *J. Soc. Chem. Ind.* **54**, 1–65.

Anon. (1959). "Sephadex." Pharmacia Fine Chemicals, A.B. Uppsala, Sweden.

Anon. (1974). "Affinity Chromatography: Principles and Methods." Pharmacia Fine Chemicals, A.B. Uppsala, Sweden.

Baranowskii, J. D., and Nagel, C. W. (1981). Isolation of L-epicatechin and D-catechin by column chromatography. *J. Agric. Food. Chem.* **29**, 63–64.

Barth, H. G. (1980). High performance gel permeation chromatography of pectins. *J. Liquid Chromatogr.* **3**, 1481–1496.

Barth, H. G. (1982). High performance size exclusion chromatography of hydrolyzed plant proteins. *Anal. Biochem.* **124**, 191–200.

Burns, B. G., Musial, C. J., and Uthe, J. F. (1981). Novel cleanup method for quantitative gas chromatographic determination of trace amounts of di-2-ethylhexylphthalate in fish lipid. *J. Assoc. Off. Anal. Chem.* **64**, 282–286.

Byrne, K. J., Reazin, G. H., and Andreasen, A. A. (1981). High pressure liquid chromatographic determination of beta sitosterol and beta sitosterol-*d*-glucoside in whiskey. *J. Assoc. Off. Anal. Chem.* **64**, 181–184.

Cantow, M. J. R. (1967). "Polymer Fractionation." Academic Press, Orlando, FL.

Cuatrecasas, P., and Anfinsen, C. B. (1971). Affinity chromatography. *Ann. Rev. Biochem.* **40**, 259–278.

Cuatrecasas, P., Wilchek, M., and Anfinsen, C. B. (1968). Selective enzyme purification by affinity chromatography. *Proc. Nat. Acad. Sci. (U.S.A.)* **61**, 636–643.

Davison, K. L. (1984). Monensin absorption and metabolism in calves and chickens. *J. Agric. Food Chem.* **32**, 1273–1277.

Droz, C., Tanner, H., and Schweiz, Z. (1982). HPLC separation and quantitative determination of organic acids in fruit juice and wines. *Obst-Weinbau* **118**, 434–438. [*Chem. Abstr.* **97**, 125805.]

Gelb, W. G. (1973). Affinity chromatography for separation of biological materials. *Amer. Lab.*, Oct., 61–67.

Gorin, N., and Heidema, F. T. (1980). Cleanup of methanolic extract of high pressure liquid chromatography of fructose, glucose, and sucrose in onion powder. *J. Agric. Food Chem.* **28**, 1340–1342.

Guilford, H. (1973). Chemical aspects of affinity chromatography. *Chem. Soc. Rev.* **2**, 249–270.

Havery, D. C., Hotchkiss, J. H., and Fazio, T. (1982). Rapid determination of volatile *N*-nitrosamines in nonfat dry milk. *J. Dairy Sci.* **65**(2), 182–185.

Heftman, E. (1967). "Chromatography." Reinhold Publishing Co., New York.

Hellwig, E., Gombocz, E. Frischenschlager, S., and Petuely, F. (1981). Detection and identification of caramel color by gel permeation chromatography. *Dtsch. Lebensm.-Rundsch.* **77**, 165–174. [*Chem. Abstr.* **95**, 78,560.]

Holaday, C. E. (1981). Minicolumn chromatograph: State of the art. *J. Am. Oil Chem. Soc.* **58**, 931A–934A.

Inui, T., Terada, S., Tamura, H., and Ichinose, N. (1982). Determination of selenium by hydride generation with reducing tube followed by graphite furnace atomic absorption spectrometry. *Fresenius' Z. Anal. Chem.* **311**(5), 494–495. [*Anal. Abstr.* **43**, 4F50.]

Kundu, M. K., and Deb, A. T. (1981). Column chromatographic determination of unsaponifiable matter in fats and oils. *Fette, Seifen, Anstrichm.* **83**(2), 73–76.

Maher, W. A. (1981). Determination of inorganic and methylated arsenic species in marine organisms and sediments. *Anal. Chim. Acta.* **126**, 157–165. [*Anal. Abstr.* **41**, 5D59.]

McKinney, J. D. (1981). Rapid analysis for aflatoxins in cottonseed products with silica gel cartridge cleanup. *J. Am. Oil. Chem. Soc.* **58**, 935–937.

Nagayama, T., Nishijima, M., Kamimura, H., Yasuda, K., Saito, K., Ibe, A., Ushiyama, H., and Naoi, Y. (1982). Analytical procedure for bromate in bread. *Shokukin Eiseigaku Zasshi* **23**, 253–258. [*Chem. Abstr.* **97**, 90539.]

Nelboeck, M., and Jaworek, D. (1975). Immobilized enzymes and principles for their use in analytical and preparative processes. *Chimia* **29**, 109–123.

Nose, N., Hoshino, Y., Kikuchi, Y., and Kawauchi, S. (1982). Studies on residue of synthetic antimicrobacterials in livestock products. VIII. Simultaneous analysis of synthetic antimicrobials in livestock products. *Shokuhin Eiseigaku Zasshi* **23,** 176–183. [*Chem. Abstr.* **97,** 54108.]

Robin, J. P., and Tollier, M. T. (1981). Automatic assay of reducing sugars using tetrazolium blue. Applications of gel permeation chromatography. *Sci. Aliments.* **1**(2), 233–246. [*Anal. Abstr.* 1982, **43,** 2G12.]

Royer, G. P. (1974). Supports for immobilized enzymes and affinity chromatography. *Chemtech* 694–700.

Rutschmann, M., Kuehn, L., Dahlman, B., and Reinauer, H. (1982). High resolution gel chromatography of proteins. *Anal. Biochem.* **124,** 134–138.

Sakurai, K., Yamada, T., Mori, M., Masuda, N., Mori, Y., and Sudo, T. (1979). BHA and BHT analysis in fish and shell fish. *Mie-ken Eisei Kenkyusho Nempo* **25,** 27–31. [*Chem. Abstr.* **95,** 167201.]

Satyanarayana, B. J., Guruprasad, A. S., and Dinanth, V. (1981). Use of ion exchange resins in the extraction of proteins from defatted groundnut and soya flours. *J. Sci. Food Agric.* **32,** 717–722.

Schmidt, J., John M., and Wandery, C. J. (1981). Rapid separation of malto- xylo-, and cello-oligosaccharides (DP2-9) on cation exchange resin using water as eluent. *J. Chromatogr.* **213,** 151–155.

Schulte, E. (1982). Detection of olive oil adulterations by HPLC. *Lebensmittelchem. Gerichtl. Chem.* **36,** 88–89.

Sen-Jones, J. W., Caper, S. G., and O'Haver, T. C. (1982). Critical evaluation of a multi-element scheme using plasma emission and hydride evolution atomic-absorption spectrometry for the analysis of plant and animal tissues. *Analyst (London)* **107,** 353–377.

Waniska, R. D., and Kinesella, J. E. (1980). Comparison of methods for separation of oligosaccharides: Ultrafiltration, gel permeation and adsorption chromatography. *J. Food Sci.* **45,** 1259–1261.

Weetal, H. H. (1974). Immobilized enzymes: Analytical applicatoins. *Anal. Chem.* **46,** 602A–610A.

Wiggins, R. A., Zai, E. S., and Lumley, I. (1982). Determination of riboflavin and flavin mono-nucleotide in foodstuffs using high-performance liquid chromatography and a column enrichment technique. *Analyst (London)* **20,** 1103–1108.

21
High-Performance Liquid Chromatography and Ion Chromatography

We have learned that the first chromatographic technique was column chromatography. Since then, paper, thin-layer, and finally gas chromatography were developed. Now, many years later, we are going back to column chromatography! This seems like a strange sequence of events but there are good reasons for it. The original column chromatographic technique employed glass columns and either gravity flow or a slight vacuum to move the mobile phase through the column. This was also "slow" chromatography and "hard to reproduce" chromatography. It was, however, "extremely flexible" chromatography in that an almost unlimited variety of solvents and column packings could be used, neither of these completely available to paper, thin-layer or gas chromatography. It was because of this recognized flexibility that scientists re-examined column chromatography.

By going to steel columns and high pressures it was found that column chromatography could be "fast" and "reproducible" as well as flexible. The result is now usually called either *high-pressure liquid chromatography* or *high-performance liquid chromatography* (HPLC).

A basic instrument, shown in diagrammatic form in Fig. 21.1, consists of a solvent reservoir, pump, gradient chamber, injection port, column, detector, fraction collector, and recorder. Depending upon the quality of the individual components, the cost of such a combination may vary from $6,000 to $30,000. Figure 21.2 is a photograph of a commercial HPLC instrument.

Of what value is this instrument to the food chemist? This instrument permits nutrition studies to be made that could only be dreamed about a

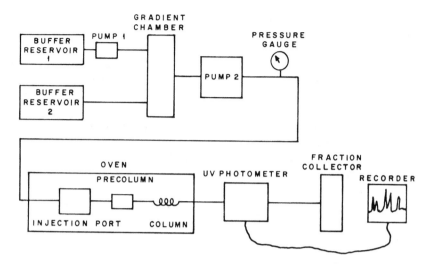

Fig. 21.1. Components of a Varian LCS 100 high-performance liquid chromatograph. (Courtesy Varian Aerograph Co.)

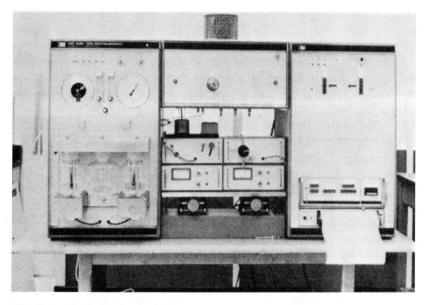

Fig. 21.2. A Hewlett-Packard high-performance liquid chromatograph. (Courtesy Hewlett-Packard Co.)

few years ago. It permits both qualitative and quantitative examination of a wide variety of nutrients, so we can now determine their biological interrelationships and just what the specific nutrient requirements are to prevent disease or to promote growth and general good health. With a capillary column filled with a high-performance ion-exchange packing it is possible to examine a wide variety of nutrients in the 1–10 nanomole range!

Ever since man realized that he had to eat to live, the more curious of his breed have wondered how the process functioned. Much progress has been made in the past, and the advent of HPLC has given scientists new tools for further understanding these vital processes and how to control them for our benefit.

In this chapter, we discuss the components and basic operation of a high-performance liquid chromatograph and describe the various types of HPLC. Ion chromatography is discussed in detail at the end of the chapter.

BASIC HPLC EQUIPMENT AND OPERATION

Solvents

The range of solvents that can be used in high-performance liquid chromatography is quite wide. Water and aqueous buffer solutions are common, as are many nonaqueous solvents of low viscosity. High-viscosity solvents are avoided because they require longer times to pass through the column, which results in peak broadening, poorer resolution, and a higher pressure to force the solvent through the column.

The solvents used must be free of suspended particles as well as chemical impurities. Suspended particles tend to plug up the column and chemical impurities produce spurious peaks at the detector. If you are serious about performing a good HPLC separation, you will purchase high-purity solvents.

The dissolved sample is also a problem. You must be sure that there are no suspended or undissolved particles in the sample before it is injected onto the column. To ensure that the sample is free of particles, it is usually passed through a Millipore-type filter such as the one shown in Fig. 21.3. A Teflon filter of about 0.45 μm is recommended when organic solvents are used, and a cellulose acetate filter is used with aqueous systems.

Degassing may also be necessary if a reciprocating pump is used. This type of pump (discussed later) has a tendency to produce gas bubbles from any dissolved gases during the intake stroke. These bubbles do not

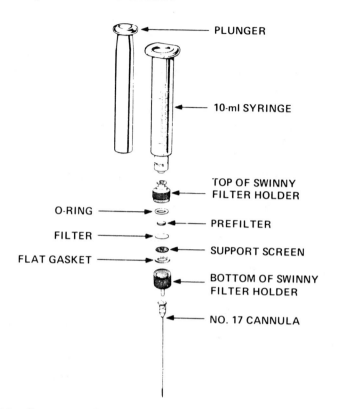

PLUNGER

10-ml SYRINGE

TOP OF SWINNY FILTER HOLDER

O-RING

PREFILTER

FILTER

SUPPORT SCREEN

FLAT GASKET

BOTTOM OF SWINNY FILTER HOLDER

NO. 17 CANNULA

Fig. 21.3. Components of apparatus for clarifying HPLC samples. (Courtesy Waters Associates.)

always redissolve and can partially block a column or create noise at the detector.

Pumps

The pump is one of the major components of a high-performance liquid chromatograph. Equation 21.1 relates several of the variables producing a pressure drop across a column.

$$P = L\eta v/\theta d^2 \qquad (21.1)$$

where L is the column length, η the solvent viscosity, v the flow rate, θ a constant, and d the particle diameter.

In order to obtain the very best separations, small particles must be packed into the column, but from equation 21.1 you can see that this greatly affects the pressure required. Pressures greater than 500 psi and sometimes more than 5000 psi are required; typical pressures are 700–1500 psi. To obtain these pressures requires good pumps, and many different types have been developed. We will discuss two major types, the reciprocating pump and positive displacement pump. A good HPLC pump should have the following characteristics:

1. High pressure to force liquid through the column.

2. Ability to obtain reproducible, stable flow rates in the 30- to 200-ml/hr range, using a variety of solvents having easy flow measurement, accurate resettability, and flow rates independent of column back pressure.

3. Nonpulsating flow to attain full sensitivity of the detectors.

4. Ability to change even immiscible solvents easily and quickly.

5. Maximum freedom in choosing solvents including those that are volatile or corrosive.

6. Gradient elution capability and ability to blend solvents in desired proportions.

7. High reliability and minimum maintenance.

Reciprocating Pumps.   Reciprocating pumps have the advantage of being able to pump solvent in unlimited quantities and are useful for long separations of preparative procedures. Figure 21.4 shows a diagram of how one type of reciprocating pump functions.

The disadvantages are that they produce a pulsating pressure, which affects most detectors; they present difficulties in providing reproducible

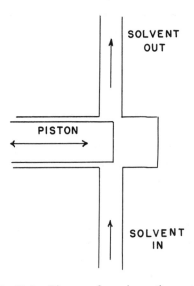

Fig. 21.4.   Diagram of a reciprocating pump.

flow control over a range of solvent types (i.e., pump volumes vary as the solvents change due to the compressibility of the solvents), and they cavitate—that is, on the inlet stroke when the pressure is released, gas bubbles can form either from dissolved gases or from solvents with a high vapor pressure. This means that solvents with a lower vapor pressure (usually also higher viscosity) must be used and the solvents should be degassed. Bubbles can cause detector noise, variable volume delivery, and "vapor lock" requiring the pump to be reprimed. The pulsating effect can be greatly reduced by using pulse suppressors. A modern reciprocating pump has very little pulsating effect.

Positive Displacement Pumps.   Figure 21.5 shows the basic design of a syringe pump, one type of positive displacement pump.

A syringe pump has the disadvantage that only a single charge of solvents can be delivered, hopefully enough to complete the analysis. Its advantages are no cavitation, easy to change solvents, reproducible and resettable flow rates at both high and low flow rates, and no check valves, thus permitting more corrosive solvents to be used for longer periods of time.

Gradient Elution

If the solvent composition is kept constant throughout the entire separation, the process is called *isocratic elution*. If, however, the solvent composition is changed in any manner (pH, buffer strength, different solvent mixtures), then the process is called *gradient elution*. Gradient elution is particularly useful when separating mixtures having widely varying characteristics because column packings generally perform well only over a narrow range of sample characteristics. However, if the solvent characteristics are progressively changed, then the behavior of the sample compounds is altered. If gradient elution is done properly, a slow separation can be speeded up or groups of compounds that would not separate in a single solvent can be made to separate.

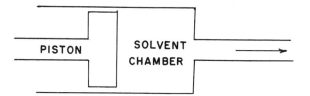

Fig. 21.5.   Diagram of a syringe pump.

One type of device used to produce a gradient (Fig. 21.6) consists of two wedge-shaped chambers, a mixing region, and proportioning valves. More elaborate gradient systems can produce either linear, exponential or step gradients to meet the operator's specifications.

Sample Injection

Samples of 5–50 μl are normal, but they must be injected onto a column that is usually under several hundred pounds of pressure. One device for doing this is shown in Fig. 21.7. This injector allows sample introduction at full column pressure without interrupting the solvent flow. All of the parts are either stainless steel or Teflon. A syringe is injected to an adjustable stop and a knurled knob is tightened, causing the Teflon cylinder to form a pressure seal around the needle. The transverse seal is then opened, the syringe pushed forward and the injection made.

More elaborate high-performance liquid chromatographs contain several columns of different characteristics. A sample injection device with 12 ports that can inject two samples at the same time onto two similar columns to obtain a differential comparison of two samples is shown in Fig. 21.8. One use of such a device is to inject a food sample extract into

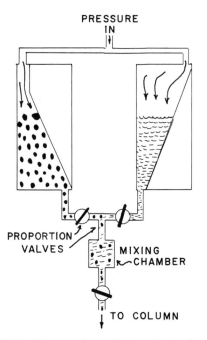

Fig. 21.6. Diagram of a gradient-producing device.

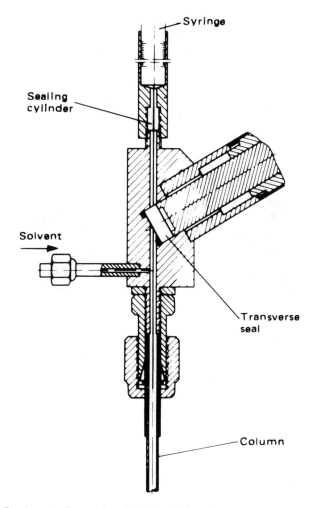

Fig. 21.7. A septumless syringe injection block. (Courtesy Hewlett-Packard Co.)

one column and simultaneously inject a food sample extract of a second food, which you suspect contains an unlisted additive, into a second column. Only the differences are detected, which greatly simplifies a very complex analysis.

Columns and Column Packing

Columns for HPLC are commonly 10–30 cm in length and 3–10 mm in diameter, the larger columns being used for preparative work. They are usually made out of stainless steel.

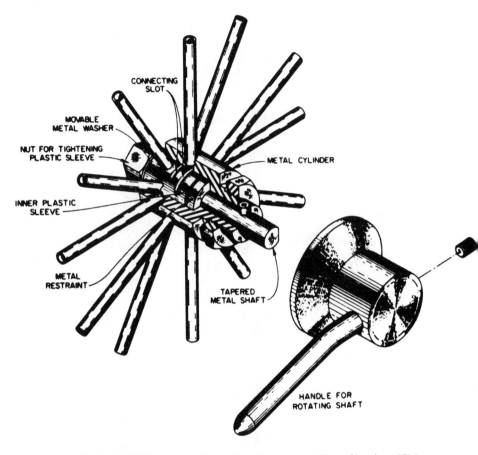

Fig. 21.8.   High-pressure 12-port injection system. (From Veening 1973.)

The high pressures employed require a very hard packing material. Furthermore, if high efficiency is to be achieved, uniform packing is necessary. Because of these requirements, column particles are usually made from silica or alumina and consist of round particles. There are currently three general types of particles: fully porous, pellicular, and microporous. These are shown diagrammatically in Fig. 21.9.

Fully Porous.   Fully porous particles come in various sizes, but a popular size for preparative work is about 50 μm in diameter. They have very large surface areas—300–500 m²/g. This large area means high column capacity, so these materials are used for preparative separations. The retention times with fully porous particles are quite long because the diffusion distances (25 μm in and 25 μm out) can be long. Examples are Porasil, Styragel, and the Durapak and Bondapak Porasils.

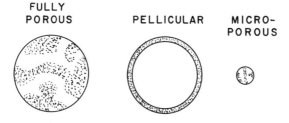

Fig. 21.9. Diagrams of the general types column packings.

Pellicular. The pellicular particles consist of a solid core with a 2- to 3-μm crust etched onto the surface. They are also about 50 μm in diameter, but their surface area (10–30 m²/g) is much less than that of fully porous particles. Thus their column capacity is much lower, but because of their small diffusion distances (2 μm in and 2 μm out), they are highly efficient and excellent for analytical separation. Examples are Corasil and the Durapak and Bondapak Corasils.

Microporous. These are approximately 10 μm in diameter and are fully porous particles. They provide a highly efficient and high-speed packing and are used for the most efficient separations. Their high porosity means heavier loading is possible than with the pellicular particles. In addition, since the diffusional distances of microporous particles are small, they have very good efficiencies. Examples of microporous materials are the micro-Porasils, Bondapaks, and Styragels.

An excellent analogy of the advantages of a micro particle over a regular-size particle has been given by Waters Associates:

Consider for a moment a sewer pipe which is packed with basketballs and filled with water. Now, between each one of these basketballs, if you stop the flow of water, there is a certain volume of liquid; if there is anything dissolved in the water, each one of these volumes of liquid acts like a large mixing chamber. Consider now the same sewer pipe, only this time packed with baseballs. The volumes between the balls are much smaller; therefore the mixing chambers are much smaller. The total volume of the water in both cases is going to be nearly the same. But because the mixing chambers are smaller and the time between mixing and interaction between the liquid and the particle is smaller, the net result is that you end up with a lot less mixing with the smaller particles and therefore higher efficiencies.

Continuing this argument it would appear that even smaller diameter particles would be the goal. Within the past ten years a 5-μm ultrasphere particle has been developed. Columns packed with these have more than 65,000 theoretical plates per meter and peak symmetry factors between 0.9 and 1.4. Because of the small particle size, greater pressures are required to maintain a reasonable flow rate; however, because of the high efficiency, a shorter column can be used.

Within the past five years, a 3-μm-diameter particle has been developed. Columns containing these particles are only 7.5 cm long, but they have 100,000 theoretical plates per meter. Figure 21.10 shows an example of the fine separations that can be obtained with such a column.

Column Cleanup. Because HPLC columns are expensive ($300–800 per column), special care must be taken to protect them. They easily become inactive due to surface contamination and must be regenerated. A good policy is to flush the column at a slow rate (0.1 ml/min) overnight.

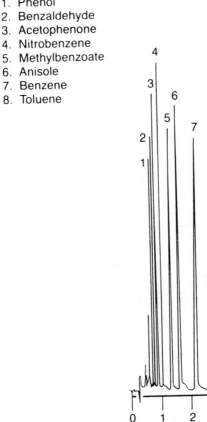

PEAK IDENTIFICATION
1. Phenol
2. Benzaldehyde
3. Acetophenone
4. Nitrobenzene
5. Methylbenzoate
6. Anisole
7. Benzene
8. Toluene

MINUTES

Fig. 21.10. Separation of substituted benzenes with very small microporous particles (3 μm diameter) in a 4.6 × 7.6 mm column. Detector: UV at 254 nm. Mobile phase: 60% methanol/40% water. Flow rate: 1 ml/min. Temperature: 30°C.

A silica column can be regenerated more quickly with 100 ml of *iso*-propanol followed by about 100 ml each of successively less polar materials such as acetone, chloroform, and finally hexane, at a rate of 2–4 ml/min.

Reversed-phase packings, such as the Bondapaks, require a water flush if buffers have been used, which is then followed with 50 ml of methanol or acetonitrile.

Proteins are removed with 8 M urea followed by water. The important concept is that columns do go bad but they can be regenerated and should not be thrown away too hastily.

Guard columns—small sections of a similar column—should always be used. If these plug or get contaminated, they can be changed without harming the main column.

Detectors

The detector is another of the critical components of a high-performance liquid chromatograph and, in fact, the practical application of HPLC had to await development of a good detector system. There are now many types of detectors on the market but we will discuss only those that are the most common.

Ultraviolet Absorption Detector. The ultraviolet absorption detector uses a low-pressure mercury lamp as a source; the cells are about 1 cm in path length and have volumes of 8–30 μl. Most are double beam. Normally, they have absorbance ranges from 0.001 to 3 absorbance units, which corresponds to a sensitivity of about 5×10^{-10} g/ml for a favorable compound.

The UV detector is particularly useful for many food components, additives, pesticides, and drugs because these compounds contain —C=C—, —C=O, —N=O, and —N=N— functional groups, which readily absorb UV radiation. Aromatic rings absorb very strongly at the 254 nm radiation of the mercury lamp emission. By using a filter, the radiation at 280 nm is also available, although it is not as sensitive for most compounds.

Multiple-wavelength detectors are now quite common, though expensive. They usually have somewhat higher detection limits at each wavelength because it is difficult to focus sufficient intensity of radiation through the cell.

Refractive Index Detector. The refractive index (RI) detector is applicable to all compounds although it is not as sensitive as the UV detector. An RI detector can detect differences of about 1×10^{-5} RI units, which means that about 5×10^{-7} g/ml must pass through the detector for a favorable sample. As a general rule the sensitivity in milligrams of sample is almost equal to the reciprocal of the difference in refractive index between the solvent and the sample.

Figure 21.11 shows the two basic types of RI detectors. In one type, the measurement is based on an optical displacement of the beam (Fig. 21.11A). A second type utilizes the Fresnel principle (Fig. 21.11B), which relates the transmittance of a dielectric interface to the refractive indices of the interface materials. Such an interface may be formed between a

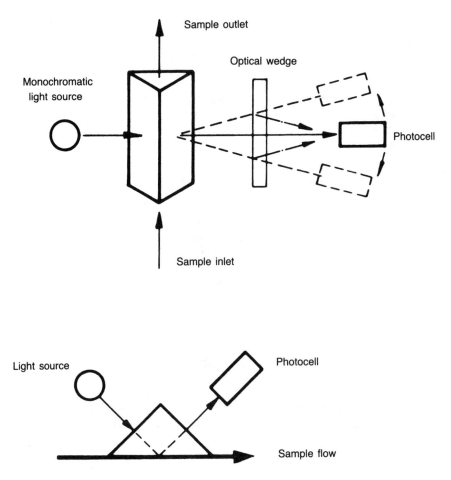

Fig. 21.11.   Two basic types of refractive index detectors. (Courtesy *Journal of Chemical Education.*)

glass prism of selected optical properties and the liquid whose refractive index is to be measured.

Refractive index detectors are difficult to use with gradient elution systems, and temperature control of the solvents is critical.

Fluorescence Detector. One way to lower the detection limits for several types of compounds is to measure their natural fluorescence. Compounds that do not fluoresce naturally can sometimes be reacted, after separation, with specific reagents to form a fluorescing product. Fluorescence detectors generally have detection limits that are 100- to 1000-fold lower than those of standard UV detectors.

Figure 21.12 is a diagram of the cell from one type of fluorescence detector. The cell volume is 5 μl. The important feature in this case is the 2π srad interceptor optics that collects the fluorescent radiation and beams it towards the S20 end-on photomultiplier tube. The detector itself has an excitation capability from 190 to 700 nm and a detection capability from 300 to 850 nm. The photometric range is 0.01–1.0 Å full scale.

To induce fluorescence in a nonfluorescing molecule, a small amount of a second component, which usually does not fluoresce, is added to the sample just as it emerges from the separation column. A postcolumn reaction takes place producing a reaction product that fluoresces. This fluorescence is then measured. Table 21.1 lists several reagents that are used for this purpose.

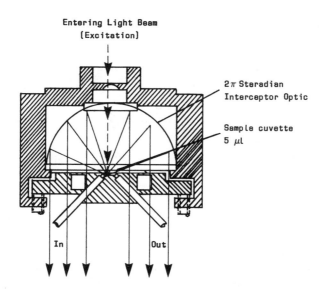

Fig. 21.12. 2 Pi steradian fluorescence cell. (Courtesy Anspec Corp.)

Table 21.1.   Some Derivatizing Reagents for HPLC

Reagent	Functional groups affected
Dansyl chloride	Primary amines, secondary amines, peptides, and amino acids
Dansyl hydrazide	Reducing sugars, aldehydes, and ketones
Fluorescamine	Primary amines and amino acids
o-Phthaldehyde	Primary amines and amino acids
4-Bromomethyl-7-methyl coumarin	Carboxylic acids

Electrochemical Detector. In Chapter 14, conductivity detectors were discussed and this is one type of electrochemical detector used with HPLC. Another type is a variation of the polarographic technique in which the potential is set at a particular value and any compound that is reduced at that potential produces a current (amperometry). The working electrode is a tubular carbon–polymer matrix, which acts as a cell. A silver–silver chloride electrode is the reference electrode and a platinum wire electrode serves as the auxillary electrode to optimize the linear dynamic range. Sensitivities from 0.1 to 200 nA are available. This corresponds to detection limits of a few tenths of a nanogram of sample.

Mass Detector. The mass detector was developed fairly recently and was designed to be a universal detector with more sensitivity than a RI detector. Figure 21.13 is a schematic diagram of one such detector. This instrument operates as follows (Anspec Corp. 1985):

> As the solvent stream leaves the column it passes into a nebulizer and an air or nitrogen supply assists in atomizing the liquid stream. The atomized solvent is sprayed into the heated evaporation column. Eluted solutes less volatile than the solvent remain as a fine cloud of particles. This particle cloud is then carried at a high speed down the evaporation column past a light source. Light from a lamp is collimated and passed through the instrument at right angles directly opposite the source of light. This eliminates internal reflections inside the instrument body. Light scattered by this particle cloud is detected by a photomultiplier located at an angle of 120° to the incident light beam.

The mass detector can be used with all solutes that are not thermally labile and not volatile. It will operate with all commonly used HPLC solvents provided they do not contain buffer salts.

MAJOR TYPES OF HPLC

There are four major types of high-performance liquid chromatography based on type of column: liquid–solid, ion-exchange, liquid–liquid, and paired-ion. Each type is described briefly and illustrated in this section.

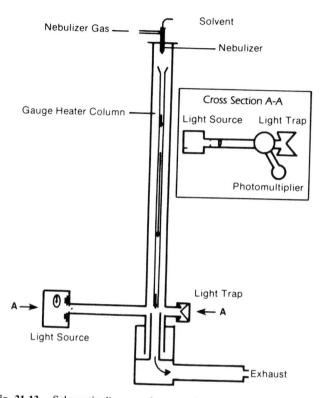

Fig. 21.13. Schematic diagram of a mass detector. (Courtesy Anspec Corp.)

Ion chromatography, a recent modification of the ion-exchange technique, is described in a separate section.

Liquid–Solid Chromatography

Although the basic packing materials in all liquid chromatography systems are solids, the outer layer may be either a liquid or a solid. In liquid–solid HPLC, the outer layer of the packing material, which comes in contact with the mobile phase and sample compounds, is a solid. This is shown in diagrammatic form in Fig. 21.14.

Since most solid column packings are clays or clay-type materials, their surfaces are aluminates or silicates and consist of large numbers of terminal —OH groups, which are highly polar. Usually a nonpolar solvent such as hexane is used as the mobile phase. When moderately polar compounds are dissolved in the mobile phase and passed over the column packing, the more polar compounds are retained more strongly than the less polar compounds and a separation results.

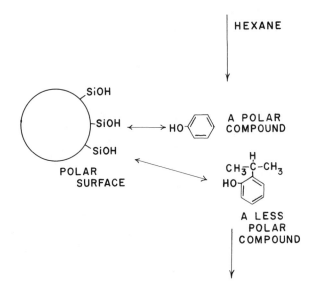

Fig. 21.14.   Liquid–solid packing system.

Reversed-Phase Liquid–Solid Chromatography.   Later on when we discuss liquid–liquid systems you will find that one of the disadvantages of that system is that the liquid coating often strips off. In addition, since the solid surfaces of the particles are polar, they are not very useful for separating nonpolar compounds, which have little affinity for the polar

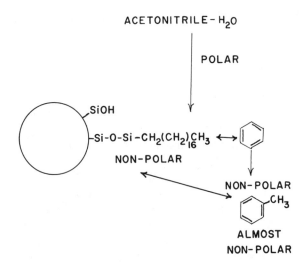

Fig. 21.15.   Reverse-phase bonded packing system.

particles. To solve this problem, we now have what is known as *reverse-phase bonded packing*. In this system, shown in Fig. 21.15, the reactive surface of the particle is changed by reaction with a nonpolar compound that is chemically bonded to the —OH group so it cannot be stripped off. A commercial system is Bondapak C_{18}.

Ion-Exchange Liquid Chromatography

Although the normal ion-exchange resins made from styrene and divinyl benzene, which were discussed in Chapter 20, can be used in HPLC, the beads are too soft, compress, and may plug up the column. What is more commonly used is a bonded packing of typical ion-exchange functional groups on a hard silica particle (Fig. 21.16).

Generally, the compounds to be separated are placed in an aqueous system buffered about 1.5 pK_a units above the highest sample pK_a. This ensures that the compounds are completely ionized and retained. They are then eluted by one of the following methods:

1. Passing the solvent buffer through the column; the compound with the highest pK_a is eluted first.

2. Using an ionic strength gradient with a low ionic strength being used first.

3. Changing the pH. With anion exchanges, the buffer is made more acidic, as this changes the sample compounds back to neutral molecules

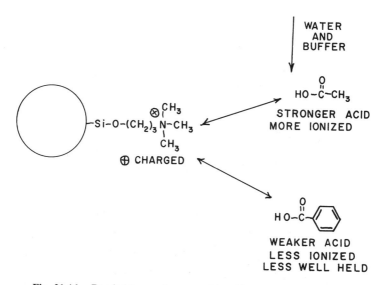

Fig. 21.16. Bonded ion-exchange packing (Corasil, Bondapak-AX).

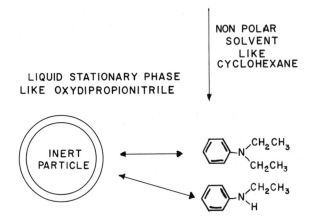

Fig. 21.17.   Liquid–liquid packing system.

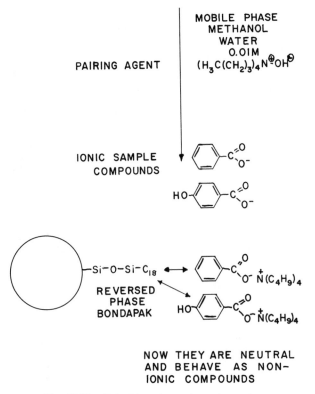

Fig. 21.18.   Paired-ion chromatography system.

and they will then elute The reverse is done with cation exchangers. Note: *The pH must be kept below 8 or the solid silica supports will begin to dissolve.*

Liquid–Liquid Chromatography

A liquid–liquid system consists of an inert phase solid support upon which is coated a liquid stationary phase (Fig. 21.17). The separation is based upon the relative solubilities of the sample compounds in the stationary phase. One difficulty is that under high pressure the stationary phase is often readily stripped from the solid support. One way to reduce this is to saturate the mobile phase with the stationary phase before the separation begins. See the precolumn arrangement in Fig. 21.1.

The amount of stationary phase used is about 1% by weight for each 15 m^2/g of solid phase surface area up to a maximum of about 15%.

Paired-Ion Chromatography

Paired-ion chromatography (PIC) in its most popular application is a modification of reversed-phase liquid–solid chromatography. It is based entirely on concentration equilibrium and can be used to separate highly polar materials with a nonpolar surface. A diagram of how this is done is shown in Fig. 21.18.

If the right system can be found, PIC usually provides better separation efficiencies than ion exchange. The ion pair reagents for cations are organic sulfonic acids like $CH_3(CH_2)_6\text{—}SO_3^-$, H^+.

DEFINITION OF TERMS

In the previous chapter we mentioned V_t, the retention volume, and V_0, the void volume or the volume required to move an unreactive compound through the column. These terms, along with W, the peak width, are identified on the typical HPLC chromtogram in Fig. 21.19.

These measurable quantities are used to determine four parameters useful in evaluating the quality of a liquid chromatographic system. These parameters are k', the *capacity factor*; α, the *separation factor*; N, the *number of theoretical plates*; and R, the *resolution*. Mathematically these

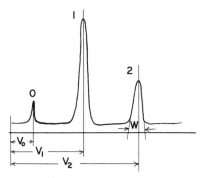

Fig. 21.19.   Typical HPLC chromatogram.

can be expressed as follows:

$$k' = \frac{V_1 - V_0}{V_0} \tag{21.2}$$

$$\alpha = \frac{V_2 - V_0}{V_1 - V_0} = \frac{k'_2}{k'_1} \tag{21.3}$$

$$N = 16 \left(\frac{V}{W}\right)^2 \tag{21.4}$$

$$R = \frac{2(V_2 - V_1)}{W_1 + W_2} \tag{21.5}$$

The k' is a measure of a compound's retention in terms of the column volume. A k' of 1 means that the sample comes through unretained, while a k' of 8 to 10 means that the sample takes a long time to come through the column. For rapid analysis, a low k' is desired, while for good separation, a high k' is needed. The compromise is a k' value of 2–6. The k' is usually controlled by changing the solvents. Table 21.2 lists several solvents and indicate their relative physical and chemical properties. We will see how this can be used later on.

The separation factor α is a ratio of the net retention time (equilibrium distribution coefficient) for any two components. If α is 1 then the separation between two components is zero. Like k', α is controlled by the solvent. Table 21.2, column 3, gives a relative measure of α.

The theoretical plate concept will be discussed in greater detail in Chapter 23 on gas chromatography. One way to increase N is to increase the column length. The separation time will be increaqsed as well, but by increasing the pressure, the flow rate can be maintained. Another means

Table 21.2. Characteristics of Solvents for HPLC[a]

Polarity index k'	Solvent	Dipole moment	α Solvent group	Viscosity
−0.3	n-Decane	0.00	0	0.92
−0.4	Isooctane	0.00	0	
−0.8	Squalane	0.00	0	
0.0	n-Hexane	0.00	0	
0.0	Cyclohexane	0.00	0	1.00
1.0	Carbon disulfide	0.00	0	0.37
1.7	Butyl ether	0.39	1	
1.7	Carbon tetrachloride	0.32	7	0.97
1.8	Triethyl amine	0.32	1	
2.2	Isopropyl ether	0.35	1	
2.3	Toluene	0.44	7	0.50
2.4	p-Xylene	0.44	7	0.70
2.7	Chlorobenzene	0.42	8	0.80
2.7	Bromobenzene	0.42	8	
2.7	Iodobenzene	0.40	8	
2.8	Phenyl ether	0.42	8	
2.9	Ethyl ether	0.34	1	0.23
2.9	Ethoxy benzene	0.44	7	
3.0	Benzene	0.43	7	0.65
3.1	Ethyl bromide	0.40	6	
3.1	Tricresyl phosphate	0.47	5	
3.2	1-Octanol	0.25	2	
3.3	Fluorobenzene	0.43	8	
3.3	Benzyl ether	0.46	7	
3.4	Methylene chloride	0.49	5	0.44
3.5	Methoxybenzene	0.41	7	
3.6	1-Pentanol	0.25	2	
3.7	Ethylene chloride	0.45	5	0.79
3.9	bis-2-Ethoxyethyl ether	0.46	5	
3.9	n-Butanal	0.26	2	
3.9	1-Butanol	0.22	2	
4.1	1-Propanol	0.27	2	2.30
4.2	Tetrahydrofuran	0.40	3	0.51
4.3	Ethyl acetate	0.42	6	0.47
4.3	2,6-Lutidine	0.35	3	
4.3	Isopropanol	0.26	2	2.30
4.3	Chloroform	0.33	9	0.57
4.4	Acetophenone	0.40	6	
4.5	Methyl ethyl ketone	0.47	6	0.43
4.5	Cyclohexanone	0.42	6	

(*continued*)

Table 21.2.   Characteristics of Solvents for HPLC[a] (*continued*)

Polarity index k'	Solvent	Dipole moment	α Solvent group	Viscosity
4.5	Nitrobenzene	0.43	7	
4.6	Benzonitrile	0.39	6	
4.8	Dioxane	0.41	6	1.54
4.8	2-Picoline	0.30	3	
5.0	Tetramethyl urea	0.40	3	
5.0	Diethylene glycol	0.33	4	
5.1	Triethyleneglycol	0.33	4	
5.2	Quinoline	0.33	3	
5.2	Ethanol	0.28	2	1.20
5.3	Pyridine	0.36	3	0.94
5.3	Nitroethane	0.42	7	
5.4	Ethylene glycol	0.30	4	19.90
5.4	Acetone	0.40	6	0.32
5.5	Benzyl alcohol	0.30	4	
5.5	Tetramethyl guanidine	0.37	1	
5.7	Methoxyethanol	0.36	4	
5.8	Triscyanoethoxypropane	0.41	6	
6.0	Propylene carbonate	0.41	7	
6.2	Aniline	0.36	6	4.40
6.2	Acetonitrile	0.41	6	0.37
6.2	Methyl formamide	0.36	3	
6.2	Acetic acid	0.30	4	1.26
6.2	Oxydipropionitrile	0.39	6	
6.3	N,N-Dimethylacetamide	0.37	3	
6.4	Dimethyl formamide	0.38	3	0.90
6.5	N-Methyl-2-Pyrolidone	0.28	3	1.65
6.5	Dimethyl sulfoxide	0.38	6	2.24
6.5	Tetrahydrothiophene-1,1-dioxide	0.38	6	
6.6	Methanol	0.30	2	0.60
6.6	Hexamethyl phosphoric acid triamide	0.36	3	
6.8	Nitromethane	0.42	7	0.67
7.0	m-Cresol	0.25	9	20.80
7.3	Formamide	0.32	4	
7.9	Dodecafluoroheptanol	0.25	9	
9.0	Water	0.26	9	1.00
9.3	Tetrafluoropropanol	0.30	9	

[a] Data from Waters Associates, Milford, MA, and from *J. Chromatogr.* **92**, 223 (1974).

of increasing N is to reduce the solvent flow rate, but the separation then takes longer. If these two methods are not sufficiently effective, then the column materials must be changed. For a given column length and flow rate, N depends upon the porosity of the particles, the size of the particles, and the homogeneity of the particle sizes. Usually the most porous, smallest, and most uniform size particles provide the highest N, all other factors being equal.

High resolution is the ultimate goal of any separation process. Equation 21.6 shows the effect of N, k', and α on R:

$$R = \frac{\sqrt{N}}{4} \left(\frac{k'}{k' + 1} \right) \left(\frac{\alpha - 1}{\alpha} \right) \qquad (21.6)$$

In practice, resolution is defined as the distance between the peak centers of two peaks divided by the average width of their respective bases, as given in equation 21.5. Figure 21.20 provides some idea of how k', α, and N are used to control resolution.

Example: The chromatogram shown in Fig. 21.21 was obtained using a 92-cm column. Calculate (A) k' for compound 1, (B) α for compounds 2 and 3, (C) N for compound 1, (D) R for compounds 2 and 3, and (E) the height equivalent to a theoretical plate (HETP) in millimeters for compound 1. $V_0 = 0.90$ ml; $W_2 = 0.80$ ml:

(A) $\quad k'_1 = \dfrac{4.10 - 0.90}{0.90} = 3.6$, $\qquad$ (B) $\quad \alpha_{2,3} = \dfrac{7.20 - 0.90}{6.10 - 0.90} = 1.2$

(C) $\quad N_1 = 16 \left(\dfrac{4.10}{0.70} \right)^2 = 549$, $\qquad$ (D) $\quad R_{2,3} = 2 \dfrac{(7.20 - 6.10)}{0.70 + 0.80} = 1.47$

(E) $\quad$ HETP $= 920$ mm/549 plates $= 1.67$ mm/plate

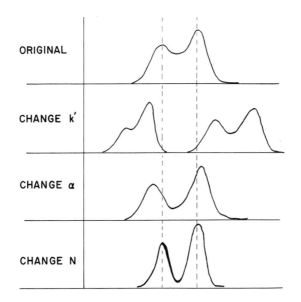

Fig. 21.20. Effects of varying k', α, and N on separation of sample components.

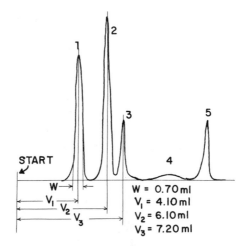

Fig. 21.21. Chromatogram obtained using a 92-cm column.

ION CHROMATOGRAPHY

Ion chromatography may be thought of as a high-performance ion-exchange separation. Essentially it is an HPLC system with an ion-exchange column, an additional *suppressor column,* and a conductivity detector. It is capable of separating and detecting nearly all cations and all anions that have pK_a's less than 7. This rapid and reproducible ion-exchange technique permits food items to be screened for their anion components. Both cation and anion systems are possible, although with foods the anions are more important because there is no other simple way to determine them.

This technique was developed by Small *et al.* (1975). A conductimetric method of detection is preferable because conductance is a universal property of ionic species and is linear with concentration. The difficulty previously was that the eluting solvents usually contain hydronium ion or hydroxide ion whose conductance overwhelms that of the ion to be measured. Small *et al.* (1975) eliminated this problem by adding a second ion-exchange column, following the separation column, that removes the ions from the eluting solution. Only the ions of interest are then present in a background of de-ionized water. Figure 21.22 is a diagram of how the apparatus is arranged.

As a example of how an ion chromatograph works, consider the analysis of a sample containing Li^+, Na^+, and K^+. The eluant, in this case dilute HCl, is pumped first through a cation exchanger in the separating column and then through a strong base resin (OH^- form) in the suppressor

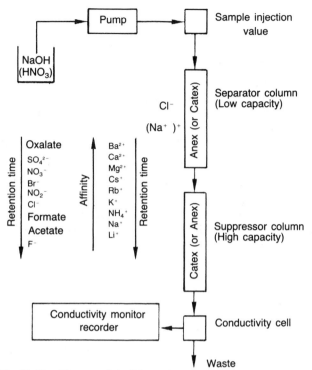

Fig. 21.22. Diagram of the Dionex ion chromatography apparatus.

column. If a sample containing Li^+, Na^+, and K^+ is injected at the head of the first column, the ions will be resolved in the separating bed and will exit at various times from the bottom of the column in a background of HCl eluent. On entering the suppressor column, two important reactions take place: HCl is removed by the strong base resin

$$HCl + resin\ OH^- = resin\ Cl^- + H_2O$$

and the alkali metal chlorides are converted to their hydroxides

$$M^+Cl^- + resin\ OH^- = M^+OH^- + resin\ Cl^-$$

which pass unretarded through the suppressor column and into the conductivity cell where they are monitored and quantified in a background of deionized water.

An analogous scheme for anion analysis can be envisioned with NaOH as the eluant, an anion-exchange resin in the separating column, and a strong acid resin in the H^+ form in the suppressor column. However, NaOH does not work well with organic anions because the selectivity ratio is not favorable, thereby requiring more neutralizing solution and a corresponding shorter life of the suppressor column. The phenate ion has been found to be quite satisfactory. Not only does it have a more favorable

Table 21.3.   Eluants for Anion Analysis by Ion Chromatography[a]

Eluant	Eluting ion	Strength	Suppressor reaction product
$Na_2B_4O_7$	$B_4O_7^{2-}$	Very weak	H_3BO_3
NaOH	OH^-	Weak	H_2O
$NaHCO_3$	HCO_3^-	Weak	H_2CO_3
Na_2CO_3	CO_3^{2-}	Medium	H_2CO_3

[a] From Fritz *et al.* (1982).

selectivity coefficient on Dowex 2 (0.14 for phenate vs. 1.5 for OH^-), but an acid resin as the suppressor would convert it to phenol, which being a very weak acid would be only feebly dissociated and contribute little to the conductivity of the effluent from the suppressor. A disadvantage of the phenate ion is that it forms oxidation products that tend

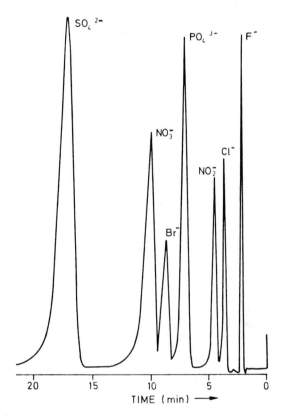

Fig. 21.23.   Separation of several inorganic anions by ion chromatography. (From Fritz *et al.* 1982.)

to poison the column and shorten its life. Table 21.3 lists some of the other common eluents for anions.

According to Fritz et al. (1982), carbonate is a very good eluant for ion chromatography because of its valence state (-2) and because it can be used at relatively low concentrations. Buffered solutions, prepared by mixing carbonate and bicarbonate, are the most commonly used eluants. The selectivity can be changed simply by varying the ratio of bicarbonate to carbonate salt, i.e., by changing the pH of the eluant. Also the eluant concentration can be varied so that faster or slower elution is obtained without affecting the elution order of analyte ions.

Ordinary ion-exchange columns prepared for HPLC should not be used for ion chromatography. The reason is that their capacity is too large, and thus they require large amounts of eluent with a corresponding reduction in the active lifetime of the suppressor column. Ion-exchange columns made specifically for the separation columns in ion chromatography have very low capacities compared to those of regular columns.

The separation of several inorganic anions by ion chromatography is shown in Fig. 21.23, and the separation of several organic anions in a coffee extract in Fig. 21.24. Mulik and Sawicki (1979) have summarized the inorganic and organic ions determined by ion chromatography. These are listed in Table 21.4.

Table 21.4. Anions Determined by Ion Chromatography[a]

Inorganic		Organic	
Arsenate	Azide	Acetate	Adipate
Borate	Bromate	Acrylate	Ascorbate
Bromide	Carbonate	Benzoate	Bromopropylsulfonate
Chlorate	Chloride	Butyrate	Butylphosphonate
Chromate	Dithionate	Butylphosphate	Citrate
Fluoride	Hypochlorite	Ethylmethyl	Formate
Iodate	Iodide	phosphonate	Fumarate
Monofluoro	Nitrate	Gluconate	Glycolate
phosphate	Nitrite	Hydroxybutyrate	Hydroxycitrate
Orthophosphate	Perchlorate	Hippurate	Itaconate
Periodate	Rhenate	Lactate	Maleate
Selenate	Sulfate	Malonate	Methacrylate
Sulfite	Tetrafluoro	Methylphosphate	Oxalate
Thiocyanate	borate	Propionate	Phthalate
Thiosulfate		Pyruvate	Succinate
		Tartrate	Trichloroacetate
		Tetraethyl	Tetramethyl
		ammonium	ammonium bromide
		bromide	

[a] From Mulik and Sawicki (1978).

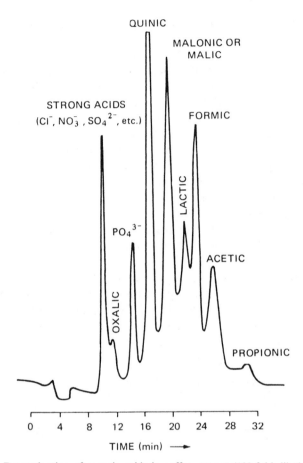

Fig. 21.24. Determination of organic acids in coffee extract (100-fold dilution). (From Dionex Corp. 1979.)

RECENT APPLICATIONS

Wheling and Wetzel (1984) used HPLC to do a simultaneous determination of pyridoxine, riboflavin, and thiamin in fortified cereal products. Morawski (1984) surveyed several HPLC methods for determining additives in dairy products and Coppola (1984) showed how HPLC can be used to detect adulteration in fruit juices. Tweeten and Euston (1980) provide a good review of how HPLC can be used in the food industry, especially for the determination of vitamins, colors, preservatives, contaminants, and pesticide residues.

Bread was analyzed for bromate by Oikawa *et al.* (1982) using ion chromatography with a borate buffer. Watanabe *et al.* (1981) also determined bromate in bread by ion chromatography utilizing column separation and concentration trapping to remove Cl^-, PO_4^{3-}, and SO_4^{2-} interference. Tateo *et al.* (1982) analyzed meat extracts with an ion chromatographic method and found high Cl^- amounts that required removal as an interference.

BIBLIOGRAPHY

Coppola, E. D. (1984). Use of HPLC to monitor juice authenticity. *Food Technol.* **38**(4), 88–91.

Dionex Corp. (1979). "Application Note 19." Sunnyvale, CA.

Fritz, J. S., Gjerde, D. T., and Pohlandt, C. (1982). "Ion Chromatography." Alfred Huthig Co., Heidelberg, Germany.

Morawski, J. (1984). Analysis of dairy products by HPLC. *Food Technol.* **38**(4), 70–78.

Mulik, J. D., and Sawicki, E. (1979). Ion chromatography. *Environ. Sci. Technol.* **13**, 804–809.

Oikawa, K., Saito, H., Sakazume, S., and Fujii, M. (1982). Ion chromatographic determination of bromate in bread. *Bunseki Kagaku* **31**, E251. [*Chem. Abstr.* **97**, 125774.]

Small, H., Stevens, T. S., and Bauman, W. C. (1975). Novel ion exchange chromatographic method using conductimetric detection. *Anal. Chem.* **47**, 1801–1809.

Tateo, F., Faleschini, M. L., and Fossati, M. (1982). Exploratory experiments on the use of ion chromatography for determination of nitrates, chlorides and phosphates in meat products, *Ind. Conserve* **57**, 30–33. [*Chem. Abstr.* **97**, 70904.]

Tweeten, T. N., and Euston, C. B. (1980). Applications of high performance liquid chromatography in the food industry. *Food Technol.* **34**, 29–37.

Veening, H. (1973). Recent developments in instrumentation for liquid chromatography. *J. Chem. Educ.* **50**, A529–A538.

Watanabe, I., Tanaka, R., and Kashimoto, T. (1981). Determination of potassium bromate (in, e.g., bread) by ion chromatography. *Shokuhin Eiseigaku Zasshi* **22**, 246–247. [*Anal. Abstr.* 1982, **43**, 1F18.]

Wheling, R. L., and Wetzel, D. L. (1984). Simultaneous determination of pyridoxine, riboflavin, and thiamine in fortified cereal products. *J. Agric. Food Chem.* **32**(6), 1326–1331.

22
Paper and Thin-Layer Chromatography

PAPER CHROMATOGRAPHY

Column chromatography was used only occasionally until 1931 when it was found that cis and trans isomers of organic compounds could be separated by this technique. From 1931 until 1944 there was a rapid development in theory and applications, but no one really challenged the basic idea of using a column.

If you have ever done a column chromatographic separation, you know that sometimes it can take a long time. If you pack the column loosely so the eluent will come through fast, then the packing is subject to cracking and the nonuniform packing causes erratic results, often producing a leading edge; however, the packing can be removed rather easily from the column. If you pack the column tightly to get a better separation, then you have to apply a vacuum to pull the eluent through the column, and it is difficult to remove the packing from the column.

Why remove the packing? Suppose you wanted to separate 10 components. Some would come out fast, but many would still be in the column. Now you could keep eluting them from the column, but this dilutes the sample, making an additional concentration step necessary. The other alternative is to push the packing out of the column after all of the components are separated and cut the packing into sections, extracting the sample from each section.

Most of these problems could be solved if the column could be disposed of. What about paper? Paper is almost pure cellulose, so the inert phase is present. Paper normally has 2–5% moisture adsorbed to its surface (can be as high as 20%), so a stationary phase is present. A sample is placed as a spot on the paper and a mobile phase passed over it, separations that take hours and days in a column can be done in minutes and hours because the paper is porous enough to allow the eluent to pass through freely, yet uniform enough to provide a good separation. Furthermore, once the sample is separated, the spots can be cut out or washed off of the paper, which is considerably easier than removing the packing from a column.

The principles of separation in paper chromatography are similar to those in column chromatography discussed previously (Consden *et al.* 1944). The main difference is that a sheet of paper is used for the inert phase.

The exact mechanism of the separation is not known. According to one theory, it is entirely an adsorption process, and the water in the paper has no effect. Another theory is that the water in the paper acts as a stationary phase and that a partition process takes place. In view of the structure of the paper, the amount of water actually present, the shape, and concentration gradient of the spots, probably a combination of both mechanisms is actually taking place.

Ascending Technique

The sample is spotted $\frac{1}{2}$ in. from the end of a paper strip. The paper is suspended vertically, in a chromatographic chamber, with the spotted end down, and allowed to dip into about $\frac{1}{4}$ in. of a solvent. The solvent rises up the paper by capillary action, effecting a separation of the components as it ascends (Fig. 22.1). The main disadvantage of the ascending technique is that the solvent will rise only 8–9 in. The main advantage is the very simple equipment required.

Remember that paper normally has 2–5% water in it, but it can adsorb as much as 20% water. What happens if a 50% water solution passes over the paper? The paper picks up water until it reaches equilibrium. The net result is that the first solvent to pass over the sample spot will be considerably different in composition than the solvent passing over it later on. The R_f value changes radically and separation may not be possible. However, if the paper is brought to equilibrium with the developing solvent at the start, a good separation is possible.

Descending Technique

The sample is spotted 2–3 in. from the end of a paper strip. The strip is suspended vertically in a chromatographic chamber with the spotted end up. The 2- to 3-in. wick is then bent over and placed in a solvent tank. After the solvent rises up the wick, it descends across the spot and

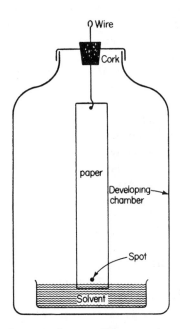

Fig. 22.1.   Apparatus for ascending paper chromatography.

down the strip. An antisiphon bar is placed at the top of the strip to prevent the solvent from simply siphoning out of the tank (Fig. 22.2). This technique permits a longer strip to be used than with ascending chromatography, and allows a corresponding increase in resolution.

Two-Dimensional Technique

The solution is spotted in the corner of a square or rectangular sheet of paper. One solvent is used to develop the chromatogram in one direction. The paper is dried, turned 90°, and a second solvent is used. An example of the high degree of separation that can be obtained in two-dimensional paper chromatography is shown in Fig. 22.3. By changing the characteristics of the solvent pair, a variety of compounds otherwise unseparable can be resolved.

Procedures and Materials

Paper.   The nature of the paper used for chromatography is most important. Ideally, it contains pure cellulose and no lignins, copper, or other impurities. Examination of cellulose under the electron microscope re-

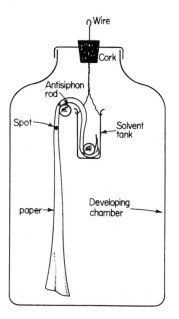

Fig. 22.2. Apparatus for descending paper chromatography.

Fig. 22.3. Two-dimensional paper chromatogram.

veals the crystallinity of the macromolecules forming closely attached fiber bundles, held together by hydrogen bonding. A small percentage of water is tightly bound. Other water molecules are adsorbed by the cellulose and fill interspaces of the fibers. It is not certain that this water participates in the partition process. The fact that the interspaces can be filled readily with a nonpolar solvent may exclude the adsorbed water as a factor in the partition process.

Chromatographic paper is usually available in sheets (18 × 22 in.) or in strips (1 in. × several hundred feet). The long direction is the machine direction and should be used for solvent development or the first direction in two-dimensional work. It is important that the sample spot be as small in diameter as possible to get good fast separations. When paper is made and pulled through the mill, the fibers tend to line up in the machine direction. A solvent moving with these fibers meets little resistance with the result that the first direction separations are fast and there is little time for spreading due to diffusion. This means that the spots are still rather small when the second direction development starts. However, if you go against the machine direction with the first solvent, greater flow resistance is encountered and the spots get large. This makes it difficult to get complete separation in the second direction.

Chromatographic paper contains organic impurities (lignin), and caution should be used when interpreting infrared or ultraviolet absorption of eluted spots of "unknown" compounds. The relative speed of several Whatman papers are listed in Table 22.1.

Preparation of the Sample. Conventional extraction procedures and evaporation techniques are applied. Depending on the nature of the compounds, different cleanup procedures are recommended: ion-exchange for amino acids and organic acids; pyridine solution for sugars. Since the final volume of the sample solution must be small, evaporations must be carried out *in vacuo* or by lyophilization. Sometimes inorganic ions (Na^+, Mg^{2+}) interfere in chromatography and the solutions must be desalted. Usually commercial (Ion Exchange, Bio-Gel) desalters are used for this purpose.

Table 22.1. Characteristics of Whatman Papers

Fast	Medium	Slow
No. 4	No. 1	No. 2
No. 54	No. 7	No. 20
No. 540		

Spotting. The principle is to spot a small amount of the solution by successive applications of small volumes of about 1–2 microliters. During solvent development some diffusion occurs, so the original spot becomes larger during the process. Spots can be kept small by applying heat or a draft of warm air from a hair drier after each application of material. Micropipets are generally available; these empty when touched to the paper. Other pipets require a syringe control. The simplest qualitative method is to use a very small loop of platinum wire sealed into a glass rod.

Chromatographic Chamber. The simplest arrangement can be made for the ascending technique. Here the upper edge of the paper may be fastened with a paper clip to a support and the bottom edge just immersed in the developing solvent. Test tubes, aquaria, biological specimen jars, pickle or large reagent jars are excellent chambers. Chrome-plated, insulated, explosion-proof commercial equipment is available for the more advanced researcher.

Solvents. By knowing the partition coefficient of the solute in two partially miscible solvents (e.g., phenol–water), one can predict the movement of the solute or suggest a better pair of solvents. Here are some general rules to follow:

1. If the substance moves too slowly, increase the solvent constituent favoring solubility of the solute. If the solute moves near the solvent front, increase the other solvent.
2. When using a two-phase system, the chromatographic chamber should be saturated with the stationary phase.
3. Some useful solvents are water–saturated phenol, butanol–NH_4OH, and acetone–water. A dependable general solvent is *n*-butanol (40%)–acetic acid (10%)–water (50%); shake and use the upper phase.

Ion-Exchange Papers

The specificity of ion exchange can be combined with the simplicity of paper chromatography in two ways. One involves impregnating the paper with ion-exchange resin, and the other involves chemical modification of the —OH groups of the cellulose to produce acidic or basic groups. Both types of papers are commercially available. Table 22.2 shows some of the common types.

Ion-exchange papers generally permit increased resolution and speed. Two-way separations can now be done in one day with these papers.

Table 22.2.   Some Commerically Available Ion-Exchange Papers

Type	Name	Ion-exchange group
	Modified cellulose	
Strong acid	Cellulose phosphate	$-OPO_3H_2$
Weak acid	Carboxymethyl cellulose	$-COOH$
	Cellulose citrate	$-COOH$
Strong base	Diethylaminoethyl cellulose (DEAE)	$-C_2H_4 \cdot NEt_2$
Weak base	Aminoethyl cellulose	$-C_2H_4 \cdot NH_2$
	Resin-loaded papers	
		(Resin incorporated)
Strong acid	Amberlite SA-2	IR-120
Weak acid	Amberlite WA-2	IRC-50
Strong base	Amberlite SB-2	IRA-500
Weak base	Amberlite WB-2	IR-4B

THIN-LAYER CHROMATOGRAPHY

Paper chromatography was a major advance over column chromatography because it removed the restriction of the column. However, the paper itself served as a restriction in that cellulose was the only inert phase. Experiments with column chromatography had shown that such packings as silica gel, alumina, diatomaceous earth, and many other materials were good chromatographic materials. The problem was, how do you make a "paper" out of these? This problem was solved in 1956 by E. Stahl in Germany who added 2–5% of plaster of Paris ($CaSO_4$) to the silica gel and "plastered" the silica gel to a glass plate (Stahl *et al.* 1956). Now the plate could be held vertical, and normal paper chromatographic techniques applied. However, the thin layer is much more uniform than paper, and sharp separations can be made much faster. Separations that require hours with paper chromatography take 15–20 min with thin-layer chromatography (TLC).

Thin-layer chromatography applications have increased at a very rapid rate since 1958. This rapid growth has been prompted by the many advantages of TLC.

1. The method is quick; generally 20–40 min are sufficient for separations.
2. Because inorganic layers are used as the sorbent, more reactive reagents can be used to visualize spots than are possible in paper chromatography.

3. The method is considerably more sensitive than paper chromatography; a lower limit of detection of 10^{-9} g is possible in some cases.
4. A wide range of sample sizes can be handled.
5. Thin-layer chromatography can be scaled up and adapted to column preparative separations.
6. The equipment for TLC is simple and readily available.
7. No special manipulative skills are required.
8. Experimental parameters are easily varied to effect separations.

General Procedures

Thin-layer adsorption chromatography involves spreading a thin layer (about 250 μm thick) of a sorbent–water slurry on a glass plate. The sorbent generally contains a binder ($CaSO_4$ or starch) to improve adhesion to the plate. After spreading, the layer is activated by drying. The activity is controlled by the time and, more importantly, the temperature of drying.

Following activation, a drop of solution containing the mixture to be separated is applied to the sorbent near one end of the plate, and the carrier solvent is allowed to evaporate. The spotted plate is placed in a closed chamber in an upright position with the lower edge (nearest the applied spot) immersed in solvent at the bottom of the chamber. When the solvent rises through the sorbent layer by capillary action, the components of the applied spot separate into individual spots in a line perpendicular to the edge of the plate. After allowing the solvent to rise to the desired distance (10–15 cm), the plate is removed from the solvent and the solvent is allowed to evaporate from the plate.

Some separated spots are visible to the naked eye. Colorless spots can be made visible by a variety of methods. The spots may be charred by spraying the plate with a mixture of sulfuric acid and an oxidizing agent ($KMnO_4$ or $K_2Cr_2O_7$), and then heating the plate. Plates also may be sprayed with specific reagents. For each separation, pure compounds should be cochromatographed for identification of components in an analyzed mixture.

Fluorescence Detection. Another way to visualize sample spots is by fluorescence. If a compound is naturally fluorescent, it can be detected by placing the developed plate in a small box and irradiating it with either short wavelength (254 nm) or long wavelength (366 nm) ultraviolet radiation. Those compounds that fluoresce produce a variety of colors. Many other compounds can be made to fluoresce by spraying the coating with acids or bases. Another fluorescence detection technique involves the use of fluorescent-coated plates. These are prepared with TLC adsorbents to which inorganic phosphors have been added. These include manganese-activated zinc silicate (254 nm), zinc/cadmium sulfide (254/366 nm), and lead–manganese-activated calcium silicate (254 nm). Occasionally organic compounds are added as phosphors, but they do not withstand treatment with strong acids and bases as well as inorganic compounds. When a fluorescent-coated plate is exposed to ultraviolet radiation, the entire plate glows with a blue-green fluorescence. Those sample compounds that do not fluoresce cover this background fluorescence and appear as dark spots. Those compounds that fluoresce naturally add their fluorescence to that on the coating and produce a colored spot.

Preparation of Plates. Commercially prepared sheets of thin-layer material already impregnated with phosphors are readily available at reasonable cost. These are so uniform that few people make their own plates.

Principles of TLC Separations

Adsorbants. Most TLC applications depend upon the adsorptive properties of the thin-layer materials. This is true even though partition chromatography is well-established in various thin-layer applications, and ion-exchange media are becoming more important. It is probably safe to say that cases of either pure adsorption or pure partitioning are rare. Rather, conditions are adjusted so that one predominates at the expense of the other. Although many materials have been tried as TLC adsorbents, the most frequently used is silica gel. Alumina follows but it is much less used than silica. Adsorption requires a high surface field strength, and activity increases with increasing lattice energy of the sorbent. Therefore, activity generally increases with hardness and melting point of the sorbent. Common TLC absorbants and their use are given in Table 22.3.

Adsorptive Processes. Valence electrons at the surface of a solid are not saturated by adjacent atoms and to a certain degree are available for bonding. Substances that are polar or polarizable undergo an electrostatic attraction to these electrons. Surface field strengths of refractory materials, as indicated by cold cathode emission of electrons, are of the order of 1,000,000 volts/cm. Therefore, polarization (field-induced charge separation) is not unexpected. The surface forces are short range and greatly diminish upon building of one to five layers of molecules on the surface. First monolayers adsorbed are bonded three to five times more strongly than succeeding layers.

It is obvious from these considerations that adsorption is a concentration dependent process and the adsorption coefficient k is not a constant. This phenomenon is in contrast to partition chromatography where the distribution coefficient is constant over a wide range. These relationships are illustrated in Fig. 22.4. Curve A represents an adsorption isotherm, and curve B represents partitioning behavior.

Table 22.3. Adsorbents for Thin-Layer Chromatography[a]

Solid	Used to separate
Silica gel	Amino acids, alkaloids, sugars, fatty acids, lipids, essential oils, inorganic anions and cations, steroids, terpenoids
Alumina	Alkaloids, food dyes, phenols, steroids, vitamins, carotenes, amino acids
Kieselguhr	Sugars, oliogosaccharides, dibasic acids, fatty acids, triglycerides, amino acids, steroids
Celite	Steroids, inorganic cations
Cellulose powder	Amino acids, food dyes, alkaloids, nucleotides
Ion exchange cellulose	Nucleotides, halide ions
Starch	Amino acids
Polyamide powder	Anthocyanins, aromatic acids, antioxidants, flavonoids, proteins
Sephadex	Amino acids, proteins

[a] From Stock and Rice (1967).

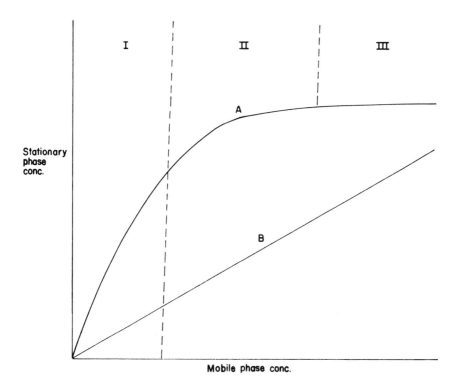

Fig. 22.4. Adsorption isotherm (A) and partitioning behavior (B).

In region I, the adsorption coefficient is approximately linear, similar to partitioning behavior. In region II, however, a rapid change occurs, and in region III, saturation is reached. A sample size exceeding the adsorptive capacity therefore would be expected to result in relatively poor separation (tailing or smearing).

In general, substrates desirable for partition chromatography are porous, rather soft materials capable of holding tenaciously a liquid phase, whereas desirable substrates for adsorption are refractory, finely divided, strongly adsorbent powders.

Sorbent–Solvent Interactions. Although adsorption is the major controlling factor in most TLC applications, simultaneous partitioning may also occur to a considerable extent. It is probably not well recognized that even in cases where active material has been prepared by long drying at elevated temperatures, the sorbent still contains appreciable quantities of water. This water may behave as a stationary liquid phase. As shown in Fig. 22.5, as much as 9% water may be retained on a silica sorbent after 3 hr activation at 150°C.

When activated at lower temperatures, the opportunity for partition of hydrophilic substances is even greater. The properties of commercial adsorbents may be modified considerably by incorporating either acids,

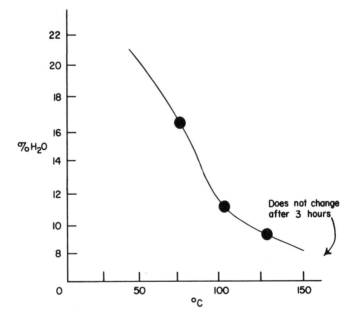

Fig. 22.5.   Effect of activation temperature on water content of a silica sorbent (0.25-mm layer).

bases, or buffer systems when the plate is prepared. In those systems, partition rather than adsorption processes predominate.

Solvents. As control of sorbent activity is somewhat restricted, adjustments needed to effect TLC separations are usually more readily accomplished by altering solvent composition. It is not surprising that, generally, solvent-eluting power correlates with dielectric constant, as adsorption chromatography relies on electrostatic attraction. Assuming other things are equal, the attractive force varies inversely with the dielectric constant of the medium. On the other hand R_f values are inversely proportional to the attractive force between the sorbent and the material adsorbed; the weaker the force the higher the compound can climb on the plate and, therefore, the higher the R_f value. Table 22.4 shows a few solvents and their dielectric constants.

Blending solvents yields a solution with a dielectric constant approximately proportional to the quantities of the individual components. Thus a single solvent system of binary mixtures can be systematically devised to effect separation. Solvent dielectric constants are strongly dependent on purity, and for reproducible results only high-purity solvents should be used.

Effect of Chemical Constitution on Adsorption. Strongly adsorbed substances require strong eluents, and weakly adsorbed materials weaker eluents. It is helpful to know which features of chemical constitution influence the strength of the adsorption bond. General rules have been summarized by Randerath (1963) as follows:

1. Saturated hydrocarbons are either adsorbed weakly or not at all. Introduction of double bonds raises adsorption affinity in proportion to the number and degree of conjugation, because the polarizability of the

Table 22.4. Dielectric Constants of Selected Solvents

Solvent	Dielectric constant (25°C)	Solvent	Dielectric constant (25°C)
n-Pentane	1.84 (20°)	Ethylene chloride	10.35
Cyclohexane	2.01	Pyridine	12.3
Carbon tetrachloride	2.23	Ammonia	16.9
Benzene	2.28	Acetone	20.70
Trichloroethylene	3.4	Ethanol	24.30
Chloroform	4.75	Methanol	32.63
Diethyl ether	5.02	Acetonitrile	37.5 (20°)
Ethyl acetate	6.02	Water	78.54
Tetrahydrofuran	8.20 (20°)	Formamide	110 (20°)

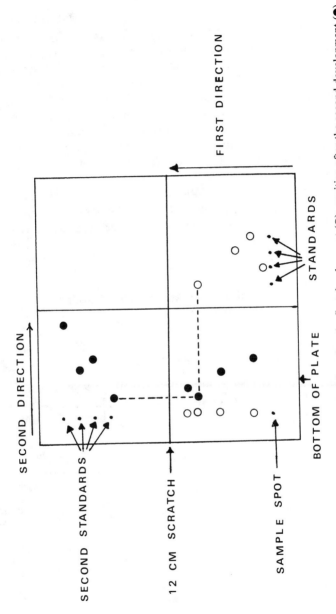

Fig. 22.6. Diagram of two-dimensional TLC. Position after the first development (○); position after the second development (●).

molecule increases and, consequently, the strength with which it is bound to the surface of the adsorbent also increases.

2. The introduction of functional groups into a hydrocarbon alters the adsorption affinity in the following order: $COOH > CONH_2 > OH > NHCOCH_3 > NH_2 > OCOCH_3 > COCH_3 > N(CH_3)_2 > NO_2 > OCH_3 > H > Cl$. These rules are derived from aromatic compounds and variations may occur for saturated compounds.

3. If there are several substituents in the same molecule, their separate influences on the adsorption affinity are only approximately additive. Steric effects are important and can greatly vary the relative activity.

Two-Dimensional Thin-Layer Chromatography

The technique of two-dimensional TLC involves first using a polar developing solvent, drying the plate, turning it 90°, and then developing the plate a second time with a solvent that is of a different polarity than the first. Two-dimensional chromatography is useful in two situations: (1) when the mixture is so complex that the components are not all separated by one solvent, and (2) when a more positive identification is required for one or more of the components. Figure 22.6 shows how the plate is scratched to prevent edge effects and to stop the solvent front. Where the sample and standards should be spotted are shown also.

The sample is spotted 2 cm in and up from one corner, and the first set of standards is spotted to the right of the vertical scratch mark, usually 1 cm apart and 2 cm from the bottom of the plate. The plate is developed as usual and in the example in Fig. 22.6 produces the separation indicated by the open circles. The plate is *not* sprayed to develop the sample spots at this time. The plate is thoroughly dried and a second set of standards is spotted in what is now the top left quadrant of the plate. The plate is turned 90° so the left side of the plate becomes the bottom, and the plate is developed with a second solvent. The plate is dried and sprayed with color-developing reagents. The results of using the second solvent are shown in Figure 22.6 by solid circles. An identification is obtained if the dashed lines from the standards intersect with a spot of the unknown.

RECENT APPLICATIONS

Penicillic acid was converted to a fluorescent derivative after two-dimensional TLC development by Ehnert *et al.* (1981) who could estimate 5 ng in a spot. Meat was assayed for traces of oleandomycin by Rutczynska-Skonieczna (1980) who separated the drug by two-dimensional TLC. Neissner (1980) described a two-dimensional TLC method for the analysis of complex mixtures of partial glycerides of castor oil fatty acids. Several phenolic glycosides of fruit including 1-*o*-hydroxycinnamyl-beta-D-glycosides were reported by Reschke and Hermann (1981) who used two-dimensional TLC. Two-dimensional TLC and HPLC have been used by Barcelon (1982) as a means of separating free amino acids as their 5-dimethylaminonaphthalene-1-sulfonyl derivatives. Puttemans *et al*. (1982) evaluated TLC and paper chromatography for the identification of food coloring dyes extracted by ion-pairing. The hydroxyflavone glycosides of some spices were isolated by Hoffman and Herrmann (1982) by paper chromatography and TLC. Twenty anions, including nitrate, nitrite, and thiocyanate, were separated by Tanabe *et al*. (1981) who measured nitrates in vegetables using paper chromatography.

A two-stage TLC method (CH_3Cl:MeOH:HAc in a ratio of 98:2:1 followed by hexane:ethyl acetate:HAc in the ratio of 470:30:1) is presented by Bitman *et al*. (1981) for the separation of lipid classes.

BIBLIOGRAPHY

Barcelon, M. D. (1982). New liquid chromatography approaches for free amino acid analysis in plants and insects. *J. Chromatogr.* **238**, 175–182.

Bitman, J., Wood, D. L., and Ruth, J. M. (1981). Two-stage, one-dimensional thin layer chromatographic method for separation of lipid classes. *J. Liq. Chromatogr.* **4**, 1007–1021. [*Anal. Abstr.* **4**, 6D146.]

Consden, R., Gordon, A. H., and Martin, A. J. P. (1944). Quantitative analysis of proteins: A partition chromatographic method using paper. *Biochem. J.* **5**, 224–232.

Ehnert, M., Popken, A. M., and Dose, K. (1981). Quantitative determination of penicillic acid in vegetables. *Z. Lebensm. Unters. Forsch.* **172**, 110–114.

Hoffman, B., and Herrmann, K. (1982). Flavonol glycosides of wormwood (*Artemisia Vulgaris* L.), tarragon (*Artemisia dracunculus* L.) and absinthe (*Artemisia absinthium* L.) 8. Phenolics of spices. *Z. Lebensm. Unters. Forsch* **174**, 211–215.

Neissner, R. (1980). Thin layer chromatographic separation of partial glycerides of castor-oil fatty acids. *Fette, Seifen, Anstrichm.* **83**, 257–262. [*Anal. Abstr.* 1981, **40**, 3F90.]

Puttemans, M., Dyron, L., and Massart, D. L. (1982). Isolation, identification, and determination of food dyes following ion-pair extraction. *J. Assoc. Off. Anal. Chem.* **65**, 737–744.

Randerrath, K. (1963). "Thin-layer Chromatography." Academic press, New York.

Reschke, A., and Herrmann, K. (1981). *Z. Lebensm. Unters. Forsch.* **172**, 110.

Rutczynska-Skonieczna, E. (1980). Detection of ethopabate residues in eggs. *Rocz. Panstw. Zakl. Hig.* **31**, 181–185. [*Anal. Abstr.* **39**, 6F44.]

Stahl, E., Schroter, G., Kraft, G., and Rentz, R. (1956). Thin layer chromatography (the method, affecting factors, and a few examples of application). *Pharmazie* **11**, 633–637.

Stock, R., and Rice, C. B. F. (1967). "Chromatographic Methods." Chapman & Hall, London.

Tanabe, S., Toida, T., Ogata, K., Taguchi, K., and Imanari, T. (1981). Densometric analysis of inorganic anions on paper chromatography by ultraviolet absorption. *Bunseki Kagaku* **30**, 30–35. [*Anal. Abstr.* **41**, 2B10.]

23
Gas–Liquid Chromatography

It was pointed out by Martin and Synge (1941) that the use of a gas as a mobile phase might have certain advantages. The big problem in achieving fast and sharp separations lies in how fast the molecules can be moved around. If a sample is moved very rapidly to the stationary phase (and hopefully very rapidly out of it), then the less time it has to diffuse in other directions, with the result that we can get narrower bands and a sharper separation.

In 1952 Martin and James published the first paper on gas chromatography (GC). In 1956 the first commercial instruments appeared, and by 1959 over 600 articles had appeared in the literature. Figure 23.1 shows the necessary components of a gas chromatograph.

INSTRUMENT COMPONENTS

Carrier Gas

The most common carrier gases (mobile phase) are helium, nitrogen, and argon. The type of carrier gas to be employed presents some choice, but is usually dictated by the type of detector used.

Pressure Differential

The pressure differential over the column plays an important role in the gas chromatograph. The gas velocity down the column is obtained by

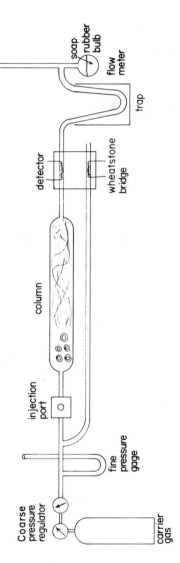

Fig. 23.1. Schematic of a gas chromatograph.

applying a pressure differential. Commonly the outlet pressure is atmospheric, and the inlet is maintained at a pressure somewhat above atmospheric. Generally a pressure ratio (p_1/p_2) of 2–3 is used. If the pressure ratio is too low, molecular diffusion remixes the separated components and efficiency decreases. If the ratio is too high, the resistance to mass transfer increases and again the efficiency of separation decreases. (See the section on the van Deemter equation later in this chapter.) For routine work the coarse pressure regulator gages on the gas cylinder can be used, but for fundamental research an auxiliary precision pressure gage is necessary.

Injection Port

The sample enters the chromatograph through the injection port, which is usually heated so that liquids can be vaporized immediately upon injection. The better instruments have a separate heater for the injection port.

Sample addition is generally done by means of a syringe, a volume of 0.5–1 ml being required for gases, and 1–100 μl for liquids. However, gas-sampling valves, backflushing valves, pyrolysis systems, inlet splitters, and solid samplers are among other sampling devices.

A *gas-sampling valve* is shown in Fig. 23.2.

Backflushing valves allow flushing out of the column the high-boiling components as a single composite peak after obtaining separation of the low-boiling components, thereby shortening the time required before the next sample can be added.

Pyrolysis systems are used for solids and very high boiling liquids (paint film, alkaloids, plastics). The object is to partially combust the sample by heating the inlet port to a high temperature. These materials are degraded to volatile components characteristic of the starting material.

Splitters are used to divide a sample into two equal parts for dual-column chromatography, or to take only a fraction of the initial sample.

A *solid sampler* is essentially a syringe within a syringe as shown in Fig. 23.3.

Column

Columns are usually made out of copper, stainless steel, aluminum, nickel, or glass, with diameters of 1/8 and 1/4 in. and lengths of 3–10 ft being the normal sizes. Columns 3/8 in. and larger are considered to be *preparative columns*; those smaller than 1/16 in. are called *capillary columns*.

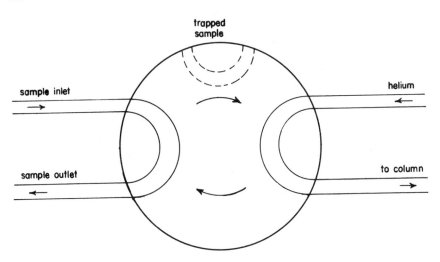

Fig. 23.2. Gas-sampling valve.

The columns must be heated to about two-thirds of the boiling point of the highest boiling material in the mixture to be separated. If one part of the column is heated to a different temperature than another, separations may be made but they will not be reproducible and good physical constant measurements cannot be made. Therefore, in order to heat the column more uniformly and at the same time save space, the column is usually coiled (3-in. diam for 1/8 in. tubing, and about 6-in. diam for 1/4 in. tubing) or bent into a U shape. The better instruments have a separate heater for the column.

Sometimes in the separation of mixtures of several components, the more volatile components come off in a short time and have nicely shaped peaks, whereas the higher boiling materials may take an hour or more to be eluted. When a material takes this long to come off of the column, it does so as a very broad band because there has been more time for diffusion. This means that quantitative analysis is hard to obtain (see Fig. 23.4A).

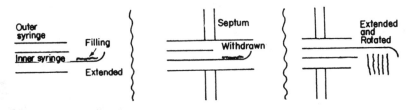

Fig. 23.3. A solid sampler.

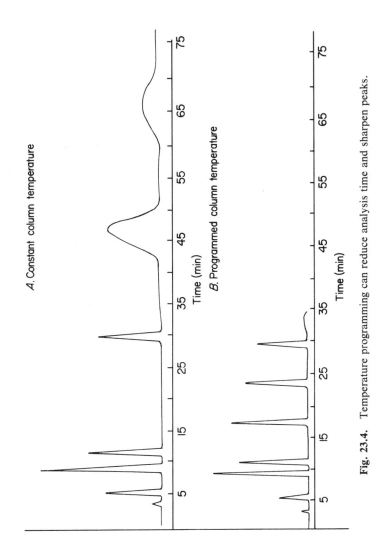

Fig. 23.4. Temperature programming can reduce analysis time and sharpen peaks.

However, once the low boiling materials have been separated, the column temperature can be raised, thus speeding up the emergence of the rest of the components; as they come out of the column, the temperature can be raised further to speed up the remaining materials. This can be done manually but usually a timing device can be installed that will increase the temperature at a set rate. The result of such temperature programming is shown in Fig. 23.4B. Note that the analysis time has been shortened and the peak shapes have been improved, so that good quantitative data can be obtained.

The disadvantages of temperature programming are (1) that the column must be cooled to the original temperature before the next sample can be added and (2) the temperature in the column lags behind the temperature in the surrounding air chamber, so reproducible results are difficult to obtain.

Capillary Columns. Capillary columns have a diameter of 0.005–0.02 in., are 100–500 ft long, and are made from copper, stainless steel, glass, or nylon. The column itself acts as the inert phase. The stationary phase is added by filling the first foot with a 10–15% solution of the stationary phase and pushing it through the column (with carrier gas) at 2–5 mm/ sec. After the column is coated, the gas flow is continued at a faster rate for an hour or so to evaporate the excess solvent.

Because the thickness of the stationary phase layer is so small (0.3–2 μm), the resistance to mass transfer decreases and highly efficient, rapid analysis can be obtained. Several hundred thousand theoretical plates are common, and there is one report of seven components being separated in 25 sec. A new type of capillary column, the fused silica column, has recently been introduced. These columns usually have about 4000 theoretical plates per meter.

Because the capacity of capillary columns is so small, it is difficult to accurately inject the required small samples (a few tenths of a microliter). A way to circumvent the problem is to use a *splitter,* which splits a large sample into two portions. The smaller portion is injected into the column, and the larger is usually discarded. Figure 23.5 shows one type of capillary inlet splitter. Detectors that have small volumes also are necessary with capillary columns.

Solid Supports

The function of the solid support is to act as an inert platform for the liquid phase in the column. The solid supports most often used are described in this section.

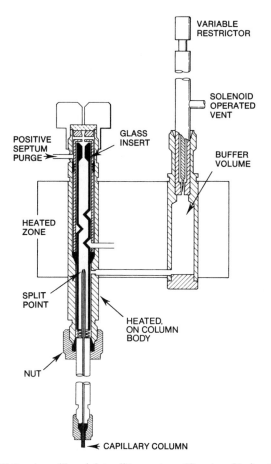

Fig. 23.5. A capillary inlet splitter system. (Courtesy Varian Corp.)

Chromosorb P is a pink diatomaceous earth material that has been carefully size-graded and calcined. The surface area is 4–6 m²/g. This material is the least inert of the Chromosorb supports but offers the highest efficiency.

Chromosorb W is a white diatomaceous earth material that has been flux-calcined with about 3% sodium carbonate and has a surface area of 1–3.5 m²/g. This material is more inert but less efficient than Chromosorb P.

Chromosorb G to a very substantial degree combines the high column efficiency and good handling characteristics of Chromosorb P with the following advantages over present white diatomite supports: greater column efficiency, less surface adsorption, harder particles, and less break-down in handling. Since Chromosorb G is about 2.4 times as heavy as Chromosorb W, a 5% liquid loading on Chromosorb G is equal to 12% on Chromosorb W. No more than 5% liquid phase should be used on Chromosorb G.

Silanized supports are Chromosorb supports that have been coated with dimethyldichlorosilane (DMCS) to reduce surface active sites of the dia-tomaceous earth material. Chromosorb supports deactivated with DMCS provide less tailing, minimized catalytic effects, and improved results with low liquid loadings. Hexamethyldisilane (HMDS) has also been used. The combination of acid washing and DMCS treatment is particularly effective in reducing adsorption.

Fluoropak 80 and *Chromosorb T* are made from fluorocarbon polymer particles. They are chemically inert and can be used, when coated, to separate very polar compounds such as water, acetic acid, ammonia, and amines, without tailing. *Chromosorb T* is made from 40/60 mesh Teflon 6 and has a surface area much greater than Fluoropak 80. This larger surface area generally results in columns with higher efficiencies.

Glass beads provide a low surface area, inert, solid support that is used with 0.25% or less stationary phase. A glass bead support is recommended for use in separating highly polar compounds of high molecular weight.

Solid Adsorbents

Several important adsorbents are used in gas–solid chromatography. Charcoal, alumina, silica gel, and molecular sieves are used generally to separate gaseous mixtures. Molecular sieves are used to separate O_2 and N_2 but do not elute CO_2 under normal conditions. Silica gel on the other hand elutes CO_2 but does not separate O_2 and N_2. Liquid-modified solid adsorbents are used for the analysis of higher molecular weight materials.

Poropak is finding many applications in gas chromatography. Poropak is a porous bead formed by the polymerization of monomers such as styrene with divinylbenzene as a cross linker. These beads serve both the function of solid support and liquid phase, although they may be mod-ified with liquid phases. Elimination of the conventional solid support removes the adsorption sites that normally cause tailing. In addition, elim-ination of the liquid phase reduces bleeding. Poropak columns appear to be particularly useful for the analysis of gases and polar compounds.

Poropak P and Q are made with ethylvinylbenzenes or styrene mon-omers. Types N, R, S, and T are similar, but have been moldified with polar monomers. This changes the elution order of polar solutes, espe-cially water.

Porous polymers also can be modified with conventional liquid phases. This reduces the absolute retention of all solutes and shifts the order of elution. The principal effect is to reduce tailing of polar materials such as amines and organic acids. A liquid coating of about 5% by weight is recommended.

Stationary Phases

Stationary phases used in gas–liquid chromatography may be classified into five types:

1. *Nonpolar:* This class consists of hydrocarbon-type liquid phases including silicone greases (e.g., SE-30, squalene) but not aromatic materials. Although many exceptions are reported, generally nonpolar liquid phases separate solutes in order of increasing boiling points.

2. *Polar:* Liquid phases containing a large proportion of polar groups are in this class (e.g., Carbowax, dimethyl sulfolane). These materials differentiate between polar and nonpolar solutes retaining only the polar materials.

3. *Intermediate:* Polar or polarizable groups on a long nonpolar skeleton are typical of the intermediate stationary phases (e.g., SE-52, diisodecyl phthalate, benzyl diphenyl). Members of this group dissolve both polar and nonpolar solutes with relative ease.

4. *Hydrogen Bonding:* This special class of polar liquid phases contains compounds with a large number of hydrogen atoms available for hydrogen bonding (e.g., the hydroxyl groups of glycerols).

5. *Specific:* Special purpose phases that rely on a specific chemical interaction between solute and solvent to perform separation (e.g., $AgNO_3$ for olefins) are termed specific phases.

Column Conditioning. Columns should be preconditioned for at least 10 hr at about 20°C above the maximum temperature at which you plan to operate the column, but below the maximum temperature limit for the stationary phase. This conditioning removes solvent and other volatile materials that will cause interference later by altering the column conditions and making reproducible results impossible. Very small amounts of carrier gas (5–10 ml/min) should be flowing through the column during the conditioning period. The columns can be conditioned in the chromatograph if the exit end is disconnected from the detector.

Common Abbreviations. Table 23.1 lists some of the common stationary phases and the abbreviations by which they generally are known.

Table 23.1.   Abbreviations of Some Stationary Phases

Abbreviation	Chemical name
Aroclor	Chlorinated biphenyls
DEGA	Diethylene glycol adipate
DEGSE	Diethylene glycol sebacate
DEGS	Diethylene glycol succinate
EGA	Ethylene glycol adipate
EGIP	Ethylene glycol isophthalate
EGSE	Ethylene glycol sebacate
EGS	Ethylene glycol succinate
HMPA	Hexamethylphosphoramide
IGEPAL	Nonyl phenoxy polyoxyethylene ethanol
NPGS	Neopentyl glycol succinate
SE-30	Methyl silicone gum rubber
THEED	Tetrahydroxyethyl ethylene diamine
TCP	Tri cresyl phosphate
TWEEN	Polyoxyethylene sorbitan monooleate

Detectors

A detector is a device that measures the change of composition of the effluent. Dozens of different kinds of detectors are used in gas chromatographs. The ones described here are thermal conductivity, hydrogen flame ionization, cross section, argon ionization, electron capture, and flame photometric. The Hall differential conductivity detector, which was described in Chapter 14, also can be used in gas chromatography.

Thermal Conductivity Detectors. If you blow air over a hot wire, it will be cooled. This phenomenon is the basis of the thermal conductivity detector. A thin filament of wire is placed at the end of the GC column and heated by passing a current through it. The carrier gas molecules strike the hot wire; as each molecule hits the wire, it takes away some heat from the wire. As the wire is cooled, its resistance changes. When a sample passes over the hot wire, the sample molecules, because of different velocities and masses, will take away a different amount of heat from the hot wire than the carrier gas. This means that the resistance of the wire will change from what it was and a signal will be produced.

The greater the difference in thermal conductivities of the carrier gas and the components of a sample, the greater the sensitivity of this type of detector. In Table 23.2 are listed the thermal conductivities of several representative materials. The two gases whose thermal conductivity (k) differs the most from that of other compounds are H_2 and He. Hydrogen would appear superior but presents safety hazards in the laboratory. Helium is nearly as satisfactory and is noncombustible.

Table 23.2. Thermal Conductivity, K, of Some Gases and Vapors

Substance	$K(\times 10^5)$	Substance	$K(\times 10^5)$
Hydrogen	41.6	i-Butane	3.32
Nitrogen	5.81	n-Pentane	3.12
Oxygen	5.89	i-Pentane	3.00
Air	5.83	n-Hexane	2.96
Ammonia	5.22	Ethylene	4.19
Water	3.5	Acetylene	4.53
Helium	34.8	Methyl chloride	2.20
Neon	11.1	Methylene chloride	1.61
Argon	3.98	Chloroform	1.58
Krypton	2.12	Methyl bromide	1.50
Xenon	1.24	Methyl iodide	1.13
CO	5.63	Freon-12	1.96
CO_2	3.52	Methanol	3.45
CS_2	3.70	Ethanol	3.5
Methane	7.21	Diethyl ether	3.6
Ethane	4.36	Acetone	3.27
Propane	3.58	Methyl acetate	1.61
n-Butane	3.22	Nitrous oxide	4.5
		Nitric oxide	5.71

The wire is heated by an electric current to a temperature T_2, which is hotter than the wall of the detector, T_1. The sensitivity of detection with these cells generally increases with a reduction in the wall temperatures (greater difference between T_2 and T_1), but one must not lower the wall temperature too greatly, or condensation of the components will occur in the cell. The better instruments have a separate heater for the detector.

Figure 23.6 show three different arrangements of thermal conductivity detectors. The through-flow type has the advantage of a fast response time constant (1 sec), but is very susceptible to small changes in flow rate. The diffusion type was designed to smooth out rapid flow rate changes, producing a more stable detector, but the response time constant is slow (20 sec). The purging type is an attempt to combine the advantages of the other two types. If one hot wire is used, the detector is unstable because it is sensitive to temperature changes, flow rate changes, and filament current changes. To remedy this, the Wheatstone bridge circuit is used.

The important characteristics of the thermal conductivity detector are as follows (courtesy Varian-Aerograph Co.):

1. Sensitive to all organic compounds (10^{-7} g/sec)
2. Linear dynamic range: 10,000
3. Sensitive to flow rate and temperature changes

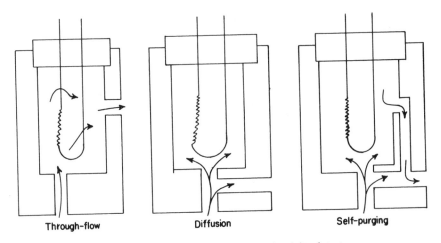

Through-flow   Diffusion   Self-purging

Fig. 23.6. Geometries of thermal conductivity detector.

4. Detector temperature limit: 500°C
5. Carrier gases: He, H_2, N_2, Ar, CO_2

Hydrogen Flame Ionization Detector. The hydrogen flame ionization detector (FID) works as follows: When carrier gas (CG) and sample emerge from the column, hydrogen (1 H_2/1 CG) and air (10 air/1 CG) are added to the carrier gas to produce a flame of about 2100°C. This flame produces + and − ions whose concentrations have been established at 10^{10} to 10^{12} ions/ml. This concentration of ions is too large to come only from the known combustion products of the flame. It has been suggested that carbon forms aggregates, since free carbon is known to polymerize, and the low work function (4.3 eV) of these carbon aggregates could explain the large ion concentrations observed. This is further substantiated by the fact that this detector's sensitivity is roughly proportional to the carbon content of the sample.

The response of the FID per gram-mole of a given compound is directly proportional to the number of C atoms bound directly to H or other C atoms. Carbons bonded to OH, amines, and X_2 provide a fractional contribution as do alkali and alkaline earth elements, but no signal is obtained from fully oxidized C atoms (COOH, C≡O) and inert gas molecules. The net result is that the FID is sensitive to all organic compounds except HCOOH and insensitive to (1) all inorganic compounds except those containing alkali and alkaline earth elements and (2) Ar, He, Ne, O_2, H_2S, CO, CO_2, NO_2, and H_2O.

The FID is about 1000 times more sensitive than a thermal conductivity cell and can easily detect 1-ng samples. As a generalization, a thermal conductivity detector is not as good as your nose, a flame ionization detector is as good as your nose, and the electron capture detector to be discussed later is better than your nose.

Two precautions should be observed with a FID: (1) A hydrogen flame is colorless, so use a piece of paper and not your finger to check if the flame is on. (2) Dust and dirt contain alkali metals, so be sure your air supply is clean or erratic results will be obtained.

The basic characteristics of the FID are as follows: (courtesy of Varian-Aerograph Co.):

1. Sensitive to all organic compounds (10^{-12} g/sec)
2. Linear dynamic range: 1,000,000
3. Insensitive to temperature change (ideal for programming)
4. Detector temperature limit: 400°C
5. Carrier gas: N_2, He, or Ar

Figure 23.7 shows the physical arrangement of a flame ionization detector.

Cross-Section Detectors. Rather than use a flame to produce ions, high-speed electrons (β particles) can be used. Strontium-90 and tritium (^3H) are used as the source of β particles. When the β particles strike the carrier gas, they can produce + ions and electrons. These electrons can be captured by other gas molecules and become − ions, or the electrons may ionize other gas molecules. The net effect is an increase in the charged particles in the gas stream with the result that the conductivity

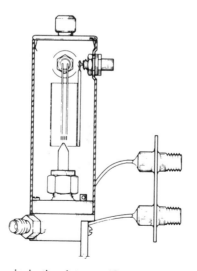

Fig. 23.7. Flame ionization detector. (Courtesy Varian-Aerograph Co.)

of the gas increases. If 300–500 V are applied to the collector plate, the charged species can be collected, producing a signal. When the sample comes along, it also will produce ions, but because of the usually larger size of the sample molecules, more electrons are utilized and the gas conductivity is different than that of the carrier gases.

The adsorption cross section is roughly proportional to the mass of the molecule, so to get maximum sensitivity. H_2 and He are usually used as the carrier gases. Because β particles do not have a high penetrating power the detector must be small in order to be efficient. The β sources are quite susceptible to high temperature; therefore, these detectors are limited to 200–225°C. A diagram of a cross-section detector is shown in Fig. 23.8.

The cross-section detector is sensitive to all organic and inorganic compounds; it is the only detector capable of measuring gas sample concentrations up to 100% within the detector. It is about as sensitive as a thermal conductivity cell.

The basic characteristics of the cross-section detector are as follows (courtesy of Varian-Aerograph Co.):

1. Sensitive to all vapors and gases; comparable to a good thermal conductivity detector (about 10^{-7} g/sec)
2. Linear dynamic range: 5×10^5 to 100% concentration
3. Insensitive to temperature change
4. Insensitive to carrier gas molecules

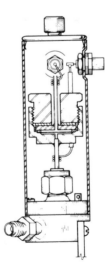

Fig. 23.8.   Micro cross-section detector. (Courtesy Varian-Aerograph Co.)

5. Detector temperature limit: 220°C

6. Carrier gases: H_2 to 100°C or He + 3 CH_4 (1:3)

Argon Ionization Detectors. The cross-section detector can be made much more efficient if argon is used as the carrier gas. In this case, the β particles produce Ar* atoms, which then transfer their energy very efficiently to the sample molecules producing ions and electrons and a very strong signal, 10–100 times more sensitive than a hydrogen flame ionization detector. The following sequence of reactions has been suggested:

Background

$$Ar + \beta^-(\text{high energy}) = Ar^+ + e^-$$

$$Ar + \beta^-(\text{low energy}) = Ar^* (10^{-4} \text{ sec half-life, } 11.6 \text{ eV})$$

$$Ar^* + \text{impurities} = Ar + M^+ + e^-$$

Signal

$$Ar^* + \text{sample} = Ar + M^+ + e^-$$

Materials that can be ionized by 11.6 eV can be detected, which means that most materials (except the noble gases, some inorganic gases, and fluorocarbons) can be detected. ^{90}Sr, ^{85}Kr, RaD, and T (^3H) are usually used to provide the β particles. The argon carrier gas must be very dry.

The output stability of the argon detector is such that no reference cell is required. Moderate variations in carrier flow rate and temperature have negligible effect on the detector base current.

Electron Capture Detector. With the cross-section detector there was an increase in the number of ions in the gas, so there was an increase in signal, and all types of compounds would respond. To make a more selective detector, use of electron capture was investigated. Suppose that the electrons striking the sample molecule had just enough energy to penetrate the electrical field of the molecule and be captured but not enough to break up the molecule into ions. If this were to happen, the original electrical signal, based on the free electrons, would now decrease because of the electrons lost due to being captured.

Halogenated compounds, conjugated carbonyls, nitriles, nitrates, and some organometallic compounds are quite sensitive to the electron capture process.

The electron capture detector has the following basic characteristics (courtesy of Varian-Aerograph Co.);

1. Variable sensitivity: 0.1 pg (10^{-13} g) for CCl_4 or 5 × 10^{-14} g/sec

2. Linear dynamic range: 500

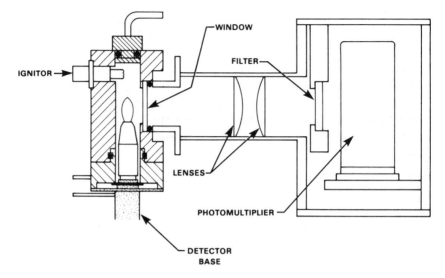

Fig. 23.9.   Flame photometric detector. (Courtesy Varian Corp.)

3. Sensitive to temperature change on column; programming not recommended

4. Insensitive to most organic compounds except those containing halogens, sulfur, nitrogen (nitriles, nitrates), and conjugated carbonyls

5. Detector temperature limit: 220°C

6. Carrier gas: N_2, must be pure and dry

7. Ionization source—250 mCi tritium

Flame Photometric Detector.   The flame photometric detector is rather specific for those compounds containing either sulfur or phosphorus. It is based on the principle that if such compounds are combusted, then the 526 nm emission band of phosphorus and the 394 nm emission band of sulfur can be detected with a phototube. Figure 23.9 is a diagram of a flame photometric detector made by Varian Corp.

The sample emerging from the column is combusted in a hydrogen–air flame, and the resulting emission is monitored. If phosphorus is desired, then a 526-nm filter is used; if sulfur is desired, then a 394-nm filter is used. This is an excellent detector to use when complex mixtures are separated and only those compounds containing sulfur or phosphorus are of interest.

The sensitivity of this detector for phosphorus is 10^{-12} g/sec and for sulfur is 10^{-10} g/sec. Its linear dynamic range is 10^5 for phosphorus and 10^3 for sulfur. The selectivity ratios compared to carbon are P/C = 10^5 and S/C = 10^3 to 10^6.

Flow-rate Devices

If retention times are to be duplicated from day to day and from laboratory to laboratory, one of the parameters that must be controlled is the carrier gas flow rate. Many precise flow meters are commercially available. However, in their absence, a buret fitted with a side arm and a rubber squeeze bulb full of liquid soap will work quite well (see Fig. 23.1). All you do is squeeze the bulb until the soap rises so that the carrier gas will form a film out of it. This film will be pushed to the top of the buret and by measuring how many milliliters it rises in a given time, you can determine the flow rate. A spot of grease at the top of the buret breaks the soap film and it runs back into the squeeze bulb.

Recently, flow programming has been suggested as an alternative to temperature programming for heat-sensitive materials. In this case the flow rate is increased as the separation progresses.

The full effect of flow rate changes is discussed in the next section.

HEIGHT EQUIVALENT TO A THEORETICAL PLATE

The height equivalent to a theoretical plate (HETP) for a column gives the operator some idea of how efficient the column is and whether or not it would be worthwhile to spend time trying to improve the separation. Two equations for obtaining the number of theoretical plates are the *Glueckauf equation*

$$N = 8(t/\beta)^2 \tag{23.1}$$

and the *van Deemter equation*

$$N = (4t/W)^2 \tag{23.2}$$

where N is the number of theoretical plates of the column, β the peak width at H/e of the peak height, and W the distance between the intersections of tangents to inflection points on the base line; H/e and W are shown schematically in Fig. 23.10. Note that t is shown as the time lapse between the injection point O and the middle of the peak; t in units of distance could also be used in determining N.

Two other equations, very similar to equation 22.2, also may be used to determine N:

$$N = \left(\frac{4t_I}{W} + 2\right)^2 \tag{23.3}$$

$$N = \left(\frac{4t_F}{W} - 2\right)^2 \tag{23.4}$$

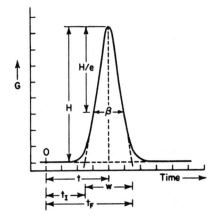

Fig. 23.10. Illustration of gas chromatographic terms.

where t_I and t_F are as defined in Fig. 23.10. Equation 23.4 is particularly useful when unsymmetrical peaks are obtained.

Example: Let us calculate the number of theoretical plates N if $H = 42$, $t = 40$, the peak area $= 540$, $\beta = 15$, $t_I = 29$, and $t_F = 51$.

(a) From equation 23.1

$$N = 8(40/15)^2 = 8 \times 7.11 = 56.9$$

(b) From equation 23.2

$$N = (4 \times 40/22)^2 = (7.273)^2 = 52.9$$

(c) From equation 23.3

$$N = \left(\frac{4 \times 29}{22} + 2\right)^2 = (5.273 + 2)^2 = 52.9$$

(d) From equation 23.4

$$N = \left(\frac{4 \times 51}{22} - 2\right)^2 = (9.273 - 2)^2 = 52.9$$

The height equivalent to one theoretical plate is defined by

$$\text{HETP} = L/N \tag{23.5}$$

where L is the length of the column. By substituting equation 23.2 we obtain

$$\text{HETP} = L(W/4t)^2 \tag{23.6}$$

where the units of W and t must be the same (i.e., either units of time or distance).

Example: Let us assume that the column described in the previous example was 4 ft in length. That is, the length, L = 4 ft × 12 in./ft × 2.54 cm/in. = 122 cm. From equation 23.5 and using an average value of N = 54 obtained in the example above, we see that

$$\text{HETP} = 122/54 = 2.26 \text{ cm}$$

Van Deemter Equation

The theoretical plate concept is useful in determining the efficiency of any given column, but it does not indicate the effects of various operational parameters. Here, however, the van Deemter rate theory proves valuable. The *van Deemter equation* is

$$\text{HETP} = A + (B/V) + CV \qquad (23.7)$$

where V is the velocity of the carrier gas and the values of A, B, and C are given by the following:

$$A = 2\lambda \bar{d} \qquad (23.8)$$

$$B = 2\gamma D_g \qquad (23.9)$$

$$C = 8kd_f/\pi^2 D_e(1 + k)^2 \qquad (23.10)$$

In these expressions, λ is a quantity characteristic of the column packing, $\bar{d}$ the average particle diameter, D_g the diffusion coefficient in the gas phase, k the fraction of the sample in the liquid phase divided by the fraction in the vapor phase, d_f the average liquid film thickness, and D_l the diffusion coefficient in the liquid phase.

The first term in equation 23.7 is due to Eddy diffusion (turbulence and the column packing), the second term to molecular diffusion, and the third term to the resistance to mass transfer. These component parts of the van Deemter equation are illustrated graphically in Fig. 23.11.

In Fig. 23.11 data are shown for the variation of the HETP of a column, prepared by coating Celite with tri-cresyl phosphate, as a function of the linear velocity of the carrier gas in cm/sec. The curve R-S is the actual variation of the HETP with velocity. The tangent to this curve, P-Q, allows the extrapolation to zero velocity. Thus the Eddy diffusion is shown as the straight line P-T. The resistance to mass transfer is represented by the contributions between lines P-T and P-Q for *n*-butane, and between P-T and P-U for air. The molecular diffusion contribution is represented by the difference between P-Q and R-S. Thus, at very low velocities of carrier gas the molecular diffusion is more important,

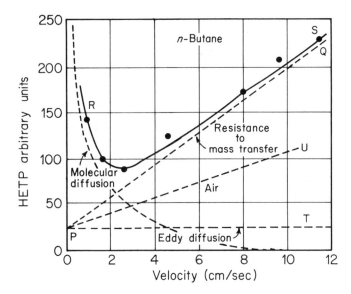

Fig. 23.11.   Variation of HETP as a function of carrier gas velocity.

whereas at high gas velocities the mass transfer resistance becomes more important.

The optimum velocity of carrier gas to be employed is at that point where the HETP–velocity curve is at a minimum.

IDENTIFICATION OF COMPOUNDS

Under a given set of conditions and with a given chromatographic column, the retention time or retention volume is characteristic of a particular component. However, in certain cases, two or more materials may have the same retention volume. They seldom have the same retention time on stationary phases varying in polarity, so if in doubt, change columns and see if other peaks appear. The result is that you can never be completely sure of a positive identification of an unknown from retention data alone. Nevertheless, the characteristic retention volume (or retention time) is extremely valuable in identifications and particularly in separations of components in mixtures. If some knowledge is already available concerning the possible constituents of the mixture, the gas chromatographic results may be sufficient for positive identification. In cases of doubt, small quantities of the separated materials may be trapped in cold traps after they emerge from the column. Mass spectrometric or fast-scan infrared measurements can be employed for positive identification.

A useful relation in gas chromatography is that the logarithm of the retention volume is often a linear function of the number of carbon groups, CH_2 groups, or the like in a series of compounds of the same general type. This is shown in Fig. 23.12 for the straight-chain hydrocarbons. This relationship may be useful in predicting retention volumes of new compounds, as explained in the following example.

Example: From the data for straight-chain hydrocarbons in Fig. 23.12, we can calculate the retention volumes for the higher homologues in the series. We may write the equations

$$\log 6 = 6x + y$$

$$\log 10.5 = 7x + y$$

$$\log 18.6 = 8x + y$$

$$\log 32.7 = 9x + y$$

Taking any two of these equations, we may solve two simultaneous equations to find x and y. Thus,

$$\log 18.6 = 8x + y$$
$$\underline{-\log 6.0 = -6x - y}$$
$$\log 18.6 - \log 6.0 = 2x$$

$$1.26951 - 0.77815 = 2x = 0.49136$$

$$x = 0.24568$$

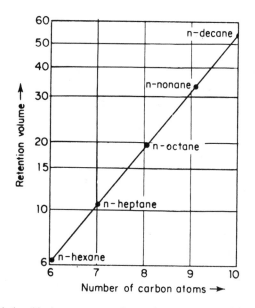

Fig. 23.12. Relationship between retention volume and number of carbon atoms in straight-chain hydrocarbons. Note that the retention volume is plotted on a log scale.

and therefore

$$y = 0.77815 - (6 \times 0.24568)$$
$$= 0.77815 - 1.47408 = -0.69593$$

As a check, we calculate the retention volume for n-decane:

$$\log V_R = 10(0.24568) - 0.6593 = 1.7975$$

Taking the antilog, we get $V_R = 57.7$, which is in good agreement with the experimental value of 57.5.

RECENT APPLICATIONS

Josephson *et al.* (1984) used a 60-m fused silica capillary column coated with Carbowax 20M to determine the amines, carbonyls, and alcohols responsible for different fish odors. Lundstrom and Racicot (1983) tested the volatile amines dimethylamine and trimethylamine, the common degradation products of trimethylamine oxide, as indicators of the quality of seafood. Larsson (1983) determined the preservatives benzoic acid and sorbic acid in almond paste, fish homogenate, and apple juice by gas chromatographic separation of their trimethylsilyl derivatives; recoveries of 99–101% were reported.

Daft used a double column (20% OV-101 and a 2:1 mixture of 20% OV-225:20% OV-17) and a ^{63}Ni electron capture detector to determine fumigant residues in stored grains. Varner *et al.* (1983), using a Chromosorb 104 column and an FID, was able to detect the styrene transferred to margarines from the packaging material down to 25 ppb. Lehtonen (1983) studies phenolics in rums, cognacs, and bourbons by forming 2,4-dinitrophenyl derivatives and separating them on a gas chromatograph with an electron capture detector. She found that Scotch, Spanish, and Japanese whiskies contained o-, m-, and p-cresols that the other whiskies did not contain.

BIBLIOGRAPHY

Daft, J. L. (1983). Gas chromatographic determination of fumigant residues in stored grains, using isooctane partitioning and dual column packings. *J. Assoc. Off. Anal. Chem.* **66**, 228–234.

Josephson, D. B., Lindsay, R. C., and Stuiber, D. A. (1984). Variations in the occurences of enzymatically derived volatile aroma compounds in salt and freshwater fish. *J. Agric. Food Chem.* **32**, 1344–1347.

Larsson, B. K. (1983). Gas–liquid chromatographic determination of benzoic acid and sorbic acid in foods. *J. Assoc. Off. Anal. Chem.* **66**, 775–781.

Lehtonen, M. (1983). Gas chromatographic determination of volatile phenols in matured distilled alcoholic beverages. *J. Assoc. Off. Anal. Chem.* **66**, 62–71.

Lundstrom, R. C., and Racicot, L. D. (1983). Gas chromatographic determination of dimethylamine and trimethylamine in seafoods. *J. Assoc. Off. Anal. Chem.* **66**, 1158–1163.

Martin, A. J. P., and Synge, R. L. M. (1941). A new form of chromatography employing two liquid phases. *Biochem. J.* **35**, 1358–1368.

Martin, A. J. P., and James A. T. (1952). Gas–liquid partition chromatography. The separation and microestimation of volatile fatty acids from formic acid to dodecanoic acid. *Biochem. J.* **50**, 679–690.

Varner, S. L., Breder, C. V., and Fazio, T. (1983). Determination of styrene migration from food container polymers into margarine using azeotropic distillation and headspace GC. *J. Assoc. Off. Anal. Chem.* **66**, 1067–1074.

24
Extraction

A large number of analytical methods for foods require an extraction as a cleanup procedure, as a concentration step, to remove a slightly soluble material, or to aid in the identification of a component.

These various procedures involve liquid–liquid and liquid–solid extractions using batch, continuous, and discontinuous countercurrent techniques.

LIQUID–LIQUID BATCH EXTRACTION

One of the commonest and simplest extraction techniques is liquid–liquid batch extraction, which is generally carried out in a separatory funnel. A *liquid–liquid* extraction is based on partition of a material between two immiscible liquids; usually one is organic and one is water. The usual parameter for expressing the extent of this partition is the *distribution ratio D*:

$$D = \frac{\text{total grams solute in the organic phase}}{\text{total grams solute in the aqueous phase}} \qquad (24.1)$$

We cannot quantitatively predict the value of D for any given system, but it is known that for a substance to be extracted it must be neutral. Ions are made neutral by chelation or ion association. For example, lead in food coloring dyes can be chelated and neutralized by diethyl dithio carbamate, whereas iron in ferbam, a plant insecticide, can form an ion association system with HCl:

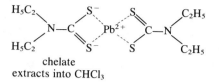

chelate
extracts into CHCl$_3$

FeCl$_4^-$H$^+$

ion association
extracts into ether

Percentage Extraction

Equation 24.2 is used to calculate the percentage of a compound that can be extracted if D and the volume of each phase is known:

$$\% \text{ extraction} = \frac{100D}{D + V_w/V_o} \qquad (24.2)$$

where V_w and V_o are the volume of the water and organic layers, respectively.

In order for two components to separate, not only must their distribution coefficients differ but they must be in the correct range. For example, suppose you have two systems: system 1 contains compound A with a D of 1 and compound B with a D of 10; system 2 contains compound X with a D of 100 and compound Y with a D of 1000. In both systems, the ratio of Ds is 10, yet system 1 is far easier to separate than system 2, as the following calculations indicate:

$$A\% = \frac{100 \times 1}{1 + 1/1} = 50\%, \qquad X\% = \frac{100 \times 100}{100 + 1/1} = 99.0\%$$

$$B\% = \frac{100 \times 10}{10 + 1/1} = 90.9\%, \qquad Y\% = \frac{100 \times 1000}{1000 + 1/1} = 99.9\%$$

B is not completely extracted but it is separating from A

Both are almost completely extracted and no separation takes place

Multiple versus Single Extractions

A common mistake is to assume that one extraction with a large volume is equivalent to several extractions with smaller volumes. Suppose, for example, that the directions call for three 10-ml extractions, but you are in a hurry and make one 30-ml extraction, assuming that the procedures are equivalent. However, they are not, as can be seen from equation 24.3:

$$W_m = W\left(\frac{V_w}{DV_o + V_w}\right)^n \qquad (24.3)$$

where n is the number of extractions, W_m the weight of solute remaining in the water layer after n extractions, W the original weight of solute in the water layer, V_w the volume of the water layer, V_o the volume of the organic layer used in *each* extraction, and D the distribution coefficient.

Example: A fat is to be removed from a meat sample by an ether extraction. The sample contains 0.1 g of fat in 1 g of meat dispersed in 30 ml of H_2O. Assume $D = 2$. Which is best, one 90-ml extraction or three 30-ml extractions?

For a single 90-ml extraction,

$$W_m = 0.1 \left(\frac{30}{(2 \times 90) + 30} \right) = 0.014 \text{ g}$$

For the three extractions,

$$W_m = 0.1 \left(\frac{30}{(2 \times 30) + 30} \right)^3 = 0.0036 \text{ g}$$

Thus, the multiple extraction leaves about four times less fat in the sample than the single extraction.

CONTINUOUS EXTRACTION

Liquid–Solid Systems

Extracting a material from a solid generally requires considerable time because it is difficult to get the extracting solvent into direct contact with the solute. Because of this, the effective distribution ratios are low and large volumes of solvent are sometimes necessary.

An efficient device for liquid–solid extraction, which eliminates the use of large solvent volumes, is the *Soxhlet extractor* (Fig. 24.1). The extracting solvent is placed in the flask at the bottom (A). When the solvent is heated, it goes up the side arm at B. C is closed off, and the siphon return (right side of B) fills with liquid almost immediately and is also closed off. The solvent vapor is condensed in the condenser and drips onto the sample. When the extractor fills up to G, the solvent will siphon back into the reservoir flask and the process will start again. The solvent is used over and over again; therefore, although a large volume of fresh solvent attacks the sample, actually only a small total volume is used.

A Soxhlet extractor is commonly used in the determination of diethylstilbestrol, a growth enhancer, in poultry meat.

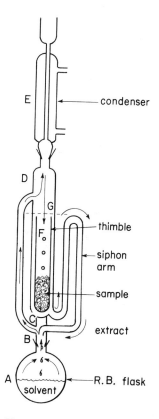

E — condenser

D

G

thimble

F

siphon
arm

sample

extract

C

B

A — R.B. flask

solvent

Fig. 24.1. Soxhlet extractor.

Liquid–Liquid Systems

The Soxhlet extractor works well with solids but not with liquids. Continuous extractors are excellent for use with materials having a low distribution ratio.

Solvent Heavier. The extractor shown in Fig. 24.2 is a continuous extractor for use with solvents heavier than water. The solvent is placed in the flask A and heated. The vapor rises to B and then up to C where it is condensed. The liquid cannot get back in the flask because of the seal at B. It runs into the extractor D, drips through the water layer E, and collects at F. The excess solvent containing the extract runs out the bottom at F and back into the flask. This is a continuous process with the extract collecting in the flask.

By using chloroform, caffeine can be extracted easily from cola beverages in about 45 min.

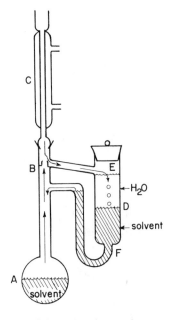

Fig. 24.2. Solvent-heavier continuous extractor.

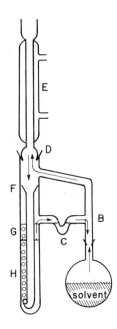

Fig. 24.3. Solvent-lighter continuous extractor.

Solvent Lighter. The coninuous extractor shown in Fig. 24.3 is used with solvents lighter than water. The solvent is placed in flask A and heated. The vapor rises past B and on to D. It cannot go to F past C because after an initial few minutes, C is filled with liquid. The vapor condenses in E and drips down into the flared tube F. The liquid will stay in tube F until it builds sufficient pressure to force its way out of the bottom opening; then it bubbles up through the sample solution H and collects at the top G. The excess runs back into the flask by tube C.

The ipecac alkaloids can be extracted from expectorants with ether by this technique in about 3 hr.

DISCONTINUOUS COUNTERCURRENT EXTRACTION

Craig and Post (1949) developed the concept of countercurrent extraction. In this technique, several extractions are performed simultaneously with the top phase shifting one tube after each extraction.

Consider a set of perhaps 100 tubes like those shown in Fig. 24.4, all placed side by side. All of the tubes are filled with the lower phase in the volume indicated. The sample is in the lower phase of the first tube; the upper phase is then added and an extraction performed. The tubes are rocked gently a few times and then stopped to allow the layers to separate. The bank of tubes is then turned to an upright position and the top layer, which contains some of the extracted sample, transfers into the transfer section. When the tube is then turned upright again, the original top layer drains out of the transfer section at C and goes to the next tube in the line. D of one tube fills the second tube at A and so forth. Fresh solvent is added to the original tube at A at the same time the other tubes transfer.

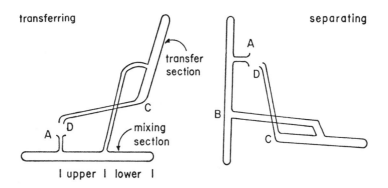

Fig. 24.4. Tubes used in countercurrent extraction process.

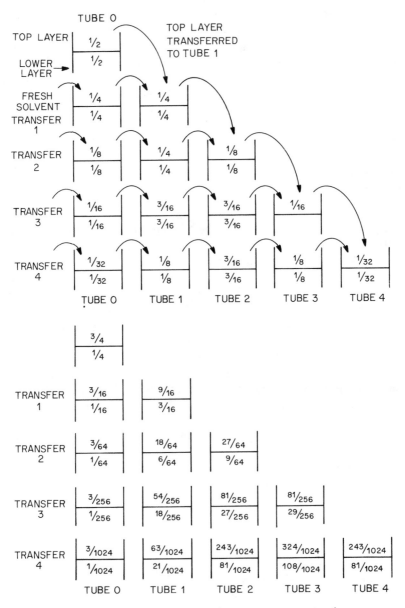

Fig. 24.5. Separation process in countercurrent extraction.

Figure 24.5 diagrams this process for five transfers with two different materials. Notice how the compounds begin to separate. If we neglect diffusion processes, this is basically the same step by step procedure that takes place in partition chromatography. Notice that at no time is a complete separation accomplished, but that at each step the separation becomes a little more complete.

Assume that a compound has a D of 1 (Fig. 24.5, top). If this is so, then $\frac{1}{2}$ goes into the top layer and $\frac{1}{2}$ goes into the lower layer. The top layer is transferred to the next tube and this $\frac{1}{2}$ is again divided in half, so we have $\frac{1}{4}$ in the top layer and $\frac{1}{4}$ in the lower layer, the total being $\frac{1}{2}$. Fresh solvent is then added to the first tube and the lower layer equilibrated with it. This process is repeated until the desired separations occur. If the second compound has a D of 3 (Fig. 24.5, bottom), then $\frac{3}{4}$ will go into the top layer and $\frac{1}{4}$ into the lower layer. In order to get an idea of how the extraction is proceeding, compare the amount of compound 1 in tube 3 with the amount of compound 2 in the same tube.

By calling the first tube 0, the mathematics of the process can be made to fit existing binomial expansion equations and the fraction in any tube after any given number of transfers has been made can be calculated.

EXTRACTION *P* VALUES

There are times when the gas chromatographic retention times of two or more materials are so close that an identification is difficult even if two different columns have been used. A thin-layer chromatographic separation can then be used, but this generally requires a lengthy cleanup and concentration because of the small (nanogram) samples involved. A much quicker method is the use of P values. This method is based on the distribution of a material between two immiscible phases.

This technique was developed primarily to help identify pesticides in foods, but there is no reason why it can not be extended to other compounds found in foods. The procedure of Beroza and Bowman (1965) is described here to illustrate this method.

A 5-ml aliquot of the upper phase containing a given pesticide is analyzed by gas chromatography. To a second 5-ml aliquot in a graduated, glass-stoppered 10-ml centrifuge tube is added an equal volume of lower phase; the tube shaken for about 1 min, and then the upper phase is analyzed exactly like the first 5-ml aliquot. The ratio of the second analysis to the first is the P value—the amount of the pesticide in the upper phase (second analysis) divided by the total amount of pesticide (first analysis). The P values in three solvent systems of o,o'-DDT and TDE, which have identical retention times, are listed in Table 24.1.

Table 24.1. *P* Values of Two Insecticides Having Identical Retention Times[a]

	Cyclohexane/ methanol	Hexane/ 90% dimethyl sulfoxide	Iso-Octane/ 80% dimethyl formamide
o,o-DDT	0.61	0.55	0.44
TDE	0.37	0.08	0.16

[a] From Beroza and Bowman (1965).

Aliquots of the equilibrated phases are measured with volumetric pipets with the phase volume being noted before and after equilibration. Emulsions are separated by centrifugation. After the distribution, the upper phase is passed through a 1-in. layer of anhydrous Na_2SO_4 to dry the solvent. The *P* value is quite reproducible (± 0.02), provided the two phases are saturated with each other before distribution.

Table 24.2 shows *P* values of several insecticides, and Table 24.3 shows the results for several foods. The 6 and 15% eluates are from a Florisil cleanup.

Table 24.2. *P* Values of Insecticides[a]

Pesticide	Hexane/ acetonitrile	Hexane/ 90% aq. dimethyl sulfoxide	Isooctane/ 85% aq. dimethyl formamide	Isooctane/ dimethyl formamide
Aldrin	0.73	0.89	0.86	0.38
Carbophenothion	0.21	0.35	0.27	0.04
Gamma chlordane	0.40	0.45	0.48	0.14
p,p'-DDE	0.56	0.73	0.65	0.16
o,o'-DDT	0.45	0.53	0.42	0.10
p,p'-DDT	0.38	0.40	0.36	0.08
Dieldrin	0.33	0.45	0.46	0.12
Endosulfan I	0.39	0.55	0.52	0.16
Endosulfan II	0.13	0.09	0.14	0.06
Endrin	0.35	0.52	0.51	0.15
Heptachlor	0.55	0.77	0.73	0.21
Heptachlor epoxide	0.29	0.35	0.39	0.10
1-Hydroxychlordene	0.07	0.03	0.06	0.03
Lindane	0.02	0.09	0.14	0.05
TDE	0.17	0.08	0.15	0.04
Telodrin	0.48	0.65	0.63	0.17

[a] From Beroza and Bowman (1965).

Table 24.3. P Values Obtained by Spiking the 6% and 15% Eluates of Various Crops[a]

Sample	Aldrin eluate		Heptaclor eluate		DDE eluate		Ronnel eluate		Kelthane eluate		Dieldrin eluate	
	6%	15%	6%	15%	6%	15%	6%	15%	6%	15%	6%	15%
Raspberries	0.72	0.86	0.50	0.62	0.56	0.65	0.22	0.30	0.15	0.20	0.37	0.42
Blueberries	0.72	0.79	0.52	0.58	0.56	0.58	0.24	0.31	0.15	0.18	0.38	0.43
Beet pulp	. .	0.81	0.57	0.59	0.59	0.50	0.23	0.25	. .	0.15	0.41	0.34
Wheat	0.75	0.80	0.53	0.57	0.59	0.60	0.26	0.29	0.16	0.17	0.39	0.39
Potatoes	0.78	0.83	0.54	0.61	0.62	0.62	0.23	0.29	0.14	0.16	0.37	0.41
Lettuce	0.81	. .	0.56	. .	0.60	. .	. .	. .	. .	. .	. .	. .
Strawberries	0.83	0.70	0.56	0.50	0.62	0.57	0.27	0.29	0.16	0.17	0.38	0.41
Pea silage	0.70	0.76	0.50	0.52	0.56	0.58	0.27	0.30	0.17	0.20	0.38	0.41
Cream	0.73	0.69	0.51	0.48	0.56	0.52	0.24	0.28	0.15	0.18	0.35	0.39
Pea silage	. .	. .	. .	. .	. .	. .	0.21	. .	0.13	. .	. .	. .
Beets	. .	. .	. .	. .	. .	. .	. .	. .	. .	. .	. .	. .
Avg.	0.75	0.78	0.53	0.56	0.58	0.58	0.25	0.29	0.15	0.18	0.38	0.40
Avg. dev.	0.04	0.05	0.02	0.04	0.02	0.04	0.02	0.01	0.01	0.01	0.01	0.02

[a] From Manske and Frasch (1966).

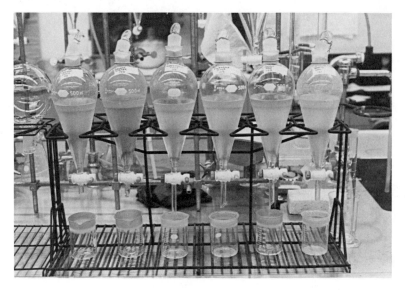

Fig. 24.6.   Extraction of fruits and vegetables for pesticide residues.

RECENT APPLICATIONS

Daughtery and Lento (1983) used a chloroform–methanol extraction method for the determination of fat in foods; they report recoveries of 93–98%. Kumpulainen *et al.* (1983) were able to remove selenium from foods by an isobutyl ketone extraction after the addition of ammonium pyrrolidine dithiocarbamate.

Dermal exposure to carbaryl by strawberry harvesters was determined by Zweig *et al.* (1984) who extracted gloves and patches for 2 hr with 50 ml of acetonitrile. A continuous extraction using isooctane in the Bleidner–Heizler apparatus was used by Hornish *et al.* (1984) to determine amitraz residues in apples and pears. Sen and Seaman (1981) described a fast liquid–liquid extraction method for volatile nitrosoamines in bacon fat.

Figure 24.6 shows a bank of standard separatory funnels used by the FDA, in total diet analyses, to extract organic pesticide residues from orange juice, winter squash, grapefruit juice, pineapple juice, raw onions, and beets.

PROBLEMS

24.1. The carbaryl in a sample of lettuce was extracted with isooctane and then concentrated to 2.0 ml. To this was added 2.0 ml of water; after mixing and separation, the isooctane layer contained 94% of the carbaryl. What is the distribution ratio for this compound in these liquids? *Ans.* 15.7.

24.2. Assume you wanted to extract 99.5% of the carbaryl in problem 24.1 into the isooctane. How many milliliters of isooctane must be used? *Ans.* 25 ml.

24.3. Lead is determined in bulk food dyes by forming the diethyl-dithiocarbamate chelate and extracting it into chloroform. Assume the *D* of the chelate is 3. How many 10-ml extractions will it take to remove 99.9% of the lead from 20 ml of aqueous digest if the original dye is assumed to contain 0.0010% Pb? *Ans.* 5.8 or 6 extractions.

24.4. Referring to Fig. 24.5 (bottom), what is the amount of the second component (*D* = 3) in the upper and lower phases in tube 2 after five transfers? *Ans.* Upper, 432/4096; lower, 144/4096.

BIBLIOGRAPHY

Beroza, M., and Bowman, M. C. (1965). Identification of pesticides at nanogram levels by extraction p-values. *Anal. Chem.* **37**, 291–292.

Brugel, P. (1981). Determination of pentachlorophenol in food gelatins. *Ann. Falsif. Expert. Chim. Toxicol.* **74**, 61. [*Chem. Abstr.* **95**, 59996u.]

Craig, L. C., and Post, O. (1949). Apparatus for countercurrent distribution. *Anal. Chem.* **21**, 500–504.

Daughtery, C. E., and Lento, H. G. (1983). A chloroform–methanol extraction method for the determination of fat in foods, *J. Assoc. Off. Anal. Chem.* **66**, 927–932.

Hornish, R. E., Closby, M. A., Nappier, J. L., Nappier, J. M., and Hoffman, G. A. (1984). Total residue analysis of amitraz in fruit and soil samples by electron capture gas chromatography. *J. Agric. Food Chem.* **32**, 1219–1223.

Kumpulainen, J., Raittila, A., Lehto, J., and Koivistainen, P. (1983). Electrochemical atomic absorption spectrometric determination of selenium in foods and diets. *J. Assoc. Off. Anal. Chem.* **66**, 1129–1135.

Manske, D., and Frasch, D. L. (1966). Identification of pesticides by extraction p-values. *Food Drug Admin. Interbureau Bylines* **4**, 171–174.

Sen, N. P., and Seaman, S. (1982). A rapid liquid–liquid extraction cleanup method for determination of volatile *N*-nitrosamines in cooked-out bacon fat. *J. Agric. Food Chem.* **30**, 364–367.

Zweig, G., Gaw, R., Witt, J. M., Popendorf, W., and Bogen, K. (1984). Dermal exposure to carbaryl by strawberry harvesters. *J. Agric. Food Chem.* **32**, 1232–1236.

25
Centrifugation

Centrifugation is used for separation of solids from liquids and from immiscible solvents, and for resolution of emulsions that are formed during extraction. Centrifugation at high speeds (*ultracentrifugation*) is useful for concentrating high-molecular-weight materials and for estimating their molecular weights.

BASIC RELATIONSHIPS

The force acting on a particle is given by

$$F = ma = m\omega^2 r \qquad (25.1)$$

where F is the force on a particle in dynes, m the mass of a particle in grams, ω the angular velocity of rotation in radians per second, and r the radial distance of a particle from the axis of rotation in cm.

We can express the centrifugal force F' (in grams) as

$$F' = m\omega^2 r/g \qquad (25.2)$$

where g is the gravitational constant (980.7 cm/sec^2) or as

$$F' = 1.118 \times 10^{-5} (mrN^2)$$

where N is the speed of rotation in rpm (revolutions per minute).

Another useful equation is

$$RCF = \omega^2 r/g \qquad (25.3)$$

where RCF is relative centrifugal force defined as the force acting on a given particle in a centrifugal field in terms of multiples of its weight in the earth's gravitational field, and

$$RCF = 1.118 \times 10^{-5} (rN^2) \qquad (25.4)$$

If the axis of rotation is horizontal, the difference between RCF at the tip and at the surface of the liquid is generally twice, or more. Because the radial distance from the axis of rotation and the depth of the liquid in a centrifuge tube vary widely with different types of centrifuges, it is inadequate to report results in terms of the speed of rotation in rpm. Reporting data in terms of centrifugal force (g) calculated from the equation 25.4 is much more meaningful. Generally, the centrifugal force at the middle of the liquid depth is reported; reporting the centrifugal force at both the tip end and the free surface of the liquid is preferable. If the axis of rotation is vertical, the downward pull of gravity on the rotating parts, is insignificant if RCF is $25g$ or above.

The effective sedimenting or centrifugal force must be corrected by the buoyancy factor. From Archimedes law,

$$m_{eff} = m - V\rho \qquad (25.5)$$

where V is the particle volume in a solution that has a density of ρ. Substituting equation 25.5 into 25.1 gives

$$F_{eff} = (m - V\rho)\omega^2 r \qquad (25.6)$$

If the particle is a sphere, then

$$F_{eff} = \left(\Delta\rho \frac{\pi}{6} D^3\right) \omega^2 r \qquad (25.7)$$

where D is the diameter of the particle and $\Delta\rho$ the difference between the density of the particle and that of the fluid in which it is suspended. This is the force that is available to move the particle through the liquid medium. The sedimenting molecules are subjected to a frictional force F exerted by the medium. For a slowly moving small particle, the frictional force is given by Stokes' law:

$$F = 3\pi\eta D V_s \qquad (25.8)$$

where η is the viscosity of the liquid and V_s the velocity of particle moving through the liquid phase.

By setting equations 25.7 and 25.8 equal, we obtain

$$V_s = \Delta\rho D^2 \omega^2 r/18\eta \qquad (25.9)$$

from which we can calculate the velocity of a particle at a distance r from the axis of rotation.

EQUIPMENT

There are two basic types of centrifuges, and a third that may be considered a combination of the two:

1. Solid-wall centrifuges in which separation or concentration is by subsidence or flotation

2. Perforated-wall centrifuges (centrifugal filters) in which the solid phase is supported on a permeable surface through which the fluid phase is free to pass.

3. Combinations of the two in which the primary concentration is effected by subsidence followed by drainage of the liquid phase away from the solid phase (Ambler and Keith 1956).

Bottle centrifuges have a motor-driven vertical spindle on which various heads or rotors can be mounted. The rotors carry metal containers into which fit glass tubes or bottles (from 2 to 16 or more) of a total capacity of up to about 1 gal. Bottle centrifuges are particularly useful for analytical and small-scale preparative separations. The time, speed, and (in some) the temperature can be closely controlled. They are generally enclosed for safety in a metal guard bowl. Most bottle centrifuges have a horizontal swinging-type rotor. The glass tubes (bottles) are placed into metal shields on a rubber cushion, and the shields or containers are supported in trunnion rings, which are set in slots in the rotor head. Since the center of gravity of the assembly is below the trunnions, the tubes hang vertically at rest. As the rotor starts to turn, they gradually swing out to a horizontal position where they remain as long as the head is rotating. The advantage of the method is fractional sedimentation across the tube length; the disadvantages are the long path of travel of some particles and hindered settling near the bottom. To overcome the disadvantages, prolonged centrifugation at high speeds is required.

In angle or conical-type centrifuges, the tubes (in metal shields) are held at a fixed angle of about 45° in holes in the rotor, both at rest and during rotation. The rotor can be used for relatively high centrifugation speeds. The particles travel in free sedimentation only a distance equal to the diameter of the tube times the secant of the angle of inclination from the vertical. The particles that strike the glass wall of the tube aggregate and slide to the bottom. The sedimenting particles deposit at an angle.

The bottle centrifuge can accomodate a large variety of tubes and bottles of various sizes: plain and graduated cylinders, plain tubes with round or conical bottoms, plain and graduated pear-shaped tubes, and separatory funnels. Special refrigerated bottle centrifuges can be used for work with low-melting-point compounds or heat-labile biochemical materials. Heated centrifuges are used in fat determination in dairy products.

The hydrostatic pressure exerted on a filled glass container in a laboratory bottle centrifuge may reach high enough values to rupture the container. This may be largely offset by filling the space between the glass tube and the metal cut with a liquid (water, glycerin, ethylene glycol) or by using semielastic, plastic containers that deform sufficiently (but do not rupture) under pressure to carry the hydrostatic pressure directly on the wall of the metal shield. All rotating parts of centrifuge are subject to stresses created by centrifugal forces; these stresses impose limitations on the maximum permissible speed of the centrifuge. The effects of those stresses can be minimized in properly designed centrifuges by careful balancing before loading, by slowly increasing the speed to that desired, and by proper maintenance and adherence to the manufacturer's instructions. To prevent remixing after centrifugation, deceleration should be slow.

ULTRACENTRIFUGATION

Separation and fractionation of macromolecules by subjecting them to a strong centrifugal force originated with T. Svedberg and co-workers, who in the early 1920s invented and developed the instrument called the *ultracentrifuge*. In 1924, Svedberg and Rinde described the use of ultracentrifugation in studies of colloidal particles. Two years later, Svedberg and Fahraens (1926) described the determination of the molecular weight of proteins from sedimentation in an ultracentrifuge by measuring the diffusion coefficient. The centrifugal forces that can be attained in ultracentrifugation are in the order of 500,000g. The high forces can be used for the determination of molecular weights of macromolecules or for preparative fractionations.

The theory and practice of ultracentrifuge measurements have been described in a number of monographs and reviews; preeminent is the classic work of Svedberg and Pedersen (1940). More recent reviews include those by Williams (1963), Schachman (1959), Fujita (1962), Bowen (1970), and Hucho and Sund (1974).

Analytical Ultracentrifuge

Commercially available analytical ultracentrifuges provide a photographic record of the migration of high-molecular-weight substances in a strong gravitational field. These instruments have automatic temperature and speed controls for the rotor, a high vacuum chamber to reduce friction, an optical system for measuring the rate at which individual peaks (representing different proteins) move towards the bottom of the cell, an automatic photographic system for recording changes in concentration at specified intervals, and special cells. Three optical systems are available: Schlieren, interference, and absorption. The most commonly used system is the astigmatic Schlieren optics. The photographic record of the sedimentation pattern using Schlieren optics gives the concentration gradient in the cell in terms of the refractive index gradient.

In using the ultracentrifuge for determining the molecular weight of proteins, two lines of approach are possible: the *sedimentation equilibrium* and the *sedimentation velocity* methods.

In sedimentation equilibrium, a relatively low centrifugal force is applied to a solution till the distribution of protein throughout the column of liquid in the centrifuge tube reaches a steady state. What one measures is the stage of equilibrium between outward sedimentation of protein and backward diffusion. At this stage, the molecular weight M can be determined from the formula

$$M = [2RT \ln(c_1/c_2)]/[\omega^2(1 - \rho/\sigma)(x_1^2 - x_2^2)] \qquad (25.10)$$

where R is the gas constant, T the absolute temperature, c_1 and c_2 the concentrations at distances x_1 and x_2 from the center of rotation, ρ the density of the solvent, σ the density of the protein, and ω the angular velocity.

The term

$$\rho/\sigma = \pi v \qquad (25.11)$$

in which v is the partial specific volume of the protein (increase in volume when 1 g of dry protein is added to a large volume of liquid), can be assumed to be 0.75. The angular velocity can be expressed as

$$\omega = V/x \qquad (25.12)$$

where V is the velocity of the centrifuged solution and x the distance from the center of the rotor. If the number of revolutions per second is z, then

$$V = 2r\pi z \qquad (25.13)$$

$$\omega = 2\pi z. \qquad (25.14)$$

It is customary to give the velocity in rpm; since $z = \text{rpm}/60$, then

$$\omega = 2 \times 3.14 \, \text{rpm}/60 = 0.105 \, \text{rpm}$$

As mentioned earlier, in the sedimentation equilibrium method relatively low centrifugal velocities are used. If very high velocities were used, there would be packing of the protein at the bottom with little difference in c_1 and c_2 or x_1 and x_2 values. The main disadvantage of the sedimentation equilibrium method is that it requires several days for a determination. In the *approach to equilibrium* technique (Trautman 1964), concentrations of protein are measured at various times at the meniscus of the solution and at the bottom of the tube. From these data, c_1 and c_2 are extrapolated. The procedure is especially useful with small protein molecules which sediment slowly. If special small cells are used, ultracentrifugation may take as little as 45–70 min.

In the sedimentation velocity method, high speeds at which the protein particles sediment at a fast rate are used. If the molecular weight of the particles is high, their rate of diffusion can be neglected. The sedimentation velocity depends on the shape and the hydration of the protein molecules, whereas the sedimentation equilibrium is independent of those factors and depends only on the molecular weight. Nevertheless, many molecular-weight determinations are made by the sedimentation velocity method, combined with diffusion measurements. The principal reason for this is the shorter time of centrifugation, which eliminates the danger of bacterial growth and decomposition of the proteins.

The rate of sedimentation is usually expressed in terms of the sedimentation constant s, the velocity per unit centrifugal field force, which has the dimensions of time. For most proteins, s ranges from 1 to 200 $\times$ 10^{-13} sec. A sedimentation constant of 1×10^{-13} is called a *Svedberg* unit (S), and sedimentation constants generally are given in Svedberg units.

Preparative Ultracentrifugation

Moving boundary ultracentrifugation is of limited use in preparative work. While the lightest protein in a mixture can be obtained in pure form, the heavier fractions contain some of the lighter ones. For preparative work several methods that eliminate diffusion of particles and make complete separations possible are used. The methods are called *density gradient ultracentrifugation*. They depend on the formation of a gradient, the density of which increases with distance from the axis of rotation. Actually, the usefulness of density gradient ultracentrifugation is threefold: it provides information on the size of the separated molecules in a mixture, permits separation on a preparative scale, and can be carried out with low concentrations of solute.

Density gradient separations can be classified into three categories according to the way in which the gradient is used (Anon. 1960). In *stabilized moving boundary centrifugation*, one uses a shallow gradient formed during the centrifugation. Its function is to stabilize against convection. Material to be fractionated is distributed throughout the solution (Pickels 1943). The sample in the preparative centrifuge is separated into fractions that are analyzed to determine the position of the boundary. If several components are present, several boundaries form and the sedimentation coefficient of each can be calculated.

In *zone centrifugation*, a solution containing particles of varied characteristics is layered on top of a steep gradient (Brakke 1951). The steep gradient keeps convective stirring at a minimum. Each substance sedimenting at its own rate forms a band or zone in the fluid column. The solute zones will be separated from one another by distances related to their sedimentation rates. After centrifugation, each substance can be drawn off separately for the determination of its sedimentation rate and for further analysis. This fractionation technique is probably the most widely used in biochemical ultracentrifugation, but it has two limitations: only small amounts of material can be separated at high speeds, and the separation is incomplete due to the wall effect (from some particles being reflected from the tube walls back into the solution, sticking to the walls, or clumping). In the zonal ultracentrifuge developed by Anderson (1962), high-speed rotors with increased capacity are used for sharp sample-zone separations under conditions that minimize wall effects. Sedimentation takes place in sector-shaped compartments in hollow rotors of capacities up to 120 times that of a high-speed swinging-bucket rotor of comparable separating capabilities. The gradient solutions are introduced and recovered while stabilized in a centrifugal field. Continuous flow zonal centrifugation is also available.

In *isopycnic gradient centrifugation* (Messelson et al. 1957), separation is based on differences in density of the macromolecules in a sample solution that is usually distributed evenly throughout the gradient column before ultracentrifugation. The gradient column must cover the entire density range of the particles, and centrifugation is continued until the particles reach positions at which the density of the surrounding liquid is equal to their own.

One should distinguish clearly between zone (or band) and isopycnic (or equilibrium) density centrifugation. Zone sedimentation is a kinetic method that depends on the sedimentation coefficient of the macromolecules (Szybalski 1968). The gradient is preformed and is usually substantially less dense than that of the macromolecules. The purpose of the gradient is to stabilize the sedimenting band of macromolecules and prevent convection. The macromolecules move continuously through the gradient and settle at the bottom of the tube if centrifugation is allowed to proceed for a very long period. Isopycnic density gradient ultracentrifugation is an equilibrium (static) method that depends on the buoyant density of the macromolecules. The gradient is self-generating in the centrifugal field, and its density range is so adjusted as to be denser at the bottom of the tube and less dense near the meniscus than the macromolecules in the column of solution. Thus, the macromolecules, most frequently nucleic acid or viruses, form a definite band at a definite level in the column and remain there irrespective of the length of centrifugation.

Materials used to form gradients should be chemically inert to the studied system, nontoxic, and soluble in water and salt solutions; they also should have a high density, high molecular weight, and low viscosity. For separation and analysis of proteins, the gradient material should contain no nitrogen. High-density materials are required to form a steep gradient; if the molecular weight is high, the osmotic pressure gradient in the tube is relatively small. Low viscosity permits easy handling during gradient formation, and rapid sedimentation and fractionation. Sucrose is the most widely used material for the gradient. Its main disadvantage is high viscosity. Ficoll is a commercially available, water-soluble, neutral colloid with properties similar to that of a polysaccharide. Its average molecular weight is about 50,000, and it is stable in nonoxidizing neutral or alkaline solutions. Its viscosity and density are lower than those of sucrose. For the separation of nucleic acids, inorganic salts (mainly cesium chloride or rubidium chloride) are used.

Gradients can be produced manually by layering sucrose (or other) solutions of decreasing density into a centrifuge tube. After 24 hr or more, the gradient becomes semicontinuous as a result of diffusion. Commercially available or mechanical devices constructed in the laboratory to produce various stable gradients (linear, exponential, concave, convex, or S-shaped) are increasingly popular. The use of a specific gradient depends on the distribution of sedimentation rates or densities. Thus, a linear gradient is most useful for a solution containing a relatively high range of densities. An S-shaped gradient would give best results for a mixture containing two components with similar densities and one with a considerably higher density.

The position of particles in the centrifuged tube can be determined visually (under ordinary or ultraviolet light) or after the contents of the tube have been separated into small fractions. This is done most commonly by carefully puncturing a hole at the bottom and removing dropwise the tube contents.

BIBLIOGRAPHY

Ambler, C. M., and Keith, F. W. (1956). Centrifuging. In "Technique of Organic Chemistry," Vol. III (A. Weissberger, ed.). Interscience Publishers, New York.

Anderson, N. G. (1962). The zonal ultracentrifuge. A new instrument for fractionating mixtures of particles. *J. Phys. Chem.* **66,** 1984–1989.

Anon. (1960). "An Introduction to Density Gradient Centrifugation." Tech. Rev. 1. Beckman Instruments, Inc., Spinco Division, Palo Alto, CA.

Bowen, T. J. (1970). "An Introduction to Ultracentrifugation." Wiley-Interscience, New York and London.

Brakke, M. K. (1951). Density gradient centrifugation: A new separation technique. *J. Am. Chem. Soc.* **73,** 1847–1848.

Fujita, H. (1962). "Mathematic Theory of Sedimentation Analysis." Academic Press, New York.

Hucho, F., and Sund, H. (1974). Ultracentrifugation. *In* "Clinical Biochemistry: Principles and Methods" (H. C. Curtius and M. Roth, eds.) pp. 148–194. Walter de Gruyter, Berlin.

Messelson, M., Stahl, F. W., and Vinograd, J. (1957). Equilibrium sedimentation of macromolecules in density gradients. *Proc. Natl. Acad. Sci.* **43,** 581–594.

Pickels, E. G. (1943). Sedimentation in the angle centrifuge. *J. Gen. Physiol.* **26,** 341–365.

Schachman, H. K. (1959). "Ultracentrifugation in Biochemistry." Academic Press, New York.

Svedberg, T., and Fahraens, R. (1926). A new method for the determination of the molecular weight of the proteins. *J. Am. Chem. Soc.* **48,** 430–438.

Svedberg, T., and Pedersen, K. O. (1940). "The Ultracentrifuge." Clarendon Press, Oxford, England.

Svedberg, T., and Rinde, H. (1924). The ultracentrifuge, a new instrument for the determination of size and distribution of size of particle in amicroscopic colloids. *J. Am. Chem. Soc.* **46,** 2677–2693.

Szybalski, W. (1968). Equilibrium sedimentation of viruses, nucleic acids, and other macromolecules in density gradients. *Fractions* **1,** 1–15.

Trautman, R. (1964). Ultracentrifugation. *In* "Instrumental Methods in Experimental Biology," (D. W. Newman ed.). MacMillan, New York.

Williams, J. W. (1963). "Ultracentrifugal Analysis in Theory and Experiment." Academic Press, New York.

26
Densimetry

Determination of density is one of the most common simple measurements in food analyses. It is made primarily on liquids, though it can also be used in analyses of solids.

The determination in liquids provides a useful parameter whenever the density of a mixture of two compounds is a function of its composition, and when the composition can be read off calibration graphs or tables. This procedure is used to determine the sugar or alcohol content of aqueous solutions.

The smaller the change in volume on mixing the components of a solution, the more closely is the concentration related to density. If in addition, changes in density per unit change in concentration are large, densimetric methods yield an accurate as well as rapid method of analysis.

Densimetry also can be used in analyses of more complex systems, such as tomato juice or milk. In those instances, density is an index of total solids contents.

The density of many substances is a characteristic physical property and serves for identification purposes. Density is a function of both the length of the carbon chain of the glyceride fatty acids and the degree of unsaturation of glyceride fatty acids. In addition, several rapid methods are available to determine oil content from measuring density of extracts of lipids obtained under specified conditions.

Density is an important criterion of seed purity; texture and softness of fruits; maturity of such products as peas, sweet corn, and lima beans; and has been proposed as an index of soundness of dried prunes or plumpness and dryness of raisins.

425

DEFINITIONS AND BASIC RELATIONSHIPS

Density d is defined as

$$d = \text{mass/volume} = m/v \qquad (26.1)$$

In accordance with the cgs system, *absolute density* d^t at temperature t is defined as

$$d^t = m \text{ (g)}/V_{cm^3} \text{ (cm}^3) \qquad (26.2)$$

In practice, density measurements are expressed in grams of weight per milliliter, which under most conditions is equivalent to grams of mass per milliliter. The *relative density* at $t°C$ is defined as

$$d_4^t = m \text{ (g)}/V_{ml} \text{ (ml)} \qquad (26.3)$$

By definition, 1 ml = 0.001 part of the volume of 1 kg of pure water at 3.98°C. Therefore, d_4^t gives the ratio of absolute density at $t°C$ to the absolute density of water at 3.98°; it is dimensionless.

Originally, it was intended that 1 cm^3 should equal 1 ml, but precise measurements have shown that

$$1 \text{ cm}^3 = 0.999973 \text{ ml,} \qquad (26.4)$$

and therefore

$$V_{cm^3} = 1.000027 \, V_{ml} \qquad (26.5)$$

For most food analyses, the difference between d^t (g/cm^3) and d_4^t (g/ml) is negligible.

The weight W is related to the mass m by the equation

$$W = mg \qquad (26.6)$$

where g is the acceleration due to gravity (980.665 cm/sec^2) at a latitude of 45° and at sea level.

Because of the proportionality between the mass and weight of calibrated weights and mass and the weight of an analyzed sample, at a given geographical location, there is no change in the numerical value of density by using grams of weight instead of grams of mass. In practice, d_4^t is used because it can be determined accurately by comparing the weights of equal volumes of the substance at 4°C and of water at 3.98°C. Sometimes the term *specific gravity*, d_t^t, is used. It is a dimensionless number defined as the mass m of a substance at $t°C$ relative to the mass m_o of an equal volume of water at $t°C$.

$$d_t^t = mV/m_o V \qquad (26.7)$$

For the calibration of instruments used in density determinations, pure water or mercury is used.

Density is affected by temperature. For water at room temperature, the density decreases by about 0.03% per °C rise in temperature, and the temperature coefficient of cubical expansion is

$$\beta_t = 3 \times 10^{-4} \, \text{deg}^{-1} \tag{26.8}$$

Some organic solvents (aliphatic hydrocarbons) have β_t values up to five times greater than water; the density of organic solids usually changes somewhat less rapidly with temperature than that of liquids.

If an error of not more than ± 0.001 in density is desired, temperature control must be within 1°C (Bauer 1945). For greater accuracy, correspondingly better control is needed. However, the temperature control required depends also on the method used. When pycnometric determinations are made, the temperature effect is somewhat reduced by the expansion of the glass container.

Atmospheric pressure differences are important if they deviate considerably from the standard one. Ordinary fluctuations in pressure encountered in the laboratory have little effect on density.

During mixing of two or more liquids, the volume increases or decreases above the theoretical one. The decrease is as high as 9.5% in mixing equimolar amounts of water and sulfuric acid, and 2.56% with water and ethanol. In the case of acetone–disulfide there is an increase of 1.2%. The changes are concentration-dependent and the resultant effects on density must be established empirically (Mahling 1965).

MEASUREMENT OF THE DENSITY OF LIQUIDS

Pycnometric Methods

The most common method of density determination consists of measuring the weight of a known volume of liquid in a vessel, the volume of which was calibrated in terms of the weight of pure water that the vessel holds. This is best done by using a pycnometer. If the weighing is done with an error of not more than 0.1 mg, for a vessel up to 30 ml, an accuracy of $\pm 5 \times 10^{-6}$ can be attained (Bauer 1945). Increasing the size of the vessel theoretically increases the precision, but is accompanied by compensating errors resulting from such factors as insensitivity of the balance and difficulty in maintaining a uniform temperature of a large volume of liquid.

Pycnometers should preferably be made from resistant glass with a

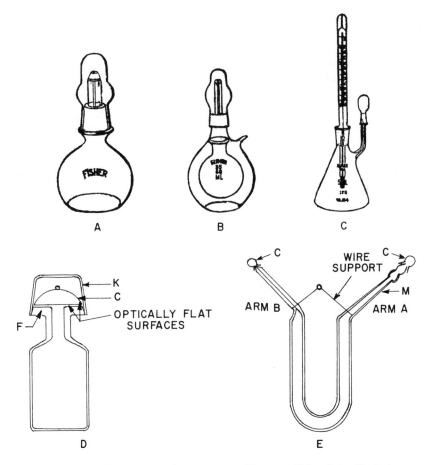

Fig. 26.1. Different types of pycnometers. (Courtesy Fisher Scientific Co.)

low coefficient of thermal expansion such as Pyrex or Vycor, or fused quartz. Figure 26.1 shows five types of commercially available pycnometers. A Gay–Lussac bottle (A) has a ground-in perforated glass stopper and outer ground-on cap to reduce evaporation. Type B is similar to A, but has an evacuated jacket to stabilize the temperature of the material being weighed. Type C has a thermometer connected to the flask by a ground-glass joint. Type D has a wide neck opening and can be used for both solids and liquids; the outside cap minimizes evaporation and leakage. A standard Sprengel–Ostwald pycnometer (E) has two openings and is easy to fill and clean even though the side arms are made of small-bore capillaries. It can be filled without danger of loss by evaporation. A simple, though less precise, Sprengel pycnometer can be made from an ordinary pipet.

Before use, the pycnometer must be thoroughly cleaned with a mixture of potassium dichromate and concentrated sulfuric acid, rinsed thoroughly with water, and dried. To avoid changes in volume, the pycnometer must not be subjected to excessive temperature or pressure changes.

Pycnometers can give reproducible results with an accuracy to the fourth decimal place. For fifth-place accuracy, the weights must be checked against one another to obtain their relative values. Metal-plated weights are preferable to lacquered weights as the latter are more susceptible to humidity effects. An error of about 2×10^{-5} in density can result from inequalities in the lengths of the balance arms. This error can be eliminated by determining the average of two weighings, one on each balance pan. A pycnometer dried at room temperature under vacuum can adsorb several milligrams of water by standing in a humid atmosphere. The adsorbed moisture can be removed by carefully wiping the glass surface with a lintless cloth and then allowing the vessel to stand in the balance case for about 15 min before weighing (Bauer 1945).

Bubbles are the most common sources of relatively large errors. Whenever possible the liquid should be boiled and cooled shortly before filling. It is advisable to cool the liquid about 1°C below the thermostat temperature just before filling to ensure an excess of liquid when equilibrium is reached. The excess liquid is removed by fine capillaries or strips of filter paper. Special filling devices are available to accurately fill the pycnometer and remove all air.

Buoyancy Methods

Several methods for measuring density are based on the Archimedes principle, which states that the upward buoyant force exerted on a body immersed in a liquid is equal to the weight of the displaced liquid. This principle can be applied in several ways to measure density.

An ordinary gravimetric balance can be adapted for making precise density measurements by removing one of the pans and attaching to the balance arm a fine, freshly platinized (black) platinum wire at the end of which is suspended a sinker (cylinder of glass or metal). The sinker is immersed in a column of liquid below the balance case. The combined volume (V) of the sinker and wire (up to the point of immersion) is determined by calibration with water. If the weight in air is W_o and in the liquid W_L, the apparent weight loss, $W_o - W_L$, is equal to the mass of displaced liquid, and the density d is given by

$$d = (W_o - W_L)/V \qquad (26.9)$$

The Mohr balance, as improved by Westphal, is a commercially available direct-reading instrument based on this principle. This balance can be used to determine the specific gravity of liquids heavier or lighter than water to the fourth decimal.

A Westphal balance is shown in Fig. 26.2. The glass plummet displaces 5 g of water at 15°C; the largest rider weighs 5 g and equals 1 when placed on the same hook with the plummet. The other riders weigh 0.5, 0.05, and 0.005 g, respectively. A 15-g weight is supplied for initially balancing the beam in the air. The riders are placed on the beam in succession until it balances. The sum of the settings shows the specific gravity. If the liquid is lighter than water, the position of the heaviest rider shows tenths of a unit; the smaller riders give respectively the second, third, and fourth decimal places. If the liquid is heavier than water, a second 5-g rider is placed on the plummet hook, where it has the value of 1. The Westphal balance is widely used for rapid determination of density of nonvolatile solutions and nonhygroscopic liquids.

Another buoyancy method was described by Wagner *et al.* (1942) who used a quartz float attached to an elastic quartz helix mounted in a temperature-controlled tube containing the investigated liquid. The float is free to move and the helix is stretched by the buoyant force acting on the float. The density is determined from the change in length of the helix. The method is rapid, can be used to determine density of liquids in a closed container, and is very precise provided temperature control is adequate. However, calibration of the helix is quite time-consuming.

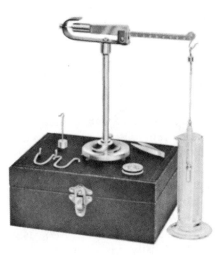

Fig. 26.2.   The Westphal balance. (Courtesy A. H. Thomas Co.)

Hydrometry is based on the principle that the same body displaces equal weights of all liquids in which it floats. If V_1 and V_2 are the volumes of two liquids displaced by the same floating body, and D_1 and D_2 their respective densities, then

$$V_1 D_1 = V_2 D_2 \qquad (26.10)$$

and

$$D_1/D_2 = V_2/V_1 \qquad (26.11)$$

Thus, the volumes of different liquids displaced by the same floating body are inversely proportional to the densities of the liquids. If the floating body is an upright cylinder of uniform diameter, the volumes displaced are proportional to the depths (H_1, H_2) to which the body sinks:

$$D_1/D_2 = H_2/H_1 \qquad (26.12)$$

Hydrometers (Fig. 26.3) are hollow glass bodies that have a broad and heavy bottom and a narrow upper stem. The lower part is loaded by a metallic insert of appropriate weight so that the whole hydrometer sinks in the tested solution to such a depth that the upper calibrated stem is in part above the liquid. The total weight of the hydrometer must be smaller than that of the liquid it displaces. The deeper the hydrometer sinks, the

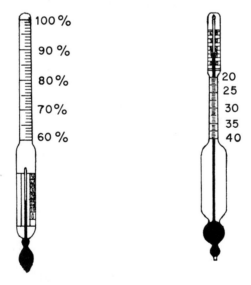

Fig. 26.3. Alcoholometer (left) and lactodensimeter (right) are hydrometers whose scales are calibrated in percentage composition.

lower the density of the solution. The calibrated stem shows increasing density values from top to the bottom of the stem. For a wide range of densities, a rather long glass stem that might break easily would be required. Generally, for orientation purposes a wide-range and less precise hydrometer is used. This is followed by more precise determinations with a set of overlapping narrow-range hydrometers.

The precision of hydrometers is generally limited to three decimal places. The container, generally a glass cylinder, should be at least twice as wide as the diameter of the bottom part of the hydrometer. A density determination with a hydrometer must be corrected for deviation from the standard temperature. Some hydrometers have a built-in thermometer and correction scale. The scale can be calibrated in density units or in percentage composition as related to density.

Alcoholometers are used for determining the percentage of alcohol by volume (Tralle scale) and deviations from "proof" (50% on Tralle scale = 0 on proof scale). The Tralle scale is graduated from 0 to 100% in 1% divisions, and the proof scale from 100 under proof to 100 over proof in single proof divisions.

Baumé hydrometers are of two kinds: heavy Bé for liquids heavier than water, and light Bé for liquids lighter than water. In the first, 0° corresponds to a density of 1.000 and 66° to a density of 1.842. In the lighter than water scale, 0° Bé is equivalent to the density of a 10% solution of NaCl, and 60° Bé corresponds to a density of 0.745. For Bé degrees on a scale of densities above 1.000,

$$\text{density} = \frac{145}{145 - \text{Bé reading}} \tag{26.13}$$

Saccharometers can have the Balling scale (% of sugar) or the Brix scale (% of sugar by weight at 20°C). The use of saccharometers to determine the composition of various sugar blends (sucrose, corn syrups, invert sugar, high-fructose syrups) in solution was described by Wartman *et al.* (1984) and by Maxwell *et al.* (1984). Kruger and Oliver-Daumen (1980) reviewed various batch, continuous, and automated methods for the determination of density in beer production. Flexural resonators, radiometric, hydrostatic, and gravimetric methods were described.

Salometers are used to determine the percentage of saturation of salt brines.

Oleometers for vegetable and sperm oils have a scale of 50–0° that corresponds to densities of 0.870–0.897.

Lactometers are used to determine dry milk solids (and adulteration by water addition) of milk from density determinations. The Soxhlet lactometer has a scale of 25–35 (density 1.025–1.035) subdivided into suitable divisions. The Quevenne scale has a range of 14–42° Quevenne in 1°

divisions, and corresponding scales to indicate percentage of water in whole or skim milk. A thermometer scale above the Quevenne scale indicates corrections that are to be made when temperatures are above or below 60°F. The density of milk also can be estimated from the sinking of plastic colored beads of specified density (Erb and Manus 1963).

Sartor and Sherry (1983) developed an automated method for the determination of alcohol (by a thermometric method) and the apparent and real extract, caloric value, original gravity, and degree of fermentation of wort and beer. A commercial instrument for continuous determination of density, specific gravity, and percentage of solids in liquids and slurries (independent of ambient temperature, viscosity, pressure, or flow velocity) is available (Anon. 1984).

MEASUREMENT OF THE DENSITY OF SOLIDS

Methods for measuring the density of solids are considerably less precise than those for liquids mainly due to nonhomogeneity of solid foods, partial solubility, and presence of occluded air bubbles. Most methods depend on immersing the solid in an inert liquid of known density. The volume of a known weight of solid can be determined from the volume of a fluid that is displaced by a submerged solid. In the pycnometric method, the volume of the solid is determined from the change in weight when the vessel is successively filled with a liquid of known density, with solid in air, and with solid in liquid.

When a solid neither rises nor falls through a liquid in which it is submerged, the density of the solid and liquid are equal. This is the principle used in determining the density of milk with calibrated beads. By preparing a series of liquid mixtures of known density, the density of solid foods can be determined.

In many instances, apparent (rather than absolute) density is determined. For example, volume of bread (determined by displacement of dwarf rapeseed) is an important criterion of the breadmaking potentialities of a wheat flour. For many years, plumpness has been considered an important characteristic of good grain. This quality is indicated in a general way by the test weight of a grain. Minimum test weight is specified by standards for cereal grains and some oilseeds (flax and soybeans) (U.S. Dep. Agric. 1947).

The equipment used in test weight determinations is shown in Fig. 26.4. In the United States and Canada, test weight is determined in pounds per bushel; in most other countries using the metric system in kilograms per hectoliter.

Fig. 26.4.   Apparatus for determination of test weight of cereal grains. (Courtesy Henry Simon Ltd.)

Test weight is influenced mainly by kernel shape and size, as they govern the packing of grain in a container. Kernel size for a specific grain is of less significance. The other important factor is the density of grain, which depends on the structure of the grain and its chemical composition including moisture content. Also wetting, subsequent drying, and even mechanical handling in grain elevators influence the test weight. Determinations of test weight must be made under rigidly controlled conditions to obtain reliable results. Theoretically, conversion factors can be applied to convert test weight values from one system to another, provided tests are made in containers of comparable size.

BIBLIOGRAPHY

Anon. (1984). "Dynatrol Bulletin *J-67*." DA Automation Products, Inc., Houston, TX.
Bauer, N. (1945). Determination of density. *In* "Physical Methods of Organic Chemistry," Vol. I (A. Weissberger, ed.). Interscience Publishers, New York.

Erb, R. E., and Manus, L. J. (1963). Estimating solids non-fat in herd milk using the plastic bead method of Golding. *J. Dairy Sci.* **46**, 1373–1379.

Kruger, E. and Oliver-Daumen, B. (1980). Present state of continuous beer analysis. *Mschr. Brauerei* **33**(1), 12–21.

Mahling, A. (1965). Density. *In* "Handbook of Food Chemistry, Analysis of Foods, Physical and Physico-Chemical Assay Methods," Vol. II, Part I (J. Schormuller, ed.). Springer-Verlag, Berlin.

Maxwell, J. L., Kurtz, F. A., and Strelka, B. J. (1984). Specific volume (density) of saccharide solutions (corn syrups and blends) and partial specific volumes of saccharide–water mixtures. *J. Agric. Food Chem.* **32**, 974–979.

Sartor, D. J., and Sherry, N. B. (1983). Automated alcohol and original gravity analyzer: A new aid to brewing. *MBAA Tech. Quart.* **20**, 160–163.

U.S. Dep. Agric. (1947). "Grain Grading Primer." Misc. Publ. *740.* U.S. Dep. Agric., Washington, DC.

Wagner, G. H., Bailey, G. C., and Eversole, W. G. (1942). Determining liquid and vapor densities in closed systems. A precise method. *Ind. Eng. Chem., Anal. Ed.* **14**, 129–131.

Wartman, A. M., Spawn, T. D., and Eliason, M. A. (1984). Relationship between density, temperature and dry substance of commercial corn syrups, high-fructose corn syrups, and blends with sucrose and invert sugar. *J. Agric. Food Chem.* **32**, 971–974.

27
Refractometry and Polarimetry

REFRACTOMETRY

When a ray of electromagnetic radiation strikes a flat surface at an angle, the ray may be bent upward (*reflected*) or bend downward (*refracted*) as illustrated in Fig. 27.1. Notice that the ray does not go straight through the material.

The amount of refraction is a characteristic of every substance. It is commonly expressed by the refractive index n, defined as

$$n = \sin i/\sin r \qquad (27.1)$$

Refractive index measurements have long been used for the qualitative identification of unknown compounds by comparing the refractive index of the unknown with literature values of various known substances. When density measurements also are considered, a certain amount of information about the structure of a compound can be obtained. Each atom, bond, or group contributes to the overall refractive index. When these individual contributions (atomic refractions) are taken together, the result is the *specific refractivity*. If the specific refractivity is multiplied by the molecular weight of a compound, then the *molecular refractivity*, R, is obtained. The Lorentz–Lorenz equation shows this relation:

$$R = \frac{(n^2 - 1)}{(n^2 + 2)} \frac{M}{d} \qquad (27.2)$$

where n is the refractive index, M the molecular weight, and d the density.

The relation is particularly useful in the identification of unknown ma-

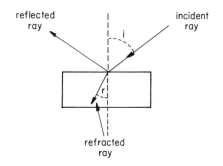

Fig. 27.1. Reflection and refraction.

terials. Suppose you have isolated a new compound and suspect that it may be one of several possible structures. Using the atomic refractions listed in Table 27.1, you can calculate the expected refractive index for each possible structure and then see which matches the experimental refractive index.

Example: We calculate the refractive index of dimethoxymethane ($CH_3OCH_2OCH_3$; $M = 76.10$, $d = 0.8560$) as follows. First, we calculate the molecular refractivity from the atomic refractions:

$$
\begin{array}{lrr}
3C: & 3 \times 2.42 = & 7.26 \\
8H: & 8 \times 1.10 = & 8.80 \\
2 \text{ ether O:} & 2 \times 1.64 = & \underline{3.28} \\
& R = & 19.34
\end{array}
$$

Then, we substitute into equation 27.2:

$$19.34 = \left(\frac{n^2 - 1}{n^2 + 2}\right) \times \frac{76.10}{0.8560}$$

$$(n^2 - 1)/(n^2 + 2) = 0.21754$$

$$n^2 - 1 = 0.2175n^2 + 0.4351$$

$$0.7825n^2 = 1.4351$$

$$n^2 = 1.834$$

$$n = 1.354$$

The value reported in the literature is 1.3534.

Refractive Index of Mixtures

This determination is very important because it can be used to determine the concentration of eluents from paper, column, and thin-layer chromatography where the sample is mixed with the solvent.

Table 27.1.   Atomic Refractions[a]

Element structural unit	Na_D	Element structural unit	Na_D
F	1.0	Nitro group in	
C	2.42	aromatic nitro compounds	7.30
H	1.10	nitroamines	7.51
O in OH	1.52	nitroparaffins	6.72
O in ester OR	1.64	alkyl nitrites	7.44
O^{2-}	2.21	alkyl nitrates	7.59
Cl	5.97		
Br	8.86	Nitroso group in	
I	13.90	nitrates	5.91
S in SH	7.69	nitrosoamines	5.37
S in R_2S	7.97		
S in RCNS	7.91	Double bond	
S in R_2S_2	8.11	no radicals	1.51
O in ether	1.64	$RCH{=}CH_2$	1.60
N		$RCH{=}CHR$	1.75
in I° aliphatic amines	2.32	$R_2C{=}CHR$	1.88
in II° aliphatic amines	2.49	$R_2C{=}CR_2$	2.00
in III° aliphatic amines	2.84		
in I° aromatic amines	3.21	Triple bond	2.40
in II° aromatic amines	3.59	3-membered ring	0.71
in III° aromatic amines	4.36	4-membered ring	0.48
in hydroxylamines	2.48		
in hydrazines	2.47	Diazo group	
in aliphatic cyanides	3.05		
in aromatic cyanides	3.79		
in aliphatic oximes	3.93		
in amides	2.65		8.43
in II° amides	2.27		
in III° amides	2.71		
			7.47

[a] From Eisendohr (1910, 1912).

The refractive index of mixtures varies linearly with the mole fraction of the components. Refractive index measurements provide one way of measuring mixtures of salts that would otherwise be difficult to determine.

Because specific and molecular refractivities are an additive property of substances, it is possible to set up an equation for calculating the refractivity of a mixture from the refractivity of the solvent and the solute. The equation is

$$Zr_m = Xr_a + Yr_b \qquad (27.3)$$

where X is the weight of compound A and Y the weight of compound B with specific refractivities r_a and r_b respectively, and Z the total weight of the mixture with specific refractivity r_m.

Example: Twenty milliliters of a mixture of xylene and carbon tetrachloride had a density of 1.2156 and $n_D^{25} = 1.4338$. Pure xylene has a density of 0.8570 and $n_D^{25} = 1.4915$; pure carbon tetrachloride has a density of 1.5816 and $n_D^{25} = 1.4562$. Calculate the wt % of this mixture.

First calculate the specific refractivity r of the pure compounds and the mixture from

$$r = \left(\frac{n^2 - 1}{n^2 + 2}\right) \frac{1}{d} \tag{27.4}$$

For xylene,

$$r = \frac{(1.4915)^2 - 1}{(1.4915)^2 + 2} \frac{1}{0.8570} = 0.3382$$

For carbon tetrachloride,

$$r = \frac{(1.4562)^2 - 1}{(1.4562)^2 + 2} \frac{1}{1.5816} = 0.1719$$

For the mixture,

$$r = \frac{(1.4738)^2 - 1}{(1.4738)^2 + 2} \frac{1}{1.2156} = 0.2311$$

Then calculate the weight of the mixture, Z, in grams from $Z =$ volume times density:

$$Z = (20 \text{ ml}) (1.2156 \text{ g/ml}) = 24.31 \text{ g}$$

Let X be the weight of carbon tetrachloride and Y the weight of xylene. Using equation 27.3, set up the following two equations:

$$X + Y = 24.31 \text{ g}$$
$$0.1719X + 0.3382Y = (24.31)(0.2311) = 5.618 \text{ g}$$

Solving these two equations gives

$$X = 15.66 \text{ g}, \qquad Y = 8.65 \text{ g}$$

Thus, the mixture is 64.4% by weight carbon tetrachloride and 35.6% by weight xylene.

Refractometers

Now that some of the applications of refractive index measurements have been discussed, the instruments for obtaining refractive indices will be examined. In order to standardize measurements, the refractive index is usually determined at 20°C with the sodium D line (the yellow doublet at 589 nm). The symbol used to indicate these standard conditions is n_D^{20}. Since a change in temperature of 1°C changes the refractive index by 0.00045, the temperature must be maintained within ±0.2°C.

Several refractometers use the critical ray approach, illustrated in Fig. 27.2. Three rays of monochromatic radiation strike a medium of different density. Two of these rays are refracted and would produce light on the other side of the medium. However, the third ray and all other rays having an angle of incidence equal to or greater than λ_2 are not refracted. Thus no light gets through the medium at this point and a dark field is produced. This *critical ray* is utilized in refractometers to measure the refractive index of various substances, since the critical angle is different for each substance. Each wavelength of incident radiation has a critical angle, and if white light were used, no sharp division would occur between the light and dark fields, and there would be a light field followed by a rainbow of colors and the dark field. We can eliminate the rainbow of colors by the *Amici prism* discussed next.

An instrument is constructed that measures the critical angle of the sodium D line. The rays of other wavelengths are then disposed of by means of an Amici prism (Fig. 27.3). Notice that only the sodium D line comes through in the same direction as the incident light. This permits the use of white light since all the rays, other than the one of interest, are scattered out of the optical path.

Abbe Refractometer.   The Abbe refractometer is probably the most common type employed today. A schematic diagram of it is shown in Fig. 27.4. This refractometer covers the refractive index range from 1.3 to 1.7 with a precision of ±0.0003 units; it can be used for direct reading and requires only 1 or 2 drops of sample.

To use an Abbe refractometer, place the sample between the two lower

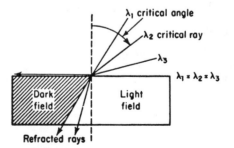

Fig. 27.2.   Illustration of the critical ray and critical angle concepts.

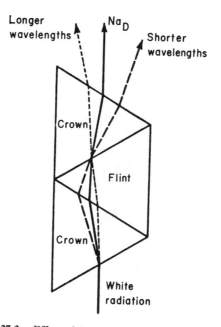

Fig. 27.3. Effect of Amici prism on white radiation.

prisms, and then rotate the connecting arm until the critical ray is centered in the eyepiece. If the division line between the light and dark field is colored, turn the Amici prism (the compensator) until a sharp division appears. If possible, do not use ether or acetone to clean off the sample from the prisms; these solvents evaporate quickly and in that process change the temperature of the prism sufficiently to cause a significant error in the measurement. It is best to use ethanol, but if you must use

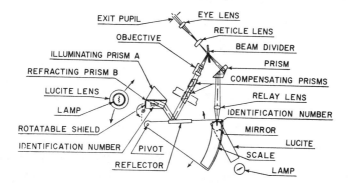

Fig. 27.4. Components of an Abbe refractometer.

ether or acetone, wait at least 10 min for temperature equilibrium to be reestablished.

Refractive Index of Solids

Probably the easiest method for determining the refractive index of solids is to use a microscope and solutions of different refractive index. A crystal of the solid is placed on a slide and a few drops of a liquid of known refractive index are placed around the crystal. A slide cover is placed on top to prevent the solution from evaporating.

A thin band of visible light outlines the crystal when a narrow axial illuminating core and an objective of low aperature are employed. This line around the crystal is called the *Becke line*. This line moves toward the medium of higher refractive index if the focus is raised, and toward the medium of lower refractive index if the focus is lowered. By changing the surrounding solutions, the refractive index can be determined to less than 0.1 of a RI unit.

Differential Refractometry

The Abbe refractometer can determine values of n to ± 0.0003 units, but for many cases this is not accurate enough, particularly when determining column eluates. The differential refractometer produces more accurate readings by using two hollow prisms set opposite to each other, so that the image displacement by one prism tends to offset the displacement by the other. This keeps the net shift small, and when this shift is compared to a fixed reference point (observed by a high-power microscope) differences of the order of ± 0.000005 RI units can be obtained. Figure 27.5 is a schematic diagram of such an instrument. A monochromatic source and very close temperature control are necessary.

Interferometry

Interferometry is a technique based upon the interference of light waves discussed previously in connection with the interference filter, the diffraction grating, and X-ray diffraction. Figure 27.6 shows the design of an interferometer.

Let us take two rays of radiation coming from the same source and have them "race" to the detector plate. The lens is used to get the rays in the direction we want them to go, and the baffle is used to ensure that only the two rays of interest continue on with the race. Assume for the moment that both the reference cell and the sample cell are empty. The

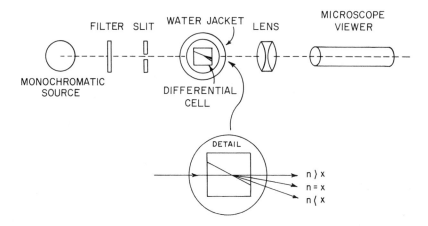

Fig. 27.5. Design of a differential refractometer.

two rays continue through the cells, are focused by the lens onto the detector plate, and (since both rays have covered the same distance) they are in phase when they hit the detector plate; consequently, each ray reinforces the other and a bright spot is formed.

If, however, some sample is added to the sample cell, that ray may find the going a bit tougher and be slowed down. If it is slowed down by one-half a wavelength, when it hits the detector plate it will completely interfere with the other ray and be destroyed, thus producing a dark spot. If the sample ray is slowed down by one full wavelength, it will again be in phase with the reference ray and again a bright spot will be produced. This bright spot will be offset from the first bright spot because the slower ray is focused differently by the lens. The net result is a series of bright and dark spots (*fringes*) if monochromatic radiation is used, and a series of rainbows if white light is used. Equation 27.5 is used to calculate the

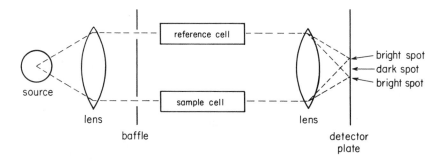

Fig. 27.6. Schematic diagram of an interferometer.

results:

$$N = b(n - n_o)/\lambda \qquad (27.5)$$

where N is the number of fringes, b the cell thickness, n the refractive index of the reference, n_o the refractive index of the sample, and λ the wavelength of radiation used.

How good is an interferometer? Using a 1-m cell it is possible to determine differences in n of ± 0.0000001 for liquids and ± 0.00000001 for gases.

Recent Applications

Sugars in sugar refinery products have been determined by Charles (1981) by HPLC using refractometric detection. Pinitol and other cyclitols (carbohydrates) have been analyzed by Ghias-Ud-Din *et al.* (1981) by HPLC with refractive index detection. HPLC methods are reported for the analysis of triglycerides with methanol–acetone as mobile phase and RI detection (Podlaha *et al.* 1982). Svensson *et al.* (1982) described the analysis of geometric and positional isomers of long-chain monounsaturated fatty acids by reversed-phase HPLC using an interference RI detector. Fruit juices were analyzed for tartaric, malic, and citric acids by reversed-phase HPLC with RI detection (Bigliardi *et al.* 1979). Automated refractive index measurements were used by Bitner *et al.* (1982) to determine the I value to follow the progress of vegetable oil hydrogenation.

POLARIMETRY

Let us examine in more detail the light coming from the lamp you are using to read this page. Recall from Chapter 5 that we said that this radiation behaved as if it had an electric component and a magnetic component, and that they acted as if they were at right angles to each other. One such ray might be represented as in Fig. 27.7.

However, there are millions of rays coming from your lamp and the direction of the electric and magnetic components are purely random and may look like Fig. 27.8(A). This radiation is said to be unpolarized. If, however, by some means we can get all of the rays to have their electric and magnetic components all in the same direction, then the radiation is *plane polarized*, as shown in Fig. 27.8(B).

Of what value is polarized light to us? Consider the following molecule

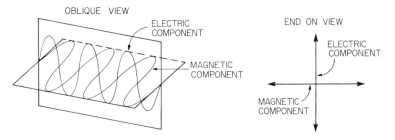

Fig. 27.7. Components of radiation.

of 1-chloro-1-ethanol:

$$\begin{array}{c} Cl \\ | \\ H\!-\!\!C\!-\!OH \\ | \\ CH_3 \end{array}$$

We see that it is unsymmetrical. No matter where we placed a mirror in this molecule, we could not get an exact mirror image. If unpolarized radiation strikes this molecule, and the electric component of one ray interacts more with one side than the other, we might notice an effect; however, since the radiation is unpolarized, there are always some rays oriented opposite to the first ray that react in just the opposite manner and cancel out any initial effect we might have seen. On the other hand, if polarized radiation strikes the molecule, there is no way for an opposite effect to take place, and as shown in Fig. 27.9 the radiation is "rotated."

Compounds that can rotate polarized light are said to be *optically active*. In the old system of nomenclature for optical rotation, if you were looking toward the light source and the rotation was clockwise, then the compound was said to be *dextro* (+); if rotation was counter clockwise, then

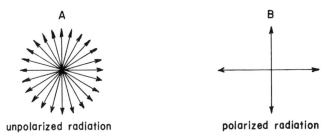

Fig. 27.8. Vector diagram of polarized and unpolarized radiation.

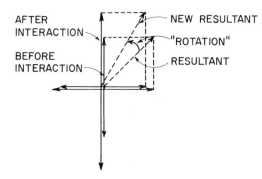

Fig. 27.9. Rotation of radiation.

the compound was *levo* (−). The difficulty was that not all dextro compounds had + rotations and not all levo compounds had − rotations.

The new system of nomenclature, the Cahn–Ingold–Prelog system, is not based on rotation at all but on the configuration of atoms around a particular center. As an example of how an optically active compound is designated in this system, consider the molecule of 1-chloro-1-ethanol, shown previously:

1. Select the "center" of the asymmetric atom to be examined. In this case, it is the C with the H, Cl, OH, and CH_3 groups attached.

2. Assign the groups in their order of importance. The atom directly bonded to the center atom with the highest atomic number is most important. In this case, $Cl > O > C > H$. In cases of a tie, then mass is the tie breaker (e.g., ethyl > methyl).

3. Hold the molecule so that the lowest priority atom is directly away from you and behind the central atom.

4. Start at the next lowest priority number and go around to the highest. If you go to the right, then the compound is designated *R*, if to the left, then it is designated *S*. In this case, we would go C to O to Cl in a counterclockwise direction, therefore this is an *S* configuration.

The new system is often called the *RS* system (*R* = restus, right; *S* = sinister, left).

If a compound has a plane of symmetry, it cannot rotate polarized radiation, because what effect occurs on one side of the molecule is canceled out by an opposite effect on the other side of the molecule.

When we have an optically active atom in a molecule, it is generally starred and is called an *asymmetric center*:

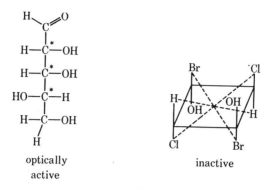

optically
active

inactive

Polarimeters

The instrument for measuring optical rotation is called a *polarimeter*. In order to see how it works, let us imagine this book and two picket fences with vertical pickets; the fences are about 10 ft apart. The book will represent a ray of unpolarized radiation initially. Suppose I ask you to throw the book through the first picket fence. You know that it will only go through the fence when the length of the book lines up with the pickets. Thus, the first picket fence will polarize our book, i.e., line it up in one direction as it comes through, and is called a *polarizer*. If nothing happens to the book as it crosses the 10 ft to the second picket fence, it will be lined up properly and go through it also. But suppose that a gust of wind tipped the book slightly after it left the first picket fence. In other words, it has been rotated just like a polarized ray would be rotated by an asymmetrical center of a molecule. Now the book would not get through the second picket fence. If we picked up one end of the second fence (rotated it) until the book lined up with the openings, the book could then pass through. Suppose we put a big wheel on this second picket fence and marked degrees of rotation on it. It would now be a simple matter to see how many degrees we had to rotate the fence to allow the book (radiation) to get through. This second picket fence would then be an *analyzer*.

A polarimeter does just this. A monochromatic source is used; a polarizer polarizes the radiation; the sample rotates the radiation; and an analyzer measures how much the rotation is. The components of a polarimeter are diagrammed in Fig. 27.10.

Sources. Optical rotation depends upon the wavelength of radiation used. For this reason it is easier to interpret the data if monochromatic radiation is used. The sodium vapor lamp and the mercury vapor lamp are the two most common sources.

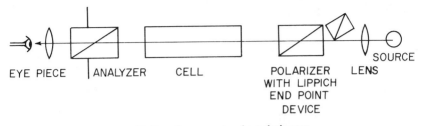

Fig. 27.10. Components of a polarimeter.

Polarizers and Analyzers. The two most common means of producing polarized radiation are the Nicol and the Glan-Thompson prisms (Fig. 27.11). These prisms are made from quartz or calcite, and are cut in such a way that the ordinary ray is totally reflected, permitting the extraordinary ray to emerge polarized. The Glan-Thompson prism has the advantage that the emerging radiation is going in the same direction as the entering radiation, and the prism can be turned without distorting the beam intensity.

The analyzer can be turned in two directions to measure the optical rotation. If we turn it one way, we get a very bright field, and if we turn it in the other direction, we get a black field. The problem is that our eyes are not able to tell where the brightest and blackest field is since they have no reference. This difficulty is overcome by use of an *end-point device* to provide a reference.

The human eye is very good at comparing shades of colors, so the end-point device is designed to produce a gray shade over half of the viewing eyepiece, which is then matched by rotating the analyzer. The most common type of end-point device is the *Lippich prism*, which is a small Glan-Thompson prism, tipped slightly, covering half of the field. This is shown in Fig. 27.10 along with the other components of a polarimeter.

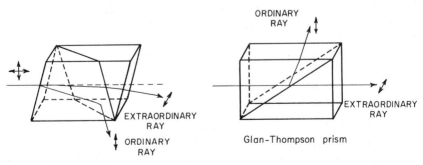

Fig. 27.11. Prisms used as polarizers.

Basic Relationships

The basic equation used for analytical polarimetry is

$$[\alpha]_\lambda^t = \alpha/lc \tag{27.6}$$

where $[\alpha]_\lambda^t$ is the specific rotation in degrees, t the temperature in °C, λ the wavelength of radiation employed, α the observed angle of rotation in degrees, l the length of the tube in decimeters, and c the concentration in g/ml (for a pure liquid, density d replaces c).

Example: A solution of an organic compound is placed in a 20-cm tube and produces a rotation of 38.73°. A tube containing only solvent has a rotation of 1.46°. If the specific rotation is 62.12°, what is the concentration of the solute?

Since $[\alpha]_\lambda^t = 62.12°$, $\alpha = (38.73° - 1.46°) = 37.27°$ corrected, and $l = 20$ cm or 2 dn., then substituting into equation 27.6 gives $62.12 = 37.27/2c$ and $c = 0.30$ g/ml.

Effect of Temperature. Equation 27.6 is valid providing a number of variables are controlled, one of which is the temperature. When the temperature of the solution in the polarimeter tube is different than the temperature at which the specific rotation was obtained, a correction is necessary. The equation for this is

$$[\alpha]_\lambda^{t_1} = [\alpha]^{t_2} + n(t_1 - t_2) \tag{27.7}$$

where t is the temperature of the solution in °C and n a constant.

Example: A solution of sucrose was placed in a polarimeter and the specific rotation of sucrose was determined at three different temperatures. From the following data, determine the constant n and the value of $[\alpha]_D^{20}$:

t (°C)	$[\alpha]_D^t$
14.0	66.57
22.0	66.375
30.0	66.18

First plot $[\alpha]_D^t$ vs. t (Fig. 27.12). From this group, the slope (n) is found to be -0.0244, over the temperature range studied. Then, to obtain $[\alpha]_D^{20}$, substitute into equation 27.7 as follows:

$$[\alpha]_D^{20} = [\alpha]_D^{22} - 0.0244\,(20-22)$$
$$= 66.375 + 0.0488 = 66.42$$

Effect of Concentration. A second correction must be applied in cases where the specific rotation changes with concentration. Biot's equations

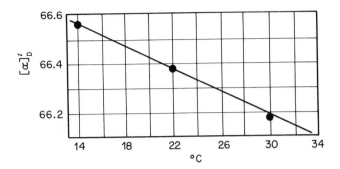

Fig. 27.12. Effect of temperature on specific rotation.

may be used to make a suitable correction:

$$[\alpha] = \begin{cases} A + Bq & \text{(linear)} & (27.8) \\ A + Bq + Cq^2 & \text{(parabolic)} & (27.9) \\ A\,[Bq/(C + q)] & \text{(hyperbolic)} & (27.10) \end{cases}$$

where A, B, and C are constants and q the weight fraction of solvent in the solution. $[\alpha]$ is generally replaced with (α/lpd), where p is the weight fraction of solute, d the density of the solution, and l the tube length in decimeters. To determine which equation to use, plot $[\alpha]$ vs. q, and see if the resulting curve more closely approaches a parabola, hyperbola, or a straight line.

Example: Three solutions of a saccharide containing 20, 50, and 80% solute were prepared. The solutions were placed in a 20-cm tube and the following data obtained:

p	q	d	α	$[\alpha]$
0.80	0.20	1.116	18.26	10.22
0.50	0.50	1.290	13.74	10.65
0.20	0.80	1.464	6.47	11.04

To determine which of the Biot equations can be used and to evaluate the constants, plot $[\alpha]$ vs. q (Fig. 27.13). Since the plot is a straight line, equation 27.8 applies:

(1) $10.22 = A + 0.20B$

(2) $10.65 = A + 0.50B$

(3) $11.04 = A + 0.80B$

Solve any pair of these equations simultaneously; in this case $A = 9.955$ and $B = 1.37$.

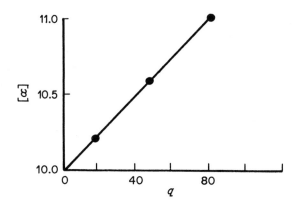

Fig. 27.13. Effect of concentration on specific rotation.

Effect of Wavelength. Specific rotation also varies with wavelength. This relationship is expressed by the Drude equation

$$[\alpha] = \frac{k_1}{\lambda_0^2 - \lambda_1^2} + \frac{k_2}{\lambda_0^2 - \lambda_2^2} + \frac{k_3}{\lambda_0^2 - \lambda_3^2} \qquad (27.11)$$

in which k_1, k_2, and k_3 along with λ_1, λ_2, and λ_3 are constants; λ_0 is the wavelength employed, generally expressed in angstroms. For most purposes, when λ is far removed from an optically active band, the simplified equation (27.12) can be used:

$$[\alpha] = k/(\lambda_0^2 - \lambda_1^2) \qquad (27.12)$$

If λ is very far removed, so that $\lambda_0 \gg \lambda_1$ then

$$[\alpha] = k/\lambda_0^2 \qquad (27.13)$$

Example: An organic compound had $[\alpha] = 14.63°$ when radiation of 589 nm was employed, and 15.27° when radiation of 436 nm was used. Calculate the Drude equation constants and predict the $[\alpha]$ when a λ_0 of 707 nm is used:

$$14.63 = \frac{k}{(0.589)^2 - \lambda_1^2}, \qquad 15.27 = \frac{k}{(0.436)^2 - \lambda_1^2}$$

These can be further reduced to

$$5.08 - 14.63\,\lambda_1^2 = k$$
$$\underline{2.90 - 15.27\,\lambda_1^2 = k}$$
$$2.18 + 0.64\,\lambda_1^2 = 0$$

POLARIZER   SOURCE

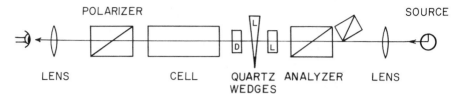

LENS   CELL   QUARTZ ANALYZER   LENS
                                        WEDGES

Fig. 27.14.   Components of a saccharimeter.

Thus, $\lambda_1^2 = -3.41$, and $k = 54.97$. These values can be substituted into equation 27.12 to obtain the specific rotation at 707 nm:

$$[\alpha] = 54.97/[(0.707)^2 + 3.41] = 14.06$$

Saccharimeters

A modification of a polarimeter used extensively by the sugar industry is the quartz wedge saccharimeter (Fig. 27.14). The *rotatory dispersion* of quartz and most sugars are about the same. This means that white light can be used rather than monochromatic radiation because any wavelength effects can be canceled out. When a sugar sample is added to the cell, a dextrorotation is usually obtained. The levo quartz wedge is then moved to exactly compensate this rotation. The smaller wedge of levo quartz is to eliminate refractive dispersion. The piece of dextro quartz is used for those sugars that produce levo readings.

PROBLEMS

27.1.   Calculate the molecular refractivity for diethylamine if $n_D = 1.3873$ and $d = 0.7180$. *Ans.* 23.98.

27.2.   The molecular refractivity of 6,8-dimethyl quinoline is 41.73. What is its refractive index if its density is 1.0660? *Ans.* 1.4776.

27.3.   Mixtures of acetic acid ($n_D = 1.3718$) and cetyl acetate ($n_D = 1.4358$) produce a linear relationship between refractive index and mole fraction. What would be the refractive index of a compound that would seem to disappear when placed in a solution containing 0.152 mole fraction of acetic acid? *Ans.* 1.4261.

27.4.   A solution of $l(-)$-glutamic acid containing 1.085 g/20 ml of 0.37 N HCl had a rotation of $-1.63°$ when placed in a 100-mm tube. Calculate the specific rotation of the glutamic acid. *Ans.* $-30.04°$.

27.5. $l(-)$-diiodotyrosine has a specific rotation of $+2.890$. What would be the expected optical rotation of a solution containing 4.41 g/60 ml of 1.1 N HCl using a 5-cm tube? *Ans*. 1.062°.

27.6. The specific rotation of $l(+)$-alanine is $+14.7°$. If 20-cm tubes are used and an optical rotation of $+1.70°$ is observed, what is the concentration of alanine in the solution? *Ans*. 0.057 g/ml.

27.7. $l(-)$-cystine has a specific rotation of $-214.0°$ at 24.35°C. At 20.0°C, 0.997 g/200 ml gave a rotation of -2.0 using a 20-cm tube. What would be the specific rotation at 30°C? *Ans*. 224.4°.

27.8. Determine which of Biot's equations best expresses the following data and calculate the constants.

p	q	$[\alpha]$
0.97	0.03	15.84
0.71	0.29	15.00
0.50	0.50	14.50
0.35	0.65	14.26
0.24	0.76	14.13
0.10	0.90	14.06

Ans. $A = 14.6$, $B = -3.78$, $C = 1.89$.

27.9. Nicotine has a specific rotation of $-162°$ using radiation of 5993 Å, and $-126°$ when radiation of 6563 Å is used. What is its specific rotation when radiation of 4861 Å is used? *Ans*. $-318°$.

27.10. A solution of santonin dissolved in alcohol produced a specific rotation of $+442°$ using radiation of 686.7 nm, and 991° using radiation of 526.9 nm. What would be the wavelength required to have a specific rotation of $+1323°$? *Ans*. 488.3 nm.

BIBLIOGRAPHY

Bigliardi, D., Gherardi, S., and Poli, M. (1979). Determination of tartaric, malic and citric acids in fruit juices by high-pressure liquid chromatography. *Ind. Conserve* **54**, 209–212. [*Anal. Abstr*. 1980 **39**, 5F69.]

Bitner, E. D., Snyder, J. M., Mounts, T. L., Dutton, H. J., and Baker, G. (1982). Analytical chain to monitor vegetable oil hydrogenation. *J. Am. Oil Chem. Soc.* **59**, 286–287. [*Chem. Abstr.* **97**, 70909e.]

Charles, D. F. (1981). Analysis of sugars and organic acids. Cane-sugar refinery experience with liquid chromatography on a sulphonic acid cation-exchange resin. *Int. Sugar J.* **83**, 169–172, 195. [*Anal. Abstr.* 1982, **42**, 1F22.]

Eisendohr, F. (1910). A new calculation of atomic refractions. Part 1. *Z. Physik. Chem.* **75**, 585–607.

Eisendohr, F. (1912). A new calculation of atomic refractions. Part 2. *Z. Physik. Chem.* **79**, 129–146.

Ghias-Ud-Din, M., Smith, A. E., and Phillips, D. V. (1981). Separation of pinitol and some other cyclitols by high-performance liquid chromatography. *J. Chromatogr.* **211**, 295–298.

454 27 REFRACTOMETRY AND POLARIMETRY

Podlaha, O., Toregard, B., and Petersson, B. (1982). Analysis of triglycerides. *Fette, Seifen, Anstrichm.* **84**, 17. [*Anal. Abstr.* **43**, 1F85.]

Svensson, L., Sisfontes, L., Nyborg, G., and Blomstrand, R. (1982). High performance liquid chromatography and glass capillary gas chromatography of geometric and positional isomers of long chain monounsaturated fatty acids. *Lipids* **17**, 50–59.

28
Rheology

RHEOLOGICAL PARAMETERS

Rheology is concerned with stress–strain relations of materials that show a behavior intermediate between those of solids and liquids. *Stress* can be compressive, tensile, or shear. The passage of time does not itself cause changes in materials. Time is, however, often introduced in measuring rates of changes of forces and deformations. Chemical changes in foodstuffs often occur in time and may be studied by rheological methods. Temperature is also important and often appears in rheological equations.

Strain is measured by deformation. Figure 28.1 classifies deformations according to a procedure proposed by the British Rheologist Club, as modified by W. Bushuk. This is a useful classification, though the limits of each category are arbitrary. For convenience, all deformations are divided into *elastic deformations* and *flow*. An elastic deformation is one that can be recovered, irrespective of whether the recovery is spontaneous or not. Elastic deformations can be divided into those that are ideal, which show no time effect, and those that are nonideal, which show a time effect. Ideal deformations may be Hookean or non-Hookean; in the latter strain may increase more or less rapidly than stress. Nonideal deformations may be completely recoverable, and the strain may vary more rapidly than, proportionally to, or less rapidly than stress. The incompletely recoverable deformations are linked with flow, and also lead to plastoelastic and plastoinelastic groups. On the flow side, the figure divides into plastic and viscous groups. Plastic deformations may be New-

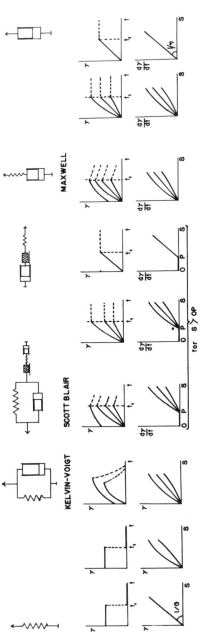

Fig. 28.1. Classification of deformation. (Courtesy W. Bushuk)

tonian or non-Newtonian; the latter is subsdivided into viscoelastic and viscoinelastic groups. The Newtonian and Bingham systems are listed as separate subcategories on account of their exceptional importance.

Figure 28.2 classifies fluids according to their behavior under stress. Viscosity is given by the slope of plots of shearing stress vs. shear velocity (top graph) and is plotted in the bottom graph as a function of shear rate, which affects the viscosity of non-Newtonian fluids (Hallikainen 1962). Viscosity terminology and equations are summarized in Table 28.1.

Texture can be regarded as a manifestation of the rheological properties of a food. It is an important attribute in that it affects processing and handling (Charm 1962), influences food habits, and affects shelf life and consumer acceptance of foods (Matz 1962). Research indicates that the consumer is highly conscious of food texture, and that in certain foods texture may be even more important than flavor (Szczesniak and Kleyn 1963). The physical and mechanical properties of foods are discussed in detail in books by Rha (1975), Mohsenin (1980), and Peleg and Bagley (1983). Muller (1973), Sone (1982), Dea *et al.* (1983), and Mitchell (1979, 1984) have published books or reviews on rheology and consistency. Food texture and its measurement were the subject of reviews and books by Prentice (1972), Sherman (1979), deMan *et al.* (1976), Kramer and Szczesniak (1973), and Brennan (1980).

Analytical versus Integral Approach

Measurement of the rheological properties of foods can be based on either the analytical or the integral approach. In the first, the properties of a material are related to such simple systems as Newtonian fluids or Hookean solids. When the material approximates any such system, the

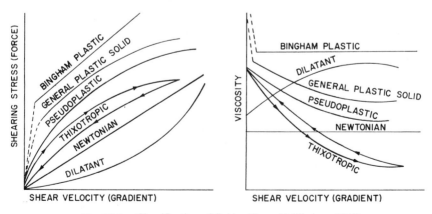

Fig. 28.2.   Classification of fluids. (From Hallikainen 1962)

Table 28.1. Viscosity Terminology and Definitions

Term	Definition
Stress	Force/area; F/A
Velocity gradient (shear)	Rate of change of liquid velocity across the stream; V/L for linear velocity profile; dV/dL for nonlinear profile (feet/sec^2)
Absolute (dynamic) viscosity	Constant of proportionality between applied stress and resulting shear velocity (Newton's hypothesis):

$$\frac{F}{A} = \frac{\eta V}{L} \quad \eta = \frac{F/A}{V/L}$$

Poise (P)	Unit of dynamic or absolute viscosity (dyne-sec/cm^2)
Kinematic viscosity V	Dynamic viscosity/density, η/ρ
Stoke (S)	Unit of kinematic viscosity
Hagen–Poiseuille law (flow through a capillary)	

$$Q = \frac{\pi R^4}{8\eta L}(P_1 - P_2)$$

Saybolt viscometer (Universal, Furol)	Measures time (sec) for a given volume of fluid to flow through a standard orifice
Fluidity	Reciprocal of absolute viscosity (rhe, $= P^{-1}$)
Specific viscosity	Ratio of absolute viscosity of a fluid at any temperature to that of water at 20°C (68°F); since for water at this temperature $\eta = 1.002$ cP, the relative viscosity of a fluid is approximately equal to its absolute viscosity; as the density of water is 1, kinematic viscosity of water is 1.002 cS at 20°C)
Apparent viscosity	Viscosity of a non-Newtonian fluid under given conditions; same as consistency
Consistency	Resistance of a substance to deformation; the same as viscosity for a Newtonian fluid and the same as apparent viscosity for a non-Newtonian fluid
Saybolt Universal seconds (SUS)	Time units referring to the Saybold viscosimeter
Saybolt Furol seconds (SFS)	Time units referring to the Saybolt viscosimeter, with a Furol capillary, which is larger than a Universal capillary
Shear viscosimeter	Viscometer that measures viscosity of a non-Newtonian fluid at several different shear rates; viscosity is extrapolated to zero shear rate by connecting the measured points and extending curve to zero shear rate

[a] From Hallikainen (1962).

appropriate equations are used with a suitable correction. For a material that does not exhibit perfect elastic or true viscous behavior, the hypothesis is made that the material consists of two or more parts, and that the effects are additive.

In the integral approach, a simple and initially empirical relation between stress, strain, and time is sought. This approach is often dictated by the fact that its results correlate better with sensory evaluation than descriptions based on rheological models or simple dimensional terms. Foods seldom have simple rheological properties. In addition, most rheological measurements refer to the arbitrary conditions imposed by a particular instrument. What we are generally measuring is not a pure rheological parameter, but the way in which the properties vary under some standardized system of applied forces.

Yet, the relationship between stress–strain systems and their time derivatives are often expressed in terms of models—a pictorial presentation of the analytical approach. The models sometimes have no more than a symbolic significance; in other cases they throw light on existing configurations and are a very valuable rheological tool (Scott-Blair 1953).

Rheological Models

Rheological models are useful because they represent, in an easily comprehensible way, the mechanical behavior of materials and facilitate mathematical description of their behavior. In this section, we describe various types of rheological models, their respective deformation–time curves, and the rheological shorthand frequently used.

In Fig. 28.3 of elementary and composite rheological models, the four basic components include a dashpot (piston sliding in a cylinder filled with oil) depicting Newtonian viscosity (N); the spring depicting the Hookean elasticity element (H); the spring clip or St. Venant body (St. V); and the shear pin (SP). The dotted lines above the deformation–time plots represent the timing of loading and unloading. Upon loading, the spring (H) lengthens immediately by an amount dependent on the load, whereas in the dashpot (N) it is the speed (or "rate") of movement that is proportional to the load. When the dashpot (N) is unloaded, the deformation remains; when the spring (H) is unloaded, there is instantaneous and complete recovery.

The St. Venant element (St. V) is shown by a friction weight. It describes ideal plasticity, i.e., an irrecoverable deformation caused at a specific stress above the yield value. Below this, no deformation takes place; above it, deformation continues at a constant rate as long as the yield stress persists. The shear pin (SP) allows presentation of the rupture of an element under a specific stress.

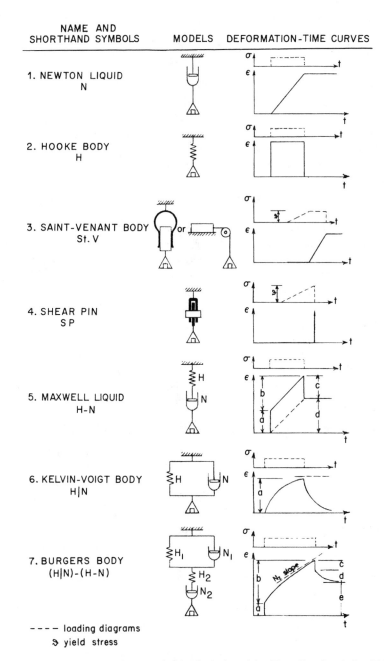

Fig. 28.3.   Elementary and composite rheological models. (From Lerchenthal and Muller 1967.)

A dashpot and a spring combined in a series (H–N) give a model of a Maxwell body (Fig. 28.3). On loading, there is at first an immediate deformation of (a) due to the spring followed by a slow deformation (b) due to the dashpot. On unloading, there is an immediate recovery of (c) equivalent to (a) and an irrecoverable residual deformation of (d) equal to (b).

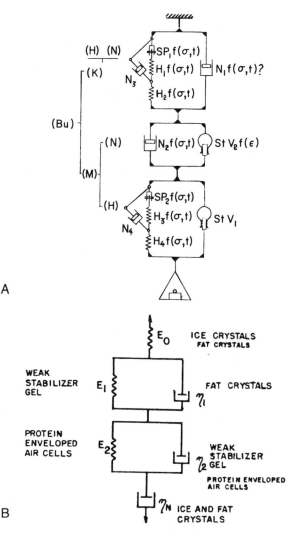

Fig. 28.4. (A) Rheological model of wheat flour dough. (From Lerchenthal and Muller 1967.) (B) Six-element model for frozen ice cream showing rheological association with structural components. (From Shama and Sherman 1966.)

When the spring and dashpot are combined in parallel (H/N), the model is referred to as the Kelvin or Voigt type. When applied, the load is gradually transferred from the dashpot to the spring, which eventually tends to hold the load entirely. This causes the deformation–time curve to level out asymptotically. On unloading, there is slow but complete recovery as the spring recovers against the friction of the dashpot.

When a Maxwell and a Kelvin–Voigt body are placed in series [(H₁/ N₁)–(N₂ − H₂)], we have a Burger's body. The deformation–time curve shows at first an immediate response (a) due to the free spring H_2 of the Maxwell element. The curve then rises (b), and due to the free dashpot N_2 does not level out as it would in the Kelvin–Voigt model. On unloading, there is an immediate response (c) due to H_2, followed by a delayed elastic response (d) caused by H_1 working against the resistance of N_1. The residual deformation (e) is due to the free Newtonian element N_2 of the Maxwell component.

The rheological model of wheat flour dough in Fig. 28.4A illustrates the complexity of model systems in describing actual food materials. A six-element model for frozen ice cream showing rheological association with structural components is shown in Fig. 28.4B.

RHEOLOGICAL METHODS AND INSTRUMENTS

Empirical Methods

Numerous schemes have been proposed for classifying the available rheological methods used in food evaluation. Scott-Blair (1958) classified rheological methods of texture measurement as fundamental, empirical, and imitative. According to Szczesniak (1966), fundamental tests measure basic rheological parameters and relate the nature of the tested product to basic rheological models. Because of the complexity of food products and the necessity to relate the obtained measurements to functional properties, fundamental tests have only limited use. However, they can serve to define a system and provide a more scientific basis for empirical and imitative tests. Thus, fundamental tests are seldom used for quality control in either flour mills or bakeries, but the information gained from them has led to the development of a number of useful empirical tests. Similarly, fundamental research on the relation between concentration and viscosity of emulsions was useful in the manufacture of condensed milk. Basic parameters determined in working molten chocolate and studies on the rheology of hydrogenated fats are additional examples of fundamental studies that have led to applied empirical tests. However, most empirical tests measure poorly defined properties that have been shown by experience to be related to textural quality.

The most common instruments used in indirect empirical tests are *penetrometers*. They are used to determine (1) rigidity of gels in terms of strain produced by a load; (2) the force required just to penetrate the material; and (3) the consistency of the material as measured by resistance to further penetration either by the rate of sinking of a column or needle or by the total depth of penetration following impact. Penetrometers have been much used for measuring the firmness of dairy products and the tenderness of fruit, vegetables, and fish. Much work has been published on the Bloom "gelometer" and its various modifications. Sinkers and line-spread consistometers are also useful tools in indirect tests. They include "plummets" and "bobs" of all kinds, and are used to determine the consistency of chocolate, creams, spreads, sauces, and fillings.

Imitative Methods

Imitative tests measure various properties under conditions similar to those to which the food is subjected in practice, that is, the properties of the material during handling and the properties of the food during consumption. Instruments to measure the first group of properties include butter spreaders, which give a measure of the spreading properties, and various mixing and load-extension meters, which measure dough-handling properties. In the second group are various tenderometers, which simulate the chewing action of teeth.

Scott-Blair (1958) pointed out that imitative tests do not necessarily correlate better with actual performance than tests somewhat removed from the practical situation. Thus, in testing butter, an instrument has been devised in which a small cube of butter is sheared by a knife edge at standard loads and speeds; the amount of shear measured indicates "spreadability." Yet, the correlation with spreadability scores of an experienced panel were higher with simple viscosity measurements than with the test imitating the conditions to which the material is subjected in practice. The reason is that the imitation is far from perfect, that a "standard" surface (differing substantially from that of bread crumb) is used, and that spreadability is not related only to "softness" assessed by determining shearing forces.

Other Methods

In addition to the classification of Scott-Blair (1958), instruments used to measure textural quality have been classified according to the foods they evaluate (Matz 1962; Amerine *et al.* 1965). The weakness in this kind of classification is that many instruments are used on more than one of the group of foods. Drake (1961) developed a classification system based on the geometry of the rheological apparatus. This classification contains the following types: (1) rectilinear motion (parallel, divergent, and convergent); (2) circular motion (rotation, torsion); (3) axially symmetric motion (unlimited, limited); (4) defined other motions (bending, transversal); and (5) undefined motions (mechanical treatment, muscular treatment). Each subheading in Drake's classification is subdivided further on the basis of the geometry of the apparatus.

The classification system proposed by Bourne (1966) is based upon the variable (or variables) that constitute the basis of the measurement. This classification system, which stresses the nature and dimensional units of the property being measured, is given in Table 28.2.

Force-measuring instruments are the basis of the most common method used in measuring food texture. The measured variable is force, usually maximum force, and distance and time are held either constant or replicatable. The group includes instruments measuring compression or tensile strength. In distance-measuring instruments, force and time are held con-

Table 28.2. Classification of Objective Methods for Measuring Texture and Consistency[a]

Method	Measured variable	Dimensional units	Examples
Force-measuring	Force F	mlt^{-2}	Tenderometer
Distance-measuring	(a) Distance	l	Penetrometers
	(b) Area	l^2	Grawemeyer consistometer
	(c) Volume	l^3	Seed displacement (bread volume)
Time-measuring	Time T	t	Ostwald viscometer
Energy-measuring	Work $F \times D$	ml^2t^{-2}	Farinograph
Ratio-measuring	F, or D, or T, or $F \times D$, measured twice	Dimensionless	Cohesiveness (G. F. texture profile)
Multiple-measuring	F, D, and T, and $F \times D$	mlt^{-2}, l, t, ml^2t^{-2}	G. F. texturometer
Multiple-variable	F, or D, or T (all vary)	Unclear	Durometer
Chemical analysis	Concentration	Dimensionless (% or ppm)	Alcohol-insoluble solids
Miscellaneous	Anything	Anything	Optical density of fish homogenate

[a] From Bourne (1966).

stant or replicatable, and distance, area, or volume is measured. The instruments measure depth of penetration, compression and flow, or volume of either end product (e.g., bread) or expressed liquid (e.g., juice).

Viscometers are the main time-measuring instruments. Energy-measuring instruments determine work or energy and include recording dough mixers or meat grinders. If a force–distance curve is drawn during a test, the area under the force–distance curve is a measure of work.

For ratio-measuring methods, at least two measurements of the same variable must be taken. Thus, cohesiveness can be computed from the ratio of the work done during the first and second bites of food.

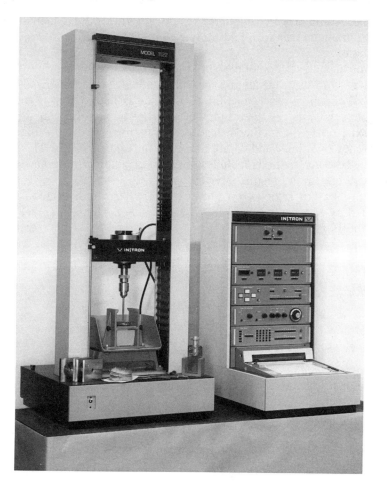

Fig. 28.5. Instron model 1122 testing equipment with punch probe set up for analyzing cooked meat. Additional punches or dies are shown in the foreground. (Courtesy Instron Corp., Canton, MA.)

Instruments in the multiple-measuring group can measure various forces, distances, areas, etc., and record the results. The instruments can be used for a large number of products. They include the Allo-Kramer Shear Press, the General Foods Texturometer, and the Instron (Fig. 28.5). Such instruments are highly versatile and useful. Generally, a special test cell for each type of product is necessary for meaningful evaluation (Kramer and Backinger 1959). Commercial universal testing machines are expensive; however, construction of a simple apparatus for food research was described by Voisey et al. (1967). Multiple-variable instruments have more than one uncontrolled variable, but only one variable is measured. Although sometimes these instruments correlate well with sensory evaluation, it is difficult to evaluate the results in terms of fundamental rheological parameters or to relate the measurements with data from other instruments.

Chemical analyses do not measure texture directly, but are often highly correlated with physical texture measurements or subjective panel tests. A typical chemical determination of this kind is the content of alcohol-insoluble solids, an accepted index of green peas' maturity. In addition, there exist objective methods that are highly correlated with texture or consistency of a food, and do not fit any of the above groups. This group of miscellaneous methods includes various refractometric, polarimetric, electrical, and sound-testing devices.

Comprehensive reviews on physical characterization of foods were published by Mohsenin (1968, 1980), Morrow and Mohsenin (1968), and Peleg and Bagley (1983).

MEASURING THE COMPONENTS OF FOOD TEXTURE

According to Szczesniak (1966), all food texture-measuring devices have five essential elements:

1. *Driving mechanism* varying from a simple weight and pulley arrangement to a more sophisticated variable-drive electric motor or hydraulic system.

2. *Probe element* in contact with food (flat plunger, shearing jaws, tooth-shaped attachment, piercing rod, spindle, or cutting blade).

3. *Force* (simple or composite) applied in a vertical, horizontal, or levered manner of the cutting, piercing, puncturing, compressing, grinding, shearing, or pulling type.

4. *Sensing element* ranging from a simple spring to a sophisticated strain gage; proving ring dynamometers and transducers may also be used.

5. *Read-out system* (maximum-force dial, an oscilloscope, or a recorder tracing the force–distance relationship).

Texture Profiles

In selecting a rheological instrument, it is essential to establish the type of information sought and its potential usefulness. For intelligent interpretation of results, it is generally desirable also to know the expected precision and applicability of results to various conditions. For quality control, generally one textural parameter determined on a composite sample may be useful. For research purposes, a *texture profile* in terms of several parameters determined on a small homogeneous sample may be desirable. The mechanical textural characteristics of foods that govern, to a large extent, the selection of a rheological procedure and instrument can be divided into the primary parameters of hardness, cohesiveness, viscosity, elasticity, and adhesiveness, and into the secondary (or derived) parameters of brittleness, chewiness, and gumminess (Szczesniak 1966).

The basic measurement in a hardness determination involves the load–deformation relationship. Cohesiveness may be measured as the rate at which the material disintegrates under mechanical action. Tensile strength is a manifestation of cohesiveness. Cohesiveness is usually tested in terms of the secondary parameters brittleness, chewiness, and gumminess. Brittleness, crunchiness, and crumbliness, which can be placed on a continuum, can be measured as the ease with which the material yields under an increasing compression load; the smaller the deformation under a given load, the lower the cohesiveness and the greater the "snappability" of the product. Tenderness, chewiness, and toughness are measured in terms of the energy required to masticate a solid food. They are the characteristics most difficult to measure precisely, because mastication involves compressing, shearing, piercing, grinding, tearing, and cutting, along with adequate lubrication by saliva at body temperatures.

Gumminess is characteristic of semisolid foods with a low degree of hardness and a high degree of cohesivenes. The rate at which a food returns to its original condition after removal of a deforming force is an index of the food's elasticity. Adhesiveness is measured in terms of the work required to overcome the attractive forces between the surface of a food and the surface of other materials with which the food comes into contact.

Sherman (1969) examined the texture profile of Szczesniak and proposed several modifications. The modified profile is shown in Fig. 28.6A. The proposed scheme consists of primary, secondary, and tertiary categories. Primary attributes are analytical composition, particle size and size distribution, particle shape, etc. The three secondary attributes are elasticity (E), viscosity (η), and adhesion (N). The tertiary characteristics are the responses most often used in sensory analyses of texture. Panel responses associated with masticatory tertiary characteristics are summarized in Fig. 28.6B. Tertiary characteristics are derived from a complex blending of two or more sensory attributes, and can be regarded as falling within a three-dimensional continuum, which has the secondary attributes as coordinate axes. All tertiary attributes can be represented by rectangular coordinates of the form (αE, $\beta \eta$, γN) in which α, β, and γ represent the respective magnitudes of the secondary attributes. Sherman (1969) postulated that since solids, semisolids, and fluids have characteristic values of these attributes, it should be possible to predict panel responses from mechanical strain–time tests, which are carried out at approximately constant rate of shear operative during mastication and adhesion tests.

Crispness has been the subject of several investigations by Vickers and coworkers (Vickers and Bourne 1976A,B; Vickers 1980, 1981, 1984; Christensen and Vickers 1981), who have proposed that crispness is primarily an acoustical sensation. The frequency spectra and the amplitude–time characteristics of sounds produced by biting crisp foods cover a wide range and show irregular variations in loudness with time. These workers have proposed a model system, involving a generalized cellular structure, in which the sounds produced as crisp cellular foods are crushed result from the rupture of single cells or cell walls. Foods vary widely in the recognizability of their crushing sounds. The recognizability does not depend on the familiarity or class of a food. The relationship between oral and auditory judgments of crispness, crunchiness, and hardness are highly correlated. It was also proposed that vibrations produced by fracturing crisp foods may be the basis of perception of crispness. Available instruments fail to measure crispness in a variety of foods.

Classification of Texture Measurements

Kramer and Twigg (1959) used the following classification as the basis for mechanical measurement of texture:

Finger feel	Mouth feel
Firmness	Chewiness
Softness or yielding quality	Fibrousness
Juiciness	Grittiness
	Mealiness
	Stickiness
	Oiliness

The parameters measured to determine these sensory characterists are compression, shearing, cutting, tensile strength, and shear pressure.

Table 28.3. Relation of Kinesthetic Characteristics to Physical Methods of Application of Force

Sensory reaction	Physical test	Instruments or procedures for measurement
Firmness	Compression	Pressure tester; shear-press
Yielding quality	Compression	Pressure tester; shear-press; ball compressor
Juiciness	Compression (juice extraction)	Puncture tester; succulometer; shear-press; moisture tests
Chewiness	Shear-pressure	Tenderometer; texturemeter; shear-press; specific gravity; solids
Fibrousness	Cutting; comminuting	Fibrometer; shear-press; fiber analysis
Grittiness	—	Comminution; elution; sedimentation
Mealiness	—	Starch, and/or gum analysis
Stickiness	Tensile strength	Jelly strength, pectin, and/or gum analysis

Table 28.3 summarizes the physical tests and procedures used to measure the sensory reactions.

According to Scott-Blair (1945), there are eight basic rheological properties measured in textural examinations:

1. Viscosity
2. Elastic modulus
3. Elastic hysteresis (or fore- and aftereffects)
4. Structural viscosity (i.e., immediately recoverable fall in viscosity with rise in shearing stress)
5. Sharpness of curvature of flow curve (i.e., extent to which material approximates yield-value behavior)
6. Extent of breakdown under shear not immediately recovered (includes irreversible breakdown and thixotrophy)
7. Extent of work hardening or dilatancy produced by strain
8. Strength factor—generally tensile strength, but sometimes also comprehensive

To measure these properties, the following methods and instruments are available:

- capillary tube viscometers or plastometers (Ostwald)
- jet or orifice viscometers (Saybolt or Engler viscosimeters)
- rotating bodies (spheres, discs, cylinders—MacMichael)
- falling sphere viscometers
- rising sphere viscometers
- displaced sphere viscometers
- hot wire anemometers

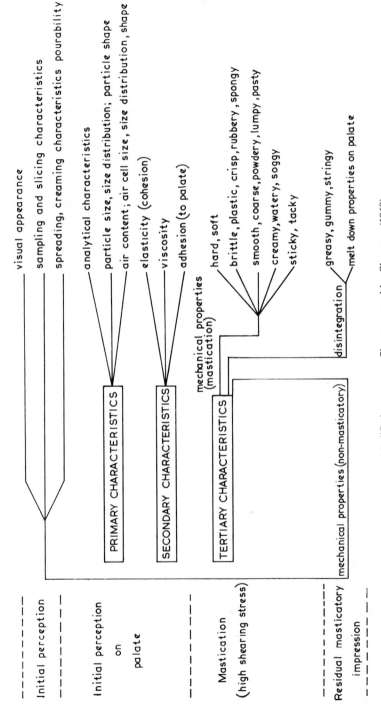

Fig. 28.6A. Modified texture profiles proposed by Sherman (1969).

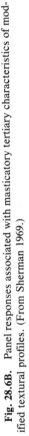

Fig. 28.6B. Panel responses associated with masticatory tertiary characteristics of modified textural profiles. (From Sherman 1969.)

- ergometer and extensimeter (Chopin)
- mixographs measuring work input in stirring
- falling solid cylinder in large hollow concentric cylinder (i.e., sinker viscosimeter)
- extensimeters (load-extension meters)
- stretching of spheres into bands
- compression or extension of cylinders or spheres and extension of strips (plastometers and parallel plate consistometers)
- compression of materials in bulk
- cutting tests (curd-o-meters, sectilometers)
- penetrometers (also used for compression without penetration)
- bending bars and rods (torque and flexion)
- tests by dropping weights on sample
- fatigue measurements of failure
- indentation hardness testers
- resilience (bounce) and pendulum methods
- vibration (without failure)
- tensile strength (direct tension)
- subjective testing by handling

A partial list of the properties and methods used in the textural evaluation of selected foods is given in Table 28.4.

According to Bourne (1982), four commonly used fundamental tests are applied to solid foods.

1. Young's modulus of elasticity:

$$E = \frac{\text{stress}}{\text{strain}} = \frac{F/A}{\Delta L/L}$$

(in compression or tension)

2. Shear modulus:

$$G = \frac{\text{shearing stress}}{\text{shearing strain}} = \frac{F/A}{\gamma/L}$$

3. Bulk modulus:

$$K = \frac{\text{hydrostatic pressure}}{\text{volume strain}} = \frac{P}{\Delta V/V}$$

4. Poisson's ratio:

$$\mu = \frac{\text{change in width per unit width}}{\text{change in length per unit length}} = \frac{\Delta D/D}{\Delta L/L}$$

In these expressions, F is the applied force, A the cross-sectional area, L the unstressed length, ΔL the change in length caused by the application

Table 28.4. Textural Evaluation of Selected Foods[a]

Food	Properties	Methods
Milk	Viscosity	Capillary tube viscometer, rotating bodies, falling solid cylinder
Honey	Viscosity	Capillary tube viscometer, rotating bodies, falling sphere cylinder, vibration
Butter	Viscosity, structural viscosity, unrecoverable breakdown, extent of work hardening due to strain	Jet viscometer, compression/extension of cylinders/spheres, cutting, penetrometers
Hard cheese	Viscosity, elastic modulus, elastic hysteresis, structural viscosity, sharpness of flow curve, extent of work hardening due to strain	Compression/extension of cylinders/spheres, cutting, penetrometers subjective testing
Soft cheese	Viscosity, elastic modulus, extent of work hardening due to strain	Compression/extension of cylinders/spheres, cutting, penetrometers subjective testing
Cream	Viscosity, unrecoverable breakdown under shear	Capillary tube viscometer, jet viscometer
Dough	Viscosity, elastic modulus, elastic hysteresis, structural viscosity, extent of work hardening due to strain	Jet viscometer, ergometer, mixographs, extensimeters, stretching of spheres into bands, compression/extension of cylinders/spheres, penetrometers, fatigue measurement, resilience/pendulum, tensile strength, subjective testing

[a] Adapted from Scott-Blair 1945.

of force F, γ the displacement (shear modulus), P the pressure, V the volume, and D the diameter. In the case of fluids, viscosity is commonly measured as viscosity $= \sigma/\dot{\gamma}$, where σ is shear stress and $\dot{\gamma}$ the shear rate.

APPLICATIONS

Cereals

Cereal technologists are faced with a variety of rheological problems. Starting with the milling process, the brittleness of the bran and the pliability of the germ are of major importance in separating the endosperm from the outer layers. Viscosity, elastic modulus, and tensile strength are the outstanding factors in determining the behavior of wheat flour dough; compressibility, crispness, and breaking strength of the baked or fried product play a role in acceptability by the consumer. The use of rheological parameters and methods in cereal chemistry and technology was the subject of a special issue of *Cereal Chemistry* (Bushuk 1975). Rheological properties of dough are of particular importance for several reasons (Bloksma and Hlynka 1964). Dough is one of the main stages in breadmaking. With increased mechanization and automation in the baking industry, dough properties are important from the mechanical viewpoint. A number of common laboratory tests for determining flour quality depend upon the measurement of the physical properties of dough. A study of the physical properties of dough contributes to our understanding of baking quality. Finally, in dough, the chemical, physical, and biological aspects interact in bringing out the full potential of the system.

The history of the rheological studies of dough was reviewed by Muller (1964) and of instrumentation by Brabender (1965). Basic considerations of dough properties were discussed by Greup and Hintzer (1953) and by Bloksma and Hlynka (1964). A review of major factors governing rheological properties and performance of wheat flour doughs was presented by Miller and Johnson (1954), Kent-Jones and Amos (1967), and by Hlynka (1967).

Kernel Hardness.   Objective hardness measurements are mainly useful in differentiating between soft or hard wheats in plant breeding programs (Miller and Johnson 1954). Extremely hard milling characteristics are usually reflected in increased power requirements and reduced yields of flours of acceptable quality (e.g., ash content or color). Extensive softness interferes with efficient bolting and increases the requirements for sieving space.

The most commonly used methods to determine wheat hardness include determination of the power to grind a sample, the time to grind under specified conditions on a burr-type mill, the resistance to grinding, particle size of ground wheat, and near infrared reflectance of ground wheat as an index of particle size. The methods were reviewed by Miller *et al.* (1982) and by Pomeranz and Miller (1982).

Physical Dough Testing.   Flow occurs in dough only if the stress exceeds a certain limiting value or yield value. In this respect dough behaves as a Bingham body. At lower stresses the deformations disappear completely after the stress is released (Bloksma and Hlynka 1964). Whether the elastic deformation is accompanied by permanent viscous deformation (at higher stresses) or not (at lower stresses), the attainment of the equilibrium shape after the release of stress requires some time. Dough shows retarded elasticity (elastic aftereffect) like that of a Kelvin body.

The viscosity of dough and, generally, its resistance against deformation increase with increasing shear; its rate of shear decreases correspondingly. The changes in properties are temporary, and the phenomenon is termed "structural activation." Apart from its dependence on the amount of shear, the apparent viscosity decreases with increasing stress. Gluten behaves essentially like dough. It does not show an appreciable yield value, and its retardation time is shorter. Satisfactory bread doughs are characterized by a high viscosity:modulus quotient and a small decrease in apparent viscosity with increasing stress. The quotient corresponds to the relaxation time for a simple Maxwell body. Bread doughs seem to differ from durum flour doughs in that deviations from linear behaviour are more pronounced in elastic deformations, whereas durum flour doughs show those deviations most clearly in viscous deformations.

Hlynka and associates conducted systematic studies of doughs rested for various periods between rounding and shaping of a dough and its stretch. By plotting extensigram heights (see section on load–extension instruments) at an arbitrarily selected fixed sample extension against rest period t, they obtained a structural relaxation curve (Fig. 28.7). The curve can be described by an equation of a hyperbola asymptotic to the load axis, and to a line parallel to the time axis and at a distance L_a above it (Fig. 28.8):

$$L = L_a + c/t \qquad (28.1)$$

where L is the load, L_a the asymptotic load, c the structural relaxation constant, and t the test period. The semiaxis constant a, which equals

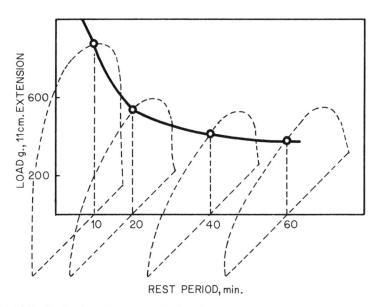

Fig. 28.7. Derivation of a structural relaxation curve. (From Dempster *et al.* 1953.)

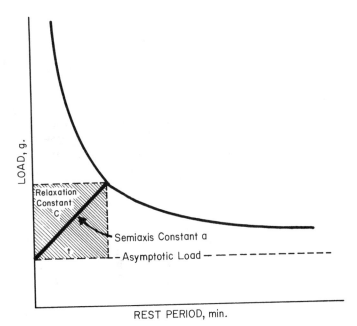

Fig. 28.8. Hyperbolic representation of the structural relaxation curve. (From Hlynka and Matsuo 1959.)

$\sqrt{2c}$, is useful in studies of the effects of oxidants in breadmaking. The parameters c and L_a can be found graphically by plotting the product of L and t vs. t (Fig. 28.9); this results in a transformation to a straight line given by

$$Lt = L_a t + c \qquad (28.2)$$

The intercept and slope of this line are c and L_a, respectively.

Muller *et al.* (1961, 1962, and 1963) have analyzed the stretching process in the Brabender extensograph. They converted the load–extension curve recorded into a stress–strain curve in which both stress and strain were expressed in cgs units. By interrupting the stretching process and allowing the test piece to recover, they divided the total extension into a recoverable elastic part and an irrecoverable viscous one.

Assuming that the rheological behavior of dough can be described by a generalized Maxwell body, one can similarly divide the work performed on a test piece into an elastic and viscous part. Bloksma (1967) analyzed the above "work technique" and developed a theory that leads to an alternative presentation of experimental results, and which permits, under the conditions of some simplifying assumptions, direct conclusions on changes in modulus and viscosity.

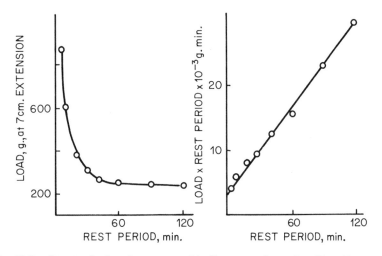

Fig. 28.9. Structural relaxation curve and its linear transformation (From Dempster *et al.* 1955.)

The main value of physical dough testing is that it provides excellent supplementary information on changes occurring during a short period of the breadmaking process. This concept is illustrated in Fig. 28.10. Physical dough testing devices are most useful in the evaluation and prediction of plant breeder samples, in quality control during flour milling and breadmaking, and in basic rheological studies. They provide information on specific properties that cannot be obtained by other means. Physical dough testing methods have many limitations, which should be realized in the interpretation of results. Thus, rheological properties may differ in flours that vary little in their breadmaking performance. Physical dough testing curves may be affected considerably by modification of flour constituents or by certain additives, which have little effect on actual breadmaking.

Most rheological measurements are made on unyeasted doughs; fermentation is known to modify considerably both rheological properties and breadmaking characteristics. Finally, although physical dough tests are useful analytical tools because they measure changes at certain stages of the breadmaking process, they fail to measure the effects of interaction between various factors on bread quality.

Recording Dough Mixers. The two most widely used types of recording dough mixers are the Brabender farinograph and the mixograph. Both instruments record the power that is needed for mixing dough at a constant speed. The record consists of (1) an initial rising part, which shows an

	FARINOGRAPH	EXTENSOGRAPH	AMLYOGRAPH
Nature of Method	dynamic	static	dynamic
Type of Indication	*Farinogram* 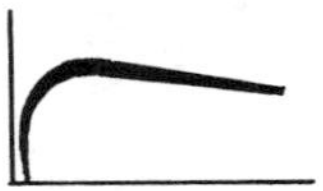 Resulting curve of above instrument	*Extensogram* 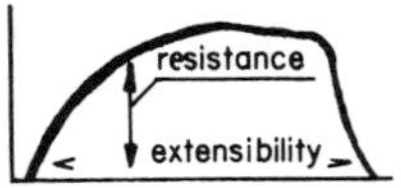 Resulting curve of above instrument	*Amylogram*  Resulting curve of above instrument
Information Obtained	Absorption Mixing time Mixing tolerance (same as general strength)	$\dfrac{\text{Resistance}}{\text{Extensibility}}$ = degree of maturing (indication of physical dough properties)	Crumb formation characteristics (gelatinization properties)
Correction Possibilities in a flour mill	changes in wheat blend	adjustment in "bleach" (chemical maturing agents)	mill stream switching and/or malt additions (diastasing)
in a bakery	changes in absorption and mixing	adjustment of yeast food	malt additions

Fig. 28.10. The three-phase concept of breadmaking. (Courtesy C. W. Brabender Instruments, Inc.)

increase in resistance with mixing time and is interpreted as dough development time; (2) a point of maximum resistance (or minimum mobility), generally identified with optimum dough development, and (3) a third part of more or less rapid decrease in consistency and resistance to mixing.

A picture and diagram of the Brabender farinograph are given in Fig. 28.11. The instrument measures plasticity and mobility of dough subjected to a prolonged, relatively gentle mixing action at constant temperature. Resistance offered by the dough to mixing blades is transmitted through a dynamometer to a pen that traces a curve on a kymograph chart. The general practice has been a determine in a preliminary "titration curve" the amount of water required to give a standard consistency (generally 500 BU). The farinograph also provides information on optimum mixing time and dough stability. Figure 28.12 compares farinograms of three types of flours and shows the main parameters calculated. Many investigators calculate the valorimeter value, which is determined by the length of time required to mix the flour to minimum mobility and the descending slope. The larger the valorimeter value, the stronger—in terms of breadmaking potential—the flour.

The mixograph is a miniature high-speed recording dough mixer (Swanson and Working 1933). The mixer (Fig. 28.13) has four vertical pins revolving about three stationary pins in the bottom of the bowl. As the gluten develops, a gradually increasing force is required to push the revolving pins through the dough. The increased force is measured by the tendency to rotate the bowl, which is placed in the center of a lever system. A record of the torque produced on the lever system is made on a chart moving at a constant rate of speed. Mixograms of a weak and a strong flour are shown in the upper part of Fig. 28.13.

Load–Extension Instruments. In a number of commercially available instruments, dough mixed and shaped under standardized conditions is stretched until rupture, and a curve of load versus elongation is recorded. From the curve, resistance to deformation, extensibility, and energy needed to rupture the dough is computed. The instruments are particularly useful in studying the effects of oxidants on dough properties.

The Brabender extensograph (Fig. 28.14) was introduced in 1936 to supplement the information provided by the farinograph. The tested flour is mixed in the farinograph into a dough reaching its maximum consistency at 500 BU. It is taken out of the mixer and two pieces are shaped into balls by means of a rounder-homogenizer and then into cylinders by means of a roller. The dough cylinders are clamped in cradles and allowed to rest for 45 min at 30°C. After stretching, the dough is shaped, allowed to rest, and stretched again. Generally, three stretching curves are obtained at 45-min intervals. It is usual to make the following measurements: (1) extensibility—length of the curve in mm, (2) resistance to extension—

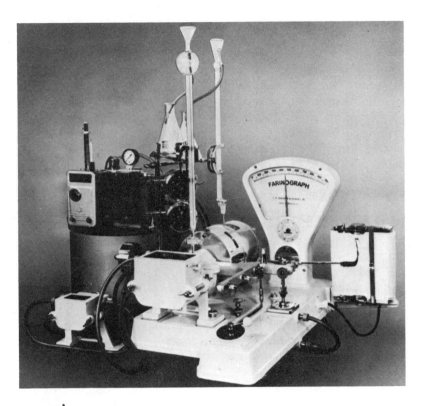

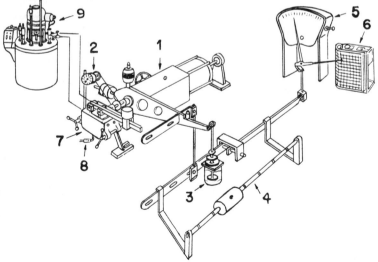

Fig. 28.11. Brabender farinograph (Courtesy C. W. Brabender Instruments, Inc.)

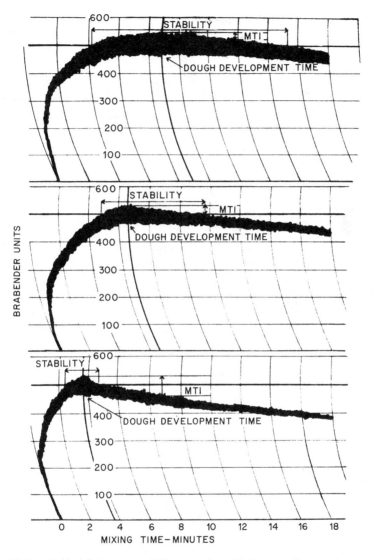

Fig. 28.12. Typical farinograms of flour samples with long, medium, and short development times, showing values for stability and mixing tolerance index (MTI). (From Miller and Johnson 1954.)

height of the extensogram in BU measured 50 mm after the curve has started, (3) strength value—area of the curve, and (4) proportionality figure—ratio between resistance and extensibility.

The research extensometer designed by Halton and his associates (Halton 1938, 1949) from the Research Association of British Flour Millers is similar in many respects to the Brabender extensometer. The extensom-

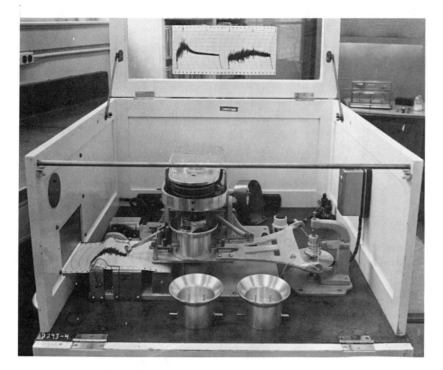

Fig. 28.13. The mixograph. (Courtesy K. F. Finney.)

eter is part of a four-unit device (Fig. 28.15) that includes a water absorption meter, mixer, shaper, and stretching unit. The water absorption meter measures the extrusion times of yeasted doughs. For load–extension curves, the doughs are mixed at optimum absorption, shaped, and impalled on a two-part peg of the research extensometer. The upper half of the peg is fixed; the lower half moves at a fixed rate and stretches the impalled dough. The force that is exerted on the upper peg is transmitted and recorded on a graph paper.

The alveograph (Chopin extensometer) consists of three parts: a mixer, bubble-blowing portion, and recording manometer. The procedure involves using air pressure to blow a bubble from a disk of a flour–water–salt dough. The bubble is expanded to the breaking point, and the recording manometer, which is operated hydraulically, records a curve from which three basic measurements are taken: distance (in mm) that the dough stretches before it ruptures; resistance to stretching at peak height; and curve area. The alveograph is quite popular in several European countries and is the most commonly used physical dough testing device in France.

Starch. For the evaluation of starch products, three aspects of starch rheology are of paramount importance (Hofstee and de Willigen 1953): (1) the alterations of rheological properties during pasting; (2) hot-paste viscosity and its variations with time; and (3) change in rheological properties during and after cooling of the paste (cold-paste viscosity; jelly strength). In all cases, the properties of the pastes are governed by the pasting or cooking method employed. On heating with water, a series of changes occurs, the most important being loss in birefringence, change in translucence, change in X-ray pattern, swelling of the grain, and abrupt change in viscosity. The changes do not occur at the same time nor at a sharply defined temperature. They take place at a small interval of temperature, conveniently called the gelatinization point.

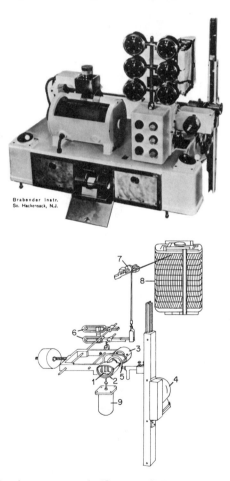

Fig. 28.14.   The Brabender extensograph. (Courtesy C. W. Brabender Instruments, Inc.)

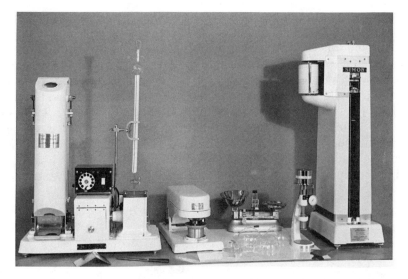

Fig. 28.15. The extensometer. (Courtesy Simon Engineering, Ltd.)

To study the complex rheological changes in starch pasting, it is essential to heat the paste under standardized conditions and record continuously the changes. Several instruments have been designed to meet these requirements including the Brabender amylograph, Corn Industries recording viscometer, and the VI Viscometer. The amylograph and its modifications (i.e., the viscograph) are most commonly used. All hot-paste recording viscometers work on a similar general principle. From the curve obtained the following parameters are generally recorded: (1) the time (temperature) of initial rise in viscosity; (2) temperature and height of maximum viscosity; and (3) viscosity after prolonged heating beyond the point of peak viscosity. There is a linear relation between the logarithm of maximum viscosity (η_{max}) and logarithm of concentration (c):

$$\log \eta_{max} = a + b \log c \qquad (28.3)$$

Although the reproducibility of the data from all recording viscometers is satisfactory, they differ in their performance and utility. Corn Industries recording viscometer allows for varying the rate of stirring. Its disadvantage is the large amount of starch (100 g) and the large sample beaker (1000 ml) required. The VI Viscograph is cheap and simple, but the determination of paste temperature is difficult and no provision is made for the condensation of evaporated water or for keeping a fixed temperature. It is commonly used in textile and paper industries.

The Brabender amylograph (Fig. 28.16) is a torsion viscometer that provides a continuous automatic record of changes in viscosity of starch, as the temperature is raised at a constant rate of approximately 1.5°C/ min (Anker and Geddes 1944). The instrument consists of a cylindrical stainless bowl that holds a suspension of 100 g flour in 460 ml of a buffered solution. The bowl is rotated at 75 rpm in an electrically heated air bath by a synchronous motor, which also operates the recording and temperature-control devices. A steel arm that dips into the bowl is connected through a shaft to a pen, which records changes in viscosity of the heated flour suspension in the bowl. Depending on the change in viscosity of the heated suspension, a torque is exerted on the steel arm and is recorded on an arbitrary scale. Viscosity of the slurry is generally recorded as the temperature rises from 30 to 95°C. Modifications of the amylograph that record viscosity changes at a uniformly rising or falling temperature are useful in studies of gelatinization characteristics of various cereal starches. In Europe, the amylograph is used widely to predict the baking performance of rye flours and to detect excessive amounts of flours from sprouted grain. In the United States, the amylograph is used primarily to control malt supplementation.

Consumer Goods.   Because many consumers often use softness as a

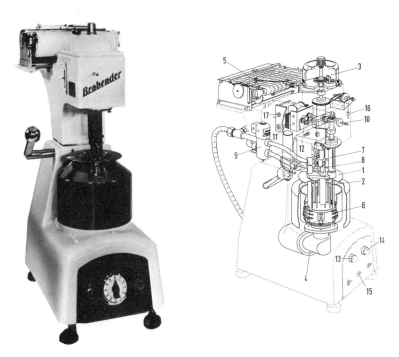

Fig. 28.16.   The Brabender amylograph. (Courtesy C. W. Brabender Instruments, Inc.)

criterion of bread freshness, there is justification for using crumb compressibility in evaluating staleness. Load softness is determined in three ways: (1) The crumb is subjected to a constant load for a fixed time, and the deformation is measured to find a softness index; (2) the force required to give a fixed deformation is measured and crumb firmness is determined; and (3) the crumb firmness is subjected to shearing or squeezing forces.

The latest edition of *Cereal Laboratory Methods* (AACC 1983) gives two tests for measuring staleness of bread, one based on organoleptic evaluation by a trained panel and one using a compressimeter, which measures crumb firmness. An instrument that quantitatively measures the effects of squeezing a load of bread was described by Hlynka and Van Eschen (1965).

The various physical tests are widely used as a convenient means of obtaining measurements of changes in crumb properties as bread ages. It should be realized, however, that the various tests record different rates of change, and that they are not entirely consistent with one another. It has also been shown that such methods frequently do not agree with human judgment of freshness.

Mechanical strength of prepacked cereal foods (cookies, macaroni, breakfast cereals, snack foods, cereal bars, and various fried or baked crisp or flaky products) is of great economic importance. Measurements of the texture of such food items are generally made by methods that measure the force necessary to break a piece of the material and that determine changes in particle size resulting from subjecting the food to the impact of crushing forces.

Dairy Products

Most liquid foods are essentially Newtonian fluids and may be considered to have their texture or " mouth feel" adequately described by their viscosity.

Measurements of the viscosity of milk were made before the end of the last century, mainly with a view to finding a property that might prove useful as an index of quality, thus avoiding lengthy analyses. This hope was never fulfilled (Scott-Blair 1953). Although milk deviates little from Newtonian behavior, its viscosity is not easily measured accurately because of creaming. Cream shows a variety of rheological anomalies, and it is difficult to define any suitable set of properties by which to describe what the consumer means by "body." The flow properties of condensed milk are important during manufacture but are commonly assessed subjectively. Rheological properties of frozen ice cream, mix, and melted ice cream by a coaxial cylinder viscosimeter were described by Shama and Sherman (1966), and by Sherman (1966).

The rheological properties of butter are important in relation to quality. The whole complex of those properties is so characteristic in butter that other substances are often said to have a butter-like consistency (Mulder 1953). Sectility, spreadability, and eating texture and consistency affect consumer evaluation of butter (and margarine). Butter should have good spreadability accompanied by some elasticity. It should be firm enough not to collapse, but not so firm that it breaks into pieces; it should be moderately extensible, long, and tough; it should not be oily or sticky or have a mealy or sandy taste.

The various methods and instruments used to estimate rheological properties of butter include penetrometers, plastometers (deformation under load), cutting methods, the sagging beam methods, and various spreadability procedures. In most cases, the properties determined by the various methods differ widely. It is unlikely that a single method would serve to evaluate all the important parameters or express consistency as it is judged by butter graders by organoleptic tests.

Davis (1937) calculated the viscosity of butter by dividing shearing stress by the rate of deformation of a cylinder of butter subjected to various loads. The elastic modulus was computed from deformation changes in samples after the loads were removed. Kruisheer and den Herder (1938) described a mechanically driven plastometer devised for the determination of permanent deformation and "yield values"—important parameters of butter consistency. An apparatus constructed to measure the force required to cut with a wire a sample of butter was described by Dolby (1941), and a penetrometer by Mohr and Wellm (1948). Kapsalis et al. (1960) developed an instrument called the "consistometer" for measuring both the spreadability and the hardness of butter samples. The instrument consists of a constant-speed motor driving a pendulum that carries the knife or wire used to measure, respectively, spreadability and hardness. The butter is held in a stainless steel frame mounted at the lowest point of an arc described by the pendulum. The sample is raised $\frac{1}{16}$ in. for the knife, and $\frac{1}{8}$ in. for the wire before each determination. The resistance offered by the sample is equivalent to the sum of the forces exerted by fixed weights and the motor. The weights act through a pulley system and can be measured directly, while the force exerted by the motor is registered on a torque meter.

The great variety of cheeses and their tremendous variations in consistency and composition stimulated development of a large number of compressimeters, penetrometers, and sectimeters. From the earliest times, the assessing of "body" of cheese and of curd (partly made cheese) by subjective means has been regarded as important. Faults in body at certain stages in manufacture effect considerably the quality of the finished cheese. Poor body in "green" (unripe cheese) foreshadows inferior keeping quality and even serious deterioration during ripening (Baron and Scott-Blair 1953).

Consistency of rennetted cheese is partly defined by the nature of the surface in rupture (clean or flocculated) and partly by the resistance of the curd to compression. The latter is best expressed by the shear modulus of the gel. For displacements under small loads applied for short times, curd gels behave approximately as Hookean elastic solids. If measurements are repeated on the same surface of the gel, there is an irreversible deformation due to some elastic fatigue. Thus readings should be taken at different spots on the surface. Numerous methods have been devised to measure the rigidity of rennetted curd by drawing a cutter through the coagulum and recording the resistance to cutting as the tension exerted on an attached spring.

An instrument manufactured under the trade name American Curd-O-Meter (and its modifications) combines features of an automatic compression type of spring balance recorder with an instrument that measures the curd tension of milk in terms of the pull in grams required to draw a 10-pronged star-shaped knife through 100 ml of coagulated milk at 35°C.

Viscosity of rennetted milk can be measured by various rotating and oscillating viscosimeters.

Meat

Tenderness is probably the most important factor affecting consumer evaluation of meat quality and acceptability. Although fibrousness is a very important textural factor in meats, the toughness of the fascia, the fat content, and the quality and amount of collagen are some additional factors that might be expected to influence tenderness.

Mechanical methods of measuring tenderness have been discussed by Pearson (1963), Schultz (1957), Kaufmann (1959), and Matz (1962). Sale (1960) divided meat tenderness devices into four types depending on their mode of action: (1) shear, (2) penetration, (3) biting, and (4) mincing. The historical development of a wide range of shear, penetration, mastication, compression, tensile, and mincing devices was reviewed by DeFelice *et al.* (1961).

The Warner–Bratzler shear press (Bratzler 1949) is one of the most commonly used and accepted mechanical devices to evaluate consumer acceptance of meat (Schultz 1957). Sensory evaluation of meat tenderness has been compared with mechanical shear measurements obtained with the Warner–Bratzler and L.E.E.–Kramer instruments (Bailey *et al.* 1962; Sharrah *et al.* 1965A,B; Voisey and Hansen 1967). Evaluation of frankfurter texture by a penetrometer was described and compared with taste panel evaluation by Simon *et al.* (1965). Kulwich *et al.* (1963) developed and evaluated a device for estimating the tenderness of slices of cooked meat. Their device consists of a sample holder, in which the slice of cooked meat is mounted, and a penetrator, which first punctures and then shears a $\frac{3}{8}$-in.-diameter circular portion of the sample. The evaluator was used in conjunction with a commercial materials-testing instrument that provided continuous recording of the force–penetration curve. The tenderness press, developed by Sperring et al. (1959), uses a cam and gear-reduction box to operate the hydraulic pump of a laboratory press. In this device, a meat sample is compressed between a plunger and cylinder base until a portion of the sample extrudes through a small orifice in the cyl-inder base. The hydraulic pressure required to extrude the first portion through the orifice is used as a measure of meat tenderness. The Warner–Bratzler shear press, slice-tenderness evaluator, and improved tenderness press were evaluated and compared with panel tenderness scores by Als-meyer et al. (1966).

Estimating meat tenderness by measuring the energy consumed in grinding a sample of food is the basis of a procedure described by Miyada and Tappel (1956). Proctor et al. (1955) devised a denture tenderometer simulating actual chewing conditions. Carpenter et al. (1965) evaluated pork samples by objective and subjective methods. The data from the Warner–Bratzler shear press, grinder tenderometer, and denture tender-ometer indicated that all three instruments possess a satisfactory potential for use in prediction of pork tenderness.

Fruits and Vegetables

Instruments used for evaluating the texture of fruits and vegetables include (1) compressimeters; (2) penetrometer; (3) shear-testing devices; and (4) to a smaller extent instruments measuring energy consumption during grinding and sonic techniques.

The Magness–Taylor apparatus (1925) has served as the basis for sev-eral pressure testers. The apparatus is composed of a plunger surrounded by a metal collar and enclosed in a tubular handle. The plunger is con-nected to the handle by a calibrated steel spring. The spring is extended to an amount proportional to the force applied to the handle when the plunger tip is in contact with the test object. As the plunger is forced into the sample, the fruit surface pushes the collar into contact with an elec-trical junction. A light bulb energized by this contact indicates the end point. The pressure is indicated on a scale attached to the handle, which is part of the plunger. Plunger-type pressure testers are simple to use and give readings that in foods of homogeneous consistency are highly cor-related with overall texture as determined organoleptically.

The Kattan Firm-O-Meter (Kattan 1957) and a commercial device based on the same principle (Garrett *et al.* 1960) crush whole tomatoes; the decrease in fruit diameter is an indication of fruit firmness. The instrument measures attributes similar to those estimated by squeezing the fruit by hand. Results correlated at a highly significant level with panel rankings of sample firmness.

Virtually all fruits and vegetables contain some components that are perceived as fibers when the food is eaten (Matz 1962). Fibrousness is a particularly serious problem with flat beans, but may also be encountered with round varieties harvested at late maturity. Although sometimes used on products such as asparagus and snap beans, pressure testers incorporating a narrow pointed or rounded plunger probably cannot give an accurate measurement of fibrousness. Most investigators use instruments based on shear-type actions for measuring fibrousness. The fibrometer devised by Wilder (1948) tests individual asparagus spears by a standardized cutting device. A wire held rigidly in a frame is arranged in such a way that the wire may be lowered upon the tested spear and allowed to press down with a definite shearing force. The presence of excessive fibrousness is indicated by the failure to cut through the spear. Wilder's asparagus fibrometer was modified by Gould (1951) to evaluate maturation of snap beans.

The most important single factor determining the quality of commercially processed green peas is their harvest maturity. This has been recognized in the U.S. standards for grades of canned and frozen peas. The unsatisfactory nature of personal assessment of grade has led to the development of a number of objective tests, which can be classified as chemical, physical, and mechanical (Lynch *et al.* 1959).

Most of the mechanical tests measure the force required to crush, puncture, or shear samples of peas. Research instruments to test single peas or small pea samples include (1) a denture tenderometer, which uses artificial dentures operated mechanically to stimulate frequency and motion of chewing; (2) various crushing instruments, which measure the force required to penetrate completely a test piece by a standard cylindrical rod, or to crush a test material to a specified fraction of its diameter; and (3) an apparatus that measures the force required to puncture a single pea with a cylindrical pin. Commercial instruments include (1) a hand-operated texturemeter that measures the force required to drive 25 steel pins through a sample of peas in a cylinder; (2) a tenderometer, which measures the force required to shear a sample of peas; (3) a shear press designed as a multipurpose instrument for measuring the force required to cut, shear, and puncture different foods (the mode of action of the shear press is similar to that of the tenderometer, and results from both are highly correlated); (4) a maturometer that measures the resistance to puncture 143 peas by pins; (5) a hardness meter in which compression by a spring-loaded piston is recorded; and (6) a succulometer that presses juice from a sample by application of thrust developed hydraulically and then measures the volume of expressed liquid.

Lynch *et al.* (1959) reported that correlations of most instrument readings with determinations of the AIS (alcohol-insoluble solids) content— a widely accepted standard against which other methods are compared— were generally high and significant. Changes in AIS equivalent to the standard deviations or replicate measurements with a tenderometer and a maturometer were similar and almost identical with the standard deviation calculated from replicate determinations of AIS on the same peas. Angel *et al.* (1965) reported that the correlations with taste panel scores of tenderness were respectively 0.93 for the AIS method and 0.91 for a shear press equipped with an extrusion cell to measure the kinesthetic properties of canned peas. The application of the extrusion principle in texture measurement of fresh peas was described by Bourne and Moyer (1968).

Johnson *et al.* (1965) and Sistrunk and Moore (1967) found that texture measurements by shear press correlated with sensory analyses of fresh, frozen, and gamma-irradiated strawberries.

Estimating firmness of an entire fruit or vegetable is preferable to determining the texture of parts of the sample or of individual tissues. Estimating fruit firmness by the thumb test has probably been in existence ever since people began eating fruit. The thumb test is more useful for evaluating maturity of fruits that soften considerably as they ripen (e.g., tomatoes) than for determining maturity of firm fruits (e.g., apples). Objective evaluations of fruit hardness have been performed for over 50 yr in the United States by the previously described Magness–Taylor pressure tester (Magness and Taylor 1925) and its modifications (Haller 1941). The Delaware jelly-strength tester was adapted to measure the resistance to crushing of cylinders cut from apples (Whittenberger 1951) or carrots (Powrie and Asselbergs 1957).

Bourne (1965) made punch tests on apples with Magness–Taylor pressure tips mounted in a commercial testing machine that draws a force–distance curve for each punch. The force increased rapidly with little deformation until the yield point was reached when the pressure tip began to penetrate the fruit tissue. The results indicated that depth of penetration of the tip should be greatly reduced to obtain a consistent measurement of yield point pressure tests. Skin on apples raised the pressure test by adding the shear strength of the skin to the pressure test. The speed of travel of the tip affected the pressure test to only a small degree.

Since the softness of the whole potato is used as a measure of quality, there is a demand for an objective method of measuring the softness of potatoes and a need to determine the factors that affect potato firmness.

Sawyer and Collins (1960) used an instrument that was designed for measuring rubber and similar products. The instrument (durometer) consists of a spring-loaded indentor that protrudes through a metal anvil. A circular dial calibrated from 0 to 100 indicates the force exerted on the spring. Finney *et al.* (1964) reported on the deformation of potatoes that were ruptured under the load of a small cylindrical punch. Bourne and Mondy (1967) evaluated firmness of potatoes by measuring in a commercial testing machine the deformation of cylinders cut from potato tissues with a cork borer and of whole potatoes under a metal punch using a constant force. Firmness measurements were highly correlated with sensory scores. Somers (1965) evaluated the usefulness of a commercial instrument in providing fundamental rheological parameters and applied information on the viscoelastic properties of storage tissues from potato, apple, and pear.

Abbott *et al.* (1968) described two ways of using sonic energy to measure the inner texture of fruits and vegetables. In one, sections cut from the fruit or vegetable were vibrated at their natural frequencies; from the data obtained, internal friction and the elastic modulus (ratio of stress to strain) were calculated. In the second method, sonic energy was applied to the whole fruit or vegetable, and the frequency of a resulting series of resonances was measured. When the frequency of one of the resonances was squared and multiplied by the mass of the food, a parameter called "stiffness coefficient" that measured stiffness or firmness was obtained. Finney *et al.* (1967) used a sonic technique to measure elasticity of cylindrical sections of fresh, firm bananas. Softening of the banana during ripening was associated with a decrease in modulus of elasticity from 272 to 85×10^5 dynes/cm^2. Modulus of elasticity was positively correlated with starch content, and negatively correlated with luminous reflectance and the log of reducing sugars.

Miscellaneous Food Products

The rheology of miscellaneous food products has been discussed by Harvey (1953), Bourne (1982), deMan *et al.* (1976), Kramer and Szczesniak (1973), Peleg and Bagley (1983), Sone (1982), and Sherman (1979).

Sugars and Honey. Sucrose, invert sugar, and confectioners' glucose are three raw materials basic to the sugar confectionery industry, and the rheological properties of their syrups and mixed syrups are of prime importance. The confectionery syrups are typical Newtonian liquids and solutions of sucrose are considered suitable for calibrations of viscometers. Sucrose solutions show the characteristic viscous behavior of nonelectrolyte solutions in that their relative viscosity falls with rising temperature. For sucrose, invert sugar, and confectioners' glucose, a simple linear relation exists between the log of viscosity and the concentration of soluble solids (expressed as the ratio of sugar dissolved per unit weight of water).

Honeys differ widely in composition and display a wide range of rheological properties. In general, honeys are Newtonian liquids, but heather honey and various honeys derived from eucalypti exhibit non-Newtonian properties.

Addition of corn syrup to "Newtonian" honeys markedly increases the viscosity. The viscosity range for various honeys is, however, very wide and markedly affected by water contents. For most "Newtonian" honeys, there is a linear inverse relationship between the logarithm of viscosity (using a falling ball apparatus) and logarithm of water content between the limits of 12.8% water (475 poises) and 22.4% water (50 poises). For honey samples ranging in water content from 12.4 to 19.7%, moisture w, viscosity η_t (determined by a falling ball viscosimeter with a tube of 21.2 mm internal diameter and steel ball of 16 mm diameter), and absolute temperature T are related as follows:

$$w = (62,500 - 156.7T) [T(\log \eta_t + 1) - 2.287(313 - T)] \quad (28.4)$$

A comparison of the results obtained by viscosity measurements and the official vacuum drying method showed that they did not differ on the average by more than 0.2%.

Good-quality heather honey sets into a rigid gel when left to rest, but flows again on stirring. Thus, this honey shows properties of *thixotropy*, that is, an isothermal reversible gel–sol–gel transformation induced by shearing and subsequent rest. Thixotropy of honey largely depends on the protein content of the sample, and its true thixotropic nature has been questioned.

The term *dilatancy* was used originally to describe the property of a granular system that increases in volume when sheared laterally. The meaning has been expanded to include any system whose viscosity increases with increasing rate of shear beyond a critical minimum value. Several honeys from eucalypti display the property of "Spinnbarkeit." They can be drawn out into long strings and show dilatancy. Dilatant honeys contain 6–7% dextrans, which can be removed by precipitation with acetone. After the honey has been restored to its original concentration, it behaves like a true Newtonian liquid. The peculiar rheological properties of honey from eucalypti can be simulated by adding dextran to clover or sainfoin honey.

Gelatin. For mixtures containing a constant proportion of gelatin by weight and constant proportion of nongelatin dissolved solids, there exists, at constant temperature, a simple linear relationship between the logarithm of viscosity and the concentration of total solids (including gelatin), when the latter is expressed in terms of weight of total solids per unit weight of water. At constant total solids concentration, there exists a simple relationship between the logarithm of viscosity and the percentage of gelatin by weight.

The ability of gelatin solutions to form a gel is the characteristic that accounts for the major uses of gelatin. The standard measure of gel strength (*Bloom*) is the main rheological property considered in grading gelatin for food uses. Bloom is measured with an instrument called the Bloom gelometer (Bloom 1925). This instrument measures the force required to depress a $\frac{1}{2}$-in.-diam plunger 4 mm into the top center of a gel prepared under standardized conditions. The force is provided by lead shot flowing into a cup on top of the plunger. Determination of Bloom in gelatin solutions at nonstandard concentration was discussed by Kramer and Rosenthal (1965).

Kramer and Hawbecker (1966) reviewed methods for measuring and recording rheological properties of gels. Most of the instruments measure the force required to break a gel. The instruments include (1) the jelmeter (Baker 1934), a simplified version of an Ostwald viscosimeter used to determine correct proportions of sugar, pectin, and acid in making jellies, jams, and marmalades; (2) the Bloom gelometer (Fellers and Griffith 1928); and (3) the penetrometer (Underwood and Keller 1948), which provides a measure of the force required to penetrate into a gel. In addition to the determination of gel strength, it may be desirable to measure uniformity, adhesiveness (stickiness), and deformation prior to breakdown of gel structure. The latter three parameters can be measured by using special extrusion-type cells in the Allo–Kramer shear press (Kramer and Hawbecker 1966). Charm and McComis (1965) reviewed critical physical measurements of gum solutions and gave examples of use of the information obtained.

Chocolate and Candy. Most studies of the rheological properties of chocolate have involved measurements made on the melted substance. The most obvious usefulness of such data is in processing. For example, the physical behavior of molten chocolate is of particular importance when it is to be used for coating purposes. The instrument most frequently used in the United States for measuring the viscosity of molten chocolate is the MacMichael viscosimeter. Capillary devices and viscosimeters of the falling sphere type have also been used (Matz 1962).

Heiss (1959) described an apparatus for measuring surface stickiness of candies. The instrument is intended to quantitate the evaluation obtained by pressing for a short time a "dry and clean finger" on to the candy. An upper plate cushioned with rubber is connected to a calibrated spring. A lower movable plate carries the candy piece, which is made flat, on the upper surface. The upper plate is initially pressed to the candy with a weight of 11.5 oz for 10 sec, and then the lower plate is gradually withdrawn at a constant rate of speed until the candy surface and the rubber surface are pulled apart. The distance traveled before the release occurs is a measure of the degree of stickiness of the candy surface.

BIBLIOGRAPHY

Abbot, J. A., *et al.* (1968). Sonic techniques for measuring texture of fruits and vegetables. *Food Technol.* **22,** 101–112.

Alsmeyer, R. H., Thornton, J. W., Hiner, R. L., and Bollinger, N. C. (1966). Beef and pork tenderness measured by the press, Warner–Bratzler, and STE methods. *Food Technol.* **20,** 115–117.

AACC. (1983). "Cereal Laboratory Methods," 8th ed. Am. Assoc. Cereal Chemists, St. Paul, MN.

Amerine, M. A., Pangborn, R. M., and Roessler, E. B. (1965). "Principles of Sensory Evaluation of Food." Academic Press, New York.

Angel, S., Kramer, A., and Yeatman, J. N. (1965). Physical methods of measuring quality of canned peas. *Food Technol.* **19,** 96–98.

Anker, C. A., and Geddes, W. F. (1944). Gelatinization studies upon wheat and other starches with the amylograph. *Cereal Chem.* **21,** 335–360.

Bailey, M. E., Hedrick, H. B., Parrish, F. C., and Naumann, H. D. (1962). L. E. E.–Kramer shear force as a tenderness measure of beef stock. *Food Technol.* **16**(12), 99–101.

Baker, G. L. (1934). New methods for determining the jelly power of fruit juice extraction. *Food Ind.* **6,** 305–307.

Baron, M., and Scott-Blair, G. W. (1953). Rheology of cheese and curd. *In* "Foodstuffs, Their Plasticity, Fluidity, and Consistency" (G. W. Scott-Blair, ed.). Interscience Publishers, New York.

Bloksma, A. H. (1967). Detection of changes in modulus and viscosity of wheat flour doughs by the "work technique" of Muller *et al. J. Sci. Food Agric.* **18,** 49–51.

Bloksma, A. H., and Hlynka, I. (1964). Basic considerations of dough properties. *In* "Wheat Chemistry and Technology" (I. Hlynka, ed.). Am. Assoc. Cereal Chemists, St. Paul, MN.

Bloom, O. T. (1925). Penetrometer for testing jelly strength of glues, gelatins, etc. U.S. Pat. 1,540,979.

Bourne, M. C. (1965). Studies on punch testing of apples. *Food Technol.* **19,** 113–115.

Bourne, M. C. (1966). A classification of objective methods for measuring texture and consistency of foods. *J. Food Sci.* **31,** 1011–1015.

Bourne, M. C. (1982). "Food Texture and Viscosity; Concept and Measurement." Academic Press, New York.

Bourne, M. C., and Mondy, N. (1967). Measurement of whole potato firmness with a universal testing machine. *Food Technol.* **21,** 97–100.

Bourne, M. C., and Moyer, J. C. (1968). The extrusion principle in texture measurement of fresh peas. *Food Technol.* **22,** 1013–1018.

Bourne, M. C., Moyer, J. C., and Hand, D. B. (1966). Measurement of food texture by a universal testing machine. *Food Technol.* **20,** 170–174.

Brabender, C. W. (1965). Physical dough testing—past, present, and future. *Cereal Sci. Today* **10**, 291–304.

Bratzler, L. J. (1949). Determining the tenderness of meat by use of the Warner–Bratzler method. *Proc. 2nd Annu. Reciprocal Meat Conf.* Nat. Livestock Meat Board.

Brennan, J. G. (1980). Food texture measurement. *In* "Developments in Food Analysis Techniques," Vol 2 (R. D. King, ed.). Applied Science Publishers, London.

Bushuk, W. (ed.). (1975). Rheology of wheat products. A symposium. *Cereal Chem.* **52**(3), Part II, 1r–183r.

Carpenter, Z. L., Kaufmann, R. G., Bray, R. W., and Weckel, K. G. (1965). Objective and subjective measures of pork quality. *Food Technol.* **19**, 118–120.

Charm, S. E. (1962). The nature and role of fluid consistency in food engineering applications. *Advan. Food Res.* **11**, 356–435.

Charm, S. E., and McComis, W. (1965). Physical measurements of gums. *Food Technol.* **19**, 58–63.

Christensen, C. M., and Vickers, Z. M. (1981). Relationship of chewing sounds to judgments of food crispness. *J. Food Sci.* **46**, 574–578.

Davis, J. G. (1937). The rheology of cheese, butter, and other milk products. *J. Dairy Res.* **8**, 245–264.

Dea, I. C. M., Richardson, R. K., and Ross-Murphy, S. B. (1983). Characterization of rheological changes during the processing of food materials. *In* "Gums and Stabilizers for the Food Industry. Applications of Hydrocolloids" (G. O. Phillips, D. J. Wedlock, and P. R. Williams, eds.). Pergamon Press, Oxford.

DeFelice, D. *et al.* (1961). "Fundamental Aspects of Meat Texture." Quartermaster Contract Res. Project Rept. 2. Project 7-84-13-002.

deMan, J. M., Voisey, P. W., Rasper, V. F., and Stanley, D. W. (eds.). (1976). "Rheology and Texture in Food Quality." AVI Publishing Co., Westport, CT.

Dempster, C. J., Hlynka, I., and Anderson, J. A. (1953). Extensograph studies of structural relaxation in bromated and unbromated doughs mixed in nitrogen. *Cereal Chem.* **30**, 492–503.

Dempster, C. J., Hlynka, I., and Anderson, J. A. (1955). Influence of temperature on structural relaxation in bromated and unbromated doughs mixed in nitrogen. *Cereal Chem.* **32**, 241–254.

Dolby, R. M. (1941). The rheology of butter. I. Methods of measuring the hardness of butter. *J. Dairy Res.* **12**, 329–336.

Drake, B. K. (1961). An attempt at a geometrical classification of rheological apparatus. (unpublished). Cited by Bourne (1966).

Fellers, C. R., and Griffiths, F. P. (1928). Jelly strength measurements of fruit jellies by Bloom gelometer. *Ind. Eng. Chem.* **20**, 857–859.

Finney, E. E., Ben-Gera, I., and Massie, D. R. (1967). An objective evaluation of changes in firmness of ripening bananas using a sonic technique. *Food Technol.* **32**, 642–646.

Finney, E. E., Hall, C. H., and Thompson, N. R. (1964). Influence of variety and time upon the resistance of potatoes to mechanical handling. *Am. Potato J.* **41**, 178.

Garrett, A. W., Desrosier, N. W., Kuhn, G. D., and Fields, M. L. (1960). Evaluation of instruments to measure firmness of tomatoes. *Food Technol.* **14**, 562–564.

Gould, W. A. (1951). "Quality Evaluation of Fresh, Frozen, and Canned Snap Beans." Ohio Agr. Expt. Sta. Res. Bull. *701*.

Greup, D. H., and Hintzer, H. M. R. (1953). Cereals. *In* "Foodstuffs, Their Plasticity, Fluidity, and Consistency" (G. W. Scott-Blair, ed.). Interscience Publishers, New York.

Haller, M. H. (1941). "Fruit Pressure Testers and Their Practical Applications." U.S. Dep. Agric. Circ. *627*.

Hallikainen, K. E. (1962). Viscosimetry. *Instrum. Control Syst.* **35**(11), 82–84.

Halton, P. (1938). Relation of water absorption to the physical properties and baking quality of flour doughs. *Cereal Chem.* **15**, 282–294.

Halton, P. (1949). Significance of load extension tests in assessing the baking quality of wheat flour doughs. *Cereal Chem.* **26**, 24–45.

Harvey, J. G. (1953). The rheology of certain miscellaneous food products. *In* "Foodstuffs, Their Plasticity, Fluidity, and Consistency" (G. W. Scott-Blair, ed.). Interscience Publishers, New York.

Heiss, R. (1959). Prevention of stickiness and graining in stored hard candies. *Food Technol.* **13**, 433–440.

Hlynka, I. (1967). Progress in the area of dough rheology. *Brot Gebaeck* **21**, 125–130.

Hlynka, I., and Matsuo, R. R. (1959). Quantitative relation between structural relaxation and bromate in dough. *Cereal Chem.* **36**, 312–317.

Hlynka, K., and Van Eschen, E. L. (1965). Studies with an improved load softness tester. *Cereal Sci. Today* **10**, 84–87.

Hofstee, J., and de Willigen, A. H. A. (1953). Starch. *In* "Foodstuffs, Their Plasticity, Fluidity, and Consistency" (G. W. Scott-Blair, ed.). Interscience Publishers, New York.

Johnson, C. F., Maxie, E. C., and Elbert, E. M. (1965). Physical and sensory tests on fresh strawberries subjected to gamma radiation. *Food Technol.* **19**, 119–123.

Kapsalis, J. G., Bettscher, J. J., Kristoffersen, T., and Gould, I. A. (1960). Effect of chemical additives on the spreading quality of butter. I. The consistency of butter as determined by mechanical and consumer panel evaluation methods. *J. Dairy Sci.* **43**, 1560–1569.

Kattan, A. A. (1957). Changes in color and firmness during ripening of detached tomatoes, and the use of a new instrument for measuring firmness. *Proc. Am. Soc. Hort. Sci.* **70**, 379–386.

Kaufmann, R. G. (1959). Techniques of measuring some quality characteristics of pork. *Proc. 12th Annu. Reciprocal Meat Conf.* Nat. Livestock Meat Board.

Kent-Jones, D. W., and Amos, A. J. (1967). "Modern Cereal Chemistry," 6th ed. Food Trade Press, London.

Kovats, L. T., and Lasztity, R. (1965). New developments in dough rheology. *Periodica Polytech.* (Budapest) **9**(1), 57–67.

Kramer, A., and Backinger, G. (1959). Textural measurement of foods. *Food* **28**, 85–86, 95.

Kramer, A., and Hawbecker, J. V. (1966). Measuring and recording rheological properties of gels. *Food Technol.* **29**, 111–115.

Kramer, A., and Rosenthal, H. (1965). Determination of bloom of gelatin in solutions of nonstandard concentration. *Food Technol.* **19**, 111–114.

Kramer, A., and Szczesniak, A. S. (eds.). (1973). "Texture Measurements of Foods." D. Reidel Publishing Co., Hingham, MA.

Kramer, A., and Twigg, B. A. (1959). Principles and instrumentation for the physical measurement of food quality with special reference to fruit and vegetable products. *Advan. Food Res.* **9**, 153–220.

Kruisheer, C. I., and den Herder, P. C. (1938). Investigations on the consistency of butter. *Chem. Weekblad* **35**, 719–730.

Kulwich, R., Decker, R. W., and Alsmeyer, R. H. (1963). Use of a slice-tenderness evaluation device with pork. *Food Technol.* **17**, 83–85.

Lerchenthal, C. H., and Muller, H. G. (1967). Research in dough rheology at the Israel Institute of Technology. *Cereal Sci. Today* **12**, 185–187, 190–192.

Lynch, L. J., Mitchel, R. S., and Casimir, D. J. (1959). The chemistry and technology of the preservation of green peas. *Advan. Food Res.* **9**, 61–152.

Magness, J. R., and Taylor, G. F. (1925). "An Improved Type of Pressure Tester for the Determination of Fruit Maturity." U.S. Dept. Agric. Circ. *350*.

Matz, S. A. (1962). "Food Texture." AVI Publishing Co., Westport, CT.

Miller, B. S., Afework, S., Pomeranz, Y., Bruinsma, B. L., and Booth, G. D. (1982). Measuring the hardness of wheat. *Cereal Foods World* **27**(2), 61–64.

Miller, B. S., and Johnson, J. A. (1954). "A Review of Methods for Determining the Quality of Wheat and Flour for Breadmaking." Kansas Agr. Expt. Sta. Tech. Bull. *76*.

Mitchell, J. R. (1979). Rheology of polysaccharide solutions and gels. *In* "Polysaccharides in Foods," (J. M. V. Blanshard and J. R. Mitchell, eds.). Butterworths, London.

Mitchell, J. R. (1984). Rheological techniques. *In* "Food Analysis, Principles and Techniques," Vol. I (D. W. Gruenwedel and J. R. Whitaker, eds.). Marcel Dekker, New York and Basel.

Miyada, D. S., and Tappel, A. L. (1956). Meat tenderization. I. Two mechanical devices for measuring texture. *Food Technol.* **10**, 142–145.

Mohr, W., and Wellm, J. (1948). Viscosity measurement of butter. *Milchwissenschaft* **3**, 181–185.

Mohsenin, N. N. (1968). "Physical Properties of Plant and Animal Materials. Part II." Dept. Agr. Eng., Pennsylvania State Univ., College Park, PA.

Mohsenin, N. N. (1980). "Physical Properties of Plant and Animal Materials." Gordon & Breach Science Publishers, New York.

Morrow, C. T., and Mohsenin, N. N. (1968). Dynamic viscoelastic characterization of solid food materials. *Food Technol.* **23**, 646–651.

Mulder, H. (1953). The consistency of butter. *In* "Foodstuffs, Their Plasticity, Fluidity, and Consistency" (G. W. Scott-Blair, ed.). Interscience Publishers, New York.

Muller, H. G. (1964). Dough rheology, I. Early developments before 1900. *Brot Gebaeck* **18**, 117–121.

Muller, H. G. (1973). "An Introduction to Food Rheology." Crane, Russak & Co., New York.

Muller, H. G., and Hlynka, I. (1964). Brabender extensigraph techniques. *Cereal Sci. Today* **9**, 422–424, 426, 430.

Muller, H. G., Williams, M. V., Russell-Eggitt, P. W., and Coppock, J. B. M. (1961). Fundamental studies on dough with the Brabender extensograph. I. Determination of stress–strain curves. *J. Sci. Food Agric.* **12**, 513–523.

Muller, H. G., Williams, M. V., Russell-Eggitt, P. W., and Coppock, J. B. M. (1962). Fundamental studies on dough with the Brabender extensograph. II. Determination of the apparent elastic modulus and coefficient of viscosity of wheat flour dough. *J. Sci. Food Agric.* **13**, 572–580.

Muller, H. G., Williams, M. V., Russell-Eggitt, P. W., and Coppock, J. B. M. (1963). Fundamental studies on dough with the Brabender extensograph. III. The work technique. *J. Sci. Food Agric.* **14**, 663–672.

Pearson, A. M. (1963). Objective and subjective measurement of meat tenderness. *Proc. Meat Tenderness Symp.*, Campbell Soup Co., Camden, NJ.

Peleg, M. (1983). The semantics of rheology and texture. Food Technol. 37(11), 54–61.

Peleg, M., and Bagley, E. B. (eds.). (1983). "Physical Properties of Foods." AVI Publishing Co., Westport, CT.

Pomeranz, Y., and Miller, B. S. (1982). Wheat hardness: Its significance and determination. *Proc. 7th World Cereal and Bread Congress*, 399–404.

Powrie, W. D., and Asselbergs, E. A. (1957). A study of canned syrup-pack whole carrots. *Food Technol.* **11**, 257–277.

Prentice, J. A. (1972). Rheology and texture of dairy products. *J. Texture Studies* **3**, 415–458.

Proctor, B. E., Davison, S., Malecki, G. J., and Welch, M. (1955). A recording strain gage denture tenderometer for foods. I. Instrument evaluation and initial tests. *Food Technol.* **9**, 471.

Pryce-Jones, J. (1953). The rheology of honey. *In* "Foodstuffs, Their Plasticity, Fluidity, and Consistency" (G. W. Scott-Blair, ed.). Interscience Publishers, New York.

Rha, C. K. (ed.). (1975). "Theory, Determination and Control of Physical Properties of Food Materials." D. Reidel Publishing Co., Dordrecht, The Netherlands.

Sale, A. J. H. (1960). Measurement of meat tenderness. *In* "Texture in Foods," Monograph 7. Soc. Chem. Ind., London.

Sawyer, R. L., and Collins, G. H. (1960). Black spot of potatoes. *Am. Potato J.* **37**, 115.

Schultz, H. W. (1957). An evaluation of the methods of measuring tenderness. *Proc. 10th Annu. Reciprocal Meat Conf.*, Nat. Livestock Meat Board.

Scott-Blair, G. W. (1945). "A Survey of General and Applied Rheology." Sir Isaac Pitman and Sons, London.

Scott-Blair, G. W. (1953). "Foodstuffs, Their Plasticity, Fluidity, and Consistency." Interscience Publishers, New York.

Scott-Blair, G. W. (1958). Rheology in food research. *Adv. Food Res.* **8**, 1–61.

Shama, F., and Sherman, P. (1966). The texture of ice cream. II. Rheological properties of frozen ice cream. *J. Food Sci.* **31**, 699–706.

Sharrah, N. Kunze, M. S., and Pangborn, R. M. (1965A). Beef tenderness: Sensory and mechanical evaluation of animals of different breeds. *Food Technol.* **19**, 131–136.

Sharrah, N., Kunze, M. S., and Pangborn, R. M. (1965B). Beef tenderness: Comparison of sensory methods with the Warner–Bratzler and L. E. E. Kramer shear press. *Food Technol.* **19**, 136–143.

Sherman, P. (1966). The texture of ice cream. III. Rheological properties of mix and melted ice cream. *J. Food Sci.* **31**, 707–716.

Sherman, P. (1969). A texture profile of foodstuffs based upon well defined rheological properties. *J. Food Sci.* **34**, 458–462.

Sherman, P. (ed.). (1979). "Food Texture and Rheology." Academic Press, London.

Simon, S., Kramlich, W. E., and Tauber, F. W. (1965). Factors affecting frankfurter texture and a method of measurement. *Food Technol.* **19**, 110–113.

Sistrunk, W. A., and Moore, J. N. (1967). Assessment of strawberry quality—fresh and frozen. *Food Technol.* **21**, 131A–135A.

Somers, G. F. (1965). Viscoelastic properties of storage tissues from potato, apple, and pear. *J. Food Sci.* **30**, 922–929.

Sone, T. (1982). "Consistency of Foodstuffs." D. Reidel Publishing Co., Dordrecht, The Netherlands.

Sperring, D. D., Platt, W. T., and Hiner, P. L. (1959). Tenderness of beef muscle as measured by pressure. *Food Technol.* **8**, 155–158.

Swanson, C. O., and Working, E. B. (1933). Testing the quality of flour by the recording dough mixer. *Cereal Chem.* **10**, 1–29.

Szczesniak, A. S. (1966). Texture measurements. *Food Technol.* **20**, 52, 55–58.

Szczesniak, A. S., and Kleyn, D. H. (1963). Consumer awareness of texture and other food attributes. *Food Technol.* **17**, 74–77.

Underwood, J. C., and Keller, G. J. (1948). A method of measuring the consistency of tomato paste. *Fruit Prod. J.* **28**, 103.

Vickers, Z. M. (1980). Food sounds: How much information do they contain? *J. Food Sci.* **45**, 1494–1496.

Vickers, Z. M. (1981). Relationship of chewing sounds to judgments of crispness, crunchiness, and hardness. *J. Food Sci.* **47**, 121–124.

Vickers, Z. M. (1984). Crackliness: Relationships of auditory judgments to tactile judgments and instrumental acoustical measurements. *J. Texture Studies* **15**, 49–58.

Vickers, Z. M., and Bourne, M. C. (1976A). Crispness in foods—a review. *J. Food Sci.* **41**, 1153–1157.

Vickers, Z., and Bourne, M. C. (1976B). A psychoacoustical theory of crispness. *J. Food Sci.* **41**, 1158–1164.

Voisey, P. W., and Hansen, H. (1967). A shear apparatus for meat tenderness evaluation. *Food Techol.* **21**, 37A–42A.

Voisey, P. W., MacDonald, D. C., and Foster, W. (1967). An apparatus for measuring the mechanical properties of foods. *Food Technol.* **21**, 43A–47A.

Whittenberger, R. T. (1951). Measuring the firmness of cooked apple tissues. *Food Technol.* **5**, 17–20.

Wilder, H. K. (1948). "Instructions for Use of the Fibrometer in the Measurement of Fiber Content in Canned Asparagus." Natl. Canners Assoc. Res. Lab. Rept. *12313-C*, San Francisco.

29

Serology, Immunochemistry, and Immunoelectrophoresis

Immune phenomena play a vital role in the health and disease of higher animals. Among these phenomena are (1) the protection conferred on higher animals against certain viral- and bacterial-induced diseases, either by their survival of an initial attack of the disease or by the deliberate injection of a modified form of the disease-producing agent (immunization); (2) the rejection of tissue or organ grafts from one individual to another of the same species; and (3) the occurrence of various diseases such as certain types of food allergy (Singer 1965).

The importance of an immunological approach in preventive medicine, in diagnosis, and in therapy is well established. Less appreciated, however, is that similar procedures can be used to yield valuable information concerning the nature of chemical components of biological materials. In the study of proteins, these contributions have been of several kinds. Immunological methods have helped to discover a number of proteins with unique properties. As microanalytical tools, serological methods

In this chapter, we discuss the possibilities that immunochemical methods offer to the food analyst. No attempt has been made to survey the extensive and rapidly growing literature. The review is limited to a short outline of basic considerations followed by a description of some of the useful procedures of particular interest to food analysts.

BASIC CONSIDERATIONS AND DEFINITIONS

When an animal receives one or more parenteral injections of certain foreign materials, there generally appear in the serum within a few days substances that have the property of reacting with the material injected. These substances are called *antibodies,* and the materials that stimulate their production are called *antigens.* Most known antigens are proteins; some are polysaccharides or lipid–carbohydrate–protein complexes. Although antibodies generally react only with the antigen used (the homologous antigen), certain exceptions, termed *cross reactions,* have been noted. Antigens are sometimes referred to as *foreign* proteins because they are not normally found in the body they invade.

Antibodies

The antibodies that form in an organism in response to the introduction of an antigen are proteins. The antibodies are globulins, have molecular weights appropriate to globulins, and are called *immunoglobulins.* The nomenclature of immunoglobulins and their genetic factors have been recommended by committees of the World Health Organization (1964).

There are three main classes of immunoglobulins present in all mammalian sera that have been studied, though their relative amounts vary considerably from one species to the other. These classes have been named IgG (gamma), IgM (macro), and IgA; in human serum their relative amounts are 71, 7, and 22%, respectively. Their common characteristics are that they all may have the same four-peptide chain structure (Porter 1967). Antigenically, they are related but not identical. The molecular weight of IgG is 150,000, of IgM 900,000, and of IgA about 400,000, but the latter has a tendency to polymerize and dissociate. Immunoglobulins of all types and of all species tested contain carbohydrates (Boyd 1966). The carbohydrate content of human IgG is 3%, of IgM 12%, and of IgA 10%.

Antigens

To elicit an antibody response, antigens must be administered parenterally to organisms of another species. Oral administration is generally not satisfactory, since the ingested proteins are digested by proteolytic enzymes of the gastrointestinal tract and the breakdown products are generally not antigenic (Haurowitz 1960).

Certain proteins of various species exist in more than one antigenic form, called *allotypes.* For example, rabbit gamma globulin exists in at least six different forms, and the differences seem to be hereditary. There is also evidence for the existence of such allotypes in man. Although we do not have the complete explanation of what makes a substance antigenic, we know that antigens must be foreign to the circulation of the experimental animal, and the more foreign—i.e., the more remote the source taxonomically—the more antigenic a protein will be. In addition, antigens must have more than a certain minimal degree of complexity and a certain molecular size. This latter requirement is supported by the non-antigenicity of the protamines and of the relatively simple molecule of gelatin, and by the high antigenicity of the large molecule of the hemocyanins.

It seems that a minimum molecular weight of approximately 5,000 to 10,000 is a prerequisite for antigenicity (Haurowitz 1960, 1961). Formation of antibodies is sometimes observed after the administration of small but highly reactive substances, such as iodine. It is generally assumed that such reactive substances combine with proteins at the site of injection and that the conjugated proteins are the true antigens. Denatured proteins may show reduced reactivity or fail to react with antibodies to native proteins, depending on the extent of denaturation.

As mentioned before, the specificity of antibodies formed in response to an antigen varies. Although antibodies generally react only with the antigen used, cross reactions, due to structural similarities between antigens, sometimes occur. For example, antibodies to horse serum protein also reacts with donkey serum protein, and ovalbumin from duck eggs reacts with antibodies to albumin from hen eggs. Serum albumins of different mammalian species are evidently antigenically similar, though not identical. The more closely related any two species are, the greater the serological likeness of their corresponding proteins. For example, there is only a slight relation, serologically, between hemoglobin of beef and man. The closer the phylogenetic relation between two protein antigens, the more extensive the cross reaction. Thus if antibodies for bovine serum albumin are reacted with a variety of other albumins, the maximum amount of precipitation is obtained in the homologous interaction. There is, however, considerable cross reaction with albumins from sheep and several other animals.

Proteins from the same species that have different functions in the body usually differ widely in specificity. Blood hemoglobin differs serologically from proteins of the kidney tissue, and the serum globulins are different from the serum albumins. There is evidence that most organs possess special proteins or carbohydrates peculiar to them, and organ-specific antisera have been obtained (Boyd 1966).

The specificity of antigens resides in the structural pecularities of their molecules. The dependence of specificity on chemical structure is shown by several lines of evidence: (1) purified proteins that exhibit chemical differences can generally be differentiated serologically; (2) simple chemical substances and carbohydrates, when chemically similar, give serological cross reactions; (3) chemical alteration of antigens generally alters their specificity; and (4) corresponding proteins from different species that are functionally, and thus probably structurally, related generally cross-react. Available evidence indicates that the reactivity of an antigen resides primarily in a small prosthetic group called *hapten*. The word "hapten" was first used by Landsteiner (1921) to describe simple organic residues that react specifically with antibodies. The word is derived from a Greek verb, the meaning of which is to touch, grasp, or fasten. The term has been used to describe substances which, in themselves, are incapable of inducing antibody formation, but which when attached to ordinary immunogens, such as proteins and polysaccharides, can induce antibodies against themselves. Day (1966) defined hapten as the specific chemical grouping to which a single antibody site conforms and with which it reacts. As long as the hapten occupies an antibody site, no other hapten can occupy the same site. A hapten, whether free or attached to a carrier in an exposed position, will react with an antibody site.

IMMUNOCHEMICAL METHODS

There are several *in vivo* and *in vitro* methods of observing the results of the combination of an antigen with its specific antibodies. If the antigen is a macromolecule, combination with its antibodies under appropriate conditions results in precipitation. If the antigen is part of a cellular surface, agglutination of the cells is observed. The formation of a specific antigen–antibody aggregate usually results in coprecipitation of certain components of the serum called *complement*. The extent of complement fixation is directly related to the extent of the antigen–antibody combination. If the antigen is on a cellular surface, reaction with antibody in the presence of complement is often cytotoxic, resulting in cell lysis.

When a crude extract containing a mixture of antigens is injected, several antibodies specific for each antigen are formed. The various antibodies formed can be removed separately from the serum by adding each antigen singly. Quantitative studies have indicated that even antibodies to a single antigen, such as crystalline egg albumin, exhibit microheterogeneity. They may still be considered to behave as a single substance when comparisons are made between a number of totally unrelated antigens, since the antibody molecules to each antigen can react specifically with their own antigen (Kabat 1961).

Available immunological methods vary in sensitivity and usefulness. It is possible to observe a positive precipitin reaction with a few micrograms of antibody per milliliter, and a passive hemagglutination reaction can be obtained with concentrations 1000 times smaller. Despite the greater sensitivity of the passive hemagglutination reaction, the precipitin reaction is commonly used because it is directly visible and can be quantitated. It has the disadvantage of being limited to antigens in aqueous solutions. The *in vitro* methods for measuring antibody potency can be divided into several types:

1. *Dilution methods:* The classical methods use as the end point the highest dilution of serum added to a constant amount of antigen at which a detectable reaction such as precipitation, agglutination, or lysis occurs. The serum dilution giving this endpoint is known as the *titer*.

2. *Optimal proportions method:* The procedure determines the ratio of antigen and serum at which flocculation is most rapid if the relative proportions are optimal. Various dilutions of antigens are mixed with a constant volume of serum in a series of tubes, the total volume in each tube being constant. The tube in which flocculation first occurs is noted, and the ratio of antigen dilution to serum dilution is calculated.

3. *Quantitative chemical methods:* They are generally based on determining the protein-N content in a precipitate or agglutination product.

The dilution methods are semiquantitative and are useful in estimating the order of magnitude of an antigen in a mixture. Greater precision may be obtained by the optimal proportions method. The most precise method of estimating amounts of antigens in unknown solutions involves using a calibration curve prepared by analyzing a series of washed specific precipitates formed by adding known volumes of antigen to a measured volume of antiserum.

The quantitative precipitin method is capable of widespread application in the analysis of mixtures containing immunologically reactive substances. It offers several advantages over the usual analytical chemical methods. It is highly specific and permits estimation of a given constituent of a mixture without chemical fractionation or purification. It requires small amounts of material for analysis, since the amount of specific precipitate analyzed is several times the amount of antigen in the sample, except for antigens of very high molecular weight. With the use of sensitive methods of protein determination (e.g., the Folin–Ciocalteau colorimetric method), samples containing as little as 1–5 μg of antigen are adequate. The method is capable of considerable precision, ± 2 to 5% under suitable conditions.

If small amounts of a chemically pure and serologically active product suitable for the preparation of a calibration curve with antiserum are obtainable, it is possible to estimate the amount of the substance in crude starting materials. This procedure can be used in estimating specific polysaccharides and proteins. Immunochemical precipitation methods can be used to supplement or to replace less precise bioassays for estimation of proteins with biological activity. In the estimation of such substances (enzymes, protein hormones, toxins in mixtures), estimations of biological activity are compared with quantitative immunochemical analysis. In many instances, inactivated material or the precursor of the active substances may still precipitate with antiserum. In such cases, use of both methods of estimation may provide evidence for the existence of altered products. Immunochemical criteria are useful in determining the homogeneity and purity of proteins and carbohydrates. Such determinations are generally used in conjunction with physicochemical methods and, whenever applicable, biological assays.

One of the main hazards in carrying out the quantitative precipitin reaction is the possibility of nonspecific precipitation, particularly with crude extracts. Nonspecific precipitation can sometimes be avoided by choice of conditions of precipitation or by fractionation of antisera. In certain cases, it occurs only at protein concentrations much higher than those used in the specific reaction. If control tests indicate that nonspecific precipitation cannot be avoided, a control can be set up for each mixture and this value subtracted from that obtained with the sample. Results under those conditions are at best rough approximations.

Immunodiffusion

The use of gelified media for the precipitin reaction was introduced at the turn of the century. As knowledge of the principles governing the precipitin reaction accumulated, several methods of specific precipitation in gels were developed (Crowle 1960, 1961). These are called *immunodiffusion*.

Oudin (1946, 1952) devised a method by which complex antigen–antibody systems could be analyzed by allowing them to react in a capillary tube filled with agar. The antibody solution is mixed with warm agar, which is then allowed to harden in the tube. When the antigen solution is added, the reaction of antigen and antibody forms a precipitation zone. The number of such zones is less than or equal to the number of independent precipitation systems (i.e., antigen–antibody reactions) present in the mixture examined (Fig. 29.1).

The technique of Ouchterlony (called the *double-diffusion method*, (1967) is similar, but differs from Oudin's in that it permits both antigen and antibody to diffuse into an agar-filled glass dish that initially contains neither reagent. A few drops each of antigen and antibody solution are placed separately in small wells cut into the agar. Antigen and antibody diffuse outward toward each other at a rate related to their concentration and their diffusion coefficients. A line of precipitate forms where an antigen interacts with its antibody. Because of differences in diffusion rates, the lines are distinctly separated (Fig. 29.2). The clean separation of lines

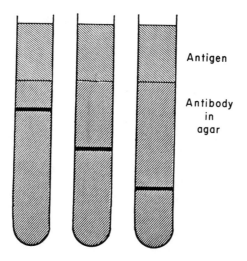

Fig. 29.1. Immunodiffusion in one dimension by the Oudin procedure

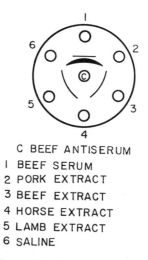

C BEEF ANTISERUM
1 BEEF SERUM
2 PORK EXTRACT
3 BEEF EXTRACT
4 HORSE EXTRACT
5 LAMB EXTRACT
6 SALINE

Fig. 29.2. Precipitation bands of unabsorbed beef antiserum on Ouchterlony gel diffusion plates. (From Warnecke and Saffle 1968.)

on Ouchterlony plates makes it possible to distinguish more reactions with them than with Oudin tubes. Consequently, the plates are more useful in studies of complex systems (Williams 1960). The line of precipitation formed by an individual reaction can be identified if either the antigen or the antibody is available in relatively pure form.

Immunoelectrophoresis

To improve the resolution and interpretation of the double-diffusion method, the *immunoelectrophoresis* was developed by Grabar and co-workers (Grabar and Williams 1953; Grabar 1950, 1957; Grabar *et al.* 1965). The principle of this method is illustrated in Fig. 29.3. The mixture to be studied is subjected to electrophoresis in a transparent gel. At the end of electrophoresis, the components of the analyzed mixture have been separated by the electrical current in the gel and occupy different positions depending on their relative mobilities. An immune serum, rich in precipitating antibodies, is then poured into elongated troughs made in the gel and parallel to the axis of migration. The antibodies diffuse perpendicularly to this axis, and as in the double-diffusion method of Ouchterlony, when antibodies and antigens meet in suitable proportions, there is specific precipitation in the form of arcs.

Immunoelectrophoresis permits one to separate and determine antigenic (or heptagenic) constituents of a liquid. In addition to characterization by precipitation, colorimetric methods and enzymic reactions may be employed. The most commonly used gel material is agar. For certain purposes, pectin or starch can be used. However, starch is opaque and the concentration needed to form a gel is, sometimes, excessively high.

The advantages and disadvantages of immunoelectrophoresis were summarized by Grabar (1959). The method can be used in studies of natural mixtures without any previous treatment. Conditions can be chosen (pH, temperature, etc.) to minimize loss of a particular biological activity

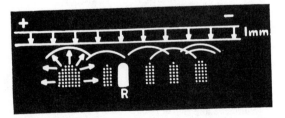

Fig. 29.3. Schematic representation of immunoelectrophoretic analysis. R indicates the hole for the antigen in which the protein mixture is poured before the run. The white, spotted areas represent the location of the different antigen constituents after their electrophoretic separation. Imm indicates the trough for the immune serum (to be filled after the run). The arrows indicate the diffusion of proteins. (From Grabar and Nummi 1967.)

of the substance to be studied. Small amounts are necessary to perform a complete analysis. As the gel is rich in liquid (98–99%), the electrophoresis resembles free electrophoresis in liquid and there is less danger of encountering effects of liquid–solid interfaces as in the case of electrophoresis on paper, starch, or cellulose. At the same time, electrophoresis in a gel has the advantage over electrophoresis in a liquid medium of slowing down the free diffusion of the macromolecular substances at the end of the electrophoresis, thus facilitating their detection.

The use of the specific precipitation reaction allows fine and specific identification of components of a liquid. Substances possessing even identical mobilities can be distinguished and impurities of less than 0.1% can be detected. Enumeration of the constituents of a mixture by the precipitation reaction is greatly facilitated by their preliminary dispersion during the electrophoresis. The components can be characterized and defined by two or three methods: electrophoretic mobility, immunological specificity, and other characteristics (color, enzymatic activity).

Comparisons of Gel Diffusion Methods

Practical applications of gel diffusion methods—radial immunodiffusion, electroimmunodiffusion, immunoelectrophoresis (IEP), Ouchterlony double diffusion, crossed electrophoresis, two-dimensional IEP, antibody ring, radioimmunodiffusion, and agar plaque techniques—were reviewed by Daniels (1979). He reached the following conclusions: the single radial immunodiffusion is a rapid and convenient method for quantitation of specific proteins; the more rapid electroimmunodiffusion technique has similar applications; IEP can be applied to specific problems of protein identification and immunochemical purity; the Ouchterlony double-diffusion test in agar provides a basis for semiquantitative identification of proteins at low concentrations and, with proper controls, can be used to establish immunochemical identify; and the other gel diffusion techniques are applicable to rather specific situations. Chait and Ebersole (1981) compared the power and limitations of various chromatographic and immunoassay methods with regard to sensitivity and specificity.

The immune sera used in all immunochemical procedures are biological products and are difficult to standardize (Maaloe and Jerne 1952). The methods are time-consuming and require the facilities of a serological laboratory. The methods are qualitative and, generally, can give only rough quantitative information. Yang et al. (1966) described conditions required for improved quantitation of immunoelectrophoretic analyses.

Labeling of Antigens and Antibodies

Methods involving labeling of antigens and antibodies include fluorescence (FIA), radio (RIA), and enzyme (EIA) immunoassays (Daussant and Bureau 1984A,B; Monroe 1984; Strecker and Eckert 1978). They are compared in Table 29.1. Radioimmunoassay was discussed in Chapter 19.

In FIA, an ultraviolet fluorescent dye marker is attached to the antibody; in RIA, a radioactive marker is attached to the antigen or antibody. FIA is difficult to standardize or quantitate. RIA is highly sensitive and specific, but the isotope labels may decay rapidly, complex and expensive equipment is required for quantitative results, and assays must be carried out by highly trained personnel.

Blake and Gould (1984) reviewed the use of enzymes with radiolabels in immunoassay. Enzyme labels have the following advantages: no radiation hazards, relatively long shelf life, inexpensive and readily available equipment, tests can be rapid and automated, and multiple simultaneous assays are possible. The disadvantages of enzyme labels are that plasma constituents may affect enzymatic activity, the assay is relatively complex, enzymic reactions are difficult to control, and some reactions have relatively low sensitivity. The two main EIA methods—ELISA (enzyme-linked immunosorbent assay) and EMIT (enzyme-multiplied immunoassay technique—are compared in Table 29.2. EMIT is used widely for rapidly assaying microamounts of drugs and biological substances in human fluids. According to Blake and Gould (1984), the following enzymes are commonly used as labels: (1) for heterogeneous EIA—acetylcholinesterase, adenosine deaminase, alkaline phosphatase, catalase, β-galactosidase, glucose oxidase, peroxidase, and urease; and (2) for homogenous EIA—acetylcholinesterase, β-galactosidase, glucose-6-phosphate dehydrogenase, lysozyme, and malate dehydrogenase.

ELISA has found many applications in analyses of foods (e.g., in the detection of pollutants and toxins in fish, of diseases in slaughter animals, and of environmental contaminants). Numerous variants of ELISA have been developed and are used in food analyses. The assay is predicated

Table 29.1. Immunoassay Comparison[a]

Characteristic	RIA	FIA	EIA
Sensitivity	High (ng-pg)	High (ng-pg)	High (ng-pg)
Specificity	High	High	High
Speed	Days	Days	Hours
Reagents	Short shelf life	Reasonable shelf life	Long shelf life
Equipment required	Scintillation or gamma counter	UV monitor	Spectrophotometer
Personnel	Skilled with license	Skilled	Minimal training
Cost	$7/test	$5/test	$2/test

[a] From Monroe (1984).

Table 29.2. EIA Comparison[a]

ELISA	EMIT
Heterogeneous assay	Homogeneous assay
Reagent separation required (centrifugation or filtration)	Reagent separation not required
Reagent washings required	Step washings not required
Slower than EMIT	Faster than ELISA
Sensitivity greater than EMIT	Sensitivity less than ELISA
Macromolecules measured (antigens, antibodies)	Measures small molecules (haptens)
For diagnosing infectious diseases; immunoglobulins	For drug, hormone, metabolite determinations
Solid phase assay	Liquid phase assay

[a] From Monroe (1984).

on the assumptions that (1) an antigen or antibody can be linked to an insoluble carrier surface and retain activity, and (2) an enzyme marker can be attached to an antibody or antigen with retention of both immunological and enzyme activity. The following steps are involved in enzyme immune assays:

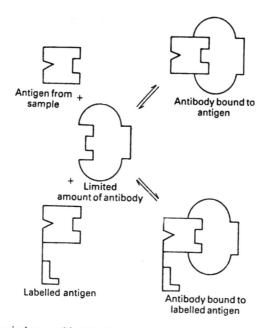

Fig. 29.4. Classical competitive EIA for antigen. All components are mixed with a limited amount of antibody specific for the antigen. Separation of bound enzyme label is necessary before measurement of enzyme activity. (From Blake and Gould 1984.)

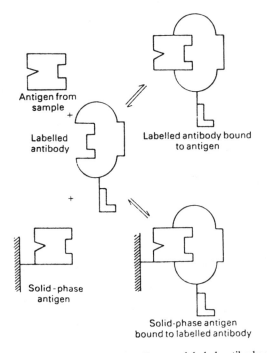

Antigen from
sample
+

Labelled
antibody

Labelled antibody bound
to antigen

+

Solid-phase
antigen

Solid-phase antigen
bound to labelled antibody

Fig. 29.5. Competitive ELISA for antigen. Enzyme-labeled antibody reacts specifically with antigen in the sample and is then added to excess of solid-phase antigen. After washing, the enzyme label still attached to the solid phase is measured. (From Blake and Gould 1984.)

1. An antigen or antibody is linked to the inside of the well of a carrier plate.

2. The sensitized carrier surface binds the specific antibody or antigen from a test solution.

3. The enzyme-labeled antiglobulin is attached to the antigen or antibody complex.

4. The complex is detected by the enzyme label and changes the color of an added substrate.

5. The optical density of the final color can be quantitated and is proportional to the amount of antibody or antigen in the original test solution.

An ideal enzyme label should show high enzyme activity at low substrate concentration; have good stability at the pH required for good antibody–antigen binding; provide an inexpensive, accurate, and sensitive assay method (spectrophotometric end points are best); have reactive groups for covalent linkage with minimum loss of activity; yield stable enzyme-labeled conjugates; be available in soluble, purified form at low

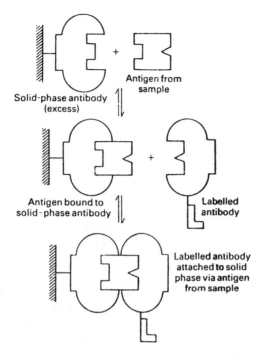

Fig. 29.6. Sandwich assay for antigen. Antigen in the sample is mixed with excess of solid-phase antibody. After washing of the solid phase enzyme-labeled antibody, which is specific for another site on the antigen, is added. The enzyme label, which remains bound after washing, is measured. (From Blake and Gould 1984.)

cost; create no health hazards; and not be affected in its activity by sample components.

Figures 29.4–29.6 illustrate classical competitive EIA, competitive ELISA, and sandwich ELISA for antigen determination.

APPLICATIONS

Precipitin formation in food extracts is affected by all factors that modify proteins. Lipids may interact with proteins and form lipoproteins. Under certain conditions, tannins can prevent precipitin formation. High concentrations of sucrose and low concentrations of glucose may reduce precipitin formation. Similar interference occurs in the presence of urea and saponins.

Intensity of precipitation depends, in addition to ratios of the various components, on the presence and concentration of electrolytes. Precipitation normally occurs in the neutral range; it is somewhat enhanced in slightly acid media, but prevented in the presence of excess inorganic acids and bases. Once formed, the precipitins are soluble in dilute alkali or acid. Antigen–antibody combination increases with temperature over the range 0 to 30°C, though the optimum range varies with the interacting components. Precipitation itself may decrease at elevated temperatures.

Serological differentiation of plant proteins is relatively simple, provided interfering materials (lipids, tannins, etc.) are removed. Quantitative serological methods have been applied for many years to taxonomic investigations. Liuzzi and Angeletti (1969) described the use of immunodiffusion in detecting 5% barley flour in wheat flour. Serological examinations are useful in distinguishing between proteins from meat, milk, or eggs. For many years, the commonly used procedure to determine the source of meat and possible adulteration by meat from other sources or other protein-rich materials was the precipitation test and its modifications (Warnecke and Saffle 1968). It can be used in studying fresh, frozen, processed (dried, smoked, and even heated at low temperatures), and damaged meat. Heating at elevated temperatures may reduce or eliminate the value of a precipitation test. It is not clear whether such heat treatment modifies and destroys serologically reactive groups or decreases the solubility and availability of the reactive groups. This reactivity can be restored, in part at least, by defatting, homogenization, and lyophilization.

Heating of milk impairs little its capacity to give reactive antigens. Similarly, manufacture of yogurt or cultured milk and processing into cheese or dried milk do not affect adversely the results. The precipitation reaction can distinguish between proteins of the white and of the yolk of an egg. It is rather difficult to establish, serologically, the presence of egg yolks in baked products, but not in margarine or alcoholic beverages. Admixture of fish eggs to caviar, determination of honey purity (depending on the presence of specific protein), and source of nuts in sweetened goods or of proteins in bread are some examples of the use of serological tests in food evaluation (Kotter 1967; Lietz, 1969).

The use of immunoassay techniques (primarily radio- and enzymeimmunoassays) for measuring veterinary drug residues in farm animals, meat, and meat products was reviewed in detail by Heitzman (1984), and their use for analysis of mycotoxins by Chu (1984). Crowther and Holbrook (1976) reviewed trends and practices in immunological detection of food-poisoning toxins. The relatively sensitive methods (0.15–1.0 μg/ml) for detecting enterotoxins are double diffusion, electroimmunodiffusion, and countercurrent immunoelectrophoresis, in decreasing order of sensitivity. For maximum sensitivity, however, the toxins must be extracted from the food and concentrated. The more sensitive methods (about 0.001 μg/ml) include ELISA, reverse passive hemagglutination, and radioimmunoassay. According to Kuo and Silverman (1980), the ad-

vantages of radioimmunoassays for detection of staphylococcal enterotoxins in foods are specificity, sensitivity, speed, and nonspecific adsorption. The advantages of ELISA are that no radioactive compounds
are required, only commonly used laboratory equipment is needed, the
enzyme conjugates are stable, and a positive reaction can be evaluated
visually. Pestka *et al.* (1980) developed a specific ELISA test for the rapid
quantitation of aflatoxin B_1 at levels as low as 25 pg per assay. Although
ELISA and RIA are highly sensitive, they cannot be used for assays of
insoluble proteins; however, such materials can be assayed by rocket and
crossed immunoelectrophoresis (Plumley and Schmidt 1983).

Baudner (1977) reviewed analyses of plant proteins using immunological techniques based on the antigen–antibody precipitation techniques.
Immunoseparation of plant proteins (by immunoaffinity chromatography)
was described by Daussant and Bureau (1984A,B). Specificity, capacity,
and quality of immunosorbents were examined, and the use of low-affin-
ity, polyclonal, and monoclonal antibodies was described in studies on
the recovery of biologically active plant proteins. Daussant (1982) discussed the potential of immunoprecipitation in gel, immunodiffusion, passive agglutination inhibition, ELISA, and immunoaffinity chromatography for the quantitative determination of active and denatured proteins.

The application of immunochemical methods in the analysis of foods
was described by Hitchcock (1984), in analyses of cereals by Baudner
(1978), and in beer production by Schurr *et al.* (1977) and Gunter and
Baudner (1979). Sinell and Mentz (1977) described the use of electroimmunodiffusion methods for quantitative determination of nonmeat proteins added to meat products, Kurth and Shaw (1983) reviewed the identification of the species of origin of meat by immunochemical methods.

Whittaker *et al.* (1983) developed an ELISA method to differentiate
between unprocessed beef, sheep, kangaroo, pig, and camel meats. The
assay detected 10% contamination and was comparable to the Ouchterlony immunodiffusion test. It is rapid (3 hr), sensitive (in terms of material
required and sampling), requires small amounts of species-specific antisera, which can be mixed for simple screening, and can be semi-auto-
mated. Subsequently Patterson *et al.* (1984) developed an improved
ELISA method to differentiate less than 1% levels of unprocessed meats.
In this method, which is a double-sandwich ELISA system, species-specific capture antibodies are coated onto the inside of microtiter plates.
The antibody-coated plates are used to immunoextract soluble protein
from meat samples. The capture antibody method simplifies sample preparation and makes possible differentiation of closely related species. Sample preparation is less critical because color development is approximately
constant between sample dilutions over a 100-fold range. The increased
sensitivity and selectivity of this method permits bulking (combining) sam-

ples for screening purposes. The test can be completed in less than 2 hr. Griffith *et al.* (1984) recommended the use of a commercial ELISA system to measure levels of soya protein in meat products. The method is particularly useful in nonspecialized laboratories.

Although species identification of animal bloods can be achieved readily by immunological tests, differentiation among fish species is more difficult. The differentiation of fish species among the Salmonidae family and some coarse fish families on the basis of an immunochemical test and electrofocusing patterns of the enzyme superoxide dismutase from the red cell was described by Sutton *et al.* (1983). The immunochemical test is based on the development of a specific anti-trout (Salmonidae) serum, which is used to differentiate the blood of the Salmonidae from that of other freshwater fish. For differentiation among the Salmonidae, the difference in polymorphic forms of the isoenzymes separated by electrofocusing in a pH 2.5 to 8 gradient is utilized. With this technique, it is possible to differentiate among salmon, sea brown trout, char, cheeta trout, and a number of varieties of brown trout.

BIBLIOGRAPHY

Baudner, S. (1977). Analysis of plant protein using immunological techniques based on the antigen–antibody precipitation. *Ann. Nutr. Aliment.* **31**, 165–178.

Baudner, S. (1978). Immunology in food analysis: Concepts, methodology, and examples from research in cereal chemistry (in German). *Getreide, Mehl, Brot* **32**, 330–337.

Blake, C., and Gould, B. J. (1984). Use of enzymes in immunoassay techniques. A review. *Analyst* **109**, 533–547.

Boyd, W. C. (1966) "Fundamentals of Immunology," 4th ed. Interscience Publishers, New York.

Chait, E. M., and Ebersole, R. C. (1981). Chemical analysis: A perspective on chromatographic and immunoassay technology. *Anal. Chem.* **53**, 682A–684A, 687A, 688A, 690A, 692A.

Chu, F. S. (1984). Immunoassays for analysis of mycotoxins. *J. Food Protection* **47**, 562–569.

Crowle, A. J. (1960), Interpretation of immunodiffusion tests. *Annu. Rev. Microbiol.* **14**, 161–176.

Crowle, A. J. (1961), "Immunodiffusion." Academic Press, New York.

Crowther, J. S., and Holbrook, R. (1976). Trends in methods for detecting foodpoisoning toxins produced by *Clostridium botulinum* and *Staphylococcus aureus. In* "Microbiology in Agriculture, Fisheries and Food" (F. A. Skinner and J. G. Carr, (eds.). Academic Press, London.

Daniels, J. C. (1979). Practical applications of gel diffusion tests. *In* "Immunoassays in the Clinical Laboratory," pp. 23–61. A. R. Liss, Inc. New York.

Daussant, J. (1982). Immunochemistry in protein analysis. *In* "Recent Developments in Food Analysis" (W. Baltes, P. B. Czedik-Eysenberg, and W. Pfannhauser, (eds.), pp. 215–228. Verlag Chemie, Weinheim, W. Germany.

Dausant, J., and Bureau, D. (1984A). Immunoseparation of plant proteins. *In* "Electrophoresis' 84" (V. Neuhoff, ed.), pp. 62–77. Verlag Chemie, Weinheim, W. Germany.

Daussant, J., and Bureau, D. (1984B). Immunochemical methods in food analysis. *In* "Developments in Food Analysis Techniques" (R. D. King, ed.), pp. 175–210. Elsevier Applied Science Publishers, London.

Day, E. D. (1966) "Foundations of Immunochemistry." Williams and Wilkins Co., Baltimore.

Grabar, P. (1950). Immunochemistry. *Annu. Rev. Biochem.* **19**, 453–486.

Grabar, P. (1957). Agar-gel diffusion and immunoelectrophoretic analysis. *Ann. N.Y. Acad. Sci.* **69**, 591–607.

Grabar, P. (1958). The use of immunochemical methods in studies of proteins. *Adv. Protein Chem.* **13**, 1–33.

Grabar, P. (1959). Immunoelectrophoretic analysis. *Methods Biochem. Anal.* **7**, 1–38.

Grabar, P., Escribano, M. J., Benhamou, N., and Daussant, J. (1965). Immunochemical study of wheat, barley and malt proteins. *Agric. Food Chem.* **13**, 392–398.

Grabar, P., and Nummi, M. (1967). Recent immunoelectrophoretic studies on soluble proteins in their transformation from barley to beer. *Brewers' Dig.* **42**(3), 68–74.

Grabar, P., and Williams, C. A. (1953). Methode permettant l'etude conjugee des prorpietes electroproretiques et immunochimiques dun melange de proteines. Application au serum sanguin. *Biochim. Biophys. Acta* **10**, 193–194.

Griffith, N. M., Billington, M. J., Crimes, A. A., and Hitchcock, C. H. S. (1984). An assessment of commercially available reagents for an enzyme-linked immunosorbent assay (ELISA) of soya proteins in meat products. *J. Sci. Food Agric.* **35**, 1255–1260.

Gunther, H. O., and Baudner, S. (1979). Immunological methods of identification of proteolytic enzymes and umalted materials in beer (in German). *Brauwissenschaft* **32**, 200–205.

Haurowitz, F. (1960). Immunochemistry. *Annu. Rev. Biochem.* **29**, 609–634.

Haurowitz, F. (1961). Use of radioisotopes in immunochemical research. *Ergeb. Mikrobiol., Immunitatsforsch. Exp. Therapie.* **34**, 1–26.

Heitzman, R. J. (1984). Immunoassay techniques for measuring veterinary drug residues in farm animals, meat, and meat products. *In* "Analysis of Food Contaminants" (J. Gilbert, ed.), pp. 73–115. Elsevier Applied Science Publishers, London.

Hitchcock, C. H. S. (1984). Immunological methods. *In* "Control of Food Quality and Food Analysis" (G. G. Birch and K. J. Parker, ed.), pp. 117–133. Elsevier Applied Science Publishers, London.

Kabat, E. A. (1961). "Kabat and Meyer's Experimental Immunochemistry," 2nd ed. Thomas Publishing Co., Springfield, IL.

Kotter, L. (1967). Serological methods in protein investigations. *In* "Handbook of Food Chemistry, Analysis of Foods, Detection and Determination of Food Components" (J. Schormuller, ed.). Springer-Verlag, Berlin.

Kuo, J. K. S., and Silverman, G. J. (1980). Application of enzyme-linked immunosorbent assay for detection of staphylococcal enterotoxins in food. *J. Food Protection* **43**, 404–407.

Kurth, L., and Shaw, F. D. (1983). Identification of the species of origin of meat by electrophoretic and immunological methods. *Food Technol. Aust.* **35**, 328–331.

Landsteiner, K. (1921). Heterogenic antigens and haptens. XV. Antigens. *Biochem. Z.* **119**, 294–306.

Lawrence, M. (1964). The techniques of immunoelectrophoresis. *Am. J. Med. Meth.* **30**, 209–221.

Lietz, A. (1969). Quantitation of food adulterants by multiple radial immuno-diffusion. II. Wheat in bread sold as "wheat-free" for the use of allergy patients. *J. Assoc. Off. Anal. Chem.* **12**, 995–998.

Liuzzi, A., and Angeletti, P. A. (1969). Application of immunodiffusion in detecting the presence of barley in wheat flour. *J. Sci. Food Agric.* **20**, 207–209.

Maaloe, O., and Jearne, N. K. (1952). The standardization of immunological substances. *Annu. Rev. Microbiol.* **6**, 349–366.

Monroe, D. (1984). Enzyme immunoassay. *Anal. Chem.* **56**, 920A, 922A, 924A, 926A, 928A, 930A.

Ouchterlony, O. (1967). "Handook of Immunodiffusion and Immunoelectrophoresis." Ann Arbor Science Publishers, Ann Arbor, MI.

Oudin, J. (1946). Immunochemical methods of analysis involving selective precipitation in solidified media. *Compt. Rend. Acad. Sci.* **222**, 115–116.

Oudin, M. (1952). Specific precipitation in gels and its application to immunochemical analysis. *Methods Med. Res.* **5**, 335–378.

Patterson, R. M., Whittaker, R. G., and Spencer, T. L. (1984). Improved species identification of raw meat by double sandwich enzyme-linked immunosorbent assay. *J. Sci. Food Agric.* **35**, 1018–1023.

Pestka, J. J., Gurr, P. K., and Chu, F. S. (1980). Quantitation of aflatoxin B_1 and aflatoxin B_1 antibody by an enzyme-linked immunosorbent microassay. Appl. *Environ. Microbiol.* **40**, 1027–1031.

Plumley, F. G., and Schmidt, G. W. (1983). Rocket and crossed immunoelectrophoresis of proteins solubilized with sodium dodecyl sulfate. *Anal. Biochem.* **134**, 86–95.

Porter, R. R. (1967). The structure of immunoglobulins. Essays in Biochem. **3**, 1–24.

Schurr, F., Anderegg, P., and Pfenninger, H. (1977). Serological detection of maize and rice in beer production (in German). *Mitt. Gebb. Lebensm. Hyg.* **68**, 538–545.

Singer, S. J. (1965). Structure and function of antigen and antibody proteins. *In* "The Proteins—Composition, Structure and Function" (H. Neurath, ed.). Academic Press, New York.

Sinnell, H.-J., and Mentz, I. (1977). Use of electroimmunodiffusion for quantitative determination of nonmeat proteins added to meat products. *Folia Vet. Latina* **VII**, 41–54.

Strecker, H., and Eckert, G. (1978). Radioimmunoassay—a revolution in analysis (in German). *In* "GIT Labor-Medizin," No. 2, pp. 106–116. Ernst Gieberler Verlag, Darmstadt, W. Germany.

Sutton, J. G., Goodwin, J., Horscroft, G., Stockdale, R. E., and Frake, A. (1983). Identification of trout and salmon bloods by simple immunological technique and by electrofocussing patterns of red cell enzyme superoxide dismutase. *J. Assoc. Off. Anal. Chem.* **66**, 1164–1174.

Warnecke, M. O., and Saffle, R. L. (1968). Serological identification of animal proteins. I. Mode of injection and protein extracts for antibody production. *J. Food Sci.* **33**, 131–135.

Whittaker, R. G., Spencer, T. L., and Copland, J. W. (1983). An enzyme-linked immunosorbent assay for species identification of raw meat. *J. Sci. Food Agric.* **34**, 1143–1148.

Williams, C. A. (1960). Immunoelectrophoresis. *Sci. Am.* **202**(3), 130–134, 136, 138–140.

World Health Organization. (1964). Nomenclature for human immunoglobulins. *Bull. World Health Organ.* **30**, 447–450.

Yang, W. K., Yaeger, R. G., and Miller, O. N. (1966). Quantitative considerations on immunoelectrophoretic analysis. *Bull. Tulane Univ. Med. Fac.* **25**, 73–84.

30
Enzymatic Methods

GENERAL CONSIDERATIONS AND USES

An analytical method is of value when its specificity, reproducibility, and sensitivity are high and when the expenditure of labor, time, and material are low. Theoretically, most of these requirements can be met admirably by enzymatic analysis (Bergmeyer 1983). The term *enzymatic analysis* is generally understood to mean analysis with the aid of enzymes. The major advantages of enzymes in analysis lie in their ability to react specifically with individual components of a mixture. This avoids lengthy separations of the components and reduces the time needed for an analysis. The amount of substrate (sample) required for analysis is small, and because of the mild conditions employed, enzymes often allow the detection and determination of labile substances that can only be estimated rather inaccurately by other methods.

For many years, enzymatic techniques in analytical chemistry were limited to biochemical and clinical laboratories because of the difficulty in preparing the enzymes. Today enzymes are used much more widely as analytical tools in analysis of agricultural and food products than in the past. New methods have been developed for the use of commercially available pure preparations. Use of enzymes in analytical chemistry admittedly has certain limitations. Reliability depends on meticulous adherence to, and satisfactory control of, standardized assay conditions. In all enzymatic analyses, the presence of contaminants in the sample, which can partially or totally inhibit the enzyme-catalyzed reactions, is a serious problem. Heavy metals, oxidants, and SH-blocking reagents may affect results. Compounds structurally similar to the enzyme but without biological activity are competitive inhibitors. And finally, the purity of a preparation is crucial. Some enzymatic impurities may enhance, others reverse a reaction. Instability of enzymes may limit their usefulness; it can, however, be minimized by using proper conditions.

The relatively high cost of using purified preparations can be reduced by use of immobilized-insolubilized enzymes (Gilbault 1970, 1976; Zaborsky 1973; Weetal 1975; Carr and Bowers 1980; Bowers and Carr 1980). When properly prepared, immobilized enzymes have nearly the same activity as free enzymes. The analytical advantages of immobilized enzymes are their high stability, reusability, ease of removal from the reaction without contamination, long half-life with predictable decay rate, and easy preparation of unstable or expensive reagents (Bowers and Carr 1980).

Enzymes can be immobilized by physical or chemical methods. The four methods used to prepare immobilized enzymes for analytical uses include (a) adsorption, (b) entrapment within a gel matrix, (c) covalent crosslinking to itself or to a second protein, and (d) covalent attachment to an insoluble carrier such as polysaccharides, Sepharose, Sephadex, polyacrylamide, several polymers and copolymers, and coated glass. Microencapsulation is used to a very limited extent for immobilization of analytical reagents. Analytical applications include determination of sugars, amino acids, cholesterol, urea and uric acids, penicillin, phosphate, organic pesticides, proteolytic enzymes, and ethanol.

The use of immobilized enzymes as analytical reagents has increased tremendously since the introduction of enzyme electrodes by Updike and Hicks (1967). Some active enzyme electrodes can be stored in the dry state (Bardelletti and Coulet 1984). Most of the assays involving immobilized enzymes have been used in the development of selective enzyme electrodes. Early enzyme electrodes were made from an electrochemical sensor, a semipermeable membrane, and an intermediate layer that contained the enzyme. Their response was slow, as they required diffusion of the substrate through the membrane. Today, many enzyme electrodes use enzymes immobilized on easily accessible surfaces of inorganic supports.

The glass pH electrode is limited in its use for enzymatic assays because most enzyme reactions are not linear over a wide pH range, and for a wide range the reaction medium must have a low buffering capacity (Mercier 1981). Specific glass electrodes (for NH_4^+, CN^-, PO_4^{3-}) can be constructed by varying the glass composition. In constant-potential polarographic measurements (amperometry), the current is proportional to the amount of reduced or oxidized species at a fixed potential. Such electrodes (e.g., the Clark pCO_2 electrode) have applications in enzymatic assays. Electrodes based on the determination of consumed oxygen or produced hydrogen peroxide are attractive for assays that involve glucose or amino acid oxidases. Amperometric determination of peroxides has been used to determine alcohol, uric acid, xanthine, galactose, and ascorbic acid. Urease and specific amino acid decarboxylase electrodes are available. Glucose in foods was determined with free glucose oxidase and with glucose oxidase bound to a glycidylmetacrylate polymer (Valentova et al. 1983). Glucose oxidase immobilized on a nylon net was used to prepare a glucose-specific electrode. The enzymic methods yielded results that were comparable to those obtained by the ferricyanide and ion-exchange chromatography methods. The use of hexokinase and glucose-6-phosphate dehydrogenase immobilized on a nylon coil for glucose analysis in a small laboratory was described by Wease et al. (1979).

The immobilization concept has been extended by attaching live microorganisms to polymers and keeping them alive while covering suitable electrodes (Townsend 1981). For example, yeast have been bound covalently to a porous polymer and placed over an oxygen electrode. The electrode can be dipped into a solution of an antibiotic. As the concentration of the antibiotic increases, more yeast are killed and less oxygen is consumed. Alternatively, the electrode can be dipped into a solution of acetate, which stimulates yeast growth and increases oxygen consumption. A variation of this concept involves incorporating thin slices of pig kidney containing glutamine oxidase into an electrode to determine glutamine in biological liquids. The highly selective glutamine–glutamine oxidase reaction liberates ammonia, which can be detected by an ammonia-sensitive detector. The kidney slices remain effective for about one month. Bacteria that cause caries (by converting dietary carbohydrates to acid) can be immobilized and used in combination with a glass electrode to determine the amount of carbohydrates from the response of the electrode to the amount of generated acid (Townsend 1981).

Methods of enzymatic analysis were comprehensively reviewed in books written or edited by Bergmeyer (1983), Bergmeyer and Gawehn (1978), Guilbault (1970, 1976), Whitaker (1972, 1974A,B), and Boehringer Mannheim (1980). Excellent review articles are those by Wiseman (1984), Mercier (1981), and Townsend (1981). Volumes 19(6) and 20(1) of Alimenta (1980 and 1981) contain in a series of seven articles in German (available from Boehringer Mannheim, Switzerland) the basis of many enzymatic tests, potentials and limitations of enzymatic assays, sample preparation, verification of results, a new enzymatic method for starch determination in foods, and determination of ascorbic acid in foods. Boehringer Mannheim (1980) has published a manual that contains detailed instructions for the determination of about 40 compounds in foods.

Enzyme assays can be used to determine (1) enzymic activity of a sample, (2) amounts of substrates, (3) amounts of enzyme activators or inhibitors, (4) isomeric configuration, (5) structures of complex molecules (carbohydrates, proteins, nucleic acids, lipids), (6) conformation of complex molecules, and (7) structures of cells and subcellular organelles. Mercier (1981) classified enzymatic assays on the basis of their action on specific substrates, activation and inhibition, use as tools in structural analyses, use as specific disintegrating agents (for breakdown of cell walls or release of bound forms) prior to actual analyses, and *in vitro* digestibility tests. Uses of enzymes to determine activators, inhibitors, toxins, sugars and polysaccharides, organic acids and bases, alcohols, lipids, and amino acids were described by Wiseman (1984).

DETERMINATION OF SUBSTRATES

According to the Michaelis–Menten equation:

$$E + S \underset{K_{-1}}{\overset{K_1}{\rightleftharpoons}} ES \overset{K_2}{\longrightarrow} E + P \tag{30.1}$$

the substrate S combines with the enzyme E to form an intermediate complex ES, which subsequently breaks down to product P and free enzyme. The equilibrium constant for the formation of the complex is called the Michaelis constant, and is denoted K_m. It is defined as $(K_2 + K_{-1})/K_1$. The rate of reaction v is a function of the enzyme and substrate concentration, and is affected by the presence of activators or inhibitors. At a fixed enzyme concentration, the initial velocity is

$$v_0 = (V_{max}[S]_0)/(K_m + [S]_0) \tag{30.2}$$

The initial rate increases with substrate until a nonlimiting excess of substrate is reached, after which additional substrate causes no increase in rate.

General Assay Procedures

The concentration of a substance modified by an enzymatic reaction can be determined in two ways. The most useful enzymatic method is the determination of a product of an enzyme-catalyzed reaction that is formed in a definite stoichiometric relation from the compound to be assayed. The ideal condition for this type of assay is a reaction that goes virtually to completion. If the substrate is consumed only partly, quantitative utilization of the substrate can be achieved by increasing the concentration of some reactants or by removing one of the products of enzymatic action. If it is not feasible to alter the reaction so that all of the substrate is consumed, the equilibrium mixture of reactants and products can be determined and compared to a standard curve prepared under defined conditions.

For determination of total change either direct measurements or measurements with the aid of coupled reactions are made. If the substrate has a characteristic absorption spectrum, the decrease in absorbance at a selected wavelength can be measured, (e.g., measurement of absorbance at 293 nm in the determination of uric acid with uricase (Plesner and Kalckar 1956). In the determination of ethanol with alcohol dehydrogenase, NAD (nicotinamide-adenine dinucleotide) is reduced to NADH, which has an absorbance peak at 340 nm. The transfer of the hydrogen of the substrate to the pyridine ring of NAD or NADP (nicotinamide-adenine dinucleotide phosphate) and measurement of the absorbance peak at 340 nm are the basis of many enzymatic assays. For example, in the oxidation of ethanol, NAD accepts the substrate hydrogen, and in the reduction of pyruvate to lactate, NADH donates hydrogen. Use of direct enzymatic methods in the determination of organic acids, carbonyls, alcohols, and sugars in wine was reviewed by Peynaud *et al.* (1966). Talalay (1960) reviewed use of enzymatic methods that utilize purified hydroxysteroid dehydrogenases for the specific sensitive microdetermination of steroid hormones. Another example of direct measurement of the enzymatically modified substrate is the titration (or manometric measurement of CO_2 released from a bicarbonate buffer) of glucose with gluconic acid formed in the determination of glucose with glucose oxidase (Free 1963).

In assays that use a coupled reaction, an auxiliary and an indicator step are involved. Dyes can serve as indicators in coupled reactions by reacting with a product of the auxiliary reaction. An example is the determination of glucose with glucose oxidase and peroxidase using *o*-dianisidine as the indicator, which is discussed later.

The second general method involving enzymes as analytical reagents is based on measuring the rate of the enzyme-catalyzed reaction. With limiting quantities of substrate and under closely controlled conditions, the initial rate of an enzyme-catalyzed reaction may be used to determine the concentration of the substrate. The region in which linearity is achieved, and in which an analytical determination of substrate concentration can be based on the rate of reaction, lies below $0.1K_m$. The rate of an enzyme-catalyzed reaction depends on pH, temperature, and ionic composition of the assay medium. Small variations in the assay conditions cause serious errors in the determinations.

The temperature coefficient of most enzyme reaction rates is roughly 10% per degree, and a 10°C rise in temperature may result in a twofold increase in the reaction rate. Hence, constant temperature is essential in the assay of enzyme activity. In enzyme reaction rate studies, standards are usually run with each test. If normal precautions are taken, the error in measurement of enzyme activity is usually less than ±5%. In automated systems, better accuracy and replicability can be attained.

A great many substrates can be assayed by enzymatic methods. Determination of some of the compounds of particular interest in food analyses is reviewed here.

Polysaccharides

Starch. Lee and Whelan (1966) determined starch and glycogen concentration after hydrolysis with amyloglucosidase. The enzyme hydrolyzes both α-1 → 4 and α-1 → 6 linkages of the polysaccharide and gives glucose. The latter can be assayed with glucose oxidase. Ettel (1981) presented a new enzymatic method for the determination of starch in foods. Total starch was defined as the amount of α-glucans that are insoluble in 80% ethanol and can be hydrolyzed by amyloglucosidase to glucose. The extracted starch is solubilized with 0.5 N NaOH and hydrolyzed enzymatically to glucose, which is determined by a hexokinase UV assay. Samples containing up to 400 mg starch per 2 g were analyzed reproducibly with a coefficient of variation of 0.5% in pure starch and 1.4% in wheat germ.

An automated photometric enzyme assay of starch content of cereals was described by Holz (1977). Lind (1977) found enzymatic assay adequate for routine starch determination in most meat products. The usefulness of the enzymatic method was confirmed by Beutler (1978) for a variety of foods. The latter two methods used dimethyl sulfoxide extracts. Skrede (1983) determined the starch content of meat products as glucose (using glucose oxidase, peroxidase, and o-dianisidine hydrochloride) in enzyme-hydrolyzed samples (using thermostable α-amylase and amyloglucosidase). The recovery in meat products containing up to 12% starch averaged 103% in a collaborative study. The average coefficient of variation was 2%. The results were unaffected by the presence of lactose. The random and systematic errors were about 8% of the starch content. Haissig and Dickson (1979) reported that less than 5% starch can be determined in 5 mg of lyophilized plant tissue by an enzymatic method with minimum manipulation. Their method is simple and reliable but rather time-consuming.

Cellulose and Hemicellulose. All forms of cellulose can be described as β-1,4 polyglucosides. As a result of the stability of native undegraded cellulose, there is no standard method for the enzymatic determination of this kind of substrate. However, partly degraded forms of cellulose are rapidly and completely hydrolyzed by enzymes (Halliwell 1959). If cellulose is treated with phosphoric acid, it swells and is transformed into a more reactive form without any great change in the degree of polymerization. Swollen cellulose or swollen cotton fibers are rendered 97% soluble in 22 hr by the action of cell-free culture filtrates from *Myrothecium verrucaria*. The extent of the enzymatic hydrolysis is determined by the loss of weight of the insoluble substrate. The loss can be determined directly or indirectly.

In the direct procedure (Halliwell 1958), the cellulase-rich filtrate is allowed to act on defatted, insoluble cellulose. The reaction mixture is filtered through sintered glass; the residue is oxidized with a dichromate–sulfuric acid mixture, and the color is measured at 430 nm. The difference between the color of the control and of the digested residue can be correlated with the amount of cellulose hydrolyzed by the enzyme. The amount of cellulose is computed from a calibration curve with glucose, and by multiplying the glucose equivalent by 0.9 to obtain cellulose content. In the indirect procedure (Halliwell 1960), the soluble carbohydrates formed on enzymatic hydrolysis are determined instead of the insoluble residue. The procedure is more rapid, but the tested solution must be protein-free or low in soluble proteins.

Hemicelluloses are plant cell-wall polysaccharides that are insoluble in water and ammonium oxalate solution, but are soluble in dilute alkali. They are bound with cellulose and lignin, and must be isolated prior to assay. A further complication in the determination of hemicelluloses is that they exist in a variety of chemical structures. Because of this structural complexity, no standard methods for the determination of hemicelluloses are available. Certain simple soluble hemicelluloses of the arabino-xylan or xylan type are completely hydrolyzed to their constituent pentoses, (i.e., arabinose and xylose) by washed suspensions of toluene-treated bacteria from rumen of sheep or by a cell-free hemicellulase preparation obtained after treatment of bacteria from sheep rumen with aqueous butanol and centrifugation. After deproteinization of the hydrolysate, the reducing value of the carbohydrates is determined (Halliwell 1957A,B).

Mono- and Oligosaccharides

Raffinose.   Raffinose is a trisaccharide of glucose, galactose, and fructose. It is present in sugar beet, beet juice, crude beet sugar, and less pure syrups containing sugar beet molasses. Raffinose can be hydrolyzed by invertase to fructose and the disaccharide melibiose. Melibiase hydrolyzes melibiose to glucose and galactose. The optical rotation and the reducing value change during this hydrolysis, and both changes can be used to determine raffinose.

The optical rotation method in sugar assays is recognized by the International Commission for Uniform Methods of Sugar Analysis. The rotation is measured (1) before the hydrolysis; (2) after the action of invertase; and (3) after the action of melibiase. Analogous to the optical rotation method, the reducing value is determined (e.g., titrimetrically with Fehling's solution) before and after the action of invertase, and invertase plus melibiase. In the reducing value method, it is not necessary to carry out a preliminary lead precipitation to remove color substances that interfere with the measurement of optical rotation.

Sucrose and Lactose.   The most common method of sucrose determination is by measuring optical rotation. The method is precise and rapid in samples free of interfering optically active compounds and relatively low in sugars other than sucrose. Sucrose is hydrolyzed by invertase to glucose and fructose. The reducing sugar formed on inversion can be determined chemically if the sample contains virtually no other reducing substances. The two hexoses are phosphorylated by ATP to the corresponding hexose-6-phosphates in a reaction catalyzed by hexokinase. Fructose-6-phosphate is isomerized to glucose-6-phosphate by phosphoglucoisomerase, and glucose-6-phosphate is oxidized by NADP and glucose-6-phosphate dehydrogenase to 6-phosphogluconate. The increase in optical density at 340 nm due to the formation of NADPH is a measure of the overall reaction. The reaction proceeds rapidly and quantitatively if the measurements are made at the pH optima of invertase (4.6) and hexose determination (7.6). The reaction can be carried out on any type of sample and requires 10 μg sucrose or less.

A lactose determination involves using β-D-galactosidase, which catalyzes the hydrolysis of lactose into glucose and galactose. In the presence of ATP and hexokinase, the glucose is phosphorylated. Oxidation of the formed glucose-6-phosphate with glucose-6-phosphate dehydrogenase in the presence of NADP yields NADPH. The NADPH formed is measured by an increase of absorbance at 340 nm.

Frank and Christen (1984) developed a simple method for determining the lactose and sucrose contents of ice cream mixes. The method is based on measurement of the freezing point depression resulting from incubation of the disaccharide with invertase and β-D-galactosidase. Sucrose and lactose concentrations are determined separately; each determination requires a 2-hr incubation at 37°C.

Glucose.   One of the most common methods for determining glucose is with glucose oxidase. At 20°C, the enzyme oxidizes β-D-glucose 150 times faster than the α isomer. However, the enzyme can be used to determine glucose in solutions containing an equilibrium mixture of the isomers because even highly purified preparations of glucose oxidase contain mutarotase and the reaction time selected for analytical assays is long enough to oxidize all of the α-glucose. The determination is specific for D-glucose. Among the commonly occurring sugars, only mannose and galactose react and give about 1% of the value with glucose. The measurement of glucose with glucose oxidase can be made manometrically, but is generally made colorimetrically. For the analysis of mixtures of sugars, it is necessary to use highly purified enzyme preparations. The procedure has been used to assay glucose in biological fluids and in various foods (corn syrup, hydrolysates of polysaccharides, fermentation liquors, etc).

As indicated previously, glucose oxidase catalyzes the reaction

$$\text{glucose} + H_2O + O_2 \rightarrow \text{gluconic acid} + H_2O_2$$

The hydrogen peroxide is decomposed in the presence of peroxidase and the oxygen liberated oxidizes a hydrogen donor DH (e.g., dianisidine) to a colored derivative D:

$$H_2O_2 + DH_2 \rightarrow 2H_2O + D$$

The amount of dye D formed from DH_2 is a measure of the glucose oxidized. The dye has maximum absorbance around 460 nm. The extinction coefficient depends on the experimental conditions, and the measured absorbance is related to a glucose standard. In addition to o-dianisidine, o-toluidine, 2,6-dinitrophenolindophenol, and other dyes have been used in glucose assays.

Many modifications of the basic glucose oxidase method have been suggested. The reaction can be accelerated by using higher temperatures (e.g., 37°C for 20 min) provided rapid color fading is prevented. The method has been adapted for automated assays. The measurements are made after 1 min, and the error is only 2%. Filter paper strips soaked with the enzymes and dye (o-toluidine) can be used for semi-quantitative determination of glucose in biological fluids.

Alcohols and Acids

Nonenzymic determination of glycerol in biological materials, foods, and industrial products usually requires an extensive purification of the samples in order to remove contaminants. Because enzymic determination of glycerol is specific, there is no need to purify the sample. Glycerol is phosphorylated by glycerokinase and ATP to give L(−)-glycerol-1-phosphate; the latter is oxidized with glycerol-1-phosphate dehydrogenase in the presence of NAD to give dihydroxyacetone phosphate and NADH.

The amount of NADH formed is equivalent to the amount of glycerol present. For quantitative oxidation in the second step, the reaction is carried out at pH 9.8 and the dihydroxyacetone phosphate formed is trapped with hydrazine.

Muscle lactate dehydrogenase catalyzes the oxidation of L-lactate by NAD to pyruvate and NADH. To obtain quantitative oxidation, the reaction products must be removed from the reaction mixture. This is done by making the medium alkaline (pH 9.5) and trapping the pyruvate as a hydrazone derivative. The course of the reaction is followed spectrophotometrically by the increase in absorbance due to the formation of NADH. Determination of lactate with yeast lactic dehydrogenase—unlike the assay with the enzyme from muscles—is not NAD-linked. It can be used to determine lactic acid in the presence of a large excess of pyruvate and is a useful routine biological assay. Yeast lactic dehydrogenase, a flavocytochrome, transfers hydrogen from lactate to potassium ferricyanide. The decrease in color on reduction of the ferricyanide ion can be followed at 405 nm.

The isolation from yeast and mammalian liver in pure form of enzymes that specifically catalyze the dehydrogenation of primary alcohols to aldehydes (alcohol dehydrogenase) opened the possibility of determining alcohols colorimetrically, since the enzymes from both sources require NAD as coenzyme. NADH has an absorbance maximum at 340 nm, and NAD has virtually no absorbance at that wavelength. At neutral reaction conditions and with the amount of NAD ordinarily employed, less than 1% of the alcohol is oxidized to acetaldehyde. It is, however, possible to drive the reaction nearly to completion by using a high pH and removing the aldehyde formed by reaction with semicarbazide. Use of the 3-acetylpyridine analog of NAD also changes the equilibrium. With this substance as the reactant, the equilibrium constant increases about 200 times. Generally, only butanol might affect the enzymatic assay of ethanol. NADH may be also measured by fluorescence, and the sensitivity of the assay is substantially improved.

In addition to the compounds discussed, methods are available to determine acetate, formate, citrate, malate, fumarate, and other organic acids present in foods. The use of enzymatic methods in the determination of organic acids (acetic, citric, lactic, and malic), glycerol, ethanol, and sugars (sucrose, glucose, and fructose) in beer, apple juice, and lemon concentrates was described by Krueger and Nordmann (1981).

Amino Acids

Certain bacteria, grown under suitable conditions, produce specific L-amino acid decarboxylases. In most cases, the pH optima for the decarboxylases are in the acid range, so the carbon dioxide produced can be measured manometrically. Decarboxylases catalyze the following type of reaction:

$$
\begin{array}{c}
\text{NH}_2 \\
| \\
\text{R—CH—COOH} \rightarrow \text{R—CH}_2\text{—NH}_2 + \text{CO}_2
\end{array}
$$

The carbon dixoide produced is a measure of the amino acid content of the sample.

Specific decarboxylases are available for the determination of lysine, ornithine, tyrosine, histidine, aspartic acid, glutamic acid, glutamine, β-hydroxyglutamic acid, phenylalanine, diaminopimelic acid, valine, and leucine. In the determination of amino acids that have a pH of about 5.8 at the end of decarboxylation (lysine, ornithine, and aspartic acid), manometer vessels with double side arms are used with a Warburg apparatus (Umbreit *et al.* 1957). The second side arm contains dilute acid, which at the end of the decarboxylation is tipped into the main compartment to release the retained carbon dioxide. Spectrophotometric assays were described by Dickerman and Carter (1962).

Amino acid decarboxylases attack free amino acids only. If an excess of decarboxylase and limiting amounts of peptidase are added to a peptide preparation, the carbon dioxide formed is an index of peptidase activity. The decarboxylases can be used to fractionate D and L amino acids by destroying the L isomer with decarboxylase, which does not attack the D isomer. Free amino acids can be determined in the presence of proteins. Many transamination reactions involve glutamic acid as one of the components. Estimation of glutamic acid by the decarboxylases can be used to follow transaminases. The decarboxylases possess pyridoxal phosphate as a prosthetic group and in some cases, e.g., tyrosine decarboxylase, it is possible to remove the prosthetic group leaving an inactive apodecarboxylase preparation. The apo-enzyme is activated by pyridoxal phosphate and the rate of decarboxylation is proportional over a limited range to the concentration of pyridoxal phosphate.

Several amino acids can be determined after dehydrogenation by measuring disappearance of NADH. In some cases, the first step involves transamination. Thus, L-alanine is converted to pyruvate by glutamate-pyruvate transaminase and α-oxoglutarate. Similar assay procedures are available for the determination of aspartate and glutamate. With threonine, the amino acid is first oxidized by periodate at neutral pH to acetaldehyde and glyoxylate, the latter undergoing a relatively rapid oxidation to formate and carbon dioxide. After destruction of excess periodate with a mercaptan, the acetaldehyde is determined with alcohol dehydrogenase and reduced NADH.

Miscellaneous Compounds

Urease catalyzes the hydrolysis of urea to carbon dioxide and ammonia.

$$O=C\begin{array}{c} \diagup NH_2 \\ \diagdown NH_2 \end{array} + H_2O \rightarrow CO_2 + 2NH_3$$

Both reaction products can be used to determine the urea content of a sample. The carbon dioxide can be determined gasometrically or color-imetrically; the ammonia can be determined directly after distillation into a receiver, or by titration after diffusion. The ammonia can also be determined colorimetrically with Nesslers' reagent after isolation by distillation or by ion exchange chromatography.

A reaction scheme for the analysis of triglycerides using enzymatic hydrolysis was described by Higgins (1984):

triglyceride $\xrightarrow{\text{lipase}}$ glycerol + free fatty acids

glycerol + ATP $\xrightarrow{\text{glycerol kinase}}$ α-glycerol phosphate + ADP

α-glycerol phosphate + NAD $\xrightarrow[\text{dehydrogenase}]{\alpha\text{-glycerol phosphate}}$ dihydroxyacetone
phosphate + NADH

NADH + indicator $\longrightarrow$ reduced indicator (colored) + NAD

Several enzymatic methods are available for the determination of lipids. The ultraviolet absorbance of conjugated diene hydroperoxides, arising from the lipoxidase-catalyzed oxidation of cis polyunsaturated fatty acids can be used to measure the concentration of the enzyme in the presence of excess substrate (Holman 1955). Alternatively, substrate concentration can be determined by oxidation in the presence of excess enzyme. All polyunsaturated fatty acids whose double bonds are separated by cis-methylene groups (linoleic, linolenic, and arachidonic) react quantitatively. The extinction coefficients of the diene hydroperoxides at 234 nm are the same for all the polyunsaturated fatty acids thus far examined. The method cannot differentiate the individual polyunsaturated fatty acids as only one mole of diene hydroperoxide is formed for each mole of fatty acid, irrespective of the number of double bonds. The method is sensitive and allows the determination of as little as 5 μg of linoleic acid. It has been used in foods, agricultural products, and biological materials (MacGee 1959). Fatty acid esters must be saponified before determination of the conjugated double bonds.

The determination of lecithin (e.g., as a measure of the egg content of foods) can be made by catalyzing the following reaction with lecithinase (phospholipase) D:

lecithin + $H_2O \rightarrow$ phosphatidic acid + choline

Lecithin and phosphatidic acid are soluble in ether, choline in aqueous solutions. After the enzymatic reaction, phosphatidic acid can be separated from the choline by extraction with ether, and the choline can be precipitated from the aqueous phase as a reineckate derivative. The red-violet reineckate is soluble in acetone and the amount is measured colorimetrically at 520 nm (Griffin and Casson 1961).

Several methods are available for the determination of inorganic compounds. Hydrogen peroxide is decomposed by peroxidase. The oxygen liberated in this process oxidizes a colorless hydrogen donor to a colored compound. o-Dianisidine (or other aromatic amines) can be used as a hydrogen donor and yields a red-brown dye with a broad absorption maximum around 460 nm. Nitrate reductase from certain strains of *Escherichia coli* catalyzes the reduction of nitrate to nitrite in the presence of a hydrogen donor (e.g., formic acid) and a hydrogen acceptor (e.g., methylene blue). The nitrite formed can be measured colorimetrically.

Magnesium ions activate isocitric dehydrogenase. The enzyme catalyzes the following reaction:

$$\text{isocitrate} + \text{NADP} \xrightarrow{\text{Mg}^{2+}} \text{NADPH} + \text{H}^+ + \alpha\text{-oxoglutarate} + CO_2$$

With constant amounts of enzyme, the rate is dependent on the magnesium ion concentration.

All inorganic pyrophosphatases so far described require magnesium ion for their action. The enzyme catalyzes the formation of orthophosphate, the concentration of which can be determined colorimetrically.

DETERMINATION OF ENZYMATIC ACTIVITY

Assays of enzymatic activity in foods are conducted to determine the soundness of foods, follow changes in heat processing, and establish damage. The common techniques for quantitatively following enzyme reactions include spectrophotometric, manometric, electrometric, polarimetric, chromatographic, and chemical procedures. The details of such assay methods have been described in reference books, textbooks, and manufacturers' catalogs, and will not be discussed here. This section is limited to a general outline of methods used in food analysis and a review of the principles involved in individual assays.

A general classifaction of enzymes is presented in Table 30.1. An *enzyme unit* is defined as the amount of enzyme required to modify under

Table 30.1.   Classification of Enzymes[a]

Type of enzyme	Examples	Remarks
Hydrolyzing		
Proteases, peptidases	Pepsin, rennin, trypsin, bromelin, papain, ficin	Hydrolyze —CO—NH— links in proteins, peptides, etc.
Carbohydrases	α- and β-amylases, amyloglucosidase, oligoglucosidase, invertase, maltase, cellulase	Hydrolyze glycosidic links
Esterases	Lipases, cholinesterases, pectin-esterases, phytase, phosphatases, tannase	Hydrolyze ester links in fats, alcohol esters, phosphoric esters, sulfuric esters, thioesters, and phenolic esters
Other	Arginase, urease, deaminase	
Adding		
Hydrolases	Aconitase, enolase, fumarase, glyoxalase, serine diaminase	Add or remove water
Carboxylases, thiamine pyrophosphokinase	Amino acid decarboxylase, carboxylase (in yeast), malic decarboxylase, oxaloacetic decarboxylase, pyruvic oxidase	Add or remove CO_2
Other	Aldolase, aspartase	
Transferring		
Oxidoreductases	D- and L-amino acid oxidases, 1-amino acid dehydrogenases, L-glutamate dehydrogenases, proline oxidase, Co II cytochrome reductase, alcohol dehydrogenase, glycerol dehydrogenase, glucose oxidase, aldehyde dehydrogenase, aldehyde oxidase, succinate dehydrogenase, cystine reductase, oxalate oxidase, cytochrome oxidases, peroxidases, catalases	Transfer hydrogen
Transaminases	Transaminases, glutamate transaminase, glycine transamidinase, γ-glutamyl transpeptidase, γ-glutamyl transferase	Transfer nitrogenous groups
Phosphotransferase	Hexokinase, glucokinase, ketohexokinase, fructokinase, ribokinase, thiamine pyrophosphokinase phosphoglucomutase (intramolecular transfer)	Transfer phosphate groups

Table 30.1. (*Continued*)

Type of enzyme	Examples	Remarks
Transacylases	Amino acid transacetylase, glucosamine transacetylase, choline transacylase, glyoxylate transacetase	Transfer acyl groups
Transglycosylases	Dextrin transglucosylase, maltose transglucosylase, sucrose transglucosylase, hyaluronidase (animal)	Transfer glycosyl groups
CoA transferase	CoA transferase	Transfer coenzyme A
Transmethylases	Nicotinamide transmethylase guanidinoacetate transmethylase, betaine-methionine transmethylase, etc.	Transfer methyl groups
Other	Transaldolase, transketolase, thiaminase, transoxidase, etc.	
Isomerizing		
Isomerases (stereoisomerases)	Alanine racemase, glutamate racemase, lactate racemase, mutarotase	
Miscellaneous		
Synthetases	Glutamine synthetase, asparagine synthetase, tryptophan hydroxamate synthetase	Catalyze synthetic reactions
Dehydrases	Fumarase, aconitase, D- and L-serine dehydrase, dihydroxy acid dehydrase, phosphopyruvate carboxylase, aldolase, citrase, isocitrase	Add groups to double bonds

a Adapted from de Becza (1965).

standardized conditions 1 μmole of substrate (or in the case of polysaccharides and proteins 1 μequivalent) per minute. The assay should be carried out—if possible—at 25°C, at optimum and defined pH, and the substrate concentration should be high enough to saturate the enzyme. Measurements should be made in the range of initial velocity, so that the kinetics follow a zero-order reaction and substrate modifications are related in a linear manner to enzyme concentration. This requires short reaction times. In case of lengthy measurements, the reaction velocity is extrapolated to zero time. Concentrations of enzymes should be expressed in units per cubic centimeter, and *specific activities* in units per milligram protein.

Methods of assaying proteinase activity include the determination of changes in the modified substrate or assay of fractions split off from the substrate. The first category includes the determination of rheological parameters (viscosity, dough consistency, coagulation, solubility, or light dispersion). In the split-off fraction method, the carboxylic or amino groups can be titrated, or free amino acids can be determined by specific or nonspecific (ninhydrin) reactions. Soluble proteins, not precipitated with trichloracetic acid, are measured before and after protease action, and the difference gives an estimate of enzymatically modified substrate. Proteolytic enzymes can cause milk coagulation; this property is particularly useful in testing preparations used in milk processing. Finally, chromogenic components of synthetic substrates are cleaved by certain proteolytic enzymes and measured colorimetrically.

Peptidases vary in their specificity. Their activity is assayed by microtitration or colorimetry of amino acids. The activity of certain peptidases, i.e. carboxypeptidases, can be assayed accurately by colorimetrically determining β-naphthol released from a synthetic substrate (naphthoxycarboxylphenylalanine). Urease activity, per se, is generally of little importance in food processing. It is used as an index of heat treatment of soy flour toasted to improve palatability and nutritional value. Enzymatic hydrolysis of pectins involves several enzymes varying in their effects on viscosity and cleavage of specific groups, which can be determined manometrically or titrimetrically.

Numerous methods are available to measure the enzymes that hydrolyze starch. Most methods for the saccharogenic β-amylase measure the amount of maltose formed. They are used widely in the brewing and malt industries. Determinations of the starch-liquefying α-amylase are made by viscometric, dextrinogenic nephelometric, or colorimetric, assays. Viscometric methods are discussed in Chapter 28. In determinations of dextrinogenic activity, the rate of dextrinization of a soluble starch solution by an extract of the material under test is measured. The time required to reach a stage at which the color produced on the addition of an iodine solution is reddish-brown instead of the blue associated with undegraded starch is determined. The determination of α-amylase activity is useful in detecting sprouted grain and in the control of malt production and supplementation. In several European countries, honey is tested for diastatic activity. Honey in which diastase was destroyed by heating is considered adulterated.

The phosphatases of milk are inactivated within the temperature range used for milk pasteurization. They are, therefore, commonly assayed to determine efficiency of pasteurization of milk and milk products. In the commonly used assay, free phenol hydrolyzed from phenyl phosphate is determined with 2,6-dibromoquinonechlorimide by a spectrophotometric procedure.

Fat acidity is often used as an index of soundness of stored grain. The acidity increases in stored grain especially if the cereals are stored at high moisture and elevated temperatures. Lipase activity in such grains is generally determined titrimetrically.

Testing for the inactivation of peroxidases in the fruit and vegetable industry is routinely done to determine efficiency of blanching. Milk peroxidase is inactivated by 81–83°C. The test is used to distinguish between milk heated by various processes. Sound milk contains insignificant amounts of catalase, which can be determined by manometric and electrometric methods. High catalase levels indicate milk from sick cows, presence of colostrum, or bacterial contamination. Bacterial dehydrogenases are assayed to determine the hygienic status of milk.

Lakon (1942) studied enzymatic reduction of several tetrazolium salts and found 2,3,5-triphenyltetrazolium chloride the best indicator of seed viability. A review of enzyme reactions related to seed viability was published by Linko (1960).

Enzyme methods to assess marine food quality were described by Jahns and Rand (1977). They are based on the formation of hypoxanthine (a metabolic byproduct of ATP breakdown), its conversion to uric acid by diamine oxidase, and development of diamine compounds. Rebhein (1979) compared the activities of mitochondrial and lysosomal enzymes (fumarase, glutamate dehydrogenase, and lactate dehydrogenase) in press juices of fresh and frozen and thawed fish fillets. Enzyme activity increased considerably and could be used to detect freezing and thawing.

OTHER APPLICATIONS OF ENZYMES

The discussion so far has shown how enzymes are used to determine specific substrates and how the measurement of enzymic activity can be used for quality control in many food systems. Other uses of enzymes of interest to food analysts are discussed in this section.

In Food Processing

Many enzymes are used in a variety of food processing systems to alleviate a processing difficulty or improve the product. Some of the common applications of enzymes in food processing are described in Table 30.2.

Determination of Pesticides

An inhibitor is a compound that causes a decrease in the rate of an enzymatic reaction by reacting with the enzyme or substrate to form a complex. In general, the initial rate of an enzymatic reaction decreases

Table 30.2. Commercial Applications of Enzymes in Food Production[a]

Processing difficulty or requirement	Enzyme function	Enzyme used	Enzyme source
Milling and baking			
High dough viscosity	Catalyzes hydrolysis of starch to smaller carbohydrates by liquefaction, thus reducing dough viscosity	Amylase	Fungal
Slow rate of fermentation	Accelerates process	Amylase	Fungal
Low level of sugars resulting in poor taste, poor crusts, and poor toasting characteristics	Converts starch to simple sugars such as dextrose, glucose, and maltose by saccharification, thus increasing sugar levels	Amylase	Bacterial
Staling of bread	Enables bread to retain freshness and softness longer	Amylase	Bacterial
Mixing time too long for optimum gas retention of doughs	Reduces mixing time and makes doughs more pliable by hydrolyzing gluten	Protease	Fungal
Curling of sheeted dough for soda crackers as dough enters continuous cracker ovens	Prevents curling of sheeted dough	Protease	Fungal
Poor bread flavor	Aids in flavor development	Lipoxidase Protease	Soy flour Fungal
Off-color flour	Bleaches natural flour pigments and lightens white bread crumbs	Lipoxidase	Soy flour
Low loaf volume and coarse texture	Hydrolyzes pentosans	Pentosanase	Fungal
Meats			
High fat content	Aids in removal or reduction of fat content	Lipase	Fungal
Upgrade meat	Hydrolyzes muscle protein and collagen to give more tender meat	Protease	Ficin (figs) Papain (papaya)
Serum separation of fat in meat and poultry products	Produces liquid meat products and prevents serum separation of fat in meat products and animal foods	Protease	Fungal
High viscosity of condensed fish solubles	Reduces viscosity while permitting solids levels over 50% without gel formation of the condensed fish solubles	Protease	Fungal

Table 30.2. (*Continued*)

Processing difficulty or requirement	Enzyme function	Enzyme used	Enzyme source
Protein shortage	Prepare fish protein concentrate	Protease	Fungal, bacterial
Distilled Beverages			
Thick mash	Thins mash and accelerates saccharification	Amylase	Bacterial, malt
Chill haze	Chillproofs beer	Protease	Fungal, bacterial, papain
Low runoff of wort	Assists in physical disintegration of resin and improves runoff of wort	Amylase	Fungal Malt
Fruit Products and Wines			
Apple juice haze	Clarifies apple juice	Pectinase	Fungal
High viscosity due to pectin	Reduces viscosity by hydrolyzing the pectin	Pectinase	Fungal
Slow filtration rates of wines and juices	Accelerates rate of filtration	Pectinase	Fungal
Low juice yield	Facilitates separation of juice from the fruit, thus increasing yield	Pectinase	Fungal
Gelled purée or fruit concentrate	Prevents pectin gel formations and breaks gels	Pectinase	Fungal
Poor color of grape juice	Improves color extraction from grape skins	Pectinase	Fungal
Fruit wastes	Produces fermentable sugars from apple and grape pomace	Cellulase	Fungal
Sediment in finished product	Helps prevent precipitation and improve clarity	Pectinase	Fungal
Syrups and Candies			
Controlled level of dextrose, maltose, and higher saccharides	Controls ratios of dextrose, maltose, and higher saccharides	Amylase	Bacterial, fungal, malt
High-viscosity syrups	Reduces viscosity	Amylase	Bacterial, fungal
Sugar loss in scrap candy	Facilitates sugar recovery from scrap candy by liquefaction of starch content	Amylase	Bacterial
Filterability of vanilla extracts	Improves filterability of vanilla extracts	Cellulase	Fungal
Miscellaneous			
Poor flavors in cheese and milk	Improves characteristic flavors in milk and cheese	Lipase	Fungal

(*continued*)

Table 30.2. (*Continued*)

Processing difficulty or requirement	Enzyme function	Enzyme used	Enzyme source
Tough cooked vegetables and fruits	Tenderize fruits and vegetables prior to cooking	Cellulase	Fungal
Inefficient degermination of corn	Produces efficient degermination of corn	Cellulase	Fungal
High set times in gelatins	Reduces set times of gelatin without significantly altering gel strength	Protease	Fungal
Starchy taste of sweet potato flakes	Increase conversion of sweet potato starch	Amylase	Fungal, bacterial
High viscosity of precooked cereals	Reduces viscosity and allows processing of precooked cereals at higher solid levels	Amylase	Fungal, bacterial

[a] From: Pulley (1969).

linearly with increasing inhibitor concentration at low inhibitor concentrations, and will gradually approach zero at higher concentrations. Several analytical procedures for pesticides have been proposed based on their inhibition of enzymes.

Organophosphorus insecticides inhibit the cholinesterase of animals and insects. This accounts to a large extent both for their effectiveness in the control of harmful insects in agriculture and for the toxicity of their residues in foods to warm-blooded animals. Many of these insecticides are such powerful inhibitors of cholinesterase that very sensitive enzymatic methods for their determination have been developed utilizing this inhibitory property (Augustinson 1959).

Enzymatic methods for the determination of organophosphorus insecticides can be classified as (1) electrometric, (2) titrimetric, (3) manometric, or (4) colorimetric. Use of electrometric methods in studies of enzymatic reactions was reviewed by Blaedel and Hicks (1964). Guilbault (1966) describes methods based on measurement of the changes in potential that occur during enzymatic hydrolysis of a thiocholine ester by cholinesterase. Organophosphorus compounds inhibit the enzyme and may be determined in 10^{-9} g concentrations by the technique. Some colorimetric methods use chromogenic substrates that form colored products on hydrolysis with cholinesterase or related esterases. With constant substrate concentration, the color intensity depends on the enzymatic activity.

Giang and Hall (1951) described a method of determining insecticide content based on changes in pH. The sample is extracted with an organic solvent, the solvent is evaporated, and the residue is incubated for 30 min with a known excess of standardized cholinesterase in a buffered solution. At the end of the inhibition period, a known excess of acetylcholine is added to the reaction mixture. After 60 min, the acetic acid produced by the hydrolysis of acetylcholine is measured by the change in pH. The less cholinesterase is inhibited by the insecticide, the more acetic acid will be formed in the reaction.

Very small amounts of DDT [di-(p-chlorophenyl)trichloroethane] can be measured accurately by its inhibitory effect on carbonic anhydrase. Methods of measuring carbonic anhydrase activity were reviewed by Davis (1963). The enzyme is inhibited by DDT at concentrations at which other inhibitors, with the exception of sulfonamides, are inactive. As the sample to be examined usually does not contain sulfonamides, special extraction and purification is not necessary.

Enzymes as Analytical Aids

Exogenous and endogenous enzymes have been employed to a limited extent to liberate proteins from tissues. With commercial availability of more purified enzymes, the lipases and carbohydrases should have increasing use in the liberation and purification of proteins especially from plants and bacteria.

The ability of proteolytic enzymes to hydrolyze the peptide bonds formed by specific amino acids gives these enzymes several advantages over acids as hydrolytic agents. Among these are high yields of peptides (or amino acids), little nonhydrolytic alteration of the products, and the requirement of only catalytic amounts of the enzyme. The only limitation of enzymatic hydrolysis is the production of artifacts through transpeptidation (Sela and Katchalski 1959). Generally proteolytic enzymes have a broad substrate specificity, but none are known which will hydrolyze all of the types of peptide bonds found in proteins (Hill 1965). The *Streptomyces griseus* proteinase (marketed as Pronase), papain, and the subtilisins extensively hydrolyze most proteins with the liberation of free amino acids, but each also leaves some peptide bonds intact. For total enzymatic hydrolysis of protein, it is necessary to employ mixtures of enzymes with several different specificities.

Complete enzymatic hydrolysis has several advantages over acid hydrolytic procedures. Acid-labile amino acids such as asparagine, glutamine, tryptophan, and the phospho- and sulfoesters of certain amino acids are not destroyed. The amino acids serine and threonine, which are destroyed partially by acid, as well as those released incompletely by acid hydrolysis, should be present in theoretical yields in enzymatic hydrolysates.

Acid hydrolysis cleaves amide groups and does not permit the direct determination of glutamine and asparagine, although the amount of both can be approximated from the amount of ammonia in the acid hydrolysate. The amount of ammonia is, however, affected by the conditions of acid hydrolysis, and especially by the destruction of serine and threonine. The amide content can be determined directly by using enzyme systems that hydrolyze peptide bonds in proteins but do not affect amide nitrogen. Tower et al. (1962) employed a whole pancreas preparation for enzymatic hydrolysis of proteins and peptides. The residues released into the hydrolysates were determined by assay with Clostridium perfringens L-glutamic acid decarboxylase, L-glutaminase, and L-aspartic acid decarboxylase, and with guinea pig serum L-asparaginase. Lin et al. (1964) hydrolyzed proteins with hydrochloric acid or Pronase (Nomoto et al. 1960A,B). Amino acids in both hydrolyzates were separated by two-dimensional paper chromatography and compared.

In microbiological assays, the vitamin must be extracted first from the sample in water-soluble form and in a state utilizable by the test organism. When the vitamin is stable to acid or alkali at high temperatures, chemical hydrolysis is used. If the vitamins are destroyed by chemical treatment, enzymatic digestion may be used. Enzymatic digestion is essential in the assay of folic acid and pantothenic acid by microbiological assay, and when the concentrations of the various members of the B-complex (thiamine, riboflavin, niacin, biotin, and inositol) are to be determined on the same sample.

A concentrated pectinase enzyme derived from Aspergillus niger is used in establishing the relationship between refractometer readings, specific gravity, and total solids for tomato pulp and paste. The pectinases, amylases, and proteases are useful in the isolation of filth and extraneous matter from food.

Enzymes in Structural Studies

Because of their specificity, enzymes are useful in various types of structural studies. For example, lipases from various sources differ in the site on the triglyceride or phospholipid molecule that they attack. Thus, pancreatic lipase shows a preferential hydrolysis of the 1 and 3 positions of a triglyceride molecule, the main monoglyceride formed during hydrolysis having the 2 configuration. Enzymes are also available that distinguish between fatty acids varying in saturation, in chain length, and in stereospecificity (Alford and Smith 1965).

Since the fundamental work of Sanger on the structure of insulin (Sanger 1952, 1956), an increasing amount of research has been undertaken to determine the amino acid sequences in proteins and peptides. It is useful to degrade the proteins with proteolytic enzymes, since the specificity of some enzymes permits the isolation of uniform fragments. The "spectrum" of the peptides obtained in this way is generally characteristic of a protein. The enzymic digests can be separated by ion-exchange column chromatography or characterized by "fingerprinting" (two-dimensional separation of paper; in one direction by conventional paper chromatography and in the other by high-voltage electrophoresis). The high selectivity of trypsin for specific peptide linkages makes this enzyme particularly useful. However, other enzymes (chymotrypsin, pepsin, and Pronase) have been found useful in some studies.

Enzymatic hydrolysis may be useful in elucidating the nature of the bonds involved in linkages between proteins and prosthetic groups, certain types of inhibitors, and coenzymes. Because of the specificity of most proteinases for bonds formed by amino acids of the L configuration, enzymatic hydrolysis provides a means for determining the stereochemical homogeneity of polypeptides and proteins (Hill 1965). Enzymes also can be used to specifically modify biologically active substances in such a way that valuable information relating structure to function is obtained. Enzymatic methods for structural analysis of complex carbohydrates were reviewed by Li (1979).

Determination of Nutritional Availability of Amino Acids

In vitro enzymatic studies have demonstrated that amino acid availability and the amino acid content of food may differ markedly. Comparative measurements of the enzymatic and chemical liberation of amino acids have been made by numerous investigators and were reviewed by Grau and Carroll (1958), Mauron (1961), and Morrison and Rao (1966).

Enzymic *in vitro* analytical procedures were developed by Scheffner *et al.* (1956) who determined the pattern of essential amino acids released by pepsin. The work involved is quite considerable, as ten amino acids have to be determined both in an acid hydrolysate and in the pepsin digest. The results showed good agreement with biological values obtained by feeding rats. The large amount of work involved limits the method to reasearch investigations and precludes its use in large-scale routine determinations.

By confining the evaluation to the release, by pepsin and pancreatin, of the key amino acids tryptophan, methionine, and lysine, which are most likely to be limiting factors in foods, Mauron *et al.* (1955) substantially reduced the work involved. Lysine was determined enzymatically by lysine decarboxylase, and methionine and tryptophan colorimetrically. The procedure was used in the quality control of heat-processed milk and was in good agreement with protein efficiency ratios measured on growing rats (Mauron and Mottu 1958). Faithfull (1984) reviewed the origin and development of *in vitro* techniques for the determination of apparent digestibility of ruminant feedstuffs. Special attention was given to the effect of the pH used for the pepsin digestion stage, its optimum value, and its solubilizing effect on various mineral components and protein–mineral–tannin components in the ruminant digestive system.

AUTOMATED METHODS

Automated methods for the determination of enzyme activity were reviewed by Schwartz and Bodansky (1963), Mason (1965), Frings (1966), and Guilbault (1976). The assay of many enzyme activities utilizes the measurement of 340 nm of the oxidation or reduction of nucleotides. During the initial stages of the enzyme reaction, which are of zero order, the change in absorbance per minute is a measure of the reaction velocity. Many of the experimental difficulties of using enzymes in analysis could be eliminated or lessened by the use of automation. A number of commercially available instruments have been used for automating the preparation of the reaction mixture in various enzyme activity assays. In many of the new instruments, assay results (i.e., absorbance) are transformed automatically by means of computers into units of enzyme activity, or various other parameters based on such activity.

A variety of automatic procedures are available for generating substrate, activator, inhibitor, or pH gradients for the characterization of enzymes. They have a wide range of applicability in enzyme research. One example is the activation of alkaline phosphatase by magnesium. The enzyme is sampled continuously and is mixed with substrate. A constantly increasing gradient of magnesium is prepared by means of addition of concentrated magnesium at a fixed rate. This gradient is then added to the sample–substrate reaction mixture. The resultant stream is incubated in a constant-temperature heating bath for a fixed time, followed by color development of the reaction product. The rate of reaction is measured by the increase in color intensity. Optimal activation concentration is reached when a plateau appears on the strip chart record.

In a number of methods, the natural fluorescence of NADH, when excited at 340 nm, provides an end point for the measurement of enzyme activity. A fluorimeter was designed to handle up to 80 samples per hour. It has a response time of less than $\frac{1}{2}$ sec while maintaining excellent stability from transient noise. The method is made more specific by dialyzing NADH, thereby eliminating background interferences.

In many studies a mixture of enzymes is fractionated by column chromatography. A small part of the effluent is continuously monitored and assayed; the bulk of the effluent is diverted to a fraction collector for the isolation and characterization of the enzymes. The system permits accurate, reproducible, and automatic separation, detection, quantitation, and collection of enzyme concentrates within a short time. At least 40 specific enzyme assay methods have been automated. Any of these methods can be used to monitor the effluent from a protein-separating column. Often enzymatic assays are made along with total protein determinations by the Lowry modification of the Lowry–Ciocalteau color reaction. In this way, the enzymatic activity per unit protein can be assayed.

Automatic enzyme assay methods are used widely in clinical laboratories. Many of the procedures are finding increasing applications in the food industry. The assays include the determination of glucose by glucose oxidase; assay of free lysine based on the continuous colorimetric determination of carbon dioxide formed during the enzymatic decarboxylation of lysine by L-lysine decarboxylase; determination of cholinesterase activity for the estimation of organic phosphate pesticide residues; and the determination of ethanol based on its oxidation in the presence of alcohol dehydrogenase and NAD.

Automated enzyme assays are used for quality control in the food industry. Feedback devices, based on enzyme determinations, make it possible to maintain a uniform product undergoing heat treatment that results in enzyme inactivation (e.g., in milk pasteurization, heat treatment in processing of fruits and vegetables, or toasting of soyflour).

BIBLIOGRAPHY

Alford, J. A., and Smith, J. L. (1965). Production of microbial lipases for the study of triglyceride structure. *J. Am. Oil Chem. Soc.* **42**, 1038–1040.

Amador, E., and Wacker, W. E. C. (1965). Enzymatic methods for diagnosis. *Methods Biochem. Anal.* **13**, 265–356.

Augustinson, Klas-Bertil. (1959). Assay methods of cholinesterases. *Methods Biochem. Anal.* **5**, 1–63.

Bardeletti, G., and Coulet, P. R. (1984). Air-dried enzyme electrodes. *Anal. Chem.* **56**, 591–593.

Bergmeyer, H. U. (1983). "Methods of Enzymatic Analysis." Academic Press, New York.

Bergmeyer, H. U., and Gawehn, K. (1978). "Principles of Enzymatic Analysis." Verlag Chemie, Weinheim, W. Germany.

Beutler, H.-O. (1978). Enzymatic determination of starch in foods by the hexokinase method (in German). *Starch* **30**, 309–312.

Boehringer, Mannheim. (1980). "Methods of Enzymatic Analysis Manual." Boehringer Mannheim Biochemicals, Indianapolis, IN.

Bowers, L. D., and Carr, P. W. (1980). Immobilized enzymes in analytical chemistry. *In* "Immobilized Enzymes" (A. Fiechter, ed.), pp. 90–128. Springer Verlag, Berlin.

Carr, P. W., and Bowers, L. D. (1980). "Immobilized Enzymes in Analytical and Clinical Chemistry." Wiley, New York.

Davis, R. P. (1963). The measurement of carbonic anhydrase activity. *Methods Biochem. Anal.* **11**, 307–327.

DeBecze, G. I. (1965). Enzymes—industrial. *In* "Kirk-Othmer Encyclopedia of Chemical Technology," Vol. 8. Wiley, New York.

Dickerman, H. W., and Carter, M. L. (1962). A spectrophotometric method for the determination of lysine utilizing bacterial lysine decarboxylase. *Anal. Biochem.* **3**, 195–205.

Ettel, W. (1981). A new enzymatic method for starch determination in food. *Alimenta* **20**(1), 7–11.

Faithfull, N. T. (1984). The *in vitro* digestibility of feedstuffs—a century of ferment. *J. Sci. Food Agric.* **35**, 819–826.

Frank, J. F., and Christen, G. L. (1984). Determination of lactose and sucrose contents of ice cream mix via enzymatic-cryoscopic methodology. *J. Food Sci.* **49**, 1332–1334.

Free, A. H. (1963). Enzymatic determination of glucose. *Adv. Clin. Chem.* **6**, 67–96.

Frings, C. S. (1966). "Analytical Applications of Enzymes." Ph.D. Thesis, Purdue Univ., Lafayette, IN.

Giang, P. A., and Hall, S. A. (1951). Enzymic determination of organic phosphorus insecticides. *Anal. Chem.* **23**, 1830–1834.

Grau, C. R., and Carroll, R. W. (1958). Evaluation of protein quality. *In* "Processed Plant Protein Foodstuffs" (A. M. Altschul, ed.). Academic Press, New York.

Griffin, F. J., and Casson, C. B. (1961). Enzymic hydrolysis of phospholipids as a means of determining egg in foods. *Analyst* **86**, 544.

Guilbault, G. G. (1970). "Enzymatic Methods of Analysis." Pergamon Press, Oxford.

Guilbault, G. G. (1976). "Handbook of Enzymatic Methods of Analysis." Marcel Dekker, New York.

Haissig, B. E., and Dickson, R. E. (1979). Starch measurement in plant tissue using enzymatic hydrolysis. *Physiol. Plant.* **47**, 151–157.

Halliwell, G. (1957A). Cellulolysis by rumen microorganisms. *J. Gen. Microbiol.* **17**, 153–165.

Halliwell, G. (1957B). Cellulolytic preparations from microorganisms of the rumen and from *Myrothecium verrucaria*. *J. Gen. Microbiol.* **17**, 166–183.

Halliwell, G. (1958). Microdetermination of cellulose in studies with cellulase. *Biochem. J.* **68**, 605–610.

Halliwell, G. (1959). The enzymic decomposition of cellulose. *Nutr. Abstr. Rev.* **29**, 747–759.

Halliwell, G. (1960). Microdetermination of carbohydrates and proteins. *Biochem. J.* **74**, 457–462.

Higgins, T. (1984). Evaluation of a colorimetric triglyceride method on the KDA analyzer. *J. Clin. Lab. Automation* **4**, 162–165.

Hill, R. L. (1965). Hydrolysis of proteins. *Adv. Protein Chem.* **20**, 37–107.

Holman, R. T. (1955). Measurement of lipoxidase activity. *Methods Biochem. Anal.* **2**, 113–119.

Holz, F. (1977). Automated enzymatic-photometric determination of starch in cereals (in German). *Landwirtsch. Forsch.* **33**(II), 228–249.

Jahns, F. D., and Rand, A. G., Jr. (1977). Enzyme methods to assess marine food quality. *In* "Enzymes in Food and Beverage Processing," (R. L. Ory and A. J. St.Angelo, eds.). ACS Symposium No. 47. American Chemical Society, Washington, DC.

Krueger, E., and Nordmann, A. (1981). Enzymic food analyses (in German). *Labor Praxis* **5**(1/2), 20–23.

Lakon, G. (1942). Topographical detection of viability of cereal seeds with tetrazolium salts. *Ber. Deut. Bot. Ges.* **60**, 299–305.

Lee, E. Y. C., and Whelan, W. J. (1966). Enzymic methods for the microdetermination of glycogen and amylopectin and their unit-chain lengths. *Arch. Biochem. Biophys.* **116**, 162–167.

Li, Y.-T. (1979). Enzymatic methods for structural analysis of complex carbohydrates. *In* "Glycoconjugate Research," Vol. 1, pp. 3–15. Academic Press, New York.

Lin, F. M., Pomeranz, Y., and Shellenberger, J. A. (1964). Determination of protein-bound glutamine and asparagine. *Proc. 6th Intern. Congr. Biochem.* **2**, 117.

Lind, J. (1977). Enzymatic determination of starch (in German). *Fleischwirtschaft* **57**, 1496–1498, 1501.

Linko, P. (1960). The biochemistry of grain storage. *Cereal Sci. Today* **10**, 302–306.

MacGee, J. (1959). Enzymic determination of polyunsaturated fatty acid. *Anal. Chem.* **31**, 298.

Mason, W. B. (1965). Bioanalytical techniques. Rep. 18th Annual Analytical Chemistry Summer Symposium. *Anal. Chem.* **37**, 1755–1758.

Mauron, J. (1961). The concept of amino acid availability and its bearing on protein evaluation. *In* "Progress in Meeting Needs of Infants and Preschool Children." Publ. **843**. National Academy of Science/National Research Council, Washington, DC.

Mauron, J., and Mottu, F. (1958). Relation between *in vitro* lysine availability and *in vivo* protein evaluation in milk powders. *Arch. Biochem. Biophys.* **77**, 312–327.

Mauron, J., Mottu, F., Bujard, E., and Egli, R. H. (1955). The availability of lysine, methionine, and tryptophan in condensed milk and milk powder. *In vitro* digestion studies. *Arch. Biochem. Biophys.* **59**, 433–451.

Mercier, C. (1981). Enzymes in food analysis. *Ernaehrung* **5**, 80–87.

Morrison, A. B., and Rao, N. (1966). Measurement of the nutritional availability of amino acids in foods. *Adv. Chem. Ser.* **57**, 159–177.

Natl. Canners Assoc. (1954). Bull. *27-L*. Berkeley, CA.

Nomoto, M. Narahashi, Y., and Murakami, M. (1960A). A proteolytic enzyme of *Streptomyces griseus*. VI. Hydrolysis of protein by *Streptomyces griseus* protease. *J. Biochem. (Tokyo)* **48**, 593–602.

Nomoto, M., Narashashi, Y., and Murakami, M. (1960B). A proteolytic enzyme of *Streptomyces griseus*. VII. Substrate specificity of *Streptomyces griseus* protease. *J. Biochem. (Tokyo)* **48**, 906–918.

Peynaud, E. Blouin, J., and Lafon-Lafourcade, Y. (1966). Review of applications of enzymatic methods to the determination of some organic acids in wines. *Am. J. Enol. Viticul.* **17**, 218–224.

Plesner, P., and Kalckar, H. M. (1956). Enzymic microdeterminations of uric acid, hypoxanthine, xanthine, and adenine, and xanthopterine by UV spectroscopy. *Methods of Biochem. Anal.* **3**, 97–109.

Pomeranz, Y. (1966). The role of enzyme additives in breadmaking. *Brot Geback* **20**, 40–45.

Pomeranz, Y. (ed.). (1971). "Wheat: Chemistry and Technology," 2nd ed. Am. Assoc. Cereal Chemists, St. Paul, MN.

Pulley, J. E. (1969). Enzymes simplify processing. *Food Eng.* **41**(2), 68–71.

Rebhein, H. (1979). Development of an enzymatic method to differentiate fresh and seafrozen and thawed fish fillets. *Z. Lebensm. Unters. Forsch.* **169**, 263–265.

Sanger, F. (1952). The arrangement of amino acids in proteins. *Adv. Protein Chem.* **7**, 2–67.

Sanger, F. (1956). The structure of insulin. *In* "Currents in Biochemical Research" (D. E. Green, ed.). Interscience Publishers, New York.

Scheffner, A. L., Eckfeld, G. A., and Spector, H. (1956). The pepsin-digest-residue (PDR) amino acid index of net protein utilization. *J. Nutr.* **60**, 105–120.

Schwartz, M. K., and Bodansky, O. (1963). Automated methods for determination of enzyme activity. *Methods Biochem. Anal.* **11**, 211–246.

Sela, M., and Katchalski, E. (1959). Biological properties of poly-α amino acids. *Adv. Protein Chem.* **14**, 391–478.

Skrede, G. (1983). An enzymic method for the determination of starch in meat products. *Food Chem.* **11**, 175–185.

Talalay, P. (1960). Enzymic analysis of steroid hormones. *Methods Biochem. Anal.* **8**, 119–143.

Tower, D. B., Peters, E. L., and Wherrett, J. R. (1962). Determination of protein-bound glutamine and asparagine. *J. Biol. Chem.* **237**, 1861–1869.

Townsend, A. (1981). Uses of enzymes in analytical chemistry. *J. Assoc. Public Anal.* **19**, 51–58.

Umbreit, W. W., Burris, R. H., and Stauffer, J. F. (1957). "Manometric Methods." Burgess Publishing Co., Minneapolis.

Updike, S. J., and Hicks, G. P. (1967). The enzyme electrode. *Nature* **214**, 986.

Valentova, O., Marek, M., Albrechtova, I., Albrecht, J., and Kas, J. (1983). Enzymic determination of glucose in foodstuffs. *J. Sci. Food Agric.* **34**, 748–754.

Wease, D. F., Anderson, Y. J., and Ducharme, D. M. (1979). Use of immobilized enzymes for glucose analysis in a small laboratory. *Clin. Chem.* **25**, 1346–1347.

Weetal, H. H. (1975). "Immobilized Enzymes, Antigen, Antibodies and Peptides." Marcel Dekker, Inc. New York.

Whitaker, J. R. (1972). "Principles of Enzymology for the Food Sciences." Marcel Dekker, New York.

Whitaker, J. R. (ed.). (1974A). "Food Related Enzymes." Adv. Chem. Ser. No. *136*. American Chemical Society, Washington, DC.

Whitaker, J. R. (1974B). Analytical applications of enzymes. *In* "Food Related Enzymes" (J. R. Whitaker, ed.). Adv. Chem. Ser. No. *136*. American Chemical Society, Washington, DC.

Wiseman, A. (1984). Enzymes in analysis of foods. *In* "Enzymes and Food Processing" (G. G. Birch, N. Blakebrough, and K. J. Parker, eds.), pp. 275–287. Applied Science Publishers, London.

Zaborsky, G. R. (1973). "Immobilized Enzymes." CRC Press, Cleveland.

31
Analytical Microbiology

Because of the similarity in nutritive requirements of microorganisms and experimental animals, it is possible to use microorganisms to determine quantitatively many of the substances that are known to be essential constituents of all living cells (Snell 1945). The requirements of certain microorganisms for specific nutritional factors reflect a loss of their ability to synthesize those factors (Snell 1946). At present, microorganisms are known that require each of the water-soluble vitamins (with the possible exception of ascorbic acid) and each of the amino acids required by higher animals. The fundamental similarity in the metabolic requirements of various organisms has facilitated greatly identification of the essential nutrients for both microorganisms and animals. Similarly, the use of microorganisms for the quantitative determination of the vitamins (Peterson and Peterson 1945; Koser 1948; Snell 1949) and the amino acids, purines, and pyrimidines (Snell 1945, 1949; Hendlin 1952) that they require is rapidly extending our knowledge of the distribution and importance of these substances in nature. The recognition that the effectiveness of antibiotic agents may sometimes be due to their interaction with essential cellular metabolites and that antibacterial agents can often be fashioned by varying the structure of known essential metabolites in a suitable manner has served further to intensify interest in microbial nutrition.

Analytical microbiology involves the use of microorganisms as reagents for the quantitative determination of certain chemical compounds (Gavin 1956), particularly vitamins, amino acids, growth factors, and antibiotics. Microbiological assays have been reviewed in several publications (Peterson and Peterson 1945; Snell 1945, 1952; Hendlin 1952; Kersey and

547

Fink 1954; Gavin 1956, 1957A,B, 1958, 1959; Hutner *et al.* 1958; Kavanagh 1960; Lwoff 1972; Nelson 1973; Friedman 1978). Methods used in microbiological assay of vitamins in clinical chemistry were reviewed by Baker and Sobotka (1962). Use of statistics in microbiological assays was discussed by Knudsen (1950) and Bliss (1956). A comprehensive book on theoretical and applied aspects of analytical microbiology was edited by Kavanagh (1963).

PRINCIPLES OF MICROBIOLOGICAL ASSAYS

The basic principle upon which a microbiological assay depends is that in the presence of limiting amounts of certain compounds, the amount of microbial growth is a function of the amount of these compounds. Although test conditions vary with the different assay methods, the fundamental procedures in most assays are the same. The test substance is added to a liquid or gel medium, the medium is inoculated with the assay microorganism, and a response is measured. The response measured depends on the biochemical effect of the substance on the metabolism of the organism. It may be a *growth response* (generally positive in the assay of nutrients and negative in the assay of antibiotics), which can be measured by numerical counts, optical density, weight, or area; the growth response may be a definite end point, or an all-or-none response. In the case of a *metabolic response* (either positive or negative), metabolic products or changes in some function may be measured. Among the measurable metabolic responses are acid production, carbon dioxide production, oxygen uptake, reduction of nitrates, hemolysis of red blood cells, antiluminescent activity, or inhibition of spore germination. Cancellieri and Morpurgo (1962) proposed an interesting method for the determination of amino acids, vitamins, and purine and pyrimidine bases. The growth of wild, protorophic strains of bacteria is inhibited by substances that are analogous and antagonistic to the regular metabolites. By adding graded amounts of the nutrilite to reactivate the growth, the nutrilite content can be assayed.

The Microorganism

The ideal test organism should (1) be sensitive to the substance being assayed; (2) be easily cultivated; (3) have some metabolic function or response that is readily measurable; and (4) not be susceptible during the assay to variation in either its sensitivity or phase. It is also desirable that the organism be nonpathogenic and have reasonable specificity. Rapid growth is often advantageous because it reduces the assay time and increases the precision. For convenience, an organism should be suitable for the assay of several compounds, although the most exacting organisms are often either less stable or more difficult to cultivate. For the examination of the amino acid content of unknowns, particularly unpurified materials, it is advantageous to introduce into assay tubes as little extraneous material as possible. Consequently, sensitivity of the organism for an assayed compound may be important.

The microorganisms used for assay include bacteria, yeasts, fungi, and protozoa. The use of bacteria generally poses fewer problems than the use of other microorganisms. They have been used to assay proteins, amino acids, carbohydrates, and vitamins, and to evaluate antiseptics, disinfectants, and chemotherapeutic agents. Lactic acid bacteria (including the genera *Lactobacillus*, *Streptococcus*, and *Leuconostoc*) equal or surpass all other groups of microorganisms in the complexity of their nutritional requirements. The complex nutritional needs of the lactic acid bacteria for one or more of the B vitamins, purines, pyrimidines, and an array of amino acids, as well as their ability to dissimilate carbohydrates to an easily measurable end point, make them exceptionally useful in microbiological assays.

Assay procedures using yeasts offer the advantages of simplicity and speed, and have been developed for some vitamins. Yeasts thrive under aerobic conditions and require the use of shakers. Fungi are not as widely used as assay organisms as bacteria. Several fungal methods have been developed for the assay of antibiotics, trace metals, and vitamins. However, because fungal growth is generally slow, fungi generally are used only when no other organism is available or when fungicidal or fungistatic evaluation is desired. Protozoa have a variety of nutritional requirements including amino acids, vitamins, nucleic acid derivatives, lipid factors, and hormones. The proteolytic enzymes elaborated by some protozoans make them useful in the assay of proteins and of the biological availability of proteins.

One of the perennial problems in any work with microorganisms is the preservation of the culture with a particular physiology. The culture must be maintained in such a manner that it will give a similar response each time it is used in an assay procedure. The simplest way to preserve a bacterial culture is in stabs or agar slants with transfer at monthly intervals. Another popular method of preservation is lyophilization. Over a long period, a culture preserved this way may differ from the material lyophilized, because differential death may reduce the distribution of organisms in the population.

The incubation times of microbiological assays range from several hours for turbidimetic assays of antibiotics to 7–8 days for assays with protozoans. In assays that involve short incubation periods, bacterial cultures in the log phase of their growth should be used; for longer incubation periods, the age may vary from 18 to 24 hr.

Similarly, the size of inoculum is more critical in short- than in long-term incubations. Inoculum size affects antibacterial potency of some antibiotics and is generally more important in diffusion than in other methods. Size of bacterial inoculum can be controlled by standardizing with a colorimeter the turbidity of the diluted inoculum.

The selected temperature for incubation should give good, but not necessarily maximum growth. It is essential, however, that the selected temperature be maintained with little variation.

The Medium

A culture medium for use in analytical procedures must have certain characteristics (Gavin 1956). First, it must contain all of the factors necessary to support growth. The test organism must be supplied with an energy source, nitrogen source, mineral salts, and, in synthetic media, growth factors. In assays of growth factors in which a positive growth response is obtained, the culture medium must be prepared so that all of the necessary nutrients are available to the test organism with the exception of one that will control growth. The limiting factor is the compound being assayed, and is supplied in graded amounts. In assays involving a negative growth response, lack of growth must be due only to the presence of an inhibitory substance in a medium containing all the nutrients essential for normal growth and development of the test organism.

Neither the medium nor the food extract should contain stimulants, substitutes, or antagonists to the assayed material. The water used should be demineralized, and the compounds used in preparing the medium should be pure.

Finally, the medium should have a pH that is compatible with the activity of the assayed substance and the growth of the microorganism.

MICROBIOLOGICAL METHODS

There are four main methods of microbiological assay by which the potency of samples and standard solutions can be compared: *diffusion*, *turbidity* or *dilution*, *gravimetric*, and *metabolic response* methods. Each of these methods is discussed in this section.

Diffusion Methods

In the diffusion method, the assayed substance is allowed to diffuse through solid inoculated culture media, and a zone of growth or inhibition of the test organism forms around the application point or area of the substance. The test organisms may be bacteria, bacteriophage, fungi, protozoa, or algae; the response zones may be of growth as in the assay for vitamins, amino acids, etc., or of inhibition as in the assay of antibiotics. The size of the zone is a function of the concentration, and in certain instances of the amount, of the assayed substance. This function can be expressed as a linear relationship between the size of the zone and the logarithm of the concentration of the substance. By measuring the distance the substance diffuses and comparing it with that of a known standard preparation, the potency of the sample may be assayed. Diffusion may be of two types: linear, which occurs when the substance is placed in contact with a column of seeded agar in a capillary or test tube, or radial, which occurs around a suitable reservoir containing the substance on a seeded agar plate. The linear diffusion method, although theoretically sound, is seldom used because of practical considerations.

In some horizontal diffusion methods, the diameter of the zone of inhibition or growth depends upon the concentration; in others, it depends upon the amount of substance being analyzed. Among those depending on concentration, the *cylinder* and *cup plate methods* are used most commonly. In the first, small cylinders (from glass, porcelain, or metal) are embedded to a fixed height in the agar; in the cup plate method, a depression is formed by removing a small object previously placed in the molten agar, after the agar has solidified, or by removing a slug from solidified agar with a sterile cork borer. The analyzed substance is placed in the cylinder or depression and diffuses; the diameter of the zone of inhibition or growth is measured and compared to a series of standard concentrations of the assayed substance. The methods depending upon the amount of material include the drop plate method and the paper disk method. The tested material is placed either directly on the agar or indirectly on the agar by means of a filter paper disk.

According to Gavin (1957A) the cylinder and cup method are well adapted to rapid assays and to testing dilute solutions; the drop and paper disk method are simple, convenient, rapid, and can be used in assaying very small samples. The accuracy and reproducibility are the same for all four methods.

Diffusion assays can be used only for substances that diffuse into agar and are generally less precise than other microbiological methods. Decreasing the thickness of the agar layer increases sensitivity; too moist agar will cause streaks and indistinct obscure zones; and the amount, quality, and origin of the agar may affect the results. The pH of the agar and assay solutions affects the activity and stability of the tested compound and assay results. Similarly, length and temperature of incubation must be rigidly standardized.

Diffusion methods are used widely in the assay of antibiotics. Their sensitivity is relatively low, and they are used less in the assay of amino acids and vitamins. In general, cup plate assays require 10 to 2000 times the vitamin levels required by conventional tube assays (Hendlin 1952; Bolinder *et al.* 1963A,B).

Turbidimetric and Dilution Methods

The distinction between dilution and turbidimetric methods is that the former gives an all-or-none end point in broth or agar, whereas the latter measures graded growth or metabolic responses.

The *serial dilution method* for assay of antibiotics is quite important for food analysts. It is simple in principle, easily performed, and of considerable utility. Several dilutions of the tested substance in small tubes are inoculated with a test organism, incubated, and the lowest concentration of the substance that causes apparently complete inhibition of growth of the organism is taken to be the minimum inhibitory concentration.

In a turbidimetric tube assay, graded concentrations of the tested substance are added to a series of test tubes or flasks containing a liquid nutrient medium. The medium is inoculated with the test organism and incubated for a suitable time. The response of the test organism is measured in a photometer, the scale readings of which may be converted by a calibration curve prepared with graded amounts of pure substance to determine the potency of the assayed sample. Turbidimetric assays are used for determining antibiotics, vitamins, amino acids, and other growth substances. A turbidimetric method is the most suitable one for assaying large numbers of samples of a known compound of approximately known concentration, which are uncontaminated with interfering substances.

There probably are fewer important factors to control in a turbidimetric than in a diffusion assay. Turbidimetric methods are more sensitive to low dilutions than diffusion methods and can be adapted to rapid assays; in addition, turbidimetric results obtained at several concentration levels can be analyzed statistically. The assayed solutions must be sterile—especially with long incubation tests; a standard curve must be prepared each time an assay is run. The presence of highly colored substances can affect the results of turbidimetric assays though such effects can be allowed for by running uninoculated blanks at each dilution level. Some organisms clump during growth, and it is difficult to suspend them uniformly for turbidimetric measurement. When this method is used, one must be certain that the organisms are uniformly suspended by shaking before measurement.

Sterilization of culture media can have adverse effects on turbidimetric assays. Heat sterilization may cause color changes, lower pH, and destroy certain nutrients due to amino acid–sugar interactions. Lactobacilli are relatively immune to interference from contaminants due to the rapidity of growth, essentially anaerobic conditions, and immediate acidification of the medium. With slow-growing organisms such as *Tetrahymena*, which require an incubation time of 6–8 days, special precautions must be taken to avoid contamination. Separate sterilization of glucose solutions or reduction of the glucose content will aid in overcoming the problem of color changes.

The limits of error reported in the literature for turbidimetric assays range from ±2 to ±20%. The majority of reports indicate that ±10% is average for this procedure, and that an experienced analyst exercising rigid control of the factors that contribute to variation can reduce the error to ±5%.

Gravimetric Methods

In gravimetric methods, the response of the test organism to graded concentrations of the analyzed substance is determined after a suitable incubation time by measurement of the amount of growth in terms of dry cell weight. Under the conditions of assay, this weight is proportional to the concentration of the limiting factor. The majority of gravimetric methods use mutants of *Neurospora* as test organisms. The methods are simple, precise, reliable, inexpensive, and can be used for assays of colored solutions. Results with some mutants are very specific.

The assays are, however, time-consuming, not well adapted to large numbers of sample, and require large incubation space. In addition, the requirements for nitrogen metabolism by the test organisms are complex and may be affected by a variety of nonspecific nitrogenous compounds. With *Neurospora* mutants, a separate strain is required for each particular assay, thus, requiring maintenance of a large number of test organisms. Occasional major changes in metabolism may occur as a result of a single-gene mutation with the resultant possibility of adaption.

Metabolic Response Methods

In a metabolic response assay, the response of the test organism to various concentrations of the assayed substance is evaluated, after a suitable incubation period, as a change in a specific measurable metabolic parameter. Several parameters can be measured; however, acid production is the only one used widely. The acid produced is determined by titration.

Rates of growth and acid production do not run exactly parallel courses. In general, growth (as measured turbidimetrically) reaches a maximum considerably before maximum acid production has occurred. The amount of acid eventually produced is, however, closely related to the number of total cells present.

Some of the precautions necessary with turbidimetric measurements are unnecessary if one titrates acid. Color of the sample, clumping, and turbidity development have little effect on titrimetric results. With some of the heterofermentative lactic acid bacteria and with *Escherichia coli*, which produce small amounts of acid, turbidimetric measurement of growth is normally used. With homofermentative lactobacilli, a titration is generally the preferred procedure.

Titrimetric methods require sterile samples and media; they also are somewhat time-consuming and tedious. The limiting buffering capacity of the medium poses problems of limits for optimum and maximum acid production. The precision is generally ± 10%, but ± 3% or lower can be attained with careful control of all variables.

DETERMINING RELIABILITY OF MICROBIAL ASSAYS

Four practices can increase the precision and reliability of many microbial assays: (1) selecting dosage levels and a function of the growth response that lead to a straight or log-dose response line; (2) randomizing the location of culture tubes within racks during sterilization, inoculation, and incubation; (3) following an objective probability rule in identifying and rejecting supposedly aberrant observations; and (4) basing assayed levels on the means of two or more independent assays.

The best and simplest criterion for establishing reliability of a microbiological assay method is agreement with other procedures. In running an assay, a standard curve is obtained in which growth response is plotted against concentration of the assayed compound. The sample is also assayed at several dilution levels selected to yield an amount of growth that falls upon the standard curve. The concentration of the assayed compound determined should be constant regardless of the portion of the standard curve from which it was calculated. Repeated assays should give consistent values.

When a known amount of a pure compound is added to a sample, it must be quantitatively recovered. High or low recoveries indicate stimulation or suppression of response; in either case, the assay value is suspect. If two or more organisms that vary in other nutritive requirements yield the same value on a sample, the probability that the assay value is correct is greatly increased. It is highly unlikely that two or more different organisms would respond to any interfering materials in exactly the same manner and to exactly the same extent. Finally, the specificity of response should always be investigated prior to use of a new medium or organism for assay.

ASSAYS OF VITAMINS, AMINO ACIDS, AND NUCLEIC ACIDS AND THEIR DERIVATIVES

In general, the essential steps involved in the assay are (1) preparation of media, carrying stock cultures and maintenance of the cultures; (2) preparation of the nutrient-deficient medium; (3) preparation of the inoculum medium and inoculum culture; (4) extraction of the nutrient from the sample prior to assay (generally, various concentrations of HCl at 15 lb in an autoclave); (5) setting up the assay; (6) sterilization of the assay tubes and media; (7) inoculation with the test organism; (8) incubation; (9) determination of response to the nutrient and nutrient-containing extract; and (10) calculation of results.

For lactobacilli assays of vitamins, amino acids, and nucleic acids and their derivatives, the selected test organism is carried as stab cultures by monthly transfer in a yeast–dextrose agar. Such cultures are incubated at 37°C for 24–48 hr or until good growth is visible, and are then held in the refrigerator for the remainder of the interval between transfers. About 24 hr before an assay is to be made, a transfer is made from this culture to a tube of sterile inoculum medium. In the assay of riboflavin, for example, this consists of the riboflavin-deficient basal medium supplemented with riboflavin (1 μg/10 cm^3). This inoculum culture is then incubated at 37°C until used.

The selected medium deficient in the nutrient to be assayed is prepared at twice its final concentration. The assay is generally carried out in 16 × 180 mm lipless test tubes held in metal racks, which are easily autoclaved. To one series of tubes, increasing amounts of a standard solution of the assayed nutrient are added. To a second series of tubes, increasing amounts of neutralized extracts of the various tested samples are added. Amounts and concentrations of the nutrient in the standard solutions and in the sample extracts are selected to produce a gradation in growth of the test organism between no growth and maximum possible growth. Contents of all tubes are diluted with water to the same volume (generally 5 cc), and an equal volume of the double-strength basic medium is added.

The tubes are then plugged with cotton or capped, sterilized by autoclaving, and cooled to room temperature; each tube is then inoculated with a drop of the washed inoculum culture.

The inoculated tubes are incubated at constant temperature near the optimum of the organism used. After a sufficiently long incubation period, growth present in each tube is determined by acidimetry or turbidimetry. The growth response to increasing concentrations of the sample nutrient is then plotted against the concentration of the nutrient in the standard solution. The amount of nutrient present in the various samples is determined by interpolating the growth response in the sample tubes onto the standard curve.

Assay Media

Snell and coworkers (Snell 1945) devised on the basis of studies of the nutritional requirements of the lactic acid bacteria many of the synthetic laboratory media used today in the assay of water-soluble vitamins, amino acids, and purines and pyrimidines. To prevent the pH from shifting to levels that interfere with growth, the media are buffered. Acetate is the buffer of choice; when citrate buffers are used it is necessary to increase the metal concentration of the medium because of the strong chelating action of citrate. Generally, a fixed mixture of minerals is used, though variation in the composition of the salt mixture may improve growth of some microorganisms.

Though the microorganisms do not require all the water-soluble vitamins and growth factors, even the nonessential nutrients often aid early growth. Thus, *L. arabinosus* grows well without vitamin B_1, but its development is greatly enhanced by pyridoxine, pyridoxamine, or pyridoxal. For most lactobacilli, adenine, guanine, and xanthine are interchangeable, but various organisms differ in the ease with which they utilize the individual compounds. The effect of thymine is shown only in a folic-free medium. Thymine plus purine bases replace folic acid more or less for certain organisms.

Any synthetic medium for microbiological assay of amino acids should contain a complete quota of water-soluble vitamins and growth factors and their physiological equivalents. In assays of vitamins, either a mixture of the essential amino acids or a hydrolysate of a complete protein is used. With the availability of pure amino acids, the nutritional requirements for individual amino acids were determined. This was done by omitting each acid in turn from a mixture that supports growth and noting the effect on growth or acid production.

Amino Acids

Microbiological assays of amino acids have certain advantages over other assay methods. They are highly specific and sensitive; they can differentiate between various forms, a differentiation that cannot be made by other means. They are admirably adapted to routine work, provided the equipment, techniques, and general background of a bacteriological laboratory is available. If a complete amino analysis is desired on a limited amount of material, on a more or less continuous basis, ion-exchange chromatography is the primary method of choice where the personnel and equipment are available. If, however, the content of one or several amino acids is desired (rather than a complete analysis), if the configuration of the amino acid is of interest, or if facilities for column chromatography are not available, properly selected and executed microbiological assays are quite simple, specific, and accurate.

Vitamins

Microbiological methods of vitamin assay can provide useful information on the total vitamin content, aid in the identification of a compound, and aid in establishing the presence of various forms of the nutritional cofactors. It should be realized, however, that the potency of the test material, determined by microbiological assay, may not be identical with the amount available for metabolism in man or animal.

In the development of microbiological assays for vitamins, the emphasis has been on extraction and hydrolysis procedures that yield maximal figures for the material being assayed. Certainly, autoclaving for 60 min at 15 lb and hydrolysis with 6 N H_2SO_4 (recommended for biotin assays) have no counterpart in animals. Microbiological assays, thus, measure *total and not available* vitamin contents.

The adequate liberation of vitamin B_6 from its multiple bound forms in biological materials is the crux of any valid method of determining this vitamin (Storvik *et al.* 1964). Vitamin B_6 exists in nature both in a "bound" and in a "free" state. The vitamin must first be liberated before accurate microbiological and chemical assays can be made. Biological assays (feeding tests) are not dependent on this step, since the test animals are able to utilize both bound and free forms of the vitamin. None of the several enzymic or chemical methods for hydrolysis or extraction is completely successful for liberation of vitamin B_6 from the multiple substances of complex nature encountered in foods and tissues. The estimation of the vitamin B_6 content of foods and biological materials is further complicated by the existence of several closely related active compounds (Snell 1958).

In biological assays, all the forms of vitamin B_6 (pyridoxine, pyridoxal, and pyridoxamine) either free or combined with other substances such as phosphate have similar growth-promoting effects. However, in microbiological assays, differences in organism response to the various forms are an important consideration. *Saccharomyces carlsbergensis* has generally been accepted as the organism of choice for the determination of total vitamin B_6. However, in view of the inherent nature of the various forms of the vitamin and of assay microorganisms, the inadequacies of extraction, the presence of growth-stimulating components in food extracts, and the interrelationships between vitamin B_6 and other components (i.e., thiamine and alanine in certain lactic acid bacteria), it is not possible to give an unqualified recommendation to any single microbiological method for the assay of vitamin B_6 in complex biological substances.

The microbiological assay of vitamin B_{12} has become almost a "science of its own" with a vast literature describing the search for suitable organisms, their specificities, sensitivities, and metabolism; the vitamin B_{12} forms and binding factors liberated during microbial growth; and the vitamin content of body fluids, organs, and foods (Skeggs 1966). Difficulties encountered in microbiological assays of folic acid resulting from nutritional interrelationships with vitamin B_{12} were reviewed by Girdwood (1960).

Microbiological assays for vitamins may measure related compounds that have similar activity for the microorganism but not for higher animals. On the other hand, microorganisms may fail to measure vitamin derivatives that are active for animals and may, therefore, give erroneously low estimates. Similarly, a food may contain antimetabolites that inhibit enzymatic systems in microbiological assays, but are rendered inactive in the animal.

In summary, although microbiological procedures can, if the proper organism is selected and the necessary precautions are used, give useful information on the vitamin content of foods, it should be realized that microbiolgical assays generally measure total rather than available vitamin content. Feeding trials with strains of some microorganisms as the experimental animal offer the advantages of speed, precision, and small requirements of labor, space, and materials. Actual feeding tests with animals are, however, the final criterion of specificity and potency. The validity of conclusions drawn from microbiological vitamin assays depends on the extent to which they give results in agreement with feeding tests.

ASSESSING THE NUTRITIONAL VALUE OF PROTEINS

There is little doubt today that the quality of a protein is primarily governed by its essential amino acid content. Consequently, much basic and useful information has been gained by development of chemical and microbiological methods of amino acid analysis.

Nevertheless, evaluation of the nutritive value of a protein from knowledge of its amino acid composition still leaves much to be desired. Two reasons are responsible for this. First, although we know the approximate amino acid needs of several species, these needs are variable according to the metabolic state of the subject. Second, amino acid content as revealed by classical amino acid analyses does not necessarily reflect amino acid availability to the organism. The basic assumption implied in all methods of chemical scoring for evaluation of protein value is that the total amount of amino acid as determined by classical methods is available to the organism. There is, however, good experimental evidence that in certain foods a proportion of the analytically determined amino acids may not be available for assimilation.

For example, Gupta et al. (1958) found that lysine availability to the weanling rat was only about 50% for corn, 70% for wheat, 85% for rice, 90–95% for spray-dried milk powder, and 68% for a roller-dried milk sample. Several workers found that lysine and, to a lesser extent, methionine and tryptophan, are made unavailable when proteins are submitted to severe heat treatment. Even under less severe conditions, such as those encountered in milk processing or milk storage, in extraction of oil seeds, and in industrial manufacture of fish flour, some of the amino acids essential and limiting in foods of plant origin may be made less available.

The accepted methods for measuring the nutritional value of proteins, involve the use of common laboratory animals—mice, rats, chicks, dogs, and others. A description of these methods is outside the scope of the book. For additional information on this subject, the interested reader may consult several excellent reviews and books (Allison 1949; Albanese 1959; Morrison and Rao 1966; Porter and Rolls 1973; White and Fletcher 1974; Friedman 1975; Bodwell 1977). Methods, nomenclature, and definitions used in such tests were described by Henry (1965). The following are some of the commonly used definitions:

biological value:

$$BV = \frac{\text{retained N} \times 100}{\text{absorbed N}}$$

true digestibility:

$$TD = \frac{\text{absorbed N} \times 100}{\text{N intake}}$$

protein efficiency ratio:

$$PER = \frac{\text{gain in body-weight}}{\text{protein intake}}$$

net protein utilization:

$$NPU = \frac{\begin{array}{c}\text{body N content with test protein} \\ - \text{ body N content with N-free diet}\end{array}}{\text{N intake}}$$

net protein ratio:

$$NPR = \frac{\begin{array}{c}\text{gain in body-weight with test protein} \\ + \text{ loss in body-weight with N-free diet}\end{array}}{\text{protein intake}}$$

Because of the many difficulties encountered in determining *in vivo* availability of individual amino acids, several authors have tried to measure amino acid availability *in vitro*. This may be done by chemical, enzymatic, and microbiological methods. The only useful chemical method developed to measure amino acid availability is that employing Sanger's reagent for determining available lysine in foods. Enzymic *in vitro* methods are discussed elsewhere in this book. The microbiological methods for assessing the nutritional value of proteins can be divided into those employing protozoans and those measuring growth of bacteria, yeasts, or fungi.

Protozoans

Some interesting studies have been made of the possibilities of using ciliated protozoa of the genus *Tetrahymena* in studies of the nutritional quality of proteins. These protozoans are known to possess proteolytic enzymes and are able to utilize intact proteins as sources of amino acids (Grau and Carroll 1958).

Studies of the nitrogen metabolism of the protozoan *T. geleii* H have shown that the amino acid requirements of the organism resemble in many respects those of higher animals. Their resemblance in requirements, rapid growth, small size, and ability to hydrolyze proteins suggested that this organism would be useful in protein evaluation. The protozoan *T. pyriformis* also has a general nutritional and metabolic pattern similar to that of higher animals. The similarities include protein quality ratings; protein–energy, protein–vitamin, and protein–amino acid relationships; and the ability to detect thermal damage to proteins and to grow on multicomponent protein mixtures such as compounded animal feeds (Rosen 1958).

Teunison (1961) used *T. pyriformis* W to evaluate the relative nutritional quality of ten protein concentrates and a mixture of amino acids. The reproducibility of the results at several nitrogen levels of the reference standard ("vitamin-free" purified casein) was established by statistical studies. Nutritional quality was estimated as an index based on the ratio of the regression coefficient of growth on the nitrogen content of the test material to the regression coefficient of growth on the nitrogen content of casein. These were estimated for the ranges 0–1.0 and 0–1.7 mg nitrogen/ml. These ranges were adopted because there was better growth at these concentrations for most of the nitrogen sources studied. Assays at higher nitrogen concentrations yielded further information on nutritional quality, particularly for diets containing mixed proteins.

The procedure used by Teunison (1961) was based on a modified procedure described by Rosen *et al.* (1962). The basal medium contains vitamins, minerals, carbon and energy sources, purines, and pyrimidines; the tested sample serves as the nitrogen source. Stott *et al.* (1963) introduced the following simplifications and modifications in experimental procedure: (1) the use of "count" to assess growth response and compute relative nutritive values; (2) elimination of routine ammonia nitrogen determination; (3) reduction of phosphate in the basal medium; and (4) use of the Fuchs–Rosenthal hemocytometer (2-mm depth). The simplified procedure was used to evaluate the relative nutritive value of proteins, the effect of thermal damage to protein, and the effects of amino acid composition and supplementation with essential and limiting amino acids. Similar results were described by Baum (1966A,B). The application of *Tetrahymena* assays to the measurement of available lysine, methionine, arginine, and histidine was described by Stott and Smith (1966).

Despite apparent advantages in simplicity and convenience, the microbiological methods involving protozoans have not been widely accepted. Ford (1960) pointed out that the procedures are unfamiliar and complex compared to usual microbiological methods and not entirely suitable as analytical tools. In a comprehensive study of methods, Bunyan and Price (1960) found difficulty in obtaining replicates using protozoans. There is no doubt, however, that good replicability can be obtained with protozoans, and that collaborative studies in which biological value of proteins and availability of amino acids are evaluated under comparable conditions could bring out the merits of the various assay procedures (Boyne *et al* 1967).

Rolle (1976) confirmed the usefulness of *T. pyriformis* W as a test organism in the evaluation of protein quality in about one hundred foods of plant and animal origin. The test organism responded to the addition of limiting amino acid(s). In an attempt to improve the determination of relative nutritive value (RNV) by the *T. pyriformis* method, Evans and Carruthers (1978) evaluated the growth of the organism by comparing ammonia nitrogen (NH_3-N) accumulation, media optical density (OD), and the reduction of triphenyltetrazolium chloride (TTC) with the results of a microscopic cell-counting procedure. They found that NH_3-N accumulation, OD, TTC reduction (log scale), and TTC reduction (linear) scale showed decreasing agreement with the reference method. They recommended NH_3-N and OD for preliminary screening of large numbers of samples. According to Maciejewicz-Rys (1979), predigested (by proteolytic enzymes) foods can be assayed by a modification of the *T. pyriformis* method. The modification involves determination of 2-amino ethylphosphoric acid, a protozoan metabolite. The agreement with rat and pig feeding tests was much better for single foods than for food mixtures.

Wang *et al.* (1979) compared four *T. pyriformis* methods (microscopic cell counting, TTC reduction, oxygen uptake, and ATP bioluminescence) with rat protein efficiency ratios (PER) assays of protein quality for the determination of RNV of casein, peanut meal, and gluten diets and breads made with wheat flour, soybean meals, and nonfat milk solids. Correlations with the PER values were 0.952 for the TTC, 0.949 for the ATP, 0.945 for the oxygen uptake, and only 0.909 for the direct count method. The TTC method was also effective in analyses of food mixtures (Wang *et al.* 1980). Li-chan and Nakai (1981) found *T. pyriformis* useful in evaluation of protein quality of wheat gluten enriched by lysine or lysine derivatives. Baker *et al.* (1978) devised a practical, relative-slope protein assay with *Tetrahymena* for screening the biological value of foods. The results showed excellent agreement with standard rat PER assays. Crucial for the agreement was predigestion and solubilization of the test samples with bromelain activated with mercaptosuccinic acid. The test involves determination of absorbance at 650 nm on a particulate-free supernatant of solubilized protein.

Bacteria

The use of *Streptococcus faecalis*, *S. zymogenes*, and *Leuconostoc mesenteroides* for assessing the nutrional value of proteins has been investigated. For these organisms, the test proteins require an *in vitro* predigestion with single or successive enzymes or chemical agents; the hydrolysates serve as nitrogen sources for the organism.

Halevy and Grossowicz (1953) tested 2-day pancreatin digests of various proteins as the sole nitrogen source for a strain of *S. faecalis* requiring 10 common essential amino acids. They determined the quantity of hydrolysate needed to achieve half-maximum growth (as determined by photometric measurement of culture turbidities after incubation for 48 hr). The hydrolysates were further tested to determine which of the ten amino acids believed to be essential for the organism was actually responsible for the observed limitation of growth. This information was obtained by supplementation of the basal medium with 9 of the 10 essential acids and retesting the hydrolysate with the supplemented medium. The amino acid that, on omission from the basal medium, failed to allow an increased growth response to the hydrolysate was chosen as the most deficient. Despite general agreement with rat growth tests, the validity of the microbiological test is questionable, as egg albumen was found to be deficient in essential amino acids not limiting for higher animals fed this protein. In addition, under the conditions of the test, gelatin as sole protein gave an appreciable response although it has practically no tryptophan.

L. mesenteroides P-60 was used by Horn *et al.* (1954) to determine the effects of heat treatment on the nutritive value of cottonseed proteins. The organism is known to require for growth the simultaneous presence of 15 different amino acids, including all those known to be indispensable to the growing rat or chick. *In vitro* enzyme digestions were carried out on successive days with pepsin, trypsin, and hog mucosa. The samples were also hydrolyzed with acid to obtain, for comparative purposes, hydrolysates representing complete liberation of the amino acids actually present in the sample. Abilities of the enzyme and acid hydrolysates to support growth of the organism were compared after 3 days of incubation by measurement of acid production. Resultant computations of "indices of protein value" agreed fairly well with values determined by rat growth, but the scope of the assay for other protein materials was not investigated.

Possible limitations of most bacterial assays for protein quality are the inability of test organisms to utilize intact protein, their requirement for amino acids nonessential for higher animals, and the probable influence of peptides that are stimulatory for bacteria but not for higher animals (Rosen 1958). Rogers *et al.* (1959) compared results of bacterial assays with PER estimations of several foods. With *S. faecalis* ATCC 9790, autolysis occurred in media containing hydrolysates of proteins deficient in lysine, and erratic results were encountered. Bunyan and Price (1960) compared microbiological methods with net protein utilization (NPU) determinations and found no correlations between the methods on a large series of meat meals.

Ford (1960, 1962) used *S. zymogenes*, a vigorously proteolytic microorganism, to measure the availability of methionine, leucine, isoleucine, arginine, histidine, tryptophan, and valine in different food proteins. The foods were hydrolyzed with papain before assay with the microorganism. In 12 whale-meat meals, differences in nutritive quality reflected corresponding differences in the biological availability of several of the constituent amino acids. Values obtained for available methionine, tryptophan, leucine, and arginine were closely correlated with each other and with rat-assay and the "available" lysine values obtained for these materials by Bunyan and Price (1960). The results of microbiological assays were affected by fineness of grinding and conditions of enzymic predigestion of the test samples (Ford 1964; Morrison and Rao 1966). According to Morkowska-Gluzinska (1966), there is good agreement in proteins of animal origin between PER or NPU values and microbiological assays by the Ford method. The usefulness of the microbiological method in assaying the nutritive value of several plant proteins was, however, very low. According to Campbell (1961), although microbiological assays for protein quality have been found useful in certain limited applications, they have inherent shortcomings, which limit their general applicability as screening procedures for predicting the quality of protein in a wide range of foodstuffs.

Potentially, the most useful applications of the microbiological approach lie in the determination of limiting amino acids, effects of supplementation of plant proteins, and changes in availability of specific amino acids as a result of processing. The details of assays of available methionine with *S. zymogenes* and of available lysine and methionine with *T. pyriformis* W were described by Boyne *et al.* (1967).

Collaborative studies (Boyne *et al.* 1975) have shown that the *S. zymogenes* assay is useful in determination of available methionine but not tryptophan. Ford (1977) pointed out that in some food grains part of the protein may be intrinsically of low digestibility or become indigestible during processing through interaction with polyphenols, which are widely distributed in higher plants. The interaction may affect the determination of available amino acids by microbiological methods, particularly the *S. zymogenes* method. The *S. zymogenes* assay was subsequently modified (Ford and Hewitt 1979) by fine grinding (ball milling) the samples and predigestion with Pronase, an enzyme of wide action, rather than the specific-action pepsin or papain. The method was used to estimate the relative nutritional value and available lysine in rice, sorghum, barley, and field beans. The results were found to be affected by the presence of tannins and "dye-binding lysine."

Fungi

Mohyuddin and Maza (1978A) used *Aspergillus flavus* to evaluate the protein quality of several breakfast foods containing wheat, corn, rice or skim milk, singly or in combination with milk casein and faba bean. Commercial breakfast foods containing 4.5–31.0% protein were administered to *A. flavus* in a liquid medium. The dry mycelial weight produced after 72 hr was considered as an index of protein quality. Fungal biomass increased two- to threefold when breakfast foods were supplemented with milk or methionine-enriched faba bean. The method was also used to evaluate the relative nutritive value in samples of barley, rye, wheat, and triticale (Mohyuddin *et al.* 1976, 1978B). The use of *A. niger* to evaluate the nutritional value of fodder (after hydrolysis in 2.5% sulfuric acid) was described by Nowacka and Nowacki (1975).

ASSAY OF ANTIBIOTICS AND HEAVY METALS

Although chemical methods of estimation have been worked out for most of the common antibiotics, it is often necessary to use biological methods because they are usually more sensitive and may be applied to both known and unknown antibiotics or to chemically heterogeneous materials without preliminary fractionation (Kersey and Fink 1954). Microbiological assay methods have been devised for almost every antibiotic in use (Loy and Wright 1959). Many of the procedures have been standardized into precise quantitative methods that have been described in official compendia (such as Official Methods of Analysis of AOAC, Federal Antibiotic Regulations of the Food and Drug Administration, the National Formulary, and U.S. Pharmacopeia). Kirshbaum and Arret (1959) outlined details for assaying the commonly used antibiotics. Detailed information on the procedures for performing the assays can be found in the Code of Federal Regulations (Anon. 1959) and the book by Grove and Randall (1955).

One of the main problems encountered in the analysis of antibiotics is the frequent use of mixtures of two or more antibiotics in one product. Test organisms are seldom so selective that they are susceptible to one antibiotic only. However, the level at which the organism is inhibited by one antibiotic may be such that it is not affected by commonly encountered amounts of a second antibiotic. To clarify this problem, Arret *et al* (1957) determined the "interference thresholds" for 10 antibiotics in two widely used tube methods.

Several methods are available for microbiological assaying mixtures of antibiotics. It is possible to develop in certain bacteria artificial resistance to a given antibiotic by growing it in the presence of increasing subinhibitory concentrations. Such artificial resistance is produced most easily with streptomycin. A culture made artificially resistant may lose its resistance unless it is maintained in a medium containing the particular antibiotic. However, when it is lyophilized, it cannot be cultured directly from the lyophilized state in a medium that contains the antibiotic. It must be grown first in an antibiotic-free medium before transfer to the antibiotic-containing medium (Loy and Wright 1959). In the analysis of mixtures of antibiotics, it is also possible to inactivate an interfering component or to separate the antibiotics. Penicillin is inactivated by the specific enzyme penicillinase. Other antibiotics can be inactivated selectively by chemical methods. Streptomycin can be inactivated with carbazide, dihydrostreptomycin with barium hydroxide, the tetracyclines by heating a solution at pH 8 at 100°C for 30 min, and erythromycin by heating a solution at pH 2 at 37°C for 3 hr. Separation by differential solubility is successful in some instances (Weiss *et al.* 1957).

Bitton *et al.* (1984) described a simple, rapid and inexpensive assay to test heavy metal (copper, mercury, zinc, nickel, and silver) toxicity, using commercial dry yeast as the test microorganism. The procedure is most sensitive for copper; intermediate for silver, zinc, and nickel; and least sensitive for mercury.

AUTOMATION OF MICROBIOLOGICAL ASSAYS

The use of automatic methods of analysis has made significant contributions to microbiology. Several methods have been proposed for removing the tedium involved in many microbiological techniques and increasing the objectivity and precision of assays.

By feeding continuously streams of inoculum and nutrient medium and by periodically introducing antibiotic solutions, Gerke *et al.* (1960) were able to obtain dose–response curves for several antibiotics. The assays measured turbidity and respiration, and were comparable with manual agar diffusion assays (Haney *et al.* 1962). Shaw and Duncombe (1963, 1965) described a method depending on measuring the carbon dioxide produced by free respiration of a bacterium during a fixed incubation time, and the depression of respiration by graded concentrations of antibiotic. Operated at a speed of 20 instrumental responses per hour, with duplicate recording for each sample and the necessary standards, an overall throughput of six samples per hour was attained. Usefulness of turbidimetric automated assays of antibiotics was described by Platt *et al.* (1965).

McMahan (1965) described an automatic turbidimetric assay for antibiotics that could handle 50 samples (400 tubes) per hour. DiCuollo *et al.* (1965) described a semiautomated instrumental system for performing large plate diffusion assays. The system (1) semiautomatically dilutes antibiotic samples; (2) applies simultaneously and quantitatively large numbers of sample solutions to the assay plate; (3) prepares a permanent photographic record of the assay results; and (4) automatically reads and records the zone sizes on digital printout and computer punch type for automated or manual computation.

A series of basic studies on application of automated analysis to the study of bacterial growth (Ferrari *et al.* 1965; Gerke 1965; Watson 1965) demonstrated the feasibility of using the available instruments for studying bacterial growth. The studies laid foundations for investigations of broader aspects of cell physiology as related to continuous flow analysis.

At present, many new, rapid, and fairly selective methods are available for determination of vitamins; the use of HPLC has been particularly effective. Still, microbiological assays continue to be used both for biological confirmation and for differentiation among various forms that differ in activity.

BIBLIOGRAPHY

Albanese, A. A. (1959). "Protein and Amino Acid Nutrition." Academic Press, New York.

Allison, J. B. (1949). Biological evaluation of proteins. *Adv. Protein Chem.* **5**, 155–200.

Anon. (1959). Compilation of regulations for tests and methods of assay and certification of antibiotic and antibiotic-containing drugs. *Federal Register* **141**, No. 21CFR.

Arret, B., Woodward, M. R., Wintermere, D. M., and Kirshbaum, A. (1957). Antibiotic interference thresholds of microbial assays. *Antibiot. Chemother.* **7**, 545–548.

Baker, H., and Sobotka, H. (1962). Microbiological assay methods for vitamins. *Adv. Clin. Chem.* **5**, 147–235.

Baker, H., Frank, O., Rusoff, I. I., Morck, R. A., and Hutner, S. H. (1978). Protein quality of foodstuffs determined with *Tetrahymena thermophila* and rat. *Nutr. Rep. Intl.* **17**, 525–536.

Baum, F. (1966A). Determining the biological value of proteins with *Tetrahymena pyriformis* W. II. The protein value of foods and feeds. *Die Nahrung* **10**, 453–459.

Baum, F. (1966B). Determining the biological value of proteins with *Tetrahymena pyriformis* W. III. The effects of amino acid supplementation and storage on protein value of cereal flours. *Die Nahrung* **10**, 571–580.

Bitton, G., Koopman, B., and Wang, H. D. (1984). Baker's yeast assay procedure for testing heavy metal toxicity. *Bull. Environ. Contam. Toxicol.* **32**, 80–84.

Bliss, C. I. (1956). The calculation of microbial assays. *Bacteriol. Rev.* **20**, 243–258.

Bodwell, C. E. (ed.). (1977). "Proteins for Humans: Evaluation and Factors Affecting Nutritional Value." AVI Publishing Co., Westport, CT.

Bolinder, A. E., Lie, S., and Ericson, L. E. (1963A). Plate assay methods for amino acids.

I. A. sensitive cup plate assay for lysine with *Streptococcus faecalis* ATCC 6057. *Biotechnol. Bioeng.* **5**, 131–146.

Bolinder, A. E., Lie, S., and Ericson, L. E. (1963B), Plate assay methods for amino acids. II. Factors affecting the cup plate assay for lysine with *Streptococcus faecalis* ATCC 6057. *Biotechnol. Bioeng.* **5**, 147–165.

Boyne, A. W., Ford, J. E., Hewitt, D., and Shrimpton, D. H. (1975). Protein quality of feeding stuffs. 7. Collaborative studies on the microbiological assay of available amino acids. *Br. J. Nutr.* **34**, 153–162.

Boyne, A. W., Price, S. A., Rosen, G. D., and Stott, J. A. (1967). Protein quality of feeding stuffs. IV. Progress report on collaborative studies on the microbiological assays for available amino acid. *Br. J. Nutr.* **21**, 181–206.

Bunyan, J., and Price, S. A. (1960). Studies on protein concentrates for animal feeding. *J. Sci. Food Agric.* **11**, 25–37.

Campbell, J. A. (1961). "Methodology of Protein Evaluation—A Critical Appraisal of Methods for Evaluation of Protein in Foods." Nutrition Document R. 10/Add 37 WHO/FAO/UNICEF—PAG.

Cancellieri, M. F. P., and Morpurgo, G. (1962). A new microbiological method for amino acid assay. *Sci. Rept. Inst. Super Sanita* **2**, 336–344.

DiCuollo, C. J., Guarini, J. R., and Pagano, J. F. (1965). Automation of large plate agar microbiological diffusion assays. *Ann. N.Y. Acad. Sic.* **130**, 672–679.

Evans, E., and Caruthers, S. C. (1978). Comparison of methods for estimating the growth of *Tetrahymena pyriformis* W. *J. Sci. Food Agric.* **29**, 703–707.

Ferrari, A., Gerke, J. R., Watson, R. W., and Umbreit, W. W. (1965). Application of automated analysis to the study of bacterial growth. I. The instrumental system. *Ann. N.Y. Acad. Sci.* **130**, 704–721.

Ford, J. E. (1960). A microbiological method for assessing the nutritional value of proteins. *Br. J. Nutr.* **14**, 485–497.

Ford, J. E. (1962). A microbiological method for assessing the nutritional value of proteins. II. The measurement of "available" methionine, leucine, isoleucine, arginine, histidine, tryptophan, and valine. *Br. J. Nutr.* **16**, 409–425.

Ford, J. E. (1964). A microbiological method for assessing the nutritional value of proteins. III. Further studies on the measurement of available amino acids. *Br. J. Nutr.* **18**, 449–460.

Ford, J. E. (1977). Microbiological procedures for protein quality assessment. *In* "Nutritional Evaluation of Cereal Mutants," pp. 73–85. IAEA, Vienna.

Ford. J. E., and Hewitt, D. (1979). Protein quality of cereals and pulses. 1. Application of microbiological and other *in vitro* methods in the evaluation of rice (*Oryza sativa* L.), sorghum (*Sorghum vulgare* Pers.), barley, and field beans. (*Vicia faba* L.). *Br. J. Nutr.* **41**, 341–352.

Friedman, M. (ed.). (1975). "Protein Nutritional Quality of Foods and Feeds." Marcel Dekker, New York.

Friedman, M. (ed.). (1978). "Nutritional Improvement of Food and Feed Proteins." Plenum Press, New York.

Gavin, J. J. (1956). Analytical microbiology. I. The test organisms. *Appl. Microbiol.* **4**, 323–331.

Gavin, J. J. (1957A). Analytical microbiology. II. The diffusion methods. *Appl. Microbiol.* **5**, 25–33.

Gavin, J. J. (1957B). Analytical microbiology. III. Turbidimetric methods. *Appl. Microbiol.* **5**, 235–243.

Gavin, J. J. (1958). Analytical microbiology. IV. Gravimetric methods. *Appl. Microbiol.* **6**, 80–85.

Gavin, J. J. (1959). Analytical microbiology. V. Metabolic response methods. *Appl. Microbiol.* **7**, 180–192.

Gerke, J. R. (1965). Application of automated analysis to the study of bacterial growth. II. Continuous analysis of microbial nucleic acids and proteins during normal and antibiotic inhibited growth. *Ann. N.Y. Acad. Sci.* **130**, 722–732.

Gerke, J. R., Haney, T. A., and Pagano, J. F. (1960). Automation of the microbiological assay of antibiotics with an Autoanalyzer instrumental system. *Ann. N.Y. Acad. Sci.* **87**, 782–791.

Girdwood, R. H. (1960). Folic acid, its analogs and antagonists. *Adv. Clin. Chem.* **3**, 236–297.

Grau, C. R., and Carroll, R. W. (1958). Evaluation of protein quality. *In* "Processed Plant Protein Foodstuffs" (A. M. Altschul, ed.). Academic Press, New York.

Grove, D. C., and Randall, W. A. (1955). "Assay Methods of Antibiotics." Medical Encyclopedia, New York.

Gupta, J. D., Dakroury, A. M., Harper, A. E., and Elvehjem, C. A. (1958). Biological availability of lysine. *J. Nutr.* **64**, 259–270.

Halevy, S., and Grossowicz, N. (1953). A microbiological approach to nutritional evaluation of protein. *Proc. Soc. Exp. Biol. Med.* **82**, 567–571.

Haney, T. A., Gerke, J. R., Madigan, M. E., and Pagano, J. F. (1962). Automated microbiological analyses for tetracycline and polyenes. *Ann. N.Y. Acad. Sci.* **93**, 627–639.

Hendlin, D. (1952). Symposium on lactic acid bacteria. III. Use of lactic acid bacteria in microbiological assays. *Bacteriol. Rev.* **16**, 241–246.

Henry, K. M. (1965). A comparison of biological methods with rats for describing the nutritive value of proteins. *Br. J. Nutr.* **19**, 125–135.

Horn, M. J., Blum, A. E., and Womack, M. J. (1954). Availability of amino acids to microorganisms. II. A short microbiological method of determining protein value. *J. Nutr.* **52**, 375–381.

Hutner, S. H., Cury, A., and Baker, H. (1958). Microbiological assays. *Anal. Chem.* **30**, 849–867.

Kavanagh, F. (1960). A commentary on microbiological assaying. *Adv. Appl. Microbiol.* **2**, 65–93.

Kavanagh, F. (1963). "Analytical Microbiology." Academic Press, New York.

Kersey, R. C., and Fink, F. C. (1954). Microbiological assay of antibiotics. *Methods Biochem. Anal.* **1**, 53–79.

Kirshbaum, A., and Arret, B. (1959). Outline of details for assaying the commonly used antibiotics. *Antibiot. Chemother.* **9**, 613–617.

Knudsen, L. F. (1950). Statistics in microbiological assay. *Ann. N.Y. Acad. Sci.* **52**, 889–902.

Koser, S. A. (1948). Growth factors for microorganisms. *Ann. Rev. Microbiol.* **2**, 121–142.

Li-Chan, E., and Nakai, S. (1981). Nutritional evaluation of covalently lysine enriched wheat gluten by *Tetrahymena* bioassay. *J. Food Sci.* **46**, 1840–1843, 1850.

Loy, H. W., and Wright, W. W. (1959). Microbiological assay of amino acids, vitamins, and antibiotics. *Anal. Chem.* **31**, 971–974.

Lwoff, A. (ed.). (1972). "Biochemistry and Physiology of Protozoa." Academic Press, New York.

Maciejewicz-Rys, J. (1979). Model systems to compare the nutritional value of proteins in feed mixtures by the protozoan *Tetrahymena pyriformis* W and rats (Polish). *Biuletyn CSOP* **10**, 3–36.

McMahan, J. R. (1965). A new automated system for microbiological assays. *Ann. N.Y. Acad. Sci.* **130**, 680–685.

Mohyuddin, M., and Maza, G. (1978A). Evaluation of protein quality in breakfast foods using *Aspergillus flavus*. *Qual. Plant. Plant Foods Hum. Nutr.* **28**, 251–259.

Mohyuddin, M., Sharma, T. R., Kaul, A. K., and Niemann, J. (1976). Use of *Aspergillus flavus* to evaluate the relative nutritive value in cultivars of rye, wheat, and triticale. *J. Sci. Food Agric.* **27**, 943–950.

Mohyuddin, M., Lewis, A. J., and Bowland, J. P. (1978B). Use of *Aspergillus flavus* to evaluate relative nutritive value of barley cultivars intended for incorporation in animal diets. *Qual. Plant. Plant Foods Hum. Nutr.* **28**, 19–27.

Morkowska-Gluzinska, W. (1966). Use of the *Streptomyces zymogenes* microbiological method in evaluating the nutritional value of proteins from selected products. *Roczniki PZH* **17**, 467–475.

Morrison, A. B., and Rao, M. N. (1966). Measurement of the nutritional availability of amino acids in foods. *Adv. Chem. Ser.* **57**, 159–177.

Nelson, O. E. (ed.). (1973). "Nuclear Techniques for Seed Protein Improvement." IAEA, Vienna.

Nowacka, D., and Nowacki, E. (1975). A trial to use *Aspergillus niger* in evaluation of fodder (in Polish). *Pamietnik Pulaski Prace IUNG* **64**, 71–86.

Peterson, N. H., and Peterson, M. S. (1945). Relation of bacteria to vitamins and other growth factors. *Bacteriol. Rev.* **9**, 49–109.

Platt, T. B., Gentile, J., and George, M. J. (1965). An automated turbidimetric system for antibiotic assay. *Ann. N.Y. Acad. Sci.* **130**, 664–671.

Porter, J. W. G., and Rolls, B. A. (eds.). (1973). "Proteins in Human Nutrition." Academic Press, London.

Rogers, C. G., McLaughlan, J. M., and Chapman, D. G. (1959). Evaluation of protein in foods. III. A study of bacterial methods. *Can. J. Biochem. Physiol.* **37**, 1351–1360.

Rolle, G. (1976). Research concerning the usefulness of *Tetrahymena pyriformis* W. as a test organism by the evaluation of protein quality. *Acta Agric. Scand.* **26**, 282–286.

Rosen, G. D. (1958). The microbiological assay of protein quality. *In* "Proceedings of International Symposium on Microchemistry" (D. Wilson, ed.). Pergamon Press, London.

Rosen, G. D., Stott, J. A., and Smith, A. (1962). Microbiological assays of amino acids and intact proteins. *Cereal Sci. Today* **7**, 36–39.

Shaw, W. H., and Duncombe, R. E. (1963). Continuous automatic microbiological assay of antibiotics. *Analyst* **88**, 694–701.

Shaw, W. H., and Duncombe, R. E. (1965). Continuous automatic microbiological assay of antibiotics. *Ann. N.Y. Acad. Sci.* **130**, 647–656.

Skeggs, H. R. (1966). Microbiological assay of vitamin B_{12}. *Methods Biochem. Anal.* **14**, 53–62.

Snell, E. E. (1945). The microbiological assay of amino acids. *Adv. Protein Chem.* **2**, 85–118.

Snell, E. E. (1946). Growth factors for microorganisms. *Ann. Rev. Biochem.* **15**, 375–391.

Snell, E. E. (1949). Nutritional requirements of microorganisms. *Annu. Rev. Microbiol.* **3**, 97–117.

Snell, E. E. (1952). Symposium on lactic acid bacteria. II. The nutrition of lactic acid bacteria. *Bacteriol. Rev.* **16**, 235–241.

Snell, E. E. (1958). Chemical structure in relation to biological activities of vitamin B_6. *Vitam. Horm.* **16**, 77–125.

Storvick, C. A., Benson, E. M., Edwards, M. A., and Woodring, M. J. (1964). Chemical and microbiological determination of vitamin B_6. *Methods Biochem. Anal.* **12**, 183–276.

Stott, J. A., and Smith, H. (1966). Microbiological assay of protein quality with *Tetrahymena pyriformis* W. IV. Measurement of available lysine, methionine, arginine, and histidine. *Br. J. Nutr.* **20**, 663–673.

Stott, J. A., Smith, H., and Rosen, G. D. (1963). Microbiological evaluation of protein quality with *Tetrahymena pyriformis* W. III. A simplified assay procedure. *Br. J. Nutr.* **17**, 227–232.

Teunison, D. J. (1961). Microbiological assay of intact proteins using *Tetrahymena pyriformis* W. I. Survey of protein concentrates. *Anal. Biochem.* **2**, 405–420.

Wang, Y. Y. D., Miller, J., and Beuchat, L. R. (1979). Comparison of four techniques for measuring growth of *Tetrahymena pyriformis* W. with rat bioassays in assessing protein quality. *J. Food Sci.* **44**, 540–544.

Wang, Y. Y. D., Miller, J., and Beuchat, L. R. (1980). *Tetrahymena pyriformis* as an organism for bioassay of protein quality. *Nutr. Rep. Int.* **21**, 645–650.

Watson, R. W. (1965). Application of automated analysis to the study of bacterial growth. III. Regulation of growth processes in *Escherichia coli* by sulfur compounds. *Ann. N.Y. Acad. Sci.* **130**, 733–743.

Weiss, P. J., Andrew, M. L., and Wright, W. W. (1957). Solubility of antibiotics in 24 solvents; use in analysis. *Antibiot. Chemother.* **7**, 374–377.

White, P. L., and Fletcher, D. C. (eds.). (1974). "Nutrients in Processed Foods." Am. Med. Assoc., Publ. Sci. Group, Acton, MA.

III
APPLICATIONS AND CHEMICAL COMPOSITION

32
General Remarks

The determination of food composition is fundamental to theoretical and applied investigations in food science and technology, and is often the basis of establishing the nutritional value and overall acceptance from the consumer standpoint. An analysis may establish the amounts of one, or all of the components in a food. Most of the methods described in Part III are useful for the proximate analysis of foods, i.e., the determination of the major components (moisture, minerals, carbohydrates, lipids, and proteins). These components are included in standard tables of composition of foods. The value and limitations of such determinations are discussed in Chapter 4 and in each of the chapters in Part III.

The constituents to be detected or determined in food analysis may be elements, radicals, functional groups, compounds, groups of compounds, or phases. Few of the chemical, physical, or physicochemical methods used in food analysis are completely specific or selective. Sometimes, careful adjustment of pH, oxidation or reduction, or complexing of certain groups or elements make determinations possible. However, (except for specific biological methods (e.g., enzymatic, immunological), fractionation procedures generally must be used. Advances in food analysis in the last two decades have resulted from the development of many instrumental methods and from improvements in separation methods (mainly chromatography) that, in turn, utilize more and more instrumental techniques. Those methods were described in Part II of this book.

The food analyst of today must have a knowledge of chemical, physical, and physicochemical properties of foods and of the methods used in their analyses. Such knowledge is essential to the selection or modification of analytical methods in order to meet requirements of speed, precision, and accuracy.

Generally, the food analyst knows the nature and qualitative composition of a sample. Often he or she also knows the approximate ranges of constituents to be determined. In such cases, there is no need for a qualitative analysis. Only occassionally will the analyst be asked to analyze a sample of totally unknown composition; in some cases, the range of a certain component may not be known, or the presence of a certain minor component (e.g., mineral) may have to be established. In such cases, a qualitative test, which generally gives information of a semi-quantitative nature, is required.

In order for an analysis to provide useful information, several criteria must be met. The information obtained must be arrived at by use of reproducible, empirically tested (preferably standardized and based on collaborative investigations), and scientifically sound techniques. In the selection of an appropriate technique, the analyst should consider first the information that is to be obtained to solve a specific problem. This selection will often depend on the nature of the food examined, availability of equipment, required speed and precision, and background information. The latter are, however, of secondary importance. The main goal is providing specific information for a specific purpose. Using a semiquantitative, rapid test or a simplified procedure instead of an appropriate instrumental assay cannot be justified if the substitute procedures cannot provide the information that is required. At the same time, using expensive equipment that requires costly maintenance cannot be justified if the same information can be obtained by a simple and rapid procedure. In any analysis, the number of assays should be reduced to a minimum of most informative tests. Reams of uniformative data are not only costly, but also complicate evaluation.

In determining the overall composition of a food sample or content of a specified component, four steps are involved: (1) obtaining a representative subsample for analysis, (2) converting the component(s) into a form that permits an assay, (3) performing the assay, and (4) calculating and interpreting the data.

The analyst often assumes that the sample he is about to analyze is homogeneous. It is advisable that before starting a determination, the whole sample be remixed to eliminate heterogeneity—mainly in particle size and moisture distribution. In some foods (concentrated sugar solutions, honey, etc.), the sample must be heated carefully to dissolve sugar crystals. Some foods pose special problems of sample preparation. The difficulty of obtaining a representative sample from a heterogeneous substance such as a meat carcass is well known. In addition, incorporation of fatty tissue with the lean meat to obtain a homogeneous mixture requires efficient mincing, mixing, and subdivision of the total sample. In some cases, complete homogenization is actually undesirable. For example, in determining the composition of an ice cream base mix, insoluble particles (fruit, nuts, confectionery) are removed from the melted sample on a sieve. Similarly, in analyzing chocolate lipids it is essential to remove nut particles.

In many foods the composition is determined in the edible portion and the inedible part is carefully removed, weighed, and often reported. It is often impossible or undesirable to incorporate and distribute all impurities uniformly in the whole sample. Thus, in grain analyses, large-sized impurities (straw, stones) are removed by sieving or mechanical cleaning from a large sample, and the partly purified sample is then subdivided for further analysis.

For a complete characterization of a totally unknown food sample, the following sequence of treatments is recommended: sublimation, steam distillation, and extraction with petroleum ether followed by ether, chloroform, ethanol (absolute or 80%), and water (cold, warm, acidified, and alkaline). Steam distillation is carried out on a separate sample aliquot. For fresh tissues, dehydration with ethanol may be required for efficient fat extraction. The residue from all treatments represents the insoluble that can be subsequently hydrolyzed for detailed determination. Generally, a determination of mineral components is made on the original untreated material. Every assay should include a moisture determination.

The solid and solute of most extracts are separated by filtration. To minimize retention of solute, suction, pressure, or centrifugation can be used. Extraction by percolation gives the most complete extraction of solute from the solid.

Whatever the extractant, the extract is generally evaporated to dryness or concentrated to reduce the bulk of solvent before subsequent fractionation and characterization. If the solution does not splash or froth, evaporation can be accomplished by simply boiling down the solution. If frothing is excessive, evaporation may be carried out at temperatures below the boiling point by heating from below in the usual manner or by radiant heating from above. The rate of evaporation can be increased by blowing a current of warm air over the surface of the fluid to remove the vapor-saturated air. Preheating the air is often effective, provided the flow of air is carefully controlled to avoid splashing and provided no water-impermeable layer of lipids or crystallized solids is formed. The extract can be dried at relatively low temperatures, without danger of dust contamination, by placing it into washed cellophane (dialysis) tubing that is hung up to dry in a well-aerated and dry area. The rate of evaporation is proportional to the exposed surface; consequently, narrow tubes give best results.

Distillation *in vacuo* is the most commonly used method for reducing the bulk of the extract. To get a high rate of distillation at a low temperature, it is essential to have a good vacuum, wide vapor paths, and efficient cooling of the distillate. Apparatus for continuous distillation *in vacuo* is available commercially.

When an aqueous extract freezes, only water solidifies, and the solutes are concentrated in the unfrozen phase. The water can then be removed by lyophilization. The process is particularly useful for heat-labile substances.

By selecting a suitable membrane, it is possible to concentrate a solute by ultrafiltration. If the solute has a high molecular weight and is water soluble, the separation by commercially available membranes can be quite rapid and efficient. With small molecules (molecular weight below 5000), the membranes must be so impervious that filtration is slow, and high pressures must be used to overcome the osmotic pressure caused by the retained solute.

Semipermeable membranes used for dialysis or for ultrafiltration are basically molecular sieves. The separation attained during the dialysis depends on (1) the opening in the membrane and (2) on the molecular weight, shape, and charge or charge distribution on the solute molecule. Generally, particles with molecular weights above 10,000 are retained. Almost all dialysis is now done with commercially available cellophane sheets or tubes. For cellulose-containing extracts, membranes of nitrated cellulose or protein are suitable.

With small amounts of fluids, dialysis is relatively simple. The fluid to be dialyzed is put into a bag folded from cellophane sheet or into a tube (tied on both ends) and immersed into water. To increase the rate of dialysis of dilute solutions, the fluid inside and outside the tubing is mixed mechanically. Less efficient mixing can be maintained by bubbling air slowly through the outside vessel or by a dropwise water current. If the concentration of solute is high, there is no need for mechanical mixing. A dialysis tube suspended near the top of a jar of water establishes a convective system in which the dense solution of diffusible solute moves to the bottom of the jar and remains there undisturbed. Consequently, dialysis takes place against a more dilute solution than if there had been mixing.

Characterization of extracts, following dialysis, centrifugation, or concentration is accomplished today most commonly after separation by any of the available, appropriate fractionation procedures.

In the actual assay, two basic approaches can be used. The desired constituent is isolated by physical, chemical, or physicochemical methods that separate it from interfering substances. Ideally, the substance to be assayed is isolated as a single component or its derivative in a specific fraction. In practice, it suffices if the fraction does not contain any uncontrollable source of interference with the subsequent measurement.

The selection of a fractionation procedure for a specific assay depends on many factors: nature and composition of the food; availability, speed, specificity, efficiency, and accuracy of separation; and recovery of minimally or optimally modified components. Generally, a single fractionation procedure, cannot separate or isolate a specified component in various foods, nor can a single procedure separate all components of a food. Consequently, the food analyst must be aware of the principles and limitations of the separation techniques and of the nature of the food examined so that the best separation technique can be selected or devised.

An assay that requires no fractionation is, of course, preferable. Such an assay must be highly selective or specific. Enzymatic or immunochemical methods are prime examples of inherently specific assays. Sometimes assay conditions can be selected so that only a specified component is determined, as in certain colorimetric tests for carbohydrates or amino acids. Alternatively, the selectivity of an assay can be attained by immobilization or sequestration of interfering substances. Any physical or chemical property can be used for identification or quantitative determination provided that the property is (or can be made) specific for the substance to be determined.

Once the analysis is completed, the results must be expressed in such a manner that their meaning and significance can be clearly understood and the information used for solving the specific problem. The food analyst is generally better qualified (than the recipient of the analysis) to evaluate the significance and limitations of the data. The analyst is expected to help interpret the results and draw appropriate conclusions. Such conclusions can be best drawn after the background information on the nature of the sample is known. Interpretation of analytical data has been greatly facilitated in recent years by the availability of computers for data processing and by the development of statistical techniques that are useful both in devising and designing analytical procedures and in evaluating analytical results. The analyst can be assisted in his interpretation by consulting available comprehensive books and tables on food composition. He will generally depend to a large extent on consultation with other chemists in the interpretation of results and ultimately on his own judgment and experience.

A recent review by Sawyer (1984) emphasized that while accurate assays of composition are a prerequisite to the evaluation of foods and food ingredients, the degree of accuracy that may be attained in practice is often quite illusory. Some compositional data may be based on well-authenticated findings, but the authenticity of other data may be questionable. The introduction of new analytical techniques has provided insight into the limitations of the more classical methods. In addition, collaborative studies have highlighted the inadequacies of some acceptable methodology and pointed to the need for more refined definitions of some of the determinants. Based on this line of reasoning, Sawyer discussed and evaluated the available methods to calculate the meat content of a food, to distinguish between meat and cured meat, to assess the true water content of frozen poultry, to estimate the butter content of composite products, and to measure in a meaningful way (from the physiological standpoint) dietary fiber content of a food. To attain progress in this area, three requirements must be met: (1) Stable, homogeneous, and relevant test material must be obtained; (2) agreed, harmonized protocols for testing methods and results must be developed, and (3) Ambiguities of definition in the legal (descriptive) sense and technical (analytical) sense must be eliminated in the specifications and standardization of foods.

According to Hildrum *et al.* (1984), the analysis of food quality has been characterized in the past by single-univariate measurements of single parameters. The development of new multivariate techniques (e.g., NIR reflectance spectroscopy) facilitates the estimation of several quality factors on the basis of large amounts of composition-related data. Once the mathematical relationships between the instrumental parameters and traditional quality assessments are established, it should be possible to develop rapid, low-cost, and reproducible methods for the continuous objective assessment and control of food quality. None of these developments will come about without a commitment to the pursuit of excellence in the control of food quality and in food analysis (Hawthorn 1984).

BIBLIOGRAPHY

Hawthorn, J. (1984). Control of food quality and food analysis or the pursuit of excellence. *In* "Control of Food Quality and Food Analysis" (G. G. Birch and K. J. Parker, (eds.), pp. 1–12. Elsevier Applied Science Publishers, London and New York.

Hildrum, K. I., Martens, M., and Martens, H. (1984). Research on analysis of food quality. *In* "Control of Food Quality and Food Analysis" (G. G. Birch and K. J. Parker, (eds.), pp. 65–80. Elsevier Applied Science Publishers, London and New York.

Sawyer, R. (1984). Food composition and analytical accuracy. *In* "Control of Food Quality and Food Analysis" (G. G. Birch and K. J. Parker, (eds.), pp. 39–64. Elsevier Applied Science Publishers, London and New York.

33
Determination of Moisture

Moisture determination is one of the most important and most widely used measurements in the processing and testing of foods. Since the amount of dry matter in a food is inversely related to the amount of moisture it contains, moisture content is of direct economic importance to the processor and the consumer. Of even greater significance, however, is the effect of moisture on the stability and quality of foods. Grain that contains too much water is subject to rapid deterioration from mold growth, heating, insect damage, and sprouting. Small differences in moisture content may be responsible for unexpected cases of spoilage in commercially stored grain. The rate of browning of dehydrated vegetables and fruits and of oxygen absorption by egg powders increases with an increase in moisture content.

Moisture determination is important in many industrial problems, e.g., in the evaluation of materials' balance or of processing losses. We must know the moisture content (and sometimes its distribution) for optimum processing of foods, e.g., in the milling of cereals, mixing of dough to optimum consistency, and production of bread with the best grain, texture, and freshness retention. Moisture content must be known in determining the nutritive value of a food, in expressing results of analytical determinations on a uniform basis, and in meeting compositional standards or laws. And finally, it is often desirable to weigh samples for analytical determinations on a given moisture basis (see Chapter 4). This is especially important if the measured analytical parameter does not vary in a linear or simple manner with an increase in dry matter content.

MOISTURE CONTENTS OF FOODS

The moisture contents of foods vary widely. Fluid dairy products (whole milk, nonfat milk, and buttermilk) contain 87–91% water; various dry milk powders contain about 4% water. Cheeses have intermediate water contents ranging, from about 40% in cheddar to 75% in cottage; the water content of butter is about 15%, of cream 60–70%, and of ice cream and sherbet around 65%. Pure oils and fats contain practically no water, but processed lipid-rich materials may contain substantial amounts of water (from about 15% in margarine and mayonnaise to 40% in salad dressings).

Some fresh fruits contain more than 90% water in the edible portion. Melons contain 92–94%, citrus fruits 86–89%, and various berries 81–90% water. Most raw tree and vine fruits contain 83–87% water; the water content of ripe guavas is 81%, of ripe olives 72–75%, and of avocado 65%. After commercial drying, fruits contain up to 25% water. Fresh fruit juices and nectars contain 85–93% water; the water content is lowered in concentrated or sweetened products.

Cereals are characterized by low moisture contents. Whole grains designed for long-term storage have 10–12% water. The moisture content of breakfast cereals is below 4%, of macaroni 9%, and of milled grain products (flour, grits, semolina, germ) 10–13%. Among baked cereals, pies are rich in water (43–59%); bread and rolls are intermediate (35–45%, and 28%, respectively); and crackers and pretzels are relatively dry (5–8%). Ripe raw nuts generally contain 3–5% water, or less, after roasting. Fresh chestnuts contain about 53% water.

The moisture contents of meat and fish depend primarily on the fat contents, and vary to a lesser degree with the age, source, and growth season of the animal or fish. The moisture contents range from 50 to 70%, but some organs may contain up to 80% water. The moisture content of sausages varies widely. Water in poultry meats is 50% (in geese) to 75% (in chicken). Fresh eggs have about 74%, and dried eggs about 5% water.

White sugar (cane or beet), hard candy, and plain chocolate contain 1% or less of water. In fruit jellies, jams, marmalades, and preserves up to 35%, in honey 20%, and in various syrups 20–40% is water.

Sweet potatoes contain less water (69%) than white potatoes (78%). Radishes have the most (93%) and parsnips the least (79%) water among the common root vegetables. Among other vegetables, a still wider range is found. Green lima beans have about 67%, and raw cucumbers over 96% water. Dry legumes contain 10–12% water, and the water content of commercially dried vegetables is 7–10% (preferably below 8).

BASIC CONSIDERATIONS

The rapid and accurate determination of water in foods varying widely in texture, overall composition, and moisture content continues to present many problems. Many workers have stressed the complexity of analytical procedures for the determination of water in foods. Though the literature is replete with methods of moisture determination, we have no methods that are both accurate and practical. Accurate, rapid, and simple methods of moisture determination applicable to all types of food materials are continuously sought, but it seems doubtful that such a goal will ever be attained.

In practice, the guiding principle has been to prefer the method that gives the highest moisture values, provided decomposition of organic components and volatilization of compounds other than water are negligible, or that such losses can be compensated by incomplete removal, under fixed experimental conditions, of strongly absorbed water. Generally, the reproducibility and practicability of a method (simplicity, convenience of apparatus, and rapidity) have been the important factors in the selection of an analytical method for water determination. Less emphasis is placed on the accuracy of a water determination. Admittedly, reproducibility is of major importance in the control of processing procedures and in the standardization of commercial products. Yet, accuracy is of great significance in establishing conditions that govern food stability. The usefulness and validity of simple and rapid moisture determinations depend on their calibration against standard and accurate reference methods. The difficulties encountered in developing such reference methods can best be understood by considering the manner in which water is held by various food components.

Water may occur in foods in at least three forms. A certain amount may be present as free water in the intergranular spaces and within the pores of the material. Such water retains its usual physical properties and serves as a dispersing agent for colloidal substances and as a solvent for crystalline compounds. Part of the water is absorbed on the surface of macromolecular colloids (starches, pectins, cellulose, and proteins). That water is closely associated with the absorbing macromolecules by forces of absorption, which are attributed to van der Waal's forces or to hydrogen bond formation. Finally, some of the water is in a bound form—in combination with various substances, i.e., as water of hydration.

This classification, though convenient, is quite arbitrary. Attempts to determine quantitatively the amounts of various forms of water in foods have been unsuccessful. Although some inorganic compounds show a

distinct discontinuous sorption isotherm of various levels of crystallization water, the sorption isotherm of water in foods shows a continuous
spectrum of the types of water binding. Consequently, the terms *free*,
absorbed, and *bound* are relative, and as the true moisture content is not
known, the conditions selected for moisture determination are arbitrary.
The isotherm describes the equilibrium relation of the amount of water
sorbed by the food components and relative vapor pressure (water activity, a_w). Depending on whether water is given off or taken up in ap-
proaching the equilibrium between vapor pressure and moisture content,
a desorption or an adsorption isotherm results.

Methods for the determination of moisture can be divided into drying
methods, distillation procedures, chemical assays, and physical procedures. Principles of the various methods are outlined and examples are
given in the rest of this chapter.

DRYING METHODS

The procedures for determining the moisture content specified in food
standards generally involve thermal drying methods. The material is
heated under carefully specified conditions and the loss of weight is taken
as a measure of the moisture content of the sample. The determination
of moisture from the loss of weight due to heating necessarily involves
an empirical choice of the type of oven and the temperature and length
of drying. Hence, the values obtained for moisture depend on the arbitrarily selected conditions, and some of the methods provide approximate
rather than accurate moisture values. Drying methods, however, are simple, relatively rapid, and permit the simultaneous analyses of large numbers of samples. They continue to be the preferred procedures for many
food analysts.

In an ideal drying procedure for the determination of water, weight
losses should result from quantitative and rapid volatilization of water
only. In practice, heating of a moist organic substance also causes volatilization of other absorbed materials and of gaseous products formed
by irreversible thermal decomposition of organic components. Further,
weight changes resulting from oxidation phenomena (e.g., of oils) occur.
These changes do not begin at a specified or fixed temperature, and occur
at all temperatures at widely varying rates. The rate at which moisture
can be removed from the surface of a solid phase is a function of the
water vapor pressure and of the drying temperature. Water can be determined at any temperature provided the partial vapor pressure in the
air above the solid phase is lower than the vapor pressure of the water

in the sample. For example, in lyophilization thermal drying occurs below the freezing point of water. For accurate moisture measurement, the tendency is to dry foods at the lowest possible temperatures. Practical considerations dictate selecting temperatures at which the decomposition of organic compounds is minimized, but at which the time required for quantitative drying is not unduly prolonged.

The accuracy of moisture determinations is affected by the drying temperature, temperature and relative humidity of the drying chamber, relative humidity and air movement in the drying chamber, vacuum in the chamber, depth and particle size of samples, drying oven construction, and the number and position of samples in the oven. The surface of the material being dried and the rate of diffusion of water vapor in the drying substance also affect results. Mitchell (1950 A,B) evaluated the effects of the following variables on the rate of evaporation of water in ovens: diameter, depth, and material of the container; number of containers and their position in the oven; rate of conduction of heat to container (i.e., from different types of shelf material); and ventilation and temperature of the oven. The rate of evaporation was higher in aluminum than in glass or porcelain dishes, higher in vacuum than in steam ovens, and higher in shallow than in deep dishes. The rate of heat supply to the bottom of the dish was the most important factor.

There are several sources of errors in the determination of moisture in cereals. In addition to sampling errors (of the bulk lot or at the laboratory subsampling stage), moisture may change during subsequent storage of the samples. If the material for a moisture determination must be ground, loss of moisture to, or gain of moisture from, the air may take place. For damp grain, a two-stage drying method has been recommended. A sample of 100 g or more is first allowed to equilibrate with the air, and the moisture loss of the damp wheat is determined. A subsample is then ground for the determination of the remaining moisture. In two-stage drying, total moisture = $A + (100 - A)B/100$, where A is the % moisture lost in air drying and B the % moisture in the air-dry sample as determined by oven drying.

[source of error]

The two-stage method gives generally higher results (0.2–0.6%) than the one stage procedure, but prolongs the determination. To prevent reabsorption of moisture after oven drying, moisture dishes with tight covers should be promptly closed after the oven is opened, rapidly transferred from the oven to a good desiccator, and weighed rapidly after they have attained room temperature. Common desiccators and desiccants are of limited usefulness and, unless properly maintained, can cause erratic results from pickup of moisture during cooling.

Solid materials must be pulverized under conditions that minimize compositional changes. Maximum particle size and distribution of particle size are of great significance. Hunt and Neustadt (1966) recommended grinding grain samples for moisture determination in a Wiley laboratory mill (intermediate model). The Wiley mill was recommended because the sample is subjected to a minimum of heating during grinding and because it is protected from contact with the atmosphere. With most cereal grains, the two-stage drying method is required for samples containing above 16% moisture. For soybean samples with more than 10% moisture and rough rice samples with more than 13% moisture, the two-stage method must be used. Cereals should be ground to pass an 18-mesh screen; grinding through a 40-mesh sieve is recommended for other foods; the rotor and stator blades of the mill must be sharpened periodically depending on use. Minimum pressure should be applied in feeding the sample through the mill and the sample weight should be limited to 5 g. Standard aluminum dishes are recommended for cereals (55 mm in diameter and 15 mm high, with slightly tapered sides and tightly fitted slip-in covers designed to fit under the dish when they are placed in the oven).

In drying liquids, it is essential to spread the material over a large surface. The liquid is preferably evaporated first on a water bath, and then drying is completed in an oven. The dried residue should weigh 1–2 g. Similarly, the loss of moisture in cheese is most rapid when the final drying in a water or vacuum oven is preceded by heating on a water bath for about 1 hr. To increase the area of dried cheese, it can be smeared on a glass rod, mixed with inert material, or placed between two previously weighed sheets of aluminum.

Drying Temperatures and Times

The drying temperature used in a water determination ranges—depending on the tested material—from 70 to 155°C. Yet, some moisture is retained by biological systems at least to temperatures as high as 365°C, the critical temperature of water. The average time of drying is from below 1 hr to 6 hr or more. Foods can be dried for moisture determination either for a selected period of time or until two successive weighings show a negligible loss in weight (generally less than 2 mg for a 5-g sample, at 1-hr intervals). The drying time is inversely related to drying temperature. However, increasing the temperature increases loss of weight to a level that cannot be attained even by prolonged heating at lower temperatures. Measurements of carbon dioxide evolved during drying indicate that some weight loss at temperatures over 80°C result from decomposition of the product. Consequently, the use of high temperatures to obtain rapid results can be justified on pragmatic grounds only. In foods susceptible to decomposition, drying temperatures can be reduced by using vacuum ovens, and drying times can be shortened by using drying agents or by passing dry air over the samples.

A common source of error in a moisture determination is the formation of a crust that is impervious to evaporation of moisture from the center of a dried sample. In drying samples rich in sugars, the effects of crust formation can be reduced by moistening with water and thorough mixing with sand or asbestos to increase the exposed surface or by top drying of thin layers under infrared heat lamps. In some plant materials, drying first at relatively low temperatures (to prevent crust formation) and completing the drying at higher temperatures is the solution. Losses of water through chemical reactions (dextrinization or hydrolysis) are accelerated at elevated temperatures and high moisture levels. To reduce such losses, predrying at low temperatures followed by final drying at a recommended temperature is helpful.

The sensitivity of certain sugars (especially fructose) to decomposition at elevated temperatures rules out determining moisture by drying in air ovens in such foods as honey and fruit syrups. Mitchell (1950A,B) found that the drying of fructose, glucose, or sucrose solutions was faster and the tendency to decompose was less if the pH was below 7. Fructose solutions decompose at temperatures above 70°C; glucose is relatively stable at 98°C. Sucrose solutions dry very slowly in the steam oven. In the vacuum oven, slightly acidic fructose solutions on pumice adsorbant dried in 4 hr at 70°C; and at 60°C (preferable temperature) in 7–9 hr.

Air-Oven Methods

Because of their convenience, air-oven methods of various types are widely used in control laboratories for the determination of moisture. Ovens ranging from simple water or steam-jacketed types to elaborate equipment with forced-air circulation and built-in balances are in wide use.

Either convection-type ovens or forced-draft ovens can be used for moisture determinations. Forced-draft ovens are preferred, since they generally accommodate more samples and attain the desired temperatures more rapidly after the samples are inserted in the oven. The principal criterion of the suitability of an oven is precision of control and uniformity of temperature at different positions in the oven. A good oven should have a thermoregulator of ±0.5°C or less that will maintain its setting without requiring constant adjustment. Heating units should give an on-and-off cycle of 15 min or less for an oven with a normal load. Variations of temperature (±3.0°C or more) with position in the oven, which is a major problem in convection ovens, can be somewhat reduced by the proper distribution of heating elements, by adequate insulation, and by placing limited numbers of samples on one shelf in a central, uniformly-heated area.

Variations of temperature with position in the oven can be minimized by mechanical circulation of the air in the oven. The Brabender, semi-automatic moisture tester (Fig. 33.1) uses a small fan to circulate air in the oven and has a built-in automatic balance. The oven holds ten 10-g samples in flat, tared dishes. After a specified drying period, each dish is weighed (without removing from the oven or cooling) and the moisture is indicated on an illuminated scale. Its precision, relative speed, and convenience have made this oven popular for the determination of moisture in various foods. The instrument is useful also for determining the

Fig. 33.1. Brabender rapid moisture oven. (Courtesy C. W. Brabender Instruments, Inc.)

drying rate of foods at various temperatures. Cereals are generally dried in the Brabender oven for 1 hr at 130°C.

Another common air oven is the Chopin oven, which uses drying temperatures up to 200°C. The vaporized water is passed through a calcium carbide container fitted with a flame jet on top. As long as water vapor is evolved, the generated acetylene is burning. Drying is completed when the flame is reduced to a specified size. The sample is cooled and weighed. Actual drying times are about 5 min for flour and 7 min for ground grain.

Vacuum-Oven Methods

Moisture determination by drying in a vacuum oven is a close and reproducible estimate of true moisture content. In many biological materials, most of the water can be removed with relative ease, but removing the last 1% (or so) is difficult. This residual water can be removed more easily, more rapidly, and with less overall change of the organic components by drying under vacuum than by other methods.

The rate of drying can be increased through lowering the vapor pressure in the air by using vacuum. If during drying no air were let into the vacuum oven, the pressure of water vapor in the oven would eliminate the usefulness of the vacuum oven, especially in samples with high moisture content. In addition, the efficiency of the oven is limited by the rate of water diffusion into the pump. Therefore, a small amount of air, preferably dried by passing through concentrated sulfuric acid, is let in to continuously sweep the oven.

The usual vacuum-oven method involves drying to constant weight at a pressure below 50 (preferably 25) mm of mercury. Drying to constant weight requires rather long periods for most foods. In cereals, practically constant weight can be attained within 16 hr. Drying temperatures specified for most vacuum-oven methods are 98–102°C, indicating an accuracy of ±2°C. Mercury-in-glass thermoregulators with suitable relays provide much better accuracy (±0.1°C), maintain their setting, and give trouble-free operation. Foods rich in levulose (fruits) must be dried at 70°C or below. Using a 0.5-in. aluminum shelf with good contact with the walls substantially improves the uniformity of temperature and heat transfer in vacuum ovens.

Other Drying Methods

Drying by conventional methods involves heat conductivity and convection, and drying times are long. Infrared drying is more effective, as it involves penetration of heat into the sample being dried. Infrared drying can shorten the drying time to 12–33% of that required in conventional drying. In infrared drying, a 250- to 500-W lamp is used, the filament of which develops temperature of 2000 to 2500 K (Schierbaum 1957). The distance of the infrared source from the dried material is a critical parameter because close proximity may cause substantial decomposition. A distance of about 10 cm is recommended. The thickness of the dried material should not exceed 10–15 mm. Drying times under optimum conditions are up to 20 min for meat products and up to 25 min for baked products. With ground grain, 10 min is generally satisfactory. Sample size should be 2.5–10 g, depending on food and moisture content. Infrared drying instruments are available equipped with forced ventilation, and/or connected to torsion balances with indicator scales to read moisture content directly. There have been many studies on the use of microwaves for the rapid determination of moisture. The method has undergone considerable development and gained considerable acceptance (Pettinati 1980; Kropf 1984). Pettinati (1975) described a procedure for meat and meat products that requires only 2.5 min of heating in a microwave oven, followed by drying in a forced-air oven for 1 min. In this method, the sample is dispersed with sodium chloride and ferrous oxide in a weighing bottle. The salt reduces spattering and the oxide strongly absorbs microwave radiation and accelerates drying. The method agrees well with the AOAC method. Lee and Latham (1976) reported that canned pet food spread on a tared filter paper can be dried in a domestic-type microwave oven in less than 2 min. The time rquired to determine moisture in meat, sausages, and meat pastes by the microwave-oven method was 4–6 min in studies by Steele (1976) and Kolar (1978). According to Risman (1978), 6–8 min are required to dry 10-g samples of meat in a compact microwave oven with a power input of 100 W. Analytical microwave ovens with built-in balances, digital displays, and microcomputers to calculate moisture are available commercially.They can dry 5- to 10-g samples spread on glass fiber filter pads within 3–4 min.

In the Moisterfuge (Marriott *et al.* 1975), samples can be dried within 12 min in a heated centrifuge. The results are in good agreement with reference determinations.

Foods can be dried by desiccation in an evacuated desiccator over dehydrating or water-absorbing chemicals, such as sulfuric acid, freshly ignited powdered lime, phosphorus pentoxide, or lumps of calcium carbide. However, at room temperature, such methods are very slow; some foods cannot be dried completely even after several months. The use of vacuum at somewhat higher temperatures (50°C) is generally more satisfactory.

Drying at low temperatures has been particularly popular with researchers interested in developing standard reference methods for mois-

ture determination. The method for moisture determination in dried vegetables by Makower and Nielsen (1948) involves the addition of a large
amount of water to a weighed sample, freezing and drying to the frozen
stage (lyophilization); and completion of the drying in a vacuum oven or
vacuum desiccator in the presence of an efficient water absorbent. The
last step can be completed in a relatively short time at, or slightly above,
room temperature because of a marked increase in drying rate brought
about by lyophilization.

True moisture contents have also been determined by several other
methods. The reversibility method of Sair and Fetzer (1942) measures the
extent of decomposition during drying. Makower *et al.* (1948) developed
the redrying technique to establish empirical conditions (time and temperature) for a true moisture determination. The isotopic dilution method
of Brand and Kassell (1942) is another primary method for the determination of water. A known amount of heavy water is added to the system,
and the total amount of water in the food can be computed from the heavy
water content in a portion of the total water.

DISTILLATION METHODS

Distillation methods have been used for moisture determinations for
almost 100 yr. There are two main types of distillation procedures. In
one, water is distilled from an immiscible liquid of high boiling point. The
sample suspended in a mineral oil having a flash point much above the
boiling point of water is heated to a predetermined temperature in a suitable apparatus. The water that distills off condenses and is collected in
a suitable measuring cylinder. In the second type, the mixture of water
and an immiscible solvent (e.g., xylene, toluene) distills off, and is collected in a suitable measuring apparatus in which water separates and its
volume can be measured. Distillation with an immiscible solvent under
a refluxing-type of condenser is the method most commonly used.

Distillation with a boiling liquid provides an effective means of heat
transfer, the water is removed rapidly, and the test is made in an inert
atmosphere that minimizes danger of oxidation. Distillation methods
cause less decomposition in some foods than drying at elevated temperatures. Although chemical reactions produced by heat are reduced, they
are not eliminated. Adverse effects of heat can be reduced still further
by selecting organic solvents with a boiling point below that of water,
such as benzene. Such a choice, however, lengthens the distillation time.
The liquids most commonly used for distillations of the second type are
xylene, toluene, and tetrachloroethylene.

If the boiling liquid is lighter than water, the collecting trap usually contains a tube sealed at the bottom and calibrated upward. This form requires reading only one meniscus in measuring the amount of collected water. The calibrated portion of the tube may be cooled by a water bath to a standard temperature; or as the water layer is covered by a layer of the immiscible liquid, the whole trap may be removed for temperature adjustment. If a liquid heavier than water is used, two menisci must be read, the condensing liquid must pass through the water, formation of an emulsion is enhanced, and the temperature adjustment somewhat complicated. On the other hand, recommended liquids with a high specific gravity (tetrachloroethylene, carbon tetrachloride) eliminate fire hazard and reduce the danger of overheating or charring, as the sample floats on top of the liquid.

The Brown–Duvel method was developed in 1907 and was used with certain modifications from 1912 till 1959 for moisture determinations in cereals by the Board of Grain Commissioners for Canada, a federal government agency responsible for quality control of Canadian grain. A drawing of the apparatus is shown in Fig. 33.2. For routine tests, units for

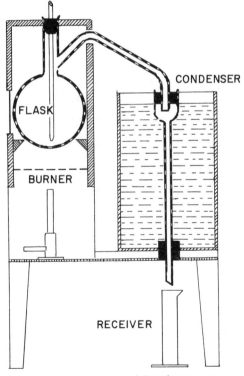

Fig. 33.2. Brown and Duvel apparatus.

heating two to six flasks simultaneously are available. A 100-g sample of whole grain is heated in the flask with 150 ml of nonvolatile oil to a specified cutoff temperature (180°C) for wheat. The amount of water that is distilled into the graduated cylinder is read in milliliters and reported as percent moisture. The determination takes about 1 hr and, with a suitable bank of equipment, one person can make about 12–18 determinations per hour. The method can be taught to inexperienced persons, and the required equipment is reasonably rugged and easily replaced.

Development of the Dean and Stark (1920) distilling receiver tube, which permits continuous refluxing and separation of water, provided a means of efficiently removing water with limited amounts of solvent. Bidwell and Sterling (1925) improved the Dean and Stark tube by placing a small reservoir above the calibrated tube. The tubes are shown in Fig. 33.3. Figure 33.4 shows an apparatus designed by the Corn Industries Research Foundation for the determination of moisture in corn products. The apparatus basically uses the Bidwell and Sterling design. Different solvents have been recommended for various corn products, and required distillation times range from 6 to 48 hr.

Many difficulties may be encountered in the determination of moisture by distillation methods. These include relatively low precision of the receiving measuring device; difficulties in reading the meniscus; adherence of moisture droplets to the glass; overboiling (especially with xylol); sol-

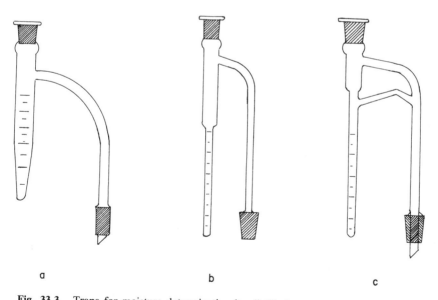

a   b   c

Fig. 33.3. Traps for moisture determination by distillation procedures. (a) Dean and Stark; (b) Bidwell and Sterling; and (c) Modified Bidwell and Sterling.

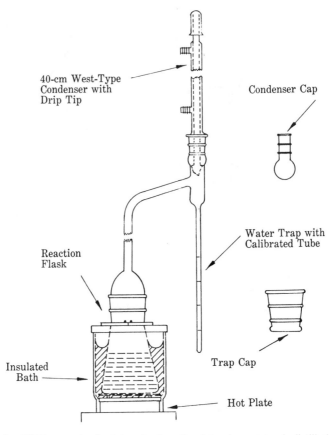

Fig. 33.4.   CIRF apparatus for determination of moisture by azeotropic distillation. (Courtesy Corn Industries Research Foundation.)

ubility of water in the distillation liquid; incomplete evaporation of water and underestimation of moisture contents; and distillation of water-soluble components. Adherence of water to the walls of the condenser tube or sides of the receiving tubes can generally be prevented by using thoroughly cleaned glassware. However, in distilling some materials a small amount of fat or wax may be carried to the reservoir trap or the drip tip of the condenser. Removing the lipid material mechanically releases water adhering to walls and gives better reading of the meniscus. Use of a small amount of wetting agent will also improve meniscus reading. Incomplete recovery of water due to the formation of an emulsion can sometimes be avoided by adding small amounts of amyl or isobutyl alcohol. To improve the moisture distillation, using wide-mouthed boiling flasks is recommended; dispersing the tested material on diatomaceous earth is useful with many viscous foods rich in sugar or protein. Foods in powder form (cereals, flours, starches) tend to bump during the distillation through overheating on the bottom of the flask. This can be largely overcome through the introduction of a small amount of dry short-fiber asbestos). If the material is heat-sensitive and decomposition of the food with the formation of water is suspected, effective techniques of food dispersion and low boiling point liquids should be used.

The main objection to distillation procedures is that they are not adaptable to routine testing. The *Official Methods of Analysis of the Association of Official Agricultural Chemists*, 13th ed. (1980) lists use of distillation with toluene only for testing of stock feeds and cheese. It is doubtful whether 48-hr distillation procedures are actually used in the corn milling industry for routine testing. A distillation procedure for rapid, approximate, determination of moisture and simultaneous removal with *n*-butyl ether of fat from meat and meat products was described by Davis *et al.* (1966).

CHEMICAL METHODS

Karl Fischer Titration

The Karl Fischer method is particularly applicable to foods that give erratic results when heated or submitted to a vacuum. The Karl Fischer titration has been found to be the method of choice for determination of water in many low-moisture foods such as dried fruits and vegetables, candies, chocolate, roasted coffee, oils, and fats. The method is superior to others for determining moisture in sugar-rich foods (sugars, honey) and in foods rich both in reducing sugars and proteins. The procedure has been applied also to foods with intermediate moisture levels (bakery doughs, baked products, fat-rich cake mixes) and to foods with high levels of volatile oils. The method is seldom used in the determination of water in structurally heterogeneous, high-moisture foods such as fresh fruits and vegetables.

The Karl Fischer method for moisture determination is based on the procedure described by Bunsen (1853) involving the reduction of iodine by sulfur dioxide in the presence of water:

$$2H_2O + SO_2 + I_2 \rightarrow H_2SO_4 + 2HI$$

Karl Fischer (1935) modified the procedure and established the conditions for quantitating the reaction. Methanol and pyridine are used in a four-component system to dissolve iodine and sulfur dioxide. The basic reaction takes place in two steps (Mitchell 1951):

$$C_5H_5N \cdot I_2 + C_5H_5N \cdot SO_2 + C_5H_5N + H_2O \rightarrow$$

$$2C_5H_5N \cdot HI + C_5H_5N \cdot SO_3$$

and

$$C_5H_5N \cdot SO_3 + CH_3OH \rightarrow C_5H_5N(H)SO_4CH_3$$

For each mole of water, 1 mole of iodine, 1 mole of sulfur dioxide, 3 moles of pyridine, and 1 mole of methanol are required. In practice, an excess of sulfur dioxide, pyridine, and methanol is used, and the strength of the reagent depends on the concentration of iodine. For general work, a methanolic solution containing the other components in the ratio of 1 iodine; 3 sulfur dioxide: 10 pyridine, and at a concentration equivalent to about 3.5 mg of water per milliliter of reagent is used. Other compositions may be preferable for various foods.

Numerous procedures have been proposed for preparation of the Fischer reagent. In laboratories that prepare and consume large amounts of the reagent, a stable stock solution of iodine dissolved in pyridine and then diluted with methanol is prepared. Liquid sulfur dioxide is added to portions of the stock solution a day or two before use. To minimize losses of active reagent from side reactions that consume iodine, many laboratory supply houses market the reagent as two solutions: (1) a solution of iodine in methanol and (2) sulfur dioxide in pyridine. The solutions are mixed shortly before use. The sample in which water is to be determined, is dispersed in an appropriate solvent (e.g., dry methanol) and the complete, four-component, reagent is added to the sample.

Alternatively, a two-mixture Karl Fischer reagent can be used. The sample is dissolved in a mixture of sulfur dioxide–pyridine–methanol and titrated with a solution of iodine in methanol. The iodine in methanol solution is stable and needs standardization only rarely, provided the solution is protected against moisture.

In the titration with the Karl Fischer reagent, iodine and sulfur dioxide are added in the appropriate form to the water-containing food. The excess of iodine that cannot react with water is in a free form. The amount required for the titration can be determined visually, till a yellow-mahagony brown color is seen. Adding to the system a few drops of methylene blue gives a green end point. Reagents, titrating flask, and buret must be protected from atmospheric moisture.

Instead of visual observations, one can use a photometric endpoint determination. In highly colored solutions, electrometric titration by the dead-stop technique is most often used. It is based on the fact that when a small potential difference is applied to two equal platinum electrodes immersed in a redox system, the potential formed by polarization is compensated and the current flow is interrupted. During the last phase of the redox titration—at the endpoint—polarization or depolarization of an electrode to a complete depolarization or polarization of both electrodes takes place. That change is accompanied by a large deflection of a galvanometer needle or change in a magic eye. Instruments based on this principle are sold by many laboratory supply houses; one such instrument is shown in Fig. 33.5. Units that automatically perform the Karl Fischer water determination by the conductometric method are also available. These units are equipped with electromagnetic, glass-sealed titrating valves for either direct or back titration. Visual determinations of end points are sensitive to differences of less than 0.5 mg water. The sensitivity is increased (to 0.2 mg) if end points are determined electrometrically.

Theoretically, the Karl Fischer reagent may be used for the determination of water in liquids, gases, and solids. Basically, the method may

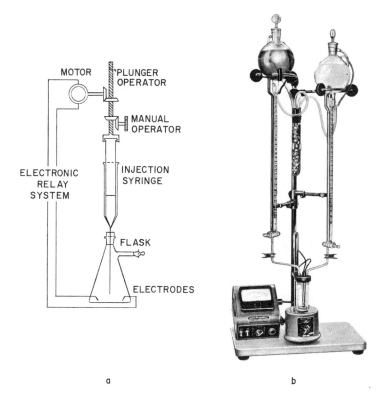

a   b

Fig. 33.5.   Apparatus for moisture determination by the Karl Fischer method. (a) Schema of apparatus; (b) Commercial equipment. (Courtesy LaPine Scientific Co.)

be considered as a primary standard method for determining water content. In most biological materials, it must be calibrated and standardized against another reference method. Most organic compounds do not interfere with the determination, but some exceptions have been observed. For example, ascorbic acid is oxidized essentially quantitatively to dehydroascorbic acid by the Fischer reagent; thus, when ascorbic acid is present, the sum of the moles of water plus ascorbic acid is determined. Aldehydes and ketones tend to react with the methanol of the Fischer reagent, forming acetals and releasing water. The interference may result in a fading end point. Similarly, mercaptans, diacylperoxides, thio acids, and hydrazines may interfere, though usually methods are available to eliminate such interferences. Inorganic compounds that affect the water titration include metal oxides, hydroxides, carbonates, bicarbonates, chromates, dichromates, borates, and sulfides.

The determination of moisture is carried out in a nonaqueous system. This requires preparing samples for analysis under conditions that minimize changes in water content. Fluids are delivered best with an automatic pipet or syringe. Viscous fluids or pastes are generally homogenized with a solvent. Solids are either homogenized with the solvents or titrated as suspensions. For granular products that must be pulverized, titration is often carried out in a grinding container. Hart and Neustadt (1957) described an accurate method for determining moisture in seeds involving an efficient grinding–extraction procedure that takes place in a completely enclosed cup. Methanol present in the cup extracts water from the grain as it is being ground. After suspended matter has settled from the extract, an aliquot is titrated with the Karl Fischer reagent. Fats and waxes are spread out or made into a paste with an appropriate fluid. Oil-rich foods (oil seeds, germ of cereals) must be ground to a fine powder. If the material is pulverized in a highspeed grinder, the temperature increases and moisture may be lost.

The simplest technique involves direct titration of the sample with the Karl Fischer reagent. To titrate all of the water in a sample, it may be necessary to extract the water by refluxing the sample with an appropriate solvent (e.g., methanol). Sometimes excess reagent is added and back-titrated with a standard water solution. For example, Fosnot and Haman (1945) found that in cereals and cereal products, completeness of the reaction between the water in the sample and the reagent varied with the kind of material and fineness of the grind. Frediani et al. (1952) found that for best results, cereals, dried yeast, etc., should be ground to pass a 40-mesh sieve. On the other hand, extraction of water from dehydrated fruits or vegetables in boiling methanol is rapid. The determination can be completed in less than $\frac{1}{2}$ hr, and the results are less dependent on the fineness of grinding than with most other common methods. In some dried vegetables, water is bound tenaciously, diffuses slowly, and the apparent moisture changes considerably with an increase in extraction time. To obtain agreement between the Karl Fischer titration and a vacuum-oven method, it is necessary to extract dry potatoes for 6 hr and carrots for less than $\frac{1}{2}$ hr.

Various modifications have been proposed for the determination of water in oils and fats. Francois and Sergent (1950) suggested boiling a large sample in methanol and titrating an aliquot of the cooled solution. Meelheim and Roark (1953) distilled water from oils and fats with toluene into a receiving titration flask containing methanol and excess Karl Fischer reagent. According to Bernetti et al. (1984), an international collaborative study of the Karl Fischer determination of water in fats and oils resulted in the following recommendations: (1) the need to exert special care in handling heterogeneous and semisolid or solid samples; (2) use of sample sizes of 5–25 g; (3) the need to adopt alternative weighed water standardization; and (4) the use of a chloroform–methanol (1:1 or 1:2) premixed solvent along with the use of reagent dilution for low-water samples.

The main difficulty in using the Karl Fischer method arises from the lack of complete water extraction. McComb and Wright (1954) found formamide to be a more rapid and versatile extractant of water from foods than methanol. Modification of the extraction procedure is exemplified by a method for water determination in dairy products in which xylene or carbon tetrachloride is employed in mixed solvent systems with alcohol (Morgareidge 1959).

Determinations of moisture by the Karl Fischer method were reviewed by Zuercher and Hadorn (1978) who emphasized the need to optimize pretreatment and test conditions for each food type. They compared the following variants: direct automatic titration, cold and warm extraction, continuous recording of the titration curve, back titrations, various extraction methods, and a heating-out method with a special kiln. For maximum values, a high ratio of methanol (i.e., about 200 ml per 1 g flour containing 15% moisture) and prolonged extraction at elevated temperatures were required. The heating-out method, which employs a special furnace, requires a long extraction time and was unsatisfactory for most foods. Best results were obtained by the use of a high-frequency disintegrator (shredder) in methanol. The addition of diethylamine prevented the splitting off of water in acid, sugar-rich foods (e.g., dry fruits) as a result of chemical decomposition. The high-frequency disintegration method provided accurate and reproducible results in the determination of water in corn starch, wheat flour, and a variety of dry fruits. The equipment for direct titration at elevated temperatures and for high-frequency disintegration are shown in Fig. 33.6.

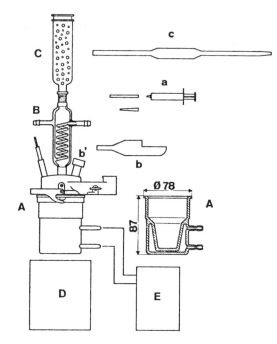

Fig. 33.6.   Equipment for direct titrations at elevated temperatures. (A) titration vessel with thermostat; (B) high-intensity cooler; (C) drying tube; (D) magnetic stirrer; (E) heater. (a) syringe, (b) weighing boat for introduction at b'; (c) pipet.

The KF Processor is a sophisticated, highly versatile instrument for the determination of water in solid and liquid foods by the Karl Fischer method (Schalch 1984). Its microprocessor performs all the necessary calculations and controls the instrument throughout all phases. Display, printer, and keyboard make the operation straightforward and effective. The instrument can be adjusted for use in foods that vary widely in water content and distribution. A special titration cell was designed for rapid and accurate results. The processor can be connected to a balance and external data system.

According to Scholz (1984A,B), there have been several important developments over the past decade in instruments, reagents, and applications of the Karl Fischer method. Pyridine has been replaced by commercially available new reagents (e.g., Hydranal) that eliminate the unpleasant smell of pyridine, speed up the reaction, and increase precision by eliminating the sluggish attainment of the end point. The use of microprocessors in new instruments allows programmed titrations. The coulometric titration is particularly useful for foods low in moisture and by titrating at elevated temperatures, extraction of water (even from high-moisture foods) can be accelerated. A whole range of microprocessor-controlled coulometric and volumetric titrators is available. An automated set up to do Karl Fischer titrations for moisture analyses is illustrated in Fig. 33.7.

Other Chemical Methods

When foods are mixed with powdered calcium carbide, the water reacts to produce acetylene. The quantity of acetylene can be measured by the loss in weight of the mixture, pressure or volume of the gas produced in a closed system, and the formation of copper acetylide (gravimetric, titrimetric, or colorimetric). Accuracy of the methods depends on the fineness of grinding of the interacting components. Gas production can be doubled by using calcium hydride in place of calcium carbide. None of the methods has found wide application.

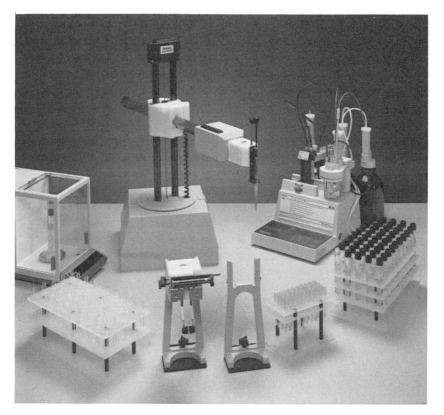

Fig. 33.7. An automated set-up to do Karl Fischer titrations for moisture analyses. (Courtesy Zymark Corp. Hopkington, MA.)

Water in sugar products has been determined by an elegant method whereby the sugar is ground with cobaltous bromide in dry chloroform (Gardiner and Keyte 1958). Combination of the salt with water causes it to precipitate from solution; it is separated subsequently by filtration and the cobaltous bromide is weighed after evaporation of the solvent. Semiquantitative estimates of free water in food can be obtained by measuring the rate of change in the color of filter paper impregnated with cobaltous chloride. The dry paper is blue and becomes pink when moist.

Mixing sulfuric acid with food results in heating, and the increase in temperature under fixed conditions is roughly proportional to the water content of a food. Launer and Tomimatsu (1952) described a procedure in which nonaqueous organic matter is determined by an oxidation with a dichromate solution in sulfuric acid, and the moisture is measured by difference. The method is rapid (5 to 10 min exclusive of sample preparation) and relatively simple. Its usefulness is somewhat limited, and the use of corrosive reagents makes it unattractive for use in routine quality control.

Determinations of water by the acetyl chloride method (Smith and Bryant 1935) is generally useful in determining moisture in products that give satisfactory results with the Karl Fischer reagent. The determination involves titration of the acid formed in the following reactions:

$$H_2O + CH_3COCl \rightarrow CH_3COOH + HCl$$

$$ROH + CH_3COCl \rightarrow CH_3COOR + HCl$$

The method is useful in determining moisture in oils, butter, margarine, dry spices, and many low-moisture foods. The acetyl chloride reagent is generally dissolved in toluol and the food is diluted or dispersed in pyridine. The method has found little acceptance.

PHYSICAL METHODS

Infrared Determination

The infrared determination of water is based on measuring the absorption at wavelengths characteristic of the molecular vibration in water (Kaye 1954). The most useful wavelengths are 3.0 and 6.1 μm, (fundamental vibrational modes of the water molecule), 1.93 μm (combination absorption band) and 1.45 μm (first overtone of the OH stretching). Measurement of the water content using infrared absorption techniques can provide analyses with a sensitivity of a few parts per million in a wide range of organic and inorganic materials. High photometric sensitivity to

1 part in 5000 is possible by use of a tungsten light source and a lead sulfide detector with carefully designed electronic circuitry. Wavelength isolation is obtained by using narrow-band interference filters or a simple grating monochromator. Use of glass or quartz windows and lens materials simplifies maintenance and permits measurements at elevated temperatures. Water absorption bands at 1.45, 1.91, and 2.83 μm have been used for water analyses at sensitivities ranging from 1 to 5 ppm to several percent (Trippeer 1965). Hart *et al.* (1962) and Norris and Hart (1965) investigated the spectral region from 0.7 to 2.4 μm for measuring the moisture content of grain and seeds. The spectral absorbance curve for a thin layer of ground wheat showed minimum interference for the 1.94-μm band. Using a 2-g sample of ground material mixed with 1.5–2.0 ml of carbon tetrachloride in a 4.4-cm-diameter cell, the transmittance values of a large number of wheat, soybean, wheat flour, and wheat bran samples are measured at 1.94 and 2.08 μm. From these data, the optical density difference ΔOD (1.94 and 2.08 μm) was computed for each sample and related to the moisture content as determined by standard procedures. Calibration curves for each of the four materials showed standard deviations from 0.28 to 0.37% moisture for the moisture range from 0 to 20%. The water absorption band at 0.97 μm was measured on individual intact peanuts and related to the moisture content. A measurement within ± 0.7% moisture content was obtained using ΔOD (0.97–0.90 μm) as the measured value.

A near infrared method has been reported for the analysis of fruit and vegetables (Gold 1964). In collaborative work the near infrared method was found to be as accurate as, and more rapid and specific than, the vacuum-oven method for the determination of moisture in dried vegetables and spices. The usefulness and limitations of the near infrared spectral absorption method for determining moisture (and fat) in meat products were described by Ben-Gera and Norris (1968) (See also infrared reflectance in Chapter 9).

Gas Chromatographic Methods

Several methods are based on extracting the moisture with an organic solvent and determining water in the extract by gas chromatography. The procedure has general application provided (1) water is extracted effectively and (2) the extracts contain no substances that coincide with either the solvent or water peaks on the chromatogram (Weise *et al.* 1965).

In a typical analysis, 15 g of sample is blended with 100 ml of absolute methanol and several milliliters of secondary butanol (depending on the expected moisture content of food). The mixture is allowed to settle for

15 sec, then a 2-μl aliquot is removed from the clear supernatant and analyzed in a gas–liquid chromatograph, which takes about 5 min. The method is useful for a variety of foods varying in moisture content from 8 to 65% (Schwecke and Nelson 1964).

Gas chromatography has been reported to be a reliable method for determining the moisture content of a number of foods such as cereals, cereal products, fruits, and fruit products (Brekke and Conrad 1965). Khayat (1974) determined moisture in 10 min by a gas chromatographic assay of an anhydrous isopropanol extract of a meat sample. The repeatability was 0.2% and the agreement with the AOAC oven-drying method within 0.2%.

Nuclear Magnetic Resonance Methods

The principles of nuclear magnetic resonance (NMR) were discussed in Chapter 18. The theory, effects of food components, physical properties, and uniformity of test specimens on moisture determination by NMR were reviewed by Shaw et al. (1953) and Rollwitz (1965). In NMR instruments, the frequency of the magnetic field is kept constant and the strength of the field is varied over a narrow range. The peak-to-peak amplitude of the signal of a sample in the field can be obtained from a calibration curve of amplitude plotted against moisture content. The method depends on the absorption of radio-frequency energy by the hydrogen nucleus and is therefore not specific for water. With most materials, however, it is easy to distinguish between the signals contributed by the hydrogen of absorbed water and that which is present in more hindered forms (Conway et al. 1957). Elsken and Kunsman (1956) reported that NMR methods are satisfactory for moisture determinations in foods with limited amounts of soluble solids, as the hydrogen nuclei of the solutes absorbed energy in the NMR spectrometer in a manner undistinguishable from those of water. In the case where the soluble solids were constant and preferably small, a calibration curve may be made.

Although the technique of NMR requires somewhat elaborate equipment, the method has sufficient advantages to have prompted extensive investigations. The usefulness of the technique for moisture determinations in food was described by Shaw et al. (1953), Conway et al. (1957), and Rubin (1958). NMR measurements are rapid (actual measurement requires 1 min), nondestructive, accurate, and for many foods more precise than drying techniques. Calibrated standards can be kept sealed in a glass tube and used repeatedly over a period of time. The relatively large sample size reduces difficulties caused by nonuniformity of specimen.

The difference in the relaxation times between free and bound water has been used in pulsed NMR to determine the relative amounts of the two types of water and the mobility of water in various foods (Leung *et al.* 1976; Brosio *et al.*, 1984).

Electrical Methods

The relatively long time required to carry out most moisture determinations has prompted the development of numerous rapid electrical methods. The most commonly used ones are based on the determination of conductivity or capacitance. Moisture determinations by electrical methods in foods are affected by many factors including the texture of the food, packing, mineral contents, temperature, and moisture distribution. The relation between the moisture content and the measured parameter is empirical and assumes that the effects of other factors have been minimized or accounted for by proper calibration.

In conductivity measurements, the conductivity or resistance of a food introduced into an electrical circuit is measured. The measurement is most commonly made on instruments using a modification of the Wheatstone bridge.

Several instruments, widely used by the industry for rapid, routine determination of moisture by electrical methods, are based on measuring resistance. These include the Universal, Marconi, and Tag–Heppenstall meters. In the Universal meter, a hand-driven megger establishes a voltage across a 20-g sample pressed in a steel cup to a specified thickness. The electrical resistance is indicated on an ohmmeter of the dynamometer type. A correction scale for temperature is provided. The instrument requires neither batteries nor power supply. The Marconi instrument is battery operated. A sample is compressed in a small cell, the bottom of which is equipped with electrodes consisting of two circular rings. The sample need not be weighed accurately, as the current penetrates the sample only to a depth of the order of the electrode separation. For many years, the Tag–Heppenstall meter (also known as the Weston moisture meter) was widely used in the determination of moisture in cereal grains. In this instrument, the electrical resistance of the grain is measured as it passes between two corrugated steel rolls (one motor driven and one stationary) that serve as electrodes. The grain is crushed to provide good electrical contact for measuring the conductance or resistance of the grain as it passes between the rolls. Galvanometer readings and the temperature corrections are used in reading moisture contents from calibration charts. Moisture determinations with the Tag–Heppenstall meter are rapid (10–20 sec), but it is difficult to maintain the calibration, as wearing of the rolls and bearings increases the spacing between the rolls, giving lower results.

In dielectric-type meters, the determination is based on the fact that the dielectric constant of water at 20°C (80.37) is higher than that of most fluids and solvents. The heart of the instrument is a capacitance cell in which two metal plates of given size are at a fixed distance. Measurements can be made over the range −20 to 180°C. The two plates have equal but opposite charges that are reversed at fixed frequencies to give an alternating current field. If a material is introduced between the plates, polarization takes place. With nonpolar substances, induced dipoles are formed in the electrical alternating field. Polar substances, with permanent dipoles, are oriented in the alternating field. This increases the charge of the plates and their capacitance. Consequently, the determination is based on measuring the capacitance, and the dielectric constant that it measures is an index of the increase in capacitance. It is determined by comparing the capacitance C of a capacitance cell filled with the tested dielectric substance to the capacitance C_0 of an empty (vacuum) condenser; thus, $C = \epsilon C_0$, where ϵ is a dimensionless constant that is affected strongly by temperature and frequency.

The dielectric-type meter most widely used in testing cereals is the Motomco. The Motomco electric moisture meter was made official in 1959 for most grain in Canada and has been used widely in the United States since 1963 (Hunt and Neustadt 1966). The principle of the instrument's operation is the balancing of two oscillating circuits as indicated by a milliammeter. One oscillator is fixed to oscillate at 18 MHz, while the other that contains the test cell for holding the grain can be adjusted to balance with the fixed oscillator by means of a variable condenser coupled to a graduated dial.

In addition to conductivity and capacitance measuring instruments, other types of electric meters are available. The signal strength from a radio transmitter falls off as the distance increases. In air, this decrease follows the inverse square law. If water or a material containing water is inserted between the transmitter and the receiver, an additional factor affects the results. The absorption of electromagnetic energy by water forms the basis of a method of measuring moisture content by microwave absorption (Watson 1965). Moisture measurements by high-frequency currents offer the advantage that they can be made by means of electrodes that do not come in contact with the material. The measurements can be made in three ways: (1) measurements of the capacitance of a flat capacitor in which the material to be measured acts as the dielectric, (2) measurement of the resistance of the material at high frequency, and (3) measurement of the impedance of an induction in which the material acts as a core (Leroy 1965).

For industrial processing, the methods of continuous moisture measurement that have been most commonly applied include resistance, capacitive reactance, power loss at microwave and radio frequencies, temperature difference due to evaporation, thermal neutron capture by hydrogen, neutron reflection by hydrogen, and nuclear magnetic resonance (Green 1965). Of these methods, capacitive reactance and resistance have been the most widely applied. The equipment employed for the two methods is generally more simple and less expensive than for the others.

Electrical instruments for moisture determination have the important advantage of speed. Some of the instruments give average accurate results if the meters are properly calibrated and checked for uniform performance. All instruments are affected by distribution of moisture in a food. For precise measurements, essentially complete equilibration of moisture is essential. Readings on all meters are also affected by temperature; in some, correction for temperature is so high that a serious error in temperature measurement introduces a serious error in moisture determination.

Moisture determinations by capacitance measurements are theoretically more satisfactory than by conductivity measurements. The latter measure free water only, and assume a constant ratio between free and bound water. Dielectric moisture meters measure both free and bound water, though the great difference in the dielectric value of the two types of water may introduce errors in the measurements when the ratio of the two shifts. Conductivity measurements are much more affected than dielectric constants measurements by the composition of soluble solids. To reduce that effect, Hancock and Burdick (1956) proposed a modified indirect conductivity method for determining water in foods and feeds. Adding a large amount of sodium chloride to an alcohol–acetone extraction mixture tends to mask the effect of other electrolytes that may be present in a sample.

Particle size and tightness of packing in the measuring cell affect all electrical measurements of moisture content. Hughes *et al.* (1965) showed that a measurement of capacitance with a correction based on the simultaneous determination of density improved correlation with moisture content. A practical guide to moisture meter calibration was presented by Gough (1983).

Miscellaneous Physical Methods

Many densimetric, refractometric, and polarimetric methods have been devised for the rapid determination of total solids in foods. In all these methods, calibration curves are available or must be prepared to correlate soluble solids with the selected physical parameter. To determine the total solids contents (or moisture by difference), the insoluble solids must be determined or assumed.

Densimetric methods (by pycnometer, specific gravity balance, or various types of hydrometers) are the most commonly used routine tests to determine dry solids in milk (in combination with the fat determination), sugar solutions (including fruit juices, syrups, concentrates), fruit products (particularly tomatoes), beverages (alcoholic and malts), and salt solutions (used by the pickling industry). Refractive index measurements give the most reliable, reproducible, and rapid values for estimating the soluble solids contents of sucrose solutions, fruits, fruit products (jellies, jams, juices, tomato products) and dried fruit (Bolin and Nury 1965), corn syrups, honey, candy, and various carbohydrate-rich products. The solids content can be determined rapidly and precisely by refractometry in milk products, protein solutions, and oil-rich foods.

The freezing point of unadulterated milk is close to $-0.550°C$; it is lowered approximately $0.0055°C$ for each 1% of added water. The percentage of added water can be determined by a cryoscopic method.

The methods described thus far were designed to measure (directly or indirectly) the total moisture content of foods. In many instances the total moisture content may be of limited significance. Moisture within a food is not distributed uniformly. Deterioration of a food might be a function of the moisture content of the component most susceptible to deterioration. Under those conditions the vapor pressure of water above (or relative humidity of the atmosphere in equilibrium with) the food may provide more meaningful information (Makower 1950). In foods, with a relatively uniform distribution of water, relative humidity elements can be used to determine moisture content rapidly and precisely (Tessem *et al.* 1965).

Measurement of water activity in foods was discussed by Roedel *et al.* (1979) and reviewed by Prior (1979). Labuza *et al.* (1976) reported on a collaborative study that compared various methods of water activity determination.

BIBLIOGRAPHY

AOAC. (1980). "Official Methods of Analysis of the Association of Official Agricultural Chemists," 13th ed. Assoc. Off. Agr. Chemists, Washington, DC.

Ben-Gera, I., and Norris, K. H. (1968). Direct spectrophotometric determination of fat and moisture in meat products. *J. Food Sci.* **33,** 64–67.

Bernetti, R., Kochan, S. J., and Pienkowski, J. J. (1984). Karl Fischer determination of water in oils and fats: International collaborative study. *J. Assoc. Off. Anal. Chem.* **67,** 299–301.

Bidwell, G. L., and Sterling, W. F. (1925). Preliminary notes on the direct determination of moisture. *J. Ind. Eng. Chem.* **17**, 147–149.

Bolin, H. R., and Nury, F. S. (1965). Rapid estimation of dried fruit moisture content by refractive index. *J. Agric. Food Chem.* **13**, 590–591.

Brand, E., and Kassell, B. (1942). Analysis and minimum molecular weight of β-lactoglobulin. *J. Biol. Chem.* **145**, 365–378.

Brekke, J., and Conrad, R. (1965). Gas liquid chromatography and vacuum oven determination of moisture in fruits and fruit products. *J. Agric. Food Chem.* **13**, 591–593.

Brosio, E., Attobelli, G., and Dinola, A. (1984). A pulsed low-resolution NMR study of water binding to milk proteins. *J. Food Technol.* **19**, 103–108.

Bunsen, R. (1853). A volumetric method of wide application. *Ann. Chem.* **86**, 265.

Conway, T. F., Cohee, R. F., and Smith, R. J. (1957). NMR moisture analyzer shows big potential. *Food Eng.* **29**(6), 80–82.

Davis, C. E., Ockerman, H. W., and Cahill, V. R. (1966). A rapid approximate analytical method for simultaneous determination of moisture and fat in meat and meat products. *Food Technol.* **20**(11), 94–98.

Dean, E. W., and Stark, D. D. (1920). A convenient method for the determination of water in petroleum and other organic emulsions. *J. Ind. Eng. Chem.* **12**, 486–490.

Elsken, R. H., and Kunsman, C. H. (1956). Further results of moisture determination of foods by hydrogen nucleic magnetic resonance. *J. Assoc. Off. Agric. Chem.* **39**, 434–444.

Fischer, K. (1935). New methods for the quantitative determination of water in fluids and solids. *Angew. Chem.* **48**, 394.

Fosnet, R. H., and Haman, R. W. (1945). A preliminary investigation of the application of Karl Fischer reagent to determination of moisture in cereals and cereal products. *Cereal Chem.* **22**, 41–49.

Francois, M. Th., and Sergent, A. (1950). Determination of moisture in oil seeds by the method of Fischer. *Bull. Mens. ITERG* **4**, 401–404.

Frediani, H. A., Owen, J. T., and Baird, J. H. (1952). Application of an automatic Karl Fischer titrator to moisture determinations in food products. *Trans. Am. Assoc. Cereal Chem.* **10**, 176–180.

Gardiner, S. D., and Keyte, H. J. (1958). A study of some methods for determining water in refined sugars, including the newly devised cobaltous bromide method. *Analyst* **83**, 150–155.

Gold, H. J. (1964). General application of near-infrared moisture analysis to fruit and vegetable materials. *Food Technol.* **18**(4), 184–185.

Gough, M. C. (1983). Moisture meter calibration: A practical guide. *Trop. Stored Prod.* **46**, 17–24.

Green, R. M. (1965). Continuous moisture measurement in solids. *In* "Humidity and Moisture, Measurement and Control in Science and Industry" (A. Wexler, ed.). Van Nostrand-Reinhold Publishing Co., New York.

Hancock, C. K., and Burdick, R. L. (1956). Modified indirect conductivity method for determining water in cottonseed meal. *Agric. Food Chem.* **4**, 800–802.

Hart, J. R., and Neustadt, M. H. (1957). Application of the Karl Fischer method to grain moisture determination. *Cereal Chem.* **34**, 26–37.

Hart, J. R., Norris, K. H., and Golumbic, C. (1962). Determination of the moisture content of seeds by near infrared spectroscopy of their methanol extracts. *Cereal Chem.* **39**, 94–99.

Hughes, E. J., Vaala, J. L., and Koch, R. B. (1965). Improvement of moisture determination by capacitance measurement through density correction. *In* "Humidity and Moisture, Measurement and Control in Science and Industry" (A. Wexler, ed.). Van Nostrand-Reinhold Publishing Co., New York.

Hunt, W. H., and Neustadt, M. H. (1966). Factors affecting the precision of moisture measurement in grain and related crops. *J. Assoc. Off. Agric. Chem* **49**, 757–763.

Kaye, W. (1954). Near infrared spectroscopy. I. Spectral identification and analytical applications. *Spectrochim. Acta* **6**, 257–287.

Khayat, A. (1974). Rapid moisture determination in meat by gas chromatography. Tuna, beef, chicken. *Can. J. Inst. Food Technol.* **7**, 25–28.

Kolar, K. (1978). A quick method for determining moisture content in meat, meat products, and ready meals. *Fleischwirtschaft* **58**, 397, 460–462.

Kropf, D. H. (1984). New rapid methods for moisture and fat analysis: A review. *J. Food Quality* **6**, 199–210.

Labuza, T. P., Acott, K., Tatini, S. R., Lee, R. Y., Flink, J., and McCall, W. (1976). Water activity determination: A collaborative study of different methods. *J. Food Sci.* **41**, 910–917.

Lang, K. (1979). Reproducibility and accuracy of measurement in determination of moisture and fat content in fresh fish using an Ultra-X-Analyzer. *Fleischwirtschaft* **59**, 413–416.

Launer, H. F., and Tomimatsu, Y. (1952). Rapid method for moisture in fruits and vegetables by oxidation with dichromate. I. Potatoes and peas. *Food Technol.* **6**, 59–64.

Lee, J. W. S., and Latham, S. D. (1976). Rapid moisture determination by a commercial-type microwave oven technique. *J. Food Sci.* **41**, 1487.

Leroy, R. P. (1965). Moisture measurements by high frequency currents. *In* "Humidity and Moisture, Measurement and Control in Science and Industry" (A. Wexler, ed.). Van Nostrand-Reinhold Publishing Co., New York.

Leung, H. K., Steinberg, M. P., Wei, L. S., and Nelson, A. I. (1976). Water binding of macromolecules determined by pulsed NMR. J. Food Sci. **41**, 297–300.

Makower, B., Chastain, S. M., and Nielsen, E. (1948). Moisture determination in dehydrated vegetables, vacuum method. *Ind. Eng. Chem.* **38**, 725–731.

Makower, B., and Nielsen, E. (1948). Use of lyophilization in determination of moisture content of dehydrated vegetables. *Anal. Chem.* **20**, 856–858.

Marriott, N. G., Smith, G. C., Carpenter, Z. L., and Dutson, T. R. (1975). Rapid moisture determination for meat samples. *J. Anim. Sci.* **41**, 296–297.

McComb, E. A., and Wright, H. W. (1954). Application of formamide as an extraction solvent with Karl Fischer reagent for the determination of moisture in some food products. *Food Technol.* **8**, 73–75.

Meelheim, R., and Roark, J. N. (1953). Determination of moisture in oils and greases. Anal. Chem. **25**, 348–349.

Mitchell, J. (1951). Karl Fischer reagent titration. *Anal. Chem.* **23**, 1069–1075.

Mitchell, T. J. (1950A). The rate of evaporation in the determination of water. *Chem. Ind.* 751.

Mitchell, T. J. (1950B). The determination of water in sugar solutions by desiccation at room temperature. *Chem. Ind.* 815.

Morgareidge, K. (1959). Food. *Anal. Chem.* **31**, 691–696.

Norris, K. H., and Hart, J. R. (1965). Direct spectrophotometric determination of moisture content of grain and seeds. *In* "Humidity and Moisture, Measurement and Control in Science and Industry" (A. Wexler, ed.). VanNostrand-Reinhold Publishing Co., New York.

Oxley, T. A., Pixton, S. W., and Howe, R. W. (1960). Determination of moisture content in cereals. I. Interaction of type of cereal and oven method. *J. Sci. Food Agric.* **11**, 18–25.

Pettinati, J. D. (1975). Microwave oven method for rapid determination of moisture in meat. *J. Assoc. Off. Anal. Chem.* **58**, 1188–1193.

Pettinati, J. D. (1980). Update: Rapid methods for the determination of fat, moisture, and protein. *Proc. 33rd Annu. Reciprocal Meat Conf.*, pp. 156–163. Natl. Live Stock and Meat Board, Chicago, Il.

Prior, B. A. (1979). Measurement of water activity in foods: A review. *J. Food Protect.* **42**, 668–674.

Risman, P. O. (1978). A microwave applicator for drying food samples. *J. Microwave Power* **13**, 297–302.

Roedel, W., Krispien, K., and Leistner, L. (1979). Measuring the water activity (a_w value) of meat and meat products. *Fleischwissenschaft* **59**, 849–851.

Rollwitz, W. L. (1965). Nuclear magnetic resonance as a technique for measuring moisture in liquids and solids. *In* "Humidity and Moisture, Measurement and Control in Science and Industry" (A. Wexler, ed.). Van Nostrand-Reinhold Publishing Co., New York.

Rubin, H. (1958). New tool for moisture analysis. Nuclear magnetic resonance. *Cereal Sci. Today* **3**, 240–243.

Sair, L., and Fetzer, W. R. (1942). The determination of moisture in the wet milling industry. *Cereal Chem.* **19**, 633–692, 714–720.

Schalch, E. (1984). A modern Karl Fischer titrator. *Am. Lab.* **16**(2), 78, 80, 85.

Schierbaum, F. (1957). Determination of moisture by infrared in food analysis. *Dtsch. Lebensm.-Rundsch.* **53**, 173–178.

Scholz, E. (1984A). New aspects in Karl Fischer titration. *Am. Lab.* **16**(10), 138–139.

Scholz, E. (1984B). "Karl Fischer Titration." Springer Verlag, New York.

Schwecke, W. M., and Nelson, J. H. (1964). Determination of moisture in foods by gas chromatography. *Anal. Chem.* **36**, 689–690.

Shaw, T. M., Elsken, R. H., and Kunsman, C. H. (1953). Moisture determination of foods by hydrogen nuclei magnetic resonance. *J. Assoc. Off. Agric. Chem.* **36**, 1070–1076.

Smith, D. M., and Bryant, W. M. D. (1935). Titrimetric determination of water in organic liquids, using acetyl chloride and pyridine. *J. Am. Chem. Soc.* **57**, 841–845.

Steele, D. J. (1976). Microwave heating applied to moisture determination. *Lab. Pract.* **25**, 515–521.

Tessem, B. M., Hughes, F. J., Pearcy, G., and Tsantir, K. (1965). Description and test evaluation of the Honeywell relative humidity flour moisture meter. *Cereal Sci. Today* **10**, 50–52, 62.

Trippeer, A. (1965). Infrared analysis of water. *In* "Humidity and Moisture, Measurement and Control in Science and Industry" (A. Wexler, ed.). Van Nostrand-Reinhold Publishing Co., New York.

Watson, A. (1965). Measurement and control of moisture content by microwave absorption. *In* "Humidity and Moisture, Measurement and Control in Science and Industry" (A. Wexler, ed.). Van Nostrand-Reinhold Publishing Co., New York.

Weise, E. L., Burke, R. W., and Taylor, J. K. (1965). Gas chromatographic determination of moisture content of grain. *In* "Humidity and Moisture, Measurement and Control in Science and Industry" (A. Wexler, ed.). Van Nostrand-Reinhold Publishing Co., New York.

Zuercher, K., and Hadorn, H. (1978). Determination of water in foods according to the Karl Fischer method. *Dtsch. Lebensm.-Rundschu.* **74**, 249–259, 287–295.

34

Ash and Minerals

ASH AND MINERAL CONTENTS

Ash is the inorganic residue from the incineration of organic matter. The amount and composition of ash in a food product depend on the nature of the food ignited and on the method of ashing.

Ash Contents

Most fluid dairy products contain 0.5–1.0% ash; the ash content increases to 1.5% in evaporated milk, and to almost 8% in nonfat dry milk. Ash in cheese depends on the water content and on the presence of mineral additives. Pure fats, oils, and shortenings contain practically no mineral components. The main mineral component in butter, margarine, mayonnaise, and salad dressing is sodium chloride.

Ash in fresh fruits ranges from 0.2 to 0.8% and is generally inversely related to moisture content. Some dried fruits (e.g., apricots) may contain as much as 3.5% ash. The ash content of most vegetables is around 1% and is generally higher than that of fruits. Pickles, sauerkraut, and other processed vegetables are rich in salt and their ash content is high. Beans contain up to 4% ash.

The white endosperm fraction obtained during milling of wheat contains less than 0.5% ash; the bran, aleurone layer, and germ are rich in minerals. The ash of baked products depends mainly on their salt content. Most nuts contain 1.5–2.5% ash.

Fresh meat and poultry contain around 1% mineral components, but the ash of processed meats may be as high as 12% (in dried and salted beef). The ash content of the edible portion of fresh fish ranges from 1 to 2%. The yolk contains almost three times as much mineral components as the white of eggs (respectively, 1.7 and 0.6%).

Pure sugar, candy, honey, and syrups contain trace amounts to 0.5% ash; but sweets made with brown sugar or chocolate contain more inorganic components.

Mineral Contents

The various minerals that compose the ash occur in different proportions in different foods. Calcium is present in relatively high concentrations in most dairy and dairy-containing products, cereals, nuts, some fish, eggs, and certain vegetables. Small concentrations of calcium are present in practically all foods, except pure sugar, starch, and oil.

Phosphorus-rich foods include most dairy products, grains and grain products, nuts, meat, fish, poultry, eggs, and legumes. Smaller concentrations of phosphorus are present in most other foods.

Iron is present in relatively high concentrations in most grains and grain products, in flours, meals and other farinaceous materials, and in baked and cooked cereals (especially enriched ones). Most nut and nut products, meat, poultry, seafoods, fish and shell fish, eggs and legumes are good sources of iron. Smaller amounts of iron are present in most dairy products, fruits and vegetables, and some sweets.

Salt is the main source of sodium in all salted foods. Most dairy products, fruits, cereals and processed cereals, nuts, meat, fish, poultry, eggs, and vegetables contain substantial amounts of potassium.

Magnesium can be found in relatively high concentrations in nuts, cereals, and legumes; manganese-rich foods include cereals, vegetables, and some fruits and meats. In addition to some seafoods and liver, cereals and vegetables are good sources of copper. Sulfur is well-distributed among most protein-rich foods and some vegetables; vegetables and fruits are good sources of cobalt. Some seafoods are particularly rich in zinc, which is present in lesser amounts in most classes of foods.

DETERMINATION OF ASH

Ash in foods is determined by weighing the dry mineral residue of organic materials heated at elevated temperatures (around 550°C). This dry-ashing procedure, which is discussed in more detail later, generally is used to determine total ash and sometimes is used before an elemental analysis for individual minerals. The form of the mineral constituents in ash differs considerably from their form in the original food. Thus, calcium oxalates are converted into carbonates, and upon further ashing to oxides. Some trace minerals linked to biologically active systems are converted to inorganic components.

In addition to the determination of total mineral content by ashing, indirect methods are available to determine the total electrolyte content of foods. Conductometric methods provide a simple, rapid, and accurate means of determining the ash content of sugars. Such foods are generally low in minerals and direct ashing requires incineration of large samples rich in strongly foaming carbohydrates. Conductometric methods are based on the principle that in a solution of sugar, the mineral matter that constitutes the ash dissociates, whereas the sucrose, a nonelectrolyte, does not dissociate. The conductance of the solution is therefore an index of the concentration of the ions present and of the mineral or ash content. Numerous papers have been published on the subject and the methods are described in detail in a number of handbooks (Browne and Zerban 1941; Bates 1942).

Results of conductometric ash determinations in sugar products are affected somewhat by nonelectrolytes, and an experimental correction factor—depending on the assayed food—is determinined. In certain foods conductivity is measured in a medium acidified to displace all the weak acids of the salts. The assay requires two conductivity measurements (before and after acid addition). In syrups or molasses, a second conductivity determination in an alkaline medium gives more precise values.

Electrolyte concentration of sugar products can be determined also by an ion-exchange method (Pomeranz and Lindner 1954). The determination is based on the fact that if a solution containing several salts is passed through a cation-exchange column (in hydrogen form), the effluent contains a quantity of acid equivalent to the original salt content. By titration of the liberated acid, the total electrolyte content of the tested solution can be determined.

Total ash is a widely accepted index of refinement of foods, such as wheat flour or sugar. Since the mineral content of the bran is about 20 times that of the endosperm, the ash test fundamentally indicates thoroughness of the separation of bran and germ from the rest of the wheat kernel. In refining cane sugar, excessive amounts of minerals interfere with processing (decolorization and crystallization). On the other hand, adequate levels of total ash are indicative of functional properties in some foods products (e.g., gelatin). In fruit jellies and marmalades, ash content is used to estimate the fruit content of the product; total electrolyte content can be used to determine adulteration of some juices and beverages.

Levels of ash and ash alkalinity are useful parameters in distinguishing fruit vinegar from synthetic vinegar. Ashing of tissue slices is useful in histological identification. Total ash content is a useful parameter of the nutritional value of some foods and feeds.

Water-soluble ash is sometimes used as an index of the fruit content of jelly and fruit preserves. *Acid-insoluble ash* is a useful index of mineral matter (dirt or sand in spices), efficiency of wheat washing prior to milling, talc in confectionery, coating of rice, or surface contamination of fruits and vegetables. Acid-insoluble ash is determined after digesting the total ash in 10% hydrochloric acid.

The ash of fruits and vegetables is alkaline in reaction; that of meat products and of certain cereals is acid. Alkaline ash is ascribed to the presence of the salts of organic acids that are converted on ashing into the corresponding carbonates. Alkalinity of ash was used for many years to analyze fruit juices. In foods rich in fruit acids or salts, ash alkalinity is an index of fruit content. The results are, however, affected by the presence of phosphates, and the effect of the latter varies with the ashing procedure. The significance of ash alkalinity determination has decreased with the availability of excellent methods for the determination of individual organic acids. The assay is occasionally made to determine the relative amounts of cream of tartar in grape products, in detecting adulteration of foods with minerals, and in determining the acid–base balance of foods.

Salt-free ash is especially important in seasoned foods. It is determined as the difference between total ash and sodium chloride in the ash (as assayed titrimetrically in ash dissolved in dilute nitric acid). Salt in foodstuffs can be determined without ashing by a potentiometric titration of chloride with silver nitrate, with a silver–silver chloride electrode as an indicator (Cole 1967).

ASHING PROCEDURES

Sampling and Contamination Problems

The problem of contamination in elemental analyses has been summarized by Thiers (1957), LaFluer (1976), and Tschopel and Tolg (1982). Contamination can be minimized by stringent precautions at all stages of the work—sampling, storage of samples, preparation for analysis, and actual assay.

Foods, especially plant materials, are subject to surface contamination because of their format and habitat. The impurities must be removed. The foods must be mixed thoroughly. Lipids are practically free of minerals, and as animal foods may vary widely in their lipid content, such variations may cause large differences in mineral content. Grinding of plant materials may result in a segregation into fractions of varying particle size and elemental composition.

Wet or liquid samples are usually dried in ovens. Conditions for corrosion are ideal in such ovens and the perforated metal shelves assist in making them a likely source of contamination. To prevent contamination, samples are often covered with fluted watch glasses so as not to interfere with drying. Such precautions have been found insufficient to prevent contamination particularly in fume hoods or in ovens with forced-air circulation. Individual small glass chambers are preferable. Contamination may come from the container in which samples are dried at elevated temperatures, (e.g., arsenic, zinc, lead, and other metals in borosilicate glass). Hardened stell mortars, widely used for powdering hard and brittle materials, result in severe contamination. It is practically impossible to make a correction for such contamination because of its extremely erratic nature. Grinding by hand with a mortar and pestle is relatively safe. However, for very precise work, a new mortar may be required to eliminate contamination from previous samples. Metallic sieves for classifying ground material may cause problems; plastic (i.e., nylon) sieves are safer. In wet digestion, contamination from reagents and containers can cause erratic results. Contamination from the interior walls and ceiling of a muffle furnace presents problems in dry ashing.

Dry Ashing

In dry ashing, the sample is weighed into a dish, and the organic matter is burned off without flaming and heated either for a fixed period of time or to constant weight. The residue must be free from carbon. The dish containing the residue is cooled in a desiccator and the amount of total ash is determined by weighing.

Crucible Materials. The selection of ashing dishes (crucibles) depends on the nature of the food analyzed, and on the analyses that are to be performed on the ash. The materials used include quartz, Vycor, porcelain, steel, nickel, platinum, and a gold–platinum alloy.

Quartz dishes, smooth on the inside, are resistant to halogens, neutral solutions, and acids (except hydrogen fluoride and phosphoric acid) at most concentrations and temperatures. Resistance to alkali is relatively poor. Quartz crucibles are stable at high temperatures (up to 1100°C for routine work) and can be cleaned with hot dilute hydrochloric acid.

Porcelain crucibles resemble quartz in chemical and physical properties. Temperature resistance is even higher—unglazed crucibles can withstand up to 1200°C for routine work. Porcelain crucibles retain their smooth surface and are easy to clean with dilute hydrochloric acid. They are widely used because of their good weight constancy and relatively low price, but they are susceptible to alkali and crack from sudden large temperature changes. This latter property makes their use somewhat troublesome.

Steel crucibles are sometimes used for ashing large samples. The low price and relatively good resistance of some alloys (e.g., steel containing 18% chromium and at least 8% nickel) to acid and alkali make their use attractive. The crucibles are cleaned mechanically with fine sand or steel wool. Nickel crucibles are used little, as they deteriorate due to nickel carbonyl formation in reducing atmospheres.

Platinum is the best widely used crucible material, but platinum crucibles are too expensive for routine ashing of large numbers of samples. Platinum has a high melting point (1773°C), good heat conductivity, and high chemical inertness. Platinum dishes can corrode and their high price makes it mandatory to eliminate conditions conducive to such corrosion. Platinum dishes should be cleaned thoroughly because dirt-containing organic matter has a reducing-corroding action. Corrosion may be caused by reduction to metals of oxides of iron, lead, and tin. Elemental lead, arsenic, antimony, silicium, and bismuth and their compounds are considered platinum poisons. Corrosion from heavy metals leads to pitting and formation of holes. Some elements give phosphides and silicides that form low-melting-point eutectic mixtures, weaken the platinum crucible or dish, and cause cracks.

Platinum crucibles should be touched only with platinum-tipped tongs and placed after ashing on clean porcelain, asbestos, or marble surfaces. They can be cleaned by boiling with water or acids, but not with aqua regia or hydrochloric acid in the presence of strong oxidants. Silicic acid residues can be removed with dilute hydrofluoric acid. Cleaning by melting (if necessary repeatedly) with potassium pyrosulfate is often effective. Mechanical cleaning should be avoided and performed, if necessary, carefully with acid-washed clean sea sand. Dishes made from gold–platinum (90:10) melt at 1100°C, but are superior to pure platinum dishes in resistance to phosphoric acid and alkali melting.

Specific Applications. For total ash determinations, where recovery and determination of individual metals is not necessary, ashing in porcelain crucibles at temperatures ranging from 400 to 700°C (most commonly around 550°C) is satisfactory.

If the ash components are to be determined, the biological material that is to be ashed and the elements that are to be determined must be considered individually. According to Grant (1951), iron, aluminum, copper, tin, silicon, and magnesium can be determined as oxides in platinum crucibles. Porcelain crucibles have been suggested for the determination of chromium as the metal. Sodium, potassium, lithium, magnesium, calcium, strontium, barium, cadmium, manganese, and lead can be determined in platinum crucibles as sulfates. In the case of lead compounds, nitric acid must be added to avoid the reduction to metallic lead and possible crucible damage.

Ashing over an open flame requires constant attention, and except in cases of ashing samples that froth and bubble excessively, only the initial stage of ashing is done with an open flame. A furnace with a rheostat for temperature control is used for most routine work.

If prolonged ashing fails to give a carbon-free ash, the residue should be moistened, dried, and reheated until a white-gray ash remains. In some cases, it may be necessary to dissolve the ash in a small amount of water, filter the carbon-containing residue through a small low-ash filter paper, dry the two parts, and ash separately. If water fails to break up the material the residue may be treated with a few drops of hydrogen peroxide, nitric acid, and/or sulfuric acid; but in the latter cases the composition of the ash is changed and special precautions must be taken to report the correct type of ash (Dunlop 1961).

Dry ashing for the destruction of organic matter prior to the determination of trace elements is not used extensively because it is generally believed that losses occur from volatilization. According to Lynch (1954), dry ashing is the most satisfactory method if no loss occurs at temperatures up to about 500°C. The method cannot be used for arsenic and mercury; its usefulness for lead is uncertain; and iron can sometimes be troublesome owing to the difficulty of getting the metal into solution after ashing is completed.

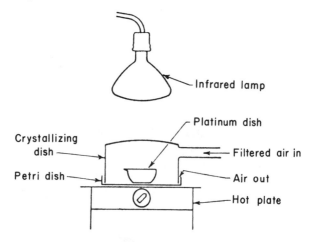

Fig. 34.1.   Chamber for drying or preashing.

Table 34.1. Summary of Reports on Losses during Dry Ashing[a]

Metal	Conditions
Arsenic	Loosely bound as in blood; may volatilize as unknown compound at 56°C
Boron	Volatilizes with steam from acid solutions
Cadmium	Volatilizes, possibly as the chloride, or metal, between 400 and 500°C
Chromium	Volatilizes as chromyl chloride at low temperatures under oxidizing conditions
Copper	Volatilizes as porphyrin compounds when petroleum samples are burned
	Volatilizes from vinegar, possibly as copper acetate at low temperatures
	Reduces to metal which is not dissolved by hydrochloric acid
Iron	Volatilizes as ferric chloride at 450°C
	Volatilizes as porphyrin compounds when petroleum samples are burned
	When materials with a high phosphorus-to-iron ratio are ashed, an unidentified compound is formed, which resists solution or hydrolysis, causing low results
Lead	Volatilizes from blood or petroleum unless sulfate is present
Mercury	Volatilizes as metal below 450°C
Nickel	Volatilizes as porphyrin compounds when petroleum samples are burned
Phosphorus	Volatilizes, presumably as one of the oxy acids, especially when sulfate is present, except in the presence of excess magnesium
Vanadium	Volatilizes as porphyrin compounds when petroleum samples are burned
	Volatilizes as the chloride below 450°C
Zinc	Volatilizes, presumably as the chloride, above 450°C

[a] From Thiers (1957).

Thiers (1957) summarized reported losses during dry ashing (Table 34.1). Most of the losses can be minimized if proper ashing conditions are used.

Thiers (1957) recommended a dry-ashing procedure for biological material that involves drying and preashing the sample in a special apparatus with the aid of a hot plate and an infrared lamp (Fig. 34.1). The temperature of the hot plate surface is gradually raised to about 300°C. This gives the sample a charred appearance. Ashing is completed in a muffle furnace with a controlled temperature that starts at 250°C and is raised within about 1 hr gradually to 450°C where it is held until the decomposition of organic matter is complete.

Gorsuch (1959) found that dry ashing of cocoa at 550°C gives satisfactory recovery of antimony, chromium, cobalt, iron, molybdenum, strontium and zinc, but recoveries of arsenic, cadmium, copper, mercury, and silver were not satisfactory. Lead was recovered completely after ashing at 450°C, but recovery was questionable at 550°C. To overcome losses at high temperature, a low-temperature dry asher has been developed and is available commercially (Anon. 1967). The instrument uses a high-purity stream of excited oxygen, produced by a high-frequency electromagnetic field as the only oxidizing agent (Gleit and Holland 1962). Figure 34.2 illustrates the principle of operation of the instrument.

Modified Dry Ashing

In addition to the simple ashing procedure, certain modifications and additives have been proposed to accelerate the process, to prevent overall losses of minerals, or to improve the retention of critical components.

Liquids and moist materials should be dried prior to ashing. The drying is generally done for moisture determination. According to Davidsohn (1948), difficulties due to spattering in determining the ash content of

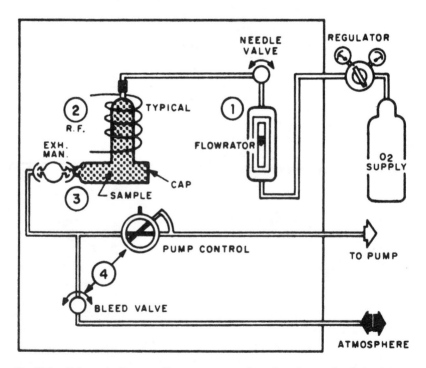

Fig. 34.2.   Schematic diagram of low-temperature dry asher. Oxygen is admitted through a flowrator (1) and passes through the radio-frequency electromagnetic field (2), which produces excited oxygen species that attack the sample in the oxidation chamber (3). Exhaust vapors are removed by a mechanical vacuum pump controlled by valves (4). (Courtesy Trapelo Lab., Waltham, MA.)

various moist materials, such as jams, may be overcome by adding about 0.2 g of pure cottonwool of known ash content. The correction to be made is very small. It has been suggested that ashing of dry materials can be better controlled when filter paper is used to line the dish.

Protein-rich materials ash slowly, particularly in salt-rich foods (e.g., salted meat). Ashing at elevated temperatures causes large losses of salt. The use of fixatives and separate ashing of the water-insoluble food and salt-rich water extract may be useful. To carbohydrate-rich materials, that swell and foam excessively, one can add a few drops of pure olive oil after drying and before ashing. The sample is heated over a small flame till foaming ends, and the ashing is completed in a furnace. Fermenting the carbohydrates prior to ashing will generally eliminate foaming difficulties, but the procedure is too lengthy and complicated for routine tests. Whenever possible, mineral constituents are estimated in carbohydrate-rich materials by indirect methods (e.g., conductivity). Lipid-rich materials ash rapidly. Schneider (1967) recommends heating such foods till they catch fire and allowing the fat to burn off. As pure fat is practically ash-free, it may be advisable to determine the ash after an extraction of the fat with an organic solvent. This is the generally recommended procedure if both ash and fat are determined.

To accelerate ashing, the addition of small amounts of pure glycerin or alcohol has been suggested. In regular dry ashing, atmospheric oxygen serves as the oxidant; chemical oxidants (e.g., hydrogen peroxide) may be added sometimes to accelerate the process. Ammonium carbonate may be useful as an aid, even if ashing is followed by ash analysis. Ammonium nitrate is not recommended as an ashing aid, as it causes puffing and ash losses (Schneider 1967). According to Zonneveld and Gersons (1966), ashing is accelerated by adding an aluminum chloride solution. An accelerated method of ash determination in cereals recommends moistening 3–5 g of the material with 5 ml of an alcoholic solution of magnesium acetate and incineration at 700°C in a muffle oven. In both the aluminum chloride and magnesium acetate methods, a blank must be subtracted from the ash.

In dry ashing of samples rich in silicon and aluminum, an insoluble residue may form. If this occurs, the sample should be fused with a small amount of sodium carbonate. In the case of silicates, the silica may be volatilized by careful treatment with sulfuric and hydrofluoric acids; special precautions should be taken to remove fluoride ions before subsequent determinations are made. Special cases of dry ashing may require the addition of a fixative; six such procedures were described by a committee on analytical methods for the destruction of organic matter (Anon. 1960). A comprehensive survey of ashing procedures was published by Middleton and Stuckey (1953). For the determination of chlorine and boron, it is necessary to ash under alkaline conditions. In the determination of fluorine in foods, the sample is ashed in the presence of calcium oxide. A mixture of aluminum nitrate and calcium nitrate is recommended as an ash aid in the determination of metals. It has been suggested that magnesium chloride be added to moisten the charred mass before final ashing of biological material rich in phosphate. As some lead losses occur in regular dry ashing above 550°C, dry ashing with sulfuric acid as an ashing aid permits increasing temperature to 650°C with little lead loss (Gorsuch 1959). The addition of acid slows down the oxidation rate and lengthens the ashing time required. Dry ashing with magnesium nitrate or acetate, or with nitric acid as an ashing aid had little advantage.

In addition to regular furnaces, special tube and combustion furnaces are available for the decomposition of organic compounds on a microscale or prior to an elemental determination. The oxygen bomb method, in which oxygen under pressure replaces air, is widely used for the determination of sulfur and halogens. Elvidge and Garratt (1954) reported that complete combustion can be achieved in a commercial bomb calorimeter in an atmosphere of oxygen without loss of the more volatile components. With light and bulky foods, preliminary compression improves combustion. Wet materials must be dried before combustion. About 3–4 g of material can be burned in one step under an initial pressure of 30 atm of oxygen.

An excellent flask combustion (Fig. 34.3) method was proposed by Schoniger (1955). The method uses a combustion procedure in which the weighed sample is rolled in a small piece of filter paper and placed in a platinum cup, which is suspended from the stopper of a 250-ml conical flask previously filled with oxygen. The filter paper is ignited and plunged back into the flask, where it burns brightly for a few seconds. The products

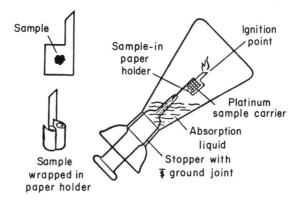

Fig. 34.3.   Equipment for oxygen flask combustion.

of combustion are absorbed in a few milliliters of water in the bottom of the flask, the final determination being made on that solution. Corner (1959) described the use of this procedure for the rapid microdetermination of organically bound halogens, arsenic, phosphorus, and boron.

A method intermediate between dry and wet ashing was proposed by Wirthle (1900). The sample is digested with sulfuric acid until a porous mass is formed. Digestion of the partly decomposed material is completed by ashing after adding soda and sodium nitrate.

Wet Ashing

Wet ashing (also called wet digestion and wet oxidation) is used primarily for the digestion of samples for determining trace elements and metallic poisons.

The use of a single acid is desirable, but usually not practical for the complete decomposition of organic material. Sulfuric acid is not a strong oxidizing agent and the time required for decomposition is long. Adding a salt, e.g., potassium sulfate, raises the boiling point of the acid and accelerates decomposition. This technique is particularly useful for samples in which adding nitric acid causes the formation of insoluble oxides.

Nitric acid alone is a good oxidant, but it usually boils away before the sample is completely oxidized. Middleton and Stuckey (1954) described a method of destroying organic matter at temperatures below 350°C by digestion with nitric acid as the only major reagent. Small amounts of sulfuric acid are added at the initial charring stage to prevent the ignition of fat-rich materials when only nitric acid is added. Gorsuch (1959) found the method generally satisfactory, though recoveries were slightly lower than with conventional wet oxidation methods. Selenium and mercury were lost almost completely. The procedure is somewhat tedious and time-consuming.

Mixed acids are the usual reagents for the decomposition of organic material in wet-ashing procedures. The use of a mixture of sulfuric and nitric acids is recommended by many workers and is the most acceptable procedure. Suggested quantities of each acid and the order and rate of addition vary with different biological materials and investigators.

To avoid excessive foaming in the digestion of fat or sugar-rich materials, it may be advisable to add sulfuric acid, allowing it to soak in, and then add nitric acid in small portions, with heating in between. In later digestion stages, hydrogen peroxide may be added to complete digestion.

The use of perchloric acid with nitric acid or with nitric–sulfuric acid mixtures has been suggested by Smith (1953) for the rapid decomposition of many organic compounds that are difficult to oxidize. Perchloric acid is an excellent oxidant at elevated temperatures and above 60%, but can be explosive and very dangerous if improperly used. It is used routinely by many laboratories. Five grams of wheat can be completely digested in 10 min with HNO_3–70% $HClO_4$ (1:2) compared with 8 hr using the usual HNO_3–H_2SO_4 method. Digestion with perchloric acid should be performed under a special hood containing no plastic ingredients and no glycerol-containing caulking substances. Figure 34.4 shows the apparatus for doing $HClO_4$ digestions in an ordinary hood.

Detailed wet-oxidation procedures using three acid mixtures (nitric and sulfuric acids, sulfuric acid and hydrogen peroxide, and mixtures containing perchloric acid) were described by Gorsuch (1975). Tolg (1974) described methods of wet oxidation applicable to various foods. The most widely used digestion reagent for material of vegetable origin is H_2SO_4–HNO_3; its use may cause volatilization of arsenic, selenium, mercury, and others. In using H_2SO_4–H_2O_2 for foods of vegetable origin, lead may be lost by coprecipitation with $CaSO_4$, and there may be a loss of germanium, arsenic, ruthenium, selenium, and others. For material of biological origin, three digestion reagents were recommended: HNO_3 (the digestion time is short at 350°C, the reagent can be purified easily, and the nitrates that form are soluble); $HClO_4$ with catalysts; and H_2SO_4–$HClO_4$ (suitable only for small samples; danger of explosion). A HNO_3–$HClO_4$ mixture is recommended for foods containing proteins and carbohydrates and no fat. Although the H_2SO_4–HNO_3–$HClO_4$ mixture is a

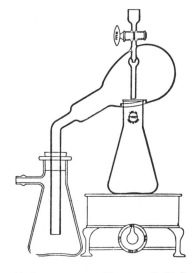

Fig. 34.4. Wet-oxidation apparatus. (Courtesy G. F. Smith Chemical Co.)

universal reagent, it requires exact temperature control and arsenic, lead, gold, iron, and some other elements may be volatilized.

Dry Ashing versus Wet Digestion

Dry ashing is the most commonly used procedure to determine the total mineral content of foods. It is used to determine water-soluble, water-insoluble, and acid-insoluble ash. Dry ashing is applicable also to the determination of most common metals. Dry ashing requires no attention, is simple, and is well suited for routinely large numbers of samples. Generally no reagents are added and no blank substraction is required. Dry ashing takes a long time, but can be accelerated by modified methods, or ashing can be done overnight. High temperatures and the relatively expensive equipment limit somewhat the usefulness of the method.

The main objection to dry ashing, however, is the occurrence of interactions between components themselves or between components and the receptacle material. In the estimation of certain trace elements in foods the use of either silica or porcelain crucibles leads to the absorption of the element by the vessels. If the temperature of ashing is not excessively high, the absorption loss may be greater than the volatilization loss. Excessive heating makes certain metallic compounds insoluble (e.g., those of tin). Foods with a high phosphorus to base ratio fuse to a dark melt in which carbon particles are trapped and will not burn. Foods with a high alkaline balance show progressive decomposition of the carbonates and the volatilization of chlorides. The high hygroscopicity, lightness, and fluffiness of the ash produced by dry ashing may sometimes present problems in determining precisely the total ash content or in handling the mineral residue for subsequent analysis.

In wet oxidation, relatively low temperatures and liquid conditions are maintained; the apparatus is simple and oxidation is rapid. On the other

Table 34.2. Comparisons of Wet-Oxidation and Dry-Ashing Techniques[a]

Wet oxidation	Dry ashing
More rapid	Rather slow
Temperature lower—less volatilization and retention	Temperature higher—more volatilization and retention
Generally less sensitive to nature of sample	Generally more sensitive to nature of sample
Relatively more supervision	Relatively less supervision
Reagent blank larger	Reagent blank smaller
Large samples inconvenient	Large samples easily handled

[a] From Gorsuch (1976).

hand, the procedures require large amounts of corrosive reagents and a correction for the reagents. Handling routinely large numbers of samples by wet digestion is difficult and the required operators' time is large.

Although significant advances have been made in elemental determination technology, little improvement has been reported in the area of sample preparation and solubilization. Some procedures for preparing samples for elemental analysis are still controversial. Gorsuch (1975, 1976) compared wet-oxidation and dry-ashing techniques. A summary comparison is presented in Table 34.2.

ELEMENTAL ANALYSIS

This chapter includes no detailed description of methods to determine individual mineral components. Such procedures are described in general textbooks of inorganic analysis, standard reference books on food analysis, and specialized textbooks on the determination of minerals in biological materials. The latter include a comprehensive treatise in 13 volumes on the chemistry and the determination of individual elements (Kolthoff and Elving 1961–1966). The principles of instrumental methods used in the determination of mineral components and trace elements also were described in previous chapters of this book. This chapter is primarily concerned with the applications of those principles to food analysis.

Developments in the measurement of trace metal components in foods were described by LaFluer (1976), Winefordner (1976), Bratter and Schramel (1980), Das (1983), Schwedt (1984) and Benton-Jones (1984). Tschoepel and Tolg (1982) reviewed the basic rules that have to be followed in trace analyses to obtain precise and accurate results at the nanogram and picogram levels. These rules are as follows:

1.   All materials used for apparatus and tools must be as pure and inert as possible. These requirements are only approximately met by quartz, platinum, glassy carbon and, to a lesser degree, polypropylene.

2.   Cleaning of the apparatus and vessels by steaming is very important to lower blanks as well as element losses by adsorption.

3.   To minimize systematic errors, microchemical techniques with small apparatus and vessels with an optimal ratio of surface to volume are recommended. All steps of the analytical procedure, such as composition, separation, pre-concentration, and determination, are best done in one vessel (single-vessel principle). If volatile elements or compounds have to be determined, the system should be closed off and the temperature should be as low as possible.

4.   Reagents, carrier gases, and auxiliary materials should be as pure

as possible. Reagents that can be purified by subboiling point distillation are preferred.

5. Contamination from laboratory air should be avoided by using clean benches and clean rooms. By this, the blanks caused by dust can be decreased by at least two to three orders of magnitude.

6. Low and constant reaction temperature should be used.

7. Manipulations and different working steps should be restricted to a minimum in order to reduce unavoidable contamination.

8. All steps of the combined procedure should be monitored, this can best be done with radiotracers.

9. All procedures have to be verified by a second independent one or, even much better, by an interlaboratory comparative analysis.

Element Enrichment

The determination of trace elements often requires enrichment of the elements, and/or the separation of many elements at the trace level from large amounts of major elements. The most useful technique for element enrichment is ion exchange. Ion exchange has proved to be a valuable tool in the concentration, isolation, and recovery of ionic materials present in a solution in trace amounts. Ion-exchange chromatography on an ion-exchange resin can be also used for fractionation, separation, and the elimination of interfering ions (e.g., phosphates). The use of ion-exchange chromatography for enrichment was described by Samuelson (1963) and is reviewed annually in *Analytical Chemistry*. A special preconcentration procedure of trace elements by precipitation ion exchange was described by Tera *et al.* (1965). Enrichment by extraction with organic solvents was described by Gorbach and Pohl (1951), and enrichment problems in microanalysis were reviewed by Schulek and Laszlovsky (1960).

Emission Spectroscopy

Emission spectroscopy is the oldest instrumental method for trace analysis. It depends on observing and measuring the radiation emitted by atoms of the various elements when planetary electrons displaced from their orbits by various means fall back to the original (or a lower) level. For each element there is a pattern of wavelengths characteristic of the element when excited in a particular way. When several elements are present, each emits its own wavelengths. By identifying the wavelengths in the spectrum, the sample can be analyzed. Emission spectroscopy is sensitive but the precision is rather low.

Flame Photometry

Early studies during the nineteenth century by J. F. Herschel, D. Alter, and G. Kirchhoff and R. Bunsen laid the foundations for the qualitative differentiation of salts depending on their emission in a flame. Later researchers developed suitable techniques and instruments for quantitative analyses based on flame photometry.

A modern flame photometer consists essentially of an atomizer, a burner, some means of isolating the desired part of the spectrum, a photosensitive detector, sometimes an amplifier, and finally a method of measuring the desired emission by a galvanometer, null meter, or chart recorder (see Chapter 10 for details). The instruments are used primarily to determine calcium, sodium, and potassium.

Atomic Absorption Spectroscopy

Within the last two decades atomic absorption spectroscopy has found enthusiastic acceptance by science and industry. Review issues of *Analytical Chemistry* indicates that hundreds of papers are published annually on basic research, instrumentation, specific analytical methods, and practical applications of atomic absorption spectroscopy.

Atomic absorption spectroscopy is not quite as free from inter-element effects as was originally expected, but it is far better in this respect than any form of emission spectrography. It is quite sensitive; the limit of detection ranges from 0.01 ppm for magnesium to 5.00 ppm for barium; and the method is rapid (about 1000 determinations can be made per week). The equipment is relatively inexpensive (about $10,000), only one-tenth the cost of X-ray fluorescence equipment. The limiting factor is the need for cathode lamps for each element or several combinations.

In atomic fluorescence spectroscopy, atoms are generated in the same way as in atomic absorption spectroscopy, except that a cylindrical flame is used. The flame is irradiated by resonance radiation from a powerful spectral source, and the fluorescence that is generated in the flame is measured at right angles to the incident beam of radiation. This is done to minimize the contamination of the fluorescence signal by light from the source.

Atomic absorption spectroscopy can be used in the ppm range (about 10^{-6} M solutions); atomic fluorescence spectroscopy in the ppb range (about 10^{-9} M solutions). For a fuller discussion of these methods, see Chapter 10.

Neutron Activation Analysis

In neutron activation analysis, a weighed sample together with a standard that contains a known weight of the element sought are exposed to nuclear bombardment. The radioactivity of the element in the sample is then compared with the radioactivity in the standard (see Chapter 19). Generally, a chemical separation is required to purify the radioisotopes of the element sought and to remove all other induced radioactivity. The quantity of the element in the sample is then calculated from the ratio of the separated activities. In some instances, the final measurement of activity can be made on the intact sample. If the background remains inactive during nuclear bombardment or if the energies of the emitted radiations differ widely, a direct measurement of trace elements is possible. Also, if the trace element has a substantially longer half life than the other induced activities, the interfering materials may be allowed to decay and the radio-assay completed when the interference is insignificant. Results obtained by neutron activation generally are within 5% of the true value, and replicate analyses under favorable conditions are within 2–3% of the mean.

The attractive features of neutron activation analyses are its wide applicability, high sensitivity (0.001–1 ppm), and satisfactory accuracy and precision. There have been numerous applications of activation analysis in botany and agriculture.

X-Ray Spectroscopy

There are three uses of X rays in chemical analysis. Absorption methods are of limited practical application because the adjustment of wavelength is most critical. X-ray diffraction is useful in crystallography and in establishing the complicated structure of biological molecules. The use of X rays for the identification of chemical components is based on emission methods, involving secondary or fluorescent emission.

Measurement of the intensity and wavelength of fluorescence radiation is a well-established method of analysis, which has been applied to the determination of elements from ^{11}Na to ^{92}U in powder, liquid, or metal samples. Coefficients of variation of about 1% in the concentration range 5–100% and of 5% in the range 0.1–1.0% can be obtained. In some instances determinations in the ppm range can be made. The method is rapid (1–4 min), independent of the chemical combination of the element, and nondestructive in the sense that the specimen examined is not destroyed, though some specimen preparation may be required. Instrumentation for X-ray spectroscopy is quite expensive and is described in Chapter 11.

Glass Electrodes

When a thin membrane of glass is interposed between two solutions, an electrical potential difference is observed across the glass. The potential depends on the ions present in the solutions. Depending on the composition of the glass, the response may be primarily to the hydrogen ion, to other cations, or even to organic cations. The electrodes are unaffected by oxidants and reducing agents, and only slightly affected by anions (except fluoride) or by high concentrations of proteins and amino acids. The design and use of ion-selective electrodes for analytical purposes are discussed in Chapter 12.

Miscellaneous Methods

Trace elements are determined in many laboratories by specific colorimetric and turbidimetric methods, by fluorescence analysis (Chapter 8), and by polarography (Chapter 16). The use of infrared spectroscopy in determining polyatomic ions was described by Miller and Wilkins (1952). Relatively simple chromatographic methods for rapid routine evaluation of trace elements in crops and foods were described by Duffield (1958) and Coulson et al. (1960). Connoly and Maguire (1963) described a quantitative paper chromatographic procedure for the determination of copper, cobalt, nickel, molybdenum, and manganese in foods. The elements are separated from the ashed materials as their chlorides on slotted chromatography papers and identified colorimetrically. The concentrations are determined by reflectance measurements on the papers and the results are calculated with the aid of prepared standard graphs. As little as 0.05 μg of the element can be determined.

Impressive advances have been made in developing instruments that permit an essentially complete elemental analysis to be performed in situ on the structures observed in the tissues of thin sections prepared by standard histological methods. The electron probe microanalyzer or electron probe X-ray scanning microscope (Birks 1960) can perform nondestructive elemental chemical analyses on localized regions with diameters as small as 1 μm and volumes of a few cubic micrometers. The limit of detectability is about 0.1%, and many inorganic elements can be measured. The method has been extended to analyses and scanning of biological specimens (Andersen 1967).

Another promising technique that has been adapted to microanalysis of inorganic elements is the laser microprobe. In this instrument, a laser beam is flashed through the optics of a regular microscope set to analyze a very small arc. The instrument is attached to a sensitive spectrograph (Glick and Rosan 1966).

Finally, mention should be made of biological (noninstrumental) methods of trace analysis. The principle of some of the procedures was described in the chapter on microbiological assays. The techniques were reviewed by Nicholas (1966).

Comparison of Methods

Bowen (1967) described the results of elemental analyses of a standard plant material (dried kale powder) analyzed for 40 elements by 29 laboratories. The techniques used were neutron activation analysis, atomic absorption spectroscopy, a catalytic technique, colorimetry, flame photometry, turbidimetry, and titrimetric analysis. Consistent results were obtained by more than one laboratory for Au, B, Ba, Br, Ca, Cl, Co, Cr, Fe, Ga, I, Mn, Mo, N, P, Rb, S, Sc, and W. Small differences in results obtained by different techniques were found for Cu, K, Mg, Na, P, Se, Sr, and Zn. For example, flame photometry gave high results for sodium, activation analysis without chemical separation was unreliable for determining potassium and magnesium, and atomic absorption spectrometry gave high results for copper and strontium. Gross discrepancies were found in the results reported for aluminum, arsenic, mercury, nickel, and titanium.

A comparison of instrumental methods for trace analysis in foods is summarized in Table 34.3.

According to Wolf and Hamly (1984), inorganic trace elements of interest in human health can be divided into those that are of nutritional and toxic interest (arsenic, cobalt, chromium, copper, fluorine, manganese, molybdenum, nickel, tin, vanadium, and zinc), those that are primarily of nutritional interest (copper, iron, iodine, and silicon), and those that are primarily of toxic interest (beryllium, cadmium, mercury, lead, and tantalum). The two techniques considered by the authors as having the required sensitivity and greatest potential for accurate trace element analysis are atomic spectroscopy and neutron activation analysis. The authors discussed in detail various procedures and instruments for atomic absorption and atomic emission spectrometry; in addition, they briefly reviewed neutron activation analysis and mass spectrometry. A major part of the review deals with statistically valid sampling, validation of analytical data (by two independent methods, use of reference materials, collaborative studies) and data handling and evaluation.

Hocquelett (1984) determined cadmium, lead, arsenic, and tin in vegetable and fish oils by atomic absorption with electrothermal atomization by an oven equipped with a graphite tube. Addition of dithiocarbamate for cadmium or of dithiocarbamate for lead and arsenic as decreased volatility of the elements. Detection limits of 0.5–3.0 ng/g and satisfactory recoveries were obtained in the 20-ppb range when samples of oil diluted in chloroform (for arsenic, cadmium, and lead) or in methylisobutyl ketone (for tin) were injected into the atomizer. This rapid (minutes compared to hours for methods with nitrosulfuric digestion) direct determination method was recommended for rapid routine testing of large numbers of samples.

Table 34.3. Comparison of Analytical Methods for Trace Element Analysis[a]

	Multielement analyses	Sensitivity (ppm)	Specificity	Accuracy	Freedom from contamination, reagent blanks, etc.	Possibility of overcoming surface contamination
Activation analysis	In some cases	Very high (0.001–1)	Good	Good	Good	Good
Atomic absorption and flame	No	0.01–5	Good	Good	Bad	Good
Emission spectroscopy	Yes	0.1	Good	Reasonable	Bad	Bad
X-ray spectroscopy	Yes	10–100	Good	Needs standards	Good	Possibly by precleaning
Mass spectrometry vacuum spark	Yes	0.01	Good	Needs standards	Good	Good
Gas analysis	Yes	Recently improved	Fair (N_2, CO)	Good	Good	Good

[a] From Ames (1966).

Noller and Bloom (1978) described an integrated analytical scheme for the determination of major (sodium, potassium, calcium, and magnesium) and minor (zinc, copper, nickel, iron, chromium, cesium, lead, tin, and mercury) elements in foods. The methods involved flame atomic absorption and flame emission spectrometry for all elements except mercury, for which flameless atomic absorption was recommended. In a collaborative study involving 13 Australian laboratories, cadmium, copper, iron, lead, tin, and zinc were determined in spiked and unspiked samples of apple puree (Steele 1984). Atomic absorption was used in the flame mode to determine copper, iron, and zinc; it was efficient and accurate and yielded low interlaboratory coefficients of variation and good recoveries. However, tin may be lost in ashing as volatile stannic chloride or as insoluble metastannic acid. Lead was determined by direct flame atomic absorption, by solvent extraction followed by flame atomic absorption, and by electrothermal atomization. The methods yielded comparable results.

BIBLIOGRAPHY

Ames, R. (1966). New instrumental procedures for determination of trace elements. *Wallerstein Lab. Commun.* **29**(100), 107–113.

Andersen, C. A. (1967). An introduction to the electron probe microanalyzer and its application to biochemistry. *Methods Biochem. Anal.* **15**, 147–270.

Anon. (1960). Methods for the destruction of organic matter. *Analyst* **85**, 643–656.

Anon. (1967). "LTA-600 Low Temperature Dry Asher." Trace-lab, Richmond, CA.

Bates, F. J. (1942). Bureau of Standards Circular *C-440*.

Benton-Jones, J. (1984). Developments in the measurement of trace metal constituents in foods. *In* "Analysis of Food Contaminants" (J. Gilbert, ed.), pp. 157–206. Elsevier Applied Science Publishers, London.

Birks, L. S. (1960). The electron probe: An added dimension in chemical analysis. *Anal. Chem.* **32**, 19A–24A.

Bowen, H. J. M. (1967). Comparative elemental analyses on standard plant material. *Analyst* **92**, 124–131.

Bratter, P., and Schramel, P. (eds.) (1980). "Trace Element Analytical Chemistry in Medicine and Biology." de Gruyter, Berlin.

Browne, C. A., and Zerban, F. W. (1941). "Sugar Analysis." Wiley, New York.

Cole, S. J. (1967). Potentiometric determination of salt in foodstuffs. *Food Technol.* **21**, 302–304.

Connoly, J. F., and Maguire, M. F. (1963). An improved chromatographic method for determining trace elements in foodstuffs. *Analyst* **88**, 125–130.

Corner, M. (1959). Rapid microdetermination of organically bound halogens, arsenic, phosphorus and boron. *Analyst* **84**, 41–46.

Coulson, C. B., Davies, R. I., and Luna, C. (1960). Quantitative paper chromatography of inorganic ions in soils and plants. *Analyst* **85**, 203–207.

Das, M. S. (ed.) (1983). "Trace Analysis and Technological Development." Wiley, New York.

Davidsohn, A. (1948). Aid for the determination of ash in matter containing high percentages of water. *Analyst* **73**, 678.

Duffield, W. D. (1958). A system for the determination of certain trace elements in foods. *Analyst* **83**, 503–508.

Dunlop, E. C. (1961). Decomposition and dissolution of samples: Organic. *In* "Treatise on Analytical Chemistry," Vol. 2 (I. M. Kolthoff, and P. J. Elving, eds.). Interscience Publishing Co., New York.

Elvidge, D. A., and Garratt, D. C. (1954). A note on a bomb technique for preparing samples for determination of lead in foodstuffs. *Analyst* **79**, 146–147.

Gleit, C. E., and Holland, W. D. (1962). Use of electrically excited oxygen for the low temperature decomposition of organic substances. *Anal. Chem.* **34**, 1454–1457.

Glick, D., and Rosan, R. C. (1966). Laser microprobe for elemental microanalysis, application in histochemistry. *Microchem. J.* **10**, 393–401.

Gorbach, G., and Pohl, F. (1951). Enrichment and spectral analytical assay of trace elements; extraction with organic solvents. *Mikrochem.* **38**, 258–267.

Gorsuch, T. T. (1959). Radiochemical investigations on the recovery for analysis of trace elements in organic and biological materials. *Analyst* **84**, 135–173.

Gorsuch, T. T. (1975). "The Destruction of Organic Matter." Pergamon Press, Oxford.

Gorsuch, T. T. (1976). "Accuracy in Trace Analysis: Sampling, Sample Handling, and Analysis." Publ. *422*. National Bureau of Standards, Washington, DC.

Grant, J. (1951). "Qualitative Organic Microanalysis," 5th ed. Blakiston Publishing Co., Philadelphia.

Hocquellet, P. (1984). Use of atomic absorption spectrometry with electrothermal atomization for the direct determination of trace elements in oils: Cadmium, lead, arsenic, and tin. *Rev. Fr. Corps Gras* **31**, 117–122.

Kolthoff, I. M., and Elving, P. J. (1961–1966). Treatise on Analytical Chemistry, Part II. *In* "Analytical Chemistry of the Elements." Interscience Publishing Co., New York.

LaFluer, F. D. (ed.) (1976). "Accuracy in Trace Analysis: Sampling, Sample Handling and Analysis," Vol. I and II. Special Publ. *422*. National Bureau of Standards, Washington, DC.

Lynch, G. R. (1954). The destruction of organic matter. *Analyst* **79**, 137.

Margoshes, M., and Vallee, B. L. (1956). Flame photometry and spectrometry. Principles and applications. *Methods Biochem. Anal.* **3**, 353–407.

Middleton, G., and Stuckey, R. E. (1953). The preparation of biological material for the determination of trace metals. I. Critical review of existing procedures. *Analyst* **78**, 532–542.

Middleton, G., and Stuckey, R. E. (1954). The preparation of biological material for the determination of trace metals. II. A method for the destruction of organic matter in biological material. *Analyst* **79**, 138–142.

Miller, F. A., and Wilkins, C. H. (1952). Infrared spectra and characteristic frequencies of inorganic ions. *Anal. Chem.* **24**, 1253–1294.

Nicholas, D. J. D. (1966). Microbiological techniques as analytical and purification tools with special reference to trace metals. *Ann. N. Y. Acad. Sci.* **137**, 217–231.

Noller, B. N., and Bloom, H. (1978). Methods of analysis of major and minor elements in foods. *Food Technol. Aust.* **30**(1), 11–19, 22–23.

Pomeranz, Y., and Lindner, C. (1954). The determination of the total electrolyte concentration of sugar products. *Anal. Chim. Acta* **11**, 2399–243.

Samuelson, O. (1963). "Ion Exchange Separations in Analytical Chemistry." Wiley, New York.

Schneider, E. (1967). Minerals; preparation of samples, determination of total ash, detection and determination of individual components. *In* "Analysis of Foods", Vol. II (W. Diemair, ed.). Springer, Berlin.

Schoniger, W. (1955). A rapid micro-analytical determination of halogen in organic substances. *Mikrochim. Acta* **1**, 123–129. [*Anal. Abstr.* **2**, 1816.]

Schulek, E., and Laszlovsky, J. (1960). Problems of destruction and enrichment in microanalysis. *Microchim. Acta*, **6**, 485–501.

Schwedt, G. (1984). Methods of multi-element trace analysis in foods (in German). *GIT Fachz. Lab.* **5**, 394–401.

Smith, G. F. (1953). The wet ashing of organic material employing hot concentrated perchloric acid. The liquid fire reaction. *Anal. Chim. Acta* **5**, 397–421.

Steele, R. J. (1984). Trace metal analysis of foods by Australian laboratories. *Food Technol. Aust.* **36**(3), 135–136, 138–139.

Tera, F., Ruch, R. R., and Morrison, G. H. (1965). Preconcentration of trace elements by precipitation exchange. *Anal. Chem.* **37**, 358–360.

Thiers, R. E. (1957). Contamination in trace element analysis and its control. *Methods Biochem. Anal.* **5**, 273–335.

Tschopel, P., and Tolg, G. J. (1982). Comments on the accuracy of analytical results in ng- and pg- trace analysis of the elements. *J. Trace Microprobe Tech.* **1**, 1–77.

Tolg, G. (1974). Wet oxidation procedures. *In* "Methodicum Clinicum," Vol. I—Analytical Methods (F. Korte, ed.), Part B, pp. 698–710. Academic Press, New York.

Walsh, A. (1955). Application of atomic absorption spectra to chemical analysis. *Spectrochim. Acta* **1**, 108–117.

Walsh, A. (1966). Some recent advances in atomic absorption spectroscopy. *J. New Zealand Instr. Chem.* **30**, 7–21.

Winefordner, J. D. (ed.). (1976). "Trace Analysis—Spectroscopic Methods for Elements." Wiley, New York.

Wirthle, F. (1900). *Cited by* Middleton and Stuckey (1954).

Wolf, W. R., and Hamly, J. M. (1984). Trace element analysis. Chapter 2, *In* "Analysis of Food Contaminants" (J. Gilbert, ed.), pp. 157–206. Elsevier Applied Science Publishers, London.

Zonneveld, H., and Gersons, L. (1966). A rapid dry ashing method. *Z. Lebensm. Unters. Forsch.* **131**, 205–207.

35
Carbohydrates

COMPOSITION AND OCCURRENCE

Carbohydrates are the most abundant and widely distributed food component. Carbohydrates include (a) *monosaccharides* (polyhydroxy aldehydes or ketones) among which are 5-carbon compounds, such as xylose or arabinose, and 6-carbon compounds, such as glucose and fructose, (b) *oligosaccharides* in which a hydroxyl group of one monosaccharide is condensed with the reducing group of another monosaccharide (if two sugar units are joined in this manner, a *disaccharide* results; a linear array of three to eight monosaccharides joined by glycosidic linkages gives *oligosaccharides*), and (c) *polysaccharides* that may be separated roughly into two broad groups, the so-called structural polysaccharides (i.e., cellulose, hemicellulose, lignin) that constitute or are part of rigid, mechanical structures in plants, and nutrient polysaccharides (i.e., starch, glycogen) that are metabolic reserves in plants and animals.

Perhaps the most important of the known monosaccharides is D-glucose. It is found as such in the blood of animals, in the sap of plants, and in many fruit juices. It also forms the structural unit of the most important polysaccharides. It is probably produced in all green plants, though its conversion into starch, cellulose, and other polysaccharides may prevent its detection. Fructose is found in fruit juices and in honey. An abundant source of both glucose and fructose is the disaccharide sucrose, which can be hydrolyzed to yield one mole of each of them. Other disaccharides include (1) the milk sugar, lactose, which constitutes about 5% of cow's milk and about 6% of human milk, and yields glucose and galactose on hydrolysis; (2) maltose, a disaccharide in which one molecule of glucose is joined through a 1,4-α-glycosidic linkage to a second molecule of glucose and which is formed abundantly by amylolytic breakdown of polysaccharides during malting or digestion in the animal body; and (3) cellobiose, a degradation product of cellulose resembling maltose except that the two glucose units are joined through a β-glycosidic linkage. The most important freely occurring trisaccharide is the sugar beet sugar, raffinose, in which galactose is linked to a sucrose unit.

In addition to their nutritional and metabolic function, carbohydrates are important as natural sweeteners, raw materials for various fermentation products including alcoholic beverages, and the main ingredient of cereals. Carbohydrates govern the rheological properties of most foods of plant origin. The involvement of carbohydrates in the *browning reaction* is known to improve or impair consumer acceptance and the nutritional value of many foods (Stadtman 1948; Coulter *et al.* 1951; Danehy and Pigman 1951; Hodge 1953).

In food composition tables, the carbohydrates content is usually given as total carbohydrates by difference, i.e., the percentage of water, protein, fat, and ash subtracted from 100. Another widely used term is nitrogen-free extract, calculated as components other than water, nitrogenous compounds, crude fiber, crude fat, and minerals. The increasing awareness that specific carbohydrates play significant metabolic and functional roles and the availability of analytical tools to determine individual components has stimulated investigations on their distribution many foods.

Fruits are a rich source of mono- and disaccharides. Dates contain up to 48.5% sucrose, and dried figs contain a mixture of 30.9% fructose and 42.0% glucose (Hardinge *et al.* 1965). The sucrose content of most fruits and fruit juices is low, though some varieties of melon, peaches, pineapple, and tangerine contain 6–9% sucrose, and mango contains 11.6% sucrose. Reducing sugars (primarily a mixture of fructose and glucose) are the main soluble carbohydrate of most fruits and account for 70% of seedless raisins. Partly ripe bananas are relatively rich in starch (8.8%), uncooked prunes in cellulose and hemicellulose, and citrus fruits in pectins. Vegetables contain substantially less glucose and fructose than fruits, and the only significant source of sucrose is sugar beets. Fresh corn and white and sweet potatoes contain about 15% or more of starch.

Practically the only carbohydrate present in unsweetened milk and milk products is lactose. Nuts are generally a poor source of mono- and disaccharides; chestnuts contain up to 33% starch. The main component of cereals and cereal products is starch. In milled products, the starch content increases with the degree of refinement, and is about 70% in white flour compared with about 60% in whole grain. The increase in starch is accompanied by a parallel decrease in cellulose, hemicellulose, and pentosans. Cloves and black pepper contain 9.0 and 38.6% reducing sugars, respectively, and the latter has as much as 34% starch.

White commercial sugar contains 99.5% (or more) sucrose; corn sugar has about 87.5% glucose; honey has about 75% reducing sugars, a mixture of fructose and glucose; and most syrups and sweets have various amounts of sucrose (up to about 65%), reducing sugars (up to 40%), and dextrins (up to 35%).

DETERMINATION OF WATER-SOLUBLE AND WATER-INSOLUBLE SOLIDS

The analyses of syrups, fruit preserves, malted products, and many other foods include the determination of the water-soluble and water-insoluble fractions. The insoluble fraction is of value in the determination of the fruit content of jams and preserves.

The determination involves heating with boiling water, removing the soluble components by filtration, and drying the insoluble fraction to constant weight. Total water-soluble solids can be determined directly by evaporating an aliquot of the extract and drying the residue in a vacuum oven at 70°C. In the indirect determination, the water-soluble solids are calculated as the difference between total solids (as determined from loss on drying) and the water-insoluble solids. If only an estimate of the water-soluble solids is required, determining the specific gravity or refractive index is rapid, simple, and reliable.

Removing Interfering Substances

In determining the composition of the carbohydrates in the water extract, it is essential to remove interfering materials. Solid foods must be ground under conditions that cause little change in moisture content and do not significantly affect the composition and properties of the foods. Lipids and chlorophyll are generally removed by extraction with petroleum ether, in which the carbohydrates are practically insoluble. Extraction is generally carried out at 40–50°C; higher temperatures may solubilize starch components. To avoid hydrolysis and inversion of sucrose by organic acids during extraction at elevated temperatures, the addition of calcium carbonate for neutralization has been recommended. This addition is inadvisable if the extract contains large amounts of reducing sugars (Streuli and Stesel 1951).

In enzymatically active extracts, it is important to prevent the hydrolysis of sugars during extraction and storage of samples. This may be accomplished by the addition of mercuric chloride (Hadorn and Jungkunz 1952). Enzymatic modifications can be eliminated by the extraction of sugars with an aqueous solution of ethanol or by dropping the finely divided material into boiling 80% ethanol. The standard procedure for fresh plant materials recommends dropping into twice distilled ethanol neutralized with calcium carbonate. The amount of ethanol is selected to give a concentration of 80% with the water extracted from the sample. The sample in ethanol is heated on a water bath for 30 min. The alcohol is generally evaporated at a low temperature, and the excess of calcium ions is removed during clarification.

Clarifying Agents

Water extracts of most foods are clarified prior to a sugar determination. Turbidity caused by proteins and soluble starch affects polarimetric assays, and end-point determinations are masked in highly colored solutions. The color of the solution may not interfere in some reductometric methods provided the coloring substances do not react with the sugar reagents. Proteins precipitate copper in copper-reduction methods.

Clarification of water extracts is based on the principle that heavy metals precipitate colloidal substances (i.e., proteins) or that precipitates formed through the action of heavy metals (e.g., zinc ferrocyanide) combine with and coprecipitate the proteins (Acker 1967). Clarification agents should remove interfering substances completely without adsorbing or modifying the sugars. A reasonable excess of clarifying agent should not affect the assay. The precipitate should be small, and the precipitation procedure should be relatively simple.

Different clarifying agents meet these criteria to varying degrees. The selection of a specific agent, therefore, depends on the analyzed food, on the kind and amount of interfering substances, and on the proposed assay method. Extracts for polarimetric assay should be clear, practically colorless, and free of optically active substances other than sugars (i.e., amino acids, tannins, glycosides) or substances that influence the optical rotation of sugars (i.e., acid salts). Ferricyanide- and copper-reducing methods are more sensitive to soluble nonsugar, reducing compounds than gravimetric copper-reducing procedures. In the latter, it is essential to remove completely colloidal matter that might be coprecipitated with Cu_2O.

Most clarifying agents also have a decolorizing action through the adsorption of coloring substances by the precipitate, or through precipitation of natural chromogens of the polyphenol type by lead salts. Lead salts, especially the basic lead acetate, also precipitate optically active organic acids and are therefore useful in polarimetric assays. Aluminum hydroxide and kieselguhr have a limited clarifying power. Generally, an increase in decolorization is accompanied by increased adsorption of reducing sugars. This correlation is particularly noticeable in activated carbons.

In the assay of sugars in colorless or slightly colored, protein-rich solutions, the Carrez precipitation method gives excellent results (Acker 1967). The precipitant is less satisfactory for solutions of plant origin that are rich in gums, pectins, and acidic colloids. The Carrez solution involves the consecutive addition of equal volumes of solutions containing per liter, respectively, 150 g $K_4Fe(CN)_6 \cdot 3H_2O$ and 300 g $ZnSO_4 \cdot 7H_2O$. The zinc is in excess but generally does not interfere, except in the complexometric determination of sugars according to Potterat and Eschman (1954), where excess zinc is removed after precipitation with 1.0 N NaOH. The excess zinc also must be removed in the determination of sugars by fermentation methods.

Neutral lead acetate is the most commonly used clarifying agent both for chemical and polarimetric determinations. Its main limitation is a low decolorizing power; consequently, it is not suited for polarimetry of dark-colored solutions. Excess lead acetate must be removed with a sodium sulfate or phosphate solution, solid sodium oxalate, or a mixture of disodium phosphate and oxalate. In the past, basic lead acetate was used widely for deproteinization, but has been largely replaced by the Carrez reagent. Its use is limited today to the clarification of highly colored solutions rich in organic acids. Whereas a slight excess of neutral lead acetate has little effect, excess basic lead acetate precipitates and occludes reducing sugars, and affects the specific rotation of sugars. Excess basic lead acetate is removed as excess of the neutral salt. Aluminum hydroxide (alumina cream) is sometimes used for clarifying slightly colored sugar solutions. It is efficient in removing flocculate colloids but not noncolloidal materials.

Additional clarifying agents include a mixture of barium hydroxide and zinc sulfate (Somogyi 1945B); dialyzed ferrioxychloride for the precipitation of slightly alkaline solutions; a mixture of copper sulfate and alkali for the precipitation of protein in milk; and mercuric nitrate in conjunction with an alkali for animal tissue extracts. Proteins can be removed by the general precipitants, trichloroacetic acid or phosphotungstic acid.

The clarifying agent is added either as a saturated aqueous solution, and the sugar solution is made to volume after clarification, or as a powder after the solution is made to volume.

Even with safe clarifying agents such as neutral lead acetate, a large excess should be avoided as it affects the polarization of sugars, and on heating, the interaction between the lead and sugar may result in some destruction of the latter.

Compounds added to remove excess clarifying agents should not be added in large excess. For certain methods of sugar determination, treatment with special reagents is mandatory. Oxalate is oxidized by ceric sulfate and must be replaced by disodium phosphate if ceric sulfate is to be used for the titration of ferrocyanide. In iodometric assays, the use of neutral lead acetate and sodium oxalate is satisfactory.

In certain assays, passing the solution through a mixed-bed ion-exchange column, for the purpose of desalting, is recommended (Wiseman *et al.* 1960). Amino acids present a special problem because they interfere with the determination of sugars and are not separated by common protein precipitants (Hadorn and Biefer 1956). Some separation of amino acids from sugars may be effected by ion-exchange chromatography.

QUALITATIVE DETECTION OF CARBOHYDRATES

Qualitative tests for sugars are based on (1) color reactions effected by the condensation of degradation products of sugars in strong mineral acids with various organic compounds; (2) the reducing properties of the carbonyl group; and (3) on oxidative cleavage of neighboring hydroxyl groups. Many qualitative tests are determined on fractions separated by paper, thin-layer, or column chromatography. Many of the qualitative tests have been adapted for quantitative determinations, which are discussed in later sections.

Color Reactions in Strong Acids

The action of strong mineral acids (sulfuric, hydrochloric, and phosphoric) on carbohydrates leads to the formation of colored decomposition products. Aldohexoses and ketohexoses give as one of the main decomposition products hydroxymethyl furfural, which in acid solution further decomposes to levulinic acid. Pentoses and hexuronic acids (after decarboxylation) give furfural; methyl pentoses produce methyl furfural.

The observations of Bandow (1937) on the reactions of carbohydrates in concentrated sulfuric acid were developed into quantitative methods by Ikawa and Niemann (1949) and Bath (1958). Scott *et al.* (1967) described a sensitive and rapid ultraviolet spectrophotometric method for the determination of hexoses, pentoses, and uronic acids after their reactions with concentrated sulfuric acid.

A more distinctive coloration is obtained when to the decomposition products of sugars in acid, some organic compounds are added (Dische 1962; Stanek *et al.* 1963; Acker 1967). The added compounds include phenols, aromatic amines, thio compounds, urea, anthrones, and others. The corresponding procedures are known as the Molisch reaction (α-naphthol), Selivanoff test (resorcinol), Bial procedure (orcinol), the naphthoresorcinol and phloroglucinol tests of Tollens, the Dische reaction (diphenylamine), and the Tauber test (benzidine). Some of the reagents are selective for certain sugars only, some show wide application, and some give various colored reactions depending on the sugar present. Anthrone and phenol have found very wide use both in qualitative and quantitative analysis, especially in determining the concentration of the fractions separated by column and paper or thin-layer chromatography (Hodge and Hofreiter 1962).

Anthrone (9,10-dihydro-9-oxoanthracene), a reduction product of anthraquinone, was first recognized by Dreywood (1946) to react specifically with many carbohydrates in concentrated sulfuric acid solutions to produce a characteristic blue-green color. The color has been attributed to the reaction product of hydroxymethyl furfural, or furfural, with anthrone. Carbohydrates and their derivatives that do not yield these substances display a wide range of different colors. The differences preclude the use of anthrone in the determination of total carbohydrate in sugar mixtures. However, in other instances this property has been used in the differential analysis of mixtures. Anthrone gives the best results when applied to pure solutions of hexose sugars or their polymers, which produce a characteristic blue-green color.

The phenol–sulfuric acid method (Dubois *et al.* 1956) is simple, rapid, sensitive, accurate, specific, and widely applicable for carbohydrates. Virtually all classes of sugars, including sugar derivatives, oligosaccharides, and polysaccharides can be determined. The reagents are inexpensive, readily available, and stable. A stable color is produced, and the results are reproducible. The method is excellent for determining sugars separated by chromatography. In the direct determination of lactose in milk and cheese, normal amounts of casein, amino acids, and organic acids do not interfere (Barnett and Tawab 1957).

Kushawaha and Kates (1981) modified the phenol–sulfuric acid method for the estimation of sugars in lipids. The dry samples are dispersed in 2 ml water and 1 ml of 5% phenol is added. After adding 5 ml H_2SO_4, the mixture is heated in a water bath. After cooling, the absorbance is read at 490 nm. The assay is linear from 0 to 80 μg. Ford (1981) described a procedure in which the ratio of hexoses to pentoses may be determined by a combination of the phenol–sulfuric acid reaction and an increase in absorbance at 312–325 nm. The absorbance of that band is related linearly to the ratio of hexoses to total sugars in the mixture in the 10–80% hexose range.

In a new method for ketose determination (Boratynski 1984), a purplish pink color developed when 100 µl of a ketose solution was mixed with 0.5 ml of a phenol–acetone–boric acid reagent (5% phenol, 2% acetone, 4% boric acid) and treated with 1.4 ml 96% sulfuric acid. The absorbance was measured at 568 nm after 60 min at 37°C. This method allowed the determination of 3–50 mmole of D-fructose with a coefficient of variation of 2.8–7.8%. Carbohydrates other than ketoses did not interfere.

Methods Based on Reducing Properties

Color reactions based on the reducing properties of monosaccharides and short-chain oligosaccharides are nonspecific and can be used only after the removal of other reducing organic compounds. The reactions are, however, sensitive and well-suited to routine assays (Dische 1962). Several types of such reactions are useful.

In the reaction of reducing sugars with arsenomolybdate, cupric (Cu^{2+}) salts are reduced to cuprous oxide (Cu_2O), which in turn reduces arsenomolybdate to molybdene blue. The absorbance of the latter is a measure of the concentration of the sugar. Ferricyanide of a pH above 10.5 can be reduced by sugars to ferrocyanide which produces Prussian blue. If an organic compound containing an easily oxidizable group (e.g., a reducing sugar) is heated with a solution of a tetrazolium salt at pH 12.5, a red, violet, or blue color—turning into a precipitate—is formed. Intensification of the color may be achieved by the addition of acetone. The test is highly sensitive. The most commonly used salt is 2,3,5-triphenyltetrazolium chloride or bromide.

In the periodate oxidation test, neighboring hydroxyl groups are oxidized and the aldehyde formed is detected by a fuchsin–sulfurous acid reagent.

In alkaline solution reducing sugars, which contain an aldehyde or keto group, can reduce copper, silver, bismuth, and mercury salts to compounds of lower valence or to a metallic state. The best known reagent, based on the reduction of copper, is Fehling's solution. It is prepared by mixing before use two solutions, one containing cupric sulfate and one containing sodium potassium tartrate and sodium hydroxide. Depending on the concentration of sugars in a solution, heating in the presence of Fehling's solution gives a yellowish orange to red solution or precipitate. Some monosaccharides (e.g., glucuronic acid) react in the cold.

The Tollens reagent is based on the oxidative effect of the complex ion $[Ag(NH_3)_2]^+$. It is the most sensitive of all reagents utilizing the reduction of metal ions by sugars. The reaction is, however, not specific and is exhibited also by other easily oxidizable organic compounds such as polyhydric phenols, amino phenols, and aldehydes. Some sugars without a free hemiacetal group (e.g., sucrose) also give a positive reaction.

The oxidation of sugars by an alkaline solution of trivalent bismuth in the presence of potassium sodium tartrate is the basis of the Nylander reaction. Crystalline phenylhydrazones with specific melting points are obtained when a cold aqueous sugar solution is treated with 1 vol of phenylhydrazine, 1 vol of 50% acetic acid, and 3 vol of water (Acker 1967). Because the phenylhydrazones are somewhat soluble in water, substituted phenylhydrazines such as bromophenyl-, nitrophenyl-, or 2,4-dinitrophenyl-hydrazine are preferred for identification. The original sugar can be recovered from the phenylhydrazones by treatment with benzaldehyde.

Osazones are obtained by the interaction of three moles of phenylhydrazine and one mole of sugar. They are less useful for identification, as epimeric sugars (glucose, mannose, and fructose) give the same osazone. Identification of sugars by their crystalline derivatives has been largely replaced by various techniques of partition chromatography.

DETERMINATION OF MONO- AND OLIGOSACCHARIDES

The available assay methods for mono- and oligosaccharides include chemical, colorimetric, chromatographic, electrophoretic, optical, and biochemical procedures. Today, more and more assay techniques involve preliminary separation by chromatographic and electrophoretic techniques prior to actual assay by classical chemical procedures or colorimetric tests. Optical tests are useful as identifying aids in the determination of total solubles and for the determination of specific sugars (generally in combination with chemical or enzymatic pretreatment). Microbiological assays of carbohydrates have found relatively little application. The use of enzymes as aids in sugar analysis or in actual assays is gaining in popularity with the commercial availability of pure, selective, and stable preparations.

Chemical Procedures

An excellent and detailed review on selected methods for determining reducing sugars was published by Hodge and Davis (1952). Comprehensive and detailed reviews on the analysis of carbohydrates include those by Bates (1942), Browne and Zerban (1941), Whistler and Wolfrom (1962–1964), and Acker (1967).

Copper Methods. Probably no other analytical method has been utilized in so many modifications as the assay of reducing sugars for their oxidation by copper ions. In all copper methods, the reduction of copper and the oxidation of sugars are not stoichiometric. Yet, the reaction conditions can be adapted to give quantitative reproducible results, so that the amount of reducing sugars can be determined from calibration tables.

The oxidation of reducing sugars by alkaline copper solutions was first proposed by Trommer in 1841. The assay procedure was improved by Barresvil, who proposed adding potassium tartrate to prevent the precipitation of cupric hydroxide. Details of the method were worked out by Fehling in 1848 and reevaluated critically by Soxhlet in 1878. Two solutions are prepared for the determination of reducing sugars by the Fehling–Soxhlet method. One contains 34.64 g of $CuSO_4 \cdot 5H_2O$ per 500 ml, and the other 173 g of Rochelle salt ($NaKC_4H_4O_6 \cdot 4H_2O$) and 50 g of NaOH per 500 ml.

When a reducing sugar is treated with alkali at elevated temperatures, the sugar is degraded, and some of the degradation products reduce the cupric ions in the solution to a cuprous oxide precipitate. For quantitative determinations the method is varied as to the composition of the alkaline copper solution and the details of the assay procedure.

Several procedures have been suggested in which the volume of sugar solution required to reduce a definite amount of alkaline copper reagent is measured. The only direct titrimetric procedure that has found wide acceptance in Europe is that proposed by Lane and Eynon (1923). It is used to a limited extent in the United States for determining the dextrose equivalent of starch syrups as the official analytical method of the Corn Industries Research Association. The *dextrose equivalent* is defined as reducing sugars, calculated as dextrose, in the dry substance of starch syrups. Use of the Lane–Eynon method in the analysis of starch hydrolysates by the International Commission for Unified Methods of Sugar Analysis was described by Heyns (1959). In principle, the sugar solution is added slowly from a buret to a vigorously boiling mixture (1:1) of the two Fehling–Soxhlet solutions. Close to the end point, the sugar solution is added dropwise in the presence of 1 ml of a 2% aqueous solution of methylene blue, which changes from blue to white by an excess of the reducing sugar. The determination is then repeated and the amount of sugar solution (less 0.5 ml) determined in the preliminary assay is added at once, and the titration is continued to the end point.

In most methods, an excess of alkaline tartrate–cupric sulfate is added to a sugar solution, the solution boiled under specified conditions, and the amount of precipitate formed is determined. The precipitate can be weighed as Cu_2O, transformed to CuO or Cu, determined by titration after dissolving the Cu_2O precipitate, or determined by measuring the unreacted cupric ion complex.

In the gravimetric methods, the precipitate is filtered through a glass or porcelain filter stick with a fritted insert or asbestos layer, washed, dried, and weighed. The method most commonly used in the United States is the modified Munson–Walker (1906) procedure of the AOAC. It uses a unified procedure for all sugars, but the results are calculated for each sugar from empirical tables. The tables were computed to allow for the presence of sucrose along with reducing sugars (glucose, fructose, maltose, lactose, and their mixtures). The Munson–Walker procedure was evaluated critically by Hammond (1940), who studied the accuracy of the procedure, developed a refined method, and prepared more concise tables for dextrose, levulose, invert sugars, and invert sugar–sucrose mixtures. The precision and errors of the Munson–Walker method were studied also by Jackson and McDonald (1941). Wise and McCammon (1945) extended the use of the Munson–Walker method by preparing tables for some of the less common sugars.

Several titrimetric methods for the indirect determination of the reduced copper are available. Titrimetric permanganate methods are based on the procedure proposed by Bertrand in 1906. The washed Cu_2O precipitate is dissolved in an acidified $Fe_2(SO_4)_3$ solution, and the amount of ferrous ions from the reduction by the cuprous ions is determined with standard permanganate. In the AOAC modification, the cuprous oxide is dissolved in neutral ferric ammonium sulfate, and the acid is added shortly before titration. The end-point determination is sharpened by adding phenanthroline as an indicator. Acid ferric sulfate dissolves cuprous oxide faster than does the neutral solution but leads to low results as some of the ferrous sulfate formed is oxidized by air under acid conditions.

In the titrimetric iodide method, the Cu_2O precipitate is filtered, washed, and oxidized to cupric nitrate with nitric acid. The excess of nitric acid is removed by boiling, and after acidification with strong acetic acid, 10 ml of a 30% potassium iodide solution is added. The liberated iodine is titrated with standard sodium thiosulfate.

Complexometric titration of cupric ions (from the Cu_2O precipitate dissolved by boiling nitric acid) with Na_2EDTA in the presence of indicators was proposed by Potterat and Eschmann (1954). The indicators change color when bound to cupric ions, but are less stable than the cupric chelates. At the end of the titration, cupric ions are removed by the chelating agent from the indicator–Cu complex and the color of the indicator changes. An indicator commonly used is murexide (ammonium salt of purpuric acid), which is blue-violet at pH 10 and yellow as a copper complex. The titration is carried out in an ammoniacal medium. The reagent (1000 ml) is prepared by adding a solution containing 25 g $CuSO_4 \cdot 5H_2O$ to 500 ml of a solution containing 286 g $Na_2CO_3 \cdot 10H_2O$ and 38 g Na_2EDTA. The reagent is stable for many months and even boiling for 24 hr affects it little.

The direct iodometric determination of Shaffer and Hartmann (1921) is based on the fact that in the presence of oxalate a complex is formed with the cupric ion in solution, and the precipitated cuprous oxide can be titrated after acidification with an iodate–iodide solution without filtering. A considerable saving of time for each determination is effected. Shaffer and Hartmann employed the alkaline copper reagent and heating conditions specified by Munson and Walker, so that the tables worked out by the latter can be used. The usefulness of the method was confirmed by Hadorn and Fellenberg (1945). Apparently, the speed is achieved at some sacrifice of accuracy.

Somogyi (1945A) modified the Shaffer–Hartmann procedure for the determination of micro quantities of reducing sugars. The alkaline reagent developed by Somogyi is buffered with phosphates and includes potassium iodate as a source of iodine for the oxidation of the cuprous ion. The inclusion of 18% sodium sulfate is claimed to eliminate back oxidation of the cuprous ions by air. The presence of inorganic halides and nitrates depresses the reducing value, apparently by partly solubilizing the cuprous oxide and enhancing its oxidation in air. The titrimetric method has a precision of $\pm 2\%$ in the range of 0.3–3.0 mg of glucose. In a colorimetric modification, reduced copper is determined by reacting it with phospho- or arsenomolybdate color-forming reagents (Nelson 1944). The colorimetric method is as precise as the titrimetric; the useful range is 5–600 μm.

Iodometric titration of excess copper sulfate has several advantages over the titration of reduced Cu_2O. The procedure is simple, requires no filtration, and one method can be used for all reducing sugars. This method, developed by Schoorl and Regenbogen (1917), employs the Fehling–Soxhlet solutions. The reduced copper is determined indirectly by iodometric titration of the unreduced copper salt remaining after the oxidation of sugars. The accuracy of the method was confirmed by collaborative studies (Flohil 1933).

The highly alkaline Fehling–Soxhlet solution causes strong and not always reproducible degradation of various sugars, and either lowers the assay precision or requires strict adherence to experimental conditions. The use of alkaline carbonate solutions has been the basis of a procedure known in Europe as the Luff–Schoorl method, and in the United States as the Benedict reagent. The latter employs sodium carbonate instead of potassium hydroxide and sodium citrate in place of tartrate. Determinations of reducing sugars in carbonate-buffered water solutions were reviewed by Heidt and Colman (1952).

As early as 1873 Barfoed proposed (Hodge and Davis 1952) using copper acetate to differentiate between reducing monosaccharides and disaccharides, as the latter are not oxidized appreciably by the reagent. The reagent was adapted for quantitative use by Steinhoff (1933) and by Sichert and Bleyer (1936), and is used for the determination of glucose in the presence of maltose and dextrin. In the modified procedures, sodium acetate is substituted for acetic acid, and the reduced copper is determined iodometrically, or by dissolving it in a ferric sulfate solution and titrating the resulting ferrous ion with ceric sulfate.

Alkaline Ferricyanide Methods. The alkaline ferricyanide method was developed as an analytical procedure for the determination of sugar in blood by Hagedorn and Jensen (1923) and modified for food analysis by Hanes (1929). The method is based on the reduction of alkaline ferricyanide to ferrocyanide in the presence of a reducing sugar, the amount of reduction being a measure of the amount of sugar the sample contains.

The amount of reduced ferrocyanide is determined as the difference between the ferricyanide added and that remaining after reduction. The changes occurring in the presence of potassium iodide are

$$2K_3Fe(CN)_6 + 2KI \rightarrow 2K_4Fe(CN)_6 + I_2$$

In the presence of zinc ions, the ferrocyanide formed is precipitated as a zinc complex and the equilibrium is shifted to the right:

$$2K_4Fe(CN)_6 + 3ZnSO_4 \rightarrow K_2Zn_3[Fe(CN)_6]_2 + 3K_2SO_4$$

The liberated iodine is titrated with standard thiosulfate. Direct titration of the ferrocyanide can be carried out with ceric sulfate and phenanthroline as the indicator (Whitmoyer 1934; Hassid 1936). Ferricyanide can also be used in the direct titration of sugars in the presence of an indicator (picric acid or methylene blue), or the ferrocyanide produced can be estimated colorimetrically as Prussian blue.

The most commonly used procedures involve the titration of excess ferricyanide. Numerous modifications of the end-point determination have been proposed, including titrimetric, colorimetric, and potentiometric procedures. Ferricyanide-reduction methods are popular because the reduction and subsequent determination of reduced ferricyanide can be carried out in one reaction vessel. The ferricyanide reagent is stable in the alkaline solution used, and the ferrocyanide formed is more stable than the Cu_2O formed in copper reduction methods. The reaction is reproducible and well suited to routine analyses. However, the oxidizing action of ferricyanide is not as specific as that of cupric reagents, since ferricyanide is reduced more easily by substances other than sugars. As in the copper-reduction methods, the oxidation yields a variety of partly unstable oxidation products. Consequently, no stoichiometric relationship exists for the sugar oxidation.

Following the investigations of Blish and Sandstedt (1933) and Kneen and Sandstedt (1941), the ferricyanide method was adapted extensively to determine reducing sugars, diastatic activity, and the β-amylase activity of wheat flour, and the saccharifying activity of enzyme preparations. It is particularly useful in the determination of small amounts of maltose in starch hydrolyzates.

Iodometric Methods. Iodine in an alkaline medium is converted rapidly into hypoiodite, which can oxidize aldoses; ketoses are oxidized little. The method, originally proposed by Willstatter and Schudel (1918), is applicable to aldoses alone or in a solution with other carbohydrates, provided no interfering iodine-consuming compounds are present. The dissolved sample is treated with an excess of dilute iodine, and sodium hydroxide is added and mixed rapidly (otherwise the iodine may be oxidized to iodate, which does not react with sugars in alkaline solutions). After the solution is acidified with hydrochloric or sulfuric acid, and left to stand for a few minutes, the excess of the standard iodine solution is titrated with a standard thiosulfate solution.

In contrast to the copper and ferricyanide methods, the oxidation of aldoses by iodine approaches the stoichiometric reaction

$$RCHO + I_2 + 3NaOH \rightarrow RCOONa + 2NaI + 2H_2O$$

Various modifications of the iodine method concern primarily the optimum alkalinity of the medium to realize the stoichiometric relation and eliminate the interference of ketoses. Ethanol, acetone, and other substances that react with iodine must be absent. Under optimum conditions only about 1% of the ketoses are oxidized. Other substances that consume small amounts of iodine include mannitol, glycerin, sodium lactate, sodium formate, urea, and sucrose (Hodge and Davis 1952). The modified procedures of Hinton and Macara (1924) and Lothrop and Holmes (1931) are accurate to within ±0.5% in samples high in aldoses. Fructose can be determined (after oxidation of glucose with iodine in an alkaline medium) by copper-reduction methods, provided the excess iodine is titrated in the glucose determination with sulfurous acid rather than with thiosulfate (Acker 1967).

By replacing the iodine solution with solutions of potassium iodide and chloramine T (the sodium salt of N-chloro-p-toluene sulfonamide), many of the limitations of the iodine method are overcome. Chloramine T hydrolyzes slowly, producing in a slightly alkaline medium sodium hypochlorite, which reacts with potassium iodide to slowly release the hypoiodite oxidizing agent. The oxidation of aldoses proceeds slowly and the danger of a side reaction in minimized.

Dextrose, maltose, lactose, and invert sugar can be analyzed satisfactorily by the iodometric method. The method has been used in the determination of lactose in milk products (Hinton and Macara 1927). The determination of lactose is unaffected by the presence of sucrose, and small amounts of fructose do not affect the determination of glucose. Substances that form iodoform with hypoiodite affect the results.

Cerimetric Methods. The cerimetric determinations of sugars can be carried out in two general ways (Stanek *et al.* 1963). Either the consumption of cerium perchlorate is estimated titrimetrically with nitroferroin as an indicator, or a sugar solution is boiled with a solution of cerium sulfate in dilute sulfuric acid and the excess of cerium salts is back-titrated with ferrous salts. Under these conditions, glucose yields formic acid as the highest oxidation product, while ketoses are partially oxidized to carbon dioxide. By adding chromic ions to the cerium sulfate solution, glucose is also oxidized to carbon dioxide.

Chromatographic and Electrophoretic Methods

Chromatographic methods have been used to fractionate, isolate, identify, and determine carbohydrates in complex mixtures. The methods range from paper chromatographic procedures for the identification of components on a microscale to large-scale column separations for the isolation of relatively large amounts of pure compounds for further identification and study.

Paper and Thin-Layer Chromatography. The separation of sugars by paper chromatography has been developed since the pioneering investigations of Partridge (1948) into one of the most versatile forms of qualitative and quantitative carbohydrate microanalysis (Cramer 1953; Hough 1954; Kowkabany 1954; Lederer and Lederer 1957; Whistler and DeMiller 1962).

Qualitative paper chromatography is the best and simplest method to distinguish between various forms of sugars present in foods along with a mixture of various other compounds. In the case of a complex mixture, separation by paper chromatography may be supplemented by the use of specific sprays, additional identification of separated spots, use of additional separation methods, and—in all cases—comparison with R_f values of mixtures of pure sugars chromatographed along with the investigated food extracts.

In the chromatography of pure sugar solutions, no special preparation is required. For food extracts, precipitation of proteins or other interfering substances by conventional sugar precipitants is advisable. To avoid tailing, the removal of inorganic compounds is recommended; this can be done by passing the solution through a mixed bed of ion exchangers prior to paper chromatography. It should be noted, however, that strongly basic anion exchangers in the hydroxyl form may adsorb or modify sugars, and that strong acidic cation exchangers may hydrolyze sensitive oligosaccharides (Jayme and Knolle 1960).

Separations of 1% sugar solutions containing a maximum of 60 μm of an individual sugar give best results. The solvent system butanol–acetic acid–water (4:1:5) gives good resolution, but requires long development times. Resolution is more rapid with phenol–water or collidin–water systems (Acker 1967). Other useful systems include acetic acid–pyridine–water and n-propanol–acetic acid–water. The selection depends also on the expected mixture of sugars. A mixture of glucose, sucrose, and lactose cannot be separated well by phenol-containing systems, but the separation is satisfactory in n-propanol–acetic acid–water systems.

Generally, the higher the sugar in a homologous series, the lower its R_f value, i.e., trisaccharides have lower R_f values than disaccharides, and monosaccharides have still higher mobilities. Pentoses have higher R_f values than hexoses; among hexoses, ketoses are faster moving than aldoses. There are exceptions to these rules. For example, in phenol-containing systems fructose moves faster than xylose. Disaccharides with 1,4-linkages migrate faster than those with 1,6-linkages; and α-D-glucosides move faster than β-D-glucosides (Isherwood and Jermyn 1951). For better resolution, two-dimensional paper chromatography can be used; generally however, separation in one dimension with several solvent systems is preferred.

For the identification of the separated components, many spray reagents are available. Aniline phthalate gives a color reaction with reducing sugars only; fructose is, however, less reactive. Spraying with m-phenylenediamine gives fluorescing acridine derivatives (Chargaff et al. 1948). Silver nitrate in ammonia is a useful general reagent for reducing sugars. For nonreducing sugars, various phenols in acid (e.g., naphthoresorcinol in syrupy phosphoric acid) are recommended. The phloroglucinol–hydrochloric acid reagent of Borenfreund and Dische (1957) gives a purple color with aldopentoses, dark green with ketopentoses, yellow brown with ketohexoses, and green with methyl pentoses.

Qualitative determinations of sugars by paper chromatographic methods are useful primarily in the separation of sugars on a microscale, and in the identification of components of a complex mixture that cannot be resolved easily by other available procedures. They are used to determine oligosaccharides in malt extracts, the composition of starch hydrolysates, or small amounts of raffinose in sucrose solutions. The two main limitations to the quantitative determinations of carbohydrates by paper chromatography are high blanks, and the lack of a linear relation between sugar concentration and a measured analytical parameter. One must prepare each time a calibration curve covering the expected range. The precision is generally only ±5% though some methods give, with proper precautions, an accuracy of ±2% (Whistler and Hickson 1955).

Quantitation of the separated carbohydrates can be accomplished in several ways. The stained spots can be determined by direct photometry (Jayme and Knolle 1960). Some colored sugar derivatives (e.g., triphenyltetrazolium) can be eluted quantitatively and the concentration measured in solution (Schoenemann et al. 1961). Most commonly, the sugars are separated by paper chromatography, their position determined, and they are eluted for subsequent determination by colorimetric methods (Dimler et al. 1952). Many methods have been developed for the quantitative determination of carbohydrates in a mixture after separation by paper chromatography. Of the microtitrimetric methods, the alkaline ferricyanide reagent of Hagedorn and Jensen (1923) is one of the more useful. Colorimetric methods are usually more sensitive than titrimetric methods. Two of the more useful are the phenol–sulfuric acid and anthrone methods. They are described earlier in this chapter. In both, it is essential that the extract be free of cellulose fibers.

Sugars also can be separated by thin-layer chromatography. The advantages of this method are speed and good resolution (Stahl and Kaltenbach 1961, 1965; Weill and Hanke 1962; Scherz et al. 1968). Huber et al. (1966) described a rapid method for the determination of saccharide distribution of corn syrups by direct densitometry of thin-layer chromatograms. The results were of comparable accuracy to those obtained by gravimetric paper chromatography.

Electrophoresis.   The application of zone electrophoresis (on a strip of filter paper or other supporting medium) for the separation of carbohydrates has found wide application for sugar determinations (Foster 1957; Weigel 1963). Separations of carbohydrates are generally made with borate derivatives. The method cannot separate borate-complex epimers such as fructose, glucose, and mannose.

Gas–Liquid Chromatography.   Gas chromatographic analysis of carbohydrates has received much attention as a result of the pioneering work of Sweeley et al. (1963) with trimethylsilyl derivatives. Other workers (Alexander and Garbutt 1965; Kagan and Mabry 1965; Sawardeker and Sloneker 1965) have followed the silyl ether derivative approach to the quantitative analysis of sugar samples. The determination of carbohydrates by gas chromatography of their alditol acetates was reviewed by Crowell and Burnett (1967). Gas–liquid chromatography of carbohydrates was reviewed by Bishop (1964) and Geyer (1965). For the separation of carbohydrates by gas–liquid partition chromatography, free reducing groups of sugars must be blocked. Excellent resolution and quantitation can be achieved.

Ion-Exchange Chromatography.   Sugars, being weak electrolytes, have little tendency to react with ion-exchange resins. However, it has been known for a long time that certain polyhydroxy compounds react with the borate ion to form complexes of negativelyy charged ions. Khym and Zill (1952) first introduced the technique of separating a mixture of sugar–borate complexes on a column of strong-base anion exchange in the borate form. Later, Zill *et al.* (1953) extended their work to include the separation of related compounds. The sugars were eluted by a stepwise gradient of borate and pH, and the elution required up to 60 hr.

Since the initial reports of Khym and Zill, a variety of ion-exchange matrices and modifications have been proposed. The procedures that were developed permitted separations of 15 sugars in an automated procedure (Green 1966). Accelerated (7.5 hr) procedures for automated separations were reported (Anon. 1966; Ohms *et al.* 1967). The sugars are eluted from a strongly anionic styrene–divinyl–benzene column with a borate buffer of gradually increasing chloride concentration and pH or with a linear gradient of borate buffer. Sugar in the eluate is determined by the orcinol–sulfuric acid method. Further development of this technique for the separation and determination of mixtures of mono-, di-, and trisaccharides was described by Kesler (1967).

Other Column Chromatographic Methods. Extrusion chromatography involves development of the column to approximately a predetermined extent, extrusion of the column from the tube, location and sectioning of the zones, and extraction of the adsorbed material with a highly polar solvent in which the material is soluble. The method is fast and eliminates the need to evaporate and handle large amounts of solvents. The method has not been used extensively for determination of carbohydrates.

Carbon column chromatography was devised by Whistler and Durso (1950) for the separation of oligosaccharides into classes according to the degree of polymerization (mono-, di-, and tri-, and oligosaccharides). Either carbon alone or carbon–Celite mixtures are used. The capacity of the columns is high and quantitative separations of large amounts of material are possible (Hoover *et al.* 1965). In automated elution and recording procedures (French *et al.* 1966), gradient elution of charcoal columns using aqueous tertiary butanol was effective in separating starch oligosaccharides in the range up to 15 D-glucose units. Branched, as well as linear, oligosaccharides could be separated. Resolution is improved if mixtures of charcoal–Celite are treated with stearic acid (Alm, 1952; Miller, 1960). Mixtures that can be resolved on paper chromatography, using a developing system that is not completely miscible with water, can usually be separated on Celite columns with the same solvent system. Cellulose column chromatography has also found wide application in preparative carbohydrate chemistry (Hough 1954; Brinkley 1955).

The HPLC determination of sugars has been the subject of numerous investigations (Hurst *et al.* 1979; Dunmire and Otto 1979; Iverson and

Bueno 1981; Reyes *et al.* 1982). An excellent, comprehensive review of HPLC of mono- and oligosaccharides was presented by Honda (1984). de Vries *et al.* (1979) evaluated an HPLC method for determining fructose, glucose, maltose, and lactose in foods. They reported a relative standard deviation of 2.8% and recoveries of 98.8% on a variety of foods.

The C_{18} reversed-phase liquid chromatographic determination of invert sugar, sucrose, and raffinose was described by Palla (1981). McFeeters *et al.* (1984) reported the routine analysis of the major substrates and products of homolactic or heterolactic acid vegetable fermentations by the use of two liquid chromatographic procedures. Sucrose, glucose, fructose, and mannitol were determined on a cation-exchange column in the lead form. The coefficient of variation of the added compounds (2–30 mM) in fermented cucumber juice ranged from 2.0 to 3.1%. Sucrose (1–12.5 mM), malic acid, lactic acid, acetic acid, and ethanol (2–25 mM) added to cucumber juice were separated on a reverse-phase column; coefficients of variation ranged between 4.3 and 5.9%. Sample preparation required only blending and filtration prior to injection.

Optical Methods

Optical methods were discussed in detail in several earlier chapters in this book. The following brief outline deals specifically with special aspects of carbohydrate analysis.

Refractometry. Refractometers are widely used in the sugar industry to measure the content of dissolved solids in sugar solutions. Generally, a sugar industry laboratory may be expected to have several refractometers, sometimes including a variety of models (Charles and Meads 1962). Refractometers are located at plant operating stations and in control laboratories to provide information for prompt process or product control. They also find application in research laboratories for process study and other investigations.

Polarimetry. Carbohydrates are optically active and can be assayed polarimetrically. Optical rotations can be measured by means of visual polarimeters, visual saccharimeters, and photoelectric spectropolarimeters (Whistler and Wolfrom 1962–1964). Visual polarimeters measure the angle of rotation directly on a circular scale. A monochromatic light source must be used. Saccharimeters are specially designed for determining the optical rotation of sucrose-containing solutions. They are generally of the quartz-wedge compensating type, which permit illumination with white light in conjunction with a dichromate filter. Photoelectric spectropolarimeters are high-precision polarimeters in which the intensity of monochromatic light is measured by photoelectric cells instead of the human eye.

Polarimetric assays of sugar are nondestructive and rapid. They are accurate provided (1) the solution is clear and colorless or only slightly colored, (2) the concentration of tested sugars is within an optimum range of the instrument, and (3) the solution contains no interfering optically active compounds. The use of clearing agents removes turbid materials and part of the coloring substances. To correct for the presence of optically active substances other than sugars, determination before and after inversion (as in sucrose analysis) is useful.

According to the Biot law, the rotating capacity of individual sugars is proportional to the concentration of the solution and to the length of the liquid column. A measure of this capacity is the specific rotation, $[\alpha]$, given by

$$[\alpha] = 100\alpha/lc = 100\alpha/lpd \tag{35.1}$$

where α is the angle of rotation of a solution of specific gravity d, containing p grams of active substance per 100 g solution (or a concentration of c grams/100 ml of solution) in a tube of length l (decimeters). The specific rotation depends on the temperature and wavelength of the rotated light. Generally, measurements are made at 20°C with a sodium light. The green mercury light is used in light-electric polarimeters because of its high intensity. (Refer to Chapter 27 for additional discussion.)

For a sucrose solution of $c = 26.016$ g (corresponding to 26.000 g weighed in air) in 100 ml, employing the sodium light, at 20°C and 760 Torr, $[\alpha] = 66.523$, and

$$c = \frac{100}{[\alpha]}\frac{\alpha}{l} = 1.5032\frac{\alpha}{l} \tag{35.2}$$

Thus, according to equation 35.2, a rotation of 1° with a 20-dm pathlength corresponds to 0.7519 g/100 ml sucrose. Corresponding values for other sugars are 0.9470, 0.5405, 0.9524, and 0.3623, respectively, for glucose, fructose (levorotatory), lactose, and maltose. The specific rotation varies with concentration, but the variation is small and can be neglected. The temperature correction is

$$[\alpha]_t^D = [\alpha]_{20}^D[1 - 0.000184(t - 20)] \tag{35.3}$$

According to decisions of the International Commission for Unified Methods of Sugar Analysis, saccharimeters should be calibrated according to the international sugar scale of ICUMSA in sugar degrees. The 100° point on the international scale is defined as the polarization of a normal solution of pure sucrose (26.000 g per 100 ml) at 20°C in a 200-mm-long tube with regular light and a dichromate filter (a solution of

$K_2Cr_2O_7$ of such concentration that the percentage of dichromate times the width of solution is 9, i.e., 6% dichromate and filter width of 1.5 cm). The details of the calibration of the saccharimeter are given in AOAC.

Products containing reducing sugars show mutarotation involving the establishment of an equilibrium state of the α and β forms, and the intermediate open-chain form. Such solutions must be allowed to stand overnight for analysis. If the determination is to be made immediately, the neutral solutions can be heated to boiling and a few drops of ammonia added before bringing to volume, or dry sodium bicarbonate can be added to solutions adjusted to final volume till clearly alkaline.

In the absence of other optically active substances, the sucrose content of food solutions can be determined by direct polarimetry. Sucrose can be determined directly by polarimetry, in the presence of reducing sugars after the latter are eliminated by heating with alkali (e.g., barium hydroxide for 1 hr at 70–80°C). The method is used in testing chocolate (Thaler 1940).

Several precise and reproducible chemical and polarimetric methods of sucrose determination are based on the fact that sucrose, a fructofuranoside, is hydrolyzed by acids faster than other disaccharides and most oligosaccharides, except those containing fructofuranosides (e.g., raffinose). On hydrolysis, one part of sucrose is hydrolyzed to 1.053 parts of an equimolar mixture of glucose and fructose known as *invert sugar*, because the hydrolysis is accompanied by a change in optical rotation (from +66.5 to −20.00 at 20°C). Thus, the sucrose content of a food solution can be determined from the change in optical rotation (or reducing power in chemical determinations) after inversion. In such methods, it is necessary that sucrose alone be hydrolyzed and that the rotation (or reducing power) of no other compounds be changed. In the optical rotation procedures, it is important to hydrolyze and polarize the samples under specified conditions. Both acids and salts affect the specific rotation, and their effects may be accounted for by blank determinations.

White and Siciliano (1980) found that honey samples containing more than 20 mg hydroxymethyl furfural (HMF) per 100 g may be adulterated and those containing about 50 mg HMF are definitely adulterated by the addition of invert syrup. The use of detailed carbohydrate analysis in detection of adulteration was described by White (1980).

Infrared Spectroscopy. Infrared spectroscopy of simple sugars is complicated by the fact that they are practically insoluble in the organic solvents commonly used in such determinations. Consequently, the potassium bromide disk and paraffin suspension methods are used to prepare samples (see Chapter 9).

Infrared spectroscopy has been widely used in studies of carbohydrate structure (Whistler and House 1953; Solms *et al.* 1954; Baker *et al.* 1956; Brock-Neely 1957; White *et al.* 1958; Spedding 1964). Investigations of Parker (1960) concerned an infrared method of studying mutarotation; Underwood *et al.* (1961) used infrared spectroscopy in investigations of browning in foods; Lin and Pomeranz (1965, 1968) found the technique useful in the detection of carbohydrates in wheat protein preparations and in investigations of the water-soluble wheat flour pentosans. The determination of lactose in milk products by infrared spectroscopy was described by Dyachenko and Samsonov (1964) and by Biggs (1967). Dull and Giangiacomo (1984) presented near infrared spectroscopic determination of glucose, fructose, and sucrose in aqueous solutions ranging from 3 to 52% (w/w). The 95% confidence limits of the method for glucose, fructose, and sucrose were ± 1.3, 1.0, and 0.9%, respectively.

Biochemical Methods

Biochemical methods of carbohydrate analysis can be classified into two types, microbiological and enzymatic. In microbiological assays, yeast strains varying in their fermentative capacity are selected. Such yeasts are particularly useful in differentiating between pentoses and hexoses (or their polymers). Bakers' yeast can also be used to differentiate between nonfermentable lactose and fermentable sucrose or maltose. In enzymatic methods, selective cleavage to monosaccharides and/or enzymatic assay of the monosaccharide is specific and widely used. The use of amylase or amyloglucosidase is an example of the first type; determination of glucose with glucose oxidase of the second type.

Microorganisms and enzymes also can be used in the pretreatment of substrate prior to assay by chemical or physical methods. For example, hydrolysis of sucrose by invertase is much more specific than acid hydrolysis, and invertase is used in various procedures requiring sucrose inversion. One such procedure, described by Blackeney and Mutton (1980), is a rapid, manual colorimetric method for determining glucose, fructose and sucrose in fruit and vegetables. The method is based on the determination of reducing sugars before and after invertase digestion using *p*-hydroxybenzoic acid hydrazide (Lever 1972). Total fructose (fructose plus fructose in sucrose) is determined using 2-thiobarbituric acid (Tanaka *et al.* 1975). Addition of sugars to fruit extracts gave recoveries of 97.4–101.8% for glucose, fructose and sucrose.

Comparison of Methods

Ugrinovits (1980) compared four methods of sugar analysis—gas chromatographic, high-performance liquid chromatographic, thin-layer chromatographic, and enzymatic—in terms of their specificity, precision, and various other characteristics. His conclusions are summarized in Table 35.1.

Copper-reduction methods are of limited value in analyses of potato sugars because they do not discriminate between glucose and fructose. Although chromatographic procedures differentiate between glucose, fructose, and sucrose, they are time-consuming and require both expensive equipment and considerable skill (Mazza 1983). The major disadvantages of using the anthrone colorimetric method for sucrose and the 3,5-dinitrosalicylate method for reducing sugars is the need for two completely separate analyses.

The Yellow Springs Instrument Co. (YSI) of Yellow Springs, Ohio, introduced an instrument that combines immobilized enzyme technology with a linear electrochemical sensor to determine sucrose, glucose, and lactose in liquid foods. The analyzer uses a thin film of oxidase immobilized within a membrane to produce H_2O_2 from the substrate of the enzyme. The H_2O_2 is measured by electrochemical oxidation at a platinum anode. Because the reaction is confined to a thin layer, equilibrium is attained quickly and steady-state readings are obtained in about 1 min. Mazza (1983) determined sucrose, glucose, and total reducing sugars of potatoes with the YSI sugar analyzer. Sucrose contents determined by the anthrone colorimetric method and the YSI were highly correlated ($r = 0.984$). Similarly, glucose levels determined with the YSI and total reducing sugars determined colorimetrically with dinitrosalicylate were positively correlated ($r = 0.956$). The YSI was recommended as capable of giving fast, simple, and reproducible assays for routine analyses and research purposes.

Table 35.1. Comparison of Analytical Methods for Sugars[a,b]

Characteristic[c]	GC	HPLC	TLC	Enzymatic
Specificity	−	−	−	+ +
Sample preparation	−	+	+	+
Assay	− −	+	−	−/+
Precision	− −	+	−	+
Sensitivity	+	+	+	+ +
Interferences	+	+	+	−
Time requirement	− −	−	−	−/+
Cost (assay)	−	+	+	−
Cost (instrument)	−	−	−	+ +
Capacity for automation	+	+	+	+

[a] Adapted from Ugrinovits (1980).
[b] GC (gas chromatography), HPLC (high-performance liquid chromatography), and TLC (thin-layer chromatography).
[c] +, Satisfactory; −, unsatisfactory.

DETERMINATION OF HEXOSAMINES

The amino sugars constitute the building stones of many biologically important substances (Balazs and Jeanloz 1965). Although many types of amino sugars are known, only a few are known to occur naturally. Of these, the most common are 2-desoxy-2-amino-D-glucose (known as D-glucosamine or chitosamine) and 2-desoxy-2-amino-D-galactose (D-galactosamine or chondrosamine). In their reactions the hexosamines have the properties of both hexoses and primary amines.

In the hydrolysis of amino sugar-containing polysaccharides, the monomers must be liberated from the polysaccharides. In general, amino-sugar glycosides are very resistant to acid hydrolysis and special precautions must be taken to secure complete hydrolysis. Gardell (1958) recommends heating a 1–2% solution for 8 hr on a boiling water bath with 6 N HCl. According to Pusztai (1965), most glucosamine-containing constituents of the seeds of kidney beans can be hydrolyzed with 0.5 N HCl for 16 hr or with 2 N HCl for 2 hr at 100°. Some acid-stable polysaccharide fractions require hydrolysis with 6 N HCl.

The isolation, crystallization, and subsequent identification of the 2-amino-sugar components of mucopolysaccharides, in which they are present in relatively large proportions, is readily accomplished (Foster and Stacey 1952). In biological materials that contain 2-amino sugars in minor proportion or only in trace amounts, other methods are required. Such methods include specific colorimetric tests, isolation of crystalline derivatives, and chromatographic procedures.

Colorimetric Tests

The most widely used of the colorimetric tests is based on the observation that glucosamine pentacetate, after being warmed with dilute alkali and then treated with p-dimethylaminobenzaldehyde (Ehrlich reagent), gives an intense reddish-purple coloration. The original procedure of Zuckerhandl and Messiner-Klebermass (1931) has undergone many modifications.

In the original procedure, free 2-amino sugars were detected and estimated colorimetrically as N-acetates after careful acetylation. The procedure was inaccurate and the color produced was not stable. Elson and Morgan (1933) and Morgan and Elson (1934) made use of the reaction noted by Pauly and Ludwig (1922) in which acetylacetone is condensed with hexosamines under alkaline conditions to yield a product that gives a stable reddish color with an acid solution of Ehrlich's reagent. In the first part of the test, the reaction of acetylacetone with glucosamine under alkaline conditions, four compounds are formed, two or three of which are chromogenic. The determination is, thus, highly empirical and the indiscriminate use of the so-called Elson–Morgan reaction may lead to confusing results. Neutral sugars (ketoses more than aldoses) and certain amino acids (mainly lysine, and to a lesser extent glycine and arginine) may react to yield products that give a color with Ehrlich's reagent indistinguishable from that produced by glucosamine. The sugar–amino acid reaction may be differentiated from the hexosamine reaction in several ways. The optimum pH for the hexosamine color is 9.5, whereas that for the amino acid–sugar complex color is 10.8–11.2. Glucosamine does not give a color with Ehrlich's reagent after being heated with a carbonate buffer alone in the first part of the test, whereas the sugar–amino acid complex gives a red color. The color intensity with glucosamine is proportional to the acetylacetone concentration, whereas a maximum intensity is obtained with the sugar–amino acid complex.

The test for N-acetyl-2-amino sugars involves warming the hexosamine-N-acetate or pentacetate with dilute alkali followed by treatment with an acid solution of Ehrlich's reagent. A reddish-purple color develops. Amino acids and neutral sugars apparently do not interfere if the test is conducted under special, rigidly controlled conditions.

The sensitivity of the original Elson–Morgan test is 20 μm. Dische and Borenfreund (1950) developed a method for determining 5 μm of amino sugar. The hexosamines are deaminated with nitrous acid, and the derivative is reacted with skatole in dilute hydrochloric acid to yield stable characteristic colors well-suited to a quantitative colorimetric determination.

The basic colorimetric procedures just described do not differentiate among the 2-amino sugars. However, several modified procedures have been proposed for assaying glucosamine and galactosamine in mixtures. One modified procedure is based on the observation that adding borate to the acetylation mixture in the Elson–Morgan method causes a depression of the color formation. With borate, glucosamine gives only 20–30% and galactosamine about 50% of the color formed without borate. Tracey (1952) used this principle in elaborating a method for the qualitative and semiqualitative determination of the glucosamine and galactosamine in mixtures. A similar method was described by Good and Bessman (1964).

Another procedure for distinguishing glucosamine and galactosamine is based on the formation of glucosamine-6-phosphate by ATP in the presence of hexokinase. This phosphorylating system, however, has little or no action on galactosamine. The glucosamine-6-phosphate can be precipitated from the reaction mixture with zinc sulfate and barium hydroxide. If the Elson–Morgan reaction is carried out before and after the enzymic reaction, the amino sugar composition of a mixture can be determined (Slein 1952). A similar procedure based on the specificity of yeast hexokinase for D-glucosamine, and of the yeast acetylating enzyme for D-glucosamine-6-phosphate was described by Luderitz *et al.* (1964).

Derivative Formation and Chromatographic Methods

The ideal means for the identification of a specific hexosamine is the isolation of a well-characterized crystalline derivative. Numerous compounds have been recorded in the literature for this purpose. The most satisfactory derivatives are the condensation products with aldehydes, the so-called Schiffs' bases (Jolles and Morgan 1940). The *p*-nitrobenzylidene, 2-hydroxy-1-naphthylidene, and 3-methoxy-4-hydroxybenzylidene derivatives have solubility properties well-suited for the isolation of small quantities of amino sugars. Large quantities of neutral sugars do not interfere, and amino acids give condensation products that can be extracted with chloroform.

Glucosamines can be separated on ion-exchange resin columns, on starch columns, and by paper or thin-layer chromatography. The most significant advance came with the introduction of 1-fluoro-2,4-dinitrobenzene as a reagent for condensation with the basic groups of amino acids to give well-defined, colored 2,4-dinitrophenyl derivatives suitable for chromatography (Annison *et al.* 1951; Kent *et al.* 1951).

Hexosamines and hexosamine uronic acids can be separated by paper electrophoresis in a borate buffer of their deamination products (Williamson and Zamenhof 1963) or of their molybdate complexes (Mayer and Westphal 1968). The separation and determination of hexosamines by gas–liquid chromatography were described by Sweeley *et al.* (1963) and Radhakrishnamurty *et al.* (1966).

DETERMINATION OF NUTRIENT POLYSACCHARIDES

Starch

Next to cellulose, starch is the most abundant and widely distributed substance in vegetable matter. Starch occurs in the form of granules as reserve food in various parts of most plants: in the seeds of cereal grains, the roots of tapioca, the tubers of potatoes, and—in small amounts—in the stem-pith of sago and fruit of the banana. The shape and size of the starch granules are characteristic for each plant. The appearance of the granules under microscopic examination is used to identify starches from various sources (Fig. 35.1). Extremes in the size of starch granules are

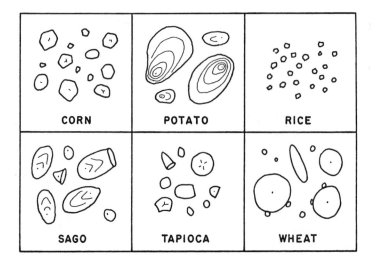

Fig. 35.1. Microscopic appearance of granular starches. (Courtesy T. J. Schoch.)

illustrated by taro or dasheen root (1 μm in diameter) and by root of the edible type of canna (150–200 μm in diameter). Under the polarizing microscope, the granular starches have a characteristic birefringence, indicating an organized spherocrystalline structure. Spherical granules give a *Maltese cross* interference pattern and others a differently shaped cross.

Starch is a natural high polymer of D-glucose units; the polymer is formed through successive condensation of glucose units involving the enzyme systems of a developing plant. Complete chemical or enzymatic hydrolysis of starch yields only glucose units.

Ordinary starches contain two types of glucose polymers. The linear fraction, called *amylose*, is a linear chain of some 500–2000 glucose units. Corn starch amylose contains an average of about 500 glucose units and has a molecular weight of about 80,000. The individual units in amylose are connected by α-1,4 linkages (i.e., from the aldehydic C_1 of each glucose to C_4 of the preceding glucose molecule). The branched fraction, *amylopectin*, contains several hundreds of short linear branches, with an average branch length of 25 glucose units. The average molecular weight of amylopectin is at least 1,000,000. The interglucose linkages in each linear portion are α-1,4 (as in amylose); the branch points are through α-1,6 linkages.

Common starches (corn, wheat, tapioca, and potato) contain amylose and amylopectin in (generally) fixed proportions. Starches in certain varieties of corn, sorghum, and rice (but not wheat or common root starches) are composed entirely of amylopectin and are known as *waxy starches*. On the other hand, wrinkled-seeded, garden-type varieties of peas and some corn varieties (amylomaize) have starches that contain from 70 to over 80% amylose. Waxy starches have certain unique properties that make them preferable for various uses. The linear fraction in cooked starch tends to deposit in the form of discrete insoluble particles as a result of starch retrogradation; amylopectin shows little tendency to retrograde.

Starches can be fractionated into amylose and amylopectin by gelatinization in water at elevated temperatures and pressures. The dissolved starch granules are cooled in the presence of polar agents such as butyl alcohol, amyl alcohol, thymol, and nitroparaffins. A complex precipitate of the polar components with amylose can be isolated by ultracentrifugation. Amylose similarly treated with fatty acids forms a complex that separates at $90 \pm 5°C$ as a pulverulent precipitate.

The intense blue color that starches give with iodine is due solely to the amylose fraction; amylopectin gives a red or violet-red color. The presence of starch can be established by the sensitive iodine reaction; 0.002 mg/ml can be positively detected.

To determine the starch content of foods, several methods are available. The starch can be extracted and dispersed into a colloidal solution, which can be separated from extraneous matter. The starch content of the dispersion can be determined by precipitation and gravimetric or titrimetric assay of the precipitate; by a polarimetric method; colorimetrically; or as glucose after chemical or enzymatic hydrolysis.

Extraction Methods. In fat- and protein-rich foods (sausages, cheese), the starch can be reacted with alkali to form alcohol-insoluble complexes that can be readily separated from other components. The method is not suitable for the determination of starch in plant materials that contain a mixture of various polysaccharides.

Although starch can be solubilized in boiling water, its complete extraction from a plant tissue is difficult to achieve because of its high molecular weight and colloidal properties. Perchloric acid is an efficient extractant. The method of Pucher et al. (1948) consists of extracting the soluble sugars from dried plant tissue with 80% ethanol. The sugar-free residue is then treated with a perchloric acid solution; the extracted starch is precipitated with iodine and the starch–iodine complex is decomposed with alkali. The liberated starch is then determined colorimetrically with the anthrone reagent.

Starch can be extracted with a hot, concentrated solution of calcium chloride, and the extract assayed by various methods. To eliminate interference from other extractable components, several modified methods have been developed (Hadorn and Doevelaar 1960; Zelenka and Sasek 1966). From the carbohydrates dissolved in the calcium chloride solution, a starch–iodine complex is precipitated. Under proper conditions, precipitation of other carbohydrates is eliminated. The starch content can be determined gravimetrically; or the amount of iodine absorbed can be measured titrimetrically or colorimetrically; or the starch–iodine complex can be decomposed with sodium thiosulfite, and the starch redissolved in calcium chloride and measured polarimetrically. Starch has a high specific optical rotation. Depending on the extractant composition and extraction method, $[\alpha]_D = +190$ to $203°$. Hemicelluloses, which may be solubilized in this procedure, are substantially less optically active. Dextrins have a high optical rotation and must be separated. Eheart and Mason (1966) have shown that in cereal products, reproducible starch determinations can be obtained by extraction of the starch with a calcium chloride solution, precipitation of proteins with uranyl acetate, and polarimetric assay of starch in the solution.

Methods based on titrimetric or colorimetric determination of the absorbed iodine are highly empirical, as the relation between color and starch concentration is not stoichiometric. The results depend on the method of starch extraction, the source of the starch, and the ratio of amylose and amylopectin.

In several methods, starch is determined polarimetrically in dilute hydrochloric extracts (Winkler and Lukov 1967). The acid extract follows the extraction of soluble carbohydrates for the determination of a blank. The method is inapplicable to foods containing pregelatinized or otherwise modified starches because they are partly soluble (Ulman and Richter 1961). Difficulties encountered in determining starch in such foods were evaluated in collaborative studies, based on the investigations of Friedman et al. (1967) and Friedman and Witt (1967). A proposed modification of the classical Maercker method consists of the extraction of nonstarchy components with isopropanol, enzymatic hydrolysis, clarification of the digest with zinc hydroxide, and determination of the starch by titration or spectrophotometry after ferricyanide reduction.

Hydrolysis Methods.   A complete acid hydrolysis of starch yields glucose, which can be determined by the usual chemical or physicochemical methods. Methods that involve acid hydrolysis are subject to error caused on the one hand by hydrolysis of nonstarch polysaccharides and on the other hand by destruction of dextrose during acid hydrolysis.

To overcome the difficulties encountered in polarimetric and acid hydrolysis methods, several procedures employing a combination of acid and enzymatic hydrolysis or hydrolysis by several enzymes have been proposed (Thivend et al. 1965; Lee and Whelan 1966; Donelson and Yamazaki 1968). For example, heat-gelatinized starch in wheat products can be partially hydrolyzed with heat-resistant bacterial alpha-amylase, and the hydrolysis to dextrose completed with amyloglucosidase. The glucose can be assayed by glucose oxidase. In all stages of the assay, only starch and its hydrolysis products are analyzed. No interference from nonstarch components is expected and the elimination of such components is therefore not required. An excellent discussion of the scope and limitations of starch determinations was presented by Richter and Schierbaum (1968).

Dextrins

Dextrins are mixtures of carbohydrates produced by *partial* acid or enzymatic hydrolysis of starch. The dextrins include glucose polymers above hexoses and below those colored blue with iodine, though the higher homologues that give a brown-yellow to brown-red color are included.

Dextrins dissolved in water give a milky suspension upon the addition of a tenfold excess of ethanol. In precipitation with alcohols, starch and proteins (and under certain conditions pectins) are coprecipitated. They can be removed by precipitation with a lead acetate, zinc acetate or potassium ferrocyanide solution. In removing the interfering substances with lead acetate, some losses of dextrins are encountered. A mixture of acetone–ethanol (1:1) containing 2 g calcium chloride per liter is a more useful precipitant (Thaler 1967). Some interference may be caused by pectins and water-soluble gums.

A relatively simple, rapid, and effective separation of dextrins can be made by thin-layer chromatography (Diemair and Kolbel 1963). In quantitative determinations, it must be realized that the assay involves an undefined mixture. Consequently, the methods are highly empirical.

Dextrins can be determined chemically. First glucose is determined by the Barfoed method, then total reducing sugars are assayed by the Fehling procedure, and finally dextrins and oligosaccharides are acid-hydrolyzed and determined as glucose by the Fehling method. Such a procedure has been developed for dextrins in syrups and bread (Rotsch 1947). A fermentation procedure with bakers' yeast was developed by Taufel and Muller (1956). If the dextrins have been separated from the starch, they can be determined polarimetrically. Van der Bij (1967) described the usefulness of thin-layer chromatography, infrared spectroscopy, and gas–liquid chromatography in testing dextrins.

Glycogen

Glycogen is a polysaccharide, generally of animal origin, that is stored in muscles and mainly in the liver. It resembles plant amylopectin, except that the number of glucose units per chain is smaller (about 12). It is water soluble and colors red-brown after iodine addition.

Most of the methods for the determination of glycogen are based on a Pfluger (1909) method as modified by Good *et al.* (1933). The glycogen is isolated by precipitating with alcohol after heating the tissue in strong potassium hydroxide. It must be emphasized (Stetten *et al.* 1958) that during alkaline extraction of glycogen its molecular weight is reduced from about 10^8 to 10^5. In addition, degradation products, which are not precipitated with 70% ethanol, are formed. After isolation, the glycogen can be measured directly with anthrone or indirectly by hydrolysis followed by a glucose determination (Seifter *et al.* 1950; Van Handel 1965).

An enzymic method for measuring glycogen has been described by Passonneau *et al.* (1967). Glycogen plus inorganic phosphate and TPN$^+$ are converted in one analytical step to 6-P-gluconolactone and TPNH. The TPNH is measured by its fluorescence or ultraviolet absorption. The method uses commercially available enzymes: phosphorylase, phosphoglucomutase, and glucose-6-P-dehydrogenase. Most commercial preparations contain enough transglucosylase and glucosidase for the complete degradation of glycogen. The method is specific, sensitive (0.05 μm can be detected), and applicable to the analyses of whole tissues.

Some Recent Studies and Evaluations of Methods

Dean (1978) analyzed several foods for dextrins, starch, glycogen, and sugars. Free sugars were extracted with 80% (v/v) ethanol. After deproteinization and deionization, an aliquot of the purified extract was injected into an autoanalyzer. The sugars were separated chromatographically as ionized borate complexes on an anion-exchange column at 53°C. Following elution, sugars were determined by the reaction with orcinol in sulfuric acid at 95°C. Starch, dextrins, and glycogen were determined in the insoluble residue after extraction of the free sugars. Glucoamylase was used for the quantitative transformation of starch to glucose by stepwise cleavage of α-D-(1 → 4) and α-D-(1 → 6) glucosidic linkages. Glucose was estimated with glucose oxidase by colorimetry at 420 nm in the autoanalyzer.

According to Kujawski *et al.* (1979), starch determination in wheat and rye flour, rye bran, potato starch, and ground potatoes by polarimetry of $CaCl_2$ extracts gave smaller standard errors than polarimetry of HCl extracts. The largest errors in determination occurred with ground potato starch. The errors in the polarimetric determination were decreased by adjusting the pH with HCl. Starch determinations employing glucoamylase were unreliable.

Starch and (1 → 3), (1 → 4)-β-D-glucan (β-glucan) of the starchy endosperm are the main polysaccharides of cereal grains. According to Ahluvalia and Ellis (1984), the methods for the determination of starch are not entirely satisfactory. This is due to difficulties in extraction or gelatinization and in assaying the extracted starch. Extraction with hot water or chloral yields low recoveries of starch. Several extractions with dimethyl sulfoxide are necessary for complete extraction. Perchloric acid (52 or 72% w/v) is the most efficient solvent for extracting starch from plant tissues and has been used for the extraction of starch from barley and wheat. Dissolution requires stirring for about 1 hr and/or inclusion of sand to facilitate extraction.

The nonspecific methods for estimating starch by precipitation with iodine and measurement of the blue value can be inaccurate because the iodine-binding capacity of amylose is reduced in the presence of other molecules and because the blue value varies with the amylose content of starch. Hydrolysis of starch and measurement of glucose by anthrone in hot concentrated sulfuric acid is inconvenient and can be inaccurate due to color reactions with other sugars. Specific enzymic hydrolysis of starch to glucose with amyloglucosidase requires elimination of β-glucanase activity. Methods in which the difference between total carbohydrate and reducing sugars is considered as starch are inaccurate because cellulose and β-glucan are not considered (Ahluvalia and Ellis 1984).

Similarly, there are no satisfactory methods to determine β-glucan. Ahluvalia and Ellis (1984) described a simple and precise method suitable for the routine determination of starch and β-glucan in barley and malt. The glucans were extracted rapidly (5 min) and exhaustively with perchloric acid (50 mM) and were measured directly from this single extract by enzymic hydrolysis of the individual glucans to glucose. The glucose was also measured enzymatically. Little or no acid hydrolysis of starch or β-glucan was observed. Most of the free glucose could be attributed to hydrolysis of sucrose. There was complete solubilization of the gum and hemicellulosis components of β-glucan. Incubation of the acid extracts with protease, prior to amyloglucosidase digestion, increased starch measurements by about 4% w/w. The method was used to measure the starch and β-glucan in five varieties of barley with contrasting malting quality, in malts, and in commercial lager and ale malts.

Lustinec et al. (1983) described methods for the accurate determination of starch, amylose, and amylopectin in plant tissues. They are based on extraction of starch with 32% perchloric acid and selective retention of the starch–iodine complex on a glass fiber disk. The retained starch is dissolved in 0.75 M sulfuric acid and estimated with phenol. For amylose and amylopectin determination, the starch on the disk is dissolved in perchloric acid, precipitated with ethanol, and retained on a 10-cm glass fiber strip. The two polysaccharides are separated by a chromatographic procedure, involving development of the strip in a mixture of ethanol and dimethyl sulfoxide and in dimethyl sulfoxide. The strip is washed in ethanol and stained with iodine or used for polysaccharide quantitation. As little as 5 μg of starch or its components can be estimated. The method is claimed to be faster and more starch-specific than enzymatic methods. It can be used to assay different starches including waxy ones (high amylopectin). It can differentiate between starch and glycogen.

Colorimetric (amylose–iodine complex at pH 4.5–4.8) and titrimetric (potentiometric and amperometric assays of iodine binding) methods of amylose determination were described by Perez and Juliano (1978) and Juliano (1979). The interfering effects of lipids in the assay were compensated (in part, at least) by defatting, determination of the color at pH 3, or use of empirical correction factors.

Scherz and Morgenthaler (1980) reviewed the analytical methodology for the determination of polysaccharides added to foods. The general methodology, which is outlined in Fig. 35.2, includes the following steps: (1) isolation and purification of polysaccharides; (2) separation into neutral and acid components by ion-exchange chromatography on DEAE cellulose (CC); (3) hydrolysis to give monosaccharides; (4) determination of the monomeric components by thin-layer chromatography (TLC), gas chromatography of derivatives (GC), or enzymatic analysis (EA); and (5)

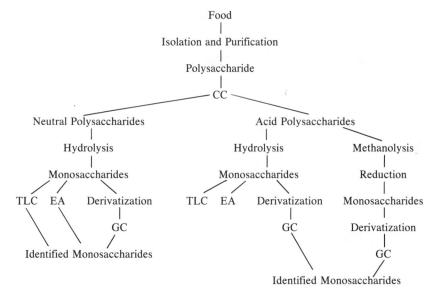

Fig. 35.2. General analytical scheme for determination of polysaccharides in foods. (Adapted from Scherz and Morgenthaler 1980.)

determination of the types of linkages between monomeric units by permethylation.

DETERMINATION OF STRUCTURAL POLYSACCHARIDES

The determination of structural polysaccharides is wrought, even today, with many difficulties, partly because the materials contain many undefined polymers varying in size and composition. In this section, we divide these complex polysaccharides into two broad groups: (1) cellulose and other cell wall components and (2) pectins.

Cellulose and Other Cell Wall Components

During isolation of the various cell wall components, modifications— including degradation—take place. Part of the cellulose is soluble in alkali. The insoluble fraction is called α-cellulose. The fraction in the alkaline extract that is precipitated upon acidification is β-cellulose, and the fraction remaining in solution, containing mainly hemicelluloses, is γ-cellulose.

Hemicelluloses comprise a mixture of alkali-soluble polysaccharides, the amount and composition of which vary with the extraction procedure. These include polymers of mannose, galactose, xylose, and arabinose as well as polymers of uronic acids and of methyl- or acetyl-substituted monoses. Cellulose and hemicelluloses are among the most abundant and widely distributed carbohydrates in plant materials. They are the main cell wall constituents of fruits, vegetables, and cereals. Their content increases with maturity; they are highly concentrated in pericarp tissues of cereal grains and present in low concentration in the starchy endosperm.

Cellulose and hemicellulose are generally determined in defatted foods following the extraction of soluble carbohydrates. To eliminate the effects of enzymes on degradation and solubilization, prior storage at low temperature and enzyme inactivation at the beginning of the analysis are important. In addition, proteins and lignin must be removed.

Jermyn (1955) described in detail a procedure for the determination of celluloses and hemicelluloses in plant material. In this procedure, after defatting the powdered material with ether and ethanol, it is boiled under reflux with 85% ethanol to extract some soluble carbohydrates. Water-soluble carbohydrates—starches and pectins—are extracted after vigorous boiling in water under reflux. Lignins are extracted by heating at

70–75°C for 4 hr with an acidified sodium chlorite solution. The residual material (*holocellulose*), which is practically lignin-free, is digested with alkali and the cellulose and hemicellulose are determined. The cellulose content in the alkali-insoluble fraction can be determined by complete hydrolysis with sulfuric acid and assay of the reducing sugars.

Mannans (e.g., in coffee or yeast) are generally determined by precipitation with an alkaline copper solution (Thaler 1967). The mannans are generally in the hemicellulose fraction from which they are extracted with alkali.

Lignin is present in trace amounts in high-quality fruits and vegetables, but is present in the bran of cereal grains and in some spices. Uncertainty regarding the detailed chemistry of lignin has precluded establishment of a generally satisfactory method for its determination. Most methods are based on the insolubility of lignin in concentrated mineral acids (i.e., 72% sulfuric acid) in which most cell constituents are soluble. The residue may, however, contain some humins as a result of the action of acids at elevated temperatures. It is, therefore, recommended to remove potentially interfering substances by an extraction with water and dilute acid. In addition, defatted, air-dried materials are generally used for assays of lignin (Freudenberg and Ploetz 1940; Ellis *et al.* 1946).

Pentosans are polymers of pentoses and methyl pentoses. The pentoses can be calculated from the pentose contents of pentosan hydrolysates. Either the reducing power after removal of fermentable hexoses or the pentose content of chromatographically fractionated hydrolysates can be determined. The pentose or pentosan content is often determined by measuring the amount of furfural formed after a reaction with hydrochloric acid. The latter method is highly empirical and significantly affected by the assay conditions and the presence of nonpentosan components that yield furfural (or hydroxymethyl furfural) under test conditions. In most methods, including the official AOAC procedure, the furfural is distilled and precipitated with phloroglucinol; in other procedures barbituric or thiobarbituric acid is used.

The furfural can also be titrated with bromine, one molecule of furfural giving two molecules of hydrogen bromide. Finally, the furfural can be determined spectrophotometrically, based on its absorbance at 275 nm (284 nm for hydroxymethyl furfural) (Thaler 1967).

Pectins

Pectins are important as jelling and thickening agents. The pectins are bound to calcium mainly in the middle lamella in growing tissues of many higher plants and to cellulose in the primary cell membrane. The pectins are generally obtained commercially by processing the parenchymatic tissues of plant materials. During fruit ripening, the cell pectins are solubilized enzymatically and are present in fruit juices. They serve as the cementing agents of the cell and regulate the water content.

Chemically, pectins are polymers of galacturonic acid connected by alpha-1,4 linkages into long chains. They can be divided and defined (Gudjons 1967) as follows:

1. *Pectic substances*—materials comprising all polygalacturonic acid containing materials.

2. *Protopectins*—water-insoluble materials, in bound form, yielding pectins upon hydrolysis.

3. *Pectin*—partly esterified polygalacturonic acids; generally methyl esters, in some (rapeseeds and beets) also esterified with acetic acid. They can be divided into low-methoxy and high-methoxy pectins, depending on whether they contain less or more than 7% methoxy-esterified galacturonic acids. A completely (100%) esterified polygalacturonide has theoretically 16.3% methoxyl groups.

4. *Pectinic acids*—have all carboxyl groups in the free form and are water insoluble. Salts of pectinic acids are water soluble.

Pectic substances isolated from plant materials often contain various arabans and galactans. The isolated substances are a heterogeneous mixture that varies (depending on the source, method of extraction, and subsequent treatment) in molecular weight, degree of esterification, and araban and galactan content. In view of the heterogeneous nature of the pectic substances, their analysis is based either on empirical procedures designed to evaluate their usefulness for specific purposes, or on the assay of specific constituents.

The solubility of pectins increases with an increase in the degree of esterification and with a decrease in molecular weight. The less soluble a pectin is in water, the easier it can be precipitated by adding an electrolyte. Pectins with an esterification grade of up to 20% are precipitated by sodium chloride solutions, with 50% by calcium chloride solutions, and with 70% by aluminum chloride or copper chloride solutions. Completely esterified pectins are not precipitated by electrolytes. The pectins can also be precipitated with organic solvents, such as acetone, methanol, ethanol, or propanol. The concentration of alcohol that is required increases with the degree of esterification.

The usefulness of pectins stems mainly from their capacity to form stable gels or films, and increase the viscosity of acidified, sugar-containing solutions. Completely esterified pectins can be jelled without acid or electrolyte addition. Pectins with a high esterification grade jell rapidly at a relatively low acidity. With a decrease in esterification, the rate of gel formation decreases and the required pH optimum decreases. The jelling capacity is greatly enhanced by calcium salts, which also increase the stability and decrease the dependence on pH and sugar concentration. Viscosity increases, at a given esterification grade, with an increase in molecular weight; decreased esterification lowers viscosity. The effects on viscosity depend also on the total concentration and the electrolytes present. The stability of gels is affected by numerous factors—including temperature, the presence of acids or alkali, enzymes, and mechanical treatment. The pectin content and its quality (determined as jelly grade, i.e., pounds of sugar that 1 lb of pectin would set to a gel under standardized conditions) are important in the manufacture of jellies, jams, and preserves.

Fuchigami and Okamoto (1984) developed a method for the fractional extraction of pectic substances of vegetable tissues using dilute HCl, acetate buffer, and Na-hexametaphosphate. The degree of esterification and neutral sugar content in HCl-soluble pectin were higher than those of acetate buffer-soluble pectin.

For the detection of pectins, the McCready and Reeve (1955) method is used. To pectin and hydroxamic acid in an alkaline solution, a ferric compound is added, and a water-insoluble red complex is formed. Numerous colorimetric and paper chromatographic methods are available to detect and estimate the components after hydrolysis with 2% hydrochloric acid or 4% sulfuric acid under reflux (Gudjons 1967).

To determine the identity and purity of a pectin preparation, the galacturonide content, neutralization equivalent, degree of esterification, and physical properties of the pectic substances are determined. The precision and significance of the determination are largely dependent on the amount and nature of nonpectic compounds, and on the difficulty of their separation.

Pectin can be determined gravimetrically, in aqueous ammonium citrate or dilute hydrochloric acid extracts, by a procedure involving precipitation with alcohol or acetone, saponification of the precipitate with cold alkali, acidification, boiling with acid, and conversion into a precipitate of pectic acid (Wichmann 1922, 1923). More commonly, the saponified pectate is precipitated as calcium pectate (Carre and Haynes 1922; Emmett and Carre 1926; Griebel and Weiss 1929; Hinton 1939).

In the titrimetric method of Deuel (1943), the pectin is saponified with sodium hydroxide, slightly acidified, precipitated with 96% ethanol, washed, dissolved with excess standardized alkali, and the excess backtitrated.

As uronic acids are decarboxylated by boiling with mineral acids, the amount of carbon dioxide evolved can be used to determine the pectin contents (Conrad 1931; McCready et al. 1940). A colorimetric procedure based on the carbazole reaction of uronic acids has been proposed by McComb and McCready (1952). The number of free carboxyl groups can be determined by titration. Numerous methods of organic analysis can be used to determine the methyl or acetyl groups linked to uronic acids in pectic substances.

Crude Fiber

In the analysis of cellulose-containing foods, the determination of crude fiber is widely used. Crude fiber includes, theoretically, materials that are indigestible in the human and animal organism. It is determined as material insoluble in dilute acid and dilute alkali under specified conditions.

The *basic* method is based on a procedure developed by Hennenberg, Stohmann, and Rautenberg in 1864 in Germany. In the determination, 2 g of material are defatted with petroleum ether and boiled under reflux for exactly 30 min with 200 ml of a solution containing 1.25 g sulfuric acid per 100 ml solution. The solution is filtered through linen or several layers of cheese cloth on a fluted funnel and washed with boiling water until the washings are no longer acid. The residue is tranferred quantitatively to a beaker, boiled for 30 min with 200 ml of a solution containing 1.25 g of carbonate-free sodium hydroxide per 100 ml. The final residue is filtered through a thin but close pad of washed and ignited absestos in a Gooch crucible, dried in an electric oven, weighed, incinerated, cooled, and weighed again. The loss in weight in incineration is taken as crude fiber.

Crude fiber determinations are greatly affected by manipulations and procedures. Particle size is very important; the finer the material is ground, the lower the determined crude fiber content. Apparent crude fiber is lowered by defatting, though the low lipid content of some foods (e.g., white flour) affects the results little. The rate of heating to the boiling point and rate of boiling must be controlled. Filtering after each digestion must be completed within a given time; delays in filtering after acid or alkali digestion generally lower the results.

To reduce evaporation losses during boiling, and concomitant changes in concentration of digestion solutions, specially constructed apparatus— with condensers—is available commercially (Fig. 35.3). Most difficulties are encountered in determining the crude fiber content of protein-rich foods; predigestion by proteolytic enzymes has been recommended to accelerate filtration after digestion.

The residue from a crude fiber determination contains about 97% cellulose and lignin. It does not represent, however, all the cellulose and lignin present initially. Thus, the crude fiber content of whole wheat is about 2%, whereas the kernel has about 7% of pericarp, most of which

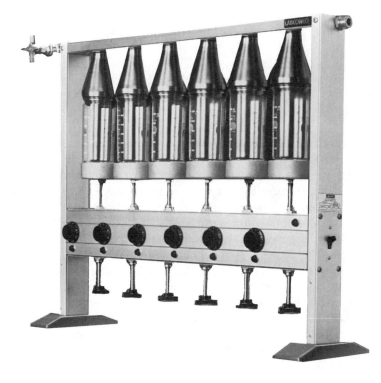

Fig. 35.3.   Crude fiber apparatus. (Courtesy Labconco Co.)

cannot be digested by humans and nonruminant animals. In addition, the crude fiber is a mixture of cellulosic materials and does not represent any specific compound or group of compounds.

Despite its nonspecific composition, crude fiber is a useful parameter in food and feed analyses. Crude fiber is commonly used as an index of the feeding value of poultry and stock feeds; seeds high in crude fiber content are low in nutritional value. A determination of crude fiber is used in evaluating the efficiency of milling and separating bran from the starchy endosperm. It is a more lengthy, laborious, and complicated but more direct index of flour purity than color or ash. Crude fiber is also useful in the chemical determination of succulence of fresh vegetables and fruits; overmature products have increased levels of crude fiber. Generally, however, substantial changes in texture and consumer acceptance are accompanied by relatively small changes in crude fiber content.

There are at least 100 modifications of the original crude fiber determination, including addition of oxidizing agents, use of special solvents (phenol or trichloroacetic acid), and various concentrations of acid and alkali. Some of the methods simplify, some shorten, and some improve reproducibility of the determination. In the Fibertec System (Tecator Inc., Herndon, VA), the weighed sample is handled in the same vessel (a glass crucible) during the whole analysis. The procedure for determination of crude fiber by the Weende method and the special crucible are illustrated diagramatically in Fig. 35.4.

The official AOAC method for the determination of crude fiber specifies the use of asbestos as a filter aid. However, because asbestos is a carcinogenic material, this method may pose health risks. A new method, which is precise, accurate, and requires less time per sample, uses the same reflux units and most of the reagents of the AOAC method but eliminates asbestos by using a new vacuum filtration unit (Holst 1978). This asbestos-free method of analysis was tested collaboratively on 10 feed samples with crude fiber contents ranging from 2 to 33% and on 6 high-protein supplement samples, of highly gelatinous material, with crude fiber contents ranging from 0.5 to 12%. The combined results of two studies showed the following for the AOAC and the asbestos-free methods, respectively; standard deviation (s_d) 0.55 and 0.49; random error (s_r) 0.26 and 0.22; bias (s_b) 0.48 and 0.43; average range between independent analyses, 0.29 and 0.22% crude fiber; and average maximum spread among laboratories 1.7 and 1.4% crude fiber.

A collaborative trial in 19 laboratories of three methods of analysis for fiber was reported by Player and Wood (1980). The two methods for crude fiber (gravimetric and volumetric) were comparable in their repeatabilities and reproducibilities. The acid-detergent method for cellulose and lignin was comparable with the crude fiber methods in reproducibility but inferior in repeatability. The content of "fiber" extracted using the acid-detergent method was significantly higher than the contents extracted by either of the other two methods.

Van Soest (1967) reviewed the limitations of the crude fiber assay in

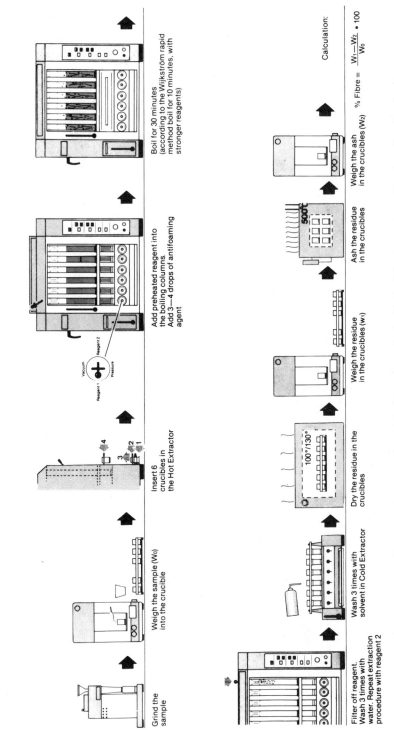

Grind the sample

Weigh the sample (W_0) into the crucible

Insert 6 crucibles in the Hot Extractor

Add preheated reagent into the boiling columns. Add 3—4 drops of antifoaming agent

Boil for 30 minutes (according to the Wijkström rapid method boil for 10 minutes, with stronger reagents)

Filter off reagent. Wash 3 times with water. Repeat extraction procedure with reagent 2

Wash 3 times with solvent in Cold Extractor

Dry the residue in the crucibles

Weigh the residue in the crucibles (w_1)

Ash the residue in the crucibles

Weigh the ash in the crucibles (W_2)

Calculation:

$$\% \text{ Fibre} = \frac{W_1 - W_2}{W_0} \cdot 100$$

Fig. 35.4 A. Crude fiber procedure according to the Weende method with the Fibertec system. (Courtesy Tecator, Inc.)

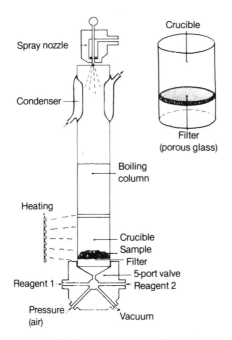

Fig. 35.4 B. The Fibertec crucible. (Courtesy Tecator, Inc.)

terms of the validity and usefulness of the information it provides. A rapid procedure was developed (Van Soest and Wine 1967) for determining the fiber insoluble in a neutral detergent. The method was compared and standardized against determinations of vegetable matter that was indigestible by proteolytic and diastatic enzymes, and that could not be utilized except by microbial fermentation in the digestive tracts of animals.

Dietary Fiber

In the last decade there has been great interest in the determination of *dietary* (rather than crude) *fiber* in our diet. Some of the recent books and reviews in this area are those by Spiller and Amen (1976), Southgate *et al.* (1978), James and Theander (1981), and Johansson (1981), Birch and Parker (1983), Asp (1983), Selvendran and DuPont (1984), and Wisker *et al.* (1985). According to Trowell *et al.* (1976) most workers define dietary fiber as nonstarchy (i.e., non-α-glucan) polysaccharides plus lignin. The main part of dietary fiber in foods originates in the plant cell wall (Asp 1983). The cell wall components and dietary fiber components are related as follows:

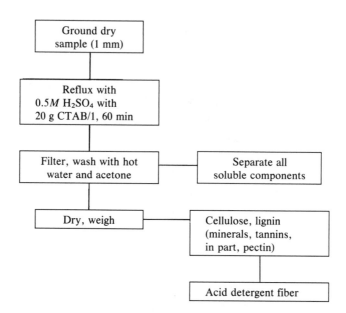

Fig. 35.5. The acid detergent fiber method. CTAB = cetylmethylammonium bromide. (From Wisker *et al.* 1985.)

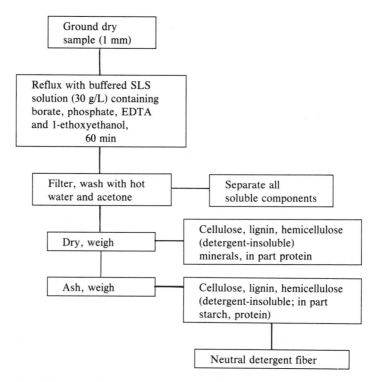

Fig. 35.6. The neutral detergent fiber method. SLS = sodium lauryl sulfate; EDTA = ethylenediaminetetraacetic acid. (From Wisker *et al.* 1985.)

Plant cell wall $\begin{cases} \text{Protein} \\ \text{Lipids} \\ \text{Inorganic constituents} \\ \text{Lignin} \\ \text{Cellulose} \\ \text{Hemicelluloses} \\ \text{Pectins} \\ \text{Gums} \\ \text{Mucilages} \\ \text{Algal polysaccharides} \\ \text{Modified cellulose} \end{cases}$ Dietary fiber

Dietary fiber determination methods were compared by Halvarson and Alstin (1984). Gravimetric methods for dietary fiber can be divided into detergent methods and enzymatic methods. The latter may determine insoluble fiber only or insoluble plus soluble fiber. The acid detergent fiber (ADF) method, which determines cellulose and lignin content, has been accepted by the AOAC as a method of testing feedstuffs. It is shown diagramatically in Fig. 35.5. The neutral detergent fiber method (NDF) is an official assay for the determination of dietary fiber in cereal grains (Fig. 35.6). It measures cellulose, hemicellulose, and lignin. Both detergent methods measure only detergent-insoluble dietary fiber. This insol-

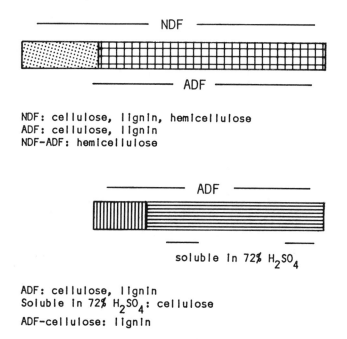

NDF: cellulose, lignin, hemicellulose
ADF: cellulose, lignin
NDF-ADF: hemicellulose

ADF: cellulose, lignin
Soluble in 72% H$_2$SO$_4$: cellulose
ADF-cellulose: lignin

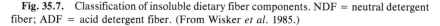

Fig. 35.7. Classification of insoluble dietary fiber components. NDF = neutral detergent fiber; ADF = acid detergent fiber. (From Wisker *et al.* 1985.)

uble dietary fiber can be separated into cellulose, hemicellulose, and lignin, as shown in Fig. 35.7. Determination of hemicellulose from the difference is not precise because of the presence of various components (in addition to cellulose, lignin, and hemicellulose) in the detergent methods. Some of the error can be reduced by a sequential analysis of NDF and ADF. Pectins and tannins are soluble in the NDF solution and hemicellulose can be estimated from the weight loss of starch and protein-free NDF residue after ADF treatment (Wisker *et al.* 1985).

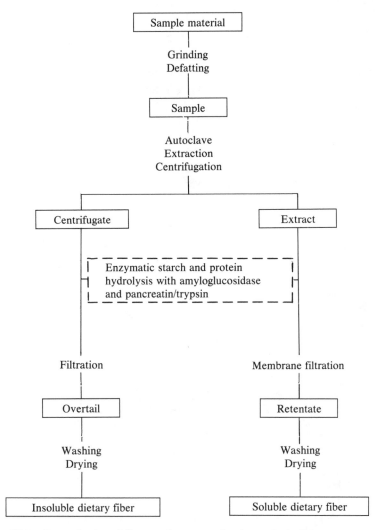

Fig. 35.8. Determination of dietary substances using the method of Meuser *et al.* (1983). (From Wisker *et al.* 1985.)

Enzymatic methods are designed to measure dietary fiber under "physiological conditions." Some enzymatic preparations may contain adventitious foreign activities and may degrade to a limited extent dietary fiber components (in addition to starch). On the other hand, some protein may be incompletely degraded and attached or bound to the separated fiber. Elchazly and Thomas (1976) developed a method for determination of enzymatically insoluble dietary fiber. Asp (1983) and Meuser *et al.* (1983) modified this procedure for determination of soluble, insoluble, and total dietary fiber (Fig. 35.8). A collaborative study to determine total dietary fiber in foods was reported by Prosky *et al.* (1984) and by Alstin (1984).

BIBLIOGRAPHY

Acker, L. (1967). Carbohydrates—water-soluble compounds (total extract); detection and determination of mono- and oligosaccharides. *In* "Handbook of Chemistry, Food Analysis," Vol. 2 (W. Diemair, ed.). Springer-Verlag, Berlin.

Ahluvalia, B., and Ellis, E. E. (1984). A rapid and simple method for the determination of starch and β-glucan in barley and malt. *J. Inst. Brew.* **90**, 254–259.

Alexander, R. J., and Garbutt, J. T. (1965). Use of sorbitol as internal standard in determination of D-glucose by gas liquid chromatography. *Anal. Chem.* **37**, 303–305.

Alm, R. S. (1952). Gradient elution analysis. II. Oligosaccharides. *Acta Chem. Scand.* **6**, 1186–1193.

Alstin, F. (1984). Experience with the Fibertec E in relation to the AOAC study on dietary fibre. *In Focus* **7**(1), 12.

Annison, E. F., James, A. T., and Morgan, W. T. J. (1951). Separation and identification of small amounts of mixed amino sugars. *Biochem. J.* **48**, 477–482.

Anon. (1966). "The Technicon Sugar Chromatography System." *1564-10-6*. Technicon Chromatography Corp., Ardsley, NY.

AOAC. (1965). "Official Methods of Analysis," 10th Edition. Assoc. Off. Agr. Chemists, Washington, DC.

Asp, N.-G. (1983). What is dietary fiber? Methods for analysis of dietary fiber. *In Focus* **6**(2), 3–7.

Asp, N.-G., and Johansson, G.-G. (1981). Techniques for measuring dietary fiber. Principal aims of methods and a comparison of results obtained by different techniques. *In* "The Analysis of Dietary Fiber in Food" (W. P. T. James and O. Theander, eds.), pp. 173–189. Marcell Dekker, New York.

Asp, N.-G., Johansson, C.-G., Hallmer, H., and Siljestroem, M. (1983). Rapid enzymatic method for assay of insoluble and soluble dietary fiber. *J. Agric. Food Chem.* **31**, 476–482.

Baker, S. A., Bourne, E. J., and Whiffen, D. H. (1956). Use of infrared analysis in the determination of carbohydrate structure. *Methods Biochem. Anal.* **3**, 213–245.

Balazs, E. A., and Jeanloz, R. W. (eds.). (1965). "The Amino Sugars—The Chemistry and Biology of Compounds Containing Amino Sugars," Vols. 1–4. Academic Press, New York.

Bandow, F. (1937). Absorption spectra of organic substances in concentrated sulfuric acid. I. Experiments on carbohydrates. *Biochem. Z.* **294**, 124–137. [*Chem Abstr.* **32**, 11819].

682   35 CARBOHYDRATES

Barnett, A. J. G., and Tawab, G. A. (1957). Determination of lactose in milk and cheese. *J. Sci. Food Agric.* **8**, 437–441.

Bates, F. J. (1942). "Polarimetry, Saccharimetry, and the Sugars." National Bureau of Standards Circ. **440**.

Bath, I. H. (1958). The ultraviolet spectrophotometric determination of sugars and uronic acids. *Analyst* **83**, 451–455.

Biggs, D. A. (1967). Milk analysis with the infrared milk analyzer. *J. Diary Sci.* **50**, 799–803.

Birch, G. G., and Parker, K. J. (eds.). (1983). "Dietary Fibre." Applied Science Publishers, London.

Bishop, C. T. (1964). Gas liquid chromatography of carbohydrate derivatives. *Adv. Carbohydr. Chem.* **19**, 95–147.

Blakeney, A. B., and Mutton, L. L. (1980). A simple colorimetric method for the determination of sugars in fruit and vegetables. *J. Sci. Food Agric.* **31**, 889–897.

Blish, M. J., and Sandstedt, R. M. (1933). An improved method for the estimation of flour diastatic value. *Cereal Chem.* **10**, 189–202.

Boratynski, J. (1984). Colorimetric method for the determination of ketoses using phenol–acetone–boric acid reagent (PABR). *Anal. Biochem.* **137**, 528–532.

Borenfreund, E., and Dische, Z. (1957). A new spray for spotting sugars on paper chromatograms. *Arch. Biochem. Biophys.* **67**, 239–240.

Brinkley, W. W. (1955). Column chromatography of sugars and their derivatives. *Adv. Carbohydr. Chem.* **10**, 55–94.

Brock-Neeley, W. (1957). Infrared spectra of carbohydrates. *Adv. Carbohydr. Chem.* **12**, 13–33.

Browne, C. A., and Zerban, F. W. (1941). "Physical and Chemical Methods of Sugar Analysis," 3rd ed. Wiley, New York.

Carre, M. H., and Haynes, D. (1922). The estimation of pectin as calcium pectate and the application of this method to the determination of the soluble pectin in apples. *Biochem. J.* **16**, 60–69.

Chargaff, E., Levine, C., and Green, C. (1948). Techniques for the demonstration by chromatography of nitrogenous lipide constituents, sulfur-containing amino acids and reducing sugars. *J. Biol. Chem.* **175**, 67–71.

Charles, D. F., and Meads, P. F. (1962). Refractive index. *In* "Methods in Carbohydrate Chemistry" (R. L. Whistler and M. L. Wolfrom, eds.). Academic Press, New York.

Conrad, C. M. (1931). Decarboxylation studies on pectins and calcium pectates. *J. Am. Chem. Soc.* **53**, 1999–2003.

Coulter, S. T., Jenness, R., and Geddes, W. F. (1951). Physical and chemical aspects of the production, storage and utility of dry milk products. *Adv. Food Res.* **3**, 45–106.

Cramer, F. (1953). "Paper Chromatography," 2nd ed. Chemie Publishing Co., Weiheim, Germany.

Crowell, E. P., and Burnett, B. B. (1967). Determination of carbohydrate composition of wood pulps by gas chromatography of alditol acetates. *Anal. Chem.* **39**, 121–124.

Danehy, J. P., and Pigman, W. W. (1951). Reactions between sugars and nitrogenous compounds and their relationship to certain food problems. *Adv. Food Res.* **3**, 241–281.

Dean, A. C. (1978). Method for the estimation of available carbohydrate in food. *Food Chem.* **3**, 241–250.

Deuel, H. (1943). Pectin as high molecular electrolyte. *Z. Lebensmittelunters. Hyg. (Bern)* **34**, 41. (German)

deVries, J., Heroff, J. C., and Egberg, D. C. (1979). High pressure liquid chromatographic determination of carbohydrates in food products: Evaluation of method. *J. Assoc. Off. Anal. Chem.* **62**, 1292–1296.

Diemair, W., and Kolbel, R. (1963). Detection and determination of dextrins. *Z. Lebens-mittelunters. Forsch.* **124**, 1–14, 157–179. (German)

Dimler, R. J., Schaefer, W. C., Wise, C. S., and Rist, C. E. (1952). Quantitative paper chromatography of D-glucose and its oligosaccharides. *Anal. Chem.* **24**, 1411–1414.

Dische, Z. (1962). Color reactions of carbohydrates. *In* "Methods in Carbohydrate Chemistry" (R. L. Whistler and M. L. Wolfrom, eds.). Academic Press, New York.

Dische, Z., and Borenfreund, E. (1950). A spectrophotometric method for the determination of hexosamines. *J. Biol. Chem.* **184**, 517–522.

Donelson, J. R., and Yamazaki, W. T. (1968). Enzymic determination of starch in wheat fractions. *Cereal Chem.* **45**, 177–182.

Dreywood, R. (1946). Qualitative test for carbohydrate material. *Ind. Eng. Chem. Anal. Ed.* **18**, 499.

Dubois, M. *et al.* (1956). Colorimetric method for determination of sugars and related substances. *Anal. Chem.* **28**, 350–356.

Dull, G. D., and Giangiacomo, R. R. (1984). Determination of individual simple sugars in aqueous solution by near infrared spectrophotometry. *J. Food Sci.* **49**, 1601–1603.

Dunmire, D. L., and Otto, S. E. (1979). HPLC determination of sugars in various food products. *J. Assoc. Off. Anal. Chem.* **62**, 176–185.

Dyachenko, P., and Samsonov, Y. (1964). Infrared spectra of dried milk and its components. *Molochn. Prom.* **25**(7), 11–13. [*Chem. Abstr.* **61**, 15263f]

Eheart, J. F., and Mason, B. S. (1966). Assay methodology studies of carbohydrate fractions of wheat products. *J. Assoc. Off. Agric. Chem.* **49**, 907–912.

Elchazly, M., and Thomas, B. (1976). A biochemical method for the determination of dietary fiber and its components in foods of plant origin (in German). *Z. Lebensm. Unters. Forsch.* **162**, 329–340.

Ellis, G. H., Matrone, G., and Maynard, L. A. (1946). A seventy-two percent sulfuric acid method for the determination of lignin and its use in animal nutrition studies. *J. Anim. Sci.* **5**, 285–297.

Elson, L. A., and Morgan, W. T. J. (1933). Colorimetric method for the determination of glucosamine and chondrosamine. *Biochem. J.* **27**, 1824–1828.

Emmett, A. M., and Carre, M. H. (1926). A modification of the calcium pectate method for the estimation of pectin. *Biochem. J.* **20**, 6–12.

Flohil, J. T. (1933). Report of the subcommittee on the development of a volumetric copper reduction method for sugar determinations. *Cereal Chem.* **10**, 470–476.

Ford, C. W. (1981). Estimation of hexose: pentose ratios in solution using the phenol–sulfuric acid method. *J. Sci. Food Agric.* **32**, 153–156.

Foster, A. B. (1957). Zone electrophoresis of carbohydrates. *Adv. Carbohydr. Chem.* **12**, 81–115.

Foster, A. B., and Stacey, M. (1952). The chemistry of the 2-amino sugars (2-amino-2-deoxy sugars). *Adv. Carbohydr. Chem.* **7**, 247–288.

French, D., Robyt, J. F., Weintraub, M., and Knock, P. (1966). Separation of maltodextrins by charcoal chromatography. *J. Chromatogr.* **24**, 68–75.

Freudenberg, K., and Ploetz, T. (1940). Quantitative assay of lignin. *Ber. Deut. Chem. Ges.* **73**, 754–757. (German)

Friedmann, T. E., and Witt, N. F. (1967). Determination of starch and soluble carbohydrates. II. Collaborative study of starch determination in cereal grains and in cereal products. *J. Assoc. Off. Agric. Chem.* **50**, 958–963.

Friedmann, T. E., Witt, N. F., and Neighbors, B. W. (1967). Determination of starch and soluble carbohydrates. I Development of method for grain stock feeds, cereal foods, fruits, and vegetables. *J. Assoc. Off. Agric. Chem.* **50**, 944–958.

Fuchigami, M., and Okamoto, K. (1984). Fractionation of pectic substances in vegetables by successive extraction with diluted hydrochloric acid and acetate buffer solutions. *Nippon Eiyo Shokuryo Gakkaishi* **37**(1), 57–64. [*Chem. Abstr.* **101**, 71125j]

Gardell, S. (1958). Determination of hexosamines. *Methods Biochem. Anal.* **6**, 289–317.

Geyer, H. V. (1965). Application of gas chromatography in analysis of sugars and starch hydrolysates. *Die Starke* **17**, 307–313. (German)

Good, C. A., Kramer, H., and Somogyi, M. (1933). The determination of glycogen. *J. Biol. Chem.* **100**, 485–499.

Good, T. A., and Bessman, S. P. (1964). Determination of glucosamine and galactosamine using borate buffers for modification of the Elson-Morgan and Morgan-Elson reactions. *Anal. Biochem.* **9**, 253–262.

Green, J. G. (1966). Automated carbohydrate analyzer. *Nat. Cancer Inst. Monograph* **21**, 447–467.

Griebel, C., and Weiss, F. (1929). The problem of pectin. *Z. Unters. Lebensm.* **58**, 189–201.

Gudjons, H. (1967). Pectins. *In* "Handbook of Food Chemistry, Food Analysis" (W. Diemair, ed.). Springer-Verlag, Berlin. (German)

Hadorn, H., and Biefer, K. W. (1956). Effects of amino acids on determination of sugars, especially by the method of Potterat and Eschmann and Hadorn and von Fellenberg. *Mitt. Lebensmitteluntersuch. Hyg. (Bern)* **47**, 4–15.

Hadorn, H., and Doevelaar, F. (1960). Systematic evaluation of titrimetric, colorimetric and polarimetric methods of starch determinations. *Mitt. Lebensmitteluntersuch. Hyg. (Bern)* **51**, 1–68. (German)

Hadorn, H., and Fellenberg, T. V. (1945). Reevaluation of iodometric sugar determinations. *Mitt. Lebensmitteluntersuch. Hyg. (Bern)* **36**, 359–367. (German)

Hadorn, H., and Jungkunz, R. (1952). Analysis and composition of chestnuts (Castanea vesca). *Z. LebensmittelUnters. Forsch.* **95**, 418–429. (German)

Hagedorn, H. C., and Jensen, B. N. (1923). Micro assay of blood sugar with ferricyanide. *Biochem. Z.* **135**, 46–58. (German)

Halvarson, H., and Alstin, F. (1984). Dietary fiber determination methods. *Cereal Foods World* **29**, 571–574.

Hammond, L. D. (1940). Redetermination of the Munson-Walker reducing-sugar values. *J. Res. Natl. Bur. Std.* **24**, 579–596 (RP 1301).

Hanes, C. S. (1929). An application of the method of Hagedorn and Jensen to the determination of larger quantities of reducing sugars. *Biochem. J.* **23**, 99–106.

Hardinge, M. G., Swarner, J. B., and Crooks, H. (1965). Carbohydrates in foods. *J. Am. Dietet. Assoc.* **46**, 197–204.

Hassid, W. Z. (1936). Determination of reducing sugars and sucrose in plant materials. *Ind. Eng. Chem. Anal. Ed.* **8**, 138–140.

Heidt, L. J., and Colman, C. M. (1952). Degradation of D-glucose, D-fructose, and invert sugar in carbonate buffered water solution. *J. Am. Chem. Soc.* **74**, 4711–4713.

Heyns, K. (1959). Standard methods of analysis of starch and starch hydrolysates. ICUMSA Methods. *Die Starke* **11**, 67–74, 215–233. (German)

Hinton, C. L., and Macara, T. (1924). Application of the iodometric method to the analysis of sugar products. *Analyst* **49**, 2–24.

Hinton, C. L., and Macara, T. (1927). The determination of aldose sugars by means of the chloramine T with special reference to the analysis of milk products. *Analyst* **52**, 668–688.

Hodge, J. E. (1953). Chemistry of browning reactions in model systems. *J. Agric. Food Chem.* **1**, 928–943.

Hodge, J. E., and Davis, H. A. (1952). "Selected Methods for Determining Reducing Sugars." U.S. Dept. Agr., Agr. Res. Serv. Publ. *AIC 333.*

Hodge, J. E., and Hofreiter, B. T. (1962). Determination of reducing sugars and carbohydrates. *In* "Methods in Carbohydrate Chemistry" (R. L. Whistler and M. L. Wolfrom, eds.). Academic Press, New York.

Holst, D. D. (1978). Asbestos-free method for determining crude fiber: Collaborative study. *J. Assoc. Off. Anal. Chem.* **61**, 154–160.

Honda, S. (1984). Review. High-performance liquid chromatography of mono- and oligo-saccharides. *Anal. Biochem.* **140**, 1–47.

Hoover, W. J., Nelson, A. I., Milner, R. T., and Wei, L. L.·(1965) Isolation and evaluation of the saccharide components of starch hydrolysis. I. Isolation. *J. Food Sci.* **30**, 248–252.

Hough, L. (1954). Analysis of mixtures of sugars by paper and cellulose column chromatography. *Methods Biochem. Anal.* **1**, 205–242.

Huber, C. M., Scobell, H., and Tai, H. (1966). Determination of saccharide distribution of corn syrup by direct densitometry of thin-layer chromatograms. *Cereal Chem.* **43**, 342–346.

Hurst, W. J., Martin, R. A., and Zoumas, B. L. (1979). Application of HPLC to characterization of individual carbohydrates in foods. *J. Food Sci.* **44**, 892–895.

Ikawa, M., and Niemann, C. (1949). A spectrophotometric study of the behavior of carbohydrates in seventy-nine percent sulfuric acid. *J. Biol. Chem.* **180**, 923–931.

Isherwood, F. A., and Jermyn, M. A. (1951). Relationship between the structure of the simple sugars and their behavior on the paper chromatogram. *Biochem. J.* **48**, 515–524.

Iverson, J. L., and Bueno, M. P. (1981). Evaluation of HPLC and GLC for quantitative determination of sugars in foods. *J. Assoc. Off. Anal. Chem.* **64**, 139–143.

Jackson, R. F., and McDonald, E. J. (1941). Errors of Munson and Walker's reducing sugar tables and the precision of their method. *J. Res. Natl. Bur. Std.* **27**, 237.

James, W. P. T., and Theander, O. (eds.). (1981). "The Analysis of Dietary Fiber in Food." Marcel Dekker, New York.

Jayme, G., and Knolle, H. (1960). Quantitative assay of paper chromatograms by direct photometry in the UV. *Z. Analyt. Chem.* **178**, 84–100.

Jermyn, M. A. (1955). Cellulose and hemicellulose. *In* "Modern Methods of Plant Analysis," Vol. 2 (K. Paech, and M. V. Tracey, eds.). Springer-Verlag, Berlin.

Jolles, Z. E., and Morgan, W. T. J. (1940). The isolation of small amounts of glucosamine and chondrosamine. *Biochem. J.* **34**, 1183.

Juliano, B. (1979). Amylose analysis in rice—a review. *Proc. Workshop Chem. Aspects Rice Grain Quality*, pp. 251–260. Intern. Rice Research Inst., Los Banos, Phillipines.

Kagan, J., and Mabry, T. J. (1965). Sugar analysis of flavonoid glycosides. *Anal. Chem.* **37**, 288–289.

Kent, P. W., Lawson, G., and Senior, A. (1951). Chromatographic separation of amino sugars and amino acids by means of the TN-(2,4-dinitrophenyl) derivatives. *Science* **113**, 354–355.

Kesler, R. B. (1967). Rapid quantitative anion-exchange chromatography. *Anal. Chem.* **39**, 1416–1422.

Khym, J. X., and Zill, L. P. (1952). Separation of sugars by ion exchange. *J. Am. Chem. Soc.* **74**, 2090–2094.

Kneen, E., and Sandstedt, R. M. (1941). Beta amylase activity and its determination in germinated and ungerminated cereals. *Cereal Chem.* **18**, 237–252.

Kowkabany, G. N. (1954). Paper chromatography of carbohydrates and related compounds. *Adv. Carbohydr. Chem.* **9**, 303–353.

Kujawski, M., Kulpinska, E., Pasternak, Z., and Scribor, T. (1979). Methods for determination of starch in high starch material (in Polish). *Przem. Spozyw.* **39**, 345–347. [*Chem. Abstr.* 1980, **92**, 162241f.]

Kushawaha, S. C., and Kates, M. (1981). Modification of phenol–sulfuric acid method for the estimation of sugars in lipids. *Lipids* **16**, 372–373.

Lane, J. H., and Eynon, L. (1923). Determination of reducing sugars by means of Fehling's solution with methylene blue as internal indicator. *J. Soc. Chem. Ind.* **42**, 32T–37T.

Lederer, E., and Lederer, M. (1957). "Chromatography." Elsevier Publishing Co., New York.

Lee, E. Y. C., and Whelan, W. J. (1966). Enzymic methods for the micro-determination of glycogen and amylopectin and their unit chain lengths. *Arch. Biochem. Biophys.* **116**, 162–167.

Lever, M. (1972). A new reaction of colorimetric determination of carbohydrates. *Anal. Biochem.* **47**, 273–279.

Lin, F. M., and Pomeranz, Y. (1965). Characterization of wheat components by infrared spectroscopy. I. Infrared spectra of major wheat flour components. *J. Assoc. Off. Agric. Chem.* **48**, 885–891.

Lin, F. M., and Pomeranz, Y. (1968). Characterization of water-soluble wheat flour pentosans. *J. Food Sci.* **33**, 599–606.

Lothrop, R. E., and Holmes, R. L. (1931). Determination of dextrose and levulose in honey by use of iodine-oxidation method. *Ind. Eng. Chem. Anal. Ed.* **3**, 334–339.

Luderitz, O., Simmons, D. A. R., Westphal, O., and Stromenger, J. L. (1964). A specific microdetermination of glucosamine and the analysis of other hexosamines in the presence of glucosamine. *Anal. Biochem.* **9**, 263–271.

Lustinec, J., Hadacova, V., Kaminek, M., and Prochazka, Z. (1983). Quantitative determination of starch, amylose, and amylopectin in plant tissues using glass fiber paper. *Anal. Biochem.* **132**, 265–271.

Mayer, H., and Westphal, O. (1968). Electrophoretic separation of hexosamine and hexatonic acid derivatives as molybdate complexes. *J. Chromatogr.* **33**, 514–525.

Mazza, G. (1983). A comparison of methods of analysis for sugars in potato tubers. *Can. Inst. Food Sci. Technol. J.* **16**, 234–236.

McComb, E. A., and McCready, R. M. (1952). Colorimetric determination of pectic substances. *Anal. Chem.* **24**, 1630–1632.

McCready, R. M., and Reeve, R. M. (1955). Test for pectin based on the reaction of hydroxamic acids with ferric ion. *J. Agric. Food Chem.* **3**, 260–262.

McCready, R. M., Swensen, H. A., and Maclay, W. D. (1940). Determination of uronic acids. *Ind. Eng. Chem. Anal. Ed.* **18**, 290–291.

McFeeters, R. F., Thompson, R. L., and Fleming, H. P. (1984). Liquid chromatographic analysis of sugars, acids, and ethanol in lactic acid vegetable fermentations. *J. Assoc. Off. Anal. Chem.* **67**, 710–714.

Meuser, F., Suckow, P., and Kulikowski, W. (1982). Analytical determination of dietary fiber in bread, fruits, and vegetables (in German). *Getreide Mehl Brot* **37**, 380–382.

Miller, G. L. (1960). Micro column chromatographic method for analysis of oligosaccharides. *Anal. Biochem.* **2**, 133–140.

Morgan, W. T. J., and Elson, L. A. (1934). A colorimetric method for the determination of N-acetylglucosamine and N-acetyl chondrosamine. *Biochem. J.* **28**, 988–995.

Munson, L. S., and Walker, P. H. (1906). The unification of reducing sugar methods. *J. Am. Chem. Soc.* **28**, 663–686.

Nelson, N. (1944). A photometric adaptation of the Somogyi method for the determination of glucose. *J. Biol. Chem.* **153**, 375–380.

Ohms, J. I., Zec, J., Benson, J. V., and Patterson, J. A. (1967). Column chromatography of neutral sugars. Operating characteristics and performance of a newly available anion-exchange resin. *Anal. Biochem.* **20**, 51–57.

Palla, G. (1981). C_{18} reversed-phase liquid chromatographic determination of invert sugar, sucrose, and raffinose. *Anal. Chem.* **53**, 1966–1967.

Parker, F. S. (1960). Infrared spectra of carbohydrates in water and a new measure of mutarotation. *Biochim. Biophys. Acta* **42**, 513–519.

Partridge, S. M. (1948). Filter-paper partition chromatography of sugars. I. General description and application to the qualitative analysis of sugars in apple juices, egg white, and foetal blood of sheep. *Biochem. J.* **42**, 238–250.

Passonneau, J. V., Gatfield, P. D., Schultz, D. W., and Lowry, O. H. (1967). An enzymic method for measurement of glycogen. *Anal. Biochem.* **19**, 315–326.

Pauly, H., and Ludwig, E. (1922). Glucosamine as a former of heterocyclic compounds. *Z. Physiol. Chem.* **121**, 170–176. [*Chem. Abstr.* **16**, 4211.]

Perez, C. M., and Juliano, B. O. (1978). Modification of the simplified amylose test for milled rice. *Starch* **30**, 424–426.

Pfluger, E. (1909). *Cited by* J. V. Passoneau *et al.* (1967).

Player, R. B., and Wood, R. (1980). Collaborative studies on methods of analysis. Part I. Determination of crude fiber in flours. *J. Assoc. Public. Anal.* **18**, 29–40.

Potterat, M., and Eschmann, H. (1954). Use of complexing agents in assay of sugars. *Mitt. Lebensmitteluntersuch. Hyg. (Bern)* **45**, 312–329. (French)

Prosky, L., Asp, N.-G., Furda, I., DeVries, J. W., Schweitzer, T. F., and Harland, B. F. (1984). Determination of total dietary fiber in foods, food products, and total diets; interlaboratory study. *J. Assoc. Off. Anal. Chem.* **67**, 1044–1052.

Pucher, G. W., Leavenworth, C. S., and Vickery, H. B. (1948). Determination of starch in plant tissues. *Anal. Chem.* **20**, 850–853.

Pusztai, A. (1965). Studies on extraction of nitrogenous and phosphorus-containing materials from the seeds of kidney beans (*Phaseolus vulgaris*). *Biochem. J.* **94**, 611–616.

Radhakrishnamurty, B., Dalferes, E. R., and Berenson, G. S. (1966). Determination of hexosamines by gas liquid chromatography. *Anal. Biochem.* **17**, 545–550.

Reyes, F. G. R., Wrolstad, R. E., and Cornwell, C. J. (1982). Comparison of enzymic, GLC, and HPLC methods for determining sugars and organic acids in strawberries at three stages of maturity. *J. Assoc. Off. Anal. Chem.* **65**, 126–131.

Richter, M., and Schierbaum, F. (1968). Possibilities and limits in starch assay in cereals. *Die Nahrung.* **12**, 189–197. (German)

Rotsch, A. (1947). Determination of low-molecular weight carbohydrates in cereals. *Getreide Mehl Brot* **1**, 10–13. (German)

Sawardeker, J. S., and Sloneker, J. H. (1965). Quantitative determination of monosaccharides by gas liquid chromatography. *Anal. Chem.* **37**, 945–947.

Scherz, H., and Morgenthaler, E. (1980). Review: Analysis of polysaccharides used as food additives (in German). *Z. Lebensm. Unters. Forsch.* **170**, 280–286.

Scherz, H., Stehlik, G., Baucher, E., and Kaindl, K. (1968). Thin-layer chromatography of carbohydrates. *Chromatogr. Rev.* **10**, 1–17. (German)

Schoenemann, K., Jeschek, G., and Frommhold, K. (1961). A new method for quantitative paper chromatographic determination of sugars and sugar alcohols. *Z. Anal. Chem.* **181**, 338–350. (German)

Schoorl, N., and Regenbogen, A. (1917). Gravimetric sugar assays. *Z. Anal. Chem.* **56**, 191–202. (German)

Scott, R. W., Moore, W. E., Effand, M. J., and Millet, M. A. (1967). Ultraviolet spectrophotometric determination of hexoses, pentoses, and uronic acids after their reactions with concentrated sulfuric acid. *Anal. Chem.* **21**, 68–80.

Seifter, S., Dayton, S., Navic, B., and Muntweyler, E. (1950). Estimation of glycogen with the anthrone reagent. *Arch. Biochem. Biophys.* **25**, 191–200.

Selvendran, R. R., and DuPont, M. S. (1984). Problems associated with the analysis of dietary fibre and some recent developments. *In* "Developments in Food Analysis Techniques," Vol. 3 (R. D. King, ed.), pp. 1–68. Elsevier Applied Science Publ., London.

Shaffer, P. A., and Hartmann, A. F. (1921). The iodometric determination of copper and its use in sugar analysis. *J. Biol. Chem.* **45**, 349–390.

Sichert, K., and Bleyer, B. (1936). Determination of dextrose, maltose, and dextrin in sugar mixtures. *Z. Anal. Chem.* **107**, 328. [*Chem. Abstr.* **31**, 13237.]

Slein, M. W. (1952). A rapid method for distinguishing D-glucosamine from galactosamine in biological preparations. *Proc. Exptl. Biol. Med.* **80**, 646–647.

Solms, J. Denzler, A., and Deuel, H. (1954). Amides of polygalacturonic acid. *Helv. Chim. Acta* **37**, 2153–2160.

Somogyi, M. (1945A). A new reagent for the determination of sugars. *J. Biol. Chem.* **160**, 61–68.

Somogyi, M. (1945B). Determination of blood sugar. *J. Biol. Chem.* **160**, 69–73.

Southgate, D. A. T., Hudson, G. J., and Englyst, H. (1978). The analysis of dietary fibre— the choices of the analyst. *J. Sci. Food Agric.* **29**, 979–988.

Spedding, H. (1964). Infrared spectroscopy and carbohydrate chemistry. *Adv. Carbohydr. Chem.* **19**, 23–49.

Spiller, G. A., and Amen, R. J. (eds.). (1976). "Fiber in Human Nutrition." Plenum Press, New York.

Stadtman, E. R. (1948). Nonenzymatic browning in fruit products. *Adv. Food Res.* **1**, 325–372.

Stahl, E., and Kaltenbach, W. (1961). Thin layer chromatography. VI. Analyses of sugar traces on silica gel plates. *J. Chromatogr.* **5**, 351–356. (German)

Stahl, E., and Kaltenbach, W. (1965). Sugars and derivatives. In "Thin-layer Chromatography, A Laboratory Handbook" (E. Stahl, ed.). Academic Press, New York.

Stanek, J., Cerny, M., Kocurek, J., and Pacak, J. (1963). "The Monosaccharides." Academic Press, New York.

Steinhoff, G. (1933). Analysis of starch sirup. Determination of glucose in the presence of maltose and dextrin. *Z. Spiritus Ind.* **56**, 64. [*Chem. Abstr.* **27**, 6004.]

Stetten, M. R., Katzen, H. M., and Stetten, D. (1958). Comparison of the glycogens isolated by acid and alkaline procedures. *J. Biol. Chem.* **232**, 475–488.

Streuli, H., and Stesel, M. (1951). Digestion of chocolate in sugar analysis. *Intern. Fachschr. Schokolade Ind.* **6**, 200–201. (German)

Streuli, H., and Stesel, M. (1952). Clarifying in assay of sugars in chocolate. *Mitt. Lebensmittelunters. Forsch.* **43**, 417–444. (German)

Sweeley, C. C., Bentley, R., Makita, M., and Wells, W. W. (1963). Gas liquid chromatography of trimethylsilyl derivatives of sugars and related substances. *J. Am. Chem. Soc.* **85**, 2497–2507.

Tanaka, M., Thananunkul, D., Lee, T. C., and Chichester, C. O. (1975). A simple method for the quantitative determination of sucrose, raffinose, and stachyose in legume seeds. *J. Food. Sci.* **40**, 1087–1088.

Taufel, K., and Muller, K. (1956). Determination of glucose, maltose, and other fermentable oligosaccharides and dextrins. *Z. Lebensmittelunters. Forsch.* **103**, 272–284. (German)

Thaler, H. (1940). Determination of lactose and sucrose in cocoa products *Z. Untersuch. Lebensm.* **80**, 439–450. (German)

Thaler, H. (1967). Detection and determination of polysaccharides. Determination of cellulose and other cell wall components. In "Handbook of Food Chemistry, Food Analysis," Vol. 2 (W. Diemair, ed.). Springer-Verlag, Berlin. (German)

Thivend, P., Mercier, Ch., and Guilbot, A. (1965). Use of glucamylase in starch determination. *Die Starke* **17**, 278–283. (German)

Tracey, M. V. (1952). Determination of glucosamine by alkaline decomposition. *Biochem. J.* **52**, 265–267.

Trowell, H., Southgate, D. A. T., Wolever, T. M. S., Leeds, A. R. L., Gussull, M. A., and Jenkins, D. A. (1976). Dietary fiber redefined. *Lancet* **1**, 967.

Ugrinovits, M. (1980). Sugar analysis by GC, HPLC, TLC, and enzymatic procedures—a comparison of methods (in German). *Chromatographia* **13**, 386–394.

Ulman, M., and Richter, M. (1961). Review of principles of polarimetric starch assays by the Ewers method. *Die Starke* **13**, 67–75. (German)

Underwood, J. C., Willits, C. O., and Lento, H. G. (1961). Browning of sugar solutions. VI. Isolation and characterization of the brown pigment in maple sirup. *J. Food Sci.* **26**, 397–400.

Van Der Bij, J. R. (1967). Modern methods of analysis of starch derivatives and starch hydrolysates. *Die Starke* **19**, 256–263. (German)

Van Handel, E. (1965). Estimation of glycogen in small amounts of tissue. *Anal. Biochem.* **11**, 256–265.

Van Soest, P. J. (1967). Development of a comprehensive system of feed analyses and its application to forages. *J. Anim. Sci.* **26**, 119–128.

Van Soest, P. J., and Wine, R. H. (1967). Use of detergents in the analysis of fibrous foods. The determination of plant cell-wall constituents. *J. Assoc. Off. Agric. Chem.* **50**, 50–55.

Weigel, H. (1963). Paper electrophoresis of carbohydrates. *Adv. Carbohydr. Chem.* **18**, 61–97.

Weill, C. E., and Hanke, P. (1962). The thin-layer chromatography of malto-oligosaccharides. *Anal. Chem.* **34**, 1736–1737.

Whistler, R. L., and DeMiller, J. N. (1962). Quantitative paper chromatographic determination of carbohydrates. *In* "Methods in Carbohydrate Chemistry" (R. L. Whistler and M. L. Wolfrom, eds.). Academic Press, New York.

Whistler, R. L., and Durso, D. F. (1950). Chromatographic separation of sugars on charcoal. *J. Am. Chem. Soc.* **72**, 677–679.

Whistler, R. L., and Hickson, J. L. (1955). Determination of components in corn sirups by quantitative paper chromatography. *Anal. Chem.* **27**, 1514–1517.

Whistler, R. L., and House, L. R. (1953). Infrared spectra of sugar anomers. *Anal. Chem.* **25**, 1463–1465.

Whistler, R. L., and Wolfrom, M. L. (eds.). (1962–1964). "Methods in Carbohydrate Chemistry," Vols. 1–5. Academic Press, New York.

White, J. W. (1980). Detection of honey adulteration by carbohydrate analysis. *J. Assoc. Off. Anal. Chem.* **63**, 11–18.

White, J. W., and Siciliano, J. (1980). Hydroxymethylfurfural in honey adulteration. *J. Assoc. Off. Anal. Chem.* **63**, 7–10.

White, J. W., Eddy, C. R., Patty, J., and Hoban, N. (1958). Infrared identification of disaccharides. *Anal. Chem.* **30**, 506.

Whitmoyer, R. B. (1934). Determination of small amounts of glucose, fructose, and invert sugar in absence and presence of sucrose. *Ind. Eng. Chem. Anal. Ed.* **6**, 268–271.

Wichmann, H. J. (1922). Report on determination of pectin in fruit and fruit products. *J. Assoc. Off. Agric. Chem.* **6**, 34–40.

Wichmann, H. J. (1923). Report on determination of pectin in fruit and fruit products. *J. Assoc. Off. Agric. Chem.* **7**, 107–112.

Williamson, A. R., and Zamenhof, S. (1963). Detection and rapid differentiation of glucosamine, galactosamine, glucosamine uronic acid and galactosamine uronic acid. *Anal. Biochem.* **5**, 47–50.

Willstatter, R., and Schudel, G. (1918). Determination of glucose with hypoiodite. *Ber. Deut. Chem. Ges.* **51**, 780–781.

Winkler, S., and Lukov, G. (1967). Standardization of polarimetric starch assays in dilute hydrochloric acid. I. Errors in the Ewers method and reproducibility of acidic digestion of starch. *Die Starke* **19**, 110–115.

Wise, L. E., and McCammon, D. C. (1945). Munson-Walker reducing values of some of the less common sugars and of sodium glucuronate. *J. Assoc. Off. Agric. Chem.* **28**, 167–174.

Wiseman, H. G., Mallack, J. C., and Jacobson, W. C. (1960). Determination of sugar in silages and forages. *J. Agric. Food Chem.* **8**, 78–80.

Wisker, E., Feldheim, W., Pomeranz, Y., and Meuser, F. (1985). Dietary fiber in cereals. *Adv. Cereal Sci. Technol.* **7**, 169–238.

Zelenka, S., and Sasek, A. (1966). Evaluation of analytical methods of starch assay. *Die Starke* **18**, 77–81.

Zill, L. P., Khym, J. X., and Chemine, G. M. (1953). The separation of the borate complexes of sugars and related compounds by ion-exchange chromatography. *J. Am. Chem. Soc.* **75**, 1339–1342.

Zuckerhandl, F., and Messiner-Klebermass, L. (1931). A method for the demonstration and determination of glucosamine. *Biochem. Z.* **236**, 19–28. [*Chem. Abstr.* **25**, 4902.]

36
Lipids

Lipids have at least three important functions in foods: culinary, physiological, and nutritional (Kummerow 1960). The ability of lipids to carry odors and flavors and their contribution to the palatability of meats, to the tenderness of baked products, and to the richness and texture of ice cream are examples of the first kind. As lipids serve as a convenient means of rapid heat transfer, they have found increasing use in commercial frying operations. Dietary lipids represent the most compact chemical energy available to man. They contain twice the caloric value of an equivalent weight of sugar. They are vital to the structure and biological function of cells. Dietary lipids provide the essential linoleic acid, which has both a structural and functional role in animal tissue, and are carriers of the nutritionally essential fat-soluble vitamins.

NOMENCLATURE AND OCCURRENCE

The term lipid is used to denote fats and fatlike substances and is synonymous with the terms lipoids or lipins, used in the earlier literature (Feldman 1967). Lipids are usually defined as food components that are insoluble in water and that are soluble in organic fat solvents. In this book, the terms lipids, fats, and oils are used interchangeably to denote such components. Solvents (and mixtures of solvents) used to extract lipids include ether, petroleum ether, acetone, chloroform, benzene, alcohols (e.g., methanol, ethanol, butanol), and water-saturated butanol.

Lipids are chemical constituents of living organisms or are derived from such constituents. Most lipids commonly possess fatty acids as part of their moiety.

This definition has, however, certain limitations. For example, sterols, squalene, and carotenoids meet the solubility criteria of lipids but contain no fatty acids. On the other side, gangliosides are soluble in water and alcohol–water mixtures but insoluble in many of the organic solvents used to extract lipids from their source. Despite these limitations, the definition is useful in describing the general characteristic of a class of compounds.

The nomenclature of lipids has been described (Anon. 1967) by a committee of the Biological Nomenclature Commission of IUPAC and the Commission of Editors of Biochemical Journals of IUB. The proposed rules concern lipids containing glycerol; sphingolipids; neuraminic acid; fatty acids, long-chain alcohols, and amino acid components of lipids; and specific generic terms such as phospholipids and others.

The classification of lipids is difficult because of their heterogeneous nature. The system most commonly used despite its limitation is that proposed by Bloor (1925) as shown in Table 36.1. Molecular structures of the major lipid classes are given in Fig. 36.1.

Foods vary widely in their lipid content and composition. Lard, shortening, and vegetable or animal cooking fats and oil contain almost 100% lipids. The fat content of butter and margarine is about 81%, and of commercial salad dressings 40–70%. Most nuts are very rich in lipids (almonds, 55; beechnuts, 50; Brazils, 67; cashews, 46; peanuts, 48; pecans, 71; and walnuts up to 64%). The main seeds used for extracting lipids on a commercial scale include (in addition to peanuts) sesame seeds (50% fat), sunflower (47%), hulled safflower seeds (60%), and soybeans (18%). Among dairy products a wide range is found. Cottage cheese contains 4%, and cream cheese 38% (on an as-is basis). Fresh, fluid cow's milk has 3.7% fat, but after drying 27.5% fat. The fat content of cream ranges from 20% in light coffee to 38% in heavy whipping cream. Ice cream contains about 12% fat. A very wide range in fat content is encountered in cereal products: grains contain only 3–5% but the germ around 30%; bread 3–6%; most cookies 15–30%; and crackers from 12% in saltines to 24% in chocolate-coated Graham crackers. Raw beef carcass trimmed to retail level contains 16–25% fat; sausages 15–50%; total edible hens and cocks 25%; and herring 11%. The fat ranges from 4% in pink to 16% in chinook salmon. Raw tuna fish contains only 4% fat, but tuna canned in oil contains 21%. The whole edible portion of eggs contains 12% lipids, the yolks alone 29%; after drying the fat content increases to 41% in commercial dried whole eggs and to 57% in dried yolks. Most fruits and vegetables contain small amounts of lipids (especially when expressed on

Table 36.1. Classification Scheme for the Lipids

Simple lipids—compounds containing two kinds of structural moieties
Glyceryl esters—these include partial glycerides as well as triglycerides, and are esters of glycerol and fatty acids
Cholesteryl esters—esters formed from cholesterol and a fatty acid
Waxes—a poorly defined group which consists of the true waxes (esters of long-chain alcohols and fatty acids), vitamin A esters, and vitamin D esters
Ceramides—amides formed from sphingosine (and its analogs) and a fatty acid linked through the amino group of the base compound. The compounds formed with sphingosine are the most common

Composite lipids—compounds with more than two kinds of structural moieties
Glyceryl phosphatides—these compounds are classified as derivatives of phosphatidic acid
Phosphatidic acid—a diglyceride esterified to phosphoric acid
Phosphatidyl choline—more descriptive term for lecithin, which consists of phosphatidic acid linked to choline
Phosphatidyl ethanolamine—often erroneously called cephalin, a term referring to phospholipids insoluble in alcohol
Phosphatidyl serine—also erroneously called cephalin
Phosphatidyl inositol—major member of a complex group of inositol-containing phosphatides including members with 2 or more phosphates
Diphosphatidyl glycerol—cardiolipin

Sphingolipids—best described as derivatives of ceramide, a unit structure common to all. However, as in the case of ceramide, the base can be any analog of sphingosine
Sphingomyelin—a phospholipid form best described as a ceramide phosphoryl choline
Cerebroside—a ceramide linked to a single sugar at the terminal hydroxyl group of the base and more accurately described as a ceramide monohexoside
Ceramide dihexosides—same structure as a cerebroside, but with a disaccharide linked to the base
Ceramide polyhexosides—same structure as a cerebroside, but with a trisaccharide or longer oligosaccharide moiety. May contain one or more amino sugars
Cerebroside sulfate—a ceramide monohexoside esterified to a sulfate group
Gangliosides—a complex group of glycolipids that are structurally similar to ceramide polyhexosides, but also contain 1 to 3 sialic acid residues. Most members contain an amino sugar in addition to the other sugars. However, not all gangliosides contain amino sugars

Derived lipids—compounds containing a single structural moiety that occur as such or are released from other lipids by hydrolysis
Fatty acids
Sterols
Fatty alcohols
Hydrocarbons—includes squalene and the carotenoids
Fat-soluble vitamins A, D, E, and K

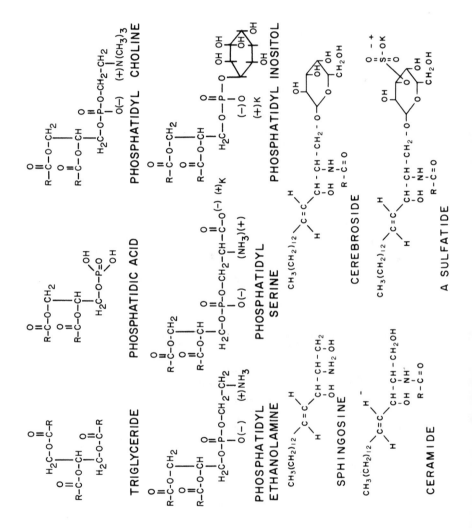

TRIGLYCERIDE

PHOSPHATIDIC ACID

PHOSPHATIDYL CHOLINE

PHOSPHATIDYL ETHANOLAMINE

PHOSPHATIDYL SERINE

PHOSPHATIDYL INOSITOL

SPHINGOSINE

CERAMIDE

CEREBROSIDE

A SULFATIDE

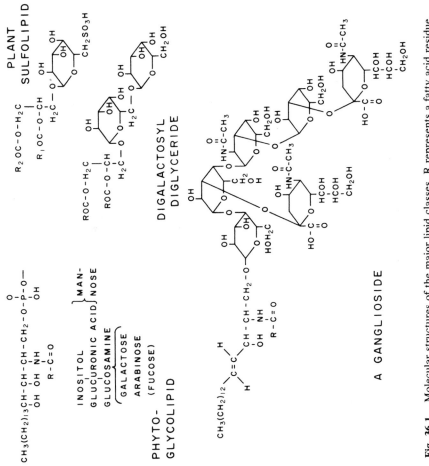

Fig. 36.1. Molecular structures of the major lipid classes. R represents a fatty acid residue.

an as-is basis), but avocado contains 16% lipids, and the lipid content ranges from 10% in giant-size pickled olives to 36% in salt-dried, oil-coated Greek-style olives. Sweet chocolate contains 35% fat, and bitter or baking chocolate 53%; in dry cocoa powders the fat ranges from 8% in low-fat powders to 24% in high-fat or breakfast-type powders.

SOURCES OF INFORMATION

The methods of analysis applied to lipids have been described in several books. Kaufmann (1958) has gathered the chemical and physical methods of analysis of fats and fat products. Mehlenbacher (1960) has collected the methods of analysis of fats and oils emphasizing those for use in industrial control laboratories. Cocks and Van Rede (1966) are the editors of a laboratory handbook for oil and fat analysis, and Boekenogen (1964) has initiated a series of volumes on the analysis and characterization of oils, fats, and fat products. Shorter reviews on methods of analysis and separation were published by Fontell et al. (1960) and by Mehlenbacher (1958). Methods of fat analysis are described in books edited by Kuksis (1978), Pryde (1979), Gurr and James (1980), Hamilton and Bhati (1980), Swern (1982), and Gunstone and Norris (1983). Falbe and Weber (1974) reviewed in detail general methods for testing fats, oils, and waxes. Fat determination was reviewed by Mattsson (1978) for feedstuffs. New and rapid (direct and indirect) methods of fat determination were reviewed by McGann (1980) for dairy products and by Pettinati (1980) and Kropf (1984) for meat and meat products.

General information on lipids is available in several textbooks including those by Williams (1950), Deuel (1951), and Hilditch and Williams (1964). A series of monographs on progress in the chemistry of lipids is published at irregular intervals. At least five scientific journals are devoted entirely or almost entirely to lipids (*Lipids; Journal of the American Oil Chemists' Society; Fette, Seifen, Anstrichmittel; Journal of Lipid Research; Chemistry and Physics of Lipids*), and practically all journals concerned with biochemistry, analysis, and food science deal occasionally with various aspects of lipid chemistry and lipid analysis. Proceedings of symposia organized by the American Oil Chemists' Society on various aspects of lipid chemistry and analysis, and published periodically in their journal, are particularly useful. The bibliography sections of several journals (including *J. Lipid Research, J. American Oil Chemists' Society*, and *J. Chromatography*) are excellent sources of information on current developments in the analysis of lipids.

The American Oil Chemists' Society has published official and tentative methods of analysis of fats, giving detailed instructions for each method (Anon. 1982) Similar collections of official methods applicable to fats have been issued by the German Society for Fat Science (Anon. 1950) and the Fat Commission of the International Union of Pure and Applied Chemistry (Anon. 1954). Standard methods of the Association of Official Analytical Chemists in the United States (Anon. 1980) include sections on fats and oils. Methods of the Society for Analytical Chemistry (England) and some methods included in British Standards are published in *Analyst*. Specific methods are included also in *Cereal Laboratory Methods* (Anon 1983) and in publications of the American Society for Testing Materials (ASTM).

Analysis of lipids involves (1) the extraction of lipids and determination of lipid content and composition and (2) assay of the extracted lipids based on their physical and chemical properties.

GENERAL PROCEDURES FOR EXTRACTING LIPIDS

As indicated, lipids are characterized by their sparing solubility in water and their considerable solubility in organic solvents—physical properties that reflect their hydrophobic, hydrocarbon nature. In practice, the wide range of relative polarities of lipids, as a result of their various structures, makes the selection of a single universal solvent impossible.

Successful extraction requires that bonds between lipids and other compounds be broken so that the lipids are freed and solubilized. Generally, such solubility is attained when polarities of the lipid and the solvent are similar. For example, the nonpolar triglycerides are dissolved in nonpolar solvents such as hexane and petroleum ether (a low boiling point distillate from petroleum). Polar compounds, such as glycolipids, are soluble in alcohols. In some cases, solubility is modified as a result of molecular interaction. Thus, phosphatidyl choline (lecithin) behaves as a base because of the quarternary ammonium group of its choline moiety. It dissolves in weakly acidic solvents such as alcohol. Phosphatidyl serine is structurally similar to phosphatidyl choline but is a polar, relatively strong acid. It is insoluble in weakly acidic alcohol but dissolves in chloroform, which readily associates with acidic polar compounds, even though it is rather a nonpolar solvent. The presence of other lipids also affects solubility. Phosphatidyl serine will be solubilized partly in the presence of phosphatidyl choline, though it is insoluble in alcohol alone. Chloroform is generally a useful solvent, but fails in quantitative extraction of compound lipids (glycolipids and proteolipids).

Preparation of Sample

There is no single standard method for lipid extraction. The method used depends on the type of analyzed material and nature of the subsequent analytical problem (Marinetti 1962). Thus, extraction of lipids from milk is relatively simple compared to the extraction of lipids from plant or animal tissue. These latter materials require some type of fragmentation, such as mechanical grinding, sonic disintegration, homogenization or compression–decompression. During these steps, it is important to keep the chemical, physical, and enzymatic degradation of lipids to a minimum. This is usually accomplished by the control of temperature, chemical environment, and time of exposure of the material to each solvent.

Entenmann (1961) stressed that if the lipids are incompletely extracted or altered during extraction, the results will be inaccurate regardless of the effort spent later or the precision of the apparatus and techniques used in the analysis. For best results (especially if the extracted lipids are to be characterized) the following requirements must be met: (1) all procedures must be carried out under an atmosphere of nitrogen; (2) the solvent(s) used should be purified, peroxide-free, and in the proper solute:solvent ratio; (3) the tissue should be removed from the source and subdivided as soon as possible; (4) heating should be minimized, (5) the extract should be purified to remove nonlipid components, and (6) the purified lipids should be stored under conditions that minimize alteration.

The moisture content is an important factor. Only part of the lipids can be extracted with ether from moist material, as the solvent cannot penetrate the tissues and the extractant becomes saturated with water and inefficient for lipid extraction. Drying at elevated temperatures is undesirable because some lipids become bound to proteins and carbohydrates and are rendered inextractable. Lyophilization affects extractability little and increases the surface area of the sample. On the other hand, more lipid will be extracted from a ground wheat sample containing above 11% moisture than from one that has a very low moisture content.

The extraction of dry materials depends on particle size; consequently, efficient grinding is very important. The classical method of determining fat in oilseeds involves extracting the ground seeds with a selected solvent after repeated grinding in a mortar with sand. Soft materials such as peanuts and copra must be grated, preferably in a grating mill without expressing oil during grinding. Cooling copra to below 0°C minimizes losses. To promote rapid extraction, the sample and solvent are mixed together in a high-speed comminuting device such as a Waring blendor. When the cutting device is operated at high speed, the particle size is rapidly reduced and the fat extraction accelerated.

Pinto and Enas (1949) described a special chopper (similar to a hammer mill but with fixed steel blades equipped with special edges) and a Waring blendor with a special blade assembly for rapid grinding–extraction of copra. Lipids are extracted with difficulty from whole yeast cells because of the limited porosity of the cell wall and its sensitivity to dehydrating agents. The extraction may be accelerated and rendered quantitative if the yeast cells are broken mechanically by shaking with glass beads (Trevelyan 1966). To reduce metabolic modification of lipids during mechanical disintegration, the yeast suspension must be preheated briefly to 90°C. To overcome the tedium involved in extracting lipids from oilseeds, Troeng (1955) proposed milling whole seeds with steel balls and sand with simultaneous extraction. This is the basis of the procedure described by Cocks and Van Rede (1966). The procedure involves preliminary extraction in a vibrating ball mill followed by final extraction for 1 hr in an efficient drip extractor.

In many processed foods; in by-products of the dairy, bread, fermentation, sugar and flour industries; and in animal products, a major part of the lipids is bound to proteins and carbohydrates and direct extraction with nonpolar solvents is inefficient. Such foods must be prepared for lipid extraction by acid hydrolysis or other methods. Two general procedures have been developed. Preliminary extraction of ground material is followed by acid hydrolysis and reextraction after acid treatment and drying. According to the procedure of Campen and Geerling (1954), which has been adopted by the Netherlands Standard Institute, the pre-extraction can be omitted. The sample is predigested by refluxing for 1 hr with 3 N hydrochloric acid; ethanol and solid hexametaphosphate—to facilitate separation—are added, and the fat is extracted with tetrachloromethane.

Solvents

Ethyl and petroleum ether are the common extraction solvents. There is a growing tendency to use petroleum ether because it is more selective toward true lipids. Ethyl ether is a better solvent for fat than petroleum ether and will dissolve oxidized lipids. It has, however, several disadvantages: it is more expensive; danger of explosion and fire hazard are somewhat greater; and it picks up water during extraction of a sample and dissolves nonlipid materials (i.e., sugars). Dried ether tends to form peroxides.

Combinations or alternate extraction with ethyl and petroleum ether are used often in extraction of lipid from dairy products. Mixtures of alcohol and ether are employed to remove fat from certain biological materials, although the extract is rich in nonlipid components that must be subsequently separated. Treatment with alcohol facilitates the removal of fat from some materials.

Of a number of solvents tried, water-saturated n-butanol was most effective in extracting liquid material from ground wheat, flour, bran, and gluten (Mecham and Mohammad 1955). Water-saturated butanol has since been used extensively to extract lipids from cereals. Bloksma (1966) extracted flours with various butanol mixtures containing 0–17.5% water. With increasing water content, the extracted lipids increased from 1.16 to 1.37%. At the same time, the extracted nonlipid components increased from 0.06 to 0.27%. Water-saturated butanol is an effective lipid extractant. Yet, some of the lipids are released only after acid hydrolysis. In addition, the solvent has a strong odor and requires relatively high temperatures for evaporation, even in vacuo.

A rapid method for lipid extraction was proposed originally by Folch *et al.* (1957) for isolation and purification of total lipids from animal tissue by means of phase partition of a ternary mixture of chloroform–methanol–water. Bligh and Dyer (1959) simplified the method. In the simplified procedure, the sample is homogenized with a mixture of chloroform and methanol in such proportions that a miscible system is formed with the water in the sample. Dilution with chloroform and water separates the homogenate into two layers, the chloroform layer containing all the lipids and the methanol layer containing all the nonlipids. A purified extract is obtained by isolating the chloroform layer. The method has been applied by the authors to fish muscle and has been adapted by Tsen *et al.* (1962) for the extraction of lipids from wheat products.

Lee *et al.* (1966) compared the extraction of lipids from fish meal by the following six methods: (1) the official AOAC method that involves a 16-hr extraction with acetone in a Soxhlet extractor, digestion of the extracted residue with dilute hydrochloric acid, filtration, drying the residue and reextraction with acetone for 16 hr; (2) extraction with ether; (3) and (4) two modifications of the Bligh–Dyer method; (5) an alkaline saponification method; and (6) the Mojonnier method. The ether and the alkaline saponification methods gave, as expected, greatly reduced yields. In the Mojonnier method the yield also was slightly reduced. One of the Bligh–Dyer methods gave practically a theoretical yield.

Two general types of lipid–protein complexes are known, *lipoproteins* and *proteolipids*, differing in the relative amounts of lipid and protein. Lipoprotein is mostly protein, whereas proteolipids contains almost equal amounts of the two and exhibit lipid solubility characteristics (Feldman 1967). Mixtures of chloroform and methanol are among the most effective and relatively mild extractants for proteolipids, which are solubilized with little damage to most proteins. The extracted proteolipids can be readily cleaved to liberate the lipids by repeated evaporation of the extract or by evaporation to dryness after forming a two-phase system with water. This cleavage is based on the ratio of chloroform to methanol during evaporation. The original ratio of 2:1 (v/v), or 1:1 on a molar basis, is useful in the extraction of intact proteolipids, as the high concentration of chloroform reduces greatly the hydrogen-bonding capacity of the methanol and prevents disruption of the electrostatic bonds of the proteolipid. During evaporation, the concentration of chloroform decreases more rapidly than that of methanol. As more methanol becomes available to attack and rupture linkages between the protein and the lipid, the protein is gradually denatured and precipitated.

Purification of Extracts

Water-soluble nonlipid substances (carbohydrates, salts, amino acids) are invariably extracted from tissues along with lipids. The nonlipids must be removed before gravimetric determination of total lipid and to prevent contamination during subsequent fractionation of the total extract. Various partition procedures have been used. The more common ones employ water alone or aqueous salt solutions. The methods fail to remove inorganic phosphate, and with phosphatide-rich extracts may produce emulsions that can be broken only by centrifugation or standing for very long times. Some lipids are appreciably soluble in water, and excessive washing may cause lipid losses.

The removal of contaminants may be accomplished (partly or completely) by evaporation of the extracts to dryness in vacuo under nitrogen, and reextraction with a nonpolar solvent. Separation by dialysis, electrodialysis, electrophoresis, and chromatography are some of the other methods. Wells and Dittmer (1963) found that a cross-linked dextran gel (Sephadex) was useful for the column chromatographic separation of lipids from nonlipid water-soluble substances. Siakotos and Rouser (1965) modified their procedure for better separation of lipid and nonlipid components.

Apparatus

Direct extraction of lipids is often carried out in a Soxhlet, which was described in Chapter 24 (see Fig. 24.1). The sample in an extraction thimble (from filter paper or Alundum), covered with defatted (by soaking in ether) cotton wool to prevent small particles from finding their way into the flask, is placed in the middle part of the apparatus. The flask is filled with solvent; the three parts of the apparatus are assembled; the condenser attached to a tap; and heating of solvent in the flask started. The condensing vapors fill the middle part containing the sample and carry

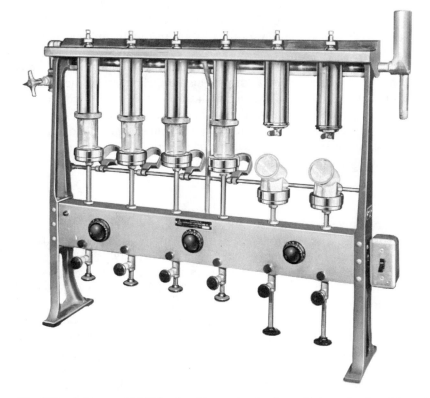

Fig. 36.2. Labconco–Goldfish solvent-type extractor for rapid determination of fat and oil content. (Courtesy Labconco Co.)

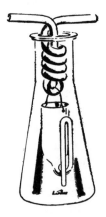

Fig. 36.3. ASTM extraction apparatus. (Courtesy LaPine Scientific Co.)

(A)

TRADITIONAL

(B)

FINAL EXTRACTION WITH
CONDENSE SOLVENT

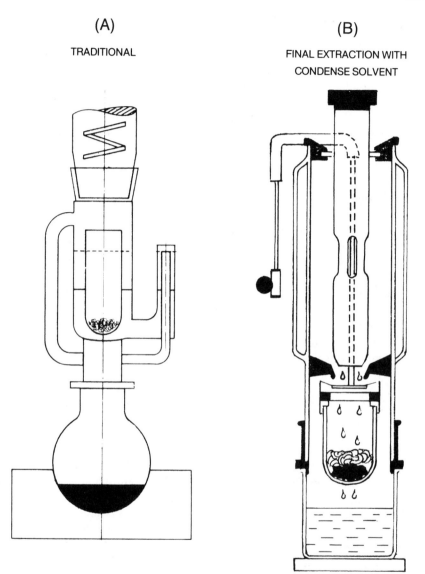

Fig. 36.4. Comparison of the Soxhlet (A) and rapid extraction apparatus (Soxtec) (B) according to Randal (1974). (Courtesy Tecator Inc.)

the dissolved lipid into the flask by a siphoning action each time the height of the siphon is attained. At the end of exhaustive extraction, the apparatus is disconnected, the solvent from the tared flask is evaporated, and the weight increases calculated.

The most efficient extractors for most materials are the continuous percolator types specified by the methods of the American Oil Chemists' Society. Commercial Goldfish extractors of the percolator type are available (Fig. 36.2). An extraction apparatus recommended by ASTM consists of a glass extraction flask, a coiled block tin condenser, polished copper cover (tinned inside), and a glass siphon cup suspended by aluminum wire (Fig. 36.3).

Randall (1974) developed a modified ether extraction apparatus that can extract fat in 30 min, instead of the 4 hr required with the conventional Soxhlet apparatus. He immersed a dried sample in a wire mesh cup directly into boiling ether in a reflux apparatus, raised the cup after 10 min, and permitted ether condensate to wash it for 20 min. A complete determination requires $2\frac{1}{4}$ hr, and 80 determinations a day can be made with a commercial multiple extractor. Precision (0.2% fat between-methods mean difference) and standard deviation (0.6% fat between methods) were equivalent to those with the Soxhlet reference method. Equipment for the procedure, available from Tecator Inc., Boulder, CO., is illustrated and compared with the Soxhlet apparatus in Fig. 36.4.

The Mojonnier extraction apparatus (Mojonnier and Troy 1925), originally devised for fat determination in dairy products, has been found to be convenient for assay of many other foods when it is desirable to make a liquid–liquid separation. The procedure utilizes a specially designed flask that permits the intimate mixture of fat and solvent essential in the extraction of fat from a liquid phase.

EXTRACTION OF LIPIDS IN SELECTED FOODS

Predried (preferably in a vacuum oven at 95–100°C under pressure not to exceed 100 Torr for about 5 hr, or in a vacuum desiccator over H_2SO_4 for 24 hr at 10 Torr) and ground grain is extracted exhaustively with dry ether in a Soxhlet, Butt-type, Goldfish, or similar extractor.

For the extraction of crude fat from bread and baked cereal products not containing fruit, the sample (2 g) is moistened with 2 ml ethanol to prevent lumping on the addition of 10 ml hydrochloric acid (mixed in a ratio of 25:11 v/v with water). The material is digested at 70–80°C in a water bath for 30–40 min. After adding 10 ml ethanol and cooling, the contents are extracted three times with ether in a Mojonnier-type apparatus, the extracts are combined and filtered, the solvent is evaporated, and the extracted lipids are weighed.

Crude fat in egg yolk and dried whole egg is extracted with a mixture of ether and petroleum ether in a sample pretreated with hydrochloric acid (mixed at a ratio of 4:1 with water) at a rather elevated temperature.

On the other hand, crude fat in cocoa can be extracted by simply passing ten 10-ml portions of petroleum ether through a sample (2 g) in a sintered glass filter (coarse porosity) attached to a suction filtering apparatus.

According to Mattsson (1978), in Sweden diethyl ether (Soxhlet) extraction is used for vegetable materials and acid hydrolysis [Schmid–Bondzynski–Ratzlaff (SBR) method] for feedstuffs rich in protein and for samples in which the fatty acids may be bound as calcium salts. In the European Economic Community methods, diethyl ether extraction is used for feedstuffs, except for those in which diethyl ether does not remove all lipids, for animal tissues, and for clover.

In milk, fat globules are present as an emulsion of oil in water and are surrounded by a thin protein film. The emulsion must be broken and the protein film removed before the fat can be separated and determined volumetrically. This is accomplished in the Babcock and Gerber methods by using sulfuric acid. In the Babcock method (Anon. 1960), 17.6 ml milk are mixed with 17.5 ml of sulfuric acid of specific gravity 1.80–1.83 in a special bottle, shaken till apparently homogeneous, centrifuged, and submerged into water at 63°C. The rising fat is determined from the height formed in a graduated neck.

In the Gerber method, 11 ml of milk followed by 1 ml isoamyl alcohol are added to a Gerber glass butyrometer containing 10 ml sulfuric acid (specific gravity 1.82). A lock stopper is inserted and the bottle is shaken until the curd disappears and the contents are homogeneous. The butyrometer is centrifuged 4 min, submerged to top of graduate stem in water at 63°C for 5 min, and the fat contents are read off.

The Gerber test is simpler and of wider general application than the Babcock test. Charring is generally avoided by the ratio of milk to sulfuric acid. Isoamyl alcohol improves the fat separation and reduces the effects of sulfuric acid.

Phospholipids are not included in the fat determined by either the Gerber or Babcock test (Levowitz 1967). This is not important in milk and cream in which the polar lipids constitute about 1% of total lipids, but is of significance in testing skim milk and buttermilk. The fat of the latter contains up to 24% phospholipids and must be determined by one of the available gravimetric methods.

In ice cream, the presence of stabilizers requires prolonged acid digestion, yet the concentration of the latter must be reduced to avoid excessive charring of the sugar-rich food. Consequently, the accuracy of the Gerber test is reduced.

The Gerber test is two to three times more rapid than the Babcock test. It is much better suited for determining fat in homogenized milk and is official in most European countries. In the United States, the Gerber test is optional in several states, but the Babcock test is used much more. The

Gerber test can be used for the assay of fat in cream and cheese provided special butyrometers are employed. The method, with various digesting reagents, has been applied satisfactorily to other fat-containing products such as processed meat and fish products. In most of those methods, speed and simplicity are achieved at some sacrifice of precision and accuracy of routine tests. Use of the Gerber method for rapid determination of the oil content of groundnut seeds was described by Shukla *et al.* (1980).

To overcome the unpleasant corrosive features of sulfuric acid in the Babcock test, Schain (1949) suggested the use of detergents. An anionic detergent, dioctyl sodium phosphate, disperses the protein layer around the fat globule and liberates the fat; a strongly hydrophilic nonionic detergent, polyoxyethylene sorbitan monolaurate, completes the separation.

Over the years, various modifications of the detergent method have been proposed. They include the DPS method of Sager *et al.* (1955) and the official AOAC Te Sa test. The Te Sa test requires a special two-neck bottle; the Schain and DPS tests are carried out in a regular Babcock milk bottle. In the Te Sa and DPS tests, fat content is read directly; in the Schain procedure, volumetric readings are converted to fat content by nomograms supplied with each batch of reagent. The rapid detergent test (Te Sa) gives significantly lower fat values in homogenized and chocolate milk than conventional gravimetric methods (Mitchell 1967).

The reference gravimetric methods for fat analysis in dairy products involve successive extractions with ether or mixed ethers after preliminary digestion with acid or alkali. Digestion with hydrochloric acid (as in the Werner–Schmidt or Schmidt–Bondzynski procedures used in Europe) are less suitable than digestion with alkali—especially in sugar-rich foods.

In the Rose–Gottlieb method, the sample is rendered alkaline with ammonium hydroxide and, if necessary, diluted with water and heated to facilitate dispersion. The digested mixture is extracted repeatedly with ethanol, ether, and petroleum ether. The combined extracts are dried, purified by extraction with petroleum ether, and weighed. In the Mojonnier test, the combined extracts are weighed without purification. Consequently, the results are slightly higher than in the Rose–Gottlieb method. The advantages and limitations of gravimetric, volumetric, and automated methods for fat estimation in milk and milk products were reviewed by Arora and Rajorbia (1978).

INDIRECT METHODS OF LIPID DETERMINATION

Various devices and techniques are available for approximate rapid determination of lipid content. Zimmerman (1962) reported a high correlation ($r = 0.96$) between flaxseed density (as determined by an air-comparison pycnometer) and oil content, and recommended seed density as a criterion for effective screening of flax lines with high oil content.

The application of wide-line nuclear magnetic resonance spectroscopy (NMR) to analysis of oil in seeds opens new opportunity for geneticists and plant breeders. The method is nondestructive, rapid, and can be used to determine the oil content of a single seed or of a bulk sample (Bauman *et al.* 1963). The measured NMR value is related to the total hydrogen in the oil fraction of the seed, independent of the hydrogen in the nonoil fraction. The oil content is calculated from calibration tables or curves. Application of the method has been initiated by Conway (1960). Although the results are affected by moisture content and distribution, the method has been applied successfully in the assay of oil in seeds such as corn (Alexander *et al.* 1967), soybeans (Collins *et al.* 1967; Fehr *et al.* 1968), and various other oil-containing seeds. The use of wide-line NMR in the determination of oil in single corn kernels was described by Wilmer *et al.* (1978) and in sunflower seed by Robertson and Morrison (1979). According to Robertson and Windham (1981), the NMR method was more precise and reproducible than the official AOCS extraction–gravimetric method and the NIR reflectance method in determining the oil content of sunflower seed.

King (1966) reported that NMR may be used to measure instantaneously and continuously moisture and fat in meat products. The two components can be determined separately with high-resolution NMR, since the resonant frequency for protons in water is slightly different than that of protons in fats. The frequency difference is adequate to permit separate analyses of the fat and water contents. The application of NMR and NIR in determining the fat content of meat and meat products was described by Pettinati (1980).

Several other instruments are also available for rapid determination of fat in meat and meat products. An instrument involving X-ray absorption is based on the fact that lean meat absorbs more X rays than fat (Kropf 1984). A unique approach to determining fat in fresh meat, large boxes of meat, and small live animals is based on differences in electrolytic properties of lean and fatty tissues. Lean meat is a 20-fold better conductor of electrical current than fat. The amount of induced current, combined with the weight of the tested sample, is used to calculate within seconds the fat percentage (Pettinati 1980). Determination of fat rendered and collected in a tube measured after a meat sample is electrically heated was described by Marriott *et al.* (1975). The method is simple, rapid, and economical and the results are in good agreement with those of the standard AOAC Soxhlet method. The 5-min HOBART FMP-1 analyzer operates on the principle of using microwave energy to separate the fat and moisture from the protein and ash of a meat sample (Bostian *et al.* 1984).

The moisture is vaporized and released and the fat is melted and collected. The sample is cooked to a point short of decomposition of the protein. The fat, moisture, and protein components are computed through microprocessing of the weight loss of a sample. The method was recommended to determine routinely analyses of many samples in meat processing.

Biggs (1967) and Dyachenko and Samsonov (1964) described the use of infrared spectroscopy in determination of fat in milk products. Fitzgerald et al. (1961) described an ultrasonic method for the determination of fat and nonfat milk solids in fluid milk. The method is based on measuring the speed of sound through milk at various temperatures. A turbidimetric procedure for the fat assay in milk was proposed by Hangaard and Pettinati (1959). The milk fat is homogenized to bring it to substantially uniform globules, the protein is chelated, light transmission is measured spectrophotometrically and converted to fat by a chart or graph. Murphy and McGann (1967) found that the turbidimetric method was quite precise and reproducible in the range 0–4% fat and was unaffected by the addition of preservatives. The method uses noncorrosive reagents and permits the testing of 75 samples per hour by an operator.

Stern and Shapiro (1953) described a colorimetric method for oil determination. The extracted lipid is treated with an alkaline solution of hydroxamic acid and allowed to react for a specified time. Upon acidification with hydrochloric acid and addition of ferric chloride, a relatively stable color with an absorbance maximum at 540 nm is formed. The use of the hydroxamic method for the determination of fat in milk was described by Katc et al. (1959).

Rapid, indirect methods for fat determination in milk were reviewed by McGann (1980). The Milko-Tester is based on determining turbidity of light scattering caused by fat globules in milk. The influence of globule size is reduced by the photometer design and use of a four-stage homogenizer. The turbidity due to protein is eliminated by diluting the sample 15-fold at 60°C with a solution of EDTA at pH 10.8 to disperse the casein molecules. The Milko-Tester and the Pro-Milk Automatic have been incorporated in the Combi-Unit, which can estimate the fat and protein contents at the rate of 180 samples/hour and provide a printout for data processing. At least three commercial instruments are available for the rapid sequential determination of gross milk composition on the basis of analyses at three specific wavelengths in a double-beam infrared spectrophotometer. The wavelengths are 5.73 μm for the carbonyl groups of fat, 6.46 μm for the amide groups of protein, and 9.60 μm for the hydroxyl groups of lactose.

The addition of fat to an organic solvent changes several properties of the mixture. Some of the changes (density and refractive index) are proportional to the concentration of the fat and large enough for analytical purposes. Standard density-measuring instruments such as hydrometers, pycnometers, and the Westphal balance have been used to a limited extent. The refractive index of an oil solution can be used as a measure of the oil content by comparison with calibration graphs or tables. The method has been adopted as an AOAC procedure for the determination of oil in flaxseed after extraction with a mixture of α-chloronaphthalene and α-bromonaphthalene. The method is used for rapid routine factory control of oilseed extraction. It can be used for cake or expeller meal of copra, palm kernels, peanuts, or soybeans (Cocks and Van Rede 1966). The precision depends primarily on the identity of the tested sample with samples used in establishing the calibration graphs or tables. In favorable cases, the precision is quite high. However, the results are affected by the free fatty acid contents of the sample, which have to be accounted for. Determinations involve extracting 10 g of the sample with 7.5 ml of tetraline in a vibrating ball mill or mortar if no mill is available. Tetraline (tetrahydronaphthalene) is used because it is a stable solvent with a high refractive index (n_D^{35} not higher than 1.5338 is recommended) and has a large refractive index difference from oils. The extract is filtered and the refractive index is determined at 35°C in a precise Abbe refractometer illuminated by a sodium lamp. The method cannot be used for the lipid determination of extracted soy meals as they absorb excessively high amounts of the solvent, and the final extract that is low in oil gives erratic results.

The Foss-Let (Foss America, Inc. Fishkill, NY) determines the fat content as a function of specific gravity of a solvent extract of a sample (Pettinati 1980). A weighed sample is extracted for 1.5–2 min in a special vibration-reaction chamber with perchloroethylene, which has a high specific gravity. The extract is filtered and its specific gravity determined in a thermostatically controlled electronic device with digital readout. The reading is converted to oil or fat percentage using a conversion chart. The method has found wide acceptance.

The refractometric method has been found useful in determining the fat content of vegetable foods, canned meat (Babicheva and Gorelik 1968), and fish and fish products (Schober 1967). The rapid refractometric method is used also for the determination of fat in chocolate-type products (Nadj and Weeden 1966).

Another method of fat determination is based on measuring the change in the dielectric properties of a selected solvent as a result of the presence of fat. The method was originally developed for estimating the oil content of soybeans (Hunt et al. 1952), but has since been applied to other oilseeds. The procedure involves grinding a sample with a solvent in a high-speed mill. The dielectric measurement is made on the solvent–solute mixture and the reading is referred to a chart that relates dielectric readings with fat content.

FRACTIONATION OF EXTRACTED LIPIDS

Extracted lipids comprise a heterogeneous mixture ranging from nonpolar hydrocarbons to the highly polar gangliosides and phytoglycolipids. This mixture is difficult to analyze, and is most commonly divided into major subgroups.

The classical method of separation by acetone precipitation of phospholipids from a solution of mixed lipids in ether is inefficient because of the ability of the triglycerides to cosolubilize polar lipids into acetone, and because some phospholipids (i.e., some unsaturated lecithins) and glycolipids are soluble in acetone (Nichols 1964). Dialysis through a rubber membrane of lipid solutions in petroleum ether, whereby the polar lipids are retained and neutral lipids pass through the membrane, is efficient but tedious if the concentration of triglycerides is high. Similarly, fractional crystallization that achieves separation based on differences in solubility is both laborious and unsatisfactory.

Separations of lipids have been tremendously improved by the introduction of newer methods. The two methods used most commonly are countercurrent distribution and silicic acid chromatography.

Countercurrent Distribution

The separation of solutes on the basis of their differential solubilities in two immiscible solvents has been known for some time. Countercurrent distribution is the name given to a particular type of liquid–liquid multiple–stage (see discussion in Chapter 24). Although this operation can be carried out as a separatory funnel procedure, the ease and labor-saving advantages of the ingenious laboratory equipment designed by Craig (1944) were most instrumental in the utilization of the method. With a modern automatic countercurrent distribution apparatus, as many as 10,000 separations are carried out in 1 hr. The technique holds particular advantage for the lipid chemist for several reasons. The conditions of fractionation are mild and well suited to the study of labile lipids. Recoveries are practically quantitative. Finally, the partition coefficient of a compound in two solutes is both a characteristic constant and a basis for isolating and characterizing structural features of a molecule.

Separations by countercurrent distribution depend upon the differences in the differential solubility of individual chemical compounds when these compounds are distributed between two immiscible solvents. The differential solubility is described by the partition coefficient K and is given by the equation

$$K = C_1/C_2 \qquad (36.1)$$

where C_1 and C_2 are the concentrations of a given solute, respectively, in two solvent layers. The partition coefficient is influenced both by chain length and degree of unsaturation. The effect of increasing the chain length by two methylene groups is opposite and nearly equal to that of one additional double bond.

Countercurrent distribution was the main separation technique used by lipid chemists prior to the availability of chromatographic procedures. Today its greatest usefulness is in preparative work. Details of countercurrent distribution apparatus and methodology were described by Ahrens and Craig (1952), Dutton (1955), and Casinovi (1963). The technique is used to separate fatty acids, glycerides, phospholipids, bile acids, pigments, fat oxidation products, and many others.

Column Chromatography

The systematic studies of Hirsch and Ahrens (1958) based on the introduction of silicic acid column chromatography by Trappe (1940) emphasized the complexity of factors involved in the separation of various components of lipid extracts. Fractionation on silicic acid columns is achieved by progressively eluting the adsorbed lipids with solvent mixtures of increasing polarity (Wren 1961). The neutral lipids (hydrocarbons, glycerides, sterols and sterol esters) are eluted with chloroform; the remaining polar lipids are eluted with methanol. Separations of compounds with widely varying properties within each group can be achieved by stepwise or gradient elutions. To make such separations reproducible, the silicic acid (moisture content, particle size, silicate content, and column preparation) and lipid sample size must be carefully controlled. If those conditions are met, separations of neutral lipids on silicic acid or Florisil (primarily magnesium silicate) columsn are highly satisfactory. Column separations of polar lipids were described by Barron and Hanahan (1958) and Lea (1956). Polar lipids are, however, partly adsorbed irreversibly, modified, and separated poorly on silicic acid (or Florisil) columns.

Recognizing that no single system was adequate for the complete fractionation of lipid classes, Rouser et al. (1965) developed multicolumn schemes for the analysis of lipids. The techniques use DEAE-cellulose, silicic acid, and silicic acid–silicate–water columns. Generally, the combination of columns was to reduce the complexity of the total extract. The DEAE-cellulose columns were introduced specifically to separate acidic lipids. The more homogeneous fractions were subsequently separated by rapid and efficient one- and two-dimensional thin-layer chromatography.

Thin-Layer Chromatography

Thin-layer chromatography (TLC) has become one of the main analytical tools in lipid research and analysis. TLC has largely replaced in lipid analyses the earlier useful but somewhat complicated methods of paper chromatography. This position of thin-layer chromatography has been attained as a result of its simplicity, speed, sensitivity, and versatility (Mangold et al. 1964; Padley 1964; Blank et al. 1964; Privett et al. 1965; Pelick et al. 1965; Pomeranz 1965). The versatility of TLC is exemplified by its application to fractionation of complex lipid mixtures, assay of purity, identification, information on structure (as related to chromatographic mobility in various solvent systems or precoated plates), use in monitoring extractions and separations on columns for preparative work, and in general preparative work.

Practically every known type of lipid can be separated from other lipids and identified by TLC. Each fraction, so separated, generally consists of a whole family of related compounds differing only in the chain length and degree of unsaturation of the component fatty acids, aldehydes, or alcohols. The composition of each lipid with respect to these functional entities can be readily determined by gas–liquid chromatography after hydrolysis or transesterification. If GLC equipment is not available, separations can be made by thin-layer or column chromatography on impregnated adsorbents. The adsorbents complex or interact selectively with specific functional groups. They include silver nitrate impregnated adsorbents for separation of compounds differing in degree or type of unsaturation, and glycol-complexing agents for the separation of isomeric polyhydroxy compounds (Morris 1964, 1966; Schmid et al. 1966; Nutter and Privett 1968).

Methods for the quantitative analysis by TLC may be divided into two main groups; (1) methods based on direct analysis of the spots on the chromatoplate, generally by photometric, reflectometric, or spectrofluorometric methods; and (2) methods involving recovery of the separated compounds from the chromatoplate, followed by analyses using conventional procedures: gravimetric, radiometric, spectroscopic for phosphorus determination; ester analysis by the hydroxamic method; oxidation by solutions of chromic acid; and determination of glycerol or carbonyl compound (Privett *et al.* 1965).

Separations of Lipid Classes

Separations of lipids by column or a combination of column and thin-layer chromatography were described by Rouser *et al.* (1965). Separations of plant lipids were reviewed by Allen *et al.* (1966). Complete separations by thin-layer chromatography alone were described by Freeman and West (1966) and Skipski *et al.* (1968).

A method of separating muscle lipids into phospholipids, free fatty acids, triglycerides, and cholesterol was developed by Hornstein *et al.* (1967). Phospholipids are separated from the total lipid extract by absorption on activated silicic acid. The free fatty acids in the supernatant are adsorbed on an anion-exchange resin. The solution, free of phospholipids and of free fatty acids, is saponified with alcoholic potassium hydroxide. After acidification, a hexane extract is obtained containing glycerides, fatty acids, and cholesterol. The fatty acids are adsorbed on an anion-exchange resin and the cholesterol remains in solution. Fatty acids are converted to methyl esters directly on the resin, and phospholipids adsorbed in silicic acid are transmethylated to produce the methyl esters of the phospholipid fatty acids. The methyl esters are analyzed by gas chromatography.

Lester (1963) developed a reproducible and sensitive anion-exchange procedure for quantitative analysis of phospholipids. The method was used by Wells and Dittmer (1966) as a basis for the quantitative determination of 24 classes of brain lipids. Selective mild and acid hydrolyses are used to obtain water-soluble phosphate esters characteristic of the diacyl phosphoglycerides and plasmalogens of brain. Those phosphate esters are separated by ion-exchange chromatography and assayed quantitatively. Phospholipids stable to hydrolysis are assayed after fractionation on silicic acid. Gangliosides, neutral lipids, and glycosphingolipids are measured by specific spectrophotometric determination of characteristic components after an initial solvent fractionation and chromatography on Florisil.

PHYSICAL ASSAY METHODS

Physical measurements of fats are useful for identification, for checking purity, and for the control of certain aspects of processing.

Color

The color of fats and oils is estimated by several methods that vary with the type of examined lipid and with the country (Mehlenbacher 1958). The Wesson method is used in the United States for most edible oils and fats. The color of a 5.25-in. column of oil is compared in a Wesson comparator under specified viewing conditions with Lovibond glasses, using specified yellow ratios. The Lovibond glasses and the Tintometer are used extensively in England and Canada.

The FAC method employs standard color solutions (prepared from solutions of inorganic salts) in 10-mm tubes, against which the tested oil in a similar tube is compared. The standards consist of 24 tubes in an odd-numbered series with 3 overlapping series; 1 normal, 1 green, and 1 red (Stillman 1955). Supplemental tubes for closer grading are available in some series. This simple and somewhat imprecise method is used widely for color-grading inedible oils, particularly tallows and greases. The official German method specifies a solution of iodine as the color standard, the comparison being made in a Pulfrich colorimeter. The Gardner color standards are specified by the methods of the American Oil Chemists' Society and the American Society for Testing Materials for drying oils. The standards are patterned along the same lines as FAC standards, but are unrelated to the latter numerically. The standards have been revised from time to time and are widely used and accepted.

A photometric method is specified for measuring the color of cotton-seed, soybean, and peanut oils. The instrument used is a wide-band spectrophotometer. The absorbance of the oil in a 25-mm cuvette is measured at 460, 550, 620, and 670 nm, and the color value calculated as follows:

$$\text{photometric color} = 1.29A_{460} + 69.7A_{550} + 41.2A_{620} - 56.4A_{670}$$

$$(36.2)$$

The photometric method was designed by the color committee of AOCS to give values identical with Lovibond color values determined by the Wesson method. There are many shortcomings to this method as it is correlated with purely arbitrary color values. A more logical approach would be the determination of the pigments responsible for oil color (Mehlenbacher 1958).

Melting Point, Solidification, and Consistency

As fats and oils are a complex mixture of compounds, they have no definite melting point and pass through a gradual softening before becoming liquid. Consequently, the melting point must be defined by the specific conditions of the method. The two methods most commonly used by the shortening and margarine industry include the capillary and Wiley melting point determinations (Smith 1955).

The capillary method is essentially the method used by the organic chemist for determining the melting point of pure organic compounds. In this method, a 1-mm (internal diameter) thin-walled capillary tube is filled to the height of 10 mm with melted fat, one end is sealed, and the fat is allowed to stand at 4–10°C for 16 hr. The tube is then attached to a thermometer and placed in a bath maintained at 8–10°C below the expected melting point. The bath is heated at a rate of 0.5°C per min. The melting point is taken as the temperature at which the fat becomes completely clear.

The Wiley melting point method is much more reproducible and reliable. In this method, a disk of the fat (⅜ in. in diameter and ⅛ in. thick) is solidified and chilled in a metal form for 2 hr or more. The disk is then suspended in an alcohol–water bath of its own density and heated slowly while being stirred with a rotating thermometer. The melting point is taken as the temperature at which the fat disk becomes completely spherical. Agreement within 0.2°C between analyses generally can be obtained. The Wiley melting point, in conjunction with the refractive index, has been used for many years in control of hydrogenation. For control purposes the sample is chilled rapidly before assay.

The titer test determines the solidification point of fatty acids. A titer tube filled with fatty acids (obtained by saponification of oil or fat with potassium hydroxide in glycerol) is suspended in an air bath surrounded by a water bath maintained at 15–20°C below the titer. The sample is stirred until the temperature begins to rise or remains constant for 30 sec, after which the stirring is stopped, and the maximum temperature that the fatty acids attain as a result of the heat of crystallization is determined. The utility of this method is limited. It is used sometimes in evaluating fats for soap manufacture.

Setting or congeal point determinations, based on the solidification point of the fat rather than that of the separated fatty acids, are used quite extensively in the margarine or shortening industries. The tests provide useful information on the consistency of a plastic fat and on the performance of a catalytic hydrogenation. In the manufacture of margarine, the solidification point (congeal) and melting point (Wiley method) are kept as close as possible (Smith 1955).

Dilatometry

Dilatometry is essentially a measurement of changes in specific volume that occur with change in temperature. It is useful in the field of fats and oils to detect or analyze phase transformations because fats expand when they melt and generally contract when they undergo polymorphic change to a more stable form (Braun 1955).

Fat dilatometers vary considerably in size and construction. Basically, they consist of a bulb that is attached to a calibrated capillary tube. The fat in the bulb is confined by a liquid such as a colored water solution or mercury. As the fat expands, the confining liquid is displaced into the capillary tube. When the volume is plotted against temperature a curve is obtained, the initial and final linear portions of which represent the completely solid and liquid states.

Nuclear Magnetic Resonance

The precise calculation of liquid and solid fats by dilatometric methods is difficult or even impossible in a complex mixture of glycerides. Low-resolution NMR appears to provide a solution to the problem (Chapman *et al.* 1960; Johnson and Shoolery 1962; Ferren and Morse 1963; Taylor *et al.* 1964).

The magnetic field strength at the center of a hydrogen resonance line always has the same value for a given frequency, but the shape of the line is influenced by the chemical and physical state of the sample. The width of the adsorption line is related to the mobility of the hydrogen in the sample or the mobility of the compound containing the hydrogen, and to the field homogeneity. The effective magnetic field strength at a nucleus is the sum of the applied magnetic field plus the field contributed by neighboring nuclei. In a solid where the nuclei are fixed rather rigidly, a nucleus is in a magnetic field significantly higher or lower than the applied field. Consequently, the line width of a solid is relatively wide. In a liquid, the molecules are in a state of thermal agitation, the field contributed by hydrogen nuclei averages out rapidly, and the absorption line is narrow. This difference in widths between solids and liquids can be used to determine the liquid content of fats. Conway and Johnson (1969) have shown that high resolution NMR can be used to determine unsaturation in single corn kernels. The procedures are useful in breeding programs to alter the fatty acid composition of corn oil.

Pulsed low resolution NMR is becoming a standard technique for the determination of solid fat content, Shukla (1983) described the critical parameters for development of a technique for the rapid, reproducible, reliable, and relatively simple determination of the solid fat content in cocoa butter equivalents. Continuous and pulsed NMR instruments were compared by Jewell (1983). They provide similar information except for complex food systems in which the continuous NMR is more informative. For routine testing of large numbers of samples, rapid pulsed NMR is more advantageous. Manufacturers of continuous-type instruments have developed a variable gate width technique to reduce the effect of interfering substances (moisture, metallic ions, etc). This is particularly applicable to measuring the fat content of cocoa where there is the risk of interference from contaminating iron.

In measuring the solid fat content, the greatest problem is how to precondition the sample before the measurement. Many workers use the dilation pretreatment in which the samples are melted to about 50–60°C cooled to 0°C, warmed to 26°C, held and cooled again to 0°C prior to starting the equilibration period at each temperature of measurement. With fats that exhibit complex polymorphism such as cocoa butter, this procedure is not acceptable and a more accurate procedure is to condition the samples at a temperature close to the minimum temperature of interest (Jewell 1983).

A new area of interest is measurement of the kinetics of fat crystallization by NMR. This involves equilibrating the sample under isothermic conditions and monitoring the solid fat build up. The assay is enhanced by using a microcomputer to collect the information and calculate it so as to obtain a printout of increasing solid content (Jewell 1983). There are significant differences in the types of polymorph, depending on the isothermal regimes. A small amount of shear significantly enhances the rate of crystallization and modifies the polymorph.

Mills and van de Voort (1981) compared a direct method of measuring solid fat content by wide-line NMR with the conventional indirect wide-line NMR procedure. The results for four fats did not differ for the two methods. The direct method required additional measurements, was more complex theoretically, and yielded more variable results because of the weak signal obtained at the wide gate setting of the instrument.

High-resolution proton NMR and natural abundance ^{13}C NMR, improved by recent developments in Fourier transform, were used to study lipid mixtures such as edible fats and oils and their hydrogenation products (Shoolery 1983). Unsaturation, linolenic acid content, fatty acid composition, and cis-trans ratios were determined, and the results were compared with those of chemical and gas chromatographic methods. Analysis of five single intact sunflower seeds for distribution of unsaturation demonstrated the usefulness of the method for single kernels. Proton NMR at 200 MHz offered a rapid way to determine the profile of types of lipids.

Refractive Index

Because the refractive index is constant, within certain limits, for each type of oil or fat, it can be used in identifying lipids. The use of a refractive index determination in the assay of lipid content was mentioned earlier in this chapter. Since the refractive index is related to unsaturation, it can be used in the determination of an iodine value. However, this correlation differs for various types of oils. A simple iodine-number refractometer for testing flaxseed and soybeans was described by Hunt *et al.* (1951).

Infrared Spectroscopy

Infrared spectroscopy has been used extensively in the analysis of lipids. In addition to structural analyses, it is useful in the identification of a lipid source; in the detection of adulteration (butterfat by plant lipids, cocoa by hydrogenated fats, durum wheat alimentary pastes by hard wheat semolina); in studies of autoxidation, rancidity, and drying properties of lipids; in following the effects of food processing on lipids; and in nutritional investigations on the effects of food composition and interaction with lipids. Extensive bibliographies on the use of infrared spectroscopy in lipid investigations were prepared by Wheeler (1954), O'Connor (1955, 1961A,B), Chapman (1960, 1965), Schwarz *et al.* (1957), Freeman (1957), Kaufman (1964), Kohn (1965), and Kohn and Laufer-Heydenreich (1966).

Ultraviolet Spectrophotometry

The common polyunsaturated fatty acids in untreated oils show no absorption peaks in the ultraviolet. If the double bonds in those oils can be rearranged to form a conjugated system, selective absorption will appear in the ultraviolet region, and can be used for analytical purposes. Polyunsaturated fatty acids are converted into conjugated isomers by heating in alkali, and the intensity of the resulting selective absorption is determined. The isomerization is enhanced at elevated temperatures. By using ethylene glycol as a solvent, the temperature can be increased and the assay performed more rapidly (Pitt and Morton 1957).

Mitchell *et al.* (1943) first published a detailed method by which both linoleic and linolenic acids can be determined. The procedure involves heating 10 ml of ethylene glycol containing 6.5% potassium hydroxide in a test tube in an oil bath. When the temperature reaches 180°C, 100 mg of fat or fatty acids is added, mixed, heated for 25 min, and cooled rapidly. After dilution with ethanol to a suitable volume, the absorbance is measured at 234 and 268 nm. A blank solution in the control cell of the spectrophotometer consists of an alkaline glycol solution heated and diluted as the assayed sample. The extinction at 268 nm is derived from linolenic acid only; the contribution of the linolenic acid at 234 nm can be computed, and the remaining extinction at 234 nm used to calculate the linoleic acid content. The method is empirical and suffers from the limitations of nonstoichiometrical analytical procedures. The ultraviolet method has undergone modifications that permit also a determination of arachidonic acid in a mixture with other unsaturated fatty acids. Use of ultraviolet absorption spectroscopy in the determination of a α-eleostearic acid in tung oil was described by O'Connor (1955).

For many years, differentiation among polyunsaturated fatty acids was based on generally nonspecific chemical reactions. The small differences in composition of these fatty acids are, however, reflected in relatively large differences in physical properties. Development of optical equipment and methods based on differences in vapor pressure, partition coefficients, adsorption, pattern of fragmentation by degradative methods, and electron resonance are some of the more powerful and specific tools that provide much more information both in scope and validity (Holman and Rahm, 1966).

Oxidation of polyunsaturated fatty acids is accompanied by increased ultraviolet absorption. The magnitude of the change is, however, not easily related to degree of oxidation because the effects vary with fatty acids. The test is, therefore, of a semiquantitative nature (Holman *et al.* 1945).

Gas–Liquid Chromatography

The use of gas–liquid chromatography (GLC) was introduced by James and Martin (1952) for the separation of normal saturated carboxylic acids up to 12 carbon atoms in chain length. Cropper and Heywood (1953) extended the method to the separation of the methyl esters of even-numbered fatty acids up to behenic acid. Since then a rapid expansion in the use of this method in lipid investigations and analyses has taken place. Today, methyl esters of fatty acids containing up to 34 carbon atoms, and minor components containing less than 0.05% of the original sample can be detected and estimated reliably. The importance of GLC has increased particularly with the development of polyester stationary liquids that enable separation of methyl esters of fatty acids varying in degree of unsaturation.

Advantages of GLC over other methods of fatty acid analysis have been so great that it has almost entirely replaced them in both research and routine determinations. The advantages include use of small quantities, relative specificity and simplicity, and adaptability to both qualitative and quantitative determinations.

In addition to the determination of total fatty acid composition, GLC can be used to determine the distribution and position of fatty acids in the lipid molecule, to study fat stability and oxidation by chemical and biological agents, to assay heat or irradiation damage to lipids, and to detect adulterants (e.g., hydrogenated fats in cocoa butter). Use of GLC in the analysis of lipids was reviewed by Kaufmann *et al.* (1961, 1962), Kohn (1964), Horning and Vandenheuvel (1964), and many others.

In most GLC determinations of fatty acids, the acids are first converted to the corresponding methyl esters before separation by GLC. Methods for obtaining the methyl esters of the fatty acids can be divided into those involving (1) transesterification of the glycerides in the presence of excess methanol and (2) saponification of the glycerides with alkali, isolation of the free fatty acids, and esterification of the acids. Transesterification methods are less time-consuming. In addition, the conditions normally employed in these procedures cause less isomerization of polyunsaturated fatty acids than occurs in saponification–esterification schemes (Jamieson and Reid 1965).

Combining GLC with Other Methods. To improve the usefulness of GLC, combined thin-layer and gas–liquid chromatographic systems (Kuksis 1966) are often used. Gas chromatography and mass spectrometry (Ryhage and Stenhagen 1960; Dutton 1961) can be used in combination for the separation of complex mixtures and for structural characterization of lipids. In both techniques, analyses of microgram quantities of samples in the vapor phase are performed. Direct combination of both techniques provides an exceptionally rapid, powerful, and versatile tool for the separation, characterization, and structural elucidation of components of complex mixtures. The ability to record rapidly several mass spectra of one emerging chromatographic peak may also be used to determine the efficiency of separation, the presence of impurities, and hydrolytic, thermal, or oxidative decomposition (Leemans and McCloskey 1967). On the other hand, interpretation of the enormous data provided by the combined instruments requires considerable experience, time, and effort.

Various additional attachments to gas chromatographs (e.g., infrared spectrophotometers) have extended the use of gas chromatography as a precise and informative technique. Alternatively, small-scale preparative chromatography can be combined with automated chemical identification. Perkins *et al.* (1977) discussed advances in the instrumental analysis of lipids, including use of HPLC, GLC, separation of geometrical isomers by GLC, mass spectrometry, and combined laser pyrolysis–gas chromatography–mass spectrometry.

Detection of Adulteration

The applications of the various physical methods discussed so far, especially gas chromatography, in the analysis of lipids in foods are too numerous to describe in detail. The usefulness of such analyses will be illustrated by a description of a few selected methods for determining fat adulteration.

If the triglyceride composition of native butter is known, adulteration with vegetable fat at the 1% level and with lard at the 3% level can be detected by GLC (Kuksis and McCarthy 1964). Adulteration of unknown butter samples can be detected at the 5–10% level due to considerable variations in chromatographic patterns of the fatty acids of butters of different origin. The ease of detection and identification depends on the fat added. Mixtures of coconut and lard, matching closely gas chromatographic patterns of butter fatty acids, can be prepared.

With the advent of modern processing techniques, animal fats can be incorporated with vegetable oils and processed into solid shortenings comparable in organoleptic and functional properties with all-vegetable shortenings. To detect such adulteration, Cannon (1964) developed a procedure for determining vegetable fat in butterfat. The method involves fat saponification followed by precipitation of the sterols with digitonin, acetylation of the digitonides, and separation of the sterol acetates by gas chromatography. The method gives satisfactory results for lard, soybean oil, and cottonseed oil. However, digitonides of beef tallow, palm kernel oil, and coconut oil are difficult to precipitate and acetylate. The method of Eisner et al. (1962) and Eisner and Firestone (1963) involves saponification, extraction of the unsaponifiable matter, separation of the latter on Florisil columns into sterols and other compounds, and gas chromatography of the sterol fraction after acetylation. This method gives satisfactory results but is time-consuming. Ettinger et al. (1965) developed a simplified and more rapid procedure. The total unsaponifiable matter (without separation on Florisil columns) is extracted with ether following saponification, acetylated, and subjected to gas chromatography analysis. The presence of 2.5% animal fat in vegetable oil could be detected.

Mani and Lakshminaryana (1976) reviewed methods for detecting adulteration of oils and fats, particularly the ones based on chromatographic separation. Included were detection of animal fats in vegetable fats, milk fat in other fats, animal fats in marine oils, lard or beef tallow in butterfat, marine oils in vegetable oils, and estimation of the composition of a great number of vegetable oil mixtures. It is a most unique compilation (text and tabulations) and list of about 200 references.

Gas chromatographic methods for the detection of shea fat in cocoa butter and chocolate on the basis of unsaponifiables, terpenic alcohols, and sterols were described by Derbesy and Richert (1979) and on the basis of sterols and steroids by Dick and Miserez (1980). Young (1984) described a new approach for interpreting triglycerol data obtained by GLC to determine cocoa butter equivalents in chocolate. It is based on the relationship between C_{50} and C_{54} triglycerides of cocoa butter of various origins. Wood (1984) presented the results of a collaborative study to evaluate a GLC method for the determination of milk fat in chocolate. The method, proposed for inclusion in the legislation of the European Economic Community, is based on the direct determination of butyric acid in the extracted fat.

The multivariate technique SIMCA was applied by Derde *et al.* (1984) to identify the origin of virgin olive oil on the basis of gas chromatographic profiles. Tentative screening for adulterants in oils and fats can be accomplished by measuring the change in the velocity of 1 MHz sound (Rao *et al.* 1980). According to Hussin and Povey (1984) and Povey (1984) the ultrasonic pulse echo method can be used to measure the solid content of oils and indicate the presence of phase transitions in the solid phase.

CHEMICAL ASSAY METHODS

Chemical characteristics of lipids are determined on the entire extract or on water-insoluble fatty acids isolated from a purified extract.

For the preparation of water-insoluble fatty acids, the fat is saponified with a potassium hydroxide solution (in ethanol) or by a glycerol–potassium hydroxide mixture, and the fatty acids in the formed soap are split by acidification with a mineral acid. The fatty acids are extracted with ether in a separatory funnel, washed, and dried cautiously. Many methods have been suggested for saponification. Some are rapid and effective, but likely to cause more or less extensive modification of the fatty acids (isomerization, oxidation, volatilization, and degradation), or formation of some mono- and diglycerides. The usefulness of the shortened methods depends on their agreement with standard procedures.

Determination of Impurities

Impurities in fats, oils, and fatty acid products are mainly moisture, volatile compounds, insoluble matter, unsaponifiable matter, trace metals, and their soaps. The term MIU (moisture, insoluble, unsaponifiable) is frequently used to designate the amount of nonfatty constituents of crude oils and other fatty acid products where settlement is on the basis of oil or acid content. It also figures predominantly in the trading rules of the various oil trading organizations (Rodeghier 1955). The total MIU is considered valueless material except to those interested in the recovery of sterols and tocopherols from the unsaponifiable fraction of fatty acids split from the soapstock of soybean oils.

The insoluble matter found in fats and oils consists of dirt, meal, and any other substances insoluble in kerosene and petroleum ether.

The unsaponifiable matter found dissolved in fats or fatty acids is that material that cannot be saponified by potassium hydroxide. The unsaponifiables include sterols, higher alcohols, and some hydrocarbons. Oil refining removes most of those substances. The analytical determination is made by weighing a 5-g sample into a 250-ml Erlenmeyer flask. Then, 30 ml of alcohol and 5 ml of a 50% potassium hydroxide solution are added, and the mixture refluxed until saponified. The saponified material is then transferred to an extraction cylinder; the flask is rinsed with alcohol and washed into the cylinder with alcohol, water, and petroleum ether. After cooling and shaking, the petroleum ether extract is siphoned off into a 500-ml separatory funnel, and the extraction repeated. The combined petroleum ether extracts are washed with 10% ethanol, dried, and weighed in a tared beaker as unsaponifiable.

Oxidized fatty acids are determined by acidifying with hydrochloric acid the saponified fat from the assay of unsaponifiable matter. Liberated free fatty acids that are insoluble in petroleum ether are considered oxidized fatty acids.

The presence of as little as 0.3 ppm iron is detrimental to the quality and stability of oils. Similarly, other metallic traces are undesirable. The determinations are made by accepted procedures.

Fatty acids react readily with alkali to produce soaps; the soaps are mainly produced during oil refinement. Levels of 5 ppm soap are detrimental to refined oil and impair significantly the quality of hydrogenated oils. In oil processing, soaps are removed by water-washing the neutral oil, followed by thorough drying before bleaching with an adsorbent. The analytical determination involves alcohol or dilute hydrochloric acid extraction of the soap from the oil followed by quantitative determination of the sodium ion. A rapid conductivity assay of soap in oil is also available.

Determination of Unsaturation

One of the most important analytical determinations that an oil chemist makes frequently is the measurement of unsaturation of an oil. This determination is important both in classification of fats and oils for trade and use, and for control of manufacturing processes. The generally accepted parameter for expressing the degree of carbon-to-carbon unsaturation of a fat, oil, or derivative is the *iodine value* (or *iodine number*), defined as the grams of iodine that add to 100 g of sample (Allen 1955). The results are expressed in terms of iodine whether iodine or some other halogen is actually used. Since all halogens add to carbon–carbon double bonds, most of the methods make use of this property.

Basically, the determination consists of adding a halogen to a weighed quantity of sample and determining the amount of reacting halogen. There are several methods for determining the iodine value, but the most important are the Wijs (developed in 1898) and the Hanus (from 1901). The Wijs method is probably the most widely used and is believed to yield results closer to theoretical values than any other method. Hanus results range from 2 to 5% below Wijs, but the Hanus reagent has the advantage of being more stable (Mehlenbacher 1958). Sources of errors in the determination of iodine values by the Wijs method were scrutinized by Oh (1982); strict adherence to the standard method is mandatory for reproducible results.

In the Wijs method, 0.1–3.0 g (depending on the expected iodine value) of sample is dissolved in 15 ml carbon tetrachloride; to this is added 25 ml of the Wijs solution (commercially available, or prepared by dissolving 9 g iodine trichloride in a mixture of 700 ml glacial acetic acid and 300 ml carbon tetrachloride). The conical flask is closed with a ground-glass stopper, mixed, and allowed to stand at about 20°C for 1 hr in the dark. Then, 20 ml of a 10% aqueous potassium iodide solution and about 150 ml water are added, and the unreacted halogen is titrated with an accurately standardized thiosulfate solution in the presence of starch towards the end of the titration.

To shorten the reaction time, a 2.5% mercuric acetate solution in glacial acetic acid is added to the Wijs reaction mixture. The addition of a catalyst shortens the reaction time from 1 hr to 3 min. For fats with a high iodine value (above 50), the difference between the two methods is negligible. The shortened method gives more meaningful results than the standard Wijs procedure in fats with hydroxy fatty acids, but is somewhat unreliable for Chinese tung oil and other highly conjugated oils (Cocks and Van Rede 1966).

The iodine value (IV) is calculated from the difference in titration of a blank and sample according to the equation

$$IV = \frac{(B - S)N \times 12.692}{\text{weight of sample}} \tag{36.3}$$

where B is the titration of blank, S the titration of sample, and N the normality of $Na_2S_2O_3$ solution.

For a *pure* oil or fat, the theoretical iodine value can be calculated as follows:

$$IV = \frac{2 \times 126.92 \times \text{no. double bonds} \times 100}{\text{molecular weight}} \tag{36.4}$$

Alternatively, if you know the IV and molecular weight, you can calculate the number of double bonds:

$$\text{no. double bonds} = \frac{\text{IV} \times \text{molecular weight}}{2 \times 126.92 \times 100} \qquad (36.5)$$

Nondrying oils have an iodine value below 100; drying oils 130 to 200; and semidrying oils have intermediate values. Theoretically, the iodine values of oleic, linoleic, and linolenic acids are respectively 89.9, 181.0, and 273.5. The iodine values of free fatty acids are higher than those of glycerides. In a mixture of free fatty acids and glycerides, for each percent of free fatty acids the iodine value increases 0.00045 × iodine value (Cocks and Van Rede 1966).

Thiocyanogen $(SCN)_2$, adds to double bonds in a manner similar to halogens. However, with polyunsaturated acids the addition is not complete. Thiocyanogen adds 1 mole per mole to dienoic acids, and 2 moles per mole to trienoic acids. This permits the calculation of oleic, linoleic, and linolenic acids from the thiocyanogen value, iodine value, and a separate assay of saturated fatty acids.

Several tests are available for specific polyunsaturated acids. These are based on the formation of crystalline polybromides varying in melting point when the fatty acids are brominated. The methods vary as to temperature, solvent used to dissolve the fatty acids, and condition of bromination. By careful standardization, the results are reproducible and provide information on composition of some fats.

Determinations of iodine value give a reasonably quantitative measure of unsaturation if the double bonds are not conjugated with each other or with a carbonyl oxygen, and if the determination is carried out under specified conditions as to the excess of halogen reagent, time of reaction, and exclusion of light (Allen 1955).

To overcome some of the limitations of the halogen methods, quantitative hydrogenation can be used. Hydrogenation is used to measure the unsaturation of acetylenic or conjugated double bonds. Although such fats do not absorb halogen readily, the addition of hydrogen can be practically quantitative. Essentially the methods consist of catalytic (nickel, palladium, or platinum) hydrogenation of a heated sample. The volume of hydrogen absorbed is determined from measurements before and after the reaction and reduced to standard conditions. The results can be expressed as mole of hydrogen per mole of sample, or can be calculated to an iodine value basis. The latter is known as the hydrogen-iodine value.

There is, as yet, no standard, rapid, and precise method for the determination of the hydrogen value. Miwa et al. (1966) developed a procedure for the quantitative determination of unsaturation in oils by using an automatic-titrating hydrogenator. The method utilizes a catalyst prepared by *in situ* treatment of platinum salts with sodium borohydride, *in situ* generation of hydrogen from sodium borohydride, and a valve at the tip of a buret that automatically introduces standardized sodium borohydride solution into the reaction mixture only as long as hydrogenation is proceeding.

Saponification Value

The *saponification value* is a measure of the amount of alkali required to saponify a definite weight of fat. It is expressed as milligrams potassium hydroxide required to saponify 1 g of fat, i.e., to neutralize the free fatty acids and the fatty acids present in the form of glycerides. The *saponification equivalent* is the amount of oil or fat saponified by 1 g-equivalent of potassium hydroxide and is equal to

$$56,108/\text{saponification value} \qquad (36.6)$$

The saponification value was originated by Koettsdorfer and is sometimes known by this name. The procedure involves saponifying under reflux 4 g filtered oil with 50 ml of a 0.5 *N* potassium hydroxide solution in 96% ethanol, for 30 min. Excess potassium hydroxide is determined by back-titration with an aqueous, standardized 0.5 *N* hydrochloric acid in the presence of phenolphthalein.

The saponification value is an indication of the average molecular weight of fat. For pure fatty acids, the saponification value equals the *acid value*. The *ester value* is the difference between the saponification value and the acid value. In oils and fats, the ester value is a measure of the amount of glycerides present.

Identification of Specific Lipids

There is no organized, foolproof scheme for the qualitative analysis of fats, and the problems of identifying individual fats and oils is quite complicated. This is particularly true in the case of mixtures and processed fats. Even with the most sophisticated instruments it is quite possible to encounter mixtures that defy identification of the source of the oil. Admittedly, the availability of more specific, meaningful, and reliable instrumental techniques has simplified identification and detection of adulteration. Some of the methods were described in an earlier section. Additional chromatographic procedures and spectrophotometry (especially infrared) are used most commonly in combination with the determination of physical and chemical constants (saponification value, iodine value, melting point, and others) that are constant and typical for individual fats and oils.

In addition, there are several specific tests for individual oils or for certain functional groups. Such tests are much more useful in detecting the presence of a specific adulterant than in ascertaining the purity of an oil. The usefulness of such tests is sometimes limited by the destruction of a specific reaction through processing. These specific tests are described in detail in the references listed at the beginning of this chapter. They are discussed briefly here.

Squalene, an unsaturated aliphatic hydrocarbon, occurs in higher concentrations in olive (and fish) oil than in most other oils. The amount of squalene, isolated from the unsaponifiable matter by column chromatography, can be used to estimate roughly the olive oil content of vegetable oils. In the Bellier test for peanut oil, insolubility of arachidic and lignoceric acids in 70% ethanol is measured. In the Halphen test, used to detect cottonseed oil, a reddish color develops on heating the oil with amyl alcohol and a solution of sulfur in carbon disulfide. Kapok oil gives a similar reaction. Processing (heating, refining, or hardening) reduces color intensity. Fats of animals fed cottonseed cake or meal may give a positive reaction.

Melting castor oil-containing oils or fats with dry potassium hydroxide, followed by precipitation of the salts from the aqueous solution of the soap and acidification, yields fatty acids from which characteristic sebacic acid crystallizes.

The difference in the melting point of the glycerides and fatty acids of lard is much greater than the difference in tallow or hardened fats. The *Bomer value* (BV) is computed as

$$BV = A + 2(A - B) \qquad (36.7)$$

where A is the melting point of glycerides and B the melting point of fatty acids. The BV is at least 73 in lard, and significantly lowered by a mixture of 10% or more of tallow and hardened fats.

Shaking a sesame oil-containing mixture with hydrochloric acid and an ethanol solution of furfural (or sucrose) gives a stable pink color that can be used for quantitative determination by the sensitive Baudouin reaction. The Fittelson test for the detection of teaseed oil in olive oil is based on the formation of an unstable deep red color after addition of acetic anhydride, chloroform, and sulfuric acid, followed by cold (5°C) ether.

The presence of adulterants in milk fat has been the subject of many investigations. The adulteration can be established in two general ways. One involves the identification of a foreign material in the fat; the other, the demonstration of an unusual concentration of specific regular milk fat components (Kurtz 1965).

The difference between the sterols of animal and vegetable fats affords a positive means of identifying adulteration by a vegetable fat. Sometimes, the identification is complicated by the presence of several phytosterols

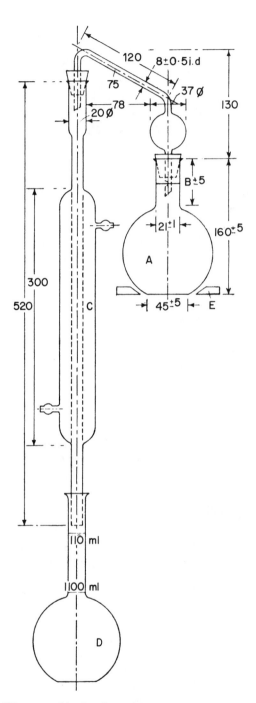

Fig. 36.5. Distilling assembly for determining Reichert–Meissl value and Polenske number.

including those that have physical and chemical properties similar to those of cholesterol. The use of one of those methods was described earlier (Eisner *et al.* 1962).

Vitamin E exists in several forms. Brown (1952) found only α-tocopherol in cow's milk, whereas the vegetable oils he examined also had β-, γ-, or σ-tocopherols.

The high proportion of the soluble, volatile fatty acids—particularly butyric acid—is the most prominent difference between milk fat and other animal fats. The Reichert–Meissl value, Polenske number, and butyric acid determination can be used to establish gross adulteration of milk fat. The Reichert–Meissl value is equivalent to the number of milliliters of a decinormal sodium hydroxide solution required to neutralize the volatile, soluble acids obtained under specified conditions (Fig. 36.5) from 5 g of fat. Milk fat contains more of these acids than any of the fats from which it might be desirable to distinguish it. The Canadian Standard for milk fat requires a Reicher–Meissl value of at least 24. The Polenske number equals the number of milliliters of a decinormal solution of sodium hydroxide required to neutralize the volatile, insoluble acids obtained from 5 g of fat. The amount of such acids in milk fat is small, but it is high in fats with a high caprylic acid content (e.g., coconut oil). The Canadian Standard requires that the Polenske number not be above 3.5 and not greater than 10% of the Reichert–Meissl value.

It is difficult to demonstrate on the basis of the Reichert–Meissl and Polenske determinations or direct butyric acid assays the adulteration of milk fat by small amounts of other fats, especially in view of some of the extreme values of those parameters in some unadulterated samples. For example, Zehren and Jackson (1956) reported that 500 samples of butterfat from 42 locations from 25 states in the United States had Reichert–Meissl values of 24.2–33.6 (avg 29.0) and Polenske values of 1.1–3.0 (avg 1.9). The refractive index varied from 1.4531 to 1.4557 (avg 1.4540). Analyses by Keeney (1956) of these samples showed a butyric acid content of 9.6–11.3 mole % (avg 10.9). The significance of the latter value is somewhat limited as high-concentrate, low-roughage diets may decrease significantly the proportion of butyric acid.

AUTOMATED LIPID ANALYSES

Fully automated systems capable of handling both sample preparation and actual analysis have been developed in clinical laboratories (Faust and Keller 1976; Burns 1981; Waiser and Bartels 1982; Dessy 1983; Roy 1984). The *Journal of Automatic Chemistry* (published quarterly by the United Trade Press, London) covers all aspects of automation and mechanization in analytical, clinical, and industrial instruments.

As in other areas, automated lipid analyses are based mainly on experience gained in clinical chemistry. Advances in automated analyses of blood cholesterol, triglycerides, and phospholipids were reviewed by Levine (1967). Cholesterol can be determined by the method of Levine and Zak (1964) that involves heating the sample at 95°C with a solution containing ferric chloride, sulfuric acid, and acetic acid, and measuring the color at 520 nm.

For the determination of triglycerides, two methods are available. In the procedure of Lofland (1964), phospholipids are removed on an ion-exchange column and the eluate is saponified with alcoholic potassium hydroxide. After evaporation and addition of sulfuric acid, the glycerol formed is oxidized to formaldehyde with periodate. Excess periodate is destroyed with sodium arsenite, and the formaldehyde reacted at 95°C with chromotropic acid to give a colored product that is measured at 570 nm. The procedure of Kessler and Lederer (1965) is based on the reaction of Hantzsch involving condensation of an amine, beta diketone, and an aldehyde. A phospholipid-free extract is prepared by adding an ion-exchange resin to the isopropanol extract. In addition, Lloyds' reagent and a copper–lime mixture are added to remove chromogens and glucose interference. The extract in isopropanol is saponified with aqueous potassium hydroxide, and mixed with periodate and the acetylacetone reagent. The formaldehyde (from the oxidation of glycerol) is condensed with acetylacetone reagent to produce 3,5-diacetyl-1,4-dihydrolutidine, which is measured in a fluorometer. Up to 20 samples can be tested per hour.

The determination of phospholipids according to Zilversmit and Davis (1960) requires precipitation with 10% trichloroacetic acid and separation of the phospholipid as a complex with protein. The precipitate is digested with perchloric acid, and the digest is assayed by a colorimetric molybdate method for phosphorus. Whitley and Alburn (1964) digested the precipitate in a continuous flow digest or with a mixture of perchloric and sulfuric acids in the presence of vanadium pentoxide as a catalyst. After neutralization, the digest is mixed with molybdate and hydrazine, and the color measured at 815 nm in a colorimeter equipped with silicon photocells.

FAT STABILITY AND RANCIDITY

The term *rancidity* is used to describe development of objectionable flavors and odors. As a result of these changes, consumer acceptance of the food is lowered. Rancidity may be caused by either hydrolytic or oxidative changes in the fat. Hydrolytic rancidity involves chemical or enzymatic hydrolysis of fats into free fatty acids and glycerol. Oxidative rancidity involves the addition of atmospheric oxygen in the presence of enzymes or certain chemicals. Hydrolytic activity is important in dairy products and coconut items, which contain glycerides of low-molecular-weight fatty acids such as butyric, caproic, caprylic, or capric. Cereal lipids, on the other hand, contain high-molecular-weight fatty acids; when hydrolyzed these do not produce the same type of off-flavors and odors as are produced by the hydrolysis of low-molecular-weight glycerides.

Fat Acidity

Deterioration of grain and milled products is accompanied by increased acidity. The acids formed include free fatty acids, acid phosphates, and amino acids; during the early stages of deterioration, fat acidity increases at a much greater rate than either of the other two types or all types of acidity combined.

The organic acidity of fats and oils can be expressed in several ways. The *acid value* is the number of milligrams of potassium hydroxide required to neutralize 1 g of fat or oil. The acidity of edible oils is sometimes expressed as milliliters N sodium hydroxide solution used to neutralize fatty acids in 100 g fat. The free fatty acid (FFA) content also is expressed as percentage of weight of a specified fatty acid (either oleic with a molecular weight of 282 or an average and specified molecular weight appropriate to the nature of the analyzed fat or oil). Acid value can be converted to FFA (expressed as oleic acid) by the following formula:

$$\text{acid value} = 1.99 \times \%\text{FFA} \qquad (36.8)$$

$$\%\text{FFA} = 0.503 \times \text{acid value} \qquad (36.9)$$

For the determination of free fatty acids, 4–20 g fat (depending on the expected acidity and normality of selected titrant) is dissolved in a neutralized solvent and titrated with a standardized aqueous solution of sodium hydroxide in the presence of phenolphthalein (or alkali-blue 6 B for dark-colored oils) as indicator. The acidity of the oil is calculated from the amount of consumed titrant.

Originally, ethanol was (and in the AOCS and British Standard Methods is) used as solvent. As many oils and fats are only slightly soluble in ethanol, the titration involves heating to 60–65°C, or even boiling. At the high temperature required, mono- and diglycerides may be saponified and some fatty acids volatilized. Using a mixture of solvents (generally ethanol and ether, 1:1) makes it possible to titrate at room temperature. On the other hand, in large laboratories that use large amounts of solvents, it is simpler to recover a single solvent. Also, the low solubility of an oil in ethanol is sometimes advantageous in the titration of dark-colored products. Pyridine-denatured alcohol cannot be used as it has a strong buffering action.

The American Association of Cereal Chemists recommends three procedures for the determination of fat acidity in grain. In the basic method, grain samples ground so that 90% or more pass a 40-mesh sieve are extracted with petroleum ether within 1 hr of grinding to forestall enzymatic breakdown of lipids. The extract is evaporated, redissolved in a benzene–ethanol mixture (1:1) containing 0.02% phenolphthalein, and titrated with 0.0178 N potassium hydroxide to the endpoint, matching color of a standard prepared from dilute solutions of potassium dichromate and potassium permanganate. In the rapid method, preground material is extracted with benzene for 4 min in a special grinder–extractor; or the ground sample is shaken mechanically for 30 min or intermittently by hand for 45 min. The extract is filtered, and an aliquot is titrated with standard potassium hydroxide. In the colorimetric method, the benzene extract (or petroleum ether extract after evaporation and dissolving in benzene) is mixed with a cupric acetate solution, and after filtration the color is determined at 640 nm. The fat acidity value (mg KOH required to neutralize FFA from 100 g grain, dry matter basis) is calculated from a calibration curve of pure oleic acid passed through the colorimetric procedure.

Detection of Oxidation Products

The *peroxide value* is the most commonly used assay of oxidation in fats and oils. Many methods have been devised for its measurement (Lea 1962). Barnard and Hargrave (1951) reviewed the value of the various methods and found the iodometric procedures were most commonly used. The original method Lea (1931) has been improved by the exclusion of oxygen from reagents and reaction flask (Lea 1946).

The peroxide value is commonly determined by measuring the amount of iodine liberated from a saturated potassium iodide solution at room temperature, by fat or oil dissolved in a mixture of glacial acetic acid and chloroform (2:1). The liberated iodine is titrated with standard sodium thiosulfate, and the peroxide value is expressed in milliequivalents of peroxide-oxygen per kilogram fat. The *Lea value*, often mentioned in the literature, is expressed in millimoles of peroxide-oxygen per kilogram fat; it is numerically half the peroxide value. The method is applicable to oils, fats, and margarine. In the latter, the oil and water phase must be separated before analysis. The peroxide value is an indicator of the products of primary oxidation. It measures rancidity or degree of oxidation but not stability of a fat.

Stability Tests

Several methods have been developed to determine the stability (or susceptibility) of an oil to the development of rancidity. The oven or Schaal Method (Pool 1931) was originally designed to evaluate the stability of shortening in baked products. The odor of a sample stored at 63°C is observed periodically until rancidity is detected organoleptically. In the accelerated test, 1 day of incubation at the elevated temperature corresponds to 6–10 days of incubation at 21°C. In the oxygen absorption test, the sample is heated in the presence of oxygen until the rate of absorption of oxygen undergoes a definite increase. The method has the advantage of being a direct measure and does not depend on the rate of formation of decomposition products (Mehlenbacher 1958).

In the active oxygen method (AOM), the fat is heated, dried, and filtered air is blown through it until the peroxide value increases to some value that has been previously shown to indicate the onset of rancidity (Fig. 36.6). The results are influenced by many factors including temperature, light, availability of oxygen, surface factors, the presence of natural and synthetic antioxidants, the nature of the fat, and the presence of impur-

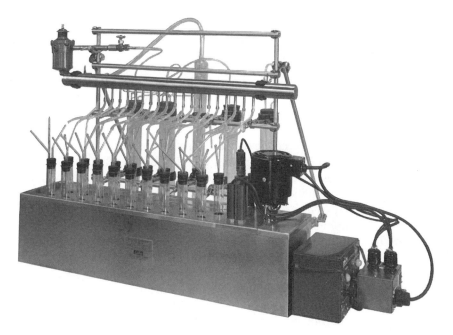

Fig. 36,6. Apparatus for determining fat stability by the active oxygen method (AOM). (Courtesy LaPine Scientific Co.)

ities, especially trace metals. The peroxide values achieved by the AOM—tested samples, at which the fat is rancid by organoleptic tests, vary with the nature of the fat and range from about 20 meq per g for lard to over 100 for vegetable oils (Dugan 1955). The AOM test is rapid; the results can be obtained within about $\frac{1}{20}$ of the time required in the oven test. The correlation with actual shelf life is, however, better in the oven test than in the AOM test, because the temperature is closer to normal storage conditions, rancidity rather than peroxide values is measured, and extreme conditions of forced-air circulation are avoided. Baumann (1959) pointed to difficulties in correlating peroxide values and chemical tests of fat stability with graders' judgments of off-flavor and off-odor in oil stored for 28 months.

Pohle et al. (1962, 1963) compared several analytical methods for the prediction of relative stability of fats and oils, and proposed a rapid oxygen bomb method for evaluating the stability of fats and shortening. A comparison of data from several accelerated laboratory tests with organoleptic evaluation of samples stored at 85°F (Pohle et al. 1964) indicated that different types of fat behaved differently. Consequently, the laboratory tests cannot be used as an index of shelf-life stability, except for a given type of formulation of fat for which the relationship between the laboratory test and shelf-life stability has been established.

Miscellaneous Tests

The odors and flavors associated with typical oxidative rancidity are mostly due to carbonyl-type compounds. The shorter-chain aldehydes and ketones isolated from rancid fats are due to oxidative fission and are associated with advanced stages of oxidation. The carbonyl-type compounds develop in low concentration early in the oxidative process.

The Kreis test is a sensitive indicator of the early stages of the oxidative process. The substance responsible for this reaction is epihydrinaldehyde. The Kreis test often indicates changes that are not necessarily consistent with fat stability as measured by other methods. Therefore, the test should be used as a supplement to (and not instead of) other tests such as peroxide value. In addition to epihydrinaldehyde, several other compounds (e.g., malonic dialdehyde and acrolein treated with hydrogen peroxide) yield a colored product with phloroglucinol. However, as the presence of these compounds has not been demonstrated in oxidized fat, they do not appear in the accepted mechanism of oxidation (Patton et al. 1951), and they seem to result from secondary oxidation products formed from the decomposition of peroxides.

Numerous workers have attempted to make quantitative evaluations of rancidity through the use of carbonyl tests. The method of Lapin and

Clark (1951) is based on the formation of a colored quinoidal ion of 2,4-dinitrophenylhydrazone in a solution of a base. An extensive study and application of the method was made by Henick *et al.* (1954). Their method involved the formation of the 2,4-dinitrophenylhydrazones in benzene solutions with trichloroacetic acid as catalyst. The absorbance at 430 and 460 nm can be used for the simultaneous determination of saturated and allenic carbonyls.

Thiobarbituric Acid Test.   A chemical test that deserves mention because of the great variety of uses to which it has been put is the thiobarbituric acid test (TBA). Kohn and Liversedge (1944) originally showed that aerobic oxidation products of animal tissues give a color reaction with thiobarbituric acid. The reaction was traced to the oxidation of unsaturated fatty acids, mainly linolenic (Bernheim *et al.* 1947; Wilbur *et al.* 1949). The pigment produced in the sensitive color reaction is a condensation product of two molecules of TBA and one of malonic dialdehyde. As proposed by Yu and Sinhuber (1967), the reaction occurs as follows:

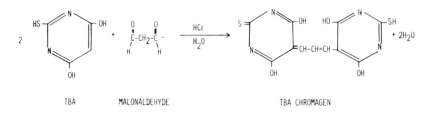

| TBA | MALONALDEHYDE | | TBA CHROMAGEN |

The TBA test was originally devised for the evaluation of dairy products (Patton and Kurz 1951) and has been found useful in testing many vegetable and animal fats.

The test is performed by treating the fat in a benzene or chloroform solution with the TBA reagent in an aqueous acetic acid solution. After shaking, the aqueous layer is separated and heated in a boiling water bath for 30 min to develop a pink chromogen with maximum absorbance at 532 nm. The intensity of the color is a measure of the degree of oxidation.

During the early stages of oxidation, the amounts of TBA-reactive substances in oxidized unsaturated fatty acids are closely correlated with peroxide value, oxygen uptake, and diene conjugation (Dahle *et al.* 1962). Kwon and Olcott (1966) showed that malonaldehyde is the principal TBA-reactive substance in oxidized methyl linolenate, fatty acid esters, and squalene. The kinds of TBA-reactive substances produced depend on the substrate and oxidation conditions. Because these compounds may undergo extensive modification at advanced stages of oxidation, the TBA test is useful as a measure of lipid oxidation only during the initial stages of oxidation. The test is most useful for detecting incipient oxidation of lipids rich in methylene interrupted, three or more, double bonds. The color formation is empirical, as the color yield from each fatty acid varies.

The test has been applied by many workers directly to a lipid-containing material without prior extraction of fat, the red pigment being subsequently extracted and measured. Such methods are, however, open to criticism as a variety of substances, other than the oxidation products of lipids, may give misleading color reactions with TBA. To overcome that difficulty, it may be advisable to purify TBA-reactive compounds by steam distillation, and then determine the oxidation products in the distillate. The TBA reagent is unstable in the presence of acid, peroxide, and heat, conditions under which it is generally used. A procedure in which an aqueous extract of a food or emulsified fat is allowed to react with TBA at room temperature for 15 hr without acid (in place of 100°C for 10–50 min) has been recommended (Tarladgis et al. 1962; Lea 1962). Several improvements of the method were described by Yu and Sinnhuber (1967).

Some Recent Investigations

Hydroxides are the major initial reaction products of fatty acids with oxygen, as shown by the following reactions:

Initiation
$$RH + O_2 \rightarrow R\cdot + \cdot OH$$

Propagation
$$R\cdot + O_2 \rightarrow ROO\cdot$$
$$ROO\cdot + RH \rightarrow ROOH + R\cdot$$

Termination
$$R\cdot + R\cdot \rightarrow RR$$
$$R\cdot + ROO\cdot \rightarrow ROOR$$
$$ROO\cdot + ROO\cdot \rightarrow ROOR + O_2$$

RH refers to any unsaturated fatty acid in which the hydrogen is labile by being on a carbon atom adjacent to a double bond. R· refers to a free radical formed by removal of a labile hydrogen (Gray 1978). Subsequent reactions control both the rate of the reaction and the nature of the formed products (Fig. 36.7). These compounds may be responsible for the development of off-flavors or for further reactions with other food components, e.g., proteins.

Rancidity in foods was reviewed in books edited by Galliard and Mercier (1975), Simic and Karel (1979), Allen and Hamilton (1983), Barnes (1983), and Gunstone and Norris (1983). Some of the subjects covered include the chemistry, measurement, and nutritional significance of ran-

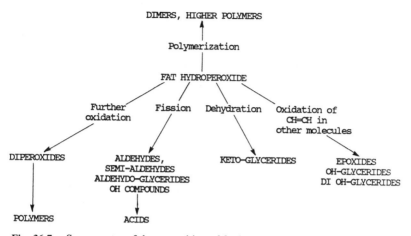

Fig. 36.7. Some routes of decomposition of fat hydroperoxide. (From Gray 1978.)

cidity. In addition, rancidity in various types of foods (cereal products, snack foods, creams and desserts, dairy products, and confectionery products) is discussed.

Grosch (1975) reviewed results and hypotheses concerning the autoxidation of unsaturated fatty acids at low temperatures. The following aspects of oxidative fat deterioration were discussed: (1) reactions during the induction period (metal catalysis, formation of singlet oxygen); (2) radical chain reactions of oxygen in the ground state (multiplication of peroxides); (3) precursors of aroma-active carbonyl compounds (peroxyl radicals, monohydroxy peroxides, polar peroxides); and (4) deductions and hypotheses based on the analyses of deterioration. The analyses included oxygen absorption, peroxide value (PV), diene absorption, thiobarbituric acid number (TBA), benzidine value, and heptanal value.

Gray (1978) listed the value and limitations of the following methods for assessing lipid oxidation: chemical (PV, TBA, Kreis test, total and volatile carbonyl compounds, and oxirane compounds) and physical (conjugated diene, fluorescence, infrared spectroscopy, polarography, gas chromatography, and refractometry). He concluded that there is no ideal method that correlates well with organoleptic changes during the entire course of autoxidation. The method of choice depends on several factors including the nature and history of the oxidized sample, the type of information required, the time available, and the test conditions (including equipment). It was postulated that the ultimate criterion for the suitability of any test is its agreement with sensory perception of rancid odors and flavors. A similar conclusion was reached by Pardun (1975) with regard to analytical methods for the assay of freshness and stability of animal fats.

Roden and Ullyot (1984) and Hung and Slinger (1981) reported on quality control in edible oil processing and on chemical methods to assess oxidative quality and storage stability of feeding oils. The PV, TBA, free fatty acids (FFA), anisidine value (AV), and carbonyl value (CV) were compared for their sensitivity and practicality in assessing the oxidative quality of salmon oil, soybean oil, canola oil, and canola soap stocks. The PV was the simplest, most sensitive, and practical indicator of changes in the four oils oxidized by bubbling air at room temperature for 792 hr. The PV, AV, TBA, and CV were sensitive to oxidation of herring oil beyond the induction period. The AV was most sensitive for detecting changes in canola oil aerated for 240 hr at 100°C. No apparent relationship was found between the TBA values and PVs of canola, corn, sunflower, and olive oil and shortening and lard tested by de Man and de Man (1984) in a home-built version of the automated, accelerated rancidity test.

The significance of temperature in accelerated shelf-life testing was reviewed by Ragnarson and Labuza (1977). Temperatures of 98–100°C may yield unpredictable variations (both under- and over-estimation) and cause formation of complicated secondary reactions.

Marcuse and Johansson (1973) investigated systematically the TBA reactivities of several aldehydes, products of lipid autoxidation. All the aldehydes studied formed a yellow 450-nm pigment, but only 2,4-alkadienals and 2-alkenals formed a red 530-nm pigment. In the case of predominant unsaturated aldehyde formation, determination of the 530-nm pigment is preferable. If alkanals are predominant, the determination of the 450-nm pigment gives higher sensitivity. Kakuda *et al.* (1981) developed a 5-min selective and sensitive HPLC test for the quantitation of malonaldehyde in aqueous distillates; the test assays were highly correlated with the standard TBA test. Methods to reduce the interference of sucrose in the detection of TBA-reactive substances were described by Shlafer and Shepard (1984). An improved method for determining TBA-reactive substances in fish tissue was described by Ke *et al.* (1984). The sample is digested in acidic media and separated by specific reflux distillation. The compounds in the distillate are estimated by spectrophotometric measurement at 538 nm after reaction with TBA.

According to Fiorti (1977), determination of oxygen absorption and PV is a satisfactory index of flavor deterioration in fats, oils, and simple model systems but not in actual food systems. In the latter, volatile (CO_2, pentane) and nonvolatile (anisidine-reactive) peroxide decomposition products yield a more reliable picture of the organoleptic status of the food.

Fritsch *et al.* (1979) described an instrument that measures changes in the dielectric constant as a measure of frying oil deterioration. Statistically significant correlations were obtained between instrument readings and increases in polar materials, color, PV, diene content, and free fatty acids, and decrease in iodine value. A portable modification of the instrument was described by Graziano (1979).

BIBLIOGRAPHY

Ahrens, E. H., and Craig, L. C. (1952). Separation of the higher fatty acids. *J. Biol. Chem.* **195**, 299–310.

Alexander, D. E., Silvela, L. S., Colins, F. I., and Rodgers, R. C. (1967). Analysis of oil content of maize by wide-line NMR. *J. Am. Oil Chem. Soc.* **44**, 555–558.

Allen, C. F., Good, P., Davis, H. F., Chisum, P., and Fowler, S. D. (1966). Methodology for separation of plant lipids and application to spinach leaf and chloroplast lamellae. *J. Am. Oil Chem. Soc.* **43**, 223–231.

Allen, J. C., and Hamilton, R. J. (eds.). (1983). "Rancidity in Foods." Applied Science Publishers, London.

Allen, R. R. (1955). Determination of unsaturation. *J. Am. Oil Chem. Soc.* **32**, 671–674.

Anon. (1950). "Deutsche Gesellschaft fur Fettwissenschaft." D. G. F. Einheitsmethoden. Munster, Germany.

Anon. (1954). "Standard Methods for the Analysis of Oils and Fats." International Union of Pure and Applied Chemistry, Paris.

Anon. (1960). "Standard Methods for the Examination of Dairy Products," 11th ed. Am. Public Health Assoc., New York.

Anon. (1967). IUPAC-IUB Commission on Biochemical Nomenclature (CBN). The nomenclature of lipids. *Eur. J. Biochem.* **2**, 127–131.

Anon. (1982). "Official and Tentative Methods of the American Oil Chemists Society," 3rd ed. Amer. Oil Chemists Soc., Chicago.

Anon. (1983). "Cereal Laboratory Methods," 8th ed. Am. Assoc. Cereal Chemists, St. Paul, MN.

AOAC. (1980). "Official Methods of Analysis of the Association of Official Analytical Chemists," 13th ed. Assoc. Off. Anal. Chemists, Washington, DC.

Arora, K. L., and Rajorbia, G. S. (1978). The volumetric determination of fat in milk and milk products—a review. *Indian J. Anim. Res.* **12**, 59–70.

Babicheva, O. I., and Gorelik, L. D. (1968). Determination of the fat content in canned meat and vegetable foods by a refractometric method. *Konserv. Ovoshchesush. Prom.* **23**, 35–37. [*Chem. Abstr.* **68**, 113376z.]

Barnard, D., and Hargrave, K. R. (1951). Analytical studies concerned with the reactions between organic peroxides and thio-esters. I. Analysis of organic peroxides. *Anal. Chim. Acta* **5**, 476–488.

Barnes, P. J. (ed.). (1983). "Lipids in Cereal Chemistry and Technology." Academic Press, London.

Barron, E. J., and Hanahan, D. J. (1958). Observations on the silicic acid chromatography of the neutral lipides of rat livers, beef liver, and yeast. *J. Biol. Chem.* **231**, 493–503.

Bauman, L. F., Conway, T. F., and Watson, S. A. (1963). Heritability of variations in oil content of individual oil kernels. *Science* **139**, 498–499.

Baumann, L. A. (1959). Evaluating refined cottonseed oils in storage. *J. Am. Oil Chem. Soc.* **36**, 28–34.

Bernheim, F., Bernheim, M. L. C., and Wilbur, K. M. (1947). The reaction between thiobarbituric acid and the oxidation products of certain lipides. *J. Biol. Chem.* **174**, 257–264.

Biggs, D. A. (1967). Milk analysis with the infrared milk analyzer. *J. Dairy Sci.* **50**, 799–803.

Blank, M. L., Schmit, J. A., and Privett, O. S. (1964). Quantitative analysis of lipids by thin-layer chromatography. *J. Am. Oil Chem. Soc.* **41,** 371–376.

Bligh, E. G., and Dyer, W. J. (1959). A rapid method of total lipid extraction and purification. *Can. J. Biochem. Physiol.* **37,** 911–917.

Bloksma, A. H. (1966). Extraction of flour by mixtures of butanol-1 and water. *Cereal Chem.* **43,** 602–622.

Bloor, J. (1925). Biochemistry of fats. *Chem. Rev.* **2,** 243–300.

Boekenoogen, H. A. (1964). "Analysis and Characterization of Oils, Fats, and Fat Products." Interscience Publishers, New York.

Bostian, M. L., Webb, N. B., and Hadden, J. P. (1984). Evaluation of rapid fat, moisture, and protein determination in meats and meat products, using an automatic meat analyzer. *J. Food Sci.* **49,** 1347–1349.

Braun, W. Q. (1955). Dilatometric measurements. *J. Am. Oil Chem. Soc.* **32,** 633–637.

Brown, F. (1952). The estimation of vitamin E. Separation of tocopherol mixtures occurring in natural products by paper chromatography. *Biochem. J.* **51,** 237–239.

Burns, D. A. (1981). Automated sample preparation. *Anal. Chem.* **53,** 1402A, 1404A, 1406A, 1408A, 1410A, 1412A, 1414A, 1417A, 1418A.

Campen, W. A. C., and Greerling, H. (1954). Fast and simple determination of the amount of crude fat and fatty acids in animal feed—a method general application. *Chem. Weekblad* **50,** 385–393. [*Chem. Abstr.* **48,** 10952i.]

Cannon, J. H. (1964). Sterol acetate test for foreign fats in dairy products. *J. Assoc. Off. Agric. Chem.* **47,** 577–580.

Casinovi, C. G. (1963). A comprehensible bibliography of organic substances by counter-current distribution. *Chromatogr. Rev.* **5,** 161–207.

Chapman, D. (1960). Infrared spectroscopic characterization of glycerides. *J. Am. Oil Chem. Soc.* **37,** 73–77.

Chapman, D. (1965). Infrared spectroscopy of lipids. *J. Am. Oil Chem. Soc.* **42,** 353–371.

Chapman, D., Richards, R. E., and Yorke, R. W. (1960). A nuclear magnetic resonance study of the liquid/solid content of margarine fat. *J. Am. Oil Chem. Soc.* **37,** 243–246.

Cocks, L. V., and Van Rede, C. (1966). "Laboratory Handbook for Oil and Fat Analysts." Academic Press, New York.

Collins, F. I., Alexander, D. E., Rodgers, R. C., and Silvella, L. S. (1967). Analysis of oil content of soybeans by wide-line NMR. *J. Am. Oil Chem. Soc.* **44,** 708–710.

Conway, T. F. (1960). Proc. Symposium on High-Oil Corn. Dept. Agronomy, Univ. Illinois. *Cited by* Alexander *et al.* (1967).

Conway, T. F., and Johnson, L. F. (1969). Nuclear magnetic resonance measurement of oil "unsaturation" in single viable corn kernels. *Science* **164,** 827–828.

Craig, L. C. (1944). Identification of small amounts of organic compounds by distribution studies. II. Separation by countercurrent distribution. *J. Biol. Chem.* **155,** 519–534.

Cropper, F. R., and Heywood, A. (1953). Analytical separation of the methyl esters of the C_{12}–C_{22} fatty acids by vapor-phase chromatography. *Nature* **172,** 1101–1102.

Dahle, L. K., Hill, E. G., and Holman, R. T. (1962). The thiobarbituric acid reaction and the autoxidation of polyunsaturated fatty acid methyl esters. *Arch. Biochem. Biophys.* **98,** 253–261.

de Man, J. M., and de Man, L. (1984). Automated AOM test for fat stability. *J. Am. Oil Chem. Soc.* **61,** 534–536.

Derbesy, M., and Richert, M. T. (1979). Detection of shea butter in cocoa butter (in French). *Oleagineaux* **34,** 405–409.

Derde, M.-P., Coomans, D., and Massart, D. L. (1984). SIMCA (Soft Independent Modelling of Class Analogy) demonstrated with characterization and classification of Italian olive oil. *J. Assoc. Off. Anal. Chem.* **67,** 721–727.

Dessy, R. (1983). Robots in the laboratory. II. *Anal. Chem.* **55,** 1232A, 1234A, 1238A, 1240A, 1242A.

Deuel, H. J., Jr. (1951). "The Lipids, Their Chemistry and Biochemistry," Vol. 1. Interscience Publishers, New York.

Dick, R., and Miserez, A. (1980). Gas chromatographic detection of adulteration of cocoa butter (in German). *Mitt. Geb. Lebensm. Hyg.* **71,** 499–508.

Dugan, L., Jr. (1955). Stability and rancidity. *J. Am. Oil Chem. Soc.* **32,** 605–609.

Dutton, H. J. (1955). The analysis of lipids by countercurrent distribution. *J. Am. Oil Chem. Soc.* **32,** 652–659.

Dutton, H. J. (1961). Some applications of mass spectrometry to lipid research. *J. Am. Oil Chem. Soc.* **38,** 660–664.

Dyachenko, P., and Samsonov, Y. (1964). Infrared spectra of dried milk and its components. *Molochn. Prom.* **25**(7), 11–13. [*Chem. Abstr.* **61,** 15263 f.]

Eisner, J., and Firestone, D. (1963). Gas chromatography of unsaponifiable matter. II. Identification of vegetable oils by their sterols. *J. Assoc. Off. Agric. Chem.* **46,** 542–550.

Eisner, J., Wong, N. P., Firestone, D., and Bond, J. (1962). Gas chromatography of unsaponifiable matter. I. Butter and margarine sterols. *J. Assoc. Off. Agric. Chem.* **45,** 337–342.

Entenmann, C. (1961). The preparation of tissue lipid extracts. *J. Am. Oil Chem. Soc.* **38,** 534–538.

Ettinger, C. L., Malanoski, A., and Kirschenbaum, H. (1965). Detection and estimation of animal fats in vegetable oils by gas chromatography. *J. Assoc. Off. Agric. Chem.* **48,** 1186–1191.

Falbe, J., and Weber, J. (1974). Determination of fats, oils, and waxes. *In* "Methodicum Chimicum," Vol. I, Part B, Analytical Methods, Micromethods, Biological Methods, Quality Control, Automatization (F. Korte, ed.), pp. 970–990. Academic Press, New York.

Faust, V., and Keller, H. (1976). Automation of analyses in the clinical-chemical laboratory (in German). *Chem. Ing. Techn.* **48,** 419–428.

Fehr, W. F., Collins, F. I., and Weber, C. R. (1968). Evaluation of methods for proteins and oil determination in soybean seed. *Crop Sci.* **8,** 47–49.

Feldman, G. D. (1967). Human occular lipids: Their analysis and distribution. *Surv. Ophthalmol.* **12,** 207–243.

Ferren, W. P., and Morse, R. E. (1963). Wide-line nuclear magnetic resonance determination of liquid–solid content of soybean oil at various degrees of hydrogenation. *Food Technol.* **17**(8), 112–114.

Fioriti, J. A. (1977). Measuring flavor deterioration of fats, oils, dried emulsions and foods. *J. Am. Oil Chem. Soc.* **54,** 450–453.

Fitzgerald, J. W., Rings, G. R., and Winder, W. C. (1961). Ultrasonic method for measurement of fluids-nonfat and milk fat in fluid milk. *J. Dairy Sci.* **44,** 1165.

Folch, J., Lees, M., and Stanley, G. H. S. (1957). A simple method for the isolation and purification of total lipides from animal tissues. *J. Biol. Chem.* **226,** 497–509.

Fontell, K., Holman, R. T., and Lambertsen, G. (1960). Some new methods for separation and analysis of fatty acids and other lipids. *J. Lipid Res.* **1,** 391–404.

Freeman, C. P., and West, D. (1966). Complete separation of lipid classes on a single thin-layer plate. *J. Lipid Res.* **7,** 324–327.

Freeman, N. K. (1957). Infrared spectroscopy of serum lipides. *Ann. N. Y. Acad. Sci.* **69,** 131–144.

Fritsch, C. W., Egberg, D. C., and Magnuson, J. S. (1979). Changes in dielectric constant as a measure of frying oil deterioration. *J. Am. Oil Chem. Soc.* **56,** 746–752.

Galliard, T., and Mercier, E. I. (eds.). (1975). "Recent Advances in the Chemistry and Biochemistry of Plant Lipids." Academic Press, London.

Gray, J. I. (1978). Measurement of lipid oxidation: A review. *J. Am. Oil Chem. Soc.* **55,** 539–546.

Graziano, V. J. (1979). Portable instrument rapidly measures quality of frying fat in food service operations. *Food Technol.* **33**(9), 50, 56, 57.

Grosch, W. (1975). Course and analysis of oxidative fat deterioration (in German). *Z. Lebensm. Unters. Forsch.* **157,** 70–83.

Gunstone, F. D., and Norris, F. A. (1983). "Lipids in Foods—Chemistry, Biochemistry and Technology." Pergamon Press, Oxford.

Gurr, M. I., and James, A. T. (1980). "Lipid Biochemistry: An Introduction," 3rd ed. Chapman & Hall, London.

Hamilton, R. J., and Bhati, A. (eds.). (1980). "Fats and Oils: Chemistry and Technology." Applied Science Publishers, London.

Hangaard, G., and Pettinati, J. D. (1959). Photometric milk fat determination. *J. Dairy Sci.* **42,** 1255–1275.

Henick, A. S., Benca, M. F., and Mitchell, J. H. (1954). Estimating carbonyl compounds in rancid fats and foods. *J. Am. Oil Chem. Soc.* **31,** 88–91.

Hilditch, T. P., and Williams, P. N. (1964). "The Chemical Constitution of Natural Fats," 4th ed. Chapman and Hall, London.

Hirsch, J., and Ahrens, E. H., Jr. (1958). The separation of complex lipide mixtures by the use of silicic acid chromatography. *J. Biol. Chem.* **233,** 311–320.

Holman, R. T., Lundberg, W. O., and Burr, G. O. (1945). Spectrophotometric studies of the oxidation of fats. III. Ultraviolet absorption spectra of oxidized octadecatrienoic acids. *J. Am. Chem. Soc.* **67,** 1390–1394.

Holman, R. T., and Rahm, J. J. (1966). Analysis and characterization of polyunsaturated fatty acids. *Prog. Chem. Fats Other Lipids* **9**(1), 15–90.

Horning, E. C., and Vandenheuvel, W. J. A. (1964). Gas chromatography in lipid investigations. *J. Am. Oil Chem. Soc.* **41,** 707–716.

Hornstein, I., Crowe, P. F., and Ruck, J. B. (1967). Separation of muscle lipids into classes by nonchromatographic techniques. *Anal. Chem.* **39,** 352–354.

Hung, S. S. O., and Slinger, S. J. (1981). Studies of chemical methods of assessing the oxidative quality and storage stability of feeding oils. *J. Am. Oil Chem. Soc.* **58,** 785–788.

Hunt, W. H., Neustadt, M. H., Hart, J. R., and Zeleny, L. (1952). A rapid dielectric method for determining the oil content of soybeans. *J. Am. Oil Chem. Soc.* **29,** 258–261.

Hunt, W. H., Neustadt, M. H., Shurkus, A. A., and Zeleny, L. (1951). A simple iodine-number refractometer for testing flaxseed and soybeans. *J. Am. Oil Chem. Soc.* **28,** 5–8.

Hussin, A. B. B. H., and Povey, M. J. W. (1984). A study of dilation and acoustic propagation in solidifying fats and oils. II. Experimental. *J. Am. Oil Chem. Soc.* **61,** 560–564.

James, A. T., and Martin, A. J. P. (1952). Gas-liquid partition chromatography; the separation and microestimation of volatile fatty acids from formic acid to dodecanoic acid. *Biochem. J.* **50,** 679–690.

Jamieson, G. R., and Reid, E. H. (1965). The analysis of oils and fats by gas chromatography. *J. Chromatogr.* **17,** 230–237.

Jewell, G. G. (1983). The application of NMR to oils and fats. *J. Sci. Food Agric.* **34,** 1024.

Johnson, L. F., and Shoolery, J. N. (1962). Determination of unsaturation and average molecular weight of natural fats by nuclear magnetic resonance. *Anal. Chem.* **34,** 1136–1139.

Kakuda, Y., Stanley, D. W., and van de Voort, F. R. (1981). Determination of TBA number by high-performance liquid chromatography. *J. Am. Oil Chem. Soc.* **58,** 773–775.

Katc, I., Keeney, M., and Bassette, R. (1959). Colorimetric determination of fat in milk and the saponification number of a fat. *J. Dairy Sci.* **42**, 903–906.

Kaufman, F. L. (1964). Infrared spectroscopy of fats and oils. *J. Am. Oil Chem. Soc.* **41**(8), 4, 6, 21, 38, 42.

Kaufmann, H. P. (1958). "Analysis of Fat and Fat Products," Vols. 1 and 2. Springer-Verlag, Berlin. (German)

Kaufmann, H. P., Mankel, G., and Lehmann, K. (1961). Gas chromatography of fatty compounds. I. General survey. *Fette, Seifen, Anstrichm.* **63**, 1109–1116. [*Chem. Abstr.* **56**, 11729h.]

Kaufmann, H. P., Seher, A., and Mankel, G. (1962). Gas chromatography of fats. II. Quantitative applications. *Fette, Seifen, Anstrichm.* **64**, 501–509. [*Chem. Abstr.* **57**, 7399b.]

Ke, P. J., Cervantes, E., and Robles-Martinez, C. (1984). Determination of thiobarbituric reactive substances (TBARS) in fish tissue by an improved distillation–spectrophotometric method. *J. Sci. Food Agric.* **35**, 1248–1254.

Keeney, M. (1956). A survey of United States butterfat constants. II. Butyric acid. *J. Assoc. Off. Agric. Chem.* **39**, 212–225.

Kessler, G., and Lederer, H. (1965). *Cited by* Levine (1967).

King, J. D. (1966). N. M. R. analysis of meat composition. *Proc. Meat Ind. Res. Conf.*, Chicago, 149–157. [*Chem. Abstr.* **67**, 115886u.]

Kohn, H. I., and Liversedge, N. (1944). A new aerobic metabolite whose production by brain is inhibited by apomorphine, emetine, ergotamine, adrenaline, and menadione. *J. Pharmacol.* **82**, 292–300.

Kohn, R. (1964). Application of gas chromatography in analyses of foods. *Qual. Plant. Mat. Veg.* **11**, 150–167. (German)

Kohn, R. (1965). Application of infrared spectroscopy in food analysis. I. Infrared spectroscopy in the intermediate range of 3 to 15 μm. *Z. Lebensm. Untersuch. Forsch.* **129**, 28–40. (German)

Kohn, R., and Laufer-Heydenreich, S. (1966). Application of infrared spectroscopy in food analysis. *Z. Lebensm. Untersuch. Forsch.* **129**, 92–97. (German)

Kropf, D. H. (1984). New rapid methods for moisture and fat analysis: A review. *J. Food Qual.* **6**, 199–210.

Kuksis, A. (1966). Quantitative lipid analysis by combined thin-layer and gas-liquid chromatographic systems. *Chromatogr. Rev.* **8**, 172–207.

Kuksis, A. (ed.). (1978). "Handbook of Lipid Research," Vol. I, Fatty Acids and Glycerides. Plenum Press, New York.

Kuksis, A., and McCarthy, M. J. (1964). Triglyceride gas chromatography as a means of detecting butterfat adulteration. *J. Am. Oil Chem. Soc.* **41**, 17–21.

Kummerow, F. A. (1960). Fats in human nutrition. *J. Am. Oil Chem. Soc.* **37**, 503–509.

Kurtz, F. E. (1965). The lipids of milk-composition and properties. *In* "Fundamentals of Dairy Chemistry" (B. H. Webb and A. H. Johnson, eds.). AVI Publishing Co., Westport, CT.

Kwon, T. W., and Olcott, H. S. (1966). Thiobarbituric-acid-reactive substances from autoxidized or ultraviolet irradiated unsaturated fatty esters and squalene. *J. Food Sci.* **31**, 552–557.

Lapin, G. R., and Clark, L. C. (1951). Colorimetric method for determination of traces of carbonyl compounds. *Anal. Chem.* **23**, 541–543.

Lea, C. H. (1931). Effect of light on the oxidation of fats. *Proc. Roy. Soc. London, Ser. B* **108**, 175–179.

Lea, C. H. (1946). The determination of the peroxide values of edible fats and oils: The iodometric method. *J. Soc. Chem. Ind.* **65**, 286–290.

Lea, C. H. (1956). "Biochemical Problems of Lipids." Butterworths Scientific Publishing Co., London.

Lea, C. H. (1962). The oxidative deterioration of food lipids. *In* "Symposium of Foods: Lipids and Their Oxidation" (H. W. Schultz, E. A. Day, and R. O. Sinnhuber, eds.). AVI Publishing Co., Westport, CT.

Lee, C. F., Ambrose, M. E., and Smith, P., Jr. (1966). Determination of lipids in fish meal. *J. Assoc. Off. Agric. Chem.* **49**, 946–949.

Leemans, F. A., and McCloskey, J. M. (1967). Combination gas chromatography–mass spectrometry. *J. Am. Oil Chem. Soc.* **44**, 11–17.

Lester, A. L. (1963). *Cited by* Wells and Dittmer (1966).

Levine, J. B. (1967). Recent advances in automated lipid analysis. *J. Am. Oil Chem. Soc.* **44**, 95–98.

Levine, J. B., and Zak, B. (1964). Automated determination of serum total cholesterol. *Clin. Chim. Acta* **10**, 381–384.

Levowitz, D. (1967). Determination of fat and total solids in dairy products. *In* "Laboratory Analysis of Milk and Milk Products." U.S. Dept. Health, Educ., Welfare, Cincinnati.

Lofland, H. B. (1964). A semiautomated procedure for the determination of triglycerides in serum. *Anal. Biochem.* **9**, 393–400.

Mangold, H. K., Schmidt, H. H. O., and Stahl, E. (1964). Thin-layer chromatography (TLC). *Methods Biochem. Anal.* **12**, 393–451.

Mani, V. V. S., and Lakshminarayana, G. (1976). Chromatographic and other methods for detection of adulteration of oils and fats. *J. Oil Technol. Assoc. India* **VIII**, 84–103.

McGann, T. C. A. (1980). Analytical chemistry in the dairy industry. *In* "Euroanalysis III. Reviews on Analytical Chemistry" (D. M. Carrol, ed.), pp. 251–270. Applied Science Publishers, London.

Marcuse, R., and Johansson, L. (1973). TBA (thiobarbituric acid) test for rancidity grading. II. TBA reactivity of different aldehyde classes. *J. Am. Oil Chem. Soc.* **50**, 387–391.

Marinetti, G. V. (1962). Chromatographic separation, identification, and analysis of phosphatides. *J. Lipid Res.* **3**, 1–20.

Marinetti, G. V. (1964). Chromatographic analysis of polar lipids on silicic acid impregnated paper. *In* "New Biochemical Separations" (A. T. James and L. J. Morris, eds.). D. Van Nostrand Co., London.

Marriott, N. G., Smith, G. C., Carpenter, Z. L., and Dutson, T. R. (1975). Rapid fat and moisture determinations for meat samples. *J. Anim. Sci.* **41**, 296–297.

Mattsson, P. (1978). Crude fat determination in feedingstuffs. Some studies of extraction and hydrolysis methods (in Swedish). St. Lantbr.–Kem. Lab. Medd. 19 pages.

Mecham, D. K., and Mohammad, A. (1955). Extraction of lipids from wheat products. *Cereal Chem.* **32**, 405–415.

Mehlenbacher, V. C. (1958). Standard methods in the fat and oil industry. *Prog. Chem. Fats Other Lipids* **5**, 1–29.

Mehlenbacher, V. C. (1960). "Analysis of Fats and Oils." The Garrard Press, Champaign, IL.

Mills, B. L., and van de Voort, F. R. (1981). Comparison of the direct and indirect wide-line nuclear magnetic resonance methods for determining solid fat content. *J. Am. Oil Chem. Soc.* **58**, 776–778.

Mitchell, D. J. (1967). Collaborative studies of methods for butterfat in homogenized and chocolate milk. *J. Assoc. Off. Anal. Chem.* **50**, 537–541.

Mitchell, J. H., Kraybill, H. R., and Zscheile, F. P. (1943). Quantitative spectrum analysis of fats. *Ind. Eng. Chem., Anal. Ed.* **15**, 1–3.

Miwa, T. K., Kwolek, W. F., and Wolff, I. A. (1966). Quantitative determination of unsaturation in oils by using an automatic-titrating hydrogenator. *Lipids* **1**, 152–157.

Mojonnier, T., and Troy, H. C. (1925). "Technical Control of Dairy Products." Mojonnier Bros. Co., Chicago.

Morris, L. J. (1964). Specific separations by chromatography on impregnated adsorbents. In "New Biochemical Separations" (A. T. James and L. J. Morris, eds.). D. Van Nostrand Co., London.

Morris, L. J. (1966). Separation of lipids by silver ion chromatography. J. Lipid Res. 7, 717–732.

Murphy, M. F., and McGann, T. C. A. (1967). Investigations on the use of the Milko-tester for routine estimation of fat content in milk. J. Dairy Res. 34, 65–72.

Nadj, L. J., and Weeden, D. G. (1966). Refractometric estimation of total fat in chocolate-type products. Anal. Chem. 38, 125–126.

Nutter, L. J., and Privett, O. S. (1968). An improved method for the quantitative analysis of lipid classes via thin-layer chromatography employing charring and densitometry. J. Chromatogr. 35, 519–525.

O'Connor, R. T. (1955). Ultraviolet absorption spectroscopy. J. Am. Oil Chem. Soc. 32, 616–624.

O'Connor, R. T. (1961A). Near-infrared absorption spectroscopy—a new tool for lipid analysis. J. Am. Oil Chem. Soc. 38, 641–648.

O'Connor, R. T. (1961B). Recent progress in the application of infrared absorption spectroscopy to lipid chemistry. J. Am. Oil Chem. Soc. 38, 648–659.

Oh, F. C. H. (1982). Sources of errors in the determination of iodine value of palm oil by Wijs method. Mardi Res. Bull. 10, 248–258.

Padley, F. B. (1964). Thin-layer chromatography of lipids. In "Thin-layer Chromatography" (G. B. Marini-Bettolo ed.). Elsevier Publishing Co., Amsterdam.

Pardun, P. (1975). Analytical methods for the assay of freshness and stability of animal fats (in German). Fette, Seifen, Anstrichm. 77, 296–305.

Patton, S., Keeney, M., and Kurtz, G. W. (1951). Compounds producing the Kreis color reaction with particular reference to oxidized milk fat. J. Am. Oil Chem. Soc. 28, 391–393.

Patton, S., and Kurtz, G. W. (1951). 2-Thiobarbituric acid as a reagent for detecting milk-fat oxidation. J. Dairy Sci. 34, 669–674.

Pelick, N., Wilson, T. L. Miller, M. E., Angeloni, F. M., and Stein, J. M. (1965). Some practical aspects of thin-layer chromatography of lipids. J. Am. Oil Chem. Soc. 42, 393–399.

Perkins, E. G., Means, J. C., and Ticciano, M. F. (1977). Recent advances in instrumental analysis of lipids. Rev. Franc. Corps Gras 24, 74–84.

Pettinati, J. D. (1980). Update: Rapid methods for the determination of fat, moisture, and protein. Proc. Recip. Meat Conf. 33, 156–163.

Pinto, A. F., and Enas, J. D. (1949). Rapid method of copra analysis and its application to the various oil seeds. J. Am. Oil Chem. Soc. 26, 723–730.

Pitt, G. A. J., and Morton, R. A. (1957). Ultraviolet spectrophotometry of fatty acids. Prog. Chem. Fats Other Lipids 4, 227–278.

Pohle, W. D., Gregory, R. L., and Taylor, J. R. (1962). A comparison of several analytical techniques for prediction of relative stability of fats and oils to oxidation. J. Am. Oil Chem. Soc. 39, 226–229.

Pohle, W. D., Gregory, R. L., and van Giessen, B. (1963). A rapid bomb method for evaluating the stability of fats and shortenings. J. Am. Oil Chem. Soc. 40, 603–605.

Pohle, W. D. et al. (1964). A study of methods for evaluation of the stability of fats and shortenings. J. Am. Oil Chem. Soc. 41, 795–798.

Pomeranz, Y. (1965). Thin layer chromatography in studies of cereal lipids. Qual. Plant. Mater. Veg. 12, 322–341.

Pool, P. O. (1931). Rancidity and stability in shortening products. *Oil Fat Ind.* **8,** 331–336.

Povey, M. J. W. (1984). A study of dilatation and acoustic propagation in solidifying fats and oils. I. Theoretical. *J. Am. Oil Chem. Soc.* **61,** 558–559.

Privett, O. S., Blank, M. L., Codding, D. W., and Nickell, E. C. (1965). Lipid analysis by quantitative thin-layer chromatography. *J. Am. Oil Chem. Soc.* **42,** 381–393.

Pryde, E. H. (ed.). (1979). "Fatty Acids." Am. Oil Chemists Soc., Champaign, IL.

Randall, E. L. (1974). Improved method for fat and oil analysis by a new process of extraction. *J. Assoc. Off. Anal. Chem.* **57,** 1165–1168.

Ragnarson, J. O., and Labuza, T. P. (1977). Accelerated shelf-life testing for oxidative rancidity in foods. A review. *Food Chem.* **2,** 291–308.

Rao, C. R., Reddy, L. C. S., and Prabhu, C. A. R. (1980). Study of adulteration in oils and fats by ultrasonic method. *Current Sci.* **49,** 185–186.

Robertson, J. A., and Morrison, W. H. (1979). Analysis of oil content of sunflower seed by wide-line NMR. *J. Am. Oil Chem. Soc.* **56,** 961–964.

Robertson, J. A., and Windham, W. R. (1981). Comparative study of methods of determining oil content of sunflower seed. *J. Am. Oil Chem. Soc.* **58,** 993–996.

Rodeghier, A. A. (1955). Determination of impurities in fats and oils. *J. Am. Oil Chem. Soc.* **32,** 578–581.

Roden, A., and Ullyot, G. (1984). Quality control in edible oil processing. *J. Am. Oil Chem. Soc.* **61,** 1109–1111.

Rouser, G., Kritchevsky, G., Galli, C., and Heller, D. (1965). Determination of polar lipids: Quantitative column and thin layer chromatography. *J. Am. Oil Chem. Soc.* **42,** 215–227.

Roy, R. B. (1984). Automated food analysis using continuous flow-analytical systems. *In* "Food Analysis—Principles and Techniques," Vol. I, Physical Characterization (D. W. Gruenwedel and J. R. Whitaker, eds.), pp. 247–294. Marcel Dekker, New York and Basel.

Ryhage, R., and Stenhagen, E. (1960). Mass spectrometry in lipid research. *J. Lipid Res.* **1,** 361–390.

Sager, O. S., Sanders, G. P., Norman, G. H., and Middleton, M. B. (1955). A detergent test for the milk fat content of dairy products. *J. Assoc. Off. Agric. Chem.* **38,** 931–940.

Schain, P. (1949). The use of detergents for quantitative fat determination. I. Determination of fat in milk. *Science* **110,** 121–122.

Schmid, H. H. O., Baumann, W. J., Cubero, J. M., and Mangold, H. K. (1966). Fractionation of lipids by successive adsorption and argentation cheomatography on adjacent layers. *Biochim. Biophys. Acta* **125,** 189–196.

Schober, B. (1967). Application of a refractometric method for lipid determination in fish and fish products. I. Determination of the fat content in herring. *Fischereiforschung* **5,** 121–124. [*Chem. Abstr.* **69,** 1825j.]

Schwarz, H. P., Dreisbach, L., Childs, R., and Mastrangelo, S. V. (1957). Infrared studies of tissue lipids. *Ann. N. Y. Acad. Sci.* **69,** 116–130.

Shlafer, M., and Shepard, B. M. (1984). A method to reduce interference by sucrose in the detection of thiobarbituric-reactive substances. *Anal. Biochem.* **137,** 269–276.

Shoolery, J. N. (1983). Applications of high resolution nuclear magnetic resonance to the study of lipids. *Am. Oil Chem. Soc. Monograph* **10,** 220–240.

Shukla, V. K. S. (1983). Studies on the crystallization behaviour of the cocoa butter equivalents by pulse nuclear magnetic resonance. Part I. *Fette, Seifen, Anstrichm.* **85,** 467–471.

Shukla, G. E., Machari, A. N., Sharma, C. K., and Murthi, T. N. (1980). A butyrometric method for rapid determination of the oil content of groundnut seeds. *J. Food Sci. Technol.* **17,** 242–244.

Siakotos, A. N., and Rouser, G. (1965). Analytical separation of nonlipid water soluble substances and gangliosides from other lipids by dextran column chromatography. *J. Am. Oil. Chem. Soc.* **42**, 913–919.

Simic, M. G., and Karel, M. (eds.). (1979). "Autoxidation in Food and Biological Systems." Plenum Press, New York.

Skipski, V. P., Good, J. J. Barclay, M., and Reggio, R. B. (1968). Quantitative analysis of simple lipid classes by thin-layer chromatography. *Biochim. Biophys. Acta* **152**, 10–19.

Smith, H. M. (1955). Melting point, solidification, and consistency. *J. Am. Oil Chem. Soc.* **32**, 593–595.

Stern, I., and Shapiro, B. (1953). A rapid and simple method for the determination of esterified fatty acids and for total fatty acids in blood. *J. Clin. Pathol.* **6**, 158–160.

Stillman, R. C. (1955). Bleach and color methods. *J. Am. Oil Chem. Soc.* **32**, 587–593.

Swern, D. (ed.). (1982). "Bailey's Industrial Oil and Fat Products." Wiley, New York.

Tarladgis, B. G., Pearson, A. M., and Dugan, L. R. (1962). The chemistry of the 2-thiobarbituric acid test for the determination of oxidative rancidity in foods. I. Some important side reactions. *J. Am. Oil Chem. Soc.* **39**, 34–39.

Taylor, J. R., Pohle, W. D., and Gregory, R. J. (1964). Measurement of solids in triglycerides by using nuclear resonance spectroscopy. *J. Am. Oil Chem. Soc.* **41**, 177–180.

Trappe, W. (1940). Separation of biological fats from natural mixtures by means of adsorption columns. I. The eluotropic series of solvents. *Biochem. Z.* **305**, 150–161.

Trevelyan, W. E. (1966). Determination of some lipid constituents of baker's yeast. *J. Inst. Brew.* **72**, 184–192.

Troeng, S. (1955). Oil determination of oilseed. Gravimetric routine method. *J. Am. Oil Chem. Soc.* **32**, 124–126.

Tsen, C. C., Levi, I., and Hlynka, I. (1962). A rapid method for the extraction of lipids from wheat products. *Cereal Chem.* **39**, 195–203.

Waiser, P. E., and Bartels, M. A. (1982). Design of a system for laboratory automation. *Am. Lab.* **14**(2), 113, 114, 116, 118–120.

Wells, M. A., and Dittmer, J. C. (1963). The use of Sephadex for the removal of nonlipid contaminants from lipid extracts. *Biochemistry* **2**, 1259–1263.

Wells, M. A., and Dittmer, J. C. (1966). A microanalytical technique for the quantitative determination of twenty-four classes of brain lipids. *Biochemistry* **5**, 3405–3408.

Wheeler, D. H. (1954). Infrared absorption spectroscopy in fats and oils. *Prog. Chem. Fats Other Lipids* **2**, 268–291.

Whitley, R. W., and Alburn, H. E. (1964). *Cited by* Levin (1967).

Wijs, J. J. A. (1929). The Wijs method as the standard for iodine absorption. *Analyst* **54**, 12–14.

Wilbur, K. M., Bernheim, F., and Shapiro, O. W. (1949). The thiobarbituric acid reagent as a test for the oxidation of unsaturated fatty acids by various reagents. *Arch. Biochem.* **24**, 305–313.

Williams, K. A. (1950). "Oils, Fats and Fatty Foods." Churchill Publishing Co., London.

Wilmer, M. C., Rettori, C., Vargas, H., Barberis, G. E., and DaSilva, W. J. (1978). Single kernel wide-line NMR oil analysis for breeding purpose. *Rev. Brasileira Fisica* **8**, 562–575.

Wood, R. (1984). Collaborative study on a potential EEC method for the determination of milk fat in cocoa and chocolate products. *Spec. Publ. R. Soc. Chem.* **49**, 56–67. [*Chem. Abstr.* **101**, 37273w.]

Wren, J. J. (1961). Chromatography of lipids on silicic acid. *Chromatogr. Rev.* **3**, 111–133.

Young, C. C. (1984). The interpretation of GLC triglyceride data for the determination of cocoa butter equivalents in chocolate: A new approach. *J. Am. Oil Chem. Soc.* **61**, 576–581.

Yu, T. C., and Sinnhuber, R. O. (1967). An improved 2-thiobarbituric acid (TBA) procedure for the measurement of autooxidation in fish oils. *J. Am. Oil Chem. Soc.* **44**, 256–258.

Zehren, V. L., and Jackson, H. C. (1956). A survey of United States butterfat constants. I. Reichert-Meissl, Polenske and refractive index values. *J. Assoc. Off. Agric. Chem.* **39**, 194–212.

Zilversmit, D. B., and Davis, A. K. (1960). Microdetermination of plasma phospholipides by trichloroacetic acid precipitation. *J. Lab. Clin. Med.* **35**, 155–160.

Zimmerman, D. C. (1962). The relationship between seed density and oil content in flax. *J. Am. Oil Chem. Soc.* **39**, 77–78.

37
Nitrogenous Compounds

PROTEINS

The problem of providing adequate protein for an expanding world population is second only to the overall food problem. Table 37.1 summarizes the protein content of selected foodstuffs. Apart from their nutritional significance, proteins play a large part in the organoleptic properties of foods (Rhodes 1963; Schultz and Anglemier 1964). Proteins exert a controlling influence on the texture of foods from animal sources. Protein content of wheat and flour is considered one of the best single indices of breadmaking quality. The protein test, although generally not included as a grading factor in grain standards, is accepted as a marketing factor (Pomeranz 1968).

Proteins often occur in foods in physical or chemical combination with carbohydrates or lipids. The glycoproteins and lipoproteins affect the rheological properties of food solutions or have technical applications as edible emulsifiers. The aging of meat is associated with chemical changes in the proteins (Whitaker 1959). Pure native proteins have little flavor. During heating (boiling, baking, roasting) the amino acid side chains are degraded or interact with other food components (e.g., lysine with reducing sugars) to give typical flavors (Danehy and Pigman 1951). Excessive heating may, on the other hand, reduce nutritive value (Rice and Beuk 1953). The role of proteins in the processing and storage of various foods was reviewed by Feeney and Hill (1960).

The food analyst most commonly wishes to know the total protein con-

Table 37.1.   Protein Content (N × 6.25%) of Selected Foodstuffs[a,b]

Animal origin	Protein	Plant origin	Protein
Milk		Rice, whole	7.5–9.0
Whole, dried	22–25	Rice, polished	5.2–7.6
Skimmed, dried	34–38	Wheat, flour	9.8–13.5
Beef		Corn, meal	7.0–9.4
Dried	81–90	Chick, pea	22–28
Roasted	72	Soybean	33–42
Egg		Peanut	25–28
Whole, dried	35	Walnut	15–21
Whole, dried,	77	Potato[c]	10–13
defatted		Tapioca[c]	1.3
Herring[c]	81	Alfalfa[c]	18–23
	69	Chlorella[c]	23–44
		Torula yeast[c]	38–55

[a] From Attschul (1962).
[b] Unless stated otherwise, on as-is basis.
[c] H_2O-free basis.

tent of a food, even though that content is made up of a complex mixture of proteins. At the present time, all methods of determining the total protein content of foods are empirical in nature. Isolation and direct weighing of the protein would provide an absolute method. Such a method is sometimes used in biochemical investigation but is completely impractical for food analysis.

The primary nutritional importance of protein is as a source of amino acids. Twenty-four amino acids are generally thought to be constituents of proteins. Some amino acids are essential to good physical and mental health. Of the amino acids in food, eight are known to be essential to man, that is, they must be supplied in the diet to maintain growth and health. Table 37.2 summarizes the contents of essential amino acids in selected proteins. Proteins from some plant sources (e.g., cereal grains) are deficient in certain amino acids (e.g., lysine). Deficient proteins must be combined with those from other sources to provide an adequate balance of the essential amino acids. Such a balance can be accomplished by a combination of wheat flour with dry skim milk or soy flour. Detailed tables on the amino acid composition of foods were presented by Block (1945), Orr and Watt (1957), and by the Food and Agriculture Organization of the United Nations (Anon. 1968).

The methods of amino acid analysis of proteins were reviewed among others by Block (1945, 1960), Dunn (1950), Martin and Synge (1945), and James and Morris (1964). The reviews give details of the preparation of samples for analysis, hydrolysis of proteins, and the separation and assay of amino acids in the hydrolysates. The latter include specific methods

Table 37.2. Essential Amino Acids in Proteins[a,b]

Amino acid	FAO reference	Skim milk	Soya	Beef	Egg	Fish	Yeast
Lysine	4.2	8.6	6.8	8.3	6.3	6.6	6.8
Tryptophan	1.4	1.5	1.4	1.0	1.5	1.6	0.8
Phenylalanine	2.8	5.5	5.3	3.5	5.7	4.1	4.5
Methionine	2.2	3.2	1.7	2.8	3.2	3.0	2.6
Threonine	2.8	4.7	3.9	4.5	4.9	4.8	5.0
Leucine	4.8	11.0	8.0	7.2	9.0	10.5	8.3
Isoleucine	4.2	7.5	6.0	4.7	6.2	7.7	5.5
Valine	4.2	7.0	5.3	5.1	7.0	5.3	5.9

[a] From Lichtfield and Sachsel (1965).
[b] Table values in grams/100 g protein.

for certain amino acids (colorimetric, enzymatic, and microbiological) and methods for assaying all or most of the amino acids after preliminary separation (by paper or thin-layer chromatography, column chromatography on ion-exchange resins, and gas chromatography). The most powerful methods of quantitative amino acid analysis in protein hydrolysates are based on ion-exchange chromatography (Spackman *et al.* 1958) gas chromatography (Gehrke and Stalling 1967) and HPLC.

The food analyst is sometimes interested in knowing the content of a particular protein in a mixture. The classical investigations of Osborne (1907) on the differences in solubility in various solvents of wheat proteins still provide a useful separation and characterization tool. Changes in protein solubility are useful in determining the length and severity of heat treatment in processing milk, soy products, or animal products. The milk proteins of chocolate can be determined from the amount of proteins precipitated from an oxalate solution with tannic acid, provided the chocolate has not been subjected in processing to elevated temperatures (Motz 1968). Finally, some of the more modern tools of fractionation such as adsorption, partition, exclusion, and exchange chromatography, and particularly electrophoresis, can help ascertain the purity of an isolated protein and identify the source of a protein.

For a detailed study, the protein generally must be extracted from natural sources by maceration (with various homogenizers, mills, or mechanical pestles), by disruption of cells (by alternate freezing and thawing, ultrasonic vibration, or fine grinding by colloidal milling), digestion of enzymes, or by solvent extraction (dilute salts, aqueous or anhydrous organic solvents, or surface-active agents). The methods of isolation and separation of proteins were described by Keller and Block (1960), James and Morris (1964), and Pfleiderer (1967).

In this chapter, various methods for determining total protein content are discussed. The most common procedure for a protein assay depends on determining a specific element or group in the protein, and calculating the protein content by using an experimentally established factor. Methods based on the analysis for constituents of proteins include those for determining carbon or nitrogen, certain amino acids, or the peptide linkage. In some proteins, certain constituents (iron in hemoglobin, iodine in thyroglobulin) can serve as a basis for protein assay. In all of these methods, it is assumed that the constituent determined is present entirely in the protein fraction. Thus, any nonprotein carbon-containing matter must be removed if the protein content is to be determined from the carbon content; and if the Kjeldahl method is used, protein nitrogen only should be measured. The common practice of estimating the protein content of foods from a total nitrogen assay is, therefore, not always correct. The content of nonprotein nitrogen compounds is, as a rule, generally small compared with the protein content of most sound foods.

Elementary Analysis

Carbon analysis has several advantages for determining the protein content of foods. The digestion can be accomplished more easily than for a nitrogen determination, and a high percentage of carbon minimizes experimental error and provides a relatively constant conversion factor. However, the difficulty of obtaining a complete and quantitative separation of protein from nonprotein carbon-containing components is practically unsurmountable.

A nitrogen determination is the most commonly used protein assay. It is generally assumed that a mixture of pure proteins will contain 16% nitrogen. Thus, the protein content of a sample is obtained by multiplying the determined nitrogen by the factor 6.25 = (100/16). For the approximate analysis of foods containing an unknown distribution of proteins of unknown composition, this is a practical and widely accepted procedure. Much confusion has resulted in reporting protein content: protein levels above 100% in pure or highly concentrated protein fractions and differences among laboratories originating from the use of various conversion factors have been reported. Much of the confusion could be eliminated by reporting the nitrogen rather than the calculated protein content. The food industry and trade will probably be reluctant to make such a change and will continue to report the calculated protein content.

Yet the inherent limitations of the procedure must be realized: the results are affected by nonprotein nitrogen; the nitrogen content of a particular protein mixture is seldom known precisely; and the methods of determining nitrogen are wrought with some difficulties. For practical purposes, unless shown to the contrary, the effects of nonprotein nitrogen can be assumed of little consequence. The general conversion factor of 6.25 is used for most foods. For wheat, milk, and gelatin, factors of 5.70, 6.38, and 5.55, respectively, are used (Jones 1931). The availability of better methods for protein isolation and characterization (including amino acid composition) provides the basis for a continuous reexamination of the conversion factors (Tkachuk 1966). Although extremes in nitrogen content range from 4.2% in beta lipoproteins to 30% in protamine, most foods contain about 16% nitrogen. Difficulties in the nitrogen determination are outlined in the following discussion of the two main methods used in food analysis, the Kjeldahl method and the Dumas procedure.

Kjeldahl Method. The Danish investigator Kjeldahl worked out in 1883 a method for determining organic nitrogen in his studies on protein changes in grain used in the brewing industry. Since the first publication of Kjeldahl, the method has undergone many changes (Bradstreet 1965). Basically, the sample is heated in sulfuric acid and digested till the carbon and hydrogen are oxidized and the protein nitrogen is reduced and transformed into ammonium sulfate. Then concentrated sodium hydroxide is added, and the digest heated to drive off the liberated ammonia into a known volume of a standard acid solution. The unreacted acid is determined and the results are transformed, by calculation, into a percentage of protein in the organic sample (for details, see later in this chapter).

Kjeldahl originally used potassium permanganate for the oxidation, but this was discontinued as the results were unsatisfactory. In 1885, Wilforth found that a digestion with sulfuric acid was accelerated by adding some catalysts. Gunning in 1889 suggested adding potassium sulfate to raise the boiling point of the digestion mixture to shorten the reaction. The procedure is, therefore, generally known as the Kjeldahl–Wilforth–Gunning method.

Various factors are known to influence the completeness and speed of the conversion of protein nitrogen into ammonia by the sulfuric acid digestion. For example, in some proteins it is more difficult to convert the organic nitrogen to ammonia. Histidine- and tryptophan-rich proteins generally require long or harsh digestion conditions. Excessive ratios of potassium or sodium sulfate (added to raise the boiling point) to acid may result in heat decomposition and the loss of ammonia. Generally, digestion temperatures of 370–410°C are best.

Nearly all of the likely elements of the periodic table have been tried as catalysts for the Kjeldahl digestion. Mercury, copper, and selenium have been widely employed. Mercury is superior to copper, though an additional step is required—precipitation of mercury with sodium thiosulfate—to decompose the mercury–ammonia complex formed during digestion. In recent years, the use of the highly toxic mercury has been largely discontinued. The most controversial catalyst is selenium. It has a more rapid effect than mercury and unlike mercury it requires no further treatment before distillation. Nevertheless, loss of nitrogen can occur if too much selenium is used or the digestion temperature is not carefully controlled; the conditions are more critical than with copper or mercury.

In commercial practice where large numbers of samples are run daily, many time-saving devices are used (Neill 1962). In wheat or flour analyses, digesting a 1-g sample, using a known amount of standardized 0.1253 N sulfuric acid, and titrating (by an inverse-reading buret) with 0.1253 N sodium hydroxide makes it possible to report percentage protein directly from the buret reading. The use of automatic pipets for dispensing the receiver acid solutions is an advantage in testing large numbers of samples. If the catalyst–digestion mixture contains mercury, a mercury precipitant is incorporated with the sodium hydroxide solution at the time of its preparation. An antibumper or pumice stone is blended with the catalyst powder mixture for a one-shot addition. Heating levels are adjusted so that digestion of 1 g with 25 ml concentrated sulfuric acid and a catalyst mixture (potassium sulfate, mercuric oxide, and copper sulfate) is completed within 35–45 min.

The boric acid modification is accurate and has the advantage that only one standard solution (of titrating acid) is required. Neither the amount (about 50 ml) nor the concentration (about 4%) of boric acid in the receiving bottle have to be precise. If small samples are available (10–30 mg), a micro-Kjeldahl modification employing the boric acid procedure and steam distillation of the liberated ammonia is used. Figures 37.1 and 37.2 show a battery for macroKjeldahl digestion and distillation, and units for digestion and distillation on a microscale.

Several methods are available to determine the ammonium sulfate in the digest. The digest may be alkalized and the liberated ammonia absorbed in acid and measured titrimetrically or colorimetrically. One colorimetric method, based on the procedure of Van Slyke and Hiller (1933), involves reacting a solution containing ammonium ions with alkaline phenol and hypochlorite. On heating the solution, an intense blue color, closely related to that of indophenol, is produced (Mann 1963; Varley 1966). Thymol hypobromite can also be used to determine nitrogen in the form of ammonia (Glebko *et al.* 1967). For the microdetermination of ammonia, the Conway microdiffusion procedure can be used. It depends on the transfer, by diffusion, of the ammonia from an alkalized solution to a standard acid, followed by titration (Conway 1957).

Because of its high sensitivity and simplicity, the Nessler colorimetric method directly applied to Kjeldahl digests can be used to determine nitrogen in foods (Hettrick and Whitney 1949; Williams 1964). However, the conditions required for the optional color reaction and stability are rigorous. The colored complex does not form a molecular solution, some cations form a precipitate with the alkali, and minute amounts of some

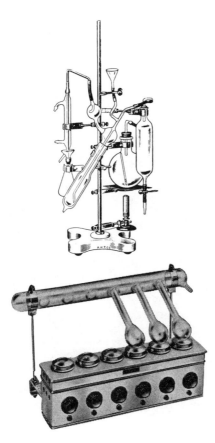

Fig. 37.1. Micro-Kjeldahl digestion (bottom) and distillation apparatus (top). (Courtesy A. H. Thomas and LaPine Scientific Co.)

heavy metals produce a considerable inhibition. The addition of potassium cyanide to the Nessler reagent avoids some of the difficulties (Minari and Zilversmit 1963). The reactions involved in a Kjeldahl digestion and direct nesslerization procedure are as follows:

Kjeldahl Reaction:

$$n{-}\overset{|}{\underset{|}{C}}{-}NH_2 + mH_2SO_4 \xrightarrow[\text{heat}]{\text{catalysts}} CO_2 + (NH_4)_2SO_4 + SO_2 \qquad (37.1)$$

protein

Nessler Reaction: $\qquad\qquad\qquad\qquad\qquad\qquad\qquad\qquad\qquad\qquad$ (37.2)

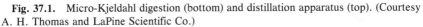

$$NH_3 + 2K_2HgI_4 + 3KOH \longrightarrow Hg_2OINH_2 + 7KI + 2H_2O$$

$\qquad\qquad$ Nessler $\qquad\qquad\qquad\qquad\qquad$ Yellow to
$\qquad\qquad$ reagent $\qquad\qquad\qquad\qquad\qquad$ reddish-brown
$\qquad\qquad\qquad\qquad\qquad\qquad\qquad\qquad\qquad\qquad$ complex

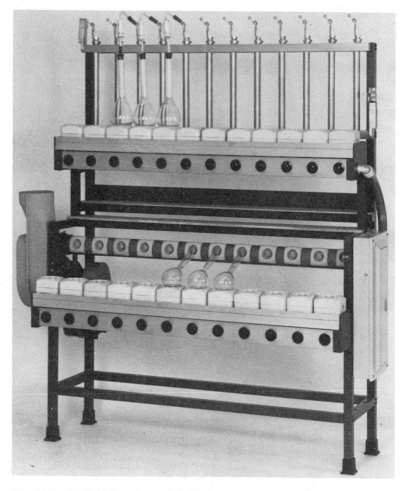

Fig. 37.2. Kjeldahl digestion and distillation apparatus. (Courtesy Labconco Co.)

The development of a continuous digestion module (Ferrari 1960) made it possible to determine nitrogen in biological fluids or suspensions within several minutes. The digestion is followed by a colorimetric determination in a neutralized digest to which alkaline phenol and sodium hypochlorite are added. The use of an automated Kjeldahl analyzer for the determination of nitrogen in biological material was described by Siriwardene *et al.* (1966). The procedure is capable of handling 20 samples an hour, reduces the labor involved to a minimum, and still maintains a high degree of accuracy and reproducibility.

Lakin (1978) emphasized that the Kjeldahl procedure continues to be the most common method for the determination of nitrogen and protein

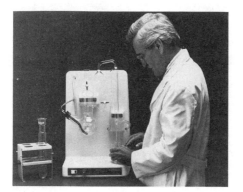

Fig. 37.3. RapidStill I for 5-min distillations in Kjeldahl procedure. (Courtesy Labconco Corp., Kansas City, MO.)

in foods. Most of the recent developments involve improvements in block digestors and in the continuous flow analysis of ammonia in digests. The RapidStill I (Fig. 37.3) features steam production by a heating element immersed in distilled water. The one-piece borosilicate still is mounted on a tough, chemical-resistant console that contains all operating controls. The instrument is designed to meet the needs of laboratories that run only a few analyses a day. Each distillation takes 5 min. A 4-sample rapid digestor and RapidStill II are shown in Fig. 37.4. A 25-sample rapid digestor and RapidStill III with many automatic features for rapid digestion (25–45 min at 450°C) and automatic distillation of micro- or macro- samples within 4 min are shown in Fig. 37.5.

The use of hydrogen peroxide, instead of sulfuric acid, for the digestion in the determination of protein by the Kjeldahl method was evaluated by

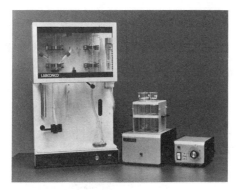

Fig. 37.4. RapidStill II with 4-sample digester in the middle. (Courtesy Labconco Corp., Kansas City, MO.)

Fig. 37.5. RapidStill III (right) and 25-place digestor (middle) for digestion of proteins and distillation of digests by the Kjeldahl procedure on micro- and macroscale. (Courtesy Labconco Corp., Kansas City, MO.)

Singh *et al.* (1984). The procedure is preferable to those employing highly toxic catalysts (mercury or selenium). Hydrogen peroxide is used in the Digestahl® (Anon. 1982) in which the digestate is analyzed for ammonia-N by direct nesslerization. This makes possible "ten minute Kjeldahls."

Dumas Method. In the classical Dumas procedure, first described in 1831, nitrogen is freed by pyrolysis, and the freed elemental nitrogen determined volumetrically. Major improvements in both the pyrolysis and nitrogen determination have ensured precise and accurate analyses of nitrogen in organic materials by the method (Sternglanz and Kollig 1962). Improved catalysts and rapid gas chromatographic methods for the nitrogen determination make it possible to conduct an assay on a microscale in 2 min. The Dumas method is used more widely because large samples can be handled by new instruments.

Biuret Method

The biuret method, proposed first by Riegler (1914), is based on the observation that substances containing two or more peptide bonds form a purple complex with copper salts in alkaline solutions. The biuret procedure is simple, rapid, and inexpensive. Because the biuret method involves a reaction with the peptide linkage, it furnishes an accurate estimate of protein. In contrast, the Kjeldahl procedure measures total nitrogen and does not distinguish between protein and nonprotein nitrogen.

The biuret method seems less subject to criticism as the basis of a colorimetric method for protein analysis than nearly any other such method (Kirk 1947). There are practically no substances other than protein normally present in biological materials that give much interference. Although the color development with various proteins is not identical, deviations are encountered less frequently than with other colorimetric methods. The results may be, however, affected by the presence of lipids and interfering opalescence from components of biological origin.

On the other hand, more material is required for the biuret than for most other colorimetric assays of protein. The biuret procedure is not an absolute method, and the color must be standardized against known protein or against another method (e.g., a Kjeldahl analysis of nitrogen).

Many modifications of the originally published biuret method have been proposed for routine determinations of small amounts of protein in physiological fluids, plant and animal tissues, and chromatographically separated protein fractions. The use of the method for the determination of protein in meats was described by Torten and Whitaker (1964) and in oilseeds by Pomeranz (1965). Sober *et al.* (1965) pointed out that by adding sodium tetraborate, which diminished the reducing action of lactose in milk, the biuret reaction could be applied without the separation of lactose.

Use of the biuret method in testing cereals is of particular interest. Originally, Pinckney (1949) described a procedure in which aliquots of a cleared alkaline protein extract were combined with measured amounts of the alkaline copper reagent. The color was measured after a given reaction period. In a modified method (Pinckney 1961), the stabilized reagent containing potassium hydroxide, sodium potassium tartarate, and copper sulfate is applied directly to the weighed sample for the simultaneous extraction and reaction. Jennings (1961) determined protein in white cereal grains by a modified biuret method that involved the simultaneous extraction of protein and color development in an alkaline copper tartarate solution. In dark-colored cereal grains, the extract prepared by the biuret extraction procedure was treated with the Folin–Ciocalteau phenol reagent. The results agreed well with the Kjeldahl nitrogen determination. In the second procedure, the extracted brown components in dark grains did not affect the precision of the determination as they did in the first method.

Lowry Method (Phenol Reagent)

One of the most widely used methods for the determination of protein in solutions is based on the interaction of proteins with the phenol reagent and copper under alkaline conditions. The color reaction involves a copper-catalyzed oxidation of aromatic amino acids and other groups by a heteropolyphosphate (phosphotungstic–phosphomolybdic) reagent. The basis of the method was established by Wu (1922), and a number of modi-

fications appeared before the important modifications of Folin and Ciocalteau (1927) and of Lowry *et al.* (1951). Chow and Goldstein (1960)
showed that many of the functional groups found in proteins (in addition
to the well-known reactivity of tyrosine and tryptophan) are responsible
for the final blue color concentration. The procedure (generally known
as the Lowry method) is highly valued because of its sensitivity.

The method is 10–20 times as sensitive as ultraviolet absorbance methods and up to 100 times as sensitive as the biuret method. The method
is relatively specific, since few substances encountered in biological materials cause serious interference (Solecka *et al.* 1968). The results are
affected little by the turbidity of the original protein solutions. Despite
these advantages, the empirical nature of the Lowry method must be
recognized. The color intensity varies with the amino acid composition
of the protein and the analytical conditions, and the color is not strictly
proportional to protein concentration. The Lowry method is more time-
consuming than direct absorbance measurement at 280 nm, is destructive,
and requires multiple operations on each sample and incubation between
the addition of reagents. The procedure is sensitive to high concentrations
of sucrose, such as are used in sucrose gradient ultracentrifugation. Attempts to overcome the sucrose inhibition have been only partly successful (Schuel and Schuel 1967).

Direct Spectrophotometric Methods

Most proteins exhibit a distinct ultraviolet absorption maximum at 280
nm, due primarily to the presence of tyrosine, tryptophan, and phenylalanine. Since the content of these amino acids in proteins from some
sources differs within a reasonably narrow range, the absorption peak at
280 nm has been used as a rapid and fairly sensitive test of protein concentration (Warburg and Christian 1941).

The measurement of ultraviolet absorption is a rapid, easily accomplished, and generally nondestructive method for the assay of protein
contents of biological fluids and protein solutions from fractionation procedures. The determination can be made under conditions where other
methods fail (in the presence of ammonium ions and certain salts). The
results, however, must be interpreted with care. Since proteins (and especially fractionated proteins) differ considerably in their amino acid composition, absorption at 280 nm may vary by a factor of five or more for
equal concentrations of protein (Sober *et al.* 1965). Nucleic acids also
have a strong ultraviolet absorption band at 280 nm, but they absorb much
more strongly at 260 than at 280 nm. With proteins, the reverse is true.
This fact is used to eliminate, by calculation, the interference of nucleic
acids in the determination of protein. However, considerable error may

be present in such calculations, since different proteins and nucleic acids do not always have the same absorption.

Nakai *et al.* (1964) described a procedure for the protein determination in milk based on absorbance measurement at 280 nm. The method can be used in testing regular, defatted, pasteurized, and sterilized milk and cream, but is not suitable for assay of whey or casein solutions. The lengthy preparation prior to an absorbance measurement partly eliminates the advantage of the absorbance procedure. Whitaker and Granum (1980) proposed an absolute method for protein determination based on the difference in absorbance at 235 and 280 nm. Gabor (1979, 1983) described methods for the determination of protein content of meat and meat products based on ultraviolet absorption spectroscopy. Spectrophotometric determination at 280 nm of protein in agricultural products, foods, and model food systems was studied by Popov *et al.* (1983).

The peptide bond has a strong absorption band at about 180 nm. Unfortunately, the technical difficulties of measurements at such a low wavelength are almost insurmountable. A compromise is often made by measuring at 210 or 220 nm; the measurement still being up to 20 times more sensitive than at 280 nm. At the low wavelengths, many of the commonly used buffers, particularly those containing carboxyl groups, absorb strongly and the selection of buffers is restricted.

Fluorescence measurements can detect and determine some proteins on a nondestructive basis and at very low levels. As the ultraviolet fluorescence is primarily due to tryptophan residues, with lesser contributions from tyrosine and alanine, the response from different proteins varies over a considerable range. Fox *et al.* (1963) evaluated the effects of pH, temperature, fluorescing impurities, and protein aggregation on determining protein in milk by fluorescence. By use of a urea solution containing citrate and phosphate, protein aggregation was eliminated and reproducibility was improved. The direct fluorimetric method was found precise and rapid. However, it requires a high dilution, good temperature control, and fails in the presence of certain impurities.

Nephelometric or Turbidimetric Methods

The turbidity produced when a protein is mixed with low concentrations of any of the common precipitants can be used as an index of protein concentration. The commonly used precipitants are trichloroacetic acid, potassium ferricyanide, and sulfosalicylic acid.

For proteins in a solution, the methods are rapid and convenient. For solid foods, extraction of the protein into a solution both lengthens and complicates an analysis. The methods yield different values with different proteins, and they do not permit differentiation between protein and other components precipitated under the employed experimental conditions. The methods are generally highly dependent on reproduction of conditions and calibration against other empirical methods.

A light-scattering technique for estimating soluble protein was described by Tappau (1966). The automatic determination is performed by treating sample solutions with acetic acid, potassium ferricyanide, and in some cases sodium tungstate and measuring the light scattered by the precipitate. A turbidimetric method based on precipitating with sulfosalicylic acid the alkaline extract of wheat or wheat flour was described by Feinstein and Hart (1959).

Dye-Binding Methods

Fraenkel-Conrat and Cooper (1944) were the first to report that proteins bind quantitatively, under specified conditions, with certain organic dyes. The dye binding can be used to determine total acidic and basic groups of proteins. Their observations led to numerous investigations on dye-binding methods for protein determination in foods.

Earlier investigations (Udy 1954, 1956) estimated protein content from the amount of a disulfonic anionic dye, orange G, bound at pH 2.2. This dye binds specifically under acidic conditions to free amino groups, lysine, the imidazole group of histidine, and the guanidyl group of arginine. It has been subsequently shown that the estimation of protein by dye binding can be improved by using the dye acid orange 12, which is structurally identical to orange G except that it has only one sulfonic acid group.

When a food sample is treated with excess dye, the dye and protein in the food react quantitatively to form an insoluble complex that can be separated either by centrifugation or filtration. From the concentration of unbound dye (measured colorimetrically), the binding capacity can be calculated. A quantitative relationship between the amount of bound dye and the protein content of the sample permits the construction, for each food, of a conversion table from which the percentage protein is read. A commercial instrumental setup for dye-binding measurements is available. In this unit, the dye is mixed with the sample in a special reactor or on a laboratory shaker and is transferred to a squeeze-type polyethylene bottle fitted with a fiberglass filter disk in the dropper cap. Light transmittance through the filtered dye solution is determined in a special colorimeter as the filtrate is transferred dropwise into a flow-through cuvette. The dye-binding procedure is rapid and eliminates the problems of skillful manipulation and corrosive reagents of the Kjeldahl procedure.

The binding of acid orange 12 can be used to determine the protein contents of sound and normal samples of cereal grains, oil seeds, legumes, and animal and dairy products. Additional dyes that have been recommended, mainly for protein determination in meat and milk products, include cochineal red A, buffalo black, and amido black 10B.

Amido black 10B was shown (Tarassuk *et al.* 1967) to give a greater change in optical density per unit of milk protein than orange G. The binding of amido black was affected by the presence of potassium dichromate and formaldehyde but not by mercuric chloride. The dye-binding capacity of milk protein was not affected by homogenizing, condensing, or heating to 32°C for 15 min. Extensive proteolysis increases dye binding, whereas heating to browning reduces it. The dye-binding test is considered suitable for normal milk samples but not for atypical milk such as colostrum, mastitic, and very late lactation milks.

Lewis *et al.* (1980) found that a colorimetric method based on the binding of coomassie blue G-250 measured accurately the decrease in protein content after chillproofing beer. A commercial dye-binding assay is based on the observation that absorbance of coomassie blue G-250 shifts from 465 nm to 595 nm when protein binding takes place (Anon. 1979). The assay is simple (one step), rapid (5 min), requires no critical timing, produces a stable color, and is free of interferences.

Other Methods for Assaying Total Protein

Protein determination by nuclear magnetic resonance was described by Coles (1980), Rutar and Blinc (1980), and Wright *et al.* (1980). Williams *et al.* (1978) found that neutron activation analysis, proton activation analysis, and thermal decomposition analysis were all reliable tests for protein content provided the samples were representative, properly prepared for analyses, and the instruments were properly calibrated.

The use of a pyrochemiluminescent method to determine the protein content of biological materials was described by Jancar *et al.* (1983). In their method, chemically bound nitrogen is converted to nitric oxide by oxidative pyrolysis. Nitric oxide, in contact with ozone, produces metastable nitrogen dioxide, which emits a photon as it relaxes to a stable state. The emission is proportional to the amount of nitrogen in the sample and can be measured quantitatively. The method is rapid (5 min), reagent costs are minimal, and no hazardous chemicals are needed.

Several methods, in addition to the Lowry method described earlier, are based on the chemical reaction of specific amino acids (especially the aromatic amino acids) with certain reagents. In the xantoproteic acid reaction, for example, a white precipitate that slowly turns yellow is formed when a protein solution is heated in the presence of concentrated nitric acid. The yellow color results from the nitration of aromatic amino acids. On neutralization, the color turns orange. The xantoproteic acid reaction is simpler, but less sensitive and more difficult to standardize, than the

Lowry method. Another such method, the Millon reaction, is specific for tyrosine. The reagent is prepared by dissolving mercury in excess nitric acid and diluting the solution. Reaction with proteins gives, upon boiling, a brick-red color.

Proteins exert relatively large effects on certain physical properties of their solutions. Determination of these effects can be used as a basis for estimating the protein content. The basic assumption in all of these tests is that the effect of nonprotein solutes is negligible. To the extent that the assumption is valid, determination of physical properties can be used to determine protein content rapidly, accurately, and reproducibly.

A refractive index determination is a good general physical method, as the refractive index is nearly the same for most proteins. Determination of bovine plasma protein from refractive index measurements was described by Weeth and Speth (1968). The advantages of refractometric assays of protein in milk were outlined by Andrievskaya (1966). Certain amino acids, polypeptides, and proteins dissolved in a cobalt-containing buffer of suitable pH produce a catalytic reaction at the dropping mercury electrode. The reaction has been named after its discoverer, Brdicka. Application of the reaction in biochemical analysis was reviewed by Muller (1963). The use of infrared spectroscopy in the analysis of proteins in foods was discussed by Kohn (1965); features of an infrared milk analyzer were described by Biggs (1967). The use of near infrared reflectance spectroscopy for determination of protein and other food components was discussed in Chapter 9. Other physical parameters that have been proposed for protein determination include specific gravity, viscosity, surface tension, conductivity, and polarization.

Formol-binding methods have gained little popularity due to their low precision (Kirk 1947). The methods measure the increase in acidic groups that can be titrated with base, when amino acids, proteins, and polypeptides are reacted with neutral formaldehyde. However, some studies (Drux and Bauer 1964; Hill and Stone 1964) indicate that formol titration can be used in determining the protein content of natural and processed milk (including ice cream).

Christian (1966) described a direct titration method of native proteins in solution. The titrant is coulometrically generated hypobromite and the end point is detected amperometrically.

Several indirect methods of protein determination in milk were reviewed by Cavagnol (1961). Milk protein heated with sodium hydroxide releases a consistent amount of ammonia; the method is rapid and linearly related to the Kjeldahl nitrogen determination. By measuring the volume of phosphotungstic acid-precipitated serum proteins, it is possible to measure the serum protein in milk. The method correlates well with the Kjeldahl procedure provided the acid precipitable casein and denatured serum protein are removed first.

Sensitivity of Various Assay Methods

Thorne (1978) and Kresze (1983) reviewed four types of assay for determining protein concentration: (1) methods based on ultraviolet absorption; (2) methods based on chemical reactions (Folin–Ciocalteau–Lowry, biuret, and biuret–Folin assays); (3) dye-binding methods for proteins in solution and electrophoretic gel staining; and (4) other high-sensitivity methods (fluorimetric, radioactive, etc). They noted that many substances interfere with methods based on chemical reactions. The high-sensitivity methods can be used to assay less than 0.1 µg; the dye-binding methods, generally 1–100 µg; the chemical methods above 10 µg (Folin–Lowry) or above 1000 µg (biuret); and the UV methods above 10 µg (at 215 nm) or above 100 µg (at 280 nm).

NONPROTEIN NITROGENOUS SUBSTANCES

The significance of nonprotein, organic, nitrogenous compounds in foods has been appreciated only in recent years. These compounds include amino acids, amines, amides, quarternary nitrogen compounds, purines, pyrimidines, and N-nitrosoamides. They contribute to nutritional value, flavor, color (especially in baked or roasted products), and other important food attributes. They provide a source of nutrients and growth factors that are important in malting, brewing, and panary fermentation. On the other hand, excessively high levels of free amino acids may result from proteolytic degradation of proteins in cereal grains stored at elevated moisture levels and temperatures. Similarly, degradation products of animal proteins are indices of incipient deterioration, and are used to ascertain storability and soundness of foods. It has been known for over a quarter of a century that N-nitrosodimethylamine can cause liver tumors in a variety of animal species. In the last two decades researchers have found that the majority of 300 N-nitroso compounds are also toxic, mutagenic, or carcinogenic. Analytical techniques for N-nitrosoamines were reviewed by Scanlan and Reyes (1985).

Inorganic nitrogenous compounds (ammonia, nitrate, and nitrite) are determined to establish the sanitary status of foods, to follow aging and processing of cheese, or to ascertain the absence of excessive amounts of undesirable pickling components in processed meats. They are determined in many investigations of nitrogen metabolism in animals and plants.

Separation from Proteins

The least empirical procedures for separating proteins from nonprotein nitrogenous compounds are dialysis and ultrafiltration with suitable membranes. Heat coagulation is likely to free most juices or extracts from protein, though some proteins (casein, gelatin) are not heat-coagulable. Heat coagulation and some protein precipitants have the advantages of rapidly inactivating proteolytic enzymes and reducing artifacts. The use of protein precipitants such as picric acid, sulfosalicylic acid, trichloroacetic acid, etc., is empirical. The use of aqueous ethanol for obtaining protein-free extracts cannot be applied to gliadin-containing materials. In most separations, possible retention on the precipitated protein of small nonprotein molecules by adsorption, ion exchange, etc., should be considered (Synge 1955).

In a study of the methods for the determination of nonprotein nitrogen, Bell (1963) used the following methods to separate protein from nonprotein nitrogenous substances in skim milk, serum, a water extract of flour, and a water extract of bran: dialysis, heat, tungstic acid, copper hydroxide, ferric oxide hydrasol, lead acetate, trichloroacetic acid, phosphotungstic acid, metaphosphoric acid, tannic acid, sulfosalicylic acid, ethanol, mercuric chloride, a mixture of chloroform and octyl alcohol (8:1), and a mixture of phenol, acetic acid, and water (1:1:1). Nonprotein fractions obtained by the various protein separation methods differed widely. Dialysis or related techniques appeared to achieve separations most closely related to nonprotein nitrogen, as theoretically defined. The binding of amino acids to proteins was small.

Determination of Free Amino Acids

Several methods are available to determine the aggregate of free amino acids in protein-free extracts. Titrimetric methods respond to the carboxyl groups of acids and other compounds and to phenolic groups of glycosides. The formol titration is fairly specific for amino groups but does not distinguish between amino acids and other amines. Gasometric methods in which the nitrogen liberated after adding nitrous acid is measured manometrically are somewhat more specific in this respect, but they give negative results with proline and other secondary amines. There is also interference by the amide group of glutamine, and glutathione gives anomalous results.

Methods based on complex formation with copper, such as that of Pope and Stevens (1939), are useful for semiquantitative determination, as the colored complexes are fairly specific for free amino acids. The color reaction of amino acids with ninhydrin is used widely though it has two disadvantages: the color and its intensity vary for different amino acids, and the procedure is not specific for amino acids. The high specificity of the ninhydrin–carbon dioxide method (Van Slyke *et al.* 1941) makes it attractive for the determination of total free amino acids.

Some free amino acids can be determined by specific colorimetric tests. Separation of free amino acids by two-dimensional paper chromatography, elution from the paper and determination by spectrophotometry or direct absorbance measurement on the paper, are useful for semiquantitative estimations. Enzymatic methods employing specific decarboxylases can be used for some amino acids. Microbiological assays sometimes give poorer results with free amino acids than with protein hydrolysates, as the extracts of the former may contain growth inhibitors or stimulants and peptides that affect the results (Synge 1955). Best quantitative results can be obtained by fractionation on ion-exchange columns.

Nitrate and Nitrite Assays

Usher and Telling (1975) critically compared methods for analysis of nitrate and nitrite in foods. Spectrophotometric methods for nitrate determination are based on (1) reduction of nitrate to nitrite or ammonia, (2) nitration of a phenolic-type compound, or (3) oxidation of an organic compound by the nitrate. Ascorbate, sulfite, and phosphate interfere with the reduction of nitrate to nitrite. Nitration substrates include chromotropic acid, 4-methylumbelliferone, salicylic acid, phenol disulfonic acid, and several xylenols. Oxidation substrates are brucine, diphenylamine, strychnine, and strychnidine.

Methods used commonly for the determination of nitrite are based on its interaction with a primary aromatic amine in acid solution to form a diazonium salt (the Griess–Ilosway procedure). Aromatic amines that have been used are sulfanilic acid, sulfanilamide, or 1-naphthylamine. The diazonium salt is coupled with an aromatic compound to form an azo color. The coupling reagent is 1-naphthylamine, 1-naphthyl ethylene diamine, 1-naphthol, or 1-naphthol sulfonic acid. Various modifications of the colorimetric methods were described by Sen and Donaldson (1978), Sen and Lee (1979), Nijhuis et al. (1979), and Lox and Okabe (1982).

Several newer methods for determining nitrate and nitrite are based on the use of ion-selective electrodes. Use of the specific nitrate reductase for the determination of nitrate in meat and fishery products was described by Hamano et al. (1983). According to Mills (1980), the nitrate electrode is susceptible to interfering ions, particularly chloride. Collet (1983) presented a procedure for the determination of nitrate and nitrite in foods without prior nitrite oxidation or nitrite reduction. In this procedure, nitrite is determined by the Griess–Ilosway procedure (with sulfanilic acid as aromatic amine and 1-naphthylamine as coupling reagent). Nitrate is determined after nitration to nitrophenol through direct current or differential pulse polarography. Both compounds are isolated by ion-exchange chromatography. The required sample is 10 g, and the detection limit is 1 mg nitrate and 0.2 mg nitrite per 1 kg of food.

The use of ion-exchange chromatography with anion-exchange columns and an ion-pair chromatography system using reversed-phase columns to eliminate interferences in colorimetric assays of nitrites was described by Luckas (1984). According to Jackson et al. (1984), conventional spectrometric methods of nitrite and nitrate determination are unsatisfactory for cured meats. They developed a new procedure based on the use of a low-capacity anion-exchange column with chloromethane sulfonate as eluent and UV absorbance at 210–214 nm as the detection method.

Other Methods

Varner et al. (1953) described a method to determine ammonium, amide, nitrate, and nitrite nitrogen in protein-free plant extracts. The extracts are buffered at pH 10 with borate, placed in a modified semimicro Kjeldahl distillation unit, and the ammonium nitrogen is removed by vacuum distillation at 40°C. Concentrated alkali is added to the distillation flask and the amide nitrogen is removed by steam distillation at 100°C. After adding ferrous sulfate as a reducing agent, the nitrite nitrogen is removed as ammonia. The nitrate nitrogen is reduced to ammonia by ferrous sulfate after addition of silver sulfate as catalyst. The procedure requires about 20 min. To overcome the interference of glucose, an ion-exchange resin mixture may be used and the total procedure requires about 2 hr. If the nitrate concentration is very low, a colorimetric determination is recommended. Modifications to overcome some of the limitations of this procedure (amino nitrogen is not estimated and nitrate reduction is incomplete) were suggested by Barker and Volk (1964). Christianson et al. (1965A,B) described chromatographic and specific spectrographic methods to identify and determine nonprotein nitrogenous substances in corn grain and corn steep liquor.

Several methods are available to detect and determine ammonia in protein-free extracts. They are based on the basic properties of the volatile ammonia, on its oxidation to elementary nitrogen, on conversion of the ammonium ion with formaldehyde to hexamethylene-tetramine, and on the Nessler reaction. In most cases, it is essential to eliminate the interference from nitrogen-containing organic compounds that might give secondary reactions leading to the formation of ammonia.

BIBLIOGRAPHY

Altschul, A. M. (1962). Seed proteins and world food problems. Econ. Bot. 16, 2–13.
Andrievskaya, L. (1966). Refractometric analyzer for milk. Proc. 17th Intern. Dairy Congr. 2, 187–189. [Chem. Abstr. 66, 9618.]

Anon. (1968). "Amino Acid Content of Food and Biological Data on Proteins." U.N. Food Agric. Organ., Rome, Italy.

Anon. (1979). "Bio-Rad Protein Assay." Bulletin *1069*. Bio·Rad Laboratories, Richmond, CA.

Anon. (1982). Ten minute 'Kjeldahls' are here. *News and Notes for the Analyst* **6**(1), 12. Hach Co., Loveland, CO.

Barker, A. V., and Volk, R. J. (1964). Determination of ammonium, amide, amino, and nitrate nitrogen in plant extracts by a modified Kjeldahl method. *Anal. Chem.* **36**, 439–441.

Bell, P. M. (1963). A critical study of method for the determination of nonprotein nitrogen. *Anal. Biochem.* **5**, 443–451.

Biggs, D. A. (1967). Milk analysis with the infrared milk analyzer. *J. Dairy Sci.* **50**, 799–803.

Block, R. J. (1945). The amino acid composition of food proteins. *Adv. Protein Chem.* **2**, 119–154.

Block, R. J. (1960). Amino acid analysis of protein hydrolysates. *In* "A Laboratory Manual of Analytical Methods of Protein Chemistry," Vol. I, The Separation and Isolation of Proteins (P. Alexander and R. J. Block, eds.). Pergamon Press, London.

Bradstreet, R. B. (1965). "The Kjeldahl Method for Organic Nitrogen." Academic Press, New York.

Cavagnol, A. (1961). Food. *Anal. Chem.* **33**, 50R–60R.

Chow, S., and Goldstein, A. (1960). Chromogenic groups in the Lowry protein determination. *Biochem. J.* **75**, 109–115.

Christian, G. D. (1966). Coulometric titration of proteins with electrogenerated hypobromite. *Anal. Biochem.* **14**, 183–190.

Christianson, D. D., Cavins, J. F., and Wall, J. S. (1965A). Steep liquor constituents. Identification and determination of nonprotein nitrogenous substances in corn steep liquor. *Agric. Food. Chem.* **13**, 277–280.

Christianson, D. D., Wall, J. S., and Cavins, J. F. (1965B). Nutrient distribution in grain. Location of nonprotein nitrogenous substances in corn grain. *Agric. Food Chem.* **13**, 272–276.

Coles, B. A. (1980). Protein determination by nuclear magnetic resonance. *J. Am. Oil Chem. Soc.* **57**, 202–204.

Collet, P. (1983). A contribution to the determination of nitrites and nitrates in foods (in German). *Dtsch. Lebensm. Rundsch.* **79**, 370–375.

Conway, E. J. (1957). "Microdiffusion Analysis and Volumetric Error." Crosby, Lockwood, and Son, London.

Danehy, J. P., and Pigman, W. W. (1951). Reactions between sugars and nitrogenous compounds and their relationship to certain food problems. *Adv. Food Res.* **3**, 241–290.

Drux, A., and Bauer, H. J. (1964). Contribution to determination of protein content in milk by formol titration (in German). *Nahrung* **8**, 99–103.

Dunn, M. S. (1950). Determination of amino acids. *Adv. Chem. Ser.* **3**, 13–28.

Feeney, R. E., and Hill, R. M. (1960). Protein chemistry and food research. *Adv. Protein Chem.* **10**, 23–73.

Feinstein, L., and Hart, J. R. (1959). A simple method for determining the protein content of wheat and flour samples. *Cereal Chem.* **36**, 191–193.

Ferrari, A. (1960). Nitrogen determination by a continuous digestion and analysis system. *Anal. N. Y. Acad. Sci.* **87**, 792–800.

Folin, O., and Ciocalteau, V. (1927). Tyrosine and tryptophan determination of proteins. *J. Biol. Chem.* **73**, 627–650.

Fox, K. K., Holsinger, V. H., and Pallansch, M. J. (1963). Fluorimetry as a method of determining protein content of milk. *J. Dairy Sci.* **46**, 302–310.

Fraenkel-Conrat, H., and Cooper, M. (1944). The use of dyes for the determination of acid and basic groups in proteins. *J. Biol. Chem.* **154**, 239–246.

Gabor, E. (1979). Determination of protein content of certain meat products by ultraviolet absorption spectrophotometry. *Acta Aliment.* **8**, 157–167.

Gabor, E. (1983). Determination of protein in meat and meat products by ultraviolet spectroscopy. *Fleisch* **37**, 194–195.

Gehrke, C. W., and Stalling, D. L. (1967). Quantitative analysis of the twenty natural protein acids by gas–liquid chromatography. *Separ. Sci.* **2**, 101–138.

Glebko, L. I., Ulkina, J. I., and Vaskovsky, V. E. (1967). Spectrophotometrical method for determination of nitrogen in biological preparations based on thymol–hypobromite reaction. *Anal. Biochem.* **20**, 16–23.

Hamano, T., Mitsuhashi, Y., Tanaka, K., Matsuki, Y., and Oji, Y. (1983). Application of nitrate reductase for the determination of nitrate in meat and fishery products. *Agric. Biol. Chem.* **47**, 2427–2433.

Hettrick, J. H., and Whitney, R. M. (1949). Determination of nitrogen in milk by direct nesslerization of the digested sample. *J. Dairy Sci.* **32**, 111–112.

Hill, R. L., and Stone, W. K. (1964). Procedure for determination of protein in ice milk and ice cream by formol titration. *J. Dairy Sci.* **47**, 1014–1015.

Jackson, P. E., Haddad, P. R., and Dilli, S. (1984). Determination of nitrate and nitrite in cured meats using high-performance liquid chromatography. *J. Chromatogr.* **295**, 471–478.

James, A. T., and Morris, L. J. (1964). "New Biochemical Separations." D. van Nostrand Publishing Co., London.

Jancar, J. C., Constant, M. D., and Hernig, W. C. (1983). Protein content of beer and wort by pyrochemiluminescence. *J. Am. Soc. Brew. Chem.,* 158–160.

Jennings, A. C. (1961). Determination of the nitrogen content of cereal grain by colorimetric methods. *Cereal Chem.* **38**, 467–478.

Jones, D. B. (1931). Factors for converting percentages of nitrogen in foods and feeds into percentages of proteins. U.S. Dept. Agr. Circl. **183**, 1–21.

Keller, S., and Block, R. J. (1960). Separation of proteins. *In* "A Laboratory Manual of Analytical Methods of Protein Chemistry," Vol. I, The Separation and Isolation of Proteins (P. Alexander and R. J. Block, eds.). Pergamon Press, London.

Kirk, P. L. (1947). The chemical determination of proteins. *Adv. Protein Chem.* **3**, 139–167.

Kjeldahl, C. (1883). New method for determination of nitrogen in organic materials. *Z. Anal. Chem.* **22**, 366–382. (German)

Kohn, R. (1965). Application of infrared spectroscopy in food analysis. 1. Infrared spectroscopy in the intermediate range of 3 to 15 μm. Z. Lebensm. *Unters. Forsch.* **129**, 28–40. (German)

Kresze, G.-B. (1983). Methods in protein determination. *In* "Methods of Enzymatic Analysis," 3rd ed. Vol. II, Samples, Reagents, Assessment of Results, pp. 84–99. Verlag Chemie, Weinheim, W. Germany.

Lakin, A. L. (1978). Determination of nitrogen and estimation of protein in foods. *In* "Developments in Food Analysis Techniques," Vol. I (R. D. King, ed.), pp. 43–74. Applied Science Publishers, London.

Lewis, M. J., Krumland, S. C., and Muhleman, D. J. (1980). Dye-binding method for measurement of protein in wort and beer. *J. Am. Soc. Brew. Chem.,* 37–41.

Lichtfield, J. H., and Sachsel, G. F. (1965). Technology and protein malnutrition. *Cereal Sci. Today* **10**, 458, 460–462, 464, 472.

Lowry, O. H. Rosebrough, N. J., Farr, A. L., and Randall, R. J. (1951). Protein measurement with the Folin phenol reagent. *J. Biol. Chem.* **193**, 265–275.

Lox, F., and Okabe, A. (1982). Comparison of nitrate and nitrite determinations in vegetables: Suitability for accurate and automated measurements. *J. Assoc. Off. Anal. Chem.* **65**, 157–161.

Luckas, B. (1984). Determination of nitrite and nitrate in foods by chromatography (in German). *Fresenius Z. Anal. Chem.* **318**, 428–433.

Mann, L. T. (1963). Spectrophotometric determination of nitrogen in total micro-Kjeldahl digests. Application of phenol–hypochlorite reaction to microgram amounts of ammonia in total digest of biological material. *Anal. Chem.* **35**, 2179–2183.

Martin, A. J. P., and Synge, R. L. M. (1945). Analytical chemistry of the proteins. *Adv. Protein Chem.* **2**, 1–83.

Mills, H. A. (1980). Nitrogen specific ion electrodes for soil, plant, and water analysis. *J. Assoc. Off. Anal. Chem.* **63**, 797–801.

Minari, O., and Zilversmit, D. B. (1963). Use of KCN for stabilization of color in direct nesslerization of Kjeldahl digests. *Anal. Biochem.* **6**, 320–327.

Motz, R. J. (1968). A critical evaluation of the AOAC method for the determination of milk protein in milk chocolate when applied to crumb chocolate. *Analyst* **93**, 116–117.

Muller, O. H. (1963). Polarographic analysis of proteins, amino acids, and other compounds by means of the Brdicka reaction. *Methods Biochem. Anal.* **11**, 329–403.

Nakai, S., Wilson, H. K., and Herreid, E. O. (1964). Spectrophotometric determination of protein in milk. *J. Dairy Sci.* **47**, 356–358.

Neill, C. D. (1962). The Kjeldahl protein test. *Cereal Sci. Today* **7**, 6–8, 10, 12.

Nijhuis, H., Heeschen, W., and Bluthgen, A. (1979). Automated determination of nitrate and nitrite in milk and dairy products by means of the continuous flow analysis. *Milchwissenschaft* **34**, 414–416.

Orr, M. L., and Watt, B. K. (1957). "Amino Acid Content of Foods." Home Econ. Res. Rept. *4*, U.S. Dept. Agric., Washington, DC.

Osborne, T. B. (1907). "The Proteins of the Wheat Kernel." Carnegie Institute of Washington, Washington, DC.

Parnas, J. K. (1938). The Kjeldahl nitrogen determination by a modification of Parnas and Wagner. *Z. Anal. Chem.* **114**, 261–275. (German)

Pfleiderer, G. (1967). Nitrogenous compounds, proteins, amino acids, and amines. *In* "Handbuch der Lebensmittelchemie," Vol. II, Part 2, Analytik der Lebensmittel, Nachweis und Bestimmung von Lebensmittel-Inhaltstoffen (J. Schormuler, ed.). Springer-Verlag, Berlin. (German)

Pinckney, A. J. (1949). Wheat protein and the biuret reaction. *Cereal Chem.* **26**, 423–439.

Pinckney, A. J. (1961). The biuret test as applied to the estimation of wheat protein. *Cereal Chem.* **38**, 501–506.

Pomeranz, Y. (1965). Evaluation of factors affecting the determination of nitrogen in soya products by the buiret and orange G binding methods. *J. Food Sci.* **30**, 307–311.

Pomeranz, Y. (1968). Relation between chemical composition and breadmaking potentialities of wheat flour. *Adv. Food Res.* **16**, 335–455.

Pope, C. G., and Stevens, M. F. (1939). Determination of amino nitrogen using a copper method. *Biochem. J.* **33**, 1070–1077.

Popov, M. P., Ilyasov, S. G., Kortunova, E. Y., and Sofronova, O. V. (1983). Detection of protein in powdery materials by a spectrophotometric method (in Russian). *Prikl. Biokhim. Mikrobiol.* **19**, 827–831.

Rhodes, D. N. (1963). Protein biochemistry. *In* "Recent Advances in Food Science," Vol. 3 (J. M. Leitch and D. N. Rhodes, eds.). Butterworths, London.

Rice, E. E., and Beuk, J. F. (1953). The effects of heat upon the nutritive value of protein. *Adv. Food Res.* **4**, 233–279.

Riegler, E. (1914). A colorimetric method for determination of albumin. *Z. Anal. Chem.* **53**, 242–245.

Rutar, V., and Blinc, R. (1980). Nondestructive determination of protein content of viable seeds by protein enhanced ^{13}C NMR. *Z. Naturforsch.* **35C**, 12–15.

Scanlan, R. A., and Reyes, F. G. (1985). An update on analytical techniques for *N*-nitrosamines. *Food Technol.* **39**(1), 95–99.

Schober, R., Niclaus, W., and Christ, W. (1964). Possibility of using the biuret reaction for determining protein in milk. *Milchwissenschaft* **19**, 75–78. [*Anal. Abstr.* **12**, 1987.]

Schuel, H., and Schuel, R. (1967). Automated determination of protein in the presence of sucrose. *Anal. Biochem.* **20**, 86–93.

Schultz, H. W., and Anglemier, A. F. (1964). "Symposium on Foods: Proteins and Their Reactions." AVI Publishing Co., Westport, CT.

Sen, N. P., and Donaldson, B. (1978). Improved colorimetric method for determining nitrate and nitrite in foods. *J. Assoc. Off. Anal. Chem.* **61**, 1389–1394.

Sen, N. P., and Lee, Y. C. (1979). Determination of nitrate and nitrite in whey powder, *Agric. Food Chem.* **27**, 1277–1279.

Singh, V., Sahranat, K. L., Jambunathan, R., and Burford, J. R. (1984). The use of hydrogen peroxide for the digestion and determination of total nitrogen in chick pea (*Cicer arietinum* L.) and pigeon pea (*Cajanus cajan* L.). *J. Sci. Food Agric.* **35**, 640–646.

Siriwardene, J. A., Evans, R. A., Thomas, A. J., and Axford, R. F. E. (1966). Use of an automated Kjeldahl analyzer for determination of nitrogen in biological material. *Proc. Technicon Symp. Automation in Analytical Chemistry.*

Sober, H. A., Hartley, R. W., Carroll, W. R., and Peterson, E. A. (1965). Fractionation of proteins. *In* "The Proteins, Composition, Structure and Function" (H. Neurath, ed.). Academic Press, New York.

Solecka, M., Ross, J. A., and Millikan, D. F. (1968). Evidence of substances interfering with the Lowry test for protein in plant leaf tissue. *Phytochemistry* **7**, 1293–1295.

Spackman, D. H., Stein, W. H., and Moore, S. (1958). Automatic recording apparatus for use in the chromatography of amino acids. *Anal. Chem.* **30**, 1190–1206.

Sternglanz, P. D., and Kollig, H. (1962). Evaluation of an automatic nitrogen analyzer for tractable and refractory compounds. *Anal. Chem.* **34**, 544–547.

Synge, R. L. M. (1955). Peptides (bound amino acids) and free amino acids. *In* "Modern Methods of Plant Analysis," Vol. I (K. Paech and M. V. Tracey, eds.). Springer-Verlag, Berlin.

Tappau, D. V. (1966). A light scattering technique for measuring protein concentration. *Anal. Biochem.* **14**, 171–182.

Tarassuk, N. P., Abe, N., and Moats, W. A. (1967). "The Dye Binding of Milk Proteins." U.S. Dept. Agr. Tech. Publ.

Thorne, C. J. R. (1978). Techniques for determining protein concentration. *In* "Techniques in Protein and Enzyme Biochemistry," B 104, pp. 1–18. Elsevier/North Holland Biomedical Press, Amsterdam.

Tkachuk, R. (1966). Note on the nitrogen to protein conversion factor for wheat flour. *Cereal Chem.* **43**, 223–225.

Torten, J., and Whitaker, J. R. (1964). Evaluation of the biuret and dye-binding methods for protein determination in meats. *J. Food Sci.* **29**, 168–174.

Udy, D. C. (1954). Dye binding capacities of wheat flour protein fractions. *Cereal Chem.* **31**, 389.

Udy, D. C. (1956). A rapid method for estimating total protein in milk. *Nature* **178**, 314–315.

Usher, C. D., and Telling, G. M. (1975). Analysis of nitrate and nitrite in foodstuffs: A critical review. *J. Sci. Food Agric.* **26**, 1793–1805.

Van Slyke, D. D., and Hiller, A. (1933). Determination of ammonia in blood. *J. Biol. Chem.* **102**, 499–504.

Van Slyke, D. D., Dillon, R. T., MacFadyen, D. A., and Hamilton, P. (1941). Gasometric determination of carboxyl groups in free amino acids. *J. Biol. Chem.* **141**, 627–669.

Varley, J. A. (1966). Automatic methods for the determination of nitrogen, phosphorus and potassium in plant material. *Analyst* **91**, 119–126.

Varner, J. E., Bulen, W. A., Vanecko, S., and Burrell, R. C. (1953). Determination of ammonium, amide, nitrite, and nitrate nitrogen in plant extracts. *Anal. Chem.* **25**, 1528–1529.

Warburg, O., and Christian, W. (1941). Isolation and crystallization of the fermentation enzyme enolase. *Biochem. Z.* **310**, 384–421.

Weeth, H. J., and Speth, C. F. (1968). Estimation of bovine plasma protein from refractive index. *J. Anim. Sci.* **27**, 146–149.

Whitaker, J. R. (1959). Chemical changes associated with aging of meat with emphasis on the proteins. *Adv. Protein Chem.* **9**, 1–60.

Whitaker, J. R., and Granum, P. E. (1980). An absolute method for protein determination based on difference in absorbance at 235 and 280 nm. *Anal. Biochem.* **109**, 156–159.

Williams, P. C. (1964). Determination of total nitrogen in feeding stuffs. *Analyst* **89**, 276–281.

Williams, P. C., Norris, K. H., Johnsen, R. L., Standing, K., Fricioni, R., MacAffrey, D., and Mercier, R. (1978). Comparison of physicochemical methods for measuring total nitrogen in wheat. *Cereal Foods World* **23**, 544–547.

Wright, R. G., Milward, R. C., and Coles, B. A. (1980). Rapid protein analysis by low-resolution pulsed NMR. *Food Technol.* **34**(12), 47–52.

Wu, H. (1922). A new colorimetric method for the determination of plasma proteins. *J. Biol. Chem.* **51**, 33–39.

38
Objective versus Sensory Evaluation of Foods

The overall quality of food can be divided into three main categories: quantitative, hidden, and sensory (Kramer 1966). Some quantitative aspects of food quality are primarily of interest to the processor, e.g., yield of product obtained from a raw material; others are of interest both to the consumer and manufacturer, e.g., the ratio of more expensive to less expensive foods or components in a processed food. In some cases the ratio can be evaluated roughly by sensory methods. Hidden quality attributes include the nutritional value of a food or the presence of toxic compounds that, generally, cannot be determined by sensory evaluation. They include, for instance, tthe vitamin C content of juices, or the presence of trace amounts of pesticides from spraying fruits and vegetables. Sensory attributes of quality guide the consumer in his selection of foods. Such attributes are measured by the processor to determine consumer preference in order to manufacture an acceptable product at maximum production economy. Sensory attributes are measured also in determining the conformity of a food with established government or trade standards and food grades.

The selection, acceptance, and digestibility of a food are largely determined by its sensory properties. Evaluation of sensory properties, is, however, affected by personal preference, which is influenced by factors ranging from the caprices of fashion to the prevalence of dentures; social, cultural, and religious patterns; psychological factors; variations in climate and in the general physical status of the individual; availability; and nutritional education (Amerine *et al.* 1965).

To minimize the effects of such factors, different procedures for sensory evaluation have been devised and the results are evaluated by statistical methods. Large consumer groups are generally used to determine consumer reaction. Highly trained experts are employed for evaluating small differences in high-quality foods. Laboratory tests may be conducted to study the human perception of food attributes; to correlate sensory attributes with chemical and physical measurements; to evaluate raw material selection; to study processing effects and the means of maintaining uniform quality; to establish shelf-life stability; and to reduce costs. The aim of such laboratory tests may be to establish differences between samples, determine directional differences, or to determine quality-preference differences. Reasons for selecting a specific test; the experimental design; the mechanics of selection and the training of judges; and details of conducting, recording, and interpreting the results are outside the scope of this book. They can be found in several books including those published by Kramer and Twigg (1966), Amerine *et al.* (1965), and in the three ASTM publications (Anon. 1968A, B, C).

According to Kramer (1966) sensory attributes include (1) appearance (color, size, shape, absence of defects, and the consistency of liquid and semisolid products;) (2) kinesthetics (texture, consistency, and viscosity), and (3) flavor (taste and odor). Objective methods for the determination of kinesthetic attributes were discussed in detail previously.

COLOR

Color is an important appearance factor. In agriculture, color development or color changes are used in assessing the maturity of fruits and vegetables. The color of a food often affects our perception of and evaluation by other senses. Discoloration or the fading of color is often accompanied or identified by consumers as being associated with undesirable changes in texture and flavor. In addition to color, the appearance characteristics of gloss, sheen, transparency, and turbidity are often important. Many objective methods are available to measure most of the appearance characteristics. For a color determination, spectrophotometric or color standards are used. In some instances, however, objective measurements cannot evaluate correctly the composite visual appearance, and sensory tests are still used widely. See Chapter 7 for a fuller discussion of food color and its measurement.

TASTE AND ODOR

The available evidence suggests that four basic taste modalities—sweet, sour, salt, and bitter—are perceived by receptors on the tongue.

The sweet taste is produced by a variety of nonionized aliphatic hydroxy compounds, the most important of which are sugars. Saccharin, the best known synthetic sweetening agent, is 200–700 times as sweet as sucrose. Not all acids are sour; indeed some amino acids are actually sweet. The threshold perception for weak organic acids is at a pH of about 3.7–3.9; for strong organic acids at about 3.4–3.5. Sugar may enhance or depress the perception of sourness. A pure salty taste is typified in foods by sodium chloride. Generally, other salts give a mixed taste. The typical bitter stimuli are given by alkaloids such as caffeine or quinine; also some amides are bitter. Tannins (in tea and alcoholic beverages) contribute to both bitterness and astringency.

Some compounds are unique. Thus, creatine—a constituent of muscle—is tasteless to some individuals and bitter to others. Similarly, sodium benzoate—a food preservative—is variously sweet, sour, salty, bitter, or tasteless. Monosodium glutamate's flavor-enhancing properties are utilized widely by the food industry. Pure glutamates are odorless, but they have a pleasant, mild flavor with a sweet-salty taste and impart some tactile sensation. A specific, synergistic flavor-enhancing action exists between monosodium glutamate and some of the 5'-ribonucleotides.

Taste responses to many organic compounds are highly specific; thus anomers of some sweet-tasting sugars are bitter. Taste sensations and thresholds are affected by many factors, including food temperature, overall food composition, concentrations of individual components, and age and individual variations among tasters.

The use of polarimetry in establishing the purity of sugars or the identity of optically active ingredients that effect the taste of foods is well-established. Determination of refractive index is used widely to determine viscosity, sweetness, or the total solids of processed fruit and vegetable products. The refractive index has been used as an aid to detect the watering of milk or gross adulteration of butter with other fats. To determine the addition of small amounts of salt to juices or purées, actual assays of sodium chloride may be required. Objective measurements are useful in detecting the presence or development of bitter components, the chemical identity and assay of which have been established.

In many instances, several objective determinations are required; in others, computing the ratio between several parameters, e.g., the sugar–titratable acidity ratio, is required to establish a satisfactory correlation with sensory evaluation. The ratio depends, however, on the type, variety, and maturity of the raw fruit, and on the composition and concentration of the syrup.

In man, the receptors for the sensation of odor—olfaction—are restricted to a small portion of the olfactory mucosa. As many as several thousand different odors can be distinguished. Classification of odors has been the subject of many controversial investigations. Thus far, no satisfactory and widely-accepted system has been described. Odor in foods is most commonly attributed to organic compounds containing sulfur, nitrogen, and certain halogens. A functional group that imparts odor to an odorless compound is called an *osmophore*. Strong osmophores include inorganic (lead, arsenic, sulfur, chlorine, bromine) or organic (esters, carbonyls, amines, imines, lactones) entities. Double bonds, hydroxyl groups, and ring structures affect odor. Compounds with a high molecular weight (above 300) and low volatility are generally odorless. In addition, small differences in structure can significantly affect odor.

Gas–liquid chromatography (GLC) has been most useful in detecting small amounts of volatile components. Its usefulness has been increased by combining it with mass spectrometry or infrared spectroscopy for the identification of column effluents. Despite these contributions, GLC has certain important limitations. Some components that contribute to odor have a limited volatility (e.g., the boiling point of vanillin is 285°C); others are highly volatile at room temperature (e.g., hydrogen sulfide). Whereas the response of the chromatograph is linear, there is good evidence that the relationship between stimulus magnitude and response magnitude of the human nose is a power function (Wick 1965).

It has been, thus far, difficult to recombine the eluates from a GLC column or to prepare a synthetic mixture of individual components that duplicates the flavor of a natural product. Hopefully, with better separation technics (more sensitive and less degradative) and with a better understanding of the interaction between the individual components, preparation of such synthetic mixtures will be possible.

But even if GLC separated and identified clearly the components of flavor, we still have to use sensory methods to determine which of the components are indispensable to the production of desirable flavors. Such determinations must be made in the parent food, as its major (nonvolatile) components and texture affect significantly our evaluation of flavor.

Admittedly, evaluation of flavor by objective tests is a most challenging and difficult problem facing the food analyst (Anon. 1968C). According to Farber and Lerke (1958), several requirements should be met in the objective evaluation of freshness of fishery products. These include a high correlation with sensory evaluation, usefulness over a wide range of quality scores—from incipient spoilage to advanced deterioration—applicability to all kinds of fish products—including factory-processed, significant instrumental or analytical response to small quality differences, and relative speed and simplicity for use in routine testing.

How well can such requirements be met? According to Kramer and Twigg (1966) in a quality control system, where objective methods are relied upon, the use of a taste-testing panel is an admission of failure, except in a case where an objective method is being tested for conformance with human evaluation.

Casimir and Whitfield (1978) proposed using the flavor impact value (FIV) to evaluate the contribution of individual flavor components to the total flavor of a food. The FIV is the slope of the psychophysical response vs. the concentration curve and is determined by the assessment of several concentrations of the component by a taste panel. The use of the FIV was illustrated by separating 23 major flavor compounds from passion fruit juice, determining their FIV, and using the flavor profile obtained to prepare synthetic passion fruit drinks. Using this data, synthetic fruit drinks with acceptable flavor were prepared by combining proper proportions of the components.

Morrison et al. (1981) found a high correlation between pentane contents and flavor intensity scores and total volatiles and flavor intensity scores as well as flavor intensity values of sunflower salad oil. Human and Khayat (1981) developed an objective method of evaluating raw tuna quality on the basis of the volatile profile pattern obtained by gas chromatography. Five components of the volatile profile were significantly correlated to the raw tuna quality evaluated organoleptically. A chemical quality index was developed on commercially available tuna using the concentration of five compounds: ethanol, propanol, butanol, hexanal, and 1-pentene-3-ol. Quality designations based on the quality index yielded more accurate and reproducible classification than sensory tests.

The problems, pitfalls, and accomplishments encountered in objective (chemical and physical) measurements of flavor quality were reviewed by Jennings (1977) and Pangborn (1984). To start with, while flavor is an integrated response to a mixture of compounds, the gas chromatograph provides a differential response. Secondly, even today the gas chromatograph has not yet matched the sensitivity of the human nose. In addition, gas chromatography measures volatile compounds only, but taste is a major contributor to flavor. Sample preparation for gas chromatography and the conditions of chromatography may modify the compounds and introduce artifacts. Synergism and antagonism between various compounds must be considered as it affects the total perception of flavor. And finally, if the objective attributes are to be related to sensory perceptions, the latter must be properly standardized, performed, and interpreted or evaluated.

OVERALL EVALUATION

As mentioned earlier, acceptance of a food by consumers is generally affected by its various attributes. Consequently, several objective tests may be required for overall evaluation. A list of such tests is given in

Table 38.1. The main difficulty is that once all the individual variables are determined objectively, they must be integrated into an overall regression equation that correlates all the individual characteristics of quality to overall consumer preferences. Introducing the individual objective determination into the overall regression equation is wrought with many difficulties. They include a nonlinear relation between sensory and chemical or objective determinations, shifting standards in sensory appraisal, establishment of meaningful limits, difference in detection and signifi-

Table 38.1. Classification of Quality Factors and Their Measurement[a]

Factor	Fruits	Vegetables	Objective methods of measurement
Appearance			
Size	Diameter, drained weight	Sieve, size, drained weight	Scales, screens, micrometers
Shape	Height/width ratio	Straightness	Dimension ratios, displacement, angles
Wholeness	Cracked pieces	Cracked pieces	Counts, percent whole, photographs, models
Pattern defects	Blemishes, bruises, spots, extraneous matter	Blemishes, spots, bruises, extraneous matter	Photographs, drawings, models
Finish	Finish, gloss		Goniophotometers, gloss meters
Color	Color	Color	Color cards, dictionaries, reflectance and transmittance meters
Consistency	Consistency	Consistency	Consistometers, viscometers, flow meters, spread meters
Kinesthetic			
	Texture, firmness. grit, character	Texture, mealiness, succulence, fiber, maturity	Tenderometers, texture meters; compressing, penetrating, cutting instruments; tests for solids, moisture, grit, fiber
Odor and Flavor			
	Flavor, aroma, ripeness	Flavor, sweetness	Hydrometers; refractometers; pH meters; determination of sugar, sodium chloride, acid, sugar/acid ratio, enzymes, volatile substances, amines; chromatography

[a] From Kramer (1965).

cance of threshold and indifference limits in objective and subjective tests, and varying significance of specific objective factors in various foods (Pilgrim 1957). In practice, there is undoubtedly an interaction among the various components. The magnitude and direction of that interaction has been studied little. From the standpoint of chemical analysis, the difficulty lies in assigning a definite value to each of the components and their interaction products. That value is variable.

Shallenberger (1984) used three examples to illustrate some aspects of the chemical basis of fruit and vegetable quality. A heat-induced flavor defect in processed table beets was traced (among others) to the formation of pyrrolidone carboxylic acid. A flavor defect was due to a sensory response to a high concentration of natural carrot root flavor involving 3-methyl-6-methoxy-8-hydroxy-3,4-dihydroisocoumarin metabolically induced by ethylene formed by storing carrots together with apples. The inferior color, flavor, and texture in processed apple products were found to depend on suboptimal Brix/acid ratios.

Although on one hand the limitation of objective tests must be recognized, the large cost, the large amount of work involved, and the uncertainty in evaluating foods consistently and meaningfully by sensory tests are well established. In addition, objective assays are applicable at higher concentrations than subjective assessments without the danger of fatigue effects. Consequently, in practice, whenever a physical or chemical test can be run to obtain an objective measure of food quality, the objective test is preferred. As the final criterion of quality is human evaluation, the value of objective measurements must be evaluated by their correlation with sensory measurements. This correlation will generally be higher, the better our understanding of the nature of the physical and chemical parameters involved, and the better the available tools and methods for their reproducible and precise measurement. According to Kramer and Twigg (1966), a correlation of 0.90, or better, is desirable though in some cases useful information can be obtained if correlations are as low as 0.80. For predicting responses in production, very high correlations are necessary.

BIBLIOGRAPHY

Amerine, M. A., Pangborn, R. M., and Roessler, E.. B. (1965). "Principles of Sensory Evaluation of Food." Academic Press, New York.

Anon. 1968A. "Basic Principles of Sensory Evaluation." Spec. Tech. Publ. *433*. American Society for Testing and Materials, Philadelphia.

Anon. 1968B. "Manual on Sensory Testing Methods." Spec. Tech. Publ. *434*. American Society for Testing and Materials, Philadelphia.

Anon. 1968C. "Correlation of Subjective-Objective Methods in the Study of Odors and Tastes." Spec. Tech. Publ. *440*. American Society for Testing and Materials, Philadelphia.

Casimir, D. J., and Whitfield, F. B. (1978). Flavor impact values: A new concept for assigning numerical values for the potency of individual flavor components and their contribution to the overall flavor profile. *Ber. Intern. Fruchtsaft Union, Wiss. Techn. Komm.* **15,** 325–347.

Farber, L., and Lerke, P. A. (1958). A review of the value of volatile reducing substances for the chemical assessment of the freshness of fish and fish products. *Food Technol.* **12,** 677–680.

Human, J., and Khayat, A. (1981). Quality evaluation of raw tuna by gas chromatography and sensory methods. *J. Food Sci.* **46,** 868–873.

Jennings, W. G. (1977). Objective measurements of flavor quality: General approaches, problems, pitfalls, and accomplishments. *In* "Flavor Quality: Objective Measurement," (R. A. Scanlan, ed.), pp. 1–10, Symposium Series No. *51.* American Chemical Society, Washington, DC.

Kramer, A. (1965). Food quality. *In* "AAAS Publ. *77*" (G. W. Irving and S. R. Hoover, eds.). AAAS, Washington, DC.

Kramer, A. (1966). Parameters of quality. *Food Technol.* **20,** 1147–1148.

Kramer, A. and Twigg, B. A. (1966). "Fundamentals of Quality Control for the Food Industry," 2nd ed. AVI Publishing Co., Westport, CT.

Kramer, A., and Twigg, B. A. (1970). "Quality Control for the Food Industry," 3rd ed., Vol. 1. AVI Publishing Co., Westport, CT.

Morrison, W. H., Lyon, B. G., and Robertson, J. A. (1981). Correlation of gas–liquid chromatographic volatiles with flavor intensity scores of sunflower oils. *J. Am. Oil Chem. Soc.* **58,** 23–27.

Pangborn, R. M. V. (1984). Sensory techniques of food analysis. Chapter 2, *In* "Food Analysis—Principles and Techniques," Vol. I, Physical Characterization (D. W. Gruenwedel and J. R. Whitaker, eds.), pp. 37–93. Marcel Dekker, New York and Basel.

Pilgrim, F. J. (1957). The components of food acceptance and their measurement. *Am. J. Clin. Nutr.* **5,** 171–175.

Shallenberger, R. S. (1984). Some aspects of the chemical basis of fruit and vegetable quality. *In* "Control of Food Quality and Food Analysis" (G. G. Birch and K. J. Parker, eds.), pp. 311–323. Elsevier Applied Science Publishers, London.

Wick, E. L. (1965). Chemical and sensory aspects of the identification of odor constituents in foods. *Food Technol.* **19,** 827–833.

Index